PROTEIN TRANSFER AND ORGANELLE BIOGENESIS

PROTEIN TRANSFER AND ORGANELLE BIOGENESIS

Edited by

RATHINDRA C. DAS

Research & Development
Biotechnology Products Division
Miles Inc.
Elkhart, Indiana

PHILLIPS W. ROBBINS

Center for Cancer Research
Massachusetts Institute of Technology
Cambridge, Massachusetts

ACADEMIC PRESS, INC.
Harcourt Brace Jovanovich, Publishers

San Diego   New York   Berkeley   Boston
London   Sydney   Tokyo   Toronto

Academic Press, Inc.
San Diego, California 92101

United Kingdom Edition published by
ACADEMIC PRESS INC. (LONDON) LTD.
24-28 Oval Road, London NW1 7DX

Library of Congress Cataloging-in-Publication Data

Protein transfer and organelle biogenesis.

Includes index.
1. cell metabolism. 2. Cell organelles. 3. Proteins— Metabolism.
4. Protein biosynthesis. 5. Biological transport. I. Das, Rathindra
C. II. Robbins, Phillips Wesley, Date. [DNLM: 1. Cells—metabolism.
2. Proteins—metabolism. QU 55 P9690127]
QH634.5.P76 1988 574.87'61 87-19554

ISBN 0-12-203460-0 (alk. paper)

PRINTED IN THE UNITED STATES OF AMERICA
88 89 90 91   9 8 7 6 5 4 3 2 1

A000004825371

Contents

PART I.   TRANSLOCATION

1. Transport of Proteins into and across the Endoplasmic Reticulum Membrane
Eve Perara and Vishwanath R. Lingappa

PART II.   MODIFICATION, MATURATION, AND TRANSPORT

2. Role of Carbohydrate in Glycoprotein Traffic and Secretion
James B. Parent

3. Membrane Insertion and Transport of Viral Glycoproteins: A Mutational Analysis
Eric Hunter

4. Posttranslational Modification during Protein Secretion
Kurt W. Runge

5. Mannosidases in Mammalian Glycoprotein Processing
Kelley W. Moremen and Oscar Touster

PART III.   SORTING AND ORGANELLE ASSEMBLY

6. Biosynthesis and Sorting of Proteins of the Endoplasmic Reticulum
Michael Green and Richard A. Mazzarella

PART IV.   TRANSFER ACROSS BACTERIAL MEMBRANES

15. Protein Secretion across the Outer Membrane of Gram-Negative Bacteria
Anthony P. Pugsley

PART V.   MITOCHONDRIAL ASSEMBLY

16. Genetic Approaches to the Study of Mitochondrial Protein Import
Marjorie C. Brandriss

17. Synthesis and Assembly of Mitrochondrial Proteins
Donald W. Nicholson and Walter Neupert

PART VI. NUCLEAR TRANSPORT

18. Transport of Proteins into the Nucleus
Pamela A. Silver and Michael N. Hall

PART VII. APPLIED RESEARCH

19. Secretion Research in Industrial Mycology
Ramunas Bigelis and Rathindra C. Das

Contributors

Numbers in parentheses indicate the pages on which the authors' contributions begin.

Ramunas Bigelis (771), Biotechnology Products Division, Miles Inc., Elkhart, Indiana 46515

Marjorie C. Brandriss (655), Department of Microbiology and Molecular Genetics, University of Medicine and Dentistry of New Jersey, New Jersey Medical School, Newark, New Jersey 07103

James A. Cardelli (364), Department of Microbiology and Immunology, Louisiana State University Medical Center, Shreveport, Louisiana 71130

Rathindra C. Das (771), Biotechnology Products Division, Miles Inc., Elkhart, Indiana 46515

Robert B. Dickson (401), Medical Breast Cancer Section, NCI, National Institutes of Health, Bethesda, Maryland 20852

Randall L. Dimond (364), Promega-Biotec, Madison, Wisconsin 53711

Becca Fleischer (289), Department of Molecular Biology, Vanderbilt University, Nashville, Tennessee 37235

Michael Green (243), Department of Microbiology, St. Louis University School of Medicine, St. Louis, Missouri 63104

Michael N. Hall (747), Department of Biochemistry and Biophysics, University of California, San Francisco, California 94143

John A. Hanover (401), Laboratory of Biochemistry and Metabolism, NIDDK, National Institutes of Health, Bethesda, Maryland 20892

Eric Hunter (109), Department of Microbiology, The University of Alabama at Birmingham, Birmingham, Alabama 35294

Vishwanath R. Lingappa (3), Departments of Physiology & Medicine, University of California, San Francisco, California 94143-0444

Richard A. Mazzarella (243), Department of Microbiology, St. Louis University School of Medicine, St. Louis, Missouri 63104

Arthur M. Mercurio (563), Laboratory of Cancer Biology, New England Deaconess Hospital, Harvard Medical School, Boston, Massachusetts 02115

Hsiao-ping Hsu Moore (521), Department of Physiology–Anatomy, University of California, Berkeley, California 94720

Kelley W. Moremen (209), Center for Cancer Research, Massachusetts Institute of Technology, Cambridge, Massachusetts 02139

Walter Neupert (677), Institut für Physiologische Chemie der Universität München, 8000 München 2, Federal Republic of Germany

Donald W. Nicholson (677), Institut für Physiologische Chemie der Universität München, 8000 München 2, Federal Republic of Germany

Lelio Orci (521), Institute of Histology and Embryology, University of Geneva Medical School, 1211 Geneva 4, Switzerland

George F. Oster (521), Departments of Biophysics, Entomology and Zoology, University of California, Berkeley, California 94720

James B. Parent (51), Metabolic Research Branch, Naval Medical Research Institute, Bethesda, Maryland 20814

Eve Perara (3), Departments of Physiology and Medicine, University of California, San Francisco, California 94143

Anthony P. Pugsley (607), Unité de Génétique Moléculaire, Institut Pasteur, Paris 75724, France

Miguel Regue[1] (587), Department of Microbiology, Uniformed Services University of the Health Sciences, Bethesda, Maryland 20814

April R. Robbins (464), Genetics and Biochemistry Branch, National Institute of Diabetes, Digestive and Kidney Diseases, National Institutes of Health, Bethesda, Maryland 20892

Joel H. Rothman (318), Institute of Molecular Biology and Department of Biology, University of Oregon, Eugene, Oregon 97403

Kurt W. Runge (159), Department of Genetics, Fred Hutchinson Cancer Research Center, Seattle, Washington 98104

Pamela A. Silver (747), Department of Biology, Princeton University, Princeton, New Jersey 08544

Tom H. Stevens (318), Institute of Molecular Biology and Department of Chemistry, University of Oregon, Eugene, Oregon 97403

Oscar Touster (209), Department of Molecular Biology, Vanderbilt University, Nashville, Tennessee 37235

Henry C. Wu (587), Department of Microbiology, Uniformed Services University of the Health Sciences, Bethesda, Maryland 20814-4799

[1]Present address: Department of Microbiology, Facultad de Farmacia, Universidad de Barcelona, Barcelona, Spain.

Preface

Intracellular protein movement and targeting are among the most active areas of research in biology today. A major reason for this is that when molecular biology and recombinant DNA technology began to be applied to problems of cell biology and cell physiology, it became apparent that hitherto unapproachable questions concerning protein transport and organelle biogenesis could be asked; the answers were indeed exciting, interpretable, and suggestive of a multitude of new experiments. When it became clear that proteins can carry signals for secretion, mitochondrial assembly, lysosomal localizations, and so forth, it immediately became important to determine exactly how these signals function and what other factors are necessary in building and maintaining the functions of a cell. The chapters in this volume all relate to these themes.

The contributions are divided into seven sections. The first three deal with general aspects of protein translocation, modification, and sorting. Among the topics considered are transport into the endoplasmic reticulum, the role of carbohydrates in glycoprotein trafficking, as well as endo- and exocytosis. Later sections deal with localization of proteins to the mitochondrion and nucleus and with bacterial protein transport. A final section emphasizes the contribution secretion research has made to the production of proteins by the biotechnology industry.

Our special thanks go to Joan W. Bennett of Tulane University, New Orleans, for her advice and comments during the planning period of this book. One of us (R.C.D.) would like to thank Ramunas Bigelis for many discussions and constructive criticisms. Douglas A. Campbell, John F. Kaumeyer, and James G. Yarger deserve our special acknowledgment for their review of some of the chapters. Our appreciation is extended to Theresa Bollinger for her patient, diligent, and accurate secretarial work. Needless to say, we are grateful to all the authors, whose expertise, hard work, and enthusiasm have made this volume a reality. And, of course, we are so very much indebted to Sheela, Rishi, and Dorothy, our family members, for their understanding, encouragement, and support.

Rathindra C. Das
Phillips W. Robbins

I

Translocation

1

Transport of Proteins into and across the Endoplasmic Reticulum Membrane

EVE PERARA AND VISHWANATH R. LINGAPPA

I. INTRODUCTION

The specific localization of proteins to specialized membrane-bound compartments allows the segregation of various biochemical functions which characterizes eukaryotic cells. Thus, the mitochondria are the exclusive site of oxidative phosphorylation because F_1 ATPase, cytochrome oxidase, and other components are exclusively localized to the mitochondrial membranes. Likewise, oxidative detoxification occurs in the endo-

3

plasmic reticulum, degradation of endocytosed proteins in the lysosomes, and so on, as a consequence of protein targeting and localization. While the delimiting lipid bilayer(s) of subcellular compartments serves to maintain specific proteins within the organelle, it poses a distinct barrier to the initial correct segregation of these proteins. Since the synthesis of all but a few proteins (those encoded by mitochondrial or chloroplast DNA) occurs in the cytoplasm, mechanisms must exist by which proteins are efficiently and accurately sequestered into their specific membranous compartments. The subject of this chapter is the transport of proteins across the endoplasmic reticulum (ER) membrane. This is the first step in the sorting of proteins destined for the plasma membrane, Golgi complex, lysosome, or the exterior of the cell. The problem of transport of proteins across the ER membrane is essentially 2-fold: (1) How do the proteins recognize specifically the ER membrane from all the other cellular membrane systems? and (2) What is the mechanism by which these proteins are unidirectionally translocated across the hydrophobic lipid bilayer?

II. HISTORICAL BACKGROUND

In pioneering work on the secretory pathway, Palade and co-workers found that nascent secretory proteins were associated with the ER membrane via the large ribosomal subunit of the synthesizing polysomes (Sabatini *et al.,* 1966; Palade, 1975), and that the mature polypeptides were released to the lumen of the ER (Redman *et al.,* 1966). These observations raised the question of how mRNAs encoding secretory proteins could be specifically selected to be translated on ER-bound ribosomes. Very early hypotheses included models in which the specificity was proposed to reside in the translating ribosome, in untranslated regions of mRNA, or in the nascent chain itself.

An answer to this question was obtained following the development of heterologous cell-free translation systems which could be programmed with purified mRNAs encoding secretory proteins. It was found that the primary cell-free translation products of immunoglobulin light chain differed from authentic light chain by an amino-terminal extension that was not observed in translation products of nonsecretory proteins. This additional peptide segment was proposed to be involved in the segregation of secretory proteins to the ER lumen (Milstein *et al.,* 1972). This idea received additional support from experiments in which cell-free translation reactions were supplemented with microsomal membranes derived from the rough ER. It was found that secretory proteins were synthesized with amino-terminal sequences not present either in secretory products *in vivo* or *in vitro* products localized to the microsomal lumen, suggesting that the precursor protein was processed to the mature form by the membranes (Blobel and Dobberstein, 1975a,b). These findings led to the formulation of the signal hypothesis (Blobel and Dobberstein, 1975a,b) in which it was postulated that the emergence of a transient "signal sequence" of amino acid residues as part of the nascent chain directs polysomes synthesizing secretory proteins to, and facilitates their transport across, the ER membrane (see below). Indeed, nearly all eukaryotic secretory proteins examined so far have been found to be synthesized as precursors with amino-terminal, cleaved signal sequences.

The concept of "vectorial discharge" of the nascent polypeptide across the membrane was established by the work of Redman and Sabatini (1966), which demonstrated that puromycin-released nascent (incomplete) polypeptide chains were localized to the lumen of vesicles isolated from the rough ER. This idea was confirmed by the finding that nascent polypeptides emerging from the ribosome were protected from proteases by the microsomal membrane (Sabatini and Blobel, 1970). In addition, a number of posttranslational modifications of proteins known to occur in the luminal space of the ER, such as cleavage of signal sequences by signal peptidase (Blobel and Dobberstein, 1975a), transfer of core oligosaccharides from lipid-linked intermediates to asparagine residues (Lingappa *et al.*, 1978a; Glabe *et al.,* 1980), and intrachain disulfide bond formation, have been shown to be carried out on nascent polypeptides as they traverse the ER membrane. Completed secretory protein precursors are incapable of being translocated (Blobel and Dobberstein, 1975b), and, in fact, there seems to exist a brief period early in a protein's elongation during which translocation can be initiated (Rothman and Lodish, 1977). Thus a picture emerged of obligate coupling of the transport of secretory proteins to their biosynthesis, i.e., translocation occurs concomitant with translation. This feature of protein translocation across the ER membrane distinguishes it from the transport of mitochondrial proteins (Schatz and Butow, 1983) and from the export of bacterial proteins (Randall, 1983).

Two general hypotheses have been advanced regarding the mechanism by which proteins may be transported across, or assembled into, the ER membrane. One type of model (von Heijne and Blomberg, 1979; Engelman and Steitz, 1981) suggests that protein transport across the membrane is dictated solely by the thermodynamics of interactions between the protein and the lipid bilayer without the participation of specific receptors or transport proteins in the membrane aside from "targeting" proteins (see below) and signal peptidase. These "spontaneous insertion" models postulate that the signal sequence approaches the membrane as the hydrophobic limb of a helical hairpin (see Fig. 1a). The free energy

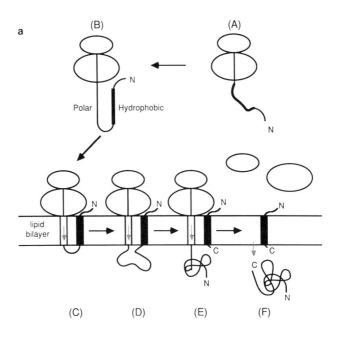

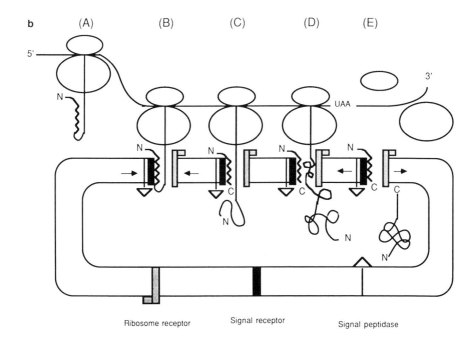

gained through the insertion of the hydrophobic signal sequence directly into the lipid bilayer is proposed to exceed the free energy "cost" of transiently burying charged residues in the hydrophilic limb of the helical hairpin. These hydrophilic sequences pulled into the bilayer would nevertheless be thermodynamically unstable and hence would move spontaneously across the bilayer by Brownian motion as protein synthesis continued. The nascent chain would thus be transported unless or until another hydrophobic sequence was encountered which would be thermodynamically stable in the lipid bilayer and thus serve to anchor the polypeptide in the membrane. Removal of signal sequences by signal peptidase is proposed to release secretory proteins to the ER lumen. Much of the appeal of this model has derived from its simplicity. Its early drawback was its difficulty in explaining the translocation of secretory proteins, such as ovalbumin, which lack cleaved signal sequences (Palmiter et al., 1978). Recently, additional evidence has emerged that is difficult to reconcile with the spontaneous insertion models (see below).

The signal hypothesis envisions a distinctly different mechanism for chain translocation (see Fig. 1b; Blobel and Dobberstein, 1975a,b; Blobel, 1980). Key features of a current version of this hypothesis (for review, see Walter and Lingappa, 1986) are as follows: (1) the information for the localization of proteins to the ER lumen is contained within a discrete portion of the nascent polypeptide, itself—the signal sequence; (2) the signal sequence facilitates translocation of the nascent chain through in-

---

Fig. 1. (a) Steps in the process of translocation according to spontaneous insertion models. Translation begins on free cytoplasmic ribosomes (A); on emergence of a sufficient length of amino acid residues, marginally stable folding begins (B). The hydrophobic limb of the hairpin inserts spontaneously into the membrane (C), pulling in the polar limb, as described in the text. As synthesis continues, the growing nascent polypeptide constitutes the relatively polar, thermodynamically unstable, limb of the hairpin and therefore passes through the membrane, folding in the extracytoplasmic space (D). Cleavage of the signal sequence occurs, releasing the amino terminus of the polypeptide into the lumen (E). On completion of translation, the C terminus passes through the membrane and is released into the ER lumen, and the cleaved signal peptide is the left in the membrane (F). (b) Steps in the process of translocation (subsequent to targeting) according to the signal hypothesis. Synthesis begins on cytoplasmic ribosomes (A). Receptor-mediated targeting of the signal sequence-bearing ribosome to the ER membrane is described in the text and depicted in Fig. 2. Once targeted correctly, the signal sequence and the ribosome interact with their respective receptors in the ER membrane, resulting in the assembly of an aqueous, proteinaceous tunnel across the membrane (B). As protein synthesis continues, the chain passes through the tunnel to the lumen of the ER, and the signal sequence is removed by signal peptidase (C). Translocation (possibly of folded polypeptide domains) continues concomitant with protein synthesis (D). On termination of protein synthesis, the ribosomal subunits dissociate, the carboxy terminus passes through the tunnel, and the tunnel components disassemble (E), restoring the integrity of the lipid bilayer.

teractions with a series of receptors both in the cytoplasm and in the ER membrane; (3) transit of the nascent chain across the membrane occurs via a proteinaceous pore or tunnel in the bilayer whose activation and assembly is catalyzed by the signal sequence; (4) the information for termination of translocation before termination of synthesis (e.g., in the case of transmembrane proteins) is encoded also in a discrete segment of the nascent chain termed the "stop transfer sequence" which also acts via particular receptors in the membrane. The appeal of the signal hypothesis has largely derived from experimental verification of its tenets. Its major weakness is that it is difficult to imagine the initiation of specific receptor-mediated events by signal sequences which have such variable primary structures (see below).

III. TARGETING

A fundamental problem for the transport of proteins across a specific intracellular membrane is the selection of the correct membrane by the protein to be transported. From the use of cell-free translation/translocation systems for the fractionation and reconstitution of translocation-associated events, a view of the molecular mechanisms of targeting of nascent secretory polypeptides to the membrane of the rough ER has emerged.

A. Signal Sequences

Most secretory proteins and many transmembrane proteins are synthesized as precursors with transient amino-terminal signal sequences as are bacterial exported proteins. In contrast to the amphipathic signal sequences of mitochondrial proteins, these signal sequences are characterized by their extreme hydrophobicity.

The most reliable definition of a signal sequence if a functional one: The ability of a sequence of amino acids within a protein to direct its translocation. Over a hundred amino-terminal signal sequences have been cataloged (Watson, 1984; von Heijne, 1985), and yet no overall homology in primary structure has been observed. Signal sequences range from 15 to 30 amino acid residues in length. They consist of a very hydrophobic core of variable length (at least six amino acids) flanked on either side by regions containing polar or hydrophilic residues. Often an amino acid with a small side chain, such as glycine, valine, or alanine, occurs at the cleavage site (von Heijne and Blomberg, 1979; von Heijne, 1984). Since no clear sequence homologies exist, the recognition features of signal sequences have been proposed to reside in their secondary structure. Bacterial and eukaryotic signal sequences are virtually indistinguishable from one another (von Heijne, 1985). In fact, eukaryotic proteins can be secreted and processed by bacteria (Talmadge *et al.*, 1980a,b), and, likewise, prokaryotic proteins are correctly segregated and processed in eukaryotic cell-free systems (Muller *et al.*, 1982).

Two protein systems have been identified with which signal sequences may interact: the targeting proteins (signal recognition particle and its receptor, see below) on the cytoplasmic side of the ER membrane and signal peptidase on the luminal aspect. Whether additional systems are involved, or whether proteins which actually span the bilayer are recognized, remains to be demonstrated. These two recognition systems may interact with distinct regions of the signal sequence (Blobel and Dobberstein, 1975b). Analysis of signal sequence mutants in bacterial export systems demonstrates that the proposed two sites are at least functionally distinguishable since mutants are observed which can be translocated but are not processed by signal peptidase (Lin *et al.*, 1978; Koshland *et al.*, 1982; Kadonaga *et al.*, 1985). Not surprisingly, the hydrophobic core which characterizes signal sequences appears to be critical for translocation since disruption of the integrity of this core, either by the introduction of charged residues or by small deletions in this hydrophobic stretch, abolishes export in bacteria (Emr and Silhavy, 1982; Bedouelle *et al.*, 1980). However, the hydrophobic core is not sufficient for translocation since deletion of the coding region for the six carboxy-terminal amino acids of the α-amylase signal sequence, leaving the hydrophobic stretch intact, abolishes secretion from *Escherichia coli* (Palva *et al.*, 1982).

While a signal sequence can be sufficient to translocate some proteins both *in vitro* and *in vivo* in eukaryotic systems (Lingappa *et al.*, 1984; Simon *et al.*, 1987), the same has not been demonstrated for export of proteins from bacteria (Moreno *et al.*, 1980; Kadonaga *et al.*, 1984). Whether this reflects differences in the passenger proteins used or differences in the mechanism of translocation between prokaryotes and eukaryotes remains to be determined. While translocation in eukaryotic cell-free systems appears to be cotranslational, translocation in *E. coli in vivo* and *in vitro* seems to be posttranslational (Randall, 1983; Randall and Hardy, 1986; Muller and Blobel, 1984). Thus, while the mechanism for translocation of proteins across cellular membranes appears to be highly conserved, critical differences do exist. It is important to note that fine structure analyses of signal sequences have been carried out exclusively on prokaryotic proteins in bacterial cells. A systematic analysis of signal sequence structure and function in eukaryotic cell-free systems has yet to be done.

Signal sequences usually occur at the extreme amino terminus of a protein and are cleaved from the nascent polypeptide before translation is complete. There are at least two exceptions to this rule: (1) Internal and uncleaved signal sequences have recently been described for a number of transmembrane proteins (Bos *et al.*, 1984; Spiess and Lodish, 1985; Friedlander and Blobel, 1985; Eble *et al.*, 1986). (2) The secretory protein, ovalbumin, does not have a cleaved signal sequence (Palmiter *et al.*, 1978) but has the functional equivalent (Lingappa *et al.*, 1978b), and location of which remains controversial (Lingappa *et al.*, 1979; Meek *et al.*, 1982; Braell and Lodish, 1982; Tabe *et al.*, 1984).

B. Signal Recognition Particle

Signal recognition particle (SRP) activity was first recognized by the ability of a microsomal membrane high salt wash to restore translocation activity to rough microsomes whose ability to translocate in the wheat germ cell-free translation system had been abolished by high salt extraction (Warren and Dobberstein, 1978). The SRP molecule has been purified to homogeneity from a high salt extract of canine pancreas rough microsomes (Walter and Blobel, 1980), and its role in translocation has been studied in detail (Walter *et al.*, 1981; Walter and Blobel, 1981a,b). Subcellular fractionation demonstrates a roughly equal distribution of SRP between a membrane-associated and cytoplasmic (ribosome-associated or free) state (Walter and Blobel, 1983b). SRP appears to act as a cytoplasmic "adaptor" for signal-bearing polysomes, targeting them to the ER membrane.

1. *Structure of Signal Recognition Particle*

SRP is a ribonucleoprotein complex composed of six nonidentical polypeptides: a 19 kDa and a 54 kDa monomer, and two heterodimers, one consisting of the 9 kDa and the 14 kDa polypeptides and the other composed of 68 kDa and 72 kDa polypeptides (Siegel and Walter, 1985) and one molecule of the small cytoplasmic 7SL RNA, 300 nucleotides in length (Walter and Blobel, 1982). When disassembled, neither the protein nor RNA fraction alone is active. Reconstitution of the active molecule is possible, however, when polypeptides and RNA are reassembled together (Walter and Blobel, 1983a; Siegel and Walter, 1985). The 7SL RNA seems to act as scaffolding, around which the polypeptide subunits assemble; the subunits have little or no affinity for one another in the absence of the RNA. Systematic reconstitution experiments reassembling SRP molecules lacking defined polypeptide or RNA domains have allowed different assayable functions of SRP to be assigned to specific structural domains of the molecule (see below; Siegel and Walter, 1985, 1986). Recent evidence demonstrates that the SRP molecule is oblong in shape (Andrews *et al.*, 1985) and that the RNA runs the length of the molecule (Andrews *et al.*, in press).

2. *Signal Sequence Recognition*

The purification of SRP has allowed detailed analysis of its role in translation and translocation of secretory proteins and its binding properties to other identified components in the translation/translocation system. While SRP was shown to bind translationally inactive ribosomes, the emergence of a signal sequence from the large ribosomal subunit results in an increase in the affinity of SRP for the ribosome by as much as four to five orders of magnitude (Walter *et al.*, 1981).

SRP binds to the signal sequence directly. This was suggested first by the finding that incorporation of the amino acid analog, β-hydroxyleucine, into leucine-rich signal sequences abolished SRP–signal sequence interactions (Walter *et al.*, 1981). Incorporation of a photoactivatable cross-linking amino acid analog into the signal sequence of nascent preprolactin has allowed cross-linking to the 54 kDa subunit of SRP, providing direct evidence for signal sequence–SRP binding (Kurzchalia *et al.*, 1986; Krieg *et al.*, 1986).

3. *Elongation Arrest*

When purified SRP is added to wheat germ cell-free translation reactions in the absence of microsomal membranes it specifically blocks elongation of signal sequence-bearing nascent chains soon after the signal emerges from the ribosome, concomitant with the increased affinity of SRP for the ribosome–nascent chain complex (Walter and Blobel, 1981b). On subsequent addition of salt-washed microsomal membranes, SRP is released from the nascent chain, translation resumes, and translocation occurs cotranslationally.

The elongation arrest function of SRP has been mapped to the 9/14 kDa protein and to 7SL RNA sequences that are homologous to repetitive Alu RNA (Siegel and Walter, 1985, 1986). Preparation of SRPs lacking either one of these domains yields a particle which can recognize signal sequences and facilitate translocation across salt-treated (SRP-free) microsomal membranes but lacks the elongation arrest activity. These partially reconstituted particles are active in promoting protein translocation, but only during a brief window of time early in nascent chain growth. Thus it appears that the role of SRP arrest may be to extend the window of time during which the nascent polypeptide is in a translocation competent state.

The mechanism by which SRP arrests translation is currently unclear.

Analysis of the structure and domain function of the SRP molecule presents an interesting possibility. The dimensions of the SRP molecule are such that the particle could span from the site at which the signal sequence emerges from the large ribosomal subunit to the elongation site between the two ribosomal subunits (Andrews *et al.*, 1985). The 54 kDa subunit binds to signal sequences directly (Kurzchalia *et al.*, 1986; Krieg, *et al.*, 1986); perhaps the 9/14 kDa subunit or 7SL RNA of SRP also bind the ribosome, hindering subsequent binding of aminoacyl-tRNAs and continued protein synthesis.

The physiological significance of the elongation arrest activity of SRP is controversial since not all signal sequence-bearing proteins experience a tight elongation arrest (Anderson *et al.*, 1983), nor is strict elongation arrest by canine SRP observed in some mammalian cell-free systems (Meyer, 1985). However, a kinetic delay in chain elongation is observed specifically for signal sequence-bearing proteins in a fractionated mammalian translation system supplemented with purified canine SRP (P. Walter, personal communication). This is consistent with the notion that an important role of SRP may be to extend the window of time during which the nascent secretory protein is in a "translocation competent state."

C. Signal Recognition Particle Receptor

The SRP receptor (also termed docking protein, Meyer *et al.*, 1982a) is an ER membrane protein localized to the cytoplasmic face of the membrane (Meyer *et al.*, 1982b) and has been purified from dog pancreas rough microsomes using SRP affinity chromatography (Gilmore *et al.*,1982a,b). A 60 kDa cytoplasmic domain of SRP receptor can be cleaved from the membrane by proteases and added back to reconstitute translocation activity (Walter *et al.*, 1979; Meyer and Dobberstein, 1980a,b), but apart from the membrane it is inactive (Gilmore *et al.*, 1982). SRP receptor has recently been shown to consist of two subunits, the previously identified 69 kDa polypeptide (now termed α subunit) and a 30 kDa β subunit (Tajima *et al.*, 1986).

The release of SRP-induced elongation arrest of secretory proteins by microsomal membranes is a function of the SRP receptor (Gilmore *et al.*, 1982a). Since SRP receptor was isolated by virtue of its affinity for SRP (Gilmore *et al.*, 1982b), it seems likely that the release of arrest may occur via a direct interaction between SRP and SRP receptor. This idea is further supported by data demonstrating that purified, detergent-solubilized SRP receptor causes SRP to lose its high affinity for signal sequence-bearing polysomes concomitant with the release of SRP-induced elongation arrest (Gilmore and Blobel, 1983).

Recently the amino acid sequence for the 69 kDa α subunit of SRP receptor has been determined from a cloned cDNA (Lauffer *et al.*, 1985). A domain consisting of clusters of predominantly basic mixed charge residues and which resembles nucleic acid binding proteins has been suggested to bind to SRP, possibly via the 7SL RNA.

The SRP receptor plays a critical role beyond releasing SRP-induced elongation arrest. Partially reconstituted SRPs which do not arrest nascent chain elongation but are capable of facilitating translocation require SRP receptor (Siegel and Walter, 1985). The requirement for SRP receptor in the absence of elongation arrest may reflect an essential role in targeting the ribosome–nascent chain to the microsomal membrane.

Quantitation of SRP and SRP receptor in pancreatic cells indicates that both are present in substoichiometric amounts relative to membrane-bound ribosomes (Gilmore *et al.*, 1982b; Walter and Blobel, 1980) and, thus, that SRP receptor is not involved directly in the process of translocation. Rather, it appears that the nascent chain–ribosome–SRP–SRP receptor interaction is transient and that the role of SRP and its receptor is primarily to target the nascent secretory polypeptide to the appropriate membrane system.

D. Summary of Targeting Events

In summary, the initial targeting events as they are currently understood have led to the model of the SRP cycle depicted in Fig. 2 (for review, see Walter *et al.*, 1984). On emergence of a signal sequence from ribosomes synthesizing secretory proteins in the cytoplasm, SRP binds the signal sequence directly, interrupting chain elongation and perhaps maintaining the nascent chain–ribosome complex in some "translocation competent state." The affinity of SRP for its receptor on the cytoplasmic face of the ER membrane targets the SRP–ribosome–nascent chain complex to that membrane system. Following interaction with its receptor in the ER membrane, SRP loses its affinity for the signal sequence-bearing ribosome and releases the complex, perhaps to another series of receptors in the membrane. Translation resumes and translocation across the membrane occurs. It is not known whether SRP initiates the translocation event or whether its role is merely to target the nascent chain to the correct location.[1]

---

[1] It should be noted that some small proteins (<10 kDa) such as M13 procoat protein (Watts, et al., 1983) and honeybee prepromelittin (Zimmermann and Mollay, 1986; Muller and Zimmermann, 1987) appear to bypass this well-characterized targeting mechanism. How these proteins are targeted to the ER membrane remains to be determined.

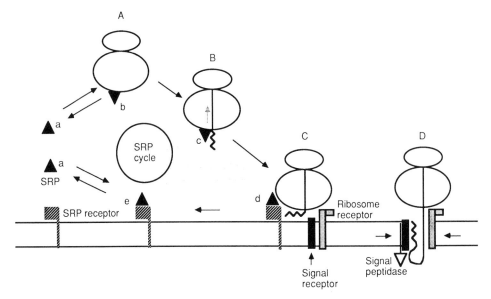

Fig. 2. Model of signal recognition particle (SRP) cycle for targeting nascent secretory and transmembrane proteins to the ER membrane. Soluble SRP (a) exists in equilibrium with a membrane-bound form, presumably bound to SRP receptor (e), and a ribosome-bound form (b). On translation of mRNA encoding a signal sequence for targeting to the ER membrane (zigzag lines), the affinity of SRP for the translating ribosome is enhanced (represented by dashed arrow, B) and SRP binds to the signal sequence directly (c), effecting elongation arrest (B–C). On interaction with ER membranes, elongation arrest is released and SRP and SRP–receptor are free to be recycled (SRP cycle, a–e), the synthesizing ribosome interacts with other transmembrane proteins, leading to formation of a functional ribosome–membrane junction, translation resumes, and translocation across the membrane occurs (D). Models for translocation are depicted in Fig. 1.

IV. MECHANISM OF TRANSLOCATION

The use of heterologous cell-free translation/translocation systems has provided a very powerful approach for the isolation of molecular components and the fractionation and reconstitution of specific activities involved in targeting of nascent chains to the ER membrane. However, the mechanism(s) by which these targeted nascent chains cross the lipid bilayer remains a mystery. This is due in large part to the obligate coupling of translocation to translation. In cell-free systems, and presumably *in vivo*, translocation can occur only during the limited time and under the fastidious conditions required for protein synthesis.

A. Altered Substrates for Translocation

One important concept regarding translocation is that information for transport across the membrane resides within the protein itself. With the availability of cloned genes and recombinant DNA technology has come the ability to manipulate this information by creating artificial substrates for translocation and thereby probe its mechanism. The analysis of altered substrates *in vitro* and *in vivo* has yielded a number of insights into the mechanism of translocation.

1. Signal Sequence Function

Initial studies involving the fusion protein consisting of the β-lactamase signal sequence fused to the cytoplasmic protein, globin, demonstrated that a signal sequence alone was sufficient to permit translocation across the ER membrane both *in vitro* (Lingappa *et al.*, 1984) and *in vivo* (Simon *et al.*, 1987). Thus, the information for translocation was contained within the signal sequence and not the protein being transported.

Expression of a fusion protein in which a cleaved amino-terminal signal sequence was engineered to an internal position raised interesting questions as to the mechanism of translocation (Perara and Lingappa, 1985). The cDNA for the normally cytoplasmic protein, globin, was engineered 5′ to that of preprolactin such that the initial 110 codons of globin were followed immediately by the entire coding region of preprolactin (see Fig. 3). The encoded hybrid protein consisted of the signal sequence of prolactin flanked at its amino terminus by globin and at its carboxy terminus by native prolactin (see Fig. 3). When the cDNA was expressed in cell-free systems by *in vitro* transcription and translation, not only was this signal sequence (formerly amino terminal, now internal) recognized by the translocation machinery of the membrane and cleaved by signal peptidase, but both the carboxy flanking domain and the amino domain (still attached to the cleaved signal at its carboxy end) were translocated to the lumen of microsomal vesicles (see Fig. 3). Extraction of the vesicles with carbonate demonstrated that neither the prolactin domain nor the globin domain with the signal attached at its carboxy terminus were integrated into the microsomal membrane. This result suggested that signal sequences do not bury into the lipid bilayer directly, as suggested by some (von Heijne and Blomberg, 1979; Engelman and Steitz, 1981; Briggs *et al.*, 1986), but rather that they facilitate translocation via associations with integral membrane proteins (see also Gilmore and Blobel, 1985). Another important implication of these findings concerned the ability of the signal sequence to facilitate translocation of an amino-terminal protein domain: Since synthesis of the amino-terminal domain proceeds before

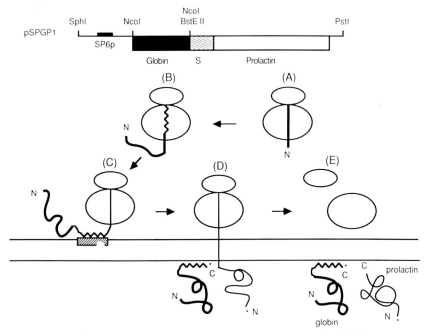

Fig. 3. Restriction map of relevant region of expression plasmid, pSPGP1, and schematic representation of translation and translocation of globin-prolactin fusion protein (GP),
encoded by pSPGP1 (for details, see Perara and Lingappa, 1985). The SP6 promoter is
represented by the small black bar denoted SP6p. The initial 109 codons of globin are
represented by the solid black bar, the 30 codon preprolactin signal sequence coding region
by the zigzag-patterned bar, and the 199 codons of mature prolactin by the open bar. In the
translocation scheme, mRNA and protein components have been omitted for simplicity. The
globin domain is represented by the heavy black line, the signal sequence by the zigzag, and
prolactin by the thin black line. The globin domain is synthesized before the signal sequence
emerges from the ribosome (A and B). The signal sequence is shown binding to a putative
signal receptor on the cytoplasmic face of the ER membrane (C); the signal sequence
facilitates translocation of both the globin and prolactin domains and is cleaved accurately
(D). Thus the globin domain, with the prolactin signal peptide now at its carboxy terminus,
and mature prolactin are localized to the microsomal lumen (E). Note: It is not known which
domain is translocated first, globin or prolactin, nor is it understood how the signal sequence
facilitates translocation of the globin (amino-terminal) domain (i.e., does it cross in an NH_2
to COOH fashion as does prolactin, COOH to NH_2, or as a folded domain?). In addition,
apparent translocation efficiency of the amino-terminal globin domain is reduced relative to
that of the processed prolactin (Perara and Lingappa, 1985).

the signal sequence emerges from the ribosome, its translocation could not be driven by the energy of its own synthesis. Thus the possibility was raised that machinery in the membrane, rather than the driving force of the synthesizing ribosome, performed work to achieve translocation.

2. Dissociation of Translocation from Translation

The use of engineered translocation substrates has allowed the processes of translation and translocation to be uncoupled for proteins whose translocation is normally strictly cotranslational (Perara *et al.*, 1986). By synthesizing truncated mRNA for a secretory protein which lacked a termination codon, arrested nascent polypeptides were generated which remained associated with the synthesizing ribosomes (see Fig. 4). When protein synthesis was blocked, such arrested chains were translocated only in the presence of nucleoside triphosphates and an ATP-regenerating system. Interestingly, the release of the "arrested" nascent polypeptide chains from the ribosome of synthesis by puromycin abolished translocation. Therefore, translocation across the ER membrane is not dependent on ongoing protein synthesis. The synthesizing ribosome does not "push" the nascent chain across the bilayer (von Heijne and Blomberg, 1979; Wickner and Lodish, 1985), nor does translocation occur "spontaneously" (Engelman and Steitz, 1981). Rather, the transport process consumes energy generated by nucleoside triphosphate hydrolysis (see also Chen and Tai, 1987; Schlenstedt and Zimmermann, 1987). In addition, it appears that the coupling of translation to translocation may reflect a role of the synthesizing ribosome in translocation independent of its role in protein synthesis. Similar findings have been demonstrated for integral membrane proteins (Perara *et al.*, 1986; Mueckler and Lodish, 1986a,b). Posttranslational translocation of human placental lactogen has also been shown to require an association of the precursor polypeptide with the ribosome (Caufield *et al.*, 1986).

It appears that this newly recognized role of the ribosome in translocation is for some aspect of targeting and not for translocation of the polypeptide per se, since ribosome-independent translocation has been observed in several instances (Hansen *et al.*, 1986; Schlenstedt and Zimmermann, 1987; Perara and Lingappa, in preparation). A small fusion protein consisting of the 23 amino acid signal sequence of β-lactamase followed by the initial 70 amino acids of chimpanzee α-globin is capable of posttranslational translocation and requires nucleoside triphosphate hydrolysis whether associated with the ribosome or not (E. Perara and V. R. Lingappa, in preparation). In most cases, however, translocation of truncated secretory and transmembrane proteins is dependent on an associa-

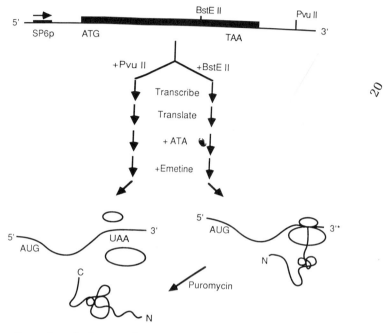

Fig. 4. Generation of substrates for dissociation of translocation from protein synthesis (see Perara *et al.*, 1986). SP6 expression plasmids encoding secretory proteins were digested with a restriction endonuclease which cut specifically either within the coding region, 5′ to the termination codon (*Bst*EII), or in the 3′noncoding region (*Pvu*II). On *in vitro* transcription, the latter treatment yields a full-length mRNA while the former treatment yields a truncated transcript lacking a termination codon. Ribosomes are able to translate the truncated transcript but do not dissociate for lack of a termination codon, thus translation yields intact polysomes with emergent nascent chains. In contrast, translation of full-length transcripts yields completed, free polypeptide chains. By treating translation reactions with aurintricarboxylic acid (ATA) and emetine, polypeptide chains were presented to microsomal membranes in the absence of protein synthesis. Ribosome-associated translation products of truncated RNAs were released from the ribosome by treatment with puromycin.

tion of the nascent chain with the ribosome of synthesis (Perara and Lingappa, unpublished observations).

Ribosome-independent translocation has also been observed for the yeast protein, prepro-α-factor (~18.5 kDa), in the recently developed homologous cell-free translation/translocation system derived from yeast (Waters and Blobel, 1986; Rothblatt and Meyer, 1986a; Hansen *et al.*, 1986). This transport was insensitive to uncouplers and ionophores but was dependent on nucleoside triphosphate hydrolysis (Hansen *et al.*, 1986; Waters and Blobel, 1986; Rothblatt and Meyer, 1986b) and required

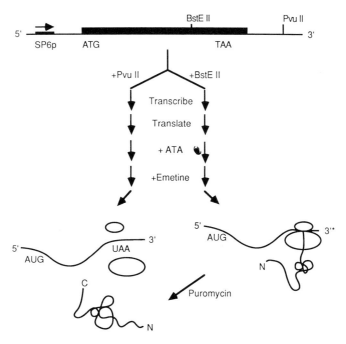

proteinaceous factors in the membrane (as demonstrated by sensitivity to the alkylating agent, N-ethylmaleimide; Hansen *et al.*, 1986). It is not clear whether the posttranslational translocation of α-factor in this system represents a peculiarity of this particular molecule or is a general characteristic of the yeast *in vitro* translocation system and/or of this organism. The tight coupling between translocation and translation has been described only for higher eukaryotic cell-free systems.

These recent findings demonstrate that the ER membrane is capable of translocating completed signal-bearing proteins and protein domains via an energy-requiring mechanism. Why, then, must some chains remain associated with the ribosome in order for translocation to occur, while others need not, and why is efficient translocation normally coupled to protein synthesis? One possible explanation is that the translocation competence of a given protein or polypeptide may reflect the accessibility of its signal sequence to signal receptors in the ER membrane and/or cytoplasmic receptors such as SRP. Thus, the signal sequence of a very short polypeptide or a protein that lacks rigid secondary structure may be readily accessible to these receptors, while longer proteins may require an association with the synthesizing ribosome, possibly in conjunction with SRP to present the signal sequence to the membrane. It appears that the ribosome serves to extend the window of time during which a given nascent polypeptide is translocation competent. The translocation competence of nascent chains varies from one protein to another, thus some nascent secretory proteins are translocation competent only during a narrow window of time early in their synthesis (Rothman and Lodish, 1977; Braell and Lodish, 1982; Siegel and Walter, 1985), while translocation of others can occur later in elongation (Ainger and Meyer, 1986) and still others, such as prepro-α-factor, may translocate even as completed polypeptides (Hansen *et al.*, 1986; Schlenstedt and Zimmermann, 1987).

A requirement for high-energy phosphate bond hydrolysis has been demonstrated for translocation (recent work suggests that ATP hydrolysis is required for co-translational translation [Chen and Tai, 1987], however, it remains to be determined how hydrolysis facilitates protein transport across the membrane. Energy may be expended to arrange and/or maintain the nascent chain in a translocation-competent state, to assemble or "activate" a protein tunnel in the membrane, or directly for movement of the polypeptide to the lumen of the ER.

B. Putative Components of Translocation Machinery

The capacity of the ER membrane to translocate already synthesized proteins (Hansen *et al.*, 1986; Mueckler and Lodish, 1986a; Müller and

Zimmermann, 1986; Schlenstedt and Zimmerman, 1987) or protein domains (Perara and Lingappa, 1985; Perara *et al.,* 1986) can be explained either by the existence of a large tunnel or pore in the membrane (through which such folded domains could diffuse or be transported), or, alternatively, by a membrane "denaturase" to effect unfolding of these domains prior to their transport. The signal hypothesis proposes that translocation occurs via an aqueous channel formed by integral membrane proteins, the assembly of which is directed by the signal sequence. Recently some evidence for this idea has emerged. Partially translocated nascent chains were generated by oligonucleotide-mediated hybrid arrest (Gilmore and Blobel, 1985). These polypeptides, spanning the membrane, with their amino termini in the lumen and their carboxy termini in the cytoplasmic space were extractable from the microsomes with protein denaturants such as urea, suggesting that the translocation process occurs in an environment accessible to aqueous perturbants and that it involves interactions of the nascent chain with proteins in the ER membrane.

It is predicted that such a putative tunnel would consist of a complex assembly of a number of integral membrane proteins essential for translocation and related processes such as signal sequence cleavage, N-linked oligosaccharide transfer, and other posttranslational modifications known to occur in the ER lumen. Candidates for components of a translocon include a ribosome receptor, signal sequence receptor, and signal peptidase.

1. Ribosome Receptor

Binding of polysomes synthesizing secretory proteins to the ER membrane has long been thought to play an important role in the vectorial transport of nascent chains across the membrane (Redman and Sabatini, 1966). Such ribosomes bind to the ER membrane both via their nascent chains (Gilmore and Blobel, 1985) and directly via their large subunits by a salt-labile interaction (Adelman *et al.,* 1973). This binding is saturable and sensitive to proteases (Hortsch *et al.,* 1986).

The ribophorins (I and II), two integral membrane glycoproteins which are present in rough microsomes but absent from smooth membranes (Kreibich *et al.,* 1978a), have been suggested as ribosome receptor(s). Several indirect lines of evidence support this idea, including cofractionation of ribosomes and ribophorins following detergent solubilization (Kreibich *et al.,* 1978a) or protein cross-linking treatment (Kreibich *et al.,* 1978b) of rough microsomes and a good stoichiometry between the number of ribophorins and the ribosome binding capacity of rough microsomes (Marcantonio *et al.,* 1984). However, controlled proteolysis of rough microsomes suggests that ribophorins do not mediate functional ribosome binding directly since ribosome binding activity of rough microsomes is lost following protease treatment to which ribophorins appear resistant (Hortsch *et al.*, 1986). In addition, translocation of some secretory proteins can occur *in vitro* across smooth microsomes which lack ribophorins (Bielinska *et al.*, 1979). Therefore, it appears that the ribophorins do not play an essential role in translocation nor in functional binding of ribosomes to the ER membrane. The role of ribophorins or other ribosome binding proteins in translocation remains to be determined.

2. *Signal Sequence Receptor*

In addition to ribosome binding sites, an independent signal sequence receptor in the ER membrane has also been suggested (Prehn *et al.*, 1980, 1981; Gilmore and Blobel, 1985; Hortsch *et al.*, 1986). Posttranslational binding of signal-bearing proteins is specific, saturable, and protease sensitive, as well as specific for rough microsomes (Prehn *et al.*, 1980, 1981). A nascent, SRP-arrested polypeptide binds to microsomal membranes in an SRP–receptor-dependent manner and remains associated with the membrane even after extraction of the ribosome with puromycin and high salt (Gilmore and Globel, 1985) but is extractable with protein denaturants such as urea or alkaline pH. Through a series of elegant cross-linking experiments, Wiedmann *et al.* (1987) have recently identified a signal sequence receptor that is an integral membrane glycoprotein of the ER.

3. *Signal Peptidase*

Signal peptidase is an integral membrane protein presumed to act on the luminal side of the ER membrane because cleavage of signal sequences takes place only on translocated secretory proteins unless the membrane is solubilized (Jackson and Blobel, 1977). Since signal peptidase removes signal sequences from nascent polypeptides as they cross the ER membrane (Blobel and Dobberstein, 1975b) it is postulated to be associated with a complex of proteins in the membrane that are involved in other aspects of translocation (e.g., tunnel proteins). Signal peptidase has recently been purified from canine pancreas rough microsomes as a relatively abundant complex of four to six polypeptides (Evans *et al.*, 1986). Bacterial leader peptidase I, which can accurately cleave eukaryotic signal sequences (Watts *et al.*, 1983), exists as a single polypeptide (Wolfe *et al.*, 1982). It is believed that eukaryotic signal peptidase is very similar to this bacterial enzyme since bacterial secretory proteins can be accurately processed by canine rough microsomes (Muller *et al.*, 1982). By analogy, it is thought that eukaryotic signal peptidase also exists as one polypeptide and that the additional copurifying proteins may be involved in other translocation-related processes. The finding that this complex exists in roughly stoichiometric amounts relative to membrane-bound ribosomes has lead to speculation that it may form a core around which still other membrane proteins assemble to form a translocation apparatus in the membrane (Evans *et al.*, 1986).

V. MEMBRANE ASSEMBLY OF INTEGRAL
TRANSMEMBRANE PROTEINS

The biogenesis of integral transmembrane proteins (ITMPs) adds an additional degree of complexity to the problem of transport of proteins across the ER membrane (Rothman and Lenard, 1977). Not only must certain polypeptide domains be translocated across the lipid bilayer but specific domains must span the membrane while others are left in the cytoplasmic space. Integral membrane proteins can be considered in two broad categories (see Fig. 5): (1) bitopic ITMPs which span the membrane only once, with their amino and carboxy termini on opposite sides of the membrane, and (2) polytopic membrane proteins which span the membrane multiple times. These can be further classified according to the disposition of the termini relative to the membrane: proteins whose amino and carboxy termini are on opposite sides of the membrane can be referred to as "trans" ITMPs, with the terminus residing in the extracytoplasmic space specified [NH_2-trans, e.g., vesicular stomatitis virus (VSV) glycoprotein or bovine rhodopsin; or COOH-trans, e.g., influenza neuraminidase]; those which span the membrane with both termini on the same side can be considered "cis" ITMPs, with the side specified (C-cis, cytoplasmic, e.g., erythrocyte Band III protein; E-cis, extracytoplasmic, e.g., hepatitis B surface antigen; see Fig. 5).

Several lines of evidence suggest that the assembly of proteins in the ER membrane is an event similar to the process of transport of secretory proteins across the membrane: both secretory and integral transmembrane proteins appear to be synthesized on membrane-bound polysomes (Morrison and Lodish, 1975), both have signal sequences that interact with SRP (Anderson *et al.*, 1982; Bos *et al.*, 1984), both compete with one another for membrane-associated components required for translocation (Lingappa *et al.*, 1978a), and both require hydrolysis of nucleoside triphosphates for translocation into or across the membrane (Waters and Blobel, 1986; Hansen *et al.*, 1986; Mueckler and Lodish, 1986b; Perara *et al.*, 1986). For a secretory protein these events result in complete translocation across the ER membrane, while in the case of ITMPs protein domains are only partially translocated. The mechanism by which pro-

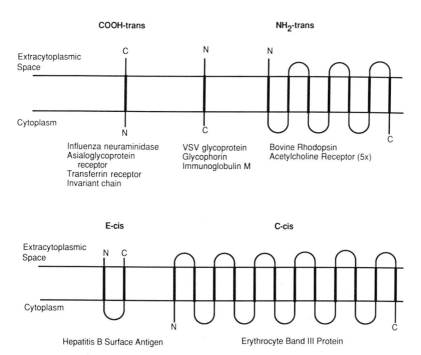

Fig. 5. Examples of orientations of integral transmembrane proteins. For definition of terminology, see text. Membrane-spanning segments are represented by heavy black lines. References for sequence and/or experimental orientation data are as follows: influenza neuraminidase (Bos *et al.,* 1984), asialoglycoprotein receptor (Spiess and Lodish, 1985, 1986), transferrin receptor (Schneider *et al.,* 1984; McClelland *et al.,* 1984; Zerial *et al.,* 1986), invariant chain (Lipp and Dobberstein, 1986), VSV glycoprotein (Katz *et al.,* 1977), glycophorin (Bretscher, 1971, 1975), immunoglobulin M (McCune *et al.,* 1980), bovine rhodopsin (Nathans and Hogness, 1983), acetylcholine receptor (Young *et al.,* 1985), hepatitis B surface antigen (Eble *et al.,* 1986), and erythrocyte Band III protein (Kopito and Lodish, 1985; Wickner and Lodish, 1985).

teins are oriented asymmetrically in the membrane operates with high fidelity—the orientation appears to be essentially identical for all copies of a given ITMP (Katz and Lodish, 1979). The orientation achieved in the ER membrane is the same as that in the final destination, e.g., the plasma membrane (Katz and Lodish, 1979). The process by which ITMPs are transported intracellularly from their site of synthesis at the ER membrane to their ultimate destination (e.g., Golgi membrane or plasma membrane) is poorly understood (for review, see Kelly, 1985).

As described for secretory proteins, two general hypotheses have been advanced to explain how the assembly of proteins in the membrane is achieved. One is a variant of the helical hairpin/spontaneous insertion

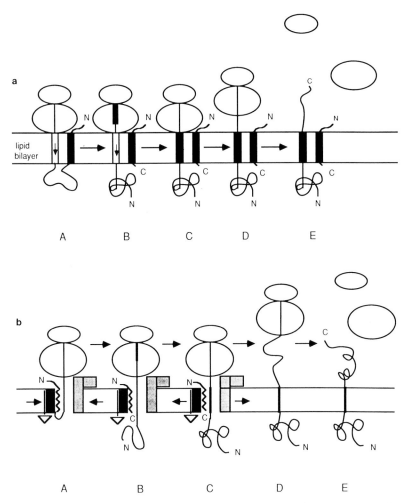

Fig. 6. (a) Integral transmembrane assembly of proteins according to spontaneous insertion/helical hairpin hypotheses. Translocation occurs as described in the text and Fig. 1a (A), until a sufficiently hydrophobic domain (stop transfer, represent by solid black bar) emerges from the ribosome (B). This region is thermodynamically stable in the lipid core of the bilayer and thus serves to halt the transfer of the nascent chain (C). The remainder of the polypeptide is synthesized in the cytoplasmic space (D, E). (b) Integral transmembrane protein assembly according to the signal hypothesis. Targeting of the nascent chain occurs as depicted in Fig. 2, and the signal sequence directs assembly of a tunnel for translocation across the membrane (A). Translocation occurs, and the signal sequence is cleaved as depicted in Fig. 1b (B). On emergence of a stop transfer sequence, components of the translocation machinery are disassembled (C) and the ribosome–membrane junction is disrupted, causing the remainder of the polypeptide to be synthesized in the cytoplasmic space (D, E).

hypothesis (Engelman and Steitz, 1981) and the direct transfer model (von Heijne and Blomberg, 1979), according to which the thermodynamics of protein–lipid interactions between the nascent chain and the ER membrane govern the protein's transport across, and integration into, the membrane. In this view, if a hydrophobic rather than a hydrophilic domain emerges as the second transmembrane region of the helical hairpin, the stability of this domain in the hydrophobic environment of the lipid bilayer will prevent further translocation and integrate the chain into the membrane (see Fig. 6a). Alternatively, according to the signal hypothesis, "topogenic" sequences in the nascent chain serve to "stitch" the nascent polypeptide in the membrane in the correct transmembrane orientation largely via interactions with receptor proteins which initiate or terminate translocation (Blobel, 1980; see Fig. 6b). According to both hypotheses the information for translocation and membrane orientation resides in particular segments of the protein such as signal and "stop transfer" sequences. Signal sequences, as described earlier, act to initiate transport of the nascent chain across the membrane, while stop transfer sequences terminate the transport process such that the subsequently synthesized polypeptide domain is maintained in the cytoplasmic space. Varying arrangements of these two types of topogenic sequences within a protein can conceivably account for any particular membrane orientation.

A. "Stop Transfer" Sequences

Early studies on transmembrane protein biogenesis carried out on VSV glycoprotein in cell-free translation systems demonstrated that this simple bitopic transmembrane protein possesses a transient amino-terminal signal sequence (Lingappa et al., 1978a) and spans the membrane with a topology indistinguishable from that observed in vivo, i.e., with a large amino-terminal extracytoplasmic domain and a small carboxy-terminal cytoplasmic domain (Katz et al., 1977; Katz and Lodish, 1979). Similar to translocation of secretory proteins in vitro, integration of VSV glycoprotein into microsomal membranes was found to occur only when microsomes were present during protein synthesis (Katz et al., 1977; Toneguzzo and Ghosh, 1977). Moreover, competition experiments revealed that nascent VSV glycoprotein competed with a nascent secretory protein for a membrane component(s) involved in transfer across the membrane (Lingappa et al., 1978a). Thus it was proposed that the biogenesis of secretory proteins and integral transmembrane proteins share early events in common. What causes some proteins to be only partially translocated?

A clue to the answer to this question is found in the immunoglobulin M (IgM) heavy chain which exists in two forms—one on the cell surface and the other secreted—which differ only in the presence on the former of a carboxy-terminal transmembrane segment and small cytoplasmic domain (Vasalli *et al.,* 1979; Kehry *et al.,* 1980; Singer and Williamson, 1980; McCune *et al.,* 1980). Apparently this carboxy-terminal extension serves to anchor the protein in the membrane. Indeed, other bitopic NH_2-trans ITMPs, which are synthesized with amino-terminal cleaved signal sequences and carboxy-terminal transmembrane segments, can be experimentally converted to secretory proteins by deletion of the transmembrane region (Boeke and Model, 1982; Gething and Sambrook, 1982; Rose and Bergmann, 1982), demonstrating that such segments are necessary for halting translocation.

That stop transfer sequences are sufficient to terminate translocation and to direct integration into the membrane was demonstrated by Yost *et al.* (1983), who engineered the IgM transmembrane segment to an internal position within a chimeric secretory protein, converting it to an integral transmembrane protein with predicted topology relative to the ER membrane (see Fig. 7). Moreover, the stop transfer sequence was able to act at other positions within the polypeptide beside the extreme carboxy terminus.

1. Structure of "Stop Transfer" Sequences

As is true for signal sequences, stop transfers are essentially functionally defined and exhibit little homology in primary sequence. Analysis of many transmembrane domains reveals that they consist of 20–30 hydrophobic and neutral amino acid residues often flanked by one or more positively charged residues on the cytoplasmic side (Sabatini *et al.,* 1982). It has been postulated that the hydrophobic stretch spans the membrane—a length of 20 amino acids is believed to be sufficient to span the 3-nm thickness of the lipid bilayer as an α-helix (Tanford, 1978). The basic residues are thought to play a role in membrane association by interactions with the negatively charged phospholipid head groups at the membrane surface (Sabatini *et al.,* 1982).

2. Structure/Function Analyses of Stop Transfer Sequences

The basic residues at the cytoplasmic boundaries of some membrane-spanning domains do not appear critical for the stop transfer function (Davis *et al.,* 1985; Cutler and Garoff, 1986; Zuniga and Hood, 1986) although they may play a role in stablizing the protein in the membrane

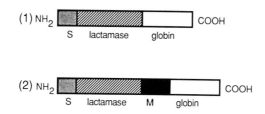

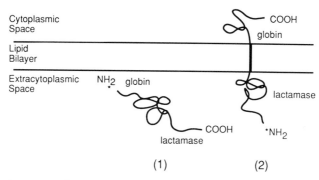

Fig. 7. A stop transfer sequence confers predictable transmembrane orientation on a previously secretory protein (see Yost *et al.*, 1983). Hybrid proteins consisting of lactamase and globin or lactamase, stop transfer, and globin are depicted as (1) and (2), respectively. The signal sequence of lactamase is denoted by S and the IgM transmembrane segment as M. The transmembrane orientation of fusion proteins 1 and 2 are depicted below. Cleavage of the amino-terminal lactamase signal sequence is indicated by an asterisk, and the IgM transmembrane segment by the bold black line spanning the bilayer. Fusion protein 1 is completely translocated across the microsomal membrane. Insertion of the M segment between the lactamase and globin domains in fusion protein 2 serves to terminate translocation at that point, leaving the lactamase domain in the lumen and the globin domain in the cytoplasmic space.

(Cutler *et al.*, 1986). Likewise, introduction of a single charged residue into the membrane-spanning domain of Semliki Forest virus protein E2 has no effect on membrane orientation or physiological function of the protein (Cutler and Garoff, 1986) but destabilizes the membrane association as determined by carbonate extraction (Cutler *et al.*, 1986), a procedure which strips membranes of all polypeptides save those integrated directly into the hydrophobic core of the lipid bilayer (Fujiki *et al.*, 1982).

Several groups have attempted to determine experimentally the minimum length of a membrane-spanning domain by carrying out gradual deletions in defined membrane-spanning regions (Adams and Rose, 1985;

Davis *et al.*, 1985). Such experiments suggest that as few as 14–17 amino acids of a stop transfer domain are sufficient for full function. When reduced to 8–12 amino acids these domains still appear able to direct proper membrane orientation, but their stability in the membrane is impaired (Davis *et al.*, 1985). A problem with these types of experiments is that the polypeptide context in which the stop transfer domains is assayed is critical, making it difficult to draw broad conclusions from a single example. For example, positioning of the potential stop transfer domains close to the carboxy terminus improves the ability of such a domain to anchor the polypeptide in the membrane (Davis and Model, 1985; Davis *et al.*, 1985; Davis and Hsu, 1986). Also, the existence of hydrophobic domains adjacent to the membrane-spanning regions being deleted [e.g., in VSV glycoprotein (Adams and Rose, 1985)] is likely to confound the interpretation of such experiments since they may compensate for the deleted hydrophobic region. Nevertheless, it appears on the basis of some of these experiments (Davis *et al.*, 1985) that stop transfer domains may have two functionally distinct roles: one to stop the translocation process and the other to engage in a stable membrane integration.

3. Role of Hydrophobicity in Stop Transfer Function

It has been suggested that topology of transmembrane proteins occurs as a result of spontaneous physical partitioning of hydrophobic protein regions into the hydrophobic core of the lipid bilayer (von Heijne and Blomberg, 1979; Engelman and Steitz, 1981). Indeed, insertion of a synthetic repetitive domain of 16 or more hydrophobic amino acid residues has been shown to be adequate to stop the translocation of a secretory protein and to confer predictable membrane topology to the resultant transmembrane protein in bacteria (Davis and Model, 1985). On the other hand, the fusion-related hydrophobic domain of the myxovirus, Sendai F protein, which consists of some 26 consecutive uncharged or hydrophobic residues is completely translocated both in the natural eukaryotic context and in a bacterial assay system (Davis and Hsu, 1986). Moreover, severely truncated membrane-spanning regions consisting of only 12–16 hydrophobic residues can direct topology similar to wild-type transmembrane sequences (Davis *et al.*, 1985).

Thus, while a "sufficiently" hydrophobic region of a translocated protein can act to halt translocation, it is not clear whether hydrophobicity is the sole feature that determines the function of authentic stop transfer sequences. Nor is it known whether critical hydrophobic interactions are between the nascent chain and the lipid bilayer or between the nascent chain and hydrophobic membrane proteins.

B. Combination Signal–Stop Transfer Sequences

Several bitopic COOH-trans ITMPs such as influenza neuraminidase (Bos *et al.*, 1984), asialoglycoprotein receptor (Spiess and Lodish, 1986), invariant chain (Lipp and Dobberstein, 1986), and transferrin receptor (Zerial *et al.*, 1986) have been identified which have internal, uncleaved signal sequences which also serve to anchor the protein in the membrane. These proteins are situated with their amino termini in the cytoplasm and their carboxy domains in the extracytoplasmic space (see Fig. 5). The hydrophobic membrane-spanning regions facilitate the transport of only the carboxy domains in an SRP-dependent fashion (Spiess and Lodish, 1986; Lipp and Dobberstein, 1986). These membrane-spanning regions may be considered simply as internal, uncleaved signal sequences. However, it is important to distinguish them from signal sequences of secretory proteins—the uncleaved signal of ovalbumin is completely translocated and secreted; moreover, when engineered to an internal position, a normally amino-terminal signal sequence can facilitate translocation of both flanking protein domains and does not integrate directly into the bilayer (Perara and Lingappa, 1985).

Signal–stop topogenic elements are not confined to this particular orientation. In the first transmembrane segment of the polytopic ITMP, bovine rhodopsin, a combined signal–stop transfer sequence has been identified which translocates the amino terminus and leaves the carboxy flanking domain in the cytoplasm (Friedlander and Blobel, 1985; Perara *et al.*, 1986). As in the case of COOH-trans ITMPs, what accounts for domain translocation specificity (i.e., whether the amino or carboxy flanking domain is translocated) of NH_2 is not clear. One possibility is that lack of translocation of a particular domain may depend on its folding in a translocation-incompetent manner. Alternatively, in the case of polytopic transmembrane proteins, adjacent topogenic domains of a protein may also influence the domain translocation specificity of a given topogenic sequence (Eble *et al.*, 1987).

We shall refer to domains which are able to facilitate both translocation and integration into the bilayer as signal–stop transfer sequences. Whether separate signal and stop transfer functions can be assigned to distinct regions of these domains remains to be determined. Until that issue is resolved it remains unclear whether such domains should be considered a special class of topogenic sequence, whether they represent two distinct elements in tandem, or whether they are distinguished from simple signal sequences merely by the inability of the cytoplasmically disposed domain to be translocated.

C. Polytopic Integral Transmembrane Proteins

Those ITMPs which span the membrane more than once are referred to as polytopic membrane proteins. The most detailed analysis of polytopic ITMPs has been done on the polytopic subunits in the photosynthetic reaction center of the bacterium, *Rhodopseudomonas viridis*. The determination of the tertiary structure by X-ray crystallography (Deisenhofer *et al.*, 1985) in conjunction with the primary amino acid sequences (Michel *et al.*, 1986) has allowed structural analysis of these proteins at nearly atomic resolution. However, the topologies of most polytopic ITMPs are predicted from their primary sequences, with very hydrophobic regions proposed to span the membrane.

Initial events in the assembly of polytopic ITMPs into the ER membrane appear to be similar to those involved in the translocation of secretory and bitopic transmembrane proteins. Some have amino-terminal cleaved signal sequences (Anderson *et al.*, 1982), while others possess uncleaved signal sequences (Anderson *et al.*, 1983; Rottier *et al.*, 1984; Friedlander and Blobel, 1985; Eble *et al.*, 1986). Their integration is dependent on SRP (Anderson *et al.*, 1982, 1983; Friedlander and Blobel, 1985; Rottier *et al.*, 1985; Mueckler and Lodish, 1986a; Eble *et al.*, 1987) although some do not experience elongation arrest (Anderson *et al.*, 1983). Thus, polytopic ITMPs utilize the same targeting system as simple secretory and bitopic transmembrane proteins. But is each transmembrane domain established by specific signal and stop transfer sequences (Blobel, 1980), or, once targeted to the ER via a signal sequence and SRP, is subsequent membrane assembly "spontaneous" (Wickner and Lodish, 1985; Mueckler and Lodish, 1986a)?

The biogenesis of only a few polytopic transmembrane proteins has been studied in any detail. The analysis of biogenesis has relied heavily on molecular genetics techniques which allow deletion of putative topogenic domains and/or analysis of their independent function in defined polypeptide contexts.

1. Multiple Signal Sequences

Bovine rhodopsin is believed to span the bilayer seven times with the amino terminus in the extracytoplasmic space and the carboxy terminus in the cytoplasm (Nathans and Hogness, 1983). Deletions of the cDNA encoding bovine opsin revealed that this protein has at least two SRP-dependent signal sequences, one of which is located in the first transmembrane segment and the other in the sixth (Friedlander and Blobel, 1985). Both signals facilitate translocation of at least the amino flanking domain and integrate into the lipid bilayer directly, thus classifying them as sig-

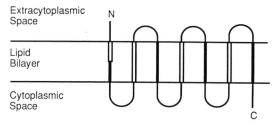

Fig. 8. Predicted arrangement of signal and stop transfer sequences to account for transmembrane orientation of bovine rhodopsin (see Friedlander and Blobel, 1985). Open rectangles represent signal sequences and solid black rectangles denote stop transfer sequences.

nal–stop transfers. The existence of two more signal sequences and four stop transfer sequences in alternating transmembrane domains has been proposed to account for the complex topology of opsin (Friedlander and Blobel, 1985; see Fig. 8). These observations must be interpreted with caution, however, in light of more recent findings (described in Section V,D) which suggest common features between signal and stop transfer sequences. Nevertheless, this study demonstrates that multiple topogenic elements, which recognize known receptors for translocation, exist in polytopic membrane proteins and suggests that the translocation of multiple protein domains of a polytopic membrane protein may occur via the same mechanism as that of a secretory or simple bitopic membrane protein.

The membrane biogenesis of hepatitis B surface antigen is somewhat more clearly understood. This is a simpler polytopic membrane protein which spans the membrane two times (Eble *et al.*, 1985, 1987). Analysis of each of the two known membrane-spanning regions in fusion proteins *in vitro* reveal that hepatitis B surface antigen contains two uncleaved signal sequences, both of which interact with SRP (Eble *et al.*, 1987). The first is located within the initial 32 amino acids of the protein and is capable of translocating both amino and carboxy flanking domains, as well as itself, completely across the membrane, although in native hepatitis B surface antigen it resides in the membrane (see Fig. 9). The second topogenic element appears to be a signal–stop transfer sequence (see Section V,B) which facilitates translocation of only the carboxy-flanking domain and integrates into the lipid bilayer directly, leaving the amino terminus in the cytoplasmic space (see Fig. 9). It is unclear exactly how these two topogenic sequences act together to achieve the ultimate transmembrane orientation, i.e., do the two signals act sequentially—first one, then the other—or do they interact with one another to specify the correct topol-

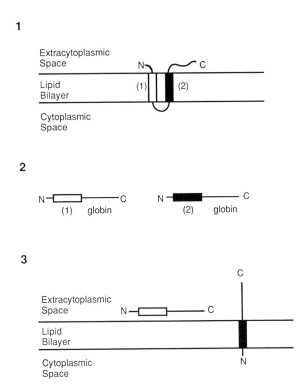

Fig. 9. Hepatitis B surface antigen possesses two types of signal sequences (see Eble *et al.*, 1986, 1987). The transmembrane orientation of hepatitis B surface antigen is represented schematically in Part 1. The two transmembrane domains are designated by open and solid black bars [(1) and (2), respectively]. Transmembrane domains (1) and (2) were analyzed independently in globin fusion proteins (shown in Part 2), with globin flanking regions (1) or (2) at their carboxy termini. Transmembrane orientations of globin fusion proteins are depicted in Part 3. Transmembrane domain 1 facilitated translocation of the flanking globin domain and was translocated itself. Transmembrane segment 2 also facilitated translocation of the globin domain but remained integrated in the membrane with a small, cytoplasmically disposed amino terminus.

ogy? It appears from this study that all "information" for residing in the bilayer need not necessarily be specified exclusively by the membrane-spanning domain itself.

2. Charged Domains May Span the Membrane

It is generally believed that membrane-spanning regions are composed of extremely hydrophobic regions of a protein, and the topology of many polytopic membrane proteins has been predicted by their amino acid sequences (Nathans and Hogness, 1983; Noda *et al.*, 1983; Devillers-Thiery *et al.*, 1983; Mueckler *et al.*, 1985). Such predictions, however, are not always correct. For example, the subunits of the acetylcholine receptor have been predicted by their primary sequence to span the membrane four times (Noda *et al.*, 1983; Devillers-Thiery *et al.*, 1983). However, immunoprecipitation and immunocytochemistry using antibodies to specific segments of the protein have provided strong evidence for a model in which the homologous receptor subunits each cross the membrane five times (Young *et al.*, 1985). Four of the predicted membrane-spanning regions are rich in hydrophobic and nonpolar residues while one (the fourth) is amphipathic and can form an α helix that is hydrophobic on one side and highly charged on the other. The homologous amphipathic helices of the five subunits are predicted to assemble in the membrane such that the charged side of each subunit faces the center, stabilizing this conformation in the lipid bilayer and forming an ion channel (Young *et al.*, 1985). It is unlikely that such amphipathic membrane-spanning segments provide "information" for membrane insertion; their localization spanning the membrane may be determined by adjacent topogenic elements in the protein.

Polytopic membrane assembly does not appear to be a simple partitioning of hydrophobic sequences into the hydrophobic core of the bilayer, since polar polypeptide domains have been shown to span the membrane (Young *et al.*, 1985) and hydrophobic regions are sometimes completely translocated (Davis and Hsu, 1986). Moreover, recent reports suggest that cytoplasmic or membrane factors, or both, may determine the orientation of a given protein relative to the membrane. A bacterial polytopic transmembrane protein integrates into *E. coli* inverted vesicles in a cell-free translation system derived from *E. coli*, but it is completely translocated across canine rough microsomes when expressed in the wheat germ system (Watanabe *et al.*, 1986). It is suggested that differences in the two types of membranes may account for the different topologies of the same protein, though differences between the *E. coli* and wheat germ extracts may also account for these observations. In fact, a similar finding has been observed in the expression of a protein in a wheat germ versus rabbit reticulocyte cell-free system, in both cases using dog pancreas rough microsomes. Expression in the wheat germ system results in a polytopic transmembrane orientation, while in reticulocyte lysate approximately equal amounts of the E-cis ITMP and a completely translocated form are observed. On expression in *Xenopus* oocytes the major product is a secreted soluble monomer (Hay *et al.*, 1987).

It is not known whether assembly of polytopic proteins into the ER membrane is receptor mediated aside from the requirement for SRP to target the nascent chain to the ER. While at least two proteins have been demonstrated to contain more than one sequence which is recognized by SRP it is not known whether SRP mediates the interactions of each sequence with the membrane or whether it is required only once to target the nascent chain to the ER membrane and that, once there, subsequent topogenic sequences in the nascent chain are able to interact directly with the membrane or membrane proteins.

While the assembly of most polytopic ITMPs appears to be "cotranslational" (Rottier *et al.*, 1984; Eble *et al.*, 1986), it is not clear whether multiple transmembrane segments insert into the membrane sequentially (i.e., as they emerge from the ribosome) or not. The observation that a fragment of the glucose transporter molecule can integrate into rough microsomes posttranslationally with an orientation similar to that observed for cotranslational integration (Mueckler and Lodish, 1986a) suggests that proper orientation need not be achieved via sequential insertion or translocation events.

D. Reevaluation of Topogenic Sequences

Information for membrane assembly appears to reside in the membrane protein itself, encoded in discrete "topogenic" sequences (Blobel, 1980; Lingappa *et al.*, 1984; Yost *et al.*, 1983). From the work discussed thus far topogenic domains can be considered as one of three basic types:

1. Conventional signal sequence. These are usually located at the amino termini of most secretory and many transmembrane proteins. They facilitate translocation of flanking protein domains in an SRP-dependent fashion and are usually (but not always) cleaved by signal peptidase in the ER membrane. These signal sequences do not appear to integrate into the bilayer directly and may themselves be translocated.
2. Stop transfer sequence. Conventional stop transfer sequences abort the translocation process initiated by a distant, preceding signal sequence and often integrate into the lipid bilayer.
3. Combined signal–stop transfer sequence. These signal–stop sequences require SRP to initiate translocation of one or the other flanking protein domain and are not themselves translocated but integrate directly into the membrane.

Recent findings suggest that the distinctions between these topogenic sequences may not be absolute. Analysis of stop transfer sequences in proteins which lack amino-terminal signal peptides reveals a number of unexpected features of these topogenic elements.

1. Domain Translocation Activity

The epidermal growth factor (EGF) receptor is a typical bitopic NH_2-trans transmembrane protein with large extracytoplasmic and cytoplasmic domains (Ullrich *et al.*, 1984). It is synthesized with an amino-terminal cleaved signal sequence which presumably facilitates the translocation of the extracytoplasmic domain. The membrane-spanning domain is believed to function as a stop transfer sequence, terminating translocation and resulting in the integration of the protein into the bilayer, with the remainder of the protein maintained in the cytoplasmic space. The retroviral oncogene, v-*erbB*, encodes a truncated version of EGF receptor which lacks the first 550 amino acids of the amino-terminal (extracytoplasmic) domain, including the signal sequence, as well as a small portion of the extreme carboxy terminus. Nevertheless, the v-*erbB* gene product spans the plasma membrane *in vivo*, as determined by immunocytochemistry, in an orientation similar to that of the EGF receptor, i.e., with an extracytoplasmic amino terminus and the carboxy terminus in the cytoplasm (Schatzman *et al.*, 1986). Thus, it appears that the membrane-spanning region of EGF receptor may possess an intrinsic capacity to direct the translocation of the truncated amino-terminal domain and to integrate itself correctly into the membrane.

In vitro analyses of a classic stop transfer sequence in chimeric proteins lacking a signal sequence also suggest that bona fide "stop transfer" sequences possess domain translocation activity (Mize *et al.*, 1986; Zerial *et al.*, 1987). The transmembrane segment at the extreme carboxy terminus of IgM (the M segment) has been shown to have intrinsic domain translocating activity in certain polypeptide contexts *in vitro* (Mize *et al.*, 1986). When an amino-terminal signal sequence was present, either in engineered proteins or in the native IgM molecule, only the domain between the signal and stop transfer sequences was translocated (Yost *et al.*, 1983; Mize *et al.*, 1986); the remainder (the domain carboxy terminal to the M segment) remained in the cytoplasmic space (see Fig. 10). However, when engineered between two antigenically distinct cytoplasmic protein domains, lacking any defined signal sequence, the M segment facilitated the transport of either the amino and carboxy flanking domains across microsomal membranes (Mize *et al.*, 1986; see Fig. 9). A normally amino-terminal signal sequence in a similar polypeptide context facilitates translocation of both flanking protein domains and is translocated itself into the lumen (Perara and Lingappa, 1985; Mize *et al.*, 1986, see Fig. 9).

2. Functional Recognition by Signal Recognition Particle

The translocation activity of "stop transfer sequences" is SRP mediated, and SRP can interact with them to arrest elongation of the nascent

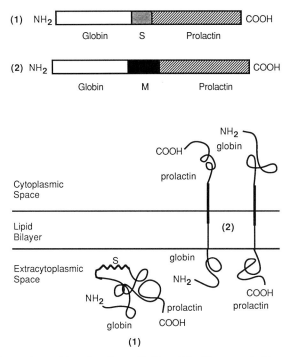

Fig. 10. Comparison of translocation activities of the M segment of immunoglobulin M and the prolactin signal sequence in similar polypeptide contexts (see Mize *et al.*, 1986). Fusion proteins 1 and 2 consisting of globin and prolactin domains flanking either a signal sequence or a transmembrane region, respectively, are shown. The dispositions of the fusion proteins are represented below. The signal sequence is represented by the zigzag line, labeled S. The stop transfer domain is represented by the heavy black line and labeled M. See text for details.

chain (Mize *et al.*, 1986; Zerial *et al.*, 1987). In this respect stop transfer sequences are qualitatively indistinguishable from classic amino-terminal signal sequences in a similar context. Thus, it appears that stop transfer and signal sequences may share at least some structural features which direct functional interactions with the same receptor, SRP, as well as with putative receptor elements on the cytoplasmic, but not the luminal (e.g., signal peptidase), aspect of the ER membrane.

Why should a transmembrane segment, such as that of IgM, possess recognition features for SRP and for initiation of translocation? Since it is located at the extreme carboxy terminus of IgM and emerges from the ribosome only after termination of protein synthesis and dissociation of the ribosomal subunits, this cotranslational translocation activity seems unlikely to play any role in its native stop transfer function. While the possibility exists that SRP may mediate normal stop transfer function, another possible explanation may be that signal and stop transfer functions are mediated by common membrane receptors which share recognition features with SRP. The observation that an artificial stretch of 23 hydrophobic amino acids interacts with SRP (Zerial *et al.*, 1987) suggests that hydrophobicity is a key recognition feature for SRP.

E. Summary of Transmembrane Protein Biogenesis

Information for translocation and membrane assembly appears to reside in the membrane protein, itself, encoded in discrete "topogenic" sequences (Lingappa *et al.*, 1984; Yost *et al.*, 1983), their position relative to one another (Coleman *et al.*, 1985; Mize *et al.*, 1986; Eble *et al.*, 1987) and their immediate polypeptide environment or context (Davis and Model, 1985; Davis and Hsu, 1986). Signals for the initiation and termination of translocation appear to share structural features for recognition by common receptors, at least for SRP. While both signals and stops may possess intrinsic domain translocation activity, they are distinguishable in that signal sequences may be translocated themselves while stop transfers are not (Perara and Lingappa, 1985; Mize *et al.*, 1986). A systematic analysis of various arrangements of identified signal and stop transfer sequences in identical polypeptide and protein synthesis contexts should allow clarification of the functions of these sequences.

Our understanding of how the process of translocation is interrupted and how complex polytopic membrane proteins are assembled is limited by our lack of understanding as to how translocation occurs in the first place. For example, is topogenic "information" in a complex polytopic membrane protein "read" in a linear order from amino to carboxy terminus as the growing polypeptide emerges from the ribosome, or do subdomains interact with cytoplasmic and ER membrane translocation components?

VI. OVERVIEW

We have reviewed here past and recent progress in the understanding of the localization of proteins to the endoplasmic reticulum. Rapid advances have been made due to the combined efforts to identify molecular components and the use of molecular genetic manipulation to create altered translocation substrates. While some questions have been answered, new ones have been raised and other long-standing issues remain unresolved.

Studies of translocation across the ER membrane initially focused on the membrane-bound ribosomes of the rough ER and the characteristic vectorial discharge (or cotranslational translocation) of the nascent chain. In the last decade the molecular mechanisms by which particular proteins are selected to be synthesized by ER-bound polysomes have been clarified, at least to one level of resolution. More recently it has been found that translocation need not be coupled to protein synthesis, rather, the characteristic coupling of translocation to translation may reflect the requirement for an association of the nascent chain with the ribosome of synthesis. Thus attention has returned to the membrane-bound ribosomes of the rough ER. What is the nature and importance of the ribosome–membrane interaction? What role does the ribosome play in the initiation and termination of translocation?

While the molecular events of targeting have been well characterized, the mechanism by which selected proteins cross the ER membrane remains a mystery. An increasing body of evidence is accumulating for the participation of membrane proteins in this process. The finding that completed protein domains can be accommodated by the translocation mechanism suggests the participation of additional proteins in the translocation process, either as components of a pore or tunnel which could accommodate folded polypeptide domains or as enzymes to unfold folded domains. In addition, translocation of completed proteins is dependent on nucleoside triphosphate hydrolysis. Thus translocation is not a "spontaneous" process as has been predicted previously (Wickner, 1979; von Heijne and Blomberg, 1979; Engelman and Steitz, 1981). While it is clear that translocation is not simply governed by thermodynamically favorable protein–lipid interactions between the nascent chain and the membrane, the molecular environment traversed by the polypeptide as it enters the ER lumen remains to be determined.

The approach of expressing engineered translocation substrates in cell-free translation/translocation systems has provided many insights into the mechanism of translocation and transmembrane integration and will continue to do so. In addition, future research will undoubtedly focus on the identification, purification, reconstitution, and mechanistic analysis of additional molecular components of the targeting and translocation apparatus. Finally, the importance of the events and components which have been defined and established in cell-free systems remains to be determined *in vivo*. Additional components may be necessary to orchestrate the complex series of molecular interactions resulting in translocation in a living cell.

ACKNOWLEDGMENTS

This work was supported by the National Institutes of Health, Grant GM 31626. We thank Russell Hirsch for helpful criticism of the manuscript.

REFERENCES

Adams, G. A., and Rose, J. K. (1985). Structural requirements of a membrane-spanning domain for protein anchoring and cell surface transport. *Cell* **41,** 1007–1015.

Adelman, M. R., Sabatini, D. D., and Blobel, G. (1973). Ribosome-membrane interaction. Nondestructive disassembly of rat liver rough microsomes into ribosomal and membranous components. *J. Cell Biol.* **56:** 206–229.

Ainger, K. I., and Meyer, D. I. (1986). Translocation of nascent secretory proteins across membranes can occur late in translation. *EMBO J.* **5,** 951–955.

Anderson, D. J., Walter, P., and Blobel, W. (1982). Signal recognition protein is required for the integration of acetylcholine receptor δ subunit, a transmembrane glycoprotein, into the endoplasmic reticulum membrane. *J. Cell Biol.* **92,** 501–506.

Anderson, D. J., Mostov, K. E., and Blobel, G. (1983). Mechanisms of integration of de novo-synthesized polypeptides into membranes: Signal-recognition particle is required for integration into microsomal membranes of calcium ATPase and of lens MP26 but not of cytochrome b_5. *Proc. Natl. Acad. Sci. U.S.A.* **80,** 7249–7253.

Andrews, D. W., Walter, P., and Ottensmeyer, F. P. (1985). Structure of the signal recognition particle by electron microscopy. *Proc. Natl. Acad. Sci. U.S.A.* **82,** 785–789.

Andrews, D. W., Walter, P., and Ottensmeyer, F. P. Evidence for an extended 75L RNA in the signal recognition particle. *EMBO J.* (in press).

Budouelle, H., Bassford, Jr., P. J., Fowler, A. V., Sabin, I., Beckwith, J., and Hofnung, M. (1980). Mutations which alter the function of the signal sequence of the maltose binding protein of *Escherichia coli. Nature (London)* **285,** 78–81.

Bergman, L. W., and Kuehl, W. M. (1979). Formation of an intrachain disulfide bond on nascent immunoglobulin light chains. *J. Biol. Chem.* **254,** 8869–8876.

Bielinska, M., Rogers, F., Rucinsky, T., and Boime, I. (1979). Processing *in vitro* of placental peptide hormones by smooth microsomes. *Proc. Natl. Acad. Sci. U.S.A.* **76,** 6152–6156.

Blobel, G. (1980). Intracellular protein topogenesis. *Proc. Natl. Acad. Sci. U.S.A.* **77,** 1496–1500.

Blobel, G., and Dobberstein, B. (1975a). Transfer of proteins across membranes I. Presence of proteolytically processed and unprocessed nascent immunoglubulin murine myeloma. *J. Cell Biol.* **67,** 835–851.

Blobel, G., and Dobberstein, B. (1975b). Transfer of proteins across membranes II. Reconstitution of functional rough microsomes from heterologous components. *J. Cell Biol.* **67,** 852–862.

Boeke, J. D., and Model, P. (1982). A procaryotic membrane anchor sequence: carboxyl terminus of bacteriophage f1 gene III protein retains it in the membrane. *Proc. Natl. Acad. Sci. U.S.A.* **79,** 5200–5204.

Bos, T. J., Davis, A. R., and Nayak, D. P. (1984). NH_2-terminal hydrophobic region of influenza virus neuraminidase provides the signal function in translocation. *Proc. Natl. Acad. Sci. U.S.A.* **81,** 2327–2331.

Braell, W. A., and Lodish, H. F. (1982). Ovalbumin utilizes an amino-terminal signal sequence. *J. Biol. Chem.* **257,** 4578–4582.

Bretscher, M. S. (1971). Major human erythrocyte glycoprotein spans the cell membrane. *Nature New Biol.* **231,** 229–232.

Bretscher, M. S. (1975). C-Terminal region of the major erythrocyte sialoglycoprotein is on the cytoplasmic side of the membrane. *J. Mol. Biol.* **98,** 831–833.

Briggs, M. S., Cornell, D. G., Dluhy, R. A., and Gierasch, L. M. (1986). Conformations of signal peptides induced by lipids suggest initial steps in protein export. *Science* **233,** 206–208.

Chen, L., and Tai, P. C. (1987). Evidence for the involvement of ATP in co-translational translocation. *Nature (London)* **328,** 164–166.

Coleman, J., Inukai, M., and Inouye, M. (1985). Dual functions of the signal peptide in protein transfer across the membrane. *Cell* **43,** 351–360.

Cutler, D. F., and Garoff, H. (1986). Mutants of the membrane-binding region of Semliki Forest virus E2 protein. I. Cell surface transport and fusogenic activity. *J. Cell Biol.* **102,** 889–901.

Cutler, D. F., Malancon, P., and Garoff, H. (1986). Mutants of the membrane-binding region of Semliki Forest virus E2 protein. II. Topology and membrane binding. *J. Cell Biol.* **102,** 902–910.

Davis, N. G., and Hsu, M. C. (1986). The fusion-related hydrophobic domain of Sendai F protein can be moved through the cytoplasmic membrane of *Escherichia coli. Proc. Natl. Acad. Sci. U.S.A.* **83,** 5091–5095.

Davis, N. G., and Model, P. (1985). An artificial anchor domain: hydrophobicity suffices to stop transfer. *Cell* **47,** 607–614.

Davis, N. G., Boeke, J. D., and Model, P. (1985). Fine structure of a membrane anchor domain. *J. Mol. Biol.* **181,** 111–121.

Deisenhofer, J., Epp, O., Miki, K., Huber, R., and Michel, H. (1985). Structure of the protein subunits in the photosynthetic reaction centre of *Rhodopseudomonas viridis* at 3 Å resolution. *Nature (London)* **318,** 618–624.

Devillers-Thiery, A., Giraudat, J., Bentaboulet, M., and Changeux, J.-P. (1983). Complete mRNA coding sequence of the acetylcholine binding α-subunit of *Torpedo marmorata* acetylcholine receptor: A model for the transmembrane organization of the polypeptide chain. *Proc. Natl. Acad. Sci. U.S.A.* **80,** 2067–2071.

Eble, B., Lingappa, V., and Ganem, D. (1986). Hepatitis B surface antigen: An unusual secreted protein initially synthesized as a transmembrane polypeptide. *Mol. Cell. Biol.* **6,** 1454–1463.

Eble, B. E., MacRae, D. R., Lingappa, V. R., and Ganem, D. (1987). Multiple topogenic sequences determine the transmembrane orientation of hepatitis B surface antigen. *Mol. Cell. Biol.* (in press).

Emr, S. D., and Silhavy, T. J. (1982). Molecular components of the signal sequence that function in the inititation of protein export. *J. Cell Biol.* **95,** 689–696.

Engelman, D. M., and Steitz, T. A. (1981). The spontaneous insertion of proteins into and across membranes: the helical hairpin hypothesis. *Cell* **23,** 411–422.

Evans, E. A., Gilmore, R., and Blobel, G. (1986). Purification of microsomal signal peptidase as a complex. *Proc. Natl. Acad. Sci. U.S.A.* **83,** 581–585.

Friedlander, M., and Blobel, G. (1985). Bovine opsin has more than one signal sequence. *Nature (London)* **318,** 338–343.

Fujiki, Y., Hubbard, A. L., Fowler, S., and Lazarow, P. B. (1982). Isolation of intracellular membranes by means of sodium carbonate treatment: application to endoplasmic reticulum. *J. Cell Biol.* **93,** 97–102.

Gething, M. J., and Sambrook, J. (1982). Construction of influenza hemagluttinin genes that code for intracellular and secreted forms of the protein. *Nature (London)* **300,** 598–603.

Gilmore, R., and Blobel, G., (1983). Transient involvement of signal recognition particle and its receptor in the microsomal membrane prior to protein translocation. *Cell* **35**, 677–685.

Gilmore, R., and Blobel, G. (1985). Translocation of secretory proteins across the microsomal membrane occurs through an environment accessible to aqueous perturbants. *Cell* **46**, 497–505.

Gilmore, R., Blobel, G., and Walter, P. (1982a). Protein translocation across the endoplasmic reticulum. I. Detection in the microsomal membrane of a receptor for the signal recognition particle. *J. Cell Biol.* **95**, 463–469.

Gilmore, R., Walter, P., and Blobel, G. (1982b). Protein translocation across the endoplasmic reticulum. II. Isolation and characterization of the signal recognition particle receptor. *J. Cell Biol.* **95**, 470–477.

Glabe, C. G., Hanover, J. A., and Lennarz, W. J. (1980). Glycosylation of ovalbumin nascent chains. *J. Biol. Chem.* **255**, 9236–9242.

Hansen, W., Garcia, P., and Walter, P. (1986). *In vitro* protein translocation across the yeast endoplasmic reticulum: ATP-dependent posttranslational translocation of prepro-α-factor. *Cell* **45**, 397–406.

Hay, B., Barry, R. A., Lieberburg, I., Prusiner, S. B., and Lingappa, V. R. (1987). Biogenesis and transmembrane orientation of the cellular isoform of the scrapie prion protein. *Mol. Cell. Biol.* **7**, 914–919.

Hay, B., Prusiner, S. B., and Lingappa, V. R. Evidence for a secretory form of the cellular prion protein. *Biochemistry* (in press).

Hortsch, M., Avossa, D., and Meyer, D. I. (1986). Characterization of secretory protein translocation: Ribosome–membrane interaction in endoplasmic reticulum. *J. Cell Biol.* **103**, 241–253.

Jackson, R. C., and Blobel, G. (1977). Posttranslational cleavage of prescretory proteins with an extract of rough microsomes from dog pancreas containing signal peptidase activity. *Proc. Natl. Acad. Sci. U.S.A.* **74**, 5598–5602.

Kadonaga, J. T., Gautier, A. E., Straus, D. R., Charles, A. D., Edge, M. D., and Knowles, J. R. (1984). The role of β-lactamase signal sequence in the secretion of proteins by E. coli. *J. Biol. Chem.* **259**, 2149–2154.

Kadonaga, J. T., Pluckthun, A., and Knowles, J. R. (1985). Signal sequence mutants of β-lactamase. *J. Biol. Chem.* **260**, 16192–16199.

Katz, F. N., and Lodish, H. F. (1979). Transmembrane biogenesis of the vesicular stomatitis virus glycoprotein. *J. Cell Biol.* **80**, 416–426.

Katz, F. N., Rothman, J. E., Lingappa, V. R., Blobel, G., and Lodish, H. F. (1977). Membrane assembly *in vitro:* Synthesis, glycosylation, and asymmetric insertion of a transmembrane protein. *Proc. Natl. Acad. Sci. U.S.A.* **74**, 3278–3282.

Kehry, M., Ewald, S., Douglas, R., Sibley, C., Raschke, W., Fambrough, D., and Hood, L. (1980). The immunoglobulin μ chains of membrane-bound and secreted IgM molecules differ in their C-terminal segments. *Cell* **21**, 393–406.

Kelly, R. B. (1985). Pathways of protein secretion in eukaryotes. *Science* **230**, 25–32.

Kopito, R. R., and Lodish, H. F. (1985). Primary structure and transmembrane orientation of the murine anion exchange protein. *Nature (London)* **316**, 234–238.

Koshland, D., Sauer, R. T., and Botstein, D. (1982). Diverse effects of mutations in the signal sequence on the secretion of β-lactamase in *Salmonella typhimurium*. *Cell* **30**, 903–914.

Kreibich, G., Ulrich, B. L., and Sabatini, D. D. (1978a). Proteins of rough microsomal membranes related to ribosome binding. I. Identification of ribophorins I and II, membrane proteins characteristic of rough microsomes. *J. Cell Biol.* **77**, 464–487.

Kreibich, G., Freienstein, C. M., Pereyra, B. N., Ulrich, B. L., and Sabatini, D. D. (1978b).

Proteins of rough microsomal membranes related to ribosome binding. II. Crosslinking of bound ribosomes to specific membrane proteins exposed at the binding sites. *J. Cell Biol.* **77**, 488–506.

Krieg, U. C., Walter, P., and Johnson, A. E. (1986). Photocross-linking of the signal sequence of nascent preprolactin to the 54 kDa polypeptide of the signal recognition particle. *Proc. Natl. Acad. Sci. U.S.A.* **83**, 8604–8608.

Kurzchalia, T. V., Wiedmann, M., Girshovich, A. S., Bochkareva, E. S., Bielka, H., and Rapoport, T. A. (1986). The signal sequence of nascent preprolactin interacts with the 54K polypeptide of the signal recognition particle. *Nature (London)* **320**, 634–636.

Lauffer, L., Garcia, P. D., Harkins, R. N., Coussens, L., Ullrich, A., and Walter, P. (1985). Topology of signal recognition particle receptor in endoplasmic reticulum membrane. *Nature (London)* **318**, 334–338.

Lin, J. J. C., Kanazawa, H., Ozols, J., and Wu, H. C. (1978). An *Escherichia coli* mutant with an amino acid alteration within the signal sequence of outer membrane prolipoprotein. *Proc. Natl. Acad. Sci. U.S.A.* **75**, 4891–4895.

Lingappa, V. R., Katz, F. N., Lodish, H. F., and Blobel, G. (1978a). A signal sequence for the insertion of a transmembrane glycoprotein. *J. Biol. Chem.* **253**, 8667–8670.

Lingappa, V. R., Shields, D., Woo, S. L. C., and Blobel, G. (1978b). Nascent chicken ovalbumin contains the functional equivalent of a signal sequence. *J. Cell Biol.* **79**, 567–572.

Lingappa, V. R., Lingappa, J. R., and Blobel, G. (1979). Chicken ovalbumin contains an internal signal sequence. *Nature (London)* **281**, 117–121.

Lingappa, V. R., Chaidez, J., Yost, C. S., and Hedgpeth, J. (1984). Determinants for protein localization: β-Lactamase signal sequence directs globin across microsomal membranes. *Proc. Natl. Acad. Aci. U.S.A.* **81**, 456–460.

Lipp, J., and Dobberstein, B. (1986). Signal recognition particle-dependent membrane insertion of mouse invariant chain: A membrane-spanning protein with a cytoplasmically exposed amino terminus. *J. Cell Biol.* **102**, 2169–2175.

McClelland, A., Kuhn, L. C., and Ruddle, F. H. (1984). The human transferrin receptor gene: Genomic organization, and the complete primary structure of the receptor deduced from a cDNA sequence. *Cell* **39**, 267–274.

McCune, J. M., Lingappa, V. R., Fu, S. M., Blobel, G., and Kunkel, H. G. (1980). Biogenesis of membrane-bound and secreted immunoglobulins. I. Two distinct translation products of human μ-chain, with identical N-termini and different C-termini. *J. Exp. Med.* **152**, 463–468.

Marcantonio, E. E., Amar-Costesec, A., and Kreibich, G. (1984). Segregation of the polypeptide translocation apparatus to regions of the endoplasmic reticulum containing ribophorins and ribosomes II. Rat liver microsomal subfractions contain equimolar amount of ribophorins and ribosomes. *J. Cell Biol.* **99**, 2254–2259.

Meek, R. L., Walsh, K. A., and Palmiter, R. D. (1982). The signal sequence of ovalbumin is located near the NH2-terminus. *J. Biol. Chem.* **257**, 12245–12251.

Meyer, D. I. (1985). Signal recognition particle (SRP) does not mediate a translational arrest of nascent secretory proteins in mammalian cell-free systems. *EMBO J.* **4**, 2031–2033.

Meyer, D. I., and Dobberstein, B. (1980a). A membrane component essential for vectorial translocation of nascent proteins across the endoplasmic reticulum: requirements for its extraction and reassociation with the membrane. *J. Cell Biol.* **87**, 498–502.

Meyer, D. I., and Dobberstein, B. (1980b). Identification and characterization of a membrane component essential for the translocation of nascent proteins across the membrane of the endoplasmic reticulum. *J. Cell Biol.* **87**, 503–508.

Meyer, D. I., Krause, E., and Dobberstein, B. (1982a). Secretory protein translocation across membranes: The role of 'docking protein.' *Nature (London)* **297,** 647–650.

Meyer, D. I., Louvard, D., and Dobberstein, B. (1982b). Characterization of molecules involved in protein translocation using a specific antibody. *J. Cell Biol.* **92,** 579–583.

Michel, H., Weyer, K. A., Gruenberg, H., Dunger, I., Oesterhelt, D., and Lottspeich, F. (1986). The 'light' and 'medium' subunits of the photosynthetic reaction centre from *Rhodopseudomonas viridis:* isolation of the genes, nucleotide and amino acid sequence. *EMBO J.* **5,** 1149–1158.

Milstein, C., Brownlee, G. G., Harrison, T. M., and Mathews, M. B. (1972). A possible precursor of immunoglobulin light chains. *Nature New Biol.* **239,** 117–120.

Mize, N. K., Andrews, D. W., and Lingappa, V. R. (1986). A stop transfer sequence recognizes receptors for nascent chain translocation across the endoplasmic reticulum membrane. *Cell* **47,** 711–719.

Moreno, F., Fowler, A. V., Hall, M., Silhavy, T. J., Zabin, I., and Schwartz, M. (1980). A signal sequence is not sufficient to lead β-galactosidase out of the cytoplasm. *Nature (London)* **286,** 356–359.

Morrison, T. G., and Lodish, H. F. (1975). The site of synthesis of membrane and nonmembrane proteins of vesicular stomatitis virus. *J. Biol. Chem.* **250,** 6955–6962.

Mueckler, M., and Lodish, H. F. (1986a). The human glucose transporter can insert posttranslationally into microsomes. *Cell* **44,** 629–637.

Mueckler, M., and Lodish, H. F. (1986b). Posttranslational insertion of a fragment of the glucose transporter into microsomes requires phosphoanhydride bond cleavage. *Nature (London)* **322,** 459–552.

Mueckler, M., Caruso, C., Baldwin, S. A., Panico, M., Blench, I., Morris, H. R., Allard, W. J., Lienhard, G. E., and Lodish, H. F. (1985). Sequence and structure of a human glucose transporter. *Science* **229,** 941–945.

Muller, M., and Blobel, G. (1984). *In vitro* translocation of bacterial proteins across the plasma membrane of *Escherichia coli. Proc. Natl. Acad. Sci. U.S.A.* **81,** 7241–7425.

Muller, G., and Zimmermann, R. (1987). Import of honeybee prepromelittin into the endoplasmic reticulum: structural basis for independence of SRP and docking protein. *EMBO J.* **6,** 2099–2107.

Muller, M., Ibrahimi, I., Chang, C. N., Walter, P., and Blobel, G. (1982). A bacterial secretory protein requires SRP for translocation across the endoplasmic reticulum. *J. Biol. Chem.* **257,** 11860–11863.

Nathans, J., and Hogness, D. S. (1983). Isolation, sequence analysis, and intron–exon arrangement of the gene encoding bovine rhodopsin. *Cell* **34,** 807–814.

Noda, M., Takahashi, H., Tanabe, T., Toyosato, M., Kikyotani, S., Furutani, Y., Hirose, T., Takashima, H., Inayama, S., Miyata, T., and Numa, S. (1983). Structural homology of *Torpedo californica* acetylcholine receptor subunits. *Nature (London)* **302,** 528–532.

Palade, G. (1975). Intracellular aspects of the process of protein synthesis. *Science* **189,** 347–358.

Palmiter, R. D., Gagnon, J., and Walsh, K. A. (1978). Ovalbumin: A secreted protein without a transient hydrophobic leader sequence. *Proc. Natl. Acad. Sci. U.S.A.* **75,** 94–98.

Palva, I., Sarvas, M., Lehtovaara, P., Sibakov, M., and Kaarianen, L. (1982). Secretion of *Escherichia coli* β-lactamase from *Bacillus subtilis* by the aid of α-amylase signal sequence. *Proc. Natl. Acad. Sci. U.S.A.* **79,** 5582–5586.

Perara, E., and Lingappa, V. R. (1985). A former amino terminal signal sequence engineered to an internal location directs translocation of both flanking protein domains. *J. Cell Biol.* **101**, 2292–2301.

Perara, E., Rothman, R. E., and Lingappa, V. R. (1986). Uncoupling translocation from translation: Implications for transport of proteins across membranes. *Science* **232**, 348–352.

Prehn, S., Tsamaloukas, A., and Rapaport, T. A. (1980). Demonstration of specific receptors of the rough endoplasmic membrane for the signal sequence of carp preproinsulin. *Eur. J. Biochem.* **107**, 185–195.

Prehn, S., Nurnberg, P., and Rapoport, T. A. (1981). A receptor for signal segments of secretory proteins in rough endoplasmic reticulum membranes. *FEBS Lett.* **123**, 79–84.

Randall, L. (1983). Translocation of domains of nascent periplasmic proteins across the cytoplasmic membrane is independent of elongation. *Cell* **33**, 231–240.

Randall, L. L., and Hardy, S. J. S. (1986). Correlation of competence for export with lack of the tertiary structure of the mature species: A study *in vivo* of maltose-binding protein in *E. coli. Cell* **46**, 921–928.

Redman, C. M., and Sabatini, D. D. (1966). Vectorial discharge of peptides released by puromycin from attached ribosomes. *Proc. Natl. Acad. Sci. U.S.A.* **56**, 608–615.

Redman, C. M., Siekevitz, P., and Palade, G. (1966). Synthesis and transfer of amylase in pigeon pancreatic microsomes. *J. Biol. Chem.* **241**, 1150–1158.

Rose, J. K., and Bergmann, J. E. (1982). Expression from cloned cDNA of cell-surface secreted forms of the glycoprotein of vesicular stomatitis virus in eurcaryotic cells. *Cell* **30**, 753–762.

Rothblatt, J. A., and Meyer, D. I. (1986a). Secretion in yeast: Reconstitution of the translocation and glycosylation of α-factor and invertase in a homologous cell-free system. *Cell* **44**, 619–628.

Rothblatt, J. A., and Meyer, D. I. (1986b). Secretion in yeast: Translocation and glycosylation of prepro-α-factor *in vitro* can occur via an ATP-dependent posttranslational mechanism. *EMBO J.* **5**, 1031–1036.

Rothman, J. E., and Lodish, H. F. (1977). Synchronized transmembrane insertion and glycosylation of a nascent membrane protein. *Nature (London)* **269**, 775–780.

Rothman, J. E., and Lenard, J. (1977). Membrane asymmetry. *Science* **195**, 743–753.

Rottier, P., Brandenburg, D., Armstrong, J., van der Zeijst, B., and Warren, G. (1984). Assembly *in vitro* of a spanning membrane protein of the endoplasmic reticulum: The E1 glycoprotein of coronavirus mouse hepatitis virus A59. *Proc. Natl. Acad. Sci. U.S.A.* **81**, 1421–1425.

Rottier, P., Armstrong, J., and Meyer, D. I. (1985). Signal recognition particle-dependent insertion of coronavirus E1, an intracellular membrane glycoprotein. *J. Cell Biol.* **260**, 4648–4652.

Sabatini, D. D., and Blobel, G. (1970). Controlled proteolysis of nascent polypeptides in rat liver cell fractions II. Location of the polypeptides in rough microsomes. *J. Cell Biol.* **45**, 146–157.

Sabatini, D. D., Tashiro, Y., and Palade, G. E. (1966). On the attachment of ribosomes to microsomal membranes. *J. Mol. Biol.* **19**, 503–524.

Sabatini, D. D., Kreibich, G., Morimoto, T., and Adesnik, M. (1982). Mechanisms for the incorporation of proteins in membranes and organelles. *J. Cell Biol.* **92**, 1–22.

Schatz, G., and Butow, R. A. (1983). How are proteins imported into mitochondria? *Cell* **32**, 316–318.

Schatzman, R. C., Ivan, G. I., Privalsky, M. L., and Bishop, M. J. (1986). Orientation of the v-*erbB* gene product in the plasma membrane. *Mol. Cell. Biol.* **6,** 1329–1333.

Schlenstedt, G., and Zimmermann, R. (1987). Import of frog prepropeptide GLa into microsomes requires ATP but does not involve docking protein or ribosomes. *EMBO J.* **6,** 699–703.

Schneider, C., Owen, M. J., Banville, D., and Williams, J. G. (1984). Primary structure of human transferrin receptor deduced from the mRNA sequence. *Nature (London)* **311,** 675–678.

Siegel, V., and Walter, P. (1985). Elongation arrest is not a prerequisite for secretory protein translocation across the microsomal membrane. *J. Cell Biol.* **100,** 1913–1921.

Siegel, V., and Walter, P. (1986). Removal of the Alu structural domain from signal recognition particle leaves its protein translocation activity intact. *Nature (London)* **320,** 81–84.

Simon, K., Perara, E., and Lingappa, V. R. (1987). Translocation of globin fusion proteins across the endoplasmic reticulum membrane in *Xenopus laevis* oocytes. *J. Cell Biol.* **104,** 1165–1172.

Singer, P. A., and Williamson, A. R. (1980) Different species of messenger RNA encode receptor and secretory IgM chains differing at their carboxyl termini. *Nature (London)* **285,** 297–299.

Spiess, M., and Lodish, H. F. (1985). Sequences of a second human asialoglycoprotein receptor: Conservation of two receptor genes during evolution. *Proc. Natl. Acad. Sci. U.S.A.* **83,** 6465–6469.

Spiess, M., and Lodish, H. F. (1986). An internal signal sequence: The asialoglycoprotein receptor membrane anchor. *Cell* **44,** 177–185.

Tabe, L., Krieg, P., Strachan, R., Jackson, D., Wallis, E., and Colman, A. (1984). Segregation of mutant ovalbumins and ovalbumin–globin fusion proteins in *Xenopus* oocytes: identification of an ovalbumin signal sequence. *J. Mol. Biol.* **180,** 645–666.

Tajima, S., Lauffer, L., Rath, V. L., and Walter, P. (1986). The signal recognition particle (SRP) receptor is a complex containing two distinct polypeptide chains: Identification of the SRP receptor β-subunit. *J. Cell Biol.* **103,** 1167–1178.

Talmadge, K., Kaufman, J., and Gilbert, W. (1980a). Bacteria mature preproinsulin to proinsulin. *Proc. Natl. Acad. Sci. U.S.A.* **77,** 3988–3992.

Talmadge, K., Stahl, S., and Gilbert, W. (1980b). Eukaryotic signal sequence transports insulin antigen in *E. coli. Proc. Natl. Acad. Sci. U.S.A.* **77,** 3369–3373.

Tanford, C. (1978). The hydrophobic effect and the organization of living matter. *Science* **200,** 1012–1018.

Toneguzzo, F., and Ghosh, H. P. (1977). Synthesis and glycosylation *in vitro* of glycoprotein of vesicular stomatitis virus. *Proc. Natl. Acad. Sci. U.S.A.* **74,** 1516–1520.

Ullrich, A., Coussens, L., Hayflick, J. S., Dull, T. H., Gray, A., Tam, A. W., Lee, J., Yarden, Y., Libermann, T. A., Schlessinger, J., Downward, J., Mayes, E. L. V., Whittle, N., Waterfield, M. D., and Seeburg, P. H. (1984). Human epidermal growth factor receptor cDNA sequence and aberrant expression of the amplified gene in A431 epidermoid carcinoma cells. *Nature (London)* **309,** 418–425.

Vasalli, P., Tedghi, R., Lisowska-Berstein, B., Tartakoff, A., and Jaton, J.-C. (1979). Evidence for hydrophobic region within heavy chains of mouse B lymphocyte membrane-bound IgM. *Proc. Natl. Acad. Sci. U.S.A.* **76,** 5515–5519.

von Heijne, G. (1984). How signal sequences maintain cleavage specificity. *J. Mol. Biol.* **173,** 243–251.

von Heijne, G. (1985). Signal sequences: The limits of variation. *J. Mol. Biol.* **184,** 99–105.

von Heijne, G., and Blomberg, C. (1979). Transmembrane translocation of proteins. The direct transfer model. *Eur. J. Biochem.* **97**, 175–181.

Walter, P., and Blobel, G. (1980). Purification of a membrane-associated protein complex required for protein translocation across the endoplasmic reticulum. *Proc. Natl. Acad. Sci. U.S.A.* **77**, 7112–7116.

Walter, P., and Blobel, G. (1981a). Translocation of proteins across the endoplasmic reticulum II. Signal recognition protein (SRP) mediates the selective binding to microsomal membranes of *in-vitro*-assembled polysomes synthesizing secretory protein. *J. Cell Biol.* **91**, 551–556.

Walter, P., and Blobel, G. (1981b). Translocation of proteins across the endoplasmic reticulum III. Signal recognition protein (SRP) causes signal sequence-dependent and site-specific arrest of chain elongation that is released by microsomal membranes. *J. Cell Biol.* **91**, 557–561.

Walter, P., and Blobel, G. (1982). SRP contains a 7S RNA essential for protein translocation across the endoplasmic reticulum. *Nature (London)* **299**, 691–698.

Walter, P., and Blobel G. (1983a). Disassembly and reconstitution of SRP. *Cell* **34**, 525–533.

Walter, P., and Blobel, G. (1983b). Subcellular distribution of signal recognition particle and 7SL RNA determined with polypeptide-specific antibodies and complementary DNA probe. *J. Cell Biol.* **97**, 1693–1699.

Walter, P., and Lingappa, V. R. (1986). Mechanism of protein translocation across the endoplasmic reticulum. *Annu. Rev. Cell Biol.* **2**, 499–516.

Walter, P., Jackson, R. C., Marcus, M. M., Lingappa, V. R., and Blobel, G. (1979). Tryptic dissection and reconsitution of translocation activity for nascent presecretory proteins across microsomal membranes. *Proc. Natl. Acad. Sci. U.S.A.* **76**, 1795–1799.

Walter, P., Ibrahimi, I., and Blobel, G. (1981). Translocation of proteins across the endoplasmic reticulum. I. Signal recognition protein (SRP) binds to in-vitro-assembled polysomes synthesizing secretory protein. *J. Cell Biol.* **91**, 545–550.

Walter, P., Gilmore, R., and Blobel, G. (1984). Protein translocation across the endoplasmic reticulum. *Cell* **38**, 5–8.

Warren, G., and Dobberstein, M. (1978). Protein transfer across microsomal membranes reassembled from separated membrane components. *Nature (London)* **273**, 569–571.

Watanabe, M., Hunt, J. F., and Blobel, G. (1986). *In vitro* synthesized bacterial outer membrane protein is integrated into bacterial inner membranes but translocated across microsomal membranes. *Nature (London)* **323**, 71–73.

Waters, M. G., and Blobel, G. (1986). Secretory protein translocation in a yeast cell-free system can occur posttranslationally and requires ATP hydrolysis. *J. Cell Biol.* **102**, 1543–1550.

Watson, M. E. E. (1984). Compilation of published signal sequences. *Nucleic Acids Res.* **12**, 5145–5164.

Watts, C., Wickner, W., and Zimmerman, R. (1983). M13 procoat and a pre-immunoglobulin share processing specificity but use different membrane receptor mechanisms. *Proc. Natl. Acad. Sci. U.S.A.* **80**, 2809–2813.

Wickner, W. (1979). The assembly of proteins into biological membranes: the membrane trigger hypothesis. *Annu. Rev. Biochem.* **48**, 23–45.

Wickner, W. T., and Lodish, H. F. (1985). Multiple mechanisms of protein insertion into and across membranes. *Science* **230**, 400–407.

Wiedmann, M., Kurzchalia, T., Hartmann, E., and Rapoport, T. A. (1987). A signal sequence receptor in the endoplasmic reticulum membrane. *Nature (London)* (in press).

Wolfe, P. B., Silver, P., and Wickner, W. (1982). The isolation of homogeneous leader peptidase from a strain of *Escherichia coli* which overproduces the enzyme. *J. Biol. Chem.* **257**, 7898–7902.

Yost, C. S., Hedgpeth, J., and Lingappa, V. R. (1983). A stop transfer confers predictable transmembrane orientation to a previously secreted protein in cell-free systems. *Cell* **34**, 759–766.

Young, E. F., Ralston, E., Blake, J., Ramachandran, J., Hall, Z. W., and Stroud, R. M. (1985). Topological mapping of acetylcholine receptor: Evidence for a model with five transmembrane segments and a cytoplasmic COOH-terminal peptide. *Proc. Natl. Acad. Sci. U.S.A.* **82**, 626–630.

Zerial, M., Melancon, P., Schneider, C., and Garoff, H. (1986). The transmembrane segment of the human transferrin receptor functions as a signal peptide. *EMBO J.* **5**, 1543–1550.

Zerial, M., Huylebroeck, D., and Garoff, H. (1987). Foreign transmembrane peptides replacing the internal signal sequence of transferrin receptor allow its translocation and membrane binding. *Cell* **48**, 147–155.

Zimmermann, R., and Mollay, C. (1986). Import of honeybee prepromelittin into the endoplasmic reticulum. *J. Biol. Chem.* **261**, 12889–12895.

Zuniga, M. C., and Hood, L. E. (1986). Clonal variation in cell surface display of an H-2 protein lacking a cytoplasmic tail. *J. Cell Biol.* **102**, 1–10.

Modification, Maturation, and Transport

2

Role of Carbohydrate in Glycoprotein Traffic and Secretion

JAMES B. PARENT

I. INTRODUCTION

This chapter reviews the evidence that carbohydrate groups may function as markers for intracellular transport and sorting of membrane and secretory glycoproteins in vertebrate cells. The receptor-mediated trafficking to lysosomes of acid hydrolases with oligosaccharide chains containing mannose 6-phosphate residues is the best understood pathway by which proteins translocated into the endoplasmic reticulum are segregated and delivered to their ultimate destination. Evidence for this pathway is presented by Robbins in Chapter 11 of this volume as well as in several recent reviews (Sly, 1982; von Figura and Hasilik, 1986; Sahagian, 1987), and I shall not focus on the mannose 6-phosphate pathway in this chapter. However, there is increasing evidence that carbyhydrate groups

51

may also be involved in the intracellular transport and/or sorting of specific plasma membrane and secretory proteins, and I shall review the literature relevant to this topic.

Several experimental approaches have given evidence that protein-bound glycan chains may have important roles in regulating intracellular glycoprotein traffic and will be discussed in this chapter. An exiciting recent approach is the use of site-directed mutagenesis of the coding sequence to insert or delete glycosylation sites in recombinant proteins. Studies of mutant cells have shown that glycosylation defects often account for a significant fraction of the mutants that are isolated after selection for absence of cell surface receptors. Numerous studies have used tunicamycin, an inhibitor of synthesis of N-asparagine-linked glycan chains, to evaluate the functional role(s) of oligosaccharide chains in glycoprotein traffic in cells. Drugs that inhibit specific steps in the major processing pathway of N-linked glycan chains have recently become available, and these drugs have been used in studies to evaluate the potential involvement of oligosaccharide intermediates in glycoprotein traffic. A number of carbohydrate-binding proteins or lectins have been isolated from vertebrate cells, and I shall also present suggestive evidence that some of these lectins may have a role in regulating glycoprotein traffic.

The interpretation of the results of studies in which carbohydrate-deficient proteins are synthesized by cells owing to either mutations or drug treatment is often difficult because in specific cases unglycosylated proteins have been shown to be more sensitive to proteases, less soluble, or abnormally folded. Therefore, in order to evaluate in the proper context the evidence that carbyhydrate groups have a direct role in regulating the intracellular traffic of specific glycoproteins, a brief overview is presented of the structure and synthesis of protein-linked glycans and the influence of glycan chains on protein solubility, structure, and stability against proteolytic degradation.

II. EVIDENCE FOR INTRACELLULAR TRANSPORT SIGNALS

Secretory proteins are synthesized on ribosomes attached to the rough endoplasmic reticulum (RER), and following vectorial discharge into the lumen of the RER they are transported in vesicles from the RER to the Golgi apparatus and finally in secretory vesicles to the cell surface (Palade, 1975; Tartakoff, 1983a; Farquhar and Palade, 1981; Farquhar, 1985). A simple model for the intracellular transport of secretory proteins is the following: The luminal volume of the RER is continually being packaged into transport vesicles for shipment first to the Golgi and then to the cell surface. Since plasma membrane proteins are also synthesized on the RER there would be a parallel flow of both secretory and plasma membrane proteins in vesicles to the cell surface. This nonspecific, passive-flow model for protein transport predicts that, since all secretory proteins are following the same pathway, the kinetics of secretion of all proteins in a single cell will be identical. Likewise, the model predicts that the kinetics of transport to the cell surface of all plasma membrane proteins will be identical. Although passive-flow models may explain the intracellular transport of some proteins (Cohen *et al.*, 1979; Cohen and Phillips, 1980), there is increasing evidence that the transport of most secretory and membrane proteins is highly regulated and that structural determinants on these proteins function as markers for intracellular transport.

Strous and Lodish (1980) were the first to report that, in the same cell, different secretory proteins are secreted at markedly different rates. They found, using cultured rat hepatoma cells, that newly synthesized albumin appeared in the medium roughly twice as fast as transferrin. These results were consistent with previous *in vivo* studies that found that albumin is secreted faster by liver than is transferrin (Morgan and Peters, 1971; Schreiber *et al.*, 1979). Asynchronous export of secretory proteins has been demonstrated in human and mouse hepatoma cell lines (Lodish *et al.*, 1983; Ledford and Davis, 1983), cultured rat hepatocytes (Fries *et al.*, 1984), and in exocrine pancreas (Scheele and Tartakoff, 1985). Export of nine secretory proteins by the human hepatoma cells in tissue culture fell into three discrete kinetic classes: (1) a rapidly secreted class with an intracellular retention half-time of 30–40 min (albumin, fibronectin, α-fetoprotein, and α_1-antitrypsin), (2) an intermediate secreted class with a half-time of 75–80 min (ceruloplasmin, α_2-macroglobulin, and plasminogen), and (3) a slowly secreted class with a half-time of 110–120 min (fibrinogen and transferrin) (Parent *et al.*, 1985).

Multiple secretory rates in single cells suggest either that there are multiple secretory pathways or that secretory proteins follow one common pathway through the cell but at different rates. Both mechanisms, however, imply that transport of secretory proteins is a selective rather than a passive-flow process and that specific receptors or carriers regulate intracellular transport of secretory proteins.

Similar results have also been reported for the kinetics of intracellular transport of newly synthesized membrane proteins. Fitting and Kabat (1982) have analyzed the synthesis and intracellular transport of two plasma membrane glycoproteins encoded by murine leukemia virus and they found that, although one viral membrane protein (gp93) was transferred quantitatively into plasma membranes of virus-infected cells within 60 min of [^{35}S]-methionine incorporation, the other glycoprotein (gp70) was still accumulating in the plasma membrane after 210 min. Williams *et al.* (1985) reported that the rates of transport of two closely related membrane glycoproteins (class 1 histocompatability antigens H-2 and H-2D) to the cell surface were remarkably different (i.e., 1 hr versus 4–5 hr). They found by subcellular fractionation that the slowly transported protein accumulated in the endoplasmic reticulum (ER). These results suggest that intracellular transport of membrane proteins is also a selective rather than a passive-flow process.

Lodish and associates reported that differences in rates of serum protein secretion by hepatoma cells were due to variability in rates of transport from the RER to the Golgi; that is, retention in the RER was primarily responsible for the overall rates of secretion. They postulated that one or more receptor proteins in the RER membrane regulates the selective transport of secretory proteins into transport vesicles en route to the Golgi (Strous and Lodish, 1980; Lodish *et al.,* 1983). Fitting and Kabat (1982) reached similar conclusions in a study of two viral glycoproteins that mature from the RER to the cell surface at different rates. In these investigations the primary assay used to measure the rate of protein transport from the RER to the Golgi depended on measurement of the time required for N-linked glycan chains to become resistant to digestion by endo-β-N-acetylglucosaminidase H (endo H). The acquisition of endo H resistance, however, does not monitor the entry of secretory glycoproteins into the Golgi since oligosaccharides become endo H resistant after processing by α-mannosidase II, an event believed to occur in the medial Golgi compartment (Dunphy and Rothman, 1985). Therefore glycoproteins may reside in the cis and medial Golgi compartments for some unknown period before becoming resistant to endo H digestion. K. T. Yeo *et al.* (1985) employed three experimental approaches, i.e., cellular fractionation, acquisition of resistance of glycan chains to endo H digestion, and binding of glycoproteins to immobilized plant lectins, to measure the rates of glycoprotein transport within the RER and Golgi of hepatoma cells. They found that the variable export rates of glycoproteins from hepatoma cells reflect heterogeneous rates of glycoprotein transport within both the RER and the Golgi.

If transport receptors with varying affinities recognize specific "markers" on different membrane proteins, then it should be possible to isolate mutants defective in cell surface expression of specific membrane proteins and not others. Several such mutants have been described. For example, studies of Thy-1 antigen synthesis in mutant mouse lymphoma cells have established that specific oligosaccharide structures are essential for maturation of Thy-1 antigen and a selected group of other glycoproteins to the cell surface (Trowbridge *et al.,* 1978a; Chapman *et al.,*

1980). Fitting and Kabat (1982) immunoselected infected cell mutants with defects in the surface expression of either one or both plasma membrane glycoproteins encoded by murine leukemia virus. In addition, treatment of cells with drugs that inhibit either the synthesis or the normal processing of N-linked glycan chains has been shown to inhibit the appearance at the cell surface of specific membrane proteins, but not others (see Sections VII and VIII).

Two types of secretory routes have been identified in cells i.e., regulative and constitutive pathways (Tartakoff and Vassalli, 1978; Kelly, 1985). In the first, which occurs in many exocrine and endocrine cells, newly synthesized secretory proteins are concentrated and stored in intracellular granules awaiting specific signaling for secretion. In the second, which occurs in most cell types including fibroblasts, muscle cells, and hepatocytes, there is an apparently continuous and nonconcentrative exocytosis of secretory products. Sorting of secretory proteins into different vesicles has been demonstrated in a pituitary tumor cell line, which has both regulative and constitutive pathways (Gumbiner and Kelly, 1982; Moore *et al.*, 1983). However, it is presently unknown if similar sorting mechanisms also function in cells, like hepatocytes, which lack regulative pathways. Albumin and transferrin have been localized in the same intracellular vesicles of hepatoma cells, suggesting a common pathway (Strous *et al.*, 1983). Yokota and Fahimi (1981), however, found that albumin and lipoprotein are separately packaged in the ER and the cis Golgi but are intermixed in vacuoles of the trans Golgi consistent with multiple intracellular transport pathways in hepatocytes.

III. OLIGOSACCHARIDE BIOSYNTHESIS

The carbohydrate chains of glycoproteins may be classified according to the type of linkage to the polypeptide backbone. N-Glycosidic chains are attached to the amide group of asparagine, whereas O-glycosidic chains are linked to the hydroxyl group of such amino acid residues as serine, threonine, and hydroxylysine (for recent reviews, see Montreuil, 1980, 1982, 1987; Kornfeld and Kornfeld, 1980; Kobata, 1984). N-Glycosidic chains are divided into two major classes: the complex-type chains containing *N*-acetylglucosamine (GlcNAc), mannose (Man), galactose (Gal), fucose (Fuc), and sialic acid (SA) and the high-mannose-type chains containing *N*-acetylglucosamine residues and variable amounts of mannose. The O-linked chains commonly contain *N*-acetylgalactosamine, galactose, fucose, and sialic acid.

The number of different oligosaccharide species that have been identified on vertebrate glycoproteins is large and numbers in the hundreds. Synthesis of this diverse array of structures requires a considerable investment of a cell's total biosynthetic capacity, and at least 9 sialyltransferases, 7 fucosyltransferases, 9 galactosyltransferases, and 14 N-acetylglucosaminyltransferases are required to account for biosynthesis of linkages commonly observed in mammalian glycoproteins (Beyer and Hill, 1982). The large diversity of oligosaccharide species found on vertebrate glycoproteins suggests that protein-linked glycans have important roles in physiological processes that require a large information content.

Although individual cells can synthesize many different protein-linked glycan chains, the process is quite specific. It is controlled so that the oligosaccharide chains at an individual glycosylation site on most glycoproteins have one or a small number of closely related structures (Swiedler et al., 1983, 1985). This implies that for many glycoproteins the specific structure of the oligosaccharide chain is important for the normal *in vivo* function of the glycoprotein.

The major pathway for biosynthesis of both complex and high-mannose N-linked glycans proceeds via a common high-mannose lipid-linked intermediate with the structure $Glc_3Man_9GlcNAc_2$-PP-dolichol (Fig. 1). The assembly of this dolichol-linked high-mannose oligosaccharide takes place in a series of reactions referred to as the lipid pathway (for reviews, see Snider, 1984; Krag, 1985; Kornfeld and Kornfeld, 1985). The first step in this pathway is the transfer of GlcNAc-1-P from UDPGlcNAc to dolichol phosphate (Dol-P) to form Dol-PP-GlcNAc. One GlcNAc, nine Man, and three Glc residues are then added by means of the corresponding sugar nucleotides or the glycolipid intermediates Dol-P-Man and Dol-P-Glc, resulting in the formation of the lipid-linked high-mannose oligosaccharide $Glc_3Man_9GlcNAc_2$. After assembly, the $Glc_3Man_9GlcNAc_2$ oligosaccharide is transferred en bloc to the nascent polypeptide chain, which is transported across or inserted into the RER membrane. Glycan transfer is carried out by oligosaccharide transferase which recognizes the acceptor amino acid asparagine (Asn) in the tripeptide sequence Asn-X-Ser (Thr), where X cannot be proline.

The $Glc_3Man_9GlcNAc_2$ oligosaccharide is the precursor of both high-mannose and complex-type oligosaccharides. In order for corresponding conversions to take place, a number of trimming reactions are required. α-Glucosidases I and II remove sequentially the outermost α-1,2-linked glucose and the two innermost α-1,3-linked glucose residues. α-Mannosidase activities then remove the four α-1,2-linked mannose residues and, after addition of GlcNAc to the $Man_5GlcNAc_2$ oligosaccharide, the α-1,3- and α-1,6-linked mannose residues to create the "core" $Man_3GlcNAc_2$ oligosaccharide intermediate. Until recently, all of the known oligosaccharide-trimming reactions involved the removal of one sugar residue in a

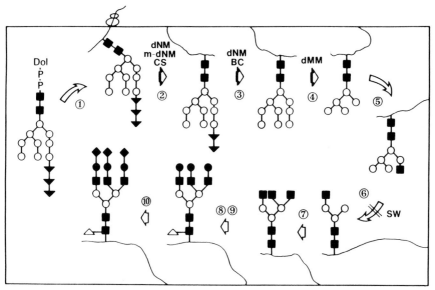

Fig. 1. Proposed sequence for the processing of peptide-bound N-linked glycans. Sites of inhibition by the following drugs are indicated by lines through arrows: dNM, 1-Deoxynorjirimycin; m-dNM, N-methyl-1-deoxynorjirimycin; CS; castanospermine; BC, bromoconduritol; dMM, 1-deoxymanojirimycin; SW, swainsonine. The reactions are catalyzed by the following enzymes: (1) oligosaccharyltransferase, (2) RER α-glucosidase I, (3) RER α-glucosidase II, (4) RER α-1,2-mannosidase, Golgi α-mannosidases 1A and 1B, (5) Golgi N-acetylglucosaminyltransferase I, (6) Golgi α-mannosidase II, (7) N-acetylglucosaminyltransferases II and IV, (8) fucosyltransferase, (9) galactosyltransferase, (10) sialyltransferase. Symbols: ■, N-acetylglucosamine; ○, mannose; ▲, glucose; △, fucose; ●, galactose; ▲, sialic acid. Adapted from Kornfeld *et al.* (1978).

single step. However, Lubas and Spiro (1987) reported that rat liver membranes contain a novel enzyme that has the capacity to trim Glc$_1$Man$_9$GlcNAc$_2$ to Man$_8$GlcNAc$_2$ by the removal of the disaccharide Glcα1-3Man. After oligosaccharide trimming, other sugars such as N-acetylglucosamine, galactose, sialic acid, and fucose may be added sequentially by the corresponding glycosyltransferases to form a glycan of the complex type.

The first processing reactions, the removal of two or three glucose residues and perhaps also removal of a single α-linked mannose residue from the Glc$_3$Man$_9$GlcNAc$_2$ oligosaccharide, occur in the RER. However, the bulk of the oligosaccharide processing reactions occur in the Golgi, and recent studies have shown that processing enzymes are localized in specific subcompartments of the Golgi stack (for recent reviews, see Dunphy and Rothman, 1985; Farquhar, 1985; Hirschberg and Snider, 1987). Using immunocytochemical techniques and thin sections of HeLa cells,

Roth and Berger (1982) showed that galactosyltransferase was localized in only two or three trans cisternae of the Golgi stacks, and Dunphy *et al.* (1985) have localized *N*-acetylglucosaminyltransferase I to the medial Golgi cisternae. It is currently thought that the sequential processing reactions of N-linked oligosaccharides occur during the stepwise transfer of glycoproteins from their site of initial assembly in the RER, through subcompartments of the Golgi apparatus, to their final destinations at the cell surface, extracellular space, or intracellular organelles. Therefore, intermediates in the N-linked oligosaccharide processing pathway are appropriate markers to follow the stepwise transport of glycoproteins through subcellular compartments (Rotundo, 1984; K. T. Yeo *et al.*, 1985; Gabel and Bergmann, 1985).

An unusual feature of the biosynthetic pathway for N-linked oligosaccharides is the specific trimming of nine sugar residues from the precursor oligosaccharide before transfer of additional GlcNAc, galactose, sialic acid, and fucose residues to form complex structures. This appears wasteful unless the oligosaccharide intermediates have important roles within the cell. Recent studies using specific inhibitors of α-glucosidases I and II have demonstrated that the trimming of three glucose residues in the ER is required for efficient transport of some secretory, plasma membrane, and lysosomal enzymes from the ER to the Golgi apparatus (Gross *et al.*, 1983a; Lodish and Kong, 1984; Lemansky *et al.*, 1984; Repp *et al.*, 1985; Parent *et al.*, 1986).

Although $Glc_3Man_9GlcNAc_2$ appears to be the major oligosaccharide transferred from lipid to protein, there is also evidence for alternative pathways for N-linked glycosylation (for a complete discussion, see Krag, 1985). This is an important point because if all N-linked oligosaccharide chains are processed along an identical pathway in the ER and cis Golgi then it is unlikely that early oligosaccharide intermediates in the pathway would have any role in routing glycoproteins to different cellular locations. However, under normal culture conditions, Chinese hamster ovary (CHO) cells synthesize low levels of lipid-linked $Glc_3Man_5GlcNAc_2$ in addition to the predominant $Glc_3Man_9GlcNAc_2$ species which raises the possibility that these cells have a second glycosylation pathway (Chapman *et al.*, 1979). Two mutant cell lines, class E Thy-1$^-$ lymphoma cells (Chapman *et al.*, 1980) and B4-2-1 CHO cells (Stoll *et al.*, 1982,) are known to transfer $Glc_3Man_5GlcNAc_2$ to protein and subsequently process this species to form complex-type oligosaccharides. Similar results are obtained in glucose-deprived CHO cells (Rearick *et al.*, 1981) and in cells treated with an inhibitor of oxidative phosphorylation (Datema and Schwarz, 1981). Yamashita *et al.* (1983) have presented evidence that the sugar chains of ovomucoid and ovalbumin are synthesized by using different lipid intermediates, and they suggest that lipid-linked $Glc_3Man_5GlcNAc_2$ may be the oligosaccharide donor for ovomucoid whereas lipid-linked $Glc_3Man_9GlcNAc_2$ is the donor for ovalbumin. Since both ovalbumin and ovomucoid are synthesized in the chicken oviduct these results imply that glycan chains on different proteins in the same cells may be processed by different pathways.

Until recently it was generally accepted that the presence of glucose in the oligosaccharide precursor is required for transfer in mammalian cells. However, Romero and Herscovics (1986) have demonstrated the transfer of a nonglucosylated oligosaccharide ($Man_7GlcNAc_2$) from dolichol pyrophosphate to protein in F9 cells, a mammalian cell line. Parodi *et al.* (1983a) have shown that glucose-free oligosaccharides are transferred in the N-glycosylation of proteins in the protozoa Trypanosoma cruzi, and this mechanism has also been described for yeast mutants which are blocked in the biosynthesis of dolichol pyrophosphate oligosaccharides (Huffaker and Robbins, 1983; Runge *et al.*, 1984).

An enzyme has been found in Triton-treated rat liver Golgi membranes which trims $Glc_1Man_9GlcNAc_2$ to $Man_8GlcNAc_2$ with the releases of $Glc\alpha1$-3Man (Lubas and Spiro, 1987). By removing a glucosylmannose disaccharide this endo-α-D-mannosidase provides a processing route alternative to the sequential actions of α-glucosidase 11 and α-mannosidase 1. These authors suggest that removal of this dissaccharide from N-linked glycans in the Golgi could serve as a signal in directing the intracellular movement and sorting of certain proteins.

Trimming of the four α-1,2-linked mannose residues from $Man_9GlcNAc_2$ oligosaccharides can be accomplished by two Golgi enzymes, α-mannosidases 1A and 1B (Tabas and Kornfeld, 1978; Tulsiani *et al.*, 1982). These two enzymes are present in comparable activities in membranes enriched in Golgi enzymes and are quite similar in substrate specificity and response to inhibitors. However, they are clearly distinguishable on the basis of behavior on cellulose phosphate chromatography, thermolability, and activity toward $Man_5GlcNAc$ and *p*-nitrophenylmannoside. As Tulsiani *et al.* (1982) have pointed out, the existence of two distinct Golgi enzymes, both of which can trim α-1,2-linked mannose residues from oligosaccharide chains, raises the possibility that different processing routes occur within the Golgi apparatus, leading to different classes of glycoproteins and/or different subsequent routing. Interestingly, mannose trimming of oligosaccharides destined to become high-mannose-type structures on mature μ chains of immunoglobulin M (IgM) was found to be much more rapid than mannose trimming on oligosaccharides destined to become complex-type structures (Brown and Hickman, 1986).

Additional evidence for multiple pathways are the findings that glucose can be directly transferred from UDPglucose to N-linked glycan chains in thyroid and liver tissue samples and that the reaction occurs in liver RER membrane fractions (Ronin and Caseti, 1981; Parodi *et al.*, 1983,a,b, 1984). Parodi *et al.* (1983a,b) have suggested that transient glucosylation of N-linked carbohydrates might be responsible for routing glycoproteins to specific cellular locations.

What factors specify the structure of the glycan at specific sites on a protein? Unlike the case of template-directed synthesis of proteins and nucleic acids, N-linked glycan chains are synthesized by the sequential activities of specific glycosyl transferases and glycosidases. Since single cells can synthesize a diverse array of glycans, the information which determines the oligosaccharide structure at an individual glycosylation site on a glycoprotein must be specified, in large part, by the primary structure of the protein. This implies that even if oligosaccharide chains function as transport markers for glycoprotein traffic in cells, perhaps by interacting with specific carbohydrate-binding receptors or carrier proteins, the information which specifies the transport marker must ultimately reside in the protein and gene which encodes the protein.

This argument is supported by recent studies on the specific recognition of lysosomal enzymes by UDP-*N*-acetylglucosamine–lysosomal-enzyme *N*-acetylglucosaminephosphotransferase (GlcNAc-P-transferase). This enzyme is responsible for the transfer of GlcNAc-1-P to selected Man residues on lysosomal enzymes. This reaction is the first in the two-step synthesis of the Man-6-P recognition marker for segregation of acid hydrolases into lysosomes (Kornfeld and Kornfeld, 1985; von Figura and Hasilik, 1986). It has been shown that GlcNAc-P-transferase selectively phosphorylates lysosomal enzymes, being able to recognize nonlysosomal glycoproteins only at very high concentrations (Reitman and Kornfeld, 1981a,b; Waheed *et al.*, 1982). Kornfeld and co-workers found that the determinant that allows the selective recognition of lysosomal enzymes by this enzyme was on the protein rather than the oligosaccharide portion of the lysosomal enzyme molecule, because deglycosylated enzymes (i.e., endo H treated) proved to be specific inhibitors of phosphorylation of intact enzymes (Lang *et al.*, 1984). Moreover, the intact protein may be required because even large proteolytic fragments did not retain the ability to be recognized.

Evidence that the primary structure of proteins has an important role in specifying the processing of N-linked chains to complex or high-mannose forms was obtained by analyzing the types of oligosaccharides present on closely related viral glycoproteins isolated from the same host cells. Since the virus utilizes the processing enzymes of the host cell, any difference in oligosaccharide structure should reflect differences in the viral glycoproteins. Differences in the processing of N-linked glycan chains on viral glycoproteins to complex or high-mannose forms were observed when different strains of either influenza virus or vesicular stomatitis virus (VSV) were grown in the same host cell (Nakamura and Compans, 1979; Schwarz and Klenk, 1981; Hunt et al., 1983) and when closely related murine leukemia viruses were grown in a number of hosts under various growth conditions (Rosner et al., 1980).

Evidence that the primary structure of proteins can also influence the finer aspects of N-linked glycan structure comes from the work of Swiedler et al. (1985), using a murine B cell lymphoma. This cell line simultaneously synthesizes two classes of major histocompatability antigens that, within each class, share a high degree of amino acid sequence homology and possess N-linked glycosylation sites at invariant positions. They found that the specific patterns of sialylation and branching at individual glycosylation sites of these structurally related glycoproteins were unique, suggesting that subtle differences in peptide structure have an important influence on both branching and sialylation of N-linked oligosaccharides. Shears and Robbins (1986) studied the expression and glycosylation of the chicken ovalbumin gene in a heterologous cell, the mouse L cell. Mature chicken ovalbumin is heterogeneous with respect to its oligosaccharide chains with approximately equal amounts of high-mannose and hybrid forms. Hybrid chains have more than three α-linked mannose residues in addition to terminal GlcNAc, Gal, and sialic acid residues, and ovalbumin represents a rare class of glycoprotein containing large amounts of hybrid chains. Interestingly, chicken ovalbumin secreted by mouse L cells is processed predominantly to hybrid structures, suggesting that it is the polypeptide chain of ovalbumin which is responsible for processing of a substantial fraction of the oligosaccharide side chains to hybrid structures.

Although much work has focused on the glycosylation of both secretory and cell surface glycoproteins, comparatively little is known about the glycosylation of resident glycoconjugates in subcellular organelles. Recently two groups have described a novel protein–saccharide linkage, O-linked GlcNAc (Torres and Hart, 1984; Schindler et al., 1987). Most of these structures are on a 62-64,000 kDa protein, which is a major component of the nuclear pore complex (Davis and Blobel, 1986; Schindler et al., 1987). Schindler and Hogan (1984) have postulated that glycosylation with O-linked GlcNAc monosaccharide residues could serve to direct the transport of proteins to the nucleus in a manner analogous to the targeting of lysosomal enzymes to lysosomes via the attachment of mannose 6-phosphate residues. However, the finding that O-linked GlcNAc moieties are also found in other subcellular compartments may argue against this model (Holt and Hart, 1986).

IV. ROLE OF CARBOHYDRATE IN PROTEIN SOLUBILITY, STRUCTURE, AND STABILITY

The presence of glycan chains can have a great effect on the solubility properties of glycoproteins (Tarentino et al., 1974). Many proteins function in aqueous environments, and this environment tends to orient hydrophobic residues in the interior of the protein and hydrophilic residues on the exterior. Oligosaccharide chains can substantially contribute to the net hydrophilicity on the exterior surface of proteins, and this influences protein solubility. The most hydrophilic sugar residues are those with a net ionic charge at physiological pH. Each sialic acid carries one net negative charge, and therefore the presence of N-linked or O-linked oligosaccharide chains containing sialic acid can substantially affect the net charge on a glycoprotein (Schauer, 1985). The effect of covalently attached carbohydrate on the solubility of a number of proteins was examined by Lawson et al. (1983), using polyethylene glycol precipitation. They found both increases and decreases in protein solubility depending on the state of glycosylation, the type of protein, and temperature. For example, ribonuclease A (nonglycosylated) was less soluble than ribonuclease B (one N-glycan chain containing $Man_6GlcNAc_2$) at temperatures between 0° and 30°C, but at higher temperatures the B form was slightly less soluble.

Protein conformation may be altered by the presence of oligosaccharide chains. However, the relative importance of carbohydrate chains on the conformation and folding of glycoproteins is unclear. Enzymatic removal of carbohydrate from glycoprotein enzymes usually has little or no effect on catalytic efficiency of the enzyme, and this has been demonstrated for numerous enzymes, including RNase B, DNase I, invertase, carboxypeptidase Y, and mung bean nuclease (Tarentino et al., 1974; Trimble and Maley, 1977).

Since N-linked glycosylation of glycoproteins is an early event and often occurs on nascent chains in the RER (Bergman and Kuehl, 1982), it is possible that N-linked oligosaccharide side chains may influence proper folding of nascent proteins. This would explain why removal of oligosaccharide units from the mature protein, which has already achieved its correct conformation, might have few deleterious effects. In recent years, however, a number of eukaryotic glycoproteins have been produced in Escherichia coli, using recombinant DNA technology, and many of these proteins are fully active despite the absence of glycosylation, which suggests that glycan chains are not required for proper folding of these proteins. Examples include fibroblast interferon (Derynck *et al.*, 1980), immune interferon (Grey *et al.*, 1982; Grey and Goeddel, 1983), plasminogen activator (Pennica *et al.*, 1983), α_1-antitrypsin (Courtney *et al.*, 1984), urokinase (Gunnzler *et al.*, 1984, and interleukin 2 (Svedersky *et al.*, 1984).

The influence of carbohydrate on the conformation and refolding of pancreatic ribonuclease has been extensively studied. Using various spectral properties (circular dichroism, optical rotatory dispersion, and UV difference spectroscopy), Puett (1973) concluded that the secondary and tertiary structures of glycosylated and nonglycosylated bovine RNase are identical. Of the known ribonucleases, the procine enzyme has the highest reported content of carbohydrate, around 35%. In a study comparing glycosylated and nonglycosylated porcine RNase, Wang and Hirs (1977) found that carbohydrate was without influence on the rate-limiting steps in the refolding of the protein from the denatured state or on the overall configurational stability of the molecule, but did influence the environment around two tyrosine side chains to effect a stabilization of the surface structure. The effects observed were relatively small, however, and two recent studies have also reported that the mechanism of folding of pancreatic ribonuclease is independent of covalently linked carbohydrate (Krebs *et al.*, 1983; Grafl *et al.*, 1987). Apparently the information for the folding of ribonuclease and perhaps most other glycoproteins is contained exclusively in the protein moiety, i.e., in the amino acid sequence.

Chu *et al.* (1978) found that the presence of carbohydrate on yeast invertase stabilized the enzyme to denaturation by mild acid, heat, or repeated cycles of freezing and thawing. Removal of the oligosaccharides of carboxypeptidase Y by digestion with endo H doubled the rate of inactivation by treatment with sodium dodecyl sulfate as compared to the glycosylated enzyme (Chu and Maley, 1982). Assuming both forms bound the same amount of detergent, this result suggests the carbohydrate may help maintain the active conformation of this enzyme.

The effects of depleting an envelope glycoprotein of carbohydrate has been extensively studied with the G protein of vesicular stomatitis virus (VSV) (Gibson *et al.*, 1979, 1980, 1981; Leavitt *et al.*, 1977). The nonglycosylated G protein of VSV (San Juan) is not transported to the cell surface and aggregates in the ER when virus is grown in cells treated with tunicamycin (TM) either at the normal temperature (38°C) or at a lower temperature (30°C). Nonglycosylated G protein from another VSV strain (Orsay), however, was partially transported to the cell surface at the lower temperature. A similar strain difference was found using *in vitro* assays designed to measure the solubility on nonionic detergent and the aggregation of nonglycosylated G proteins. The aggregation was most dramatic for the G protein (San Juan), but was also evident for the G protein (Orsay) at the higher temperature (Gibson *et al.*, 1979). This strain difference was also noted in assays for aggregation of G proteins isolated from VSV-infected mutant cell lines in which a truncated core oligosaccharide was transferred to the polypeptide chain (Gibson *et al.*, 1981). The G proteins of the San Juan and Orsay virus strains are closely related, but they contain 10 amino acid differences (Gallione and Rose, 1983). These amino acid differences must be responsible for the variation in the carbohydrate requirement for solubility.

The results of these studies using two VSV strains have been interpreted to indicate that carbohydrate has a significant influence on the folding and conformation of G proteins (Gibson *et al.*, 1980). It should be pointed out, however, that alterations in the folding and conformation of G proteins were not directly demonstrated in these studies. Rather, the influence of carbohydrate on the aggregation and solubility properties of G proteins from the two VSV strains was demonstrated. Since carbohydrate chains per se can have a large effect on the temperature-dependent solubility properties of glycoproteins, it is not clear from the results of these studies if carbohydrate significantly alters the conformation of G proteins.

Perhaps one reason why glycan chains do not appear to have a major effect on the conformation of many glycoproteins is that most glycan chains are located in or near β turns or other turn or loop structures in the protein. Beeley (1977) applied methods for predicting peptide chain conformation to amino acid sequences adjacent to the carbohydrate attachment sites of glycoproteins containing N-linked oligosaccharide chains. Of 31 glycosylated residues examined, 30 occur in sequences favoring turn or loop structures. Twenty-two of the glycosylated asparagine residues occur in tetrapeptides predicted to have the β-turn conformation. Similar conclusions were drawn in another study of 9 O-glycosidic linkages and 28 N-glycosidic linkages (Aubert *et al.*, 1976). For the O-glycosidic linkages, the serine or threonine residues that were involved in the linkages were predicted to be in a β turn. Nineteen of 28 N-glycosidic linkages occur in sequences favoring β turns. The N-glycosidic linkages that were not located in β turns were situated in a random region near a β turn, probably exposed to the solvent or near the carboxy-terminal end.

There are numerous studies that suggest that the presence of oligosaccharide chains on a protein may decrease its susceptibility to proteases (for a recent review, see Olden *et al.*, 1982). These studies include both comparisons of loss of activity during treatments with proteases *in vitro* and comparisons of turnover rates *in vivo* after treatment of cells with inhibitors of glycosylation. The reason for the increased susceptibility of nonglycosylated protein probably varies with the protein. One reason, however, may be the decreased protection from proteases of bends (β turns) in the polypeptide backbone believed to occur at glycosylation sites in glycoproteins (Bause, 1983). In some cases the rapid degradation of nonglycosylated secretory proteins in TM-treated cells may be due to the improper compartmentalization of lysosomal proteases with secretory proteins rather than to the inherent sensitivity of nonglycosylated proteins to proteolytic attack. TM treatment of cells interferes with the Man-6-P-directed transport of acid hydrolases to lysosomes, and we found that chicken embryo fibroblasts treated with TM secrete lysosomal cathepsin B-like activity (Parent *et al.*, 1982). Examples of proteins known to have an increased susceptibility to proteases in their nonglycosylated form include hemagglutinin precursor of influenza virus (Schwarz *et al.*, 1976), fibronectin in chick embryo fibroblasts (Bernard *et al.*, 1982; Hynes and Yamada, 1982), adrenocorticotropin–endorphin precursor in toad pituitaries (Loh and Gainer, 1978, 1979, 1980), alkaline phosphatase in murine cells (Firestone and Heath, 1981), and carboxypeptidase Y of yeast cells (Chu and Maley, 1982).

V. EVIDENCE FOR CARBOHYDRATE TRANSPORT SIGNALS USING SITE-DIRECTED MUTAGENESIS

The laboratory of John Rose has recently developed a novel experimental approach to study the role of carbohydrate in glycoprotein traffic by the use of site-directed mutagenesis of the coding sequence either to insert or to delete glycosylation sites in cloned proteins. Using these techniques, Machamer *et al.* (1985) investigated the role of glycosylation in intracellular transport and cell surface expression of the VSV G protein in cells expressing G protein from cloned cDNA. The individual contributions of the two N-linked oligosaccharide chains of G protein to the cell surface expression were assessed by eliminating one or the other or both of the glycosylation sites in cloned proteins. Interestingly, they found that one oligosaccharide chain at either position was sufficient for cell surface expression of G protein in transfected cells, and the apparent rates of intracellular transport of these proteins were similar to the rate observed in wild-type protein. However, the nonglycosylated G protein synthesized when both glycosylation sites were eliminated did not reach the cell surface. This protein did appear to reach a Golgi-like region, as determined by immunofluorescence microscopy, and was not subject to increased intracellular proteolytic degradation. These results suggest that carbohydrate is a direct signal for intracellular transport of G protein from the Golgi to the cell surface.

An alternative model is that carbohydrate is required for maintaining a polypeptide conformation essential for intracellular transport. This model is less appealing, however, because glycosylation at either site promotes efficient transport of G protein, and it seems unlikely that glycosylation at two different positions in the molecule would induce the same conformational changes. In addition, Machamer and Rose have introduced glycosylation sites at several new positions in G protein lacking the two normal sites. Their initial results show that at least one new glycosylation site does promote transport of G protein to the cell surface (Machamer *et al.*, 1985).

Additional evidence that carbohydrate can be a signal for intracellular transport has come from studies of a hybrid membrane protein. Guan and Rose (1984) constructed a hybrid gene encoding a membrane-anchored form of rat growth hormone. This protein is anchored in cellular membranes by a carboxy-terminal extension on the hormone composed of the transmembrane and cytoplasmic domains of the VSV G protein. The protein is transported efficiently to the Golgi but not to the cell surface. In order to determine if N-linked carbohydrate promotes surface expression of this hybrid protein, Guan *et al.* (1985) generated mutant proteins using *in vitro* mutagenesis in which single amino acids at two sites in anchored growth hormone were changed to generate consensus sequences required for addition of N-linked oligosaccharides. These mutant proteins, and a protein with both glycosylation sites, were both glycosylated and transported to the cell surface. Kalderon *et al.* (1985) have suggested that one criterion of a transport signal is that it be able to function at multiple sites in the polypeptide chain. Therefore, the findings that glycosylation at multiple sites in the G protein as well as in the growth hormone–hybrid membrane protein promotes efficient transport to the cell surface is strong evidence that the carbohydrates function as a transport signal on these proteins.

VI. MUTATIONS THAT ALTER GLYCOSYLATION AND GLYCOPROTEIN TRAFFIC

It is frequently found that mutants defective in surface appearance of membrane proteins are defective in carbohydrate processing (Stanley, 1984, 1985; Tartakoff, 1983b). Perhaps the most extensive studies have been concerned with the surface expression of the Thy-1 surface antigen by mouse lymphoma cells (Trowbridge *et al.*, 1978a,b; Trowbridge and

Hyman, 1979). The Thy-1 glycoprotein is about 30% carbohydrate by weight (Barclay, 1976), and the molecule from rat brain contains three N-linked oligosaccharides (Campbell, 1981). Thy-1 is found on the neurons and T lymphocytes of many animal species, and it is the most abundant surface protein of the mouse T cell. Mutant mouse lymphoma cells which bore less than 0.1% as much surface Thy-1 as did the wild-type cells were selected. Several clones were identified that continued to synthesize Thy-1 and apparently were defective in intracellular transport. The mutations proved to be recessive, and five complementation classes of mutant cells were characterized by somatic genetic analysis. Defects in the oligosaccharide units of Thy-1 glycoprotein were demonstrated in four of the five classes of mutants.

The specific defect in most of these mutant classes remains unknown; however, in the case of mutants belonging to complementation group E the analysis has been taken much further. The class E Thy-1$^-$ mutation has been shown to be due to a mutation in a gene that acts to transfer mannose from GDPmannose to dolichol phosphate (Chapman et al., 1980). In consequence, class E Thy-1$^-$ mutant cells are unable to synthesize a normal lipid-linked $Glc_3Man_9GlcNAc_2$ oligosaccharide but instead synthesize a smaller $Glc_3Man_5GlcNAc_2$ moiety that is transferred to the nascent polypeptide. Consequently, the cell surface glycoproteins of class E Thy-1$^-$ mutant cells show a pleiotropic alteration in their mannose-containing N-linked oligosaccharides. Interestingly, lipid-linked $Glc_3Man_5GlcNAc_2$ has also been identified as a minor species present in normal CHO cells (Chapman et al., 1979) and may be the oligosaccharide donor lipid for specific proteins in normal cells (Yamashita et al., 1983).

Many cell surface glycoproteins are expressed normally on the class E Thy-1$^-$ mutant cell. However, in addition to the Thy-1 glycoprotein, molecules bearing the Ly-6.2 and H9/25 determinants are not expressed on the cell surface of class E Thy-1$^-$ mutant cells (Horton and Hyman, 1983). It is probably due to the selectivity of the defect that the mutant cells are viable. Immunocytochemical studies show that wild-type cells have Thy-1 glycoprotein predominantly localized on the cell surface with only minimal amounts distributed in the cytoplasm. In class E mutant cells, Thy-1 glycoprotein is not expressed on the cell surface but is detectable in the ER, Golgi region, and lysosomes (Bourguignon et al., 1982).

It is not clear why the lesion in normal oligosaccharide processing characteristic of class E mutant cells prevents transport of Thy-1 and other glycoproteins to the cell surface. However, specific changes in oligosaccharide structure are necessary to block transport of Thy-1 glycoprotein. Lectin-resistant mutants with specific defects in glycosylation have been selected from mouse lymphoma cells (Trowbridge et al.,

variant allele encodes a mutant protein with a single substitution of glutamic acid by lysine at position 342 (Yoshida et al., 1976; Jeppsson, 1976). Since homozygotes for the Z allele have an 85% reduction in circulating levels of α_1-antitrypsin and substantial deposits of α_1-antitrypsin accumulate within the RER of hypatocytes, there appears to be a defect in secretion of the Z variant. The Z form of α_1-antitrypsin isolated from RER inclusions has normal protease inhibitory activity and essentially normal thermal stability (Bathurst et al., 1984). Thus, the Z variant of α_1-antitrypsin must be in its native form with the reactive center in the intact strained conformation (Lobermann et al., 1984), and it is unlikely that accumulation in RER is due to a gross misfolding of the polypeptide.

Interestingly, RER inclusions recovered from the liver of ZZ homozygotes contained α_1-antitrypsin bearing high-mannose N-linked glycan chains with attached glucose residues (Hercz et al., 1978; Hercz and Harpaz, 1980; Bathurst et al., 1984). Removal of glucose from the oligosaccharide precursor on glycoproteins usually occurs very rapidly after transfer to protein (Hubbard and Robbins, 1979; Atkinson and Lee, 1984). Treatment of rat hepatocytes and human hepatoma cells with deoxynorjirimycin, an inhibitor of processing ER glucosidases, has been shown to inhibit secretion and transport of α_1-antitrypsin from the RER to the Golgi, and it has been postulated by Lodish and Kong that oligosaccharide processing intermediates may function as part of a tag for rapid transport of α_1-antitrypsin from the RER to the Golgi (Gross et al., 1983a; Lodish and Kong, 1984; Parent et al., 1986). A plausible hypothesis is that the Z form of α_1-antitrypsin accumulates in the liver of ZZ homozygotes because the single amino acid substitution in the primary sequence alters the conformation of the molecule, thereby preventing the trimming of glucose residues from N-linked glycan chains on α_1-antitrypsin. In consequence, the rapid, receptor-mediated transport of α_1-antitrypsin from the RER to the Golgi is blocked. In this context, it is interesting that in a class of cellular transport mutants identified among immunoselected cells infected with Friend erythroleukemia virus, viral glycoprotein gp55 is not transported to the cell surface and the intracellular gp55 in these mutants contains immature forms of high-mannose oligosaccharides containing glucose (Ruta et al., 1982).

VII. TUNICAMYCIN AND GLYCOPROTEIN TRAFFIC

One of the most widely used inhibitors to evaluate the role of N-linked oligosaccharides in the transport and expression of secretory and membrane glycoproteins is tunicamycin (TM), a nucleoside analog which was isolated from Streptomyces lysosuperificus by Takatsuki et al. (1971).

This antibiotic inhibits the transfer of GlcNAc-1-P from UDPGlcNAc to dol-P (Tkacz and Lampen, 1975, Takatsuki *et al.*, 1975). Since this step is first in the lipid-linked pathway, TM treatment blocks N-linked glycosylation of protein (for reviews, see Schwarz and Datema, 1982a,b; Olden *et al.*, 1982; Elbein, 1984).

I believe there has been some confusion in the interpretation of results using this drug because it is sometimes assumed that, if N-linked oligosaccharides have an important role in the intracellular transport of secretory and membrane glycoproteins, then TM treatment should totally block secretion or surface expression of these proteins. Therefore, the appearance at the cell surface of TM-treated cells of even greatly reduced levels of secretory or membrane proteins has been used as evidence in some studies that carbohydrate does not have an important role in glycoprotein transport to the cell surface. However, this assumption is misleading because recent studies mostly of the Man-6-P-dependent transport of acid hydrolases to lysosomes have clearly demonstrated that cells can export proteins by a nonspecific pathway when the receptor-mediated pathway is inoperative. Cells defective in the Man-6-P pathway, either because they are deficient in functional Man-6-P receptors (Robbins and Myerowitz, 1981, Gonzalez-Noriega *et al.*, 1980) or lack enzymes for synthesis of the Man-6-P transport marker (Kornfeld and Kornfeld, 1985; von Figura and Hasilik, 1986), have low levels of activities of acid hydrolases in lysosomes and secrete large amounts of acid hydrolases. There is also evidence for a nonspecific pathway for protein export in yeast. Normally carboxypeptidase Y, and vacuolar enzyme in yeast analogous to a lysosomal enzyme in mammals, is efficiently localized to the vacuole. Plasmid-directed overproduction, however, results in secretion of a large fraction of the protein, apparently because the carboxypeptidase Y-sorting receptor is saturable (Stevens *et al.*, 1986).

The nonspecific pathway for protein export which operates when receptor-mediated pathways are inoperative may be analogous to passive flow of protein in vesicles from the RER to the Golgi and then to the cell surface. Therefore, it is not logical to expect that TM treatment would totally prevent the surface appearance of secretory and membrane glycoproteins even if carbohydrate does have an important functional role in the normal transport of these proteins. TM treatment, however, may substantially alter both the kinetics of intracellular transport of secretory and membrane proteins as well as the precise pathway taken by these proteins from their sites of synthesis in the RER to the cell surface.

The effect of TM on the secretion of immunoglobulins has been extensively studied. Different classes of immunoglobulins have different carbohydrate requirements for secretion. Thus, secretion of IgM and IgE was almost totally inhibited by TM, secretion of IgA was partially inhibited, and secretion of IgG was essentially unaffected (Hickman *et al.*, 1977, Hickman and Kornfeld, 1978). IgA accumulated in the RER, and no increase in intracellular degradation of nonglycosylated immunoglobulins could be demonstrated. The extent of inhibition of immunoglobulin secretion by TM in these studies correlates with the carbohydrate content of the various immunoglobulin classes, and Hickman and Kornfeld (1978) suggested that those molecules that are highly glycosylated might be most affected by inhibition of glycosylation. However, Sidman (1981) found that secretion of IgD, which is as heavily glycosylated as IgM, is not inhibited by TM, and a survey of published studies using this drug does not suggest a correlation between the extent of glycosylation of secretory proteins and the sensitivity of secretion to TM.

The level of differentiation of cells may be important in determining the sensitivity to TM. Sibley and Wagner (1981) found that secretion of IgM by a mouse B cell lymphoma (W279) was not affected by TM treatment, in contrast to the profound inhibition of secretion observed by Hickman and Kornfeld (1978) when plasmacytoma cell (104E) was treated with TM. Although both W279 and 104E cells secrete IgM, the two are models for cells at very different points along the B cell differentiation pathway and the plasmacytoma cell synthesizes about 50 times as much IgM as the B cell lymphoma. Interestingly, Ciccimarra *et al.* (1976) have described a defect in agammaglobulinemia in which the IgG secretory block was associated with a failure to glycosylate the IgG heavy chains while the incorporation of [^{3}H]mannose and [^{3}H]glucosamine into other cellular proteins was normal.

In single cells, TM inhibits secretion of specific glycoproteins only, and export of other glycoproteins is unaffected. TM treatment of human hepatoma cells significantly delayed secretion of α_2-macroglobulin, ceruloplasmin, and α_1-antitrypsin, but secretion of transferrin, fibrinogen, α-fetoprotein, fibronectin, and plasminogen was not affected by the drug (Bauer *et al.*, 1985). The most dramatic effect of TM treatment was on the secretion of α_2-macroglobulin, with more than 90% of the protein still present in the cell fraction in an undegraded form 7 hr after synthesis, while in control cells 50% of newly synthesized α_2-macroglobulin is secreted by 75–80 min. TM treatment of human hepatoma cells also markedly affected thyroxine-binding globulin (TBG) secretion, almost doubling the time required for secretion of 50% of the protein (Bartalena and Robbins, 1984).

Secretion of a major glycoprotein constituent of the mouse egg's extracellular coat (ZP2) was inhibited by TM, and nonglycosylated ZP2 accumulated intracellularly (Roller and Wassarman, 1983). However, TM did not inhibit secretion of ZP3, another mouse egg glycoprotein. TM inhibits secretion and proteolytic processing of prepro-α-factor in yeast (Julius *et al.*, 1984). The effect appears to be on transport since intracellular forms accumulate. Inhibition of glycosylation of the secreted acid phosphatase of yeast by TM resulted in synthesis of nonglycosylated, membrane-associated forms of the enzyme (Mizunaga and Noguchi, 1982). Treatment of human umbilical vein endothelial cells with TM inhibited secretion of von Willebrand factor (vWf) (Wagner *et al.*, 1986). Nonglycosylated pro-vWf monomers failed to dimerize, were not processed further, and accumulated in the RER. There was no apparent effect of TM on degradation. TM inhibition of vWf transport and processing is specific to human endothelial cells, since precursor subunits were cleaved and secreted by bovine cells in the presence of TM (Lynch *et al.*, 1983). TM inhibited both the intracellular movement and secretion of procollagen from human fibroblasts (Housley *et al.*, 1980), but the secretion of procollagen was only minimally affected by TM in chick fibroblasts (Duksin and Bornstein, 1977; Olden *et al.*, 1978). TM markedly inhibited the secretion of surfactant-associated glycoprotein A from lung epithelial cells (Whitsett *et al.*, 1985) and the secretion of thyroglobulin cells (Björkman and Ekholm, 1982; Eggo and Burrow, 1983).

TM treatment did not significantly affect ovalbumin secretion by chick oviduct (Keller and Swank, 1978), very low density lipoprotein assembly and secretion by hepatocytes (Siuta-Mangano *et al.*, 1982), processing and secretion of haptoglobin by rat hepatocytes (Hanley *et al.*, 1983), nor processing and secretion of proopiomelanocortin peptides by mouse pituitary cells (Budarf and Herbert, 1982).

Glycosylation is important for the structural and functional maturation of the insulin receptor and for its translocation to the cell membrane (Reed *et al.*, 1981; Ronnett and Lane, 1981; Ronnett *et al.*, 1984). Treatment of cells with TM rapidly depletes cell surface insulin-binding capacity and leads to the production of a 180 kDa polypeptide, precipitated by antibodies to the insulin receptor, which is not proteolytically cleaved to yield the α and β subunits, does not bind insulin, and does not reach the cell surface. Ronnett and Lane (1981) have shown by the heavy isotope density shift method that blocking glycosylation with TM caused the intracellular accumulation of the inactive, nonglycosylated receptor. Subsequent withdrawal of TM and a shift of the cells to heavy medium (^{15}N-, ^{13}C-, and ^{2}H-labeled amino acids) led to the "activation" of previously synthesized "light" receptor. Similar levels of light receptor accrued even when protein synthesis was blocked with cycloheximide. Apparently, the inactive nonglycosylated receptor accumulated during TM treatment and then reentered the glycosylation pathway on removal of the inhibitor. Since transfer of oligosaccharide from the oligosaccharide–lipid donor to protein occurs in the ER, these results suggest that N-linked oligosaccharide chains are required for transport of the receptor precursor out of the RER.

Inhibitors of N-linked glycosylation (TM and glucosamine) also prevent maturation and translocation to the cell surface of active epidermal growth factor (EGF) receptors (Soderquist and Carpenter, 1984; Slieker and Lane, 1985). The transferrin receptor, which has both high-mannose and complex-type oligosaccharide side chains, is transported to the cell surface in the presence of TM, but the levels of surface receptor are significantly reduced compared to untreated cells (Omary and Trowbridge, 1981). LDL receptor expression was reduced at the cell surface following TM treatment (Chatterjee *et al.,* 1979, 1981; Filipovic and von Figura, 1980).

Cell surface acetylcholine receptor expression is greatly decreased when muscle cells are treated with TM. Merlie *et al,* (1982) found that TM treatment of a mouse muscle cell line decreases the amount of functional acetylcholine receptor by blocking assembly of subunit precursors to mature functional molecules. They suggest that TM may prevent transport of subunits to the Golgi where assembly may occur. Prives and Bar-Sagi (1983) suggest that enhanced proteolytic degradation of nonglycosylated acetylcholine receptor in TM-treated chick muscle cells may explain the reduced receptor levels since receptor loss can be prevented, in part, by the protease inhibitor leupeptin. Rat basophilic leukemia cells grown in the presence of TM expressed fewer IgE receptors at the cell surface, as judged both by ligand binding studies and external labeling procedures (Pecoud *et al.,* 1981). TM reduced the expression of surface Na^+ channels in neuroblastoma cells (Waechter *et al.,* 1983) and in chicken muscle cells (Bar-Sagi and Prives, 1983) as measured by [^{3}H]saxitoxin binding and neurotoxin-activated ^{22}Na influx. This effect was partially reversed (30%) by leupeptin in muscle cells. TM completely blocks β-adrenergic receptors in growing human astrocytoma cells (Doss *et al.,* 1985), but the inhibitor had no influence on the steady-state level of β-receptors in S49 cells (George *et al.,* 1986).

In the presence of TM, mouse mammary tumor virus (MTV) glycoproteins synthesized by virus-infected hepatoma cells accumulate within the cells, proteolytic processing of the precursor polyprotein does not occur, and transport of MTV polypeptides to the cell surface and extracellular fractions is inhibited (Firestone, 1983). TM treatment also inhibits mumps virus particle formation (Herrler and Compans, 1983). Since mumps virus polypeptides are synthesized in the presence of the drug and assembly of nucleocapsids is observed, it is likely that TM blocks the surface expression of the viral glycoproteins.

The N-linked oligosaccharides of the H-2D^k mouse histocompatability protein heavy chain may influence its transport and cellular distribution (Le and Doyle, 1985). In mouse macrophages the H-2D^k glycoproteins exist as two distinct forms: an intracellular form with high mannose type chains and a cell surface form with complex type chains. H-2D^k glycoproteins destined for the cell surface are rapidly transported to this localization after synthesis in the RER, whereas most intracellular H-2D^k glycoproteins remain associated with intracellular membranes for prolonged periods in an undegraded form. To determine if the oligosaccharide structure on H-2D^k glycoproteins contains information important in intracellular distribution, macrophage cultures were treated with TM under conditions in which the inhibition of glycosylation is incomplete. In TM-treated cell cultures the newly synthesized nonglycosylated H-2D^k polypeptides were rapidly transported to the cell surface with no significant intracellular pool. By contast, in the same inhibitor-treated cultures, H-2D^k polypeptides containing one, two, or three oligosaccharide side chains of the high mannose type were not transported to the cell surface. These authors suggest that cell surface expression of H-2D^k polypeptides requires either processing of the oligosaccharide side chains from the high mannose form to the complex type or the absence of oligosaccharide chains.

The major human red cell glycoprotein, glycophorin A, contains an N-linked glycan chain and 15 O-linked glycan chains. The synthesis of the N-linked chain is inhibited by TM, while the O-linked chains are not affected. Incomplete glycophorin A, lacking the N-glycosidic oligosaccharide, is incorporated into the surface membrane, but the total amount of glycophorin A is decreased 2- or 3-fold compared to untreated cells (Gahmberg et al., 1980). TM treatment of hepatoma cells delayed the appearance of γ-glutamyltransferase at the cell surface (Barouki et al., 1984). TM also inhibited transport to the cell surface of specific yeast membrane proteins, suggesting that glycosylation is necessary for the correct processing and intracellular transport of some, but not all, yeast membrane proteins (Novick and Schekman, 1983). TM treatment of frog retinas blocks the incorporation of opsin into retinal rod outer segment (ROS) membranes (Fliesler and Basinger, 1985; Fliesler et al., 1985). This effect of TM may be specific to amphibian retinas, however, since Plantner et al. (1980) reported that inhibition of opsin glycosylation by TM in the bovine retina does not prevent incorporation of the visual pigment apoprotein into ROS membranes.

Epithelial cells express cell surface polarity with different glycoproteins present on apical and basolateral surfaces (for a recent review, see Simmons and Fuller, 1985). Roth et al. (1979) and Green et al. (1981) have studied viral protein expression in Madin–Darby canine kidney (MDCK) cells in the presence of TM and saw no marked effect on polarity. Similar results were obtained with mutant MDCK cells that were defective in glycosylation (Green *et al.*, 1981; Meiss *et al.*, 1982). Although N-linked glycosylation does not appear to have an important role in the polarized expression of viral proteins, it is unknown whether carbohydrate chains function as transport markers for endogenous epithelial apical and basolateral proteins.

TM does not alter expression at the cell surface of nonglycosylated HLA-A and HLA-B antigens on two lymphoblastoid cell lines (Owen *et al.*, 1980; Ploegh *et al.*, 1981) nor initial cell surface expression of human complement receptor type 2 (CR2) on human B lymphoblastoid cells (Weis and Fearon 1985). It also does not alter the intracellular routing, turnover, or function of the asialoglycoprotein receptor in a human hepatoma cell line (Breitfeld *et al.*, 1984).

VIII. INHIBITORS OF OLIGOSACCHARIDE PROCESSING AND GLYCOPROTEIN TRAFFIC

Drugs that inhibit specific steps in the processing pathway have become available, and it is now possible to study the role of oligosaccharide processing intermediates in intracellular transport of glycoproteins (Fig. 1) (for reviews, see Lalegerie *et al.*, 1982; Schwarz and Datema, 1984; Elbein, 1984; Fuhrmann *et al.*, 1985). 1-Deoxynojirimycin, N-methyl-deoxynojirimycin and castanospermine inhibit RER α-glucosidases I and II, but not Golgi endo-α-D-mannosidase and these drugs prevent removal in the RER of three glucose residues from $Glc_3Man_9GlcNAc_2$ (Saunier *et al.*, 1982; Hettkamp *et al.*, 1982; Pan *et al.*, 1983; Palamarczyk and Elbein, 1985; Lubas and Spiro, 1987); bromoconduritol prevents trimming of glucose from $GlcMan_9GlcNAc_2$ by inhibiting α-glucosidase II (Datema *et al.*, 1982). 1-Deoxymannojirimycin inhibits Golgi α-mannosidases 1A and 1B (Fuhrmann *et al.*, 1984; Burke *et al.*, 1984; Elbein *et al.*, 1984) but not the RER α-mannosidase (Bischoff and Kornfeld, 1984). Swainsonine inhibits Golgi α-mannosidase II (Tulsiani *et al.*, 1982). Inhibition of the processing of glycoproteins by swainsonine (Elbein *et al.*, 1981, 1982; Tulsiani *et al.*, 1982) results in the synthesis of glycoproteins with hybrid-type oligosaccharides, in which one branch contains sugars commonly found on complex oligosaccharides (GlcNAc, Gal, NeuAc) and the other mannose-containing branch remains unsubstituted (Gross *et al.*, 1983b; Tulsiani and Touster, 1983; Arumugham and Tanzer, 1983).

An important advantage in the use of these oligosaccharide processing inhibitors over the use of TM is that glycoproteins synthesized by treated cells do not lack N-linked glycan chains but rather retain intermediates in the normal oligosaccharide processing pathway (1-deoxynojirimycin, N-methyldeoxynorjirimycin, castanospermine, bromoconduritol, 1-deoxymannojirimycin) or hybrid oligosaccharides (swainsonine). Therefore, it is unlikely that treatment with these inhibitors will induce glycoprotein aggregation, alter protein folding, or increase sensitivity to proteolysis, and this simplifies the interpretation of studies using these drugs.

Studies using inhibitors of RER α-glucosidases have demonstrated that trimming of glucose residues from the major N-linked oligosaccharide precursor is required for the rapid maturation of specific secretory, membrane, and lysosomal enzymes from the RER to the Golgi apparatus. Gross *et al.* (1983a) were first to report that treatment of rat hepatocytes with 1-deoxynorjirimycin resulted in a long delay in the secretion of α_1-antitrypsin and that the cells accumulated α_1-antitrypsin with carbohydrate chains of the high-mannose type containing glucose, the largest species being $Glc_3Man_9GlcNAc_2$. Secretion of the nonglycoprotein albumin was unaffected by the drug. 1-Deoxynojirimycin did not completely block oligosaccharide processing, and only α_1-antitrypsin molecules processed to the complex type in one or two of their oligosaccharide chains were secreted.

Removal of three glucose residues from the $Glc_3Man_9GlcNAc_2$ precursor is believed to occur rapidly after translocation of the glycoprotein into the RER (Hubbard and Robbins, 1979) and may occur on nascent chains (Atkinson and Lee, 1984). Lodish and Kong (1984) found that glucose removal from N-linked oligosaccharides is required for efficient passage of some secretory glycoproteins from the RER to the Golgi complex. 1-Deoxynojirimycin treatment inhibited the secretion of α_1-antitrypsin and antichymotrypsin by human hepatoma cells, but it had little effect on secretion of glycoproteins C3 and transferrin or albumin. Since, after removal of 1-deoxynojirimycin, α_1-antitrypsin was secreted normally, its inhibitory effect is reversible. From cellular fractionation studies using sucrose density gradients, it was found that drug treatment caused α_1-antitrypsin and antichymotrypsin to accumulate in the RER. In untreated cells α_1-antitrypsin is rapidly transported from the RER whereas transferrin leaves the RER slowly. Lodish and Kong (1984) postulated that rapid transport of α_1-antitrypsin and antichymotrypsin from the RER to the Golgi requires that the N-linked oligosaccharide chains be processed to at least the $Man_9GlcNAc_2$ form and that this oligosaccharide forms parts of the recognition site of a transport receptor for these proteins. The slow transport of transferrin or α_1-antitrypsin synthesized in the presence of 1-deoxynojirimycin would not involve binding to a receptor, but rather bulk-phase fluid movement from the ER to the Golgi. Interestingly, in the human disease α_1-antitrypsin deficiency the Z variant of α_1-antitrypsin is poorly secreted by the liver, accumulates in the RER, and retains glucose residues (Carrell *et al.*, 1984).

In hybridoma cells producing IgD and IgM, Peyrieras *et al.* (1983) found that secretion of IgD but not IgM was blocked by 1-deoxynojirimycin. From experiments using the drug, Lemansky *et al.* (1984) found that the transport of the lysosomal enzymes cathepsin D and β-hexosaminidase from the ER to the Golgi apparatus depends on the removal of glucose residues from the carbohydrate side chains. 1-Deoxynojirimycin had no effect on surface expression of VSV G protein, influenza virus hemagglutinin, and human class I histocompatability antigens (Schlesinger *et al.*, 1984; Burke *et al.*, 1984). Glucosidase inhibitors also did not have any effect on transport of incorrectly processed *erbB* oncogene glycoprotein to the cell surface (Schmidt *et al.*, 1985).

The effect of processing inhibitors on the intracellular migration of glycoprotein E2 of mouse hepatitis virus was studied by Repp *et al.* (1985). In the presence of N-methyldeoxynojirimycin or castanospermine, glycoprotein E2 was synthesized in normal amounts but accumulated intracellularly, with transport to the cell surface being greatly delayed. Glycoprotein E2 isolated from virus particles from cells grown in the presence of N-methyldeoxynojirimycin carried predominantly Glc-$_3$Man$_9$GlcNAc$_2$ oligosaccharide side chains, and the remaining chains showed trimming of one or two mannose residues. TM treatment also prevented surface expression of glycoprotein E2, but inhibitors of trimming α-mannosidases did not affect transport. Apparently, efficient transport of glycoprotein E2 to the plasma membrane requires removal of glucose but not trimming of mannose.

Treatment of Sindbis virus-infected cells with either 1-deoxynorjirimycin or castanospermine inhibited proteolytic cleavage of the glycoprotein polyprotein precursor to glycoproteins E1 and E2 but did not prevent the migration of the precursor to the cell surface (Schlesinger *et al.*, 1985). However, treatment of Sindbus virus-infected cells with bromoconduritol, which prevents removal of the innermost glucose residue but not the outer two glucose residues, did not inhibit proteolytic cleavage of the polyprotein precursor (Datema *et al.*, 1984). Fowl plague virus shows the interesting characteristic that release of infectious virus is inhibited by bromoconduritol but not by N-methyldeoxynorjirimycin (Datema *et al.*, 1982; Romero *et al.*, 1983). Unfortunately, bromoconduritol is quite unstable and has a half-life in water of only 16 min at 37°C and pH 7.3. This, and the fact that millimolar concentrations are required for inhibition, complicates interpretation of studies using this drug.

We have found that 1-deoxynorjirimycin treatment of human hepatoma cells inhibits the maturation of N-linked glycan chains on only a selected group of secretory glycoproteins to forms resistant to digestion by endo H. (Parent *et al.,* 1986). Also, drug treatment inhibits transport from the RER to the Golgi of the same subgroup of glycoproteins, including α_1-antitrypsin, ceruloplasmin, and α_2-macroglobulin but not transferrin, α-fetoprotein, or fibronectin. Interestingly, TM treatment had a similar differential effect on secretion of these glycoproteins (Bauer *et al.,* 1985). Since trimming of three glucose residues from the α-1,3 branch is required for processing of N-linked chains to endo H-resistant forms in the major N-linked glycan processing pathway (Harpaz and Schachter, 1980), our findings that 1-deoxynorjirimycin treatment inhibits the maturation of N-linked glycans on some, but not all, secretory glycoproteins was unexpected. One possible explanation for these results is that glycan chains on some secretory glycoproteins are processed by α-glycosidases resistant to the drug. Alternatively, there may be another pathway for processing N-linked glycan side chains on some secretory glycoproteins to endo H-resistant forms that does not involve glycosylated intermediates.

Both of these explanations are consistent with previous reports in the literature. Reitman *et al.* (1982) examined the glycopeptides produced in a mouse lymphoma cell line that is severely deficient (less than 0.3% of parent) in RER α-glucosidase II. The major oligosaccharides produced in this mutant were characterized as $Glc_2Man_9GlcNAc_2$ and $Glc_2Man_8GlcNAc_2$. However, this mutant still contained 25% of the complex chains of the parent line. The authors suggest that the mutant may still have enough α-glucosidase II to allow some processing to complex chains, or there may be an alternative pathway.

Parodi *et al.* 1983a) have shown that glucose-free oligosaccharides are transferred in the N-glycosylation of proteins in the protozoa *Trypanosoma cruzi* and Romero and Herscovics (1986) have demonstrated transfer of glucose-free oligosaccharides from lipid to protein in a mammalian cell line. Krag (1979) has described a CHO cell mutant, B211, selected for increased resistance to concanavalin A, that is defective in the glucosylation of lipid-linked oligosaccharides. Interestingly, incorporation of mannose into glycoproteins occurred at near normal rates in B211, and these mutant cells also synthesized high levels of complex-type oligosaccharides (Krag and Robbins, 1982; S. Krag, personal communication). Similarly, in other studies, using several different glucosidase inhibitors (Datema *et al.,* 1982; Pan *et al.,* 1983; Gross *et al.,* 1983a; Romero *et al.,* 1985), the inhibition of synthesis of complex-type glycan chains was not complete even at high concentrations of inhibitors, suggesting either that some processing α-glucosidase activity is resistant to the inhibitors or that there are alternative pathways.

Lubas and Spiro (1987) have described a novel enzyme present in Golgi membranes which is capable of converting the $Glc_1Man_9GlcNAc_2$ unit to $Man_8GlcNAc_2$ with the release of $Glc\alpha1$-3Man. This endomannosidase is fully active in the presence of 1-deoxynojirimycin. The authors point out that this alternative pathway for the removal of glucose and mannose as a disaccharide rather than the sequential release of the monosaccharides by α-glucosidase 11 and α-mannosidase 1 could provide an explanation for the incomplete block in oligosaccharide processing which is observed in cells with inhibited or deficient α-glucosidase 11. This alternative pathway may also explain why treatment of hepatoma cells with 1-deoxynojirimycin has differential effects on both the glycan maturation and intracellular transport of specific secretory glycoproteins (Parent *et al.*, 1986).

At least four distinct exo-α-mannosidases are responsible for the trimming of six mannose residues in the conversion of $Glc_3Man_9GlcNAc_2$ oligosaccharides to complex forms in the major processing pathway of N-linked glycans (Fig. 1), and inhibitors of three of the trimming mannosidases are known. As yet no inhibitor has been reported for the α-mannosidase activity localized in the RER (Bischoff and Kornfeld, 1983). Inhibitors of trimming α-mannosidases, unlike those of the α-glycosidases, do not appear to have dramatic effects on the transport of secretory and membrane glycoproteins. 1-Deoxymannojirimycin inhibits Golgi mannosidases 1A and 1B, blocking the conversion of high-mannose oligosaccharides to complex-type oligosaccharides. 1-Deoxymannojirimycin did not prevent the secretion of either IgD or IgM (Fuhrmann *et al.*, 1984), in contrast to 1-deoxynorjirimycin, which blocked the secretion of IgD. Gross *et al.* (1985) found that high-mannose-type α_1-antitrypsin and α_1-acid glycoprotein were secreted by 1-deoxymannojirimycin-treated hepatocytes at the same rate as the normal glycoproteins.

Using another α-mannosidase inhibitor, α-mannopyranosylmethyl-*p*-nitrophenyltriazene, Docherty *et al.* (1986) found that the drug treatment did not alter the secretion of α_1-acid glycoprotein although the secreted form contained $Man_{(7-9)}GlcNAc_2$ chains rather than usual complex-type chains. 1-Deoxymannojirimycin was also without effect on the surface expression of several membrane glycoproteins tested, such as VSV G protein, influenza virus hemagglutinin, and human class I histocompatability antigens (Burke *et al.*, 1984; Elbein *et al.*, 1984).

1-Deoxymannojirimycin, which inhibits Golgi α-mannosidase 1 but not ER α-mannosidase, has been used to determine the role of the ER α-mannosidase in the processing of the N-linked oligosaccharides on glycoproteins in intact cells (Bischoff *et al.*, 1986). Analysis of the total [^{3}H]-mannose labeled N-linked oligosaccharides synthesized by control and 1-deoxymannojirimycin-treated rat hepatocytes indicates that about one-third of the oligosaccharides on secretory glycoproteins and about one-half of the oligosaccharides on cellular glycoproteins in rat hepatocyte cultures are processed by the 1-deoxymannojirimycin-resistant ER α-mannosidase. Results from similar studies using the UT-1 cell line indicate that the ER α-mannosidase is responsible for the conversion of $Man_9GlcNAc_2$ oligosaccharides to $Man_8GlcNAc_2$ units on a resident ER glycoprotein, 3-hydroxy-3-methylglutaryl-CoA reductase (Bischoff et al., 1986). The finding that only specific glycoproteins are substrates for the 1-deoxymannojirimycin-resistant ER α-mannosidase is additional evidence for multiple oligosaccharide processing pathways in cells.

Swainsonine is an inhibitor of Golgi α-mannosidase II, and T. K. Yeo et al. (1985) found that swainsonine treatment of human hepatoma cells accelerated the secretion of a number of glycoproteins (transferrin, ceruloplasmin, α_2-macroglobulin, and α_1-antitrypsin) by decreasing the lag period before secretion by 10–15 min relative to untreated cultures. Since conversion of the high-mannose precursor on α_1-antitrypsin to the hybrid form in swainsonine-treated cells occurred more rapidly (by about 10 min) than the conversion to the complex form in control cells, these results suggest that swainsonine-modified glycoproteins in hepatoma cells traverse the Golgi more rapidly than their normal counterparts. The magnitude of this effect on the rate of protein secretion is small, however, and in other studies it is reported that swainsonine treatment had little effect on the secretion of a number of proteins, including fibronectin by skin fibroblasts (Arumugham and Tanzer, 1983), α_1-antitrypsin by rat hepatocytes (Gross et al., 1983b), aminopeptidase N by organ-cultured intestinal mucosal explants (Danielsen et al., 1983), five glycoproteins by human hepatoma cells (Lodish and Kong, 1984), and surfactant-associated glycoprotein A by lung epithelial cells (Whitsett et al., 1985). Treatment of thyroid cells with both swainsonine and 1-deoxynojirimycin increased the half-time of maximal secretion of thyroglobulin from 90 to 120 min (Frank et al., 1986).

In addition, swainsonine treatment had little effect on the intracellular routing, turnover, or function of the asialoglycoprotein receptor in human hepatoma cell (Breitfeld et al., 1984). Soderquist and Carpenter (1984) demonstrated that drug treatment of A431 cells resulted in the production of incompletely processed EGF receptors which were still able to function normally. Duronio et al. (1986) found that swainsonine treatment did not significantly affect the movement of insulin and insulin-like growth factor 1 receptors through the cells, their proteolytic processing, hormone binding, or hormone-stimulated autophosphorylation. The number and affinity of β-adrenergic receptors in swainsonine-treated S49 cells were normal (George et al., 1986). Influenza particles isolated from swainsonine-treated infected cells had the same infectivity and hemagglutination titer as virus particles from control cells (Elbein et al., 1982).

IX. VERTEBRATE LECTINS AND GLYCOPROTEIN TRAFFIC

The receptor-mediated trafficking to lysosomes of acid hydrolases with oligosaccharides containing phosphorylated mannose is the best understood pathway by which proteins translated into the RER are segregated and delivered to their ultimate destination (for reviews, see Sly, 1982; von Figura and Hasilik, 1986; Robbins, Chapter 11, this volume). Currently it is thought that newly synthesized acid hydrolases bearing the Man-6-P transport marker bind to specific receptors for Man-6-P located on Golgi membranes, to be transported from these Golgi cisternae by a receptor-mediated process, similar to that which occurs at the cell surface for other receptor–ligand complexes, and to be delivered in vesicles to lysosomes where low pH causes dissociation from receptor.

Two receptors with a binding specificity for Man-6-P have been identified. The first to be isolated was a 215 kDa integral membrane protein which is found in many different cell types (Sahagian *et al.*, 1981). Recently a second Man-6-P receptor has been isolated from murine P388D1 macrophage and bovine liver that differs from the 215 kDa receptor in its subunit molecular weight (46 kDa), oligosaccharide-binding specificity, and by the requirement for divalent cations for binding (Hoflack and Kornfeld, 1985a,b). The new receptor has been named the cation-dependent Man-6-P receptor (CD Man-6-P receptor) to distinguish it from the 215 kDa or cation-independent Man-6-P receptor (CI Man-6-P receptor). In preliminary experiments, Hoflack and Kornfeld (1985b) have identified the CD Man-6-P receptor in a number of cell types, including those that contain the CI Man-6-P receptor. They suggest that the two Man-6-P receptors may be involved in targeting acid hydrolases to different populations of lysosomes in the same cells.

In addition to the two Man-6-P receptors, a number of carbohydrate-binding proteins or lectins have been identified in vertebrate cells (Table I), and, although their physiological function is generally unclear, some may have roles in the intracellular transport or targeting of glycoproteins (for reviews, see Ashwell and Harford, 1982; Barondes, 1981, 1984; Kobiler, 1987). Vertebrate lectins can be classified roughly into two groups. The first comprises integral membrane proteins generally of high molecular weight that can be solubilized by detergents (lectins 1–6, Table I). The second comprises a group of soluble or peripheral membrane proteins, low in molecular weight and often developmentally regulated (lectins 7–12, Table I).

Many of the carbohydrate-binding membrane receptors are known to bind and internalize glycoproteins, and it is often assumed that they function on the cell surface. However, one important lesson that comes from the work on Man-6-P receptors and lysosomal localization is that identification of an efficient cell surface, receptor-mediated glycoprotein uptake system does not necessarily mean that the primary *in vivo* function of the receptor is associated with the plasma membrane. Identification of the CI Man-6-P receptor resulted from earlier studies in which it was found that fibroblasts from patients with various inherited lyosomal storage diseases could take up lysosomal enzymes secreted by normal fibroblasts, and that this process was specific and saturable (Neufeld *et al.*, 1975). An early model for the delivery of acid hydrolases to lysosomes suggested by Neufeld *et al.* (1977) emphasized the role of cell surface receptor activity; they postulated that the main route for incorporation of these enzymes into lysosomes required their prior secretion and recapture by the cells. Subsequent studies have not supported this model (von Figura and Weber, 1978; Sly and Stahl, 1978; Vladutiu and Ratazzi, 1979). The demonstration that only a small fraction of the Man-6-P receptor activity is localized on the cell surface directed attention to the physiological role of intracellular receptors, and it is now generally accepted that the major pathway for transport of hydrolases to lysosomes is mediated by receptors on Golgi membranes.

Like the CI Man-6-P receptor, several other carbohydrate-binding proteins are also concentrated on intracellular membranes. Pricer and Ashwell (1976) demonstrated the binding of [³H]-asialoorosomucoid to Golgi and lysosomal membranes isolated from rat liver, and these observations were confirmed by studies on solubilized hepatocytes (Steer and Ashwell, 1980). Rat hepatocytes were found to contain a total of 860,000 asialoglycoprotein receptors of which only 80,000 reside in the plasma membrane. Similar results have been provided by Weigel and Oka (1983), using hepatocytes permeabilized with digitonin. Only 20–30% of the Man/ GlcNAc-binding activity of alveolar macrophages could be inhibited by mild trypsinization, suggesting that 70–80% of the receptor may be inside the cell inaccessible to trypsin (Stahl *et al.*, 1980). This result was subsequently confirmed by direct binding assays on cells made permeable by the presence of saponin (Wileman *et al.*, 1984). Human fibroblasts contain a large intracellular pool of binding sites specific for Man-6-P; this pool contains 80% of the total receptors found in the cell (Fisher *et al.*, 1980). In the presence of permeabilizing agents (saponin and digitonin), macrophage Man-6-P receptors showed a distribution of 15–20% on the surface and 80–85% inside (Shepherd *et al.*, 1984). Receptor distribution within cells has also been studied by electron microscopy. Antibody bound to the asialoglycoprotein receptor can be visualized when frozen sections are incubated with colloidal gold absorbed to Protein A. A quantitative analysis of gold bead distribution in sections of hepatocytes shows that

TABLE I

Vertebrate Lectins

Designation, carbohydrate-binding specificity[a]	Physical characteristics	Cellular localization	Major sources	Other characteristics
1. Cation-dependent Man-6-P receptor,[1] Man-6-P	Integral membrane glycoprotein, subunit M_r 215,000	80–90% on internal membranes (endoplasmic reticulum, Golgi, lysosomes), 10–20% on plasma membrane	Many mammalian tissues including testis, brain, spleen, liver, and lung	No divalent cation requirement for binding; receptor-mediated endocytosis
2. Cation-independent Man-6-P receptor,[2,3] Man-6-P	Integral membrane glycoprotein, subunit M_r 46,000	Plasma membranes and internal membranes	Murine P388D1 macrophages, bovine liver, and other cell types	Requires divalent cations for binding; oligomer (3 subunits)
3. Asialoglycoprotein receptor,[4,5] GalNAc > Gal, Glc	Integral membrane glycoprotein, subunit M_r 40,000–65,000	80–90% on internal membranes (endoplasmic reticulum, Golgi, lysosomes), 10–20% on plasma membrane	Mammalian liver	Requires Ca^{2+} for binding; receptor-mediated endocytosis
4. Avian N-acetylglucosamine receptor,[6,7] GlcNAc, Man, Glc	Integral membrane glycoprotein, subunit M_r 26,000	Unknown	Avian liver	Requires Ca^{2+} for binding
5. Mannose/N-acetylglucosamine receptor,[8–11] D-Man, L-Fuc > D-GlcNAc > Glc	Integral membrane glycoprotein M_r 175,000–180,000	80% on internal membranes, 20% on surface membranes	Alveolar macrophage, Kupffer cells, hepatic endothelial cells, spleen and skeletal muscle	Requires Ca^{2+} for binding; receptor-mediated endocytosis
6. Fucose receptor,[12,13] D-Fuc, D-GalNAc > D-Gal > L-Fuc ≫ D-GlcNAc	Integral membrane glycoprotein, subunit M_r 88,000	Plasma membranes and perhaps also internal membranes	Kupffer cells	Requires Ca^{2+} for binding; receptor-mediated endocytosis

7. Galactose-particle receptor,[19-21] GalNAc > Gal > Fuc	Peripheral membrane protein, subunit M_r 30,000	Plasma membranes and internal membranes	Kupffer cells	Receptor-mediated endocytosis of particulate structures with exposed Gal
8. Mannose/N-acetylglucosamine core-specific lectin,[14-18] Man, Fuc > GlcNAc > Glc	Soluble or peripheral membrane protein, subunit M_r 24,000–32,000	80% in microsomal fraction, primarily in rough microsomes	Mammalian and avian hepatocytes	Slowly secreted; requires Ca^{2+} for binding, contains collagenlike domains
9. Ligatin,[22,23] Glc-1-P and Man-6-P	Peripheral membrane protein, M_r 10,000–20,000	Cell surface	Suckling rat ileum, brain macrophages, and embryonic chick neural retina	Forms filaments 3nm in diameter in the presence of Ca^2
10. Lactose-binding lectin (Class I),[24,25] Gal (β-1,4) GlcNAc > Gal (β-1,4) Glc, Gal (β-1,3) GalNAc	Soluble dimeric protein, subunit M_r 13,000–16,500	Intracellular and on the cell surface	Many mammalian and avian tissues including heart, lung, spleen, muscle, and liver	Developmentally regulated; requires thiol-reducing groups for activity
11. Lactose-binding lectin (Class II),[26,27] Gal (β-1,4) Glc	Monomeric soluble protein, M_r 13,000–14,000	Secretory granules of intestinal goblet cells	Rabbit bone marrow, chicken intestine and kidney	
12. Heparin-binding lectin,[28] heparin, GalNAc	Soluble protein	Unknown	Rat lung, chicken muscle and liver	Developmentally regulated; secreted by embryonic muscle

[a] Key to references: 1. von Figura and Hasilik (1986); 2. Hoflack and Kornfeld (1985a); 3. Hoflack and Kornfeld (1985b); 4. Ashwell and Harford (1982); 5. Schwartz (1984); 6. Kawasaki and Ashwell (1977); 7. Drickamer (1981); 8. Maynard and Baenziger (1981); 9. Stahl et al. (1980); 10. Haltiwanger et al. (1986c); 11. Lennartz et al. (1987); 12. Lehrman et al. (1986a,b,c); 13. Haltiwanger et al. (1986a); 14. Maynard and Baenziger (1982); 15. Mori et al. (1983); 16. Brownell et al. (1984); 17. Mori et al. (1984); 18. Colley and Baenziger (1987a,b); 19. Kolb-Bachofen et al. (1982); 20. Roos et al. (1985); 21. Schlepper-Schofer et al. (1986); 22. Jakoi et al. (1981); 23. Marchase et al. (1984); 24. Barondes (1984); 25. Barondes (1981); 26. Beyer et al. (1980); 27. Harrison and Chesterton (1980); 28. Ceri et al. (1981).

85

35% of the receptors are confined to the plasma membrane; the rest are found near Golgi, smooth endoplasmic reticulum, and endosomal membranes (Gueze *et al.*, 1982, 1983).

The asialoglycoprotein receptor on mammalian hepatocytes has been extensively studied as a model system for receptor-mediated endocytosis (for reviews, see Harford and Ashwell, 1982; Schwartz, 1984; Weigel, 1987). The *in vivo* function of this receptor, however, is uncertain. Numerous investigators have shown in many systems that desialylated glycoproteins, when injected into mammals, are rapidly removed from serum by this receptor in liver. However, four laboratories have shown that under normal circumstances serum glycoproteins are not desialylated *in vivo* and are not cleared from the circulation by this receptor (Wong *et al.*, 1974; Clarenburg, 1983; Kuranda and Aronson, 1983; Lefort *et al.*, 1984).

Vertebrate lectins have binding specificities for sugars that are commonly found in membrane and secretory glycoproteins. Several are known to have binding specificities for the sugars that occur at the nonreducing termini of oligosaccharide chains (Sarkar *et al.*, 1979; Briles *et al.*, 1979; Kawasaki and Ashwell, 1977). For example, liver cells contain a surface receptor that binds glycan chains containing terminal fucose in α-1,3 linkages; glycan chains containing terminal fucose in α-1,2 or α-1,6 linkages appear not to be recognized by this lectin (Prieels *et al.*, 1978). The asialoglycoprotein receptor of mammalian liver binds Glc-terminated oligosaccharides in addition to Gal- and GalNAc-terminated oligosaccharides (Stowell and Lee, 1978; Baenziger and Maynard, 1980).

Purified human hepatic asialoglycoprotein receptor has a much higher affinity of binding for complex oligosaccharides bearing three terminal Gal residues than those with two residues (Baenziger and Maynard, 1980). Triantennary complex oligosaccharides with three terminal Gal residues are rapidly endocytosed by hepatocytes, but complex oligosaccharides with one or two terminal Gal residues are not endocytosed although they are bound by the hepatocytes (Baenziger and Fiete, 1980). Maynard and Baenziger (1981) have studied the specificity of the Man/GlcNAc receptor for binding N-linked glycan chains and found that the minimum structure required for binding and endocytosis by rat liver reticuloendothelial cells is the six residue oligosaccharide GlcNAcβ-1,6 or Manα-1,6, Manα-1,6Manα-1,3Manβ-1,4GlcNAcβ-1,4GlcNAcβAsn. Attachment of additional sugar residues to this hexasaccharide can prevent binding and endocytosis by liver reticuloendothelial cells. This lectin must have an extended binding site which accommodates several sugars and a highly restricted binding specificity.

A Man/GlcNAc-specific lectin has been isolated from rabbit and rat liver by affinity chromatography on mannan–Sepharose (Kawasaki *et al.*, 1978; Mizuno *et al.*, 1981; Townsend and Stahl, 1981). Detailed characterization of this lectin's carbohydrate specificity by Maynard and Baenziger (1982) indicated that the structural requirements for recognition of N-linked glycans include primarily the core region of oligosaccharide chains plus at least some peptide. This lectin is mostly associated with the rough microsomal fractions of liver (Mori *et al.*, 1984) and is either a soluble or peripheral membrane protein rather than an integral membrane protein since it can be extracted without detergent (Maynard and Baenziger, 1982; Mori *et al.*, 1983).

Mori *et al.* (1984) found that there was a gradient of lectin occupancy of endogenous glycoproteins from the RER to the Golgi. Most of the lectin binding sites in the rough microsomes (85%) and about one-third of the binding sites in the smooth microsomes (36%) and Golgi membranes (35%) were occupied by specifically bound (i.e., mannan-displaceable) glycoproteins. In this respect, the Man/GlcNAc lectin resembles the CI Man-6-P receptor that is also found in association with endogenous ligands (Fisher *et al.*, 1980). The Man/GlcNAc lectin is also found in the plasma of a number of mammals (Kozutsumi *et al.*, 1980, 1981). Brownell *et al.* (1984) reported that the lectin is secreted by hepatocytes and hepatoma cells; however, the kinetics of secretion are unusually slow (~50% of the lectin is secreted in 7 hr).

Although the *in vivo* function of the liver Man/GlcNAc lectin is currently unclear, its binding specificity, localization in the RER, and association with endogenous ligands have led to the suggestion that the lectin may serve as a carrier protein for the transport of biosynthetic intermediates of glycoproteins with high-mannose-type oligosaccharides from the RER to the Gogi apparatus (Mori *et al.*, 1984; Kawasaki *et al.*, 1987). Alternatively, this lectin may serve to anchor glycoproteins within the cisternal space of the ER, preventing them from moving to the Golgi apparatus as has been described by Le and Doyle (1985). The presence of high-mannose-type oligosaccharide chains on microsomal enzymes supports this latter hypothesis (Mizuochi *et al.*, 1981; Liscum *et al.*, 1983).

Soluble lectins with a binding specificity for lactose have been implicated in secretion of mucin by chicken intestinal mucosa. Both chicken lactose-lectin I (the minor component) and chicken lactose-lectin II have been localized in secretory vesicles of the goblet cells of the intestinal mucosa (Beyer *et al.*, 1979; Beyer and Barondes, 1982). Purified chicken intestinal mucin, which is also concentrated in these vesicles, can interact with these lectins since it is a potent inhibitor of their specific hemagglutination activities. These results suggest that the lectins are somehow involved in the organization and/or secretion of mucin from goblet cells into the intestinal lumen.

Marchase *et al.* (1982) have identified a distinct subclass of peripheral membrane glycoproteins whose N-linked glycan chains, unlike those on lysosomes, contained phosphodiester-linked glucose. A lectin, ligatin, has been isolated with a binding site for such a phosphooligosaccharide and seems to utilize this binding site to secure the phosphoglucose-containing glycoproteins to the external surfaces of cells. Ligatin was first identified as the filamentous baseplate for a morphologically distinctive array of β-N-acetylhexosaminidase molecules in suckling rat ileum (Jakoi *et al.*, 1976). It has since been found to bind hydrolases in rat macrophages (Jakoi *et al.*, 1981) and brain (Gaston *et al.*, 1982). In addition, it has been found in embryonic chicken neural retina (Jakoi and Marchase, 1979), where it inhibits the intercellular adhesion of dissociated retinal cells (Marchase *et al.*, 1981), and thus has been implicated in intercellular recognition.

Although not identical, ligatin preparations from all of the above sources are closely related and share the following properties: they are solubilized from membranes by high Ca^{2+} concentrations and are depolymerized by ethylenebis (oxyethylenenitrilo) tetraacetic acid; they form filaments 3 nm in diameter in the presence of Ca^{2+}; they have apparent monomeric molecular weights of 10,000–20,000 and they possess amino acid compositions characterized by relatively high levels of acidic residues and low levels of hydrophobic residues. Additionally, in affinity studies, they bind proteins that cosolubilize with them and release these proteins on treatment with specific phosphorylated sugars.

The synthetic origin of the phosphate present in the ligatin-associated oligosaccharide has been studied (Koro and Marchase, 1982; Hiller *et al.*, 1987). Similar to the results found by Reitman and Kornfeld (1981a) and Hasilik *et al.* (1980) in their studies of lysosomal enzymes, the presence of phosphate in the ligatin-associated oligosaccharides seems to arise from the transfer of sugar 1-phosphate by a phosphoglycosyltransferase to terminal mannose residues on high-mannose N-linked oligosaccharides. The reaction for UDPGlc:glycoprotein glucose-1-phosphotransferase (Glc-P-transferase) in the retinal system is UDPGlc + oligosaccharide $\rightarrow$ oligosaccharide-P-Glc + UMP. Unlike synthesis of the lysosomal transport marker, the phosphodiester linkage described for ligatin-associated retinal cell surface proteins does not seem to undergo cleavage *in vivo* to expose Man-6-P, since phosphodiester-linked glucose is present in cell surface preparations and apparently is required for binding to ligatin.

The location of ligatin and its associated phosphoglycoproteins on cell surfaces has led to the suggestion that ligatin may function in intracellular localization and that phosphodiester-linked glucose may be a tag for targeting specific glycoproteins to the surface membrane in much the same way that Man-6-P targets acid hydrolases to lysosomes (Jakoi *et al.*, 1981; Marchase *et al.*, 1982, 1984, 1987). Newly synthesized glycoprotiens destined for the cell surface may share a common feature of their peptide structure that would result in their recognition by Glc-P-transferase. The phosphodiester-linked glucose would then be recognized by a receptor distinct from the Man-6-P receptor utilized in localization of lysosomal hydrolases. This receptor could be ligatin or another protein of similar binding specificity. This recognition would ultimately lead to the segregation of these proteins in vesicles transported to the cell surface. At the cell surface these protiens could establish a stable presence through binding with cell surface baseplate, ligatin. Additional studies are needed, however, to confirm this interesting model.

X. CONCLUDING REMARKS

The stepwise transport of newly synthesized secretory and membrane glycoprotiens through the subcompartments of the RER and the Golgi apparatus to the cell surface is highly regulated, and presumably signaling and sorting mechanisms are essential at every step for proper traffic regulation. The evidence that carbohydrate groups have important functional roles in the intracellular transport and/or sorting of specific secretory and membrane glycoproteins has increased substantially in the last 5 years. The elegant studies from the laboratory of John Rose, using site-directed mutagenesis techniques, have clearly demonstrated that N-linked glycan chains promote the efficient transport of specific membrane proteins to the cell surface. The close associations between glycosylation defects and altered expression of the Thy-1 antigen, the Man-6-P receptor, and the LDL receptor emphasize the importance of carbohydrate chains for the expression and function of some cell surface glycoproteins. Many, but not all, glycoproteins fail to be secreted or to be expressed on the cell surface if normal glycosylation is inhibited with TM. In addition, studies mostly using 1-deoxynorjirimycin have demonstrated that the trimming of glucose residues from N-linked glycan chains on specific secretory, lysosomal, and membrane proteins is required for efficient transport of these glycoproteins from the RER to the Golgi.

Although these studies demonstrate that carbohydrate groups are important for the traffic of many, but not all, secretory and membrane glycoproteins, the precise role of carbohydrate in intracellular transport is unclear. In analogy with the Man-6-P pathway for transport of acid hydrolases to lysosomes, it is tempting to postulate that specific carbohydrate-binding receptors mediate the intracellular transport of specific nonlysosomal glycoproteins. This has not been clearly established, as yet, however, and efforts to identify and iolate carbohydrate-binding transport receptors distinct form the Man-6-P receptors are clearly indicated. In future studies in this area, it will also be very important to determine completely the structure of the oligosaccharide chains attached to glycoproteins being investigated. There is considerable evidence for alternative pathways for the processing of N-linked glycans, and detailed structural information will probably be required to make significant progress in understanding the presumably complex roles of carbohydrate in glycoprotein traffic.

REFERENCES

Arumugham, R. G., and Tanzer, M. L. (1983). Abnormal glycosylation of human cellular fibronectin in the presence of swainsonine. *J. Biol. Chem.* **258,** 11883–11889.

Ashwell, G., and Harford, J. (1982). Carbohydrate-specific receptors of the liver. *Annu. Rev. Biochem.* **51,** 531–554.

Atkinson, P. H., and Lee, J. T. (1984). Cotranslational excision of α-glucose and α-mannose in nascent vesicular stomatitis virus G protein. *J. Cell Biol.* **98,** 2245–2249.

Aubert, J. P., Biserte, G., and Loucheux-Lefebvre, M. H. (1976). Carbohydrate–peptide linkage in glycoproteins. *Arch. Biochem. Biophys.* **175,** 410–418.

Baenziger, J. U., and Fiete, D. (1980). Galactose and N-acetylgalactosamine-specific endocytosis of glycopeptides by isolated rat hepatocytes. *Cell* **22,** 611–620.

Baenziger, J. U., and Maynard, Y. (1980). Human hepatic lectin: Physiochemical properties and specificity. *J. Biol. Chem.* **255,** 4607–4613.

Barclay, A. N., Letarte-Muirhead, M., Williams, A. F., and Faulkes, R. A. (1976). Chemical characterization of the Thy-1 glycoproteins from the membranes of rat thymocytes and brain. *Nature (London)* **263;** 563–567.

Barondes, S. H. (1981). Lectins: Their multiple endogenous cellular functions. *Annu. Rev. Biochem.* **50,** 207–231.

Barondes, S. H. (1984). Soluble lectins: A new class of extracellular proteins. *Science* **223,** 1259–1264.

Barouki, R., Finidori, J., Chobert, M. N., Aggerbeck, M., Laperche, Y., and Hanoune, J. (1984). Biosynthesis and processing of γ-glutamyl transpeptidase in hepatoma tissue culture cells. *J. Biol. Chem.* **259,** 7970–7974.

Bar-Sagi, D., and Prives, J. (1983). Tunicamycin inhibits the expression of surface Na^+ channels in cultured muscle cells. *J. Cell. Physiol.* **114,** 77–81.

Bartalena, L., and Robbins, J. (1984). Effect of tunicamycin and monensin on secretion of thyroxine-binding globulin by cultured human hepatoma (Hep G2) cells. *J. Biol. Chem.* **259,** 13610–13614.

Bathurst, I. C., Travis, J., George, P. M., and Carrell, R. W. (1984). Structural and functional characterization of the abnormal Z α_1-antitrypsin isolated from human liver. *FEBS Lett.* **177,** 179–183.

Bauer, H. C., Parent, J. B., and Olden, K. (1985). Role of carbohydrate in glycoprotein secretion by human hepatoma cells. *Biochem. Biophys. Res. Commun.* **128,** 368–375.

Bause, E. (1983). Structural requirements of N-glycosylation of proteins. *Biochem. J.* **209,** 331–336.

Beeley, J. G. (1977). Peptide chain conformation and the glycosylation of glycoproteins. *Biochem. Biophys. Res. Commun.* **76,** 1051–1055.

Bergman, L. W., and Kuehl, W. M. (1982). *In* "The Glycoconjugates" (M. I. Horowitz, ed.), Vol. 3, pp. 81–98. Academic Press, New York.

Bernard, B. A., Yamada, K. M., and Olden, K. (1982). Carbohydrates selectively protect a specific domain of fibronectin against proteases. *J. Biol. Chem.* **257,** 8549–8554.

Beyer, E. C., and Barondes, S. H. (1982). Secretion of endogenous lectin by chicken intestinal goblet cells. *J. Cell Biol.* **92,** 28–33.

Beyer, E. C., Tokuyasu, K. T., and Barondes, S. H. (1979). Localization of an endogenous lectin in chicken liver, intestine and pancreas. *J. Cell Biol.* **82,** 565–571.

Beyer, E. C., Zweig, S. E., and Barondes, S. H. (1980). Two lactose binding lectins from chicken tissues. *J. Biol. Chem.* **255,** 4236–4239.

Beyer, T. A., and Hill, R. L. (1982). *In* "The Glycoconjugates" (M. I. Horowitz, ed.), Vol. 3, pp. 25–45. Academic Press, New York.

Bischoff, J., and Kornfeld, R. (1983). Evidence for an α-mannosidase in endoplasmic reticulum of rat liver. *J. Biol. Chem.* **258,** 7907–7910.

Bischoff, J., and Kornfeld, R. (1984). The effect of 1-deoxymannojirimycin on rat liver α-mannosidases. *Biochem. Biophys. Res. Commun.* **125,** 324–331.

Bischoff, J., Liscum, L., and Kornfeld, R. (1986). The use of 1-deoxymannojirimycin to evaluate the role of various α-mannosidases in oligosaccharide processing in intact cells. *J. Biol. Chem.* **261,** 4766–4774.

Bjorkman, U., and Ekholm, R. (1982). Effect of tunicamycin on thyroglobulin secretion. *Eur. J. Biochem.* **125,** 585–591.

Bourguignon, L. Y. M., Balazovich, K., Trowbridge, I. S., and Hyman, R. (1982). Immunoelectron microscopic localization of the Thy-1 glycoprotein in wild type and Thy-1 negative lymphoma cells. *Cell Biol. Int. Rep.* **6,** 745–755.

Breitfeld, P. P., Rup, D., and Schwartz, A. L. (1984). Influence of the N-linked oligosaccharides on the biosynthesis, intracellular routing, and function of the human asialoglycoprotein receptor. *J. Biol. Chem.* **259,** 10414–10421.

Briles, E. B., Gregory, W., Fletcher, P., and Kornfeld, S. (1979). Vertebrate lectins: Comparison of properties of β-galactoside-binding lectins from tissues of calf and chicken. *Cell Biol.* **81,** 528–537.

Brown, P. H., and Hickman, S. (1986). Oligosaccharide processing at individual glycosylation sites on MOPC 104E immunoglobulin M. Differences in α-1,2-linked mannose processing. *J. Biol. Chem.* **261,** 2575–2582.

Brownell, M. D., Colley, K. J., and Baenziger, J. U. (1984). Synthesis, processing and secretion of the core-specific lectin by rat hepatocytes and hepatoma cells. *J. Biol. Chem.* **259,** 3925–3932.

Budarf, M. L., and Herbert, E. (1982). Effect of tunicamycin on the synthesis, processing and secretion of proopiomelanocortin peptides in mouse pituitary cells. *J. Biol. Chem.* **257,** 10128–10135.

Burke, B., Matlin, K., Bause, E., Legler, G., Peyrieras, N., and Ploegh, H. (1984). Inhibition of N-linked oligosaccharide trimming does not interfere with surface expression of certain integral membrane proteins. *EMBO J.* **3,** 551–556.

Campbell, D., Gagnon, J., Reid, K., and Williams, A. (1981). Rat brain Thy-1 glycoprotein. The amino acid sequence, disulfide bonds, and an unusual hydrophobic region. *Biochem. J.* **195,** 15–30.

Carrell, R. W., Bathurst, I. C., and Brennan, S. O. (1984). The molecular pathology of α_1-antitrypsin. *Biochem. Soc. Symp.* **49**, 55–66.

Ceri, H., Kobiler, D., and Barondes, S. H. (1981). Heparin-inhibitable lectin. *J. Biol. Chem.* **256**, 390–394.

Chapman, A., Li, E., and Kornfeld, S. (1979). The biosynthesis of the major lipid-linked oligosaccharide of Chinese hamster ovary cells occurs by the ordered addition of mannose residues. *J. Biol. Chem.* **254**, 10243–10249.

Chapman, A., Fujimoto, K., and Kornfeld, S. (1980). The primary glycosylation defect in class E thy-1 negative mutant mouse lymphoma cells is an inability to synthesize dolichol-P-mannose. *J. Biol. Chem.* **255**, 4441–4446.

Chatterjee, S., Kwiterovich, Jr., P. O., and Sekerke, C. S. (1979). Effects of tunicamycin on the binding and degradation of low density lipoproteins and glycoprotein synthesis in cultured human fibroblasts. *J. Biol. Chem.* **254**, 3704–3707.

Chatterjee, S., Sekerke, C. S., and Kwiterovich, P. O., Jr. (1981). Effects of tunicamycin on the cell-surface binding, internalization, and degradation of low-density lipoproteins in human fibroblasts. *Eur. J. Biochem.* **120**, 435–441.

Chu, F. K., and Maley, F. (1982). Stabilization of the structure and activity of yeast carboxypeptidase Y by its high-mannose oligosaccharide chains. *Arch. Biochem. Biophys.* **214**, 134–139.

Chu, F. K., Trimble, R. B., and Maley, F. (1978). The effect of carbohydrate depletion on the properties of yeast external invertase. *J. Biol. Chem.* **253**, 8691–8693.

Ciccimarra, F., Rosen, F. S., Schneeberger, E., and Merler, E. (1976). Failure of heavy chain glycosylation of IgG in some patients with common, variable agammaglobulinemia. *J. Clin. Invest.* **57**, 1386–1390.

Clarenburg, R. (1983). Asialoglycoprotein receptor is uninvolved in clearing intact glycoproteins from rat blood. *Am. Phys. Soc.* **244**, G247–G253.

Cohen, B. G., and Phillips, A. H. (1980). Evidence for rapid and concerted turnover of membrane phospholipids in MOPC 41 myeloma cells and its possible relationship to secretion. *J. Biol. Chem.* **255**, 3075–3079.

Cohen, B. G., Mosler, S., and Phillips, A. H. (1979). Rapid turnover of intracellular membranes in MOPC 41 myeloma cells and its possible relationship to secretion. *J. Biol. Chem.* **254**, 4267–4275.

Colley, K. J., and Baenziger, J. U. (1987a). Identification of the post-translational modifications of the core-specific lectin. *J. Biol. Chem.* **262**, 10290–10295.

Colley, K. J., and Baenziger, J. U. (1987b). Posttranslational modifications of the core-specific lectin. *J. Biol. Chem.* **262**, 10296–10303.

Courtney, M., Buchwalder, A., Tessier, L. H., Jaye, M., Benavente, A., Balland, A., Kohli, V., Lathe, R., Tolstoshev, P., and Lecocq, J. P. (1984). High level production of biologically active human α_1-antitrypsin in *Escherichia coli*. *Proc. Natl. Acad. Sci. U.S.A.* **81**, 669–673.

Cummings, R. D., Kornfeld, S., Schneider, W. J., Hobgood, K. K., Tolleshaug, H., Brown, M. S., and Goldstein, J. L. (1983). Biosynthesis of N- and O-linked oligosaccharides of the low density lipoprotein receptor. *J. Biol. Chem.* **258**, 15261–15273.

Danielsen, E. M., Cowell, G. M., Noren, O., Sjostrom, H., and Dorling, P. R. (1983). Biosynthesis of intestinal microvillar proteins. The effect of swainsonine on post-translational processing of aminopeptidase N. *Biochem. J.* **216**, 325–331.

Datema, R., and Schwarz, R. T. (1981). Effect of energy depletion on the glycosylation of a viral glycoprotein. *J. Biol. Chem.* **256**, 11191–11198.

Datema, R., Romero, P. A., Legler, G., and Schwarz, R. T. (1982). Inhibition of formation of complex oligosaccharides by the glucosidase inhibitor bromoconduritol. *Proc. Natl. Acad. Sci. U.S.A.* **79**, 6787–6791.

Datema, R., Romero, P. A., Rott, R., and Schwarz, R. T. (1984). One the role of oligosaccharide trimming in the maturation of Sindbis and influenza virus. *Arch. Virol.* **81**, 25–39.

Davis, C. G., Elhammer, A., Russell, D. W., Schneider, W. J., Kornfeld, S., Brown, M. S., and Goldstein, J. L. (1986). Deletion of clustered O-linked carbohydrates does not impair function or low density lipoprotein receptor in transfected fibroblasts. *J. Biol. Chem.* **261**, 2828–2838.

Davis, L. I., and Blobel, G. (1986). Identification and characterization of a nuclear pore complex protein. *Cell* **45**, 699–709.

Derynck, R., Remaut, E., Saman, E., Stansseus, P., De Clercq, E., Content, J., and Fiers, W. (1980). Expression of human fibroblast interferon gene in Escherichia coli. *Nature (London)* **287**, 193–197.

Docherty, P. A., Kuranda, M. J., Aronson, N. N., Jr., BeMiller, J. N., Myers, R. W., and Bohn, J. A. (1986). Effect of α-D-mannopyranosylmethyl-p-nitrophenyltriazene on hepatic degradation and processing of the N-linked oligosaccharide chains of α_1-acid glycoprotein. *J. Biol. Chem.* **261**, 3457–3463.

Doss, R. C., Kramarcy, N. K., Harden, T. K., and Perkins, J. P. (1985). Effects of tunicamycin on the expression of β-adrenergic receptors in human astrocytoma cells during growth and recovery from agonist-induced down-regulation. *Mol. Pharmacol.* **27**, 507–516.

Drickamer, K. (1981). Complete amino acid sequence of a membrane receptor for glycoproteins. *J. Biol. Chem.* **256**, 5827–5839.

Duksin, D., and Bornstein, P. (1977). Impaired conversion of procollagen to collagen by fibroblasts and bone treated with tunicamycin, an inhibitor of protein glycosylation *J. Biol. Chem.* **252**, 955–962.

Dunphy, W. G., and Rothman, J. E. (1985). Compartmental organization of the Golgi stack. *Cell* **42**, 13–21.

Dunphy, W. G., Brands, R., and Rothman, J. E. (1985). Attachment of terminal N-acetylglucosamine to asparagine-linked oligosaccharides occurs in central cisternae of the Golgi stack. *Cell* **40**, 463–472.

Duronio, V., Jacobs, S., and Cuatrecasas, P. (1986). Complete glycosylation of the insulin and insulin-like growth factor 1 receptors is not necessary for their biosynthesis and function. *J. Biol. Chem.* **261**, 970–975.

Eggo, M. C., and Burrow, G. (1983). Glycosylation of thyroglobulin. Its role in secretion, iodination and stability. *Endocrinology* **113**, 1655–1663.

Elbein, A. D. (1984). Inhibitors of the biosynthesis and processing of N-linked oligosaccharides. *CRC Crit. Rev. Biochem.* **16**, 21–49.

Elbein, A. D. (1987). Inhibitors of the biosynthesis and processing of N-linked oligosaccharide chains. *Annu. Rev. Biochem.* **56**, 497–534.

Elbein, A. D., Solf, R., Dorling, P. R., and Vosbeck, K. (1981). Swainsonine: An inhibitor of glycoprotein processing. *Proc. Natl. Acad. Sci. U.S.A.* **78**, 7393–7397.

Elbein, A. D., Dorling, P. R., Vosbeck, K., and Horisberger, M. (1982). Swainsonine prevents the processing of the oligosaccharide chains of influenza virus hemagglutinin. *J. Biol. Chem.* **257**, 1573–1576.

Elbein, A. D., Legler, G., Tlusty, A., McDowell, W., and Schwarz, R. (1984). The effect of deoxymannojirimycin on the processing of the influenza viral glycoproteins. *Arch. Biochem. Biophys.* **235**, 579–588.

Farquhar, M. G. (1985). Progress in unraveling pathways of Golgi traffic. *Annu. Rev. Cell Biol.* **1**, 447–488.

Farquhar, M. G., and Palade, G. E. (1981). The Golgi apparatus (complex)—(1954–1981)—from artifact to center stage. *J. Cell Biol.* **91**, 77s–103s.

Filipovic, I., and von Figura, K. (1980). Effect of tunicamycin on the metabolism of low-density lipoproteins by control and low-density-lipoprotein-receptor-deficient human skin fibroblasts. *Biochem. J.* **186**, 373–375.

Firestone, G. L. (1983). The role of protein glycosylation in the compartmentalization and processing of mouse mammary tumor virus glycoproteins in mouse mammary tumor virus-infected rat hepatoma cells. *J. Biol. Chem.* **258**, 6155–6161.

Firestone, G. L., and Heath, E. C. (1981). Role of protein glycosylation in the cAMP-mediated induction of alkaline phosphatase in mouse L-cells. *J. Biol. Chem.* **256**, 1404–1411.

Fisher, H. D., Gonzalez-Noriega, A., Sly, W. S., and Morre, D. J. (1980). Phosphomannosyl–enzyme receptors in rat liver. Subcellular distribution and role in intracellular transport of lysosomal enzymes. *J. Biol. Chem.* **255**, 9608–9615.

Fitting, T., and Kabat, D. (1982). Evidence for a glycoprotein "signal" involved in transport between subcellular organelles. Two membrane glycoproteins encoded by murine leukemia virus reach the cell surface at different rates. *J. Biol. Chem.* **257**, 14011–14017.

Fitting, T., Ruta, M., and Kabat, D. (1981). Mutant cells that abnormally process plasma membrane glycoproteins encoded by murine leukemia virus. *Cell* **24**, 847–858.

Fliesler, S. J., and Basinger, S. F. (1985). Tunicamycin blocks the incorporation of opsin into retinal rod outer segment membranes. *Proc. Natl. Acad. Sci. U.S.A.* **82**, 1116–1120.

Fliesler, S. J., Rayborn, M. E., and Hollyfield, J. G. (1985). Membrane morphogenesis in retinal rod outer segments: Inhibition by tunicamycin. *J. Cell Biol.* **100**, 574–587.

Frank, J. L., Housepian, S., Fayet, G., and Bouchilloux (1986). Inhibition of N-linked oligosaccharide processing does not prevent the secretion of thyroglobulin. *Eur. J. Biochem.* **157**, 225–232.

Fries, E., Gustafsson, L., and Peterson, P. A. (1984). Four secretory proteins synthesized by hepatocytes are transported from endoplasmic reticulum to Golgi complex at different rates. *EMBO J.* **3**, 147–152.

Fuhrmann, U., Bause, E., Legler, G., and Ploegh, H. (1984). Novel mannosidase inhibitor blocking conversion of high mannose to complex oligosaccharides. *Nature (London)* **307**, 755–758.

Fuhrmann, U., Bause, E., and Ploegh, H. (1985). Inhibitors of oligosaccharide processing. *Biochim. Biophys. Acta* **825**, 95–110.

Gabel, C. A., and Bergmann, J. E. (1985). Processing of the asparagine-linked oligosaccharides of secreted and intracellular forms of the vesicular stomatitis virus G protein: *In vivo* evidence of Golgi apparatus compartmentalization. *J. Cell Biol.* **101**, 460–469.

Gabel, C. A., and Kornfeld, S. (1982). Lysosomal enzyme phosphorylation in mouse lymphoma cell lines with altered asparagine-linked oligosaccharides. *J. Biol. Chem.* **257**, 10605–10612.

Gahmberg, C. G., Jokinen, M., Karhi, K. K., and Andersson, L. C. (1980). Effect of tunicamycin on the biosynthesis of the major human red cell sialoglycoprotein, glycophorin A, in the leukemia cell line K562. *J. Biol. Chem.* **255**, 2169–2175.

Gallione, C. J., and Rose, J. K. (1983). Nucleotide sequence of a cDNA clone encoding the entire glycoprotein from the New Jersey serotype of vesicular stomatitis virus. *J. Virol.* **46**, 162–169.

Gaston, S. M., Marchase, R. B., and Jakoi, E. R. (1982). Brain ligatin: A membrane lectin that binds acetylcholinesterase. *J. Cell. Biochem.* **18**, 447–459.

George, S. T., Ruoho, A. E., and Malbou, C. C. (1986). N-glycosylation in expression and function of β-adrenergic receptors.

Geuze, H. J., Slot, J. W., Strous, G. J. A. M., Lodish, H. F., and Schwartz, A. L. (1982). Immunocytochemical localization of the receptor for asialoglycoprotein in rat liver cells. *J. Cell Biol.* **92**, 865–870.

Geuze, H. J., Slot, J. W., Strous, G. J. A. M., and Schwartz, A. L. (1983). The pathway of the asialoglycoprotein–ligand during receptor-mediated endocytosis: A morphological study with colloidal gold/ligand in the human hepatoma cell line, Hep G2. *Eur. J. Cell Biol.* **32**, 38–44.

Gibson, R., Schlesinger, S., and Kornfeld, S. (1979). The nonglycosylated glycoprotein of vesicular stomatitis virus is temperature-sensitive and undergoes intracellular aggregation at elevated temperatures. *J. Biol. Chem.* **254**, 3600–3607.

Gibson, R., Kornfeld, S., and Schlesinger, S. (1980). A role for oligosaccharides in glycoprotein biosynthesis. *Trends Biochem. Sci.* **5**, 290–293.

Gibson, R., Kornfeld, S., and Schlesinger, S. (1981). The effect of oligosaccharide chains of different sizes on the maturation and physical properties of the G protein of vesicular stomatitis virus. *J. Biol. Chem.* **256**, 456–462.

Gonzalez-Noriega, A., Grubb, J. H., Talkad, V., and Sly, W. S. (1980). Chloroquine inhibits lysosomal enzyme pinocytosis and enhances lysosomal enzyme secretion by impairing receptor recycling. *J. Cell Biol.* **85**, 839–852.

Grafl, R., Lang, K., Vogl, H., and Schmid, F. X. (1987). The mechanism of folding of pancreatic ribonuclease is independent of the presence of covalently linked carbohydrate. *J. Biol. Chem.* **262**, 10624–10629.

Green, R., Meiss, H., and Rodriguez-Boulan, E. J. (1981). Glycosylation does not determine segregation of viral envelope proteins in the plasma membrane of epithelial cells. *J. Cell Biol.* **89**, 230–239.

Grey, P. W., and Goeddel, D. V. (1983). Cloning and expression of murine immune interferon cDNA. *Proc. Natl. Acad. Sci. U.S.A.* **80**, 5842–5846.

Grey, P. W., Leung, D. W., Pennica, D., Yelverton, E., Najarian, R., Simonsen, C. C., Derynck, R., Sherwood, P. J., Wallace, D. M., Berger, S. L., Levinson, A. D., and Goeddel, D. V. (1982). Expression of human immune interferon cDNA in *E. coli* and monkey cells. *Nature (London)* **295**, 503–508.

Gross, V., Andus, T., Tran-Thi, T. A., Schwarz, R. T., Decker, K., and Heinrich, P. C. (1983a). 1-Deoxynojirimycin impairs oligosaccharide processing of α_1-proteinase inhibitors and inhibits its secretion in primary cultures of rat hepatocytes. *J. Biol. Chem.* **258**, 12203–12209.

Gross, V., Tran-Thi, T. A., Vosbeck, K., and Heinrich, P. C. (1983b). Effect of swainsonine on the processing of the asparagine-linked carbohydrate chains of α_1-antitrypsin in rat hepatocytes. *J. Biol. Chem.* **258**, 4032–4036.

Gross, V., Steube, K., Tran-Thi, T. A., McDowell, W., Schwarz, R. T., Decker, C., Gerok, W., and Heinrich, P. C. (1985). Secretion of high-mannose-type α_1-proteinase inhibitor and α_1-acid glycoprotein by primary cultures of rat hepatocytes in the presence of the mannosidase 1 inhibitor 1-deoxymannojirimycin. *Eur. J. Biochem.* **150**, 41–46.

Guan, J. L., and Rose, J. K. (1984). Conversion of a secretory protein into a transmembrane protein results in its transport to the Golgi complex but not to the cell surface. *Cell* **37**, 779–787.

Guan, J. L., Machamer, C. E., and Rose, J. K. (1985). Glycosylation allows cell-surface transport of an anchored secretory protein. *Cell* **42**, 489–496.

Kelly, R. B. (1985). Pathways of protein secretion in eukaryotes. *Science* **230**, 25–32.

Kingsley, D. M., Kozarsky, K. F., Hobbie, L., and Krieger, M. (1986a). Reversible defects in O-linked glycosylation and LDL receptor expression in a UDPGal/UDPGalNAc 4-epimerase deficient mutant. *Cell* **44**, 749–759.

Kingsley, D. M., Kozarsky, K. F., Segal, M., and Krieger, M. (1986b). Three types of low density lipoprotein receptor-deficient mutants have pleiotropic defects in the synthesis of N-linked, O-linked and lipid-linked carbohydrate chains. *J. Cell Biol.* **102**, 1576–1585.

Kobata, A. (1984). *In* "Biology of Carbohydrates" (V. Ginsburg and P. W. Robbins, eds.), Vol. 2, pp. 87–161. Wiley (Interscience), New York.

Kobiler, D. (1987). Developmentally regulated soluble lectins. *In* "Vertebrate Lectins (K. Olden and J. B. Parent, eds.) pp. 195–210. Van Nostrand Reinhold, New York.

Kolb-Bachofen, V., Schlepper-Schöfer, J., and Vogell, W. (1982). Electron microscopic evidence for an asialoglycoprotein receptor on Kupffer Cells: Localization of lectin-mediated endocytosis. *Cell* **29**, 859–866.

Kornfeld, R., and Kornfeld, S. (1980). *In* "The Biochemistry of Glycoproteins and Proteoglycans" (W. J. Lennarz, ed.), pp. 1–34. Plenum, New York.

Kornfeld, R., and Kornfeld, S. (1985). Assembly of asparagine-linked oligosaccharides. *Annu. Rev. Biochem.* **54**, 631–664.

Kornfeld, S., Li, E., and Tabas, I. (1978). The synthesis of complex-type oligosaccharides. II. Characterization of the processing intermediates in the synthesis of the complex oligosaccharide units of the vesicular stomatitis virus G protein. *J. Biol. Chem.* **253**, 7771–7778.

Kornfeld, S., Gregory, W., and Chapman, A. (1979). Class E Thy-1 negative mouse lymphoma cells utilize an alternate pathway of oligosaccharide processing to synthesize complex-type oligosaccharides. *J. Biol. Chem.* **254**, 11649–11654.

Koro, L. A., and Marchase, R. B. (1982). A UDPglucose:glycoprotein glucose-1-phosphotransferase in embryonic chicken neural retina. *Cell* **31**, 739–748.

Kozutsumi, Y., Kawasaki, T., and Yamashina, I. (1980). Isolation and characterization of a mannan-binding protein from rabbit serum. *Biochem. Biophys. Res. Commun.* **95**, 658–664.

Kozutsumi, Y., Kawasaki, T., and Yamashina, I. (1981). Kinetical properties of the serum mannan-binding protein from rabbit. A comparison with those of the liver mannan-binding protein. *J. Biochem.* **90**, 1799–1807.

Krag, S. S. (1979). A concanavalin A-resistant Chinese hamster ovary cell line is deficient in the synthesis of [³H]glucosyl oligosaccharide–lipid. *J. Biol. Chem.* **254**, 9167–9177.

Krag, S. S. (1985). Mechanisms and functional role of glycosylation in membrane protein synthesis. *Curr. Top. Membr. Transport* **24**, 181–249.

Krag, S. S., and Robbins, A. R. (1982). A Chinese hamster ovary cell mutant deficient in glucosylation of lipid-linked oligosaccharide synthesizes lysosomal enzymes of altered structure and function. *J. Biol. Chem.* **257**, 8424–8431.

Krebs, H., Schmid, F. X., and Jaenicke, R. (1983). Folding of homologous proteins. The refolding of different ribonucleases is independent of sequence variations, proline content and glycosylation. *J. Mol. Biol.* **169**, 619–635.

Krieger, M., Kingsley, D. M., Sege, R. D., Hobbie, L., and Kozarsky, K. F. (1985). Genetic analysis of receptor-mediated endocytosis. *Trends Biochem. Sci.* **10**, 447–452.

Kuranda, M. J., and Aronson, N. N., Jr. (1983). Tissue locations for the turnover of radioactively labeled rat orosomucoid *in vivo. Arch. Biochem. Biophys.* **224**, 526–533.

Lalegerie, P., Legler, G., and Yon, J. M. (1982). The use of inhibitors in the study of glycosidases. *Biochimie* **64**, 977–1000.

Lang, L., Reitman, M., Tang, J., Roberts, R. M., and Kornfeld, S. (1984). Lysosomal enzyme phosphorylation. Recognition of a protein-dependent determinant allows specific phosphorylation of oligosaccharides present on lysosomal enzymes. *J. Biol. Chem.* **259**, 14663–14671.

Lawson, E. Q., Hedlund, B. E., Ericson, M. E., Mood, D. A., Litman, G. W., and Middaugh, R. (1983). Effect of carbohydrate on protein solubility. *Arch. Biochem. Biophys.* **220**, 572–575.

Le, A. V., and Doyle, D. (1985). N-linked oligosaccharides of the H-2D^k histocompatability protein heavy chain influence its transport and cellular distribution. *Biochemistry* **24**, 6238–6245.

Leavitt, R., Schlesinger, S., and Kornfeld, S. (1977). Impaired intracellular migration and altered solubility of nonglycosylated glycoproteins of vesicular stomatitis virus and sindbis virus. *J. Biol. Chem.* **252**, 9018–9023.

Ledford, B. E., and Davis, D. F. (1983). Kinetics of serum protein secretion by cultured hepatoma cells. *J. Biol. Chem.* **258**, 3304–3308.

Lefort, G. P., Stolk, J. M., and Nisula, B. C. (1984). Evidence that desialylation and uptake by hepatic receptors for galactose-terminated glycoproteins are immaterial to the metabolism of human choriogonadotropin in the rat. *Endocrinology* **115**, 1551–1557.

Lehrman, M. A., Haltiwanger, R. S., and Hill, R. L. (1986a). The binding of fucose-containing glycoproteins by hepatic lectins. The binding specificity of the rat liver fucose lectin. *J. Biol. Chem.* **261**, 7426–7432.

Lehrman, M. A., and Hill, R. L. (1986). The binding of fucose-containing glycoproteins by hepatic lectins. Purification of a fucose-binding lectin from rat liver. *J. Biol. Chem.* **261**, 7419–7425.

Lehrman, M. A., Pizzo, S. V., Imber, M. J., and Hill, R. L. (1986b). The binding of fucose-containing glycoproteins by hepatic lectins. Examination of the clearance from blood and the binding to membrane receptors and pure lectins. *J. Biol. Chem.* **261**, 7412–7418.

Leichter, A. M., and Krieger, M. (1984). Addition of mannose 6-phosphate-containing oligosaccharides alters cellular processing of low density lipoprotein by parental and LDL-receptor-defective Chinese hamster ovary cells. *J. Cell Sci.* **68**, 183–194.

Lemansky, P., Gieselmann, V., Hasilik, A., and von Figura, K. (1984). Cathepsin D and β-hexosaminidase synthesized in the presence of 1-deoxynorjirimycin accumulate in the endoplasmic reticulum. *J. Biol. Chem.* **259**, 10129–10135.

Liscum, L., Cummings, R. D., Anderson, R. G. W., DeMartino, G. N., Goldstein, J. L., and Brown, M. S. (1983). 3-Hydroxy-3-methylglutaryl-CoA reductase: A transmembrane glycoprotein of the endoplasmic reticulum with N-linked "high mannose" oligosaccharides. *Proc. Natl. Acad. Sci. U.S.A.* **80**, 7165–7169.

Lobermann, H., Tokuoka, R., Deisenhofer, J., and Huber, R. (1984). Human α_1-proteinase inhibitor. Crystal structure analysis of two crystal modifications. *J. Mol. Biol.* **255**, 4053–4061.

Lodish, H. F., and Kong, N. (1984). Glucose removal from N-linked oligosaccharides is required for efficient maturation of certain secretory glycoproteins from the rough endoplasmic reticulum to the Golgi complex. *J. Cell Biol.* **98**, 1720–1729.

Lodish, H. F., Kong, N., Snider, M., and Strous, G. J. A. M. (1983). Hepatoma secretory proteins migrate from rough endoplasmic reticulum to Golgi at characteristic rates. *Nature (London)* **304**, 80–83.

Loh, Y. P., and Gainer, H. (1978). The role of glycosylation of the biosynthesis, degradation and secretion of the ACTH–β-lipotropin common precursor and its peptide products. *FEBS Lett.* **96**, 269–272.

Loh, Y. P., and Gainer, H. (1979). The role of carbohydrate in the stabilization, processing, and packaging of the glycosylated adrenocorticotropin–endorphin common precursor in toad pituitaries. *Endocrinology* **105**, 474–487.

Loh, Y. P., and Gainer, H. (1980). Evidence that glycosylation of proopiocortin and ACTH influences their proteolysis by trypsin and blood proteases. *Mol. Cell. Endocrinol.* **20**, 35–44.

Lubas, W. A., and Spiro, R. G. (1987). Golgi endo-α-D-mannosidase from rat liver, a novel N-linked carbohydrate unit processing enzyme. *J. Biol. Chem.* **262**, 3775–3781.

Lynch, D. C., Williams, R., Zimmerman, T. S., Kirby, E. P., and Livingston, D. M. (1983). Biosynthesis of the subunits of factor VIIIR by bovine aortic endothelial cells. *Proc. Natl. Acad. Sci. U.S.A.* **80**, 2738–2742.

Machamer, C. E., Florkiewicz, R. Z., and Rose, J. K. (1985). A single N-linked oligosaccharide at either of the two normal sites is sufficient for transport of vesicular stomatitis virus G protein to the cell surface. *Mol. Cell. Biol.* **5**, 3074–3083.

Marchase, R. B., Harges, P., and Jakoi, E. R. (1981). Ligatin from embryonic chick neural retina inhibits retinal cell adhesion. *Dev. Biol.* **86**, 250–255.

Marchase, R. B., Koro, L. A., Kelly, C. M., and McClay, D. R. (1982). Retinal ligatin recognizes glycoproteins bearing oligosaccharides terminating in phosphodiester-linked glucose. *Cell* **28**, 813–820.

Marchase, R. B., Koro, L. A., and Hiller, A. M. (1984). Receptors for glycoproteins with phosphorylated oligosaccharides. *In* "The Receptors" (P. M. Conn, ed.), Vol. 1, pp. 261–313. Academic Press, New York.

Marchase, R. B. (1987). Receptors for glycoproteins with oligosaccharides containing α-glucose-1-phosphate. *In* "Vertebrate Lectins" (K. Olden and J. B. Parent, eds.), pp. 108–123. Van Nostrand Reinhold, New York.

Maynard, Y., and Baenziger, J. U. (1981). Oligosaccharide specific endocytosis by isolated rat hepatic reticuloendothelial cells. *J. Biol. Chem.* **256**, 8063–8068.

Maynard, Y., and Baenziger, J. U. (1982). Characterization of a mannose- and N-acetylglucosamine-specific lectin present in rat hepatocytes. *J. Biol. Chem.* **257**, 3788–3794.

Mega, T., Lujan, E., and Yoshida, A. (1980). Studies of the oligosaccharide chains of human α_1-protease inhibitor. II. Structure of oligosaccharides. *J. Biol. Chem.* **255**, 4057–4061.

Meiss, H. K., Green, R. F., and Rodriguez-Boulan, E. J. (1982). Lectin-resistant mutant of polarized epithelial cells. *Mol. Cell. Biol.* **2**, 1287–1294.

Merlie, J. P., Sebbane, R., Tzartos, S., and Lindstrom, J. (1982). Inhibition of glycosylation with tunicamycin blocks assembly of newly synthesized acetylcholine receptor subunits in muscle cells. *J. Biol. Chem.* **257**, 2694–2701.

Mizunaga, T., and Noguchi, T. (1982). The role of core-oligosaccharide in formation of an active acid phosphatase and its secretion by yeast protoplasts. *J. Biochem.* **91**, 191–200.

Mizuno, Y., Kozutsumi, Y., Kawasaki, T., and Yamashina, I. (1981). Isolation and characterization of a mannan-binding protein from rat liver. *J. Biol. Chem.* **256**, 4247–4252.

Mizuochi, T., Nishimura, Y., Kato, K., and Kobata, A. (1981). Comparative studies of asparagine-linked oligosaccharide structures on rat liver microsomal and lysosomal β-glucuronidases. *Arch. Biochem. Biophys.* **209**, 298–303.

Montreuil, J. (1980). Primary structure of glycoprotein glycans. Basis for the molecular biology of glycoproteins. *Adv. Carbohydr. Chem. Biochem.* **37**, 157–223.

Montreuil, J. (1982). Glycoproteins in "Comprehensive Biochemistry" (A. Neuberger and L. L. M. Van Deenen, eds.), Vol 19B, Part II, pp. 1–188. Elsevier, Amsterdam.

Montreuil, J. (1987). Structure and conformation of glycoprotein glycans. *In* "Vertebrate

Lectins'' (K. Olden and J. B. Parent, eds.), pp. 1–26. Van Nostrand Reinhold, New York.

Moore, H. P., Gumbiner, B., and Kelly, R. B. (1983). A subclass of proteins and sulfated macromolecules secreted by AtT-20 (mouse pituitary tumor) cells is sorted with adrenocorticotropin in dense secretory granules. *J. Cell Biol.* **97,** 810–817.

Morgan, E. H., and Peters, T., Jr., (1971). Intracellular aspects of transferrin synthesis and secretion in the rat. *J. Biol. Chem.* **246,** 3508–3511.

Mori, K., Kawasaki, T., and Yamashina, I. (1983). Identification of the mannan-binding protein from rat livers as a hepatocyte protein distinct from the mannan receptor on sinusoidal cells. *Arch. Biochem. Biophys.* **222,** 542–552.

Mori, K., Kawasaki, T., and Yamashina, I. (1984). Subcellular distribution of the mannan-binding protein and its endogenous inhibitors in rat liver. *Arch. Biochem. Biophys.* **232,** 223–233.

Nakamura, K., and Compans, R. W. (1979). Host cell- and virus strain-dependent differences in oligosaccharides of hemagglutinin glycoproteins of influenza A viruses. *Virology* **95,** 8–23.

Neufeld, E. F., Lim, T. W., and Shapiro, L. J. (1975). Inherited disorders of lysosomal metabolism. *Annu. Rev. Biochem.* **44,** 357–376.

Neufeld, E. F., Sando, G. N., Garvin, A. J., and Rome, L. H. (1977). The transport of lysosomal enzymes. *J. Supramol. Struct.* **6,** 95–101.

Novick, P., and Schekman, R. (1983). Export of major cell surface proteins is blocked in yeast secretory mutants. *J. Cell Biol.* **96,** 541–547.

Olden, K., Pratt, R. M., and Yamada, K. M. (1978). Role of carbohydrates in protein secretion and turnover. Effects of tunicamycin on the major cell surface glycoprotein of chick embryo fibroblasts. *Cell* **13,** 461–473.

Olden, K., Parent, J. B., and White, S. L. (1982). Carbohydrate moieties of glycoproteins. A re-evaluation of their function. *Biochim. Biophys. Acta* **650,** 209–232.

Omary, M. B., and Trowbridge, I. S. (1981). Biosynthesis of the human transferrin receptor in cultured cells. *J. Biol. Chem.* **256,** 12888–12892.

Owen, M. J., Kissonerghis, A. M., and Lodish, H. F. (1980). Biosynthesis of HLA-A and HLA-B antigens *in vivo. J. Biol. Chem.* **255,** 9678–9684.

Palade, G. (1975). Intracellular aspects of the process of protein secretion. *Science* **189,** 347–358.

Palamarczyk, G., and Elbein, A. D. (1985). The effect of castanospermine on the oligosaccharide structures of glycoproteins from lymphoma cell lines. *Biochem. J.* **227,** 795–804.

Pan, Y. T., Hori, H., Saul, R., Sanford, B. A., Molyneux, R. J., and Elbein, A. D. (1983). Castanospermine inhibits the processing of the oligosaccharide portion of the influenza viral hemagglutinin. *Biochemistry* **22,** 3975–3984.

Parent, J. B., Bauer, H. C., and Olden, K. (1982). Tunicamycin treated fibroblasts secrete a cathepsin B-like protease. *Biochem. Biophys. Res. Commun.* **108,** 552–558.

Parent, J. B., Bauer, H. C., and Olden, K. (1985). Three secretory rates in human hepatoma cells. *Biochim. Biophys. Acta* **846,** 44–50.

Parent, J. B., Yeo, T. K., Yeo, K. T., and Olden, K. (1986). Differential effects of 1-deoxynorjirimycin on the intracellular transport of secretory glycoproteins of human hepatoma cells in culture. *Mol. Cell. Biochem.* **72,** 21–33.

Parodi, A. J., Lederkremer, G. Z., and Mendelzon, D. H. (1983a). Protein glycosylation in *Trypanosoma cruzi.* The mechanism of glycosylation and structure of protein-bound oligosaccharides. *J. Biol. Chem.* **258,** 5589–5595.

Parodi, A. J., Mendelzon, D. H., and Lederkremer, G. Z. (1983b). Transient glucosylation of protein-bound Man$_9$GlcNAc$_2$, Man$_8$GlcNAc$_2$ and Man$_7$GlcNAc$_2$ in calf thyroid cells. A possible recognition signal in processing of glycoproteins. *J. Biol. Chem.* **258**, 8260–6265.

Parodi, A. J., Mendelzon, D. H., Lederkremer, G. Z., and Martin-Barrientos, J. (1984). Evidence that transient glucosylation of protein-linked Man$_9$GlcNAc$_2$, Man$_8$GlcNAc$_2$ and Man$_7$GlcNAc$_2$ occurs in rat liver and *Phaseolus vulgaris* cells. *J. Biol. Chem.* **259**, 6351–6357.

Pecoud, A. R., Ruddy, S., and Conrad, D. H. (1981). Functional and partial characterization of the carbohydrate moieties of the IgE receptor on rat basophilic leukemia cells and rat mast cells. *J. Immunol.* **126**, 1624–1629.

Pennica, D., Holmes, W. E., Kohr, W. J., Harkins, R. N., Vehar, G. A., Ward, C. A., Bennett, W. F., Yelverton, E., Seeburg, P. H., Heyneker, H. L., Goeddel, D. V., and Collen, D. (1983). Cloning and expression of human tissue-type plasminogen activator cDNA in *E. coli. Nature* (*London*) **301**, 214–221.

Peyrieras, N., Bause, E., Legler, G., Vasilov, R., Claesson, L., Peterson, P., and Ploegh, H. (1983). Effect of the glucosidase inhibitors norjirimycin and deoxynorjirimycin on the biosynthesis of membrane and secretory glycoproteins. *EMBO J.* **2**, 823–832.

Plantner, J. J., Poncz, L., and Kean, E. L. (1980). Effect of tunicamycin on the glycosylation of rhodopsin. *Arch. Biochem. Biophys.* **201**, 527–532.

Ploegh, H. L., Orr, H. T., and Strominger, J. L. (1981). Biosynthesis and cell surface localization of nonglycosylated human histocompatability antigens. *J. Immunol.* **126**, 270–275.

Pricer, W. E., Jr., and Ashwell, G. (1976). Subcellular distribution of a mammalian hepatic binding protein specific for asialoglycoproteins. *J. Biol. Chem.* **251**, 7539–7544.

Prieels, J. P., Pizzo, S. V., Glasgow, L. R., Paulson, J. C., and Hill, R. L. (1978). Hepatic receptor that specifically binds oligosaccharides containing fucosyl-α1-3 N-acetylglucosamine linkages. *Proc. Natl. Acad. Sci. U.S.A.* **75**, 2215–2219.

Prives, J., and Bar-Sagi. D. (1983). Effect of tunicamycin, an inhibitor of protein glycosylation, on the biological properties of acetylcholine receptor in cultured muscle cells. *J. Biol. Chem.* **258**, 1775–1780.

Puett, D. (1973). Conformational studies on a glycosylated bovine pancreatic ribonuclease. *J. Biol. Chem.* **248**, 3566–3572.

Rearick, J. I., Chapman, A., and Kornfeld, S. (1981). Glucose starvation alters lipid-linked oligosaccharide biosynthesis in Chinese hamster ovary cells. *J. Biol. Chem.* **256**, 6255–6261.

Reed, B. C., Ronnett, G. V., and Lane, M. D. (1981). Role of glycosylation and protein synthesis in insulin receptor metabolism by 3T3-L1 mouse adipocytes. *Proc. Natl. Acad. Sci. U.S.A.* **78**, 2908–2912.

Reitman, M. L., and Kornfeld, S. (1981a). UDP-N-acetylglucosamine : glycoprotein N-acetylglucosamine-1-phosphotransferase. Proposed enzyme for the phosphorylation of the high mannose oligosaccharide units of lysosomal enzymes. *J. Biol. Chem.* **256**, 4257–4281.

Reitman, M. L., and Kornfeld, S. (1981b). Lysosomal enzyme targeting. N-Acetylglucosaminylphosphotransferase selectively phosphorylates native lysosomal enzymes. *J. Biol. Chem.* **256**, 11977–11980.

Reitman, M. L., Trowbridge, I. S., and Kornfeld, S. (1982). A lectin-resistant mouse lymphoma cell line is deficient in glucosidase II, a glycoprotein-processing enzyme. *J. Biol. Chem.* **257**, 10357–10363.

Repp, R., Tamura, T., Boschek, C. B., Wege, H., Schwarz, R. T., and Niemann, H. (1985). The effects of processing inhibitors of N-linked oligosaccharides on the intracellular migration of glycoprotein E2 of mouse hepatitis virus and the maturation of corona virus particles. *J. Biol. Chem.* **260**, 15873–15879.

Robbins, A. R., and Myerowitz, R. (1981). The mannose 6-phosphate receptor of Chinese hamster ovary cells. Compartmentalization of acid hydrolases in mutants with altered receptors. *J. Biol. Chem.* **256**, 10623–10627.

Robbins, A. R., Myerowitz, R., Youle, R. J., Murray, G. J., and Neville, D. M., Jr. (1981). The mannose 6-phosphate receptor of Chinese hamster ovary cells. Isolation of mutants with altered receptors. *J. Biol. Chem.* **256**, 10618–10622.

Roller, R. J., and Wassarman, P. M. (1983). Role of asparagine-linked oligosaccharides in secretion of glycoproteins of the mouse egg's extracellular coat. *J. Biol. Chem.* **258**, 13243–13249,

Romero, P. A., Datema, R., and Schwarz, R. T. (1983). N-Methyl-1-deoxynorjirimycin, a novel inhibitor of glycoprotein processing and its effect of fowl plague virus maturation. *Virology* **130**, 238–242.

Romero, P. A., Saunier, B., and Herscovics, A. (1985). Comparison between 1-deoxynorjirimycin and N-methyl-1-deoxynorjirimycin as inhibitors of oligosaccharide processing in intestinal epithelial cells. *Biochem. J.* **226**, 733–740.

Romero, P. A., and Herscovics, A. (1986). Transfer of nonglucosylated oligosaccharide from lipid to protein in a mammalian cell. *J. Biol. Chem.* **261**, 15936–15940.

Ronin, C., and Caseti, C. (1981). Transfer of glucose in the biosynthesis of thyroid glycoproteins. II. Possibility of a direct transfer of glucose from UDPglucose to proteins. *Biochim. Biophys. Acta* **674**, 58–64.

Ronnett, G. V., and Lane, M. D. (1981). Posttranslational glycosylation-induced activation of aglycoinsulin receptor accumulated during tunicamycin treatment. *J. Biol. Chem.* **256**, 4704–4707.

Ronnett, G. V., Knutson, V. P., Kohanski, R. A., Simpson, T. L., and Lane, M. D. (1984). Role of glycosylation in the processing of newly translated insulin proreceptor in 3T3-L1 adipocytes. *J. Biol. Chem.* **259**, 4566–4575.

Roos, P. H., Harman, H. J., Schlepper-Schäfer, J., Kolb, H., and Kolb-Bachofen, V. (1985). Galactose-specific receptors on liver cells. II. Characterization of the purified receptor from macrophages reveals no structural relationship to the hepatocyte receptor. *Biochim. Biophys. Acta* **847**, 115–121.

Rosner, M. R., Grinna, L. S., and Robbins, P. W. (1980). Differences in glycosylation patterns of closely related murine leukemia viruses. *Proc. Natl. Acad. Sci. U.S.A.* **77**, 67–71.

Roth, J., and Berger, E. G. (1982). Immunocytochemical localization of galactosyltransferase in HeLa cells: Codistribution with thiamine pyrophosphatase in trans-Golgi cisternae. *J. Cell Biol.* **93**, 223–229.

Roth, M. G., Fitzpatrick, J. P., and Compans, R. W. (1979). Polarity of influenza and vesicular stomatitis virus maturation in MDCK cells: Lack of a requirement for glycosylation of viral glycoproteins. *Proc. Natl. Acad. Sci. U.S.A.* **76**, 6430–6434.

Rotundo, R. L. (1984). Asymmetric acetylcholinesterase is assembled in the Golgi apparatus. *Proc. Natl. Acad. Sci. U.S.A.* **81**, 479–483.

Runge, K. W., Huflaker, T. C., and Robbins, P. W. (1984). Two yeast mutations in glycosylation steps of the asparagine glycosylation pathway. *J. Biol. Chem.* **259**, 412–417.

Ruta, M., Clarke, S., Boswell, B., and Kabat, D. (1982). Heterogeneous metabolism and subcellular localization of a potentially leukemogenic membrane glycoprotein encoded by Friend erythroleukemia virus. *J. Biol. Chem.* **257**, 126–134.

Sahagian, G. G. (1987). The mannose 6-phosphate receptor and its role in lysosomal enzyme transport. In "Vertebrate Lectins" (K. Olden and J. B. Parent, eds.), pp. 46–64. Van Nostrand Reinhold, New York.

Sahagian, G. G., Distler, J., and Jourdian, G. W. (1981). Characterization of a membrane-associated receptor from bovine liver that binds phosphomannosyl residues of bovine testicular β-galactosidase. *Proc. Natl. Acad. Sci. U.S.A.* **78,** 4289–4293.

Sarkar, M., Liao, J., Kabat, E. A., Tanabe, T., and Ashwell, G. (1979). The binding site of rabbit hepatic lectin. *J. Biol. Chem.* **254,** 3170–3174.

Saunier, B., Kilker, R. D., Jr., Tkacz, J. S., Quaroni, A., and Herscovics, A. (1982). Inhibition of N-linked complex oligosaccharide formation by 1-deoxynorjirimycin, an inhibitor of processing glucosidases. *J. Biol. Chem.* **257,** 14155–14161.

Schauer, R. (1985). Sialic acids and their roles as biological masks. *Trends Biol. Sci.* **10,** 357–360.

Scheele, G., and Tartakoff, A. (1985). Exit of nonglycosylated secretory proteins from the rough endoplasmic reticulum is asynchronous in the exocrine pancreas. *J. Biol. Chem.* **260,** 926–931.

Schindler, M., and Hogan, M. (1984). Carbohydrate moieties of nuclear glycoproteins are predominantly N-acetylglucosamine. *J. Cell biol.* **99,** 99a (abstr.).

Schindler, M., Hogan, M., Miller, R., and DeGaetano, D. (1987). A nuclear specific glyco-protein representative of a unique pattern of glycosylation. *J. Biol. Chem.* **262,** 1254–1260.

Schlepper-Schäfer, J., Hulsmann, D., Djoulcar, A., Meyer, H. E., Herbertz, L., Kolb, H., and Kolb-Bachofen, V. (1986). Endocytosis via galactose receptors in vivo. *Exp. Cell Res.* **165,** 494–506.

Schlesinger, S., Malfer, C., and Schlessinger, M. J. (1984). The formation of vesicular stomatitis virus (San Juan strain) becomes temperature-sensitive when glucose resi-dues are retained on the oligosaccharides of the glycoprotein. *J. Biol. Chem.* **259,** 7597–7601.

Schlesinger, S., Koyama, A. H., Malfer, C., Gee, S. L., and Schlesinger, M. J. (1985). The effects of inhibitors of glucosidase 1 on the formation of Sindbis virus. *Virol. Res.* **2,** 139–149.

Schmidt, J. A., Beug, H., and Hayman, M. J. (1985). Effects of inhibitors of glycoprotein processing on the synthesis and biological activity of the *erbB* oncogene. *EMBO J.* **4,** 105–112.

Schreiber, G., Dryburgh, H., Millership, A., Matsuda, Y., Inglis, A., Phillips, J., Edwards, K., and Maggs, J. (1979). The synthesis and secretion of rat transferrin. *J. Biol. Chem* **254,** 12013–12019.

Schwartz, A. L. (1984). The hepatic asialoglycoprotein receptor. *CRC Crit. Rev. Biochem.* **16,** 207–233.

Schwarz, R. T., and Datema, R. (1982a). The lipid pathway of protein glycosylation and its inhibitors: The biological significance of protein-bound carbohydrates. *Adv. Carbo-hydr. Chem. Biochem.* **40,** 287–379.

Schwarz, R. T., and Datema, R. (1982b). *In* "The Glycoconjugates" (M. I. Horowitz, ed.), Vol. 3, pp. 47–79. Academic Press, New York.

Schwarz, R. T., and Datema, R. (1984). Inhibitors of trimming: New tools in glycoprotein research. *Trends Biol. Sci.* **9,** 32–34.

Schwarz, R. T., and Klenk, H. D. (1981). Carbohydrates of influenza virus: IV. Strain-dependent variations. *Virology* **113,** 584–593.

Schwarz, R. T., Rohrschneider, J. M., and Schmidt, M. F. G. (1976). Suppression of glycoprotein formation of Semliki Forest, influenza, and avian sarcoma virus by tunicamycin. *J. Virol.* **19**, 782–791.

Sheares, B. T., and Robbins, P. W. (1986). Glycosylation of ovalbumin in a heterologous cell: Analysis of oligosaccharide chains of the cloned glycoprotein in mouse L cells. *Proc. Natl. Acad. Sci. U.S.A.* **83**, 1993–1997.

Shepherd, V. L., Freeze, H. H., Miller, A. L., and Stahl, P. D. (1984). Identification of mannose 6-phosphate receptors in rabbit alveolar macrophage. *J. Biol. Chem.* **259**, 2257–2261.

Sibley, C. H., and Wagner, R. A. (1981). Glycosylation is not required for membrane localization or secretion of IgM in a mouse B cell lymphoma. *J. Immunol.* **126**, 1868–1873.

Sidman, C. (1981). Differing requirements for glycosylation in the secretion of related glycoproteins is determined neither by the producing cell nor by the relative number of oligosaccharide units. *J. Biol. Chem.* **256**, 9374–9376.

Simmons, K., and Fuller, S. D. (1985). Cell surface polarity in epithelia. *Annu. Rev. Cell Biol.* **1**, 243–288.

Siuta-Mangano, P., Janero, D. R., and Lane, M. D. (1982). Association and assembly of triglyceride and phospholipid with glycosylated and unglycosylated apoproteins of very low density lipoprotein in the intact liver cell. *J. Biol. Chem.* **257**, 11463–11467.

Slieker, L. J., and Lane, M. D. (1985). Posttranslational processing of the epidermal growth factor receptor. Glycosylation-dependent acquisition of ligand-binding capacity. *J. Biol. Chem.* **260**, 687–690.

Sly, W. S. (1982). *In* "The Glycoconjugates" (M. I. Horowitz, ed.), Vol. 3, pp. 3–25. Academic Press, New York.

Sly, W. S., and Stahl, P. (1978). *In* "Transport of Macromolecules in Cellular Systems" (S. C. Silverstein, ed.), pp. 229–245. Dahlen Kouferenzen, Berlin.

Snider, M. D. (1984). *In* "Biology of Carbohydrates" (V. Ginsburg and P. W. Robbins, eds.), Vol. 2, pp. 163–198. Wiley (Interscience), New York.

Soderquist, A. M., and Carpenter, G. (1984). Glycosylation of the epidermal growth factor receptor in A-431 cells. The contribution of carbohydrate to receptor function. *J. Biol. Chem.* **259**, 12586–12594.

Stahl, P., Schlessinger, P. H., Sigardon, E., Rodman, J. S., and Lee, Y. C. (1980). Receptor-mediated pinocytosis of mannose glycoconjugates by macrophages: Characterization and evidence for receptor recycling. *Cell* **19**, 207–215.

Stanley, P. (1984). Glycosylation mutants of animal cells. *Annu. Rev. Genet.* **18**, 525–552.

Stanley, P. (1985). *In* "Molecular Cell Genetics: The Chinese Hamster Cell" (M. M. Gottesman, ed.), pp. 745–772. Wiley (Interscience), New York.

Steer, C. J., and Ashwell, G. (1980). Studies on a mammalian hepatic binding protein specific for asialoglycoproteins. *J. Biol. Chem.* **255**, 3008–3018.

Stevens, T. H., Rothman, J. H., Payne, G. S., and Schekman, R. (1986). Gene dosage-dependent secretion of yeast vacuolar carboxypeptidase Y. *J. Cell Biol.* **102**, 1551–1557.

Stoll, J., Robbins, A. R., and Krag, S. S. (1982). Mutant Chinese hamster ovary cells with altered mannose 6-phosphate receptor activity is unable to synthesize mannosylphosphoryldolichol. *Proc. Natl. Acad. Sci. U.S.A.* **79**, 2296–2300.

Stowell, C. P., and Lee, Y. C. (1978). The binding of D-glucosyl–neoglycoproteins to the hepatic asialoglycoprotein receptor. *J. Biol. Chem.* **253**, 6107–6110.

Strous, G. J. A. M., and Lodish, H. F. (1980). Intracellular transport of secretory and membrane proteins in hepatoma cells infected by vesicular stomatitis virus. *Cell* **22,** 709–717.

Strous, G. J. A. M., Willemsen, R., van Kerkhof, P., Slot, J. W., Geuze, H. J., and Lodish, H. (1983). Vesicular stomatitis virus glycoprotein, albumin, and transferrin are transported to the cell surface via the same Golgi vesicles. *J. Cell Biol.* **97,** 1815–1822.

Svedersky, L. P., Shepard, H. M., Spencer, S. A., Shalaby, M. R., and Palladino, M. A. (1984). Augmentation of human natural cell-mediated cytotoxicity by recombinant human interleukin 2. *J. Immunol.* **133,** 714–718.

Swiedler, S. J., Hart, G. W., Tarentino, A. L., Plummer, T. H., Jr., and Freed, J. H. (1983). Stable oligosaccharide microheterogeneity at individual glycosylation sites of a murine major histocompatability antigen derived from a B-cell lymphoma. *J. Biol. Chem.* **258,** 11515–11523.

Swiedler, S. J., Freed, J. H., Tarentino, A. L., Plummer, T. H., Jr., and Hart, G. W. (1985). Oligosaccharide microheterogeneity of the murine major histocompatibility antigens. Reproducible site-specific patterns of sialylation and branching in asparagine-linked oligosaccharides. *J. Biol. Chem.* **260,** 4046–4054.

Tabas, I., and Kornfeld, S. (1978). The synthesis of complex-type oligosaccharides. III. Identification of an α-D-mannosidase activity involved in a late stage of processing of complex-type oligosaccharides. *J. Biol. Chem.* **253,** 7779–7786.

Takatsuki, A., Arima, K., and Tamura, G. (1971). Tunicamycin, a new antibiotic. I. Isolation and characterization of tunicamycin. *J. Antibiot.* **24,** 215–223.

Takatsuki, A., Kohno, K., and Tamura, G. (1975). Inhibition of biosynthesis of polyisoprenol sugars in chick embryo microsomes by tunicamycin. *Agric. Biol. Chem.* **39,** 2089–2091.

Tarentino, A. L., Plummer, T. H., Jr., and Maley, F. (1974). The release of intact oligosaccharides from specific glycoproteins by endo-β-N-acetylglucosaminidase H. *J. Biol. Chem.* **249,** 818–824.

Tartakoff, A. M. (1983a). The confined function model of the Golgi complex: Center for ordered processing of biosynthetic products of the rough endoplasmic reticulum. *Int. Rev. Cytol.* **85,** 221–252.

Tartakoff, A. M. (1983b). Mutations that influence the secretory path in animal cells. *Biochem. J.* **216,** 1–9.

Tartakoff, A., and Vassalli, P. (1978). Comparative studies of intracellular transport of secretory proteins. *J. Cell Biol.* **79,** 694–707.

Tkacz, J. S., and Lampen, J. B. (1975). Tunicamycin inhibition of polyisoprenyl N-acetylglucosaminyl pyrophosphate formation in calf-liver microsomes. *Biochem. Biophys. Res. Commun.* **65,** 248–257.

Torres, C. R., and Hart, G. W. (1984). Topography and polypeptide distribution of terminal N-acetylglucosamine residues on the surfaces of intact lymphocytes. *J. Biol. Chem.* **259,** 3308–3317.

Townsend, R., and Stahl, P. (1981). Isolation and characterization of a mannose/N-acetylglucosamine/fucose binding protein from rat liver. *Biochem. J.* **194,** 209–214.

Trimble, R. B., and Maley, F. (1977). The use of endo-β-N-acetylglucosaminidase H in characterizing the structure and function of glycoproteins. *Biochem. Biophys. Res. Commun.* **78,** 935–944.

Trowbridge, I. S., and Hyman, R. (1979). Abnormal lipid-linked oligosaccharides in class E Thy-1-negative mutant lymphomas. *Cell* **17,** 503–508.

Trowbridge, I. S., Hyman, R., and Mazauskas, C. (1978a). The synthesis and properties of T25 glycoprotein in Thy-1 negative mutant lymphoma cells. *Cell* **14,** 21–32.

Trowbridge, I. S., Hyman, R., Ferson, T., and Mazauskas, C. (1978b). Expression of Thy-1 glycoprotein on lectin-resistant lymphoma cell lines. *Eur. J. Immunol.* **8,** 716–723.

Tulsiani, D. R. P., and Touster, O. (1983). Swainsonine causes the production of hybrid glycoproteins by human skin fibroblasts and rat liver Golgi preparations. *J. Biol. Chem.* **258,** 7578–7585.

Tulsiani, D. R. P., Hubbard, S. C., Robbins, P. W., and Touster, O. (1982a). α-D-Mannosidases of rat liver Golgi membranes. *J. Biol. Chem.* **257,** 3660–3668.

Tulsiani, D. R. P., Harris, T. M., and Touster, O. (1982b). Swainsonine inhibits the biosynthesis of complex glycoproteins by inhibition of Golgi mannosidase II. *J. Biol. Chem.* **257,** 7936–7939.

Vladutiu, G. D., and Rattazzi, M. (1979). Excretion–reuptake route of β-hexosaminidase in normal and I-cell disease cultured fibroblasts. *J. Clin. Invest.* **63,** 595–601.

von Figura, K., and Hasilik, A. (1986). Lysosomal enzymes and their receptors. *Annu. Rev. Biochem.* **55,** 167–193.

von Figura, K., and Weber, E. (1978). An alternative hypothesis of cellular transport of lysosomal enzymes in fibroblasts. *Biochem. J.* **176,** 943–950.

Waechter, C. J., Schmidt, J. W., and Catterall, W. A. (1983). Glycosylation is required for maintenance of functional sodium channels in neuroblastoma cells. *J. Biol. Chem.* **258,** 5117–5123.

Wagner, D. D., Mayadas, T., and Marder, V. J. (1986). Initial glycosylation and acidic pH in the Golgi apparatus are required for multimerization of von Willebrand factor. *J. Cell Biol.* **102,** 1320–1324.

Waheed, A., Hasilik, A., and von Figura, K. (1982). UDP-*N*-acetylglucosamine : lysosomal enzyme precursor *N*-acetylglucosamine-1-phosphotransferase. *J. Biol. Chem.* **257,** 12322–12331.

Wang, F. F. C., and Hirs, C. H. W. (1977). Influence of the heterosaccharides in porcine pancreatic ribonuclease on the conformation and stability of the protein. *J. Biol. Chem.* **252,** 8358–8364.

Weigel, P. H., and Oka, J. A. (1983). The large intracellular pool of asialoglycoprotein receptors functions during the endocytosis of asialoglycoproteins by isolated rat hepatocytes. *J. Biol. Chem.* **258,** 5095–5102.

Weigel, P. H. (1987). Receptor recycling and ligand processing mediated by hepatic glycosyl receptors: A two-pathway system. *In* "Vertebrate Lectins" (K. Olden and J. B. Parent, eds.), pp. 65–91. Van Nostrand Reinhold, New York.

Weis, J. J., and Fearon, D. T. (1985). The identification of N-linked oligosaccharides on the human CR2/Epstein–Barr virus receptor and their function in receptor metabolism, plasma membrane expression, and ligand binding. *J. Biol. Chem.* **260,** 13824–13830.

Whitsett, J. A., Ross, G., Weaver, T., Rice, W., Dion, C., and Hull, W. (1985). Glycosylation and secretion of surfactant-associated glycoprotein A. *J. Biol. Chem.* **260,** 15273–15279.

Wileman, T., Boshans, R., Schlesinger, R. H., and Stahl, P. D. (1984). Monensin inhibits recycling of macrophages' mannose–glycoprotein receptors and ligand delivery to lysosomes. *Biochem. J.* **220,** 665–675.

Williams, D. B., Swiedler, S. J., and Hart, G. W. (1985). Intracellular transport of membrane glycoproteins: Two closely related histocompatability antigens differ in their rates of transit to the cell surface. *J. Cell Biol.* **101,** 725–734.

Wong, K. L., Charlwood, P. A., Hatton, M. W. C., and Regoeczi, E. (1974). Studies of the metabolism of asialotransferrins: Evidence that transferrin does not undergo desialylation *in vivo*. *Clin. Sci. Mol. Med.* **46,** 763–774.

the secretory pathway to a common point of assembly. Since these viruses depend on the preexisting host cell processes and possess lipid envelopes that are biochemically similar to cellular membranes, they can provide ideal systems for probing the cellular mechanisms involved in glycoprotein biosynthesis and transport.

The relatively simple structure of virions, the high level of expression of viral genes, the ease of molecularly cloning and manipulating those genes, together with the availability of conditional lethal mutants with defects in viral protein transport, confer several additional advantages for such studies. We have chosen an enveloped, RNA-containing virus, Rous sarcoma virus (RSV), for an analysis of viral glycoprotein biosynthesis and transport, because in addition to the aspects delineated above, the availability of molecularly cloned, infectious DNA copies of the genome of this retrovirus has also allowed us to pose questions about the role of the glycoproteins in virus assembly and virus infectivity.

Like most simple enveloped viruses, the retroviruses consist of a host membrane-derived, lipid bilayer that surrounds (envelopes) a protein capsid structure (Fig. 1). The icosahedral capsid of RSV assembles as the virus particle buds from the plasma membrane of the cell, and so these two events are linked both temporally and spacially. Glycoprotein knobbed spikes extend from the surface of the virion, and it is generally thought that during virus assembly a specific interaction between the glycoproteins and capsid (and/or ''matrix'' protein) is required, since cell-derived polypeptides are for the most part excluded from the budding structure.

The envelope glycoproteins span the lipid bilayer and are thereby divided into three distinct domains: an external, hydrophilic receptor-bind-

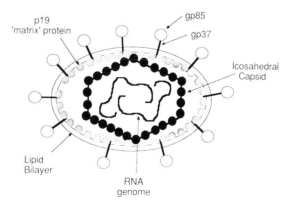

Fig. 1. Schematic representation of a Rous sarcoma virus virion.

ing domain that functions in virus–cell attachment; a hydrophobic membrane-spanning domain; and a hydrophilic cytoplasmic domain. In RSV two polypeptides, gp85 and gp37, make up this structure (Fig. 2); the external domain is primarily made up of the 341 amino acid long gp85 which contains regions that define the host-range and neutralization properties of the virus. The 198 amino acid long gp37 polypeptide, on the other hand, is a bitopic protein that anchors the envelope glycoprotein complex into the virion via disulfide linkages to gp85. It is less heavily glycosylated than gp85 (having only 2 versus 14 potential glycosylation sites) and contains two apolar regions in addition to the hydrophilic cytoplasmic domain. One apolar region is located near the amino terminus of gp37 and may be analogous to the fusion peptide of the hemagglutinin HA2 polypeptide of influenza virus that mediates viral entry into the cell. The

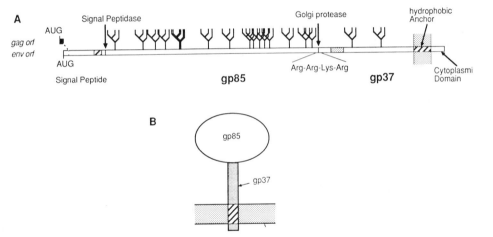

Fig. 2. Schematic representation of the Rous sarcoma virus *env* gene product. (A) The translation product of the open reading frame (orf) from the AUG codon at nucleotide 5054 to the TAG codon at nucleotide 6863 is shown as the long open box. Hatched regions represent the highly hydrophobic signal and anchor sequences found at the N and C terminus, respectively. The location of a nonpolar region that may be analogous to the fusion peptide of ortho- and paramyxoviruses is shown by the stippled box. The AUG at the start of the *env* orf was used to initiate translation of transcripts expressed in an SV40 vector (Wills *et al.*, 1984), but during a virus infection the genomic length transcript is spliced such that the AUG and first 5 codons from gag (black box) are spliced into the *env* orf (Hunter *et al.*, 1983). In both cases the signal peptide is removed during translation, and cleavage of the polyprotein precursor to gp85 and gp37 occurs in the Golgi at the basic tetrapeptide, -Arg-Arg-Lys-Arg-. Branched structures denote potential N-linked oligosaccharide addition sites (Asn-X-Ser or Asn-X-Thr). (B) Orientations of gp85 and pg37 in the viral membrane are depicted schematically. In the electron microscope this structure is seen as a spiked knob, where gp37 is the spike and gp85 is the knob.

second apolar region in gp 37 consists of a 27 amino acid long stretch of hydrophobic residues near the carboxy terminus that functions during translation to stop the movement of the protein into the lumen of the rough endoplasmic reticulum (RER) and to anchor the complex in the membrane. The general orientation and structure of the RSV glycoprotein is thus similar to that of the influenza virus hemagglutinin (HA) (Porter *et al.*, 1979; Gething *et al.*, 1980), the vesicular stomatitis virus (VSV) G protein (Gallione and Rose, 1983), and several membrane-spanning cell-encoded glycoproteins.

B. Viral Polypeptides as Probes for Intracellular Compartments

The two viral glycoproteins of RSV are encoded by a single viral gene, *env*, and are translated in the form of a heavily glycosylated precursor polypeptide, Pr95^{env}. Since several processing and maturation events occur during the transport of the *env* gene products to the plasma membrane, they provide excellent markers for the subcellular compartments of the cell. A long (62 amino acid) amino-terminal signal peptide, which mediates translocation of the *env* gene product across the rough endoplasmic reticulum (RER), is removed cotranslationally from the precursor protein, and in the lumen of the RER 15–16 high-mannose core glycosylation units are added to the nascent Pr95. The marked increase in molecular weight (MW) (approximately 40K) that results from the addition of this endo-β-N-acetylglucosaminidase H (endo H) sensitive carbohydrate provides a clear indicator for the translocation event. Removal of glucose residues and some mannose moities appears to occur prior to transport of the protein to the Golgi, where further trimming of the mannose residues and addition of glucosamine, galactose, and fucose are observed. Cleavage of Pr95 to gp85 and gp37 takes place after galactose and prior to fucose addition; it thus provides an excellent marker for trans-Golgi locations. While the transit time of Pr95 from the RER to Golgi is quite long ($t_{1/2}$ = 90 min) compared to other viral glycoproteins, movement through the Golgi appears rapid and precludes any dissection of individual compartments within this organelle. These major biochemical modifications to the *env* glycoprotein coupled with sensitive immunological probes allow a fairly accurate mapping of the cell's secretory pathway.

Several viruses assemble at points within the secretory pathway, and since in many cases the location of the viral glycoproteins defines the virus maturation point (Roth *et al.*, 1983b; Jones *et al.*, 1985; Gottlieb *et al.*, 1986; Stephens *et al.*, 1986; Gahmberg, 1984; Kabcenell and Atkinson, 1985) these systems are proving useful in investigating the signals that target proteins to specific subcellular locations. Figure 3 is a schematic summary of the assembly points for the major groups of enveloped viruses. Herpes simplex virus (HSV), a complex enveloped DNA virus that encodes several glycoproteins (gB, gC, gD, gE), buds into the nuclear envelope (Spear, 1985). Virion glycoproteins synthesized on the RER appear to be transported to the nuclear membrane in an endo H-sensitive form where they are incorporated into nascent virions (Compton and Courtney, 1984). It has been postulated that intact HSV virions traverse the secretory pathway thereby exposing the glycoproteins to the entire array of carbohydrate-modifying enzymes such that the mature virion contains glycoproteins with complex carbohydrate side chains (Compton and Courtney, 1984; Spear, 1985). A majority of the HSV glycoproteins can be found on the surface of infected cells, suggesting that targeting to the nuclear membrane is not absolute. Expression of the cloned HSV *gD* gene in the absence of other viral components resulted in more rapid

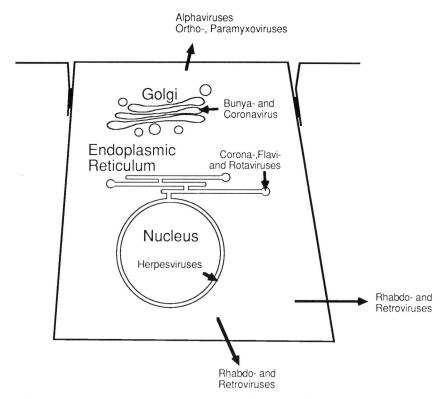

Fig. 3. Schematic representation of virus assembly sites within the secretory pathway.

transport to the plasma membrane and reduced accumulation on the nuclear membrane, indicating that interactions with other viral-encoded proteins may be required for normal nuclear membrane localization (Johnson and Smiley, 1985).

The corona-, flavi-, and rotaviruses have been reported to undergo assembly at the RER (Dubois-Dalcq *et al.*, 1984), but the rotaviruses appear to target this organelle most specifically. While the mature rotavirus is not enveloped, it contains a glycosylated capsid protein in its outer shell that is derived from a transient membrane during the assembly process. The inner and outer protein shells of these viruses form sequentially and by very different mechanisms. The inner capsids, containing the genomic RNA segments, are the equivalent of nucleocapsids of other viruses and assemble in the cytoplasm at the edge of electron-dense inclusions called "viroplasm" (Petrie *et al.*, 1982). They acquire a transient envelope, or pseudoenvelope, by budding at RER membranes adjacent to the viroplasm. Further maturation of rotaviruses occurs within the cisternae of the RER where enveloped particles are converted to mature double-shelled virions and the lipid bilayer is removed by a process that remains to be elucidated (Dubois-Dalcq *et al.*, 1984). VP7, the glycosylated protein found in the outer capsid, has been shown to have carbohydrate structures consistent with its RER location and to target specifically to this organelle when expressed in the absence of other viral proteins from a recombinant expression vector (Kabcenell and Atkinson, 1985; Poruchynsky *et al.*, 1985).

The Golgi body is the site of assembly for several viruses. Coronaviruses, for example, mature by budding into the lumina of RER or Golgi cisternae; virions form as the intracytoplasmic, helical neucleocapsids align under regions of intracellular membranes containing viral proteins. Two glycoproteins, E1 and E2, comprise these membrane-associated polypeptides. E2 forms the large peplomers or spikes characteristic of coronaviruses and is a multifunctional molecule, being responsible for virus-induced cell fusion, binding of the virion to receptors on the plasma membrane of susceptible cells, and for inducing neutralizing antibody (Dubois-Dalcq *et al.*, 1984). E1, in contrast, is an unusual polypeptide; it has only a short amino-terminal domain, which contains the glycosylation sites of the protein, and two long stretches of hydrophobic amino acids, suggesting that it may traverse the membrane more than once (Dubois-Dalcq *et al.*, 1984; Boursnell *et al.*, 1984). During infection, E2 can be transported through the secretory pathway to the plasma membrane, whereas E1 is transported only as far as the Golgi apparatus, where it accumulates during the infection cycle (Sturman and Holmes, 1983); it has been suggested that this restricted intracellular movement of E1 accounts for the intracellular budding site of coronaviruses (Sturman and Holmes, 1983).

Members of the bunyavirus family also mature intracellularly, by budding at the Golgi complex (Bishop and Shope, 1979; Dubois-Dalcq *et al.*, 1984). By immunofluorescent microscopy, Kuismanen and colleagues showed that in cells infected with Uukuniemi virus the Golgi region underwent an expansion and became vacuolized (Kuismanen *et al.*, 1984). Both glycoproteins, G1 and G2, accumulated in the Golgi region during virus infection; neither polypeptide could be chased out of the Golgi even after a 6-hr treatment with cycloheximide (Gahmberg *et al.*, 1986), conditions that would allow complete transport of the Semliki Forest virus membrane proteins from the Golgi (Green *et al.*, 1981a). Furthermore, the glycoproteins of a temperature-sensitive strain of Uukuniemi virus were retained in the Golgi even under conditions where no virus maturation took place and no nucleocapsids accumulated in the Golgi region (Gahmberg *et al.*, 1986). Thus intracellular targeting of these viral components appears to be independent of other viral components and of the assembly process itself. Moreover, it supports the concept that it is the glycoproteins themselves that dictate the cellular site of virus maturation.

For several virus groups, virion assembly does not occur until the envelope components have traversed the entire secretory pathway. Thus the ortho- and paramyxoviruses, rhabdoviruses, alphaviruses, and retroviruses mature at the plasma membrane. Even these proteins, however, possess additional membrane-targeting information such that in polarized epithelial cells, where different cell proteins are inserted in the apical and basolateral membranes, the different viruses assemble from distinct membranes; for example, the ortho- and paramyxoviruses bud from apical membranes, rhabdo- and retroviruses from basolateral membranes (Rodriguez-Boulan and Sabatini, 1978; Herrler *et al.*, 1981; Roth *et al.*, 1983a; Rindler *et al.*, 1985). As with those viruses that mature at points within the secretory pathway, it is the glycoproteins themselves that appear to specify the specific plasma membrane location for virus assembly, since glycoproteins expressed from recombinant expression vectors are transported in a polarized fashion (Roth *et al.*, 1983b; Jones *et al.*, 1985; Gottlieb *et al.*, 1986; Stephens *et al.*, 1986). The problem of polarized expression will be dealt with in more detail later in this chapter (see Section II,B,4).

From the brief outline presented above it is clear that viral envelope components provide a plethora of systems for studying intracellular protein targeting. Recombinant DNA approaches, described below, are already providing information on the role of the different glycoprotein domains in this important aspect of viral and cell biology.

C. Endocytic Pathway

In addition to utilizing the secretory pathway of vertebrate cells for transporting viral components to the point of virus maturation, several enveloped viruses take advantage of a second vesicle-mediated transport system, the endocytic pathway, to gain entry into susceptible cells. These viruses, such as orthomyxoviruses, rhabdoviruses, and togaviruses, in contrast to paramyxoviruses, such as Sendai virus, which bind and fuse with the plasma membrane of the host cell, bind to the host cell surface and are subsequently internalized by endocytosis. This latter process serves an important role in the normal uptake of nutrients and in the internalization of receptor-bound ligands such as hormones, growth factors, lipoproteins, and antibodies (Mellman *et al.*, 1986; Hopkins, 1983). Bound virions are carried into clathrin-coated pits, which form continually on the surface of the cell, and which fold inward and pinch off into the cytoplasm to form "coated vesicles." As the coated vesicle moves into the cytoplasm, it loses its clathrin and fuses with an endosome, a large acidic vacuole with a smooth outer surface. For viruses entering by this pathway, membrane fusion occurs in the endosomal compartment (Marsh, 1984; Yoshimura and Ohnishi, 1984). Fusion is triggered by the mildly acidic endosomal pH and is catalyzed by the virally encoded glycoproteins which undergo a low pH-dependent configurational change (Skehel *et al.*, 1982; Kielian and Helenius, 1985). The pH dependence of fusion varies among virus types, with the optimal pH for fusion generally falling within the range of pH 5.0–6.2 for endocytosed viruses (White *et al.*, 1983).

In order to obtain an understanding of the molecular mechanisms involved in these low pH-induced fusion reactions, virus mutants have been isolated which fuse with pH optima different from those of their respective parents. Through the use of an elegant selection scheme, in which mutagenized virus was allowed to fuse with nuclease-filled liposomes at a pH below 6.0, Kielian *et al.* (1984) isolated the first such fusion mutant of Semliki Forest virus. This virus, fus-1, fused at a pH optimum 0.7 pH units lower than that of the wild type (pH 5.5 versus 6.2). The mutant was, nevertheless, fully capable of infecting cells under standard infection conditions and even under conditions that prevent fusion of endosomes with lysosomes. On the other hand, the fus-1 mutant showed increased sensitivity to lysosomatropic agents that increase the pH in acidic vacuoles of the endocytic pathway. In addition to proving that alterations within viral structural components can significantly affect the pH at which virus-induced fusion can occur, these results showed that a pH below 5.5 exists within the endosomal compartment and thereby demonstrated the usefulness of mutant viruses as biological pH probes of this pathway.

In parallel studies, Rott and co-workers (1984) have shown that variants of the X31 strain of influenza virus, selected for their ability to undergo activation cleavage and growth in Madin–Darby canine kidney (MDCK) cells, have an elevated fusion pH threshold (approximately 0.7 pH units higher than the wild type). Similar virus variants have been selected by growth of influenza virus in the presence of amantadine, a compound that raises endosomal pH (Daniels *et al.*, 1985). In this latter study, viruses were obtained that fused at pH values 0.1–0.7 units higher than the parental strain. Analogous mutants have been reported to occur naturally within stocks of the X31 strain of influenza virus (Doms *et al.*, 1986). Such mutants should provide useful probes for elucidating the endocytic pathway.

II. GENETIC APPROACHES TO VIRAL GLYCOPROTEIN TRANSPORT

The mechanisms by which cells send membrane-bound and secreted proteins to their proper subcellular locations remain a central problem in cell biology. It has been postulated to involve the specific interaction of "sorting signals," located within the structure of the newly synthesized proteins, with membrane-bound receptors in the RER and Golgi apparatus of the cell (for review, see Sabatini *et al.*, 1982; Silhavy *et al.*, 1983). This concept is supported by the facts that cell, as well as viral, glycoproteins can be retained at or targeted to specific points in the secretory pathway and that cells can transport and secrete a variety of glycosylated and nonglycosylated proteins at distinctly different rates (Strous and Lodish, 1980; Fitting and Kabat, 1982; Gumbiner and Kelly, 1982; Ledford and Davis, 1983; Lodish *et al.*, 1983, Kelly, 1985).

Very little is known about the composition(s) or indeed the exact role of sorting signals, but it is generally thought that they are composed of protein. Clearly, the initial step that introduces polypeptides into the secretory pathway is mediated by the interaction of a sequence of amino acids (the signal sequence) within the polypeptide and the signal recognition particle (SRP)/docking protein (DP) complex (Blobel and Dobberstein, 1975; Walter and Blobel, 1980, 1981a,b; Walter *et al.*, 1981; Meyer and Dobberstein, 1980; Meyer *et al.*, 1982; Gilmore *et al.*, 1982a,b; 1984). Mutants of secreted proteins that are defective in later stages of transport have been identified that differ from the wild-type forms by one (Mosmann and Williamson, 1980; Wu *et al.*, 1983; Shida and Matsumoto, 1983) or two (Yoshida *et al.*, 1976; Hercz *et al.*, 1978) amino acid substitutions, supporting the concept that sorting signals are composed of protein. Also, several conditional transport-defective mutants of membrane-bound viral glycoproteins have been identified (for example, Knipe et al., 1977b; Zilberstein et al., 1980; Lodish and Kong, 1983; Pesonen et al., 1981).

Studies using tunicamycin, an inhibitor of glycosylation, suggest that carbohydrate moieties are not recognized directly by the sorting machinery but may be important for maintaining the proper secondary or tertiary structures of (protein-composed) sorting signals (Struck et al., 1978; Gibson et al., 1978, 1979; Leavitt et al., 1977; Hickman et al., 1977; Roth et al., 1979; Strous et al., 1983; Green et al., 1981b). The lack of a direct role for carbohydrate moieties in the sorting process is perhaps to be expected in view of the fact that many secreted proteins are not glycosylated at all (for example, Strous and Lodish, 1980; Underdown et al., 1971). Nevertheless, addition of carbohydrate to molecules that are unable to be transported through the secretory pathway can release the block to their transport (Guan et al., 1985; Machamer et al., 1985). In addition, the transport of certain hydrolases to the lysosome (and away from the secretory pathway) does appear to require the addition of a carbohydrate moiety (mannose 6-phosphate) (Hasilik and Neufeld, 1980; Sly and Fisher, 1982; Creek and Sly, 1984), but these additions in turn must require the recognition of signals within the polypeptide chains.

Even less is known about the intramolecular location(s) of sorting signals. In the case of the membrane-spanning glycoproteins, three protein domains exist which together or separately may harbor sorting signals: (1) the internal or cytoplasmic domain, (2) the hydrophobic or transmembrane domain, and (3) the extracytoplasmic or external domain. Since most secreted proteins (which may be cotransported with membrane-bound glycoproteins, Strous et al., 1983) possess only external domains, it might be reasonable to expect the transmembrane and cytoplasmic domains to be unimportant to the sorting process. We have tested this hypothesis by introducing genetic lesions into the gene encoding the envelope glycoproteins of RSV (Wills et al., 1983, 1984; Hardwick et al., 1986; Davis and Hunter, 1987; Davis et al., 1987; Perez and Hunter, 1987; Perez et al., 1987), as have others for the VSV G protein (Rose and Bergmann, 1982, 1983; Rose et al., 1984; Adams and Rose, 1985a,b), the influenza virus hemagglutinin protein (Sveda et al., 1982, 1984; Gething and Sambrook, 1982; Doyle et al., 1985, 1986; Gething et al., 1986), and the glycoproteins of Semliki Forest virus (Garoff et al., 1983; Garoff, 1985; Cutler and Garoff, 1986; Cutler et al., 1986). Genetic analyses of protein transport in prokaryotic systems have provided both support for the role of the signal peptide in protein translocation and valuable insights into the polypeptide interactions that are required for the intracellular targeting of bacterial secreted and membrane proteins (reviewed by Michaelis and Beckwith, 1982; Silhavy et al., 1983; Benson et al., 1985; Oliver, 1985). While similar experiments are more difficult to perform in eukaryotic cells with the enveloped virus systems described here, both the classic and molecular genetic approaches outlined below are providing information on the role of different protein domains in the transport process.

A. Classic Approaches

During the genetic analysis of enveloped virus replication through the isolation and biochemical characterization of spontaneous and mutagen-induced variants, complementation groups were established for several viruses that contained mutants defective in normal transport of the viral glycoproteins (Knipe et al., 1977a,b; Zilberstein et al., 1980; Lodish and Kong, 1983; Pesonen et al., 1981; Gahmberg, 1984; Ueda and Kilbourne, 1976). The existence of conditional lethal mutants that were blocked at different stages of virus glycoprotein maturation suggested that the viral polypeptides themselves might contain the signals necessary for normal sorting by the cells' transport machinery and raised the possibility that such mutants could be used to dissect the maturation pathway of a glycoprotein. Since it is impossible in this chapter to provide a detailed review of the characterization of mutants in each of these systems, and since the transport of the influenza virus glycoproteins has recently been discussed in detail by Roth et al. (1987), this section will concentrate primarily on mutants of the VSV G protein gene as an example of these approaches.

Temperature-sensitive (ts) mutants in complementation group V of this VSV have defects in the structural gene for the viral glycoprotein, G, and cells infected at the nonpermissive temperature with such mutants produce markedly reduced yields of virus like particles which are noninfectious and specifically deficient in G protein. At the nonpermissive temperature the mutant G polypeptide is synthesized normally; however, it does not accumulate on the cell surface, nor is it incorporated into virions (Knipe et al., 1977b; Zilberstein et al., 1980). The ts(V) mutants can be subdivided into two subclasses with respect to the stage of posttranslational processing at which the block occurs (Zilberstein et al., 1980). Three mutants, tsL513, tsM501, and tsO45, encode G proteins that at the nonpermissive temperature are blocked at an early, pre-Golgi step of the secretory pathway. While insertion into the ER membrane, removal of the amino-terminal hydrophobic signal sequence, and addition of the two N-linked high-mannose core oligosaccharides occur in a way that is indistinguishable from the wild type, all subsequent Golgi-mediated carbohydrate processing reactions are blocked (Zilberstein et al., 1980).

These results are consistent with subcellular fractionation and immunoelectron microscopy studies which indicated that the G protein in tsM501 or tsO45-infected cells was arrested in its transport from the RER to the Golgi complex at the nonpermissive temperature (Zilberstein et al.,

1980; Bergmann *et al.*, 1981; Bergmann and Singer, 1983). The defect in transport in these mutants is a reversible phenomenon, thereby excluding irreversible denaturation as the basis for lack of movement; proteins synthesized at the nonpermissive temperature rapidly move by stages to the plasma membrane upon shift to the permissive temperature (Lodish and Kong, 1983; Bergmann and Singer, 1983). Within 3 min after shift to 32°C, G protein of *ts*O45 could be seen by immunoelectron microscopy at high density in saccules at one face of the Golgi complex and by 3 min later was uniformly distributed through the complex (Bergmann and Singer, 1983). Movement of the mutant proteins to the cell surface occurred rapidly and was accompanied by incorporation into virions (Lodish and Kong, 1983).

The second class of *ts* mutants of VSV is represented by *ts*L511. G protein encoded by this mutant is transported normally through most of the Golgi-mediated functions involved in the processing of carbohydrate side chains, including addition of the terminal sialic acid residues. However, this molecule does not undergo two posttranslation modification reactions that take place with the wild-type G protein. The first is the addition of fucose to the *ts*L511 oligosaccharide chains, which is reduced at both permissive and nonpermissive temperatures (Zilberstein *et al.*, 1980). In the second, a molecule of palmitic acid (a 12-carbon fatty acid) is covalently attached to the wild-type G polypeptide near the membrane-spanning region (Schmidt and Schlesinger, 1979; Schmidt *et al.*, 1979; Rose *et al.*, 1984). Attachment occurs at a late stage of maturation, just before oligosaccharide processing is completed (Schmidt and Schlesinger, 1980), probably in the cis compartment of the Golgi complex (Dunphy *et al.*, 1981). This modification of the G protein does not occur at the nonpermissive temperature in cells infected with the *ts*L511 mutant. Taken together with the almost complete processing of the mutant's oligosaccharide side chains, this suggests that at the nonpermissive temperature the *ts*L511 glycoprotein accumulates at a specific region within the Golgi complex.

The mutants of VSV, together with equivalent mutants in other viral systems that were blocked at different stages of the secretory pathway (Pesonen *et al.*, 1981; Gahmberg, 1984; Kuismanen *et al.*, 1984), raised the possibility of identifying and understanding the nature of sorting signals in secreted polypeptides. With the advent of recombinant DNA and rapid nucleotide sequence techniques, it has been possible to determine at the amino acid level the basis for these defects (Gallione and Rose, 1985; Arias *et al.*, 1983), but the interpretation of this information with regard to protein transport has been less than straightforward.

Gallione and Rose (1985) determined the nucleotide sequence of the *ts*O45 mutant of VSV and compared it to that of the parent and a wild-type revertant. The mutant and revertant differed in three amino acid residues, and through the construction and expression of hybrid genes it was possible for these investigators to demonstrate that the basis of the temperature-sensitive phenotype was a single amino acid change of phenylalanine to serine. Since this polar substitution occurred within a very hydrophobic region of the G protein, it was suggested that it might significantly affect protein folding in this region such that reversible denaturation of the protein might occur at the nonpermissive temperature. This denaturation could prevent further transport to the Golgi apparatus and the cell surface. Alternatively, Gallione and Rose (1985) pointed out that the conformational change at the nonpermissive temperature might be more subtle, perhaps preventing recognition by a component of the protein transport machinery. However, since hydrophobic residues are generally buried within a protein (Kyte and Doolittle, 1982), it is unlikely that the mutated sequence itself would play a direct role in such an interaction. Nothing is known about the 3-dimensional structure of the G protein or whether accessory proteins are involved at this stage of protein transport, thus both suggestions remains viable possibilities.

A similar analysis has been carried out by Arias *et al.* (1983), who sequenced the genes encoding the viral glycoproteins of *ts*10 and *ts*23, mutants of Sindbis virus defective in the intracellular transport of their glycoproteins, and of revertants of these mutants. These investigators found *ts*23 to have a double mutation in glycoprotein E1, while *ts*10 was a single mutant in the same glycoprotein. In each case reversion to temperature insensitivity occurred by changes at the same site as the mutation, in two cases restoring the original amino acid and in the third case substituting a homologous amino acid (arginine in place of lysine). Since the three mutations were far apart from each other in the protein, these authors concluded that the 3-dimensional conformation of E1 was very important for the correct migration of the glycoproteins from the ER to the plasma membrane. Similarly, two *ts* mutants in the HA gene of influenza virus that result in HA protein transport being arrested in the RER are also caused by single point mutations that probably disrupt the tertiary structure of the molecule (Nakajima *et al.*, 1986).

In summary, several conditional mutants have been isolated by classic genetic approaches that at the nonpermissive temperature disrupt the normal transport of viral glycoproteins through the secretory pathway. While these mutants carried the promise of defining specific protein domains that might interact with components of the transport machinery, the evidence from the nucleotide sequencing experiments outlined above suggests that many or all of the mutations may exert their phenotype through distortion of the 3-dimensional shape of the molecule. While this mechanism does not preclude a role for specific protein–protein interactions in the secretory pathway, it provides no direct evidence for it at this time.

B. Recombinant DNA Approaches

The recent development of cDNA cloning, gene sequence manipulation, and gene expression technologies has opened up new approaches for localizing and characterizing those structural features of a protein that act as sorting signals. These new methodologies have allowed investigators to delete or modify potentially important structural regions of a protein at the nucleotide level and then to determine the effect of such changes by expressing modified genes in suitable eukaryotic expression vectors. Furthermore, in some instances it has been possible to test directly the functional role of a particular peptide region by fusing it to another protein and analyzing the behavior of the chimera. Since these general approaches to protein sorting have been reviewed recently (Garoff, 1985; Gething, 1985), this chapter will describe our recent recombinant DNA analyses of the biosynthesis and transport of the RSV envelope glycoproteins, within the context of similar analyses in other viral systems.

1. Mutations in the Signal Peptide Region

As we have discussed earlier, the RSV *env* gene encodes two viral glycoproteins, gp85 and gp37, that mediate recognition of, attachment to, and penetration of the susceptible target cell. These proteins are synthesized as a glycosylated precursor protein, Pr95, that is proteolytically cleaved in the Golgi complex. The coding sequences for gp85 and gp37 have been placed in an open reading frame that extends from nucleotide 5054 to nucleotide 6863, and predict sizes of 341 amino acids (40,000 MW) for gp85 and 198 amino acids (21,500 MW) for gp37 (Fig. 2) (Hunter *et al.*, 1983). Carbohydrate makes up a significant contribution to the observed molecular weights of these polypeptides—the predicted amino acid sequence contains 14 potential glycosylation sites (Asn-X-Ser/Thr) in gp85 and 2 in gp37. Experiments aimed at determining the number of carbohydrate side chains yielded results consistent with most or all of the sites being occupied (Hunter *et al.*, 1983).

Although an initiation codon is located early (codon 4) in the open reading frame, during a viral infection splicing yields an mRNA on which translation initiates at the same AUG as the *gag* gene to produce a nascent polypeptide in which gp85 is preceded by a 62 amino acid long leader (signal) peptide (Fig. 2). This peptide contains a hydrophobic sequence that we have shown (see below) is necessary for translocation across the RER and is completely removed from the *env* gene product during translation (Hunter *et al.*, 1983). It represented one of the longest signal peptides described to date, and we were therefore interested in determining the signal peptide requirements for normal biosynthesis of gp85 and gp37. For these studies the *env* open reading frame was excised from the RSV genome and inserted into an SV40 expression vector under the control of the late-region promoter (Wills *et al.*, 1983; Perez *et al.*, 1987). In this construction translation is initiated at the AUG present at the start of the open reading frame (at nucleotide 5054 of the RSV genome) and results in the synthesis of an even longer (64 amino acid) signal peptide; nevertheless, biosynthesis of Pr95 and signal peptide cleavage occur normally. Furthermore, expression of the RSV *env* gene in African Green monkey (CV-1) cells parallels that seen in a normal virus infection in avian cells (Wills *et al.*, 1984; Hardwick *et al.*, 1986), making it an excellent system for the analysis of mutant *env* genes.

   a. Deletion/Substitution of the Signal Peptide. In order to examine the role of the signal peptide in RSV glycoprotein biosynthesis we constructed a series of deletion mutations within the 5' coding region of the *env* gene using the double-stranded exonuclease *Bal*31. Oligonucleotide linkers of the sequence CATCGATG were ligated to the ends of the truncated molecules to introduce a unique restriction endonuclease cleavage site and to replace the deleted in-frame AUG. The mutants were then sized and their nucleotide sequence determined to find those with a suitable deletion and an in-frame AUG. One such mutant, A1, contained a deletion of 171 nucleotides within the *env* coding sequences and encoded an *env* product that completely lacked an aminoterminal hydrophobic sequence (Fig. 4). Expression of this gene from the SV40 vector resulted in the synthesis of a nonglycosylated, 58 kDa cytoplasmic protein that was similar in size to the nonglycosylated wild-type *env* gene product produced in the presence of the glycosylation inhibitor, tunicamycin. In contrast to the tunicamycin product, however, the A1 protein was not associated with membrane vesicles and was rapidly degraded (half-life < 5 min; E. Hunter, K. Shaw, and J. Wills, unpublished). Thus the signals for initiating translocation of the RSV *env* gene product must reside within the cotranslationally removed amino-terminal sequence, and in their absence the molecule is synthesized as an unstable cytoplasmic protein. This result is similar to those obtained by Gething and Sambrook (1982) and by Sekikawa and Lai (1983) with the influenza virus HA gene product.

	Translocation	Glycosylation	Plasma memb.
RSV env	+	+	+
env. A1	–	–	–
HA$_S$.A1	+	+	+
HA$_{Sal}$.env$_{Xho}$	+	+	+
env$_{Xho}$.HA$_{Sal}$	+	+	–
env$_S$.HA$_{12}$	+	+	–
env$_S$.HA$_{16}$	+	+	+
Influenza HA	+	+	+

RSV env   M-(42 aa)-RVN<u>YILIIGVLVLC</u>EVTGVRA▸DVHLLEQPGNLWITWANATGQTDFCLS

env. A1   MEVTGVRA▸DVHLLEQPGNLWITWANATGQTDFCLS

HA$_S$.A1   *MAIIYLILLFTAVRG*-MEVTGVRA-DVHLLEQPGNLWITWANATGQTDFCLS

HA$_{Sal}$.env$_{Xho}$ *MAIIYLILLFTAVRG*▸*DQICIGYHANNSTEKV* QPGNLWITW

env$_{Xho}$.HA$_{Sal}$ M-(42 aa)-RVNYILIIGVLVLCEVTGVRA▸DVHLLE————*DTNLERNVT*

env$_S$.HA$_{12}$ M-(42 aa)-RVNYILIIGVLVLCEVTGVRA▸DVHLLE<u>*ELGICIGYHANNSTEKVDTNLERNVT*</u>

env$_S$.HA$_{16}$ M-(42 aa)-RVNYILIIGVLVLCEVTGVRA▸*DQICIGYHANNSTEKVDTNLERNVT*

Influenza HA   *MAIIYLILLFTAVRG*▸*DQICIGYHANNSTEKVDTNLERNVT*

Fig. 4. Mutant and hybrid *env*/HA genes. RSV *env* sequences are depicted in plain text. Influenza hemagglutinin sequences are depicted by italicized text. Underlined text denotes amino acid residues encoded by oligonucleotide linkers. Arrows depict the signal peptidase cleavage site at which the signal is removed cotranslationally from each of the constructs. Plus symbols indicate that translocation, glycosylation, or transport to a plasma membrane location is observed; minus symbols mean that the properties above are not observed.

Since the signal peptide of the RSV *env* gene product is exceptionally long, it was of interest to determine whether another signal peptide could substitute for it. For these experiments we have utilized the signal sequences of the influenza virus HA gene (A/Jap/305/57; Gething *et al.,* 1980). Two constructions were made: in the first of these the A1 deletion mutant coding sequence was fused in-frame to the HA signal coding sequence at the signal peptidase cleavage site of the latter (Fig. 4) and in the second, we made use of a *Sal*I restriction enzyme site in the HA gene and an *Xho*I site in the RSV *env* gene, so that the signal sequence and 16 amino acids of HA1 were fused with *env* 6 amino acids into gp85 (Fig. 4). Expression of these hybrid genes in CV-1 cells resulted in the biosynthesis of a glycosylated Pr95 protein that was transported to the Golgi complex, cleaved to gp85 and gp37, and displayed on the cell surface (E. Hunter, K. Shaw, and J. Wills, unpublished).

To demonstrate that the Pr95 molecules expressed from the HA–A1 fusion gene had undergone signal peptide cleavage, Pr95 was immunoprecipitated from [³H]leucine, pulse-labeled cells and analyzed by sequential Edman degradation, in order to determine the amino-terminal sequence. To our surprise the signal peptidase had cleaved at the HA cleavage site, despite the fact that according to the analyses of von Heijne (1983) only the RSV cleavage sequence should have been recognized. Thus the HA–A1 fusion protein contains two potential signal peptidase cleavage sites (that from the HA and that remaining in the *env* sequences), but only the first of these is utilized. Both gene fusions, therefore, result in the synthesis of aberrant gp85 proteins—that from the HA–A1 fusion having an 8 amino acid amino-terminal extension, and that from the *Sal/Xho* fusion having lost 6 amino-terminal amino acids and gained 16 from HA1—which nevertheless can be transported to the plasma membrane (Wills *et al.,* unpublished data).

Reciprocal gene fusions, in which the *env* gene signal peptide was fused to the structural sequences of HA (Fig. 4), also resulted in translocation of the HA molecule across the RER membrane, supporting the concept that this transient sorting sequence is not polypeptide specific. However, only in the construction where the signal sequence of *env* was precisely fused to the amino terminus of HA1 was transport beyond the RER observed (Wills *et al.,* unpublished data). In constructions where the amino-terminal sequence of HA1 was perturbed, the recombinant protein was apparently prevented from assembling into trimers and its transport was blocked in the RER (Gething *et al.,* 1986). Thus, while signal peptides may be capable of mediating the translocation of foreign polypeptides across the RER, other sorting "signals" must be active for transport of the molecule to continue.

b. Mutations within the Signal Peptide and Signal Peptidase Cleavage Site. The experiments described above have been extended to determine the following: (a) whether the hydrophobic region of the signal peptide carries all the information required for transfer of the *env* gene product into the RER; (b) what the structural specificities of the signal peptide are; and (c) where the specificity for signal peptidase cleavage is located. More than 200 prokaryotic and eukaryotic signal peptides have been sequenced (Watson, 1984). Comparison shows that most extensions comprise 20–40 amino acid residues; one of the longest being that of the RSV envelope glycoprotein. There is no homology between sequences, but a character-istic distribution of amino acid chains is observed. Three structurally distinct regions have been observed so far: a positively charged amino-terminal region, a central region of 9 or more hydrophobic residues, and a more polar carboxy-terminal region that appears to define the cleavage site (von Heijne, 1983, 1984, 1985; Perlman and Halvorson, 1983). The importance of these general features has been supported by the genetic studies in prokaryotic systems (reviewed by Silhavy *et al.,* 1983; Benson *et al.,* 1985).

To investigate these questions we initially constructed a series of inter-nal deletion mutants that initiated within the amino-terminus of gp85 and extended into the signal peptide. The deletion mutations were introduced

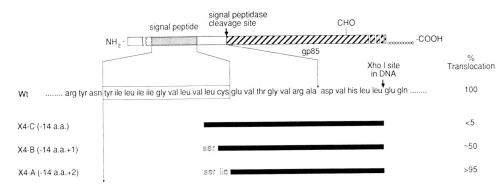

Fig. 5. Amino acid sequence deduced from DNA sequence of mutants X4-A, -B, and -C. The RSV glycoprotein is schematically represented: the location of the hydrophobic signal sequence within the long (64 amino acid) leader peptide is denoted by a stippled bar and the mature gp85 glycoprotein by a hatched bar. The signal peptidase cleavage site in both the cartoon and the amino acid sequence is denoted by a long arrow. The potential glycosylation site in the amino terminus of gp85 is shown as a CHO. The amino acid sequence of the last 21 amino acids of the signal peptide and the amino-terminal 7 amino acids of gp85 are shown for the wild-type gene product. The solid black bars show the lengths and positions of the deletions in the mutants. The outlined letters denote foreign amino acids inserted during construction of the nested set.

into the coding region for the envelope glycoprotein by digestion of a plasmid containing the *env* gene at a unique *Xho*I site located 13 base pairs (bp) from the 5' end of the coding sequence for gp85, followed by digestion with the double-stranded exonuclease *Bal*31. Potential mutants were identified by restriction enzyme analysis and DNA sequencing, and those of interest were engineered into the SV40 expression vector. These are depicted in Figs. 5 and 6. Mutants X4-A -B, and -C were derived from a single out-of-frame parent, and they represent a nested set of mutants in which the hydrophobic sequence varies from the wild-type length of 11 to only 9 amino acids. Expression of these mutant genes in CV-1 cells gave the results summarized in Fig. 5. Mutant polypeptides with the shortest hydrophobic domain (X4-C) resembled the A1 mutant polypeptides in that they had a cytoplasmic location, were nonglycosylated, and were rapidly degraded. They differed from the A1 mutant in length (65 versus 58 kDa) confirming that the mutated signal was not removed. Mutant X4-A polypeptides, on the other hand, were translocated and glycosylated with an efficiency equivalent to wild type, despite the substitution of serine and isoleucine for leucine and cysteine residues within the hydrophobic domain. Mutant X4-B expressed a phenotype intermediate between that of X4-A and X4-C, approximately 50% of the polypeptides being translocated and glycosylated. None of the mutants contain the sequences that specify the signal peptidase cleavage site, and molecules of X4-A/X4-B that were translocated across the RER retained an uncleaved signal peptide.

The data from these mutations suggest the following: (1) that the length, rather than the amino acid composition, of the hydrophobic domain of the *env* signal peptide is critical for translocation across the RER and (2) that signal peptide cleavage is not a requirement for translocation. The first of these conclusions is supported by genetic experiments in prokaryotic systems, where a requirement for secondary structure in the signal peptide was suggested (Emr and Silhavy, 1983; Bankaitis *et al.*, 1984). The second is consistent with the presence of permanent insertion sequences in secreted and membrane-spanning proteins that are translocated across the RER membrane without removal of an amino-terminal signal peptide (Palmiter *et al.*, 1978; Bos *et al.*, 1984; Markoff *et al.*, 1984; Zerial *et al.*, 1986; Spiess and Lodish, 1986). Several membrane-spanning proteins are anchored in the membrane by an amino-terminal anchor/signal domain and display what is termed group II protein topology (Garoff, 1985; Wickner and Lodish, 1985), where the amino terminus of the protein is cytoplasmic and the carboxy terminus is luminal.

After translocation of a nascent chain across the endoplasmic reticulum has been initiated, the signal peptide is removed. This cleavage is carried

signal peptidase cleavage; however, in this latter instance the procoat protein was transported inefficiently across the inner membrane (Boeke *et al.*, 1980; Russel and Model, 1981). Although more mutants will be required to properly define these systems, the prokaryotic cleavage site appears to be more sensitive to manipulation than that of eukaryotes. There is accumulating evidence that transported prokaryotic proteins, unlike those of eukaryotes, may not be transferred across membranes in a strictly cotranslational manner (Randall and Hardy, 1984). Thus, altered regions within the structural protein portion of a molecule would have the opportunity to interact and interfere with signal peptidase cleavage; such as interaction would not be possible in the cotranslational system described for eukaryotes.

Although the mutant X1 polypeptides were translocated across the RER membrane in a normal fashion, immunofluorescence experiments and posttranslational modification probes indicated that the transport and maturation of the X1 glycoprotein was halted shortly after exiting the RER, perhaps within pre- or cis-Golgi vesicles. Cells synthesizing the mutant protein showed no surface immunofluorescence, no cleavage of the Pr95 to gp85/gp37, and no terminal sugar additions (Hardwich *et al.*, 1986). The basis for this block appears to be the altered amino acid sequence rather than the loss of the carbohydrate side chain since by using a mutagenic oligonucleotide we have modified the amino terminus of the X1 gp85 from Asp-Val-His-Arg-Thr- to Asp-Val-Asn-Arg-Thr-, thereby reinserting the glycosylation site missing from this mutant. The derivative mutant, X1A, is glycosylated at this site but remains blocked at the same intracellular point in the secretory pathway (K. Shaw, K. Kervin, and E. Hunter, unpublished). A second deletion mutant of the RSV *env* gene is also blocked in intracellular transport. This mutant, C3, has an engineered deletion at the carboxy terminus of gp37 that removed the cytoplasmic tail and transmembrane region (see below and Wills *et al.*, 1984). Its transport is clearly blocked at an earlier stage than that of the X1 mutant since it was localized to the ER and never reached the Golgi apparatus, whereas by immunofluorescent staining of fixed cells the X1 protein appeared to colocalize with the Golgi complex (see Fig. 7; Hardwick *et al.*, 1986).

Although the X1 and C3 mutants contain alterations at opposite ends of the *env* gene product, they both appear to lack an element that normally signals their transport to and beyond the cis-Golgi. While there may be a specific amino acid sequence (analogous to the amino-terminal signal sequence) that is required for these transport steps, it is as likely that a correctly aligned tertiary structure is the critical factor. Just as small changes as the amino terminus of HA1 can disrupt assembly and transport

Fluorescein   Rhodamine

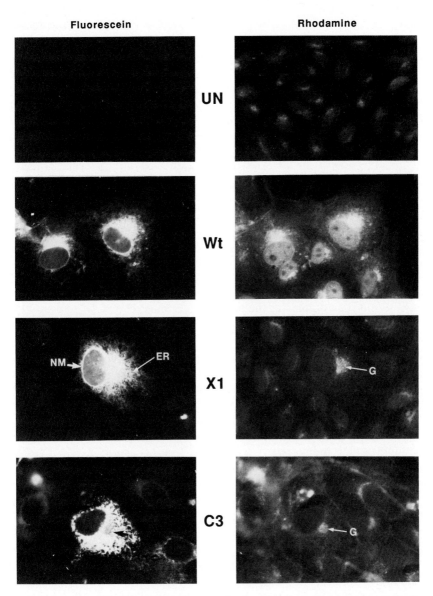

UN

Wt

NM →   ← ER
X1   ← G

C3   ← G

Fig. 7. Intracellular immunofluorescence of cells expressing wild-type and mutant poly-peptides. Fixed infected CV-1 cells were stained to detect the intracellular localization of wild-type and mutant glycoproteins. Rabbit anti-glycoprotein antibodies were tagged with fluorescein-conjugated goat anti-rabbit antibodies which in wild-type and X1-infected cells could be localized on the nuclear membrane (NM), endoplasmic reticulum (ER), and Golgi apparatus (G). The Golgi was localized by staining the same cells with rhodamine-conju-gated wheat germ agglutinin. In mutant C3-infected cells neither the nuclear membrane nor the Golgi apparatus stained with the antiglycoprotein antiserum. Magnification is 500.

131

of the HA trimer, the deleted amino acids unique to the X1 mutation may similarly play a critical role in the tertiary structure of the *env* complex, such that sorting signals required for transport through the Golgi complex are lost. Mutants X2 and X3 have deletions that begin at the amino terminus of gp85 and extend nine and six amino acids into this structural protein; thus these mutations overlap with the deletion in X1. Nevertheless, both mutant proteins were transported to the cell surface and were indistinguishable from the wild type. Mutants X2 and X3 thus indicate that the terminal nine amino acids of Pr95 are not required for normal intracellular transport and define the critical region in gp85 as the seven amino acids that are uniquely deleted in the X1 mutant.

2. Mutations within the Cytoplasmic Domain

a. Truncation of the Cytoplasmic Sequences. DNA and protein sequence studies demonstrated the presence of a 27 amino acid long hydrophobic (and presumably membrane-spanning) domain and a 22 amino acid long cytoplasmic domain at the carboxy terminus of gp37 (Hunter *et al.,* 1983; see Fig. 1). Comparison of these domains with those of other exogenous and endogenous strains of RSV has revealed that the sequence within the hydrophobic domain is highly conserved and that within the cytoplasmic domain the sequence of the first 18 amino acids (adjacent to the hydrophobic domain) is also highly conserved while those at the carboxy termini diverge greatly (Hughes, 1982; Hunter *et al.,* 1983). These results raised the possibility that the conserved region of the cytoplasmic domain might play a functional role in either transport of the *env* gene product through the secretory pathway or in virus assembly. To investigate this question we initially altered the cytoplasmic domain by introducing deletion mutations into the molecularly cloned sequences of the proviral *env* gene and examined the effects of the mutations on transport and subcellular localization in CV-1 cells. We found that replacement of the nonconserved region of the cytoplasmic domain with a longer unrelated sequence of amino acids from SV40 vector sequences (mutant C1) did not alter the rate of transport to the Golgi apparatus nor the appearance of the glycoprotein on the cell surface. Larger deletions, extending into the conserved region of the cytoplasmic domain (mutant C2), however, resulted in a 3-fold slower rate of transport to the Golgi complex, but did not prevent transport to the cell surface (Wills *et al.,* 1983, 1984). These results were thus consistent with the cytoplasmic domain of the RSV *env* gene product playing some role in transport to the Golgi complex.

Similar results were obtained by Rose and Bergmann (1983) who introduced into the cDNA clone encoding the VSV G protein a series of deletions that affected the cytoplasmic domain. These mutants fell into two classes; the first was completely arrested in their transport at a stage prior to the addition of complex oligosaccharides (presumably the RER) and the second showed severely reduced rates of transport to the Golgi complex although the proteins were ultimately transported to and expressed on the cell surface. The method by which these mutants were constructed (as with the RSV *env* mutants) meant that the truncated G proteins terminated in SV40 sequences, and in at least one case the block to transport could be alleviated by substitution of a termination codon for these "poison" sequences. Even in these constructions, however, three foreign amino acids were translated prior to termination (Rose and Bergmann, 1983).

The concept that the cytoplasmic domain might influence or govern the rate at which membrane-spanning proteins were transported to the Golgi complex was supported by similar studies of Doyle *et al.* (1985) on the HA polypeptide. While the HA cytoplasmic domain could be replaced by the equivalent region from the RSV *env* gene product without affecting the rate of transport of the hybrid HA from the RER to the Golgi complex, truncation of the HA cytoplasmic domain or addition of the 21 amino acid long cytoplasmic domain from gp37 slowed transport significantly, and addition of 16 amino acids encoded by pBR322 sequences blocked transport of the HA from the RER. On the other hand, studies on the class I histocompatibility antigens (Zuniga *et al.*, 1983; Murre *et al.*, 1984), the p62 of Semliki Forest virus (Garoff *et al.*, 1983), and additional studies on the HA of influenza (Doyle *et al.*, 1986) indicated that the cytoplasmic domains of these proteins could be truncated without affecting transport to the cell surface, although the kinetics of transport were not determined in every case.

b. Substitution Mutations. It should be noted that in most of the mutant constructions described above, the carboxy-terminal region contained one or several aberrant amino acids as a result of the recombinant DNA approach. Thus in order to determine more directly the role of the cytoplasmic domain of gp37, we have used oligonucleotide-directed mutagenesis to introduce an early termination codon in the coding sequences of gp37 such that the arginine residue that represents the first amino acid of the cytoplasmic domain is changed to an opal terminator. This mutation creates a truncated viral glycoprotein lacking specifically the cytoplasmic domain of gp37. The biosynthesis and transport of the products of this mutant viral glycoprotein gene were analyzed by expression from an SV40 late-region replacement vector, and its ability to be active in viral assembly was investigated by substitution of the mutated gene for the wild-type gene in an infectious avian retrovirus vector. In contrast to our previous results, deletion of the entire cytoplasmic domain alone had no effect on the biosynthesis or rate of intracellular transport of the *env* glycoprotein. Thus it seems unlikely that the conserved amino acids present in this region play a role in intracellular transport. Although the cytoplasmic domain contains several charged, hydrophilic residues, it does not appear, by itself, to be required for anchoring the complex in the membrane, since molecules lacking the cytoplasmic domain were expressed stably on the plasma membrane and were not shed into the cell culture medium (Perez *et al.*, 1987).

A recent study by Gething *et al.* (1986) has demonstrated that mutations within the cytoplasmic domain of the influenza virus HA can affect the conformation of the extracellular domain by preventing assembly and trimerization of the HA molecule, thereby resulting in a failure of those mutants to be efficiently transported. A similar requirement for the assembly of oligomeric forms of the VSV G protein prior to its transport to the Golgi complex has also been reported (Kreis and Lodish, 1986). The inconsistency of our previous results with the ones obtained with the opal mutant could be explained in a similar way. We cannot rule out the possibility that in our earlier experiments the extra amino acids, added as a consequence of the loss of the *env* termination codon, created a conformational change in the extracellular domain of Pr95 and slowed its transport from the RER. Our present results indicate that the cytoplasmic domain of gp37 is neither a recognition signal for transport to the plasma membrane nor a requirement for anchoring the molecule to it. These findings also support the idea that the charged amino acids present in most of the cytoplasmic domains of many transmembrane proteins (Garoff *et al.*, 1983; Sabatini *et al.*, 1982) are dispensable for anchor function (Davis *et al.*, 1984).

This latter question has also been addressed by Cutler and co-workers (Cutler and Garoff, 1986; Cutler *et al.*, 1986), who mutated the cytoplasmic domain of the p62 polypeptide of Semliki Forest virus. This region, which normally contains a charge cluster (Arg-Ser-Lys) flanking the hydrophobic domain, was changed to a neutral (Met-Ser-Gly) or an acidic (Met-Ser-Glu) one using oligonucleotide mutagenesis. Expression analyses of these mutant proteins confirmed that the basic amino acids were not required for cell surface transport since they reached the surface in a biologically active form. Nevertheless, both mutant polypeptides showed reduced stability when membranes containing them were extracted with high-pH buffer (Cutler *et al.*, 1986). Charged residues within the cytoplasmic domain may thus provide an additional measure of stability to the membrane-bound complex.

Since the conserved residues in the cytoplasmic domain of gp37 were not required for protein transport it seemed possible that this region might play a role in the process of infectious virus assembly. The fact that the mutant protein was efficiently transported to the cell surface allowed us to analyze this potential role for the cytoplasmic domain in the process of virus budding. Chemical cross-linking experiments have demonstrated an interaction between gp37 and p19, one of the *gag* gene products that structure the viral core of RSV (Gebhardt *et al.*, 1984). While it is clear that virus assembly can occur in the absence of glycoproteins, it was suggested that the p19/gp37 interaction may be part of the driving force for the process of viral assembly and budding. Furthermore, since host membrane glycoproteins are excluded from the viral membrane there must be some positive signal for inclusion of the viral *env* gene products in the budding virion. To determine whether the cytoplasmic domain is involved in this interaction and required for infectious virus assembly, we reconstructed a retrovirus genome carrying the "tail(−)" *env* gene mutation. Surprisingly, such mutant viruses were infectious on avian cells and spread through the culture with similar efficiency to those containing a native *env* glycoprotein complex. Furthermore, this truncated *env* gene complex was incorporated as efficiently into virus particles as the wild-type complex (Perez *et al.*, 1987). This fact suggests that if an interaction between gp37 and p19 is required to mediate the incorporation of the glycoproteins into the envelope of the budding viral particle, it must occur within the lipid bilayer, presumably with the hydrophobic anchor domain. It is thus unlikely that interactions between viral capsid proteins and the cytoplasmic domain of the *env* complex constitute a driving force for preferential incorporation of the viral glycoproteins in the avian retroviral envelope.

What then is the function of the cytoplasmic domain of the *env* glycoprotein? Since this segment of the viral polypeptide does show a region of conserved sequence, it is possible that it has evolved to facilitate transport to the plasma membrane without being a requirement for it; clearly, randomly inserted alterations within this domain can exert a negative effect on the transport process. While we have observed normal assembly and infection by virus encoding a "tail(−)" *env* product, it will be of interest to determine whether continued growth of the virus results in the dominant appearance of revertants that encode a functional cytoplasmic domain.

3. *Mutations in the Transmembrane Anchor Domain*

Many cell surface and membrane proteins of animal viruses are bound to the lipid bilayer by a membrane-spanning hydrophobic peptide close to the carboxy terminus of the polypeptide (reviewed by Warren, 1981; Armstrong *et al.*, 1981). Experimental evidence for this first came from deletion mutants of the influenza virus HA (Gething and Sambrook, 1982; Sveda *et al.*, 1982), the VSV G protein (Rose and Bergmann, 1982), and the minor coat protein of phage f1 (Boeke and Model, 1982) in which removal of sequences that encoded the cytoplasmic and membrane-spanning domains resulted in secretion of the protein.

The hydrophobic membrane-spanning peptide of these polypeptides is thought to be an essential component of the cotranslational signal that results in the arrest of chain transfer across the RER membrane during synthesis. These stop-translocation sequences have been proposed to be a region of the nascent protein molecule which halts insertion through the membrane by disassembling the translocation apparatus and thereby creates proteins with three topological domains (Blobel, 1980). They appear to be inseparable from the anchor sequences (Yost *et al.*, 1983; Rettenmier *et al.*, 1985) since the transfer of intact transmembrane domains to normally secreted proteins has caused translocation of the constructed hybrid molecules to stop at the added sequences (Yost *et al.*, 1983; Guan and Rose, 1984). However, the precise structural and physical properties of the stop-translocation sequences have not been defined. Wold *et al.* (1985) have suggested that the cytoplasmic domain of membrane-spanning proteins might act to interrupt translocation; however, this seems unlikely since deletion mutants lacking this domain are found to be associated with the membrane in a normal manner (see above; Garoff *et al.*, 1983; Zuniga *et al.*, 1983; Murre *et al.*, 1984; Doyle *et al.*, 1986; Perez *et al.*, 1987).

While length and sequence vary widely among regions described as transmembrane anchors, they do have characteristics in common. Most often, they are long stretches (19–30 residues) of predominantly nonpolar and hydrophobic amino acids, bounded by charged residues, at the carboxy terminus of membrane proteins. Membrane-spanning sequences have also been described, however, at the amino terminus of some viral proteins (Blok *et al.*, 1982; Palmiter *et al.*, 1978; Bos *et al.*, 1984; Markoff *et al.*, 1984; Zerial *et al.*, 1986; Spiess and Lodish, 1986) and in the middle of other proteins (Rettenmier *et al.*, 1985; Kopito and Lodish, 1985; Finer-Moore and Stroud, 1984). We have investigated the structural requirements for a functional anchor/stop-translocation sequence in the RSV *env* system by constructing both deletion and point mutations in this region.

a. Deletion of the Anchor Domain. During our studies on the role of the cytoplasmic domain in *env* product transport, we characterized a mutant (C3) in which the entire cytoplasmic and transmembrane domains were deleted. This mutant, in contrast to those described for the influenza virus HA and the VSV G protein, was arrested in its transport at the RER and thus was not secreted from the cell (Wills *et al.*, 1984). Pulse–chase experiments coupled with oligosaccharide precursor labeling experiments showed that the C3 polypeptide was not transported to the Golgi complex, even though it accumulated in a soluble, nonanchored form in the lumen of the RER; the mutant thus appeared to lack a functional sorting signal. Surprisingly, immunofluorescent labeling studies showed that the C3 protein (unlike the wild type) did not accumulate on the nuclear membrane but rather in vesicles distributed throughout the cytoplasm (Fig. 7), suggesting that movement to the nuclear membrane, blocked in C3, may require a specific transport event, even though the RER and nuclear membranes appear to be continuous. This hypothesis is supported by studies on the VSV G protein that indicate that transitional vesicles (for the transport of glycoproteins to the Golgi apparatus) may be derived from "blebs" in the nuclear membrane (Bergmann and Singer, 1983).

Although these studies raised the possibility that sorting signals might exist within the deleted region of C3, the 95 amino acid deletion in this mutant, which extends into the external domain of gp37, would be expected to prevent normal folding of the glycoprotein. To determine whether the transmembrane domain was required for intracellular transport, we have modified the *env* gene by oligonucleotide-directed mutagenesis, changing the lysine (AAA) codon, which precedes the hydrophobic domain of gp37, to an ochre nonsense codon (TAA). This modified gene thus encodes a protein consisting of the entire external domain of Pr95 and lacking precisely the hydrophobic membrane-spanning and hydrophilic cytoplasmic domain. The biosynthesis and intracellular transport of the truncated protein in CV-1 cells was not significantly different from that of the wild-type glycoprotein, suggesting that any protein signals for biosynthesis and intracellular transport of this viral glycoprotein complex must reside in its extracellular domain. In contrast to the case of the C3 mutant, this complex lacking just the transmembrane and cytoplasmic domains is secreted as a soluble molecule into the culture medium (Perez *et al.*, 1987). Since the glycoprotein complex lacking only the cytoplasmic domain of gp37 is stably expressed on the cell surface, in a manner similar to the wild-type complex, it can be concluded that the transmembrane domain alone is required for anchoring the RSV *env* complex in the cell membrane.

b. Requirements for a Functional Stop-Translocation/Anchor Sequence. We have approached the question of the compositional requirements for membrane anchoring and orientation (stop-translocation) of a membrane-spanning protein by substituting an arginine for a centrally positioned leucine in the hydrophobic anchor region of the RSV *env* gene product (Fig. 8). The arginine substitution is one of the most drastic

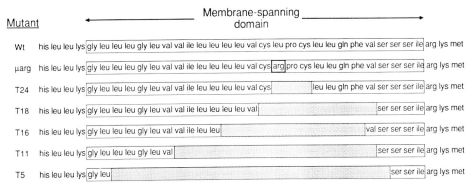

Fig. 8. Schematic summary of selected mutants within the hydrophobic domain of gp37. The amino acids within the hydrophobic membrane-spanning anchor domain of wild-type gp37 are shown within the long box. The location of the arginine mutation (μarg) is shown by the bold box. Deletions within the hydrophobic domain are depicted by stipled boxes, and the mutants are named according to the number of remaining hydrophobic residues (i.e., T24, T18, etc.). In each case, flanking external domain and cytoplasmic domain residues are identical.

compositional point mutations that could be made since it is only rarely found buried in hydrophobic environments (Kyte and Doolittle, 1982) and has a high predicted potential for terminating membrane-buried helices (Rao and Argos, 1986). The substitutions we have made fall within the conserved leucine-rich "IC" region proposed by Patarca and Haseltine (1984) and near the two cysteine residues where palmitate may be covalently added (Gebhardt *et al.*, 1984; Kaufman *et al.*, 1984).

By changing the anchor's hydrophobic integrity through the insertion of point mutations we hoped to define better what constituted a functional anchor sequence. Placing a highly charged basic side chain into the hydrophobic core of the membrane might be expected to either (1) terminate the membrane-spanning helix, thereby partitioning the charged residue to one side or the other of the membrane, or (2) destroy the stop-translocation signal, causing the protein to be secreted. The results of these experiments showed that a single amino acid substitution in the transmembrane anchor did not affect membrane association or its orientation in the membrane; unexpectedly, however, it affected targeting of the protein at a stage late in the transport pathway, such that the mutant protein was rapidly degraded in lysosmes (Davis and Hunter, 1987). The early translation products of both the arginine-mutant and wild-type genes behaved normally: they were synthesized with equal efficiency, had normal bitopic symmetry, and were glycosylated. The kinetics for the turnover of these precursors were nearly identical to those previously reported in infected chicken embryo fibroblasts (Bosch and Schwarz, 1984) and in SV40 expression vectors (Wills *et al.*, 1984). Furthermore, in the Golgi, palmitate was added to the precursors, they were cleaved to gp85-gp37, and they received terminal sugars. Only after this last stage did the presence of the charged side chain of the substituted arginine alter expression. At the level of the trans Golgi, a post-Golgi compartment (Saraste and Kuismanen, 1984), or at the cell surface, the gp85–gp37 complex was rapidly shunted to lysosomes and degraded—as shown by the protection afforded the terminally glycosylated *env* proteins by the lysosomatropic agent, chloroquine (Davis *et al.*, 1987). The exact pathway that the molecules take to the lysosome is not known. They may be transported directly from the trans Golgi or first to the surface where they are rapidly endocytosed. Discriminating between the alternate pathways has not been possible from current data.

Why the insertion of an arginine into the anchor should result in targeting to lysosomes is not obvious. A possible explanation is that we have introduced a specific sorting signal into the molecule; however, this is unlikely since other polypeptides with charged residues in the membrane-spanning domain are not so targeted (Kabcenell and Atkinson, 1985; Saito *et al.*, 1984; Hayday *et al.*, 1985). The arginine's charge is incompatible with the hydrophobic environment of the lipid bilayer; to achieve stability, the charged guanidinium group needs to be neutralized, and how this is done inside the bilayer is not clear. Parsegian (1969) has postulated that a lone charge sequestered in a membrane must form a pore or tunnel along with localized membrane thinning to acheive the lowest energy state. If the charged residue in the gp37 anchor causes the mutant molecules to aggregrate and form channels in the membrane in an analogous manner, it would likely kill the cell unless there was a mechanism to remove it rapidly. Alternatively, since the *env* protein is not an isolated entity in the membrane, it is conceivable that it aggregates with other components of the membrane to reduce net charge cooperatively, and thereby triggers the endocytotic machinery (Mellman and Plutner, 1984).

Charged residues are found in several proposed membrane-spanning helices (Kabcenell and Atkinson, 1985; Saito *et al.*, 1984; Hayday *et al.*, 1985; reviewed by Rao and Argos, 1986). The charged residues in bacteriorhodopsin membrane-spanning α helices have been suggested to be neutralized by forming ion pairs (Engelman *et al.*, 1980). This is likely a special case, however, since the energy required to bury an ion pair in the membrane is not much different from that required to bury the free charged group itself (Parsegian, 1969). Neutralization of strong charges, particularly of lysine and arginine, may occur through the formation of strong hydrogen bonds with tyrosine (Kyte and Doolittle, 1982); however, no tyrosine residues are present in the anchor domain of the *env* gene product which could participate with the arginine. The T cell α, β, and γ gene products and the rotavirus VP7 protein have a putative structure similar to that of the arginine mutant, with a lysine centered within the transmembrane anchor; however, unlike the molecule we created, they invariably have tyrosine residues adjacent to the lysine (Saito *et al.*, 1984; Kabcenell and Atkinson, 1985; Hayday *et al.*, 1985) which could stabilize the charge through hydrogen bonds (Kyte and Doolittle, 1982).

Adams and Rose (1985a) have described the similar insertion of an arginine (and glutamine) residue at the center of the transmembrane domain of the VSV G protein. Their mutant protein, like the *env* mutant we have characterized, was bitopic, could be seen localizing in the Golgi, and did not accumulate on the cell surface. Since these investigators observed a "lower level of protein expression" with their mutant G protein, it is possible that it also was rapidly degraded in lysosomes following terminal glycosylation. In contrast, Cutler and Garoff (1986) observed no alteration in the biosynthesis and transport of a mutant p62 polypeptide of Semliki Forest virus in which the hydrophobic domain was interrupted by insertion of a glutamic acid residue. In this protein, however, the outer boundary of the hydrophobic domain is not delineated by a charged residue and so it is possible that additional uncharged residues from the external domain were pulled into the membrane.

Since the insertion of a charged polar residue into the transmembrane region of the RSV *env* gene product did not interfere with its anchor/stop-translocation function, we have investigated the requirement for the long (27 amino acid) hydrophobic domain in arresting translocation and anchoring the *env* complex. A series of deletion mutations was generated by progressively removing base pairs to either side of a unique *Sph*1 restriction site that had been previously engineered into the center of the anchor coding region. This produced *env* proteins with truncated transmembrane anchors that ranged in length from 24 (T24) to a single apolar amino acid (T1) summarized in Fig. 8; Davis and Hunter, 1987).

While the effects of the deletions on the transport and subcellular localization of the *env* gene product appeared to be a complex function of the length and composition of the remaining anchor, the mutants appeared to fall into three broad phenotypic classes (summarized in Table I). Even the smallest deletion (T24), which removed only three amino acids, greatly reduced the surface expression of the mature *env* proteins. T24 and mutants T18, T17, and T16 had a normal bitopic orientation in the membrane but appeared to be cleared from the cell surface and degraded in lysosomes, since they accumulated only in the presence of choroquine, an inhibitor of lysosomal degradation. The reduced surface expression of the

TABLE I

Expression of *env* Gene Transmembrane Anchor Deletion Mutants

Name	Base pairs deleted/ anchor length[a]	Hydro-phobicity[b] Score/Mean		Processing to gp85 and gp37	gp85 and gp37: Degradation/ chloroquine protection	Surface expression
Wild type	0/27	67	2.5	+[c]	−/−	+
T24	9/24	59	2.5	+	+/+	±
T18	27/18	53	2.9	+	+/+	±
T17	30/17	31	1.9	+	+/+	±
T16	33/16	41	2.6	+	+/+	±
T14	69/14	15	1.1	−	+/−	−
T11	48/11	21	1.9	−	+/−	−
T7	105/7	14	2.0	+, SEC[d]	ND[e]	−
T5	66/5	6	1.3	+, SEC	−/(+Pr95)ϕ[f]	−
T1	87/1	5	5.0	+, SEC	ND	−
A[-g]	27/0	0	na	+, SEC	ND	−

[a] As determined by the distance between charged residues.

[b] Hydrophobicity score as determined using the hydropathicity values of Kyte and Doolittle (1982). Mean hydrophobicity is hydrophobicity score divided by the apparent anchor length.

[c] Symbols: + denotes either a wild-type or positive response; ± denotes an intermediate response; − denotes a negative response.

[d] SEC, Secreted.

[e] ND/na, Not determined/not applicable.

[f] With chloroquine treatment, Pr95 appears in the medium with gp85 and gp37.

[g] Anchorless and tailless deletion mutant (Perez *et al.*, 1987).

largest deletion mutant (T24) was surprising since the effective hydrophobic peptide remaining in this construct was as long or longer than functional anchors reportedly present in other integral membrane proteins [e.g., 19 amino acids: M_2 protein of influenza (Lamb, 1985); 20 amino acids: VSV G protein (Rose *et al.*, 1980) and adenovirus E3 protein (Wold *et al.*, 1985)]. Deletions which reduced the size of the transmembrane anchor to seven amino acids (T7) or less resulted in the secretion of mature glycoproteins into the medium. A third class of mutants with hydrophobic regions of 14 (T14) and 11 (T11) amino acids, respectively, while remaining membrane associated, no longer appeared to span the RER as bitopic proteins. Neither mutant could be found at the surface of cells, nor could their degradation be arrested by chloroquine treatment.

From the sum of the data obtained with these mutants, it would appear that bitopic insertion of the *env* gene product is possible with effective anchor domains of at least 16 amino acids; if additional amino acids are removed from the domain, the protein can no longer exist bitopically, and it either partitions monotopically to the luminal side of the RER membrane or withdraws amino acids from the cytoplasmic side of the membrane into the bilayer. Nevertheless, such sequences from the cytoplasmic domain are not able to stabilize the shortest anchors (T1, T5, and T7) since these are secreted from the cell.

Davis and Model (1985) have investigated the requirements of a functional anchor domain by inserting artificial hydrophobic peptides of varying length into the membrane-associated pIII protein of the bacteriophage f1. Their results show that 17 hydrophobic amino acids are sufficient to maintain the protein in a bitopic configuration; however, the 17 amino acid anchor was "deleterious to the cell" presumably because it was too short to assume a stable conformation compatible with existence in the bilayer and thereby destabilized the membrane. A construct with an anchor of 12 hydrophobic amino acids was membrane associated but showed an intermediate phenotype in its sensitivity to solubilization by alkali. In contrast, a construct with only 8 hydrophobic amino acids—also membrane associated—was completely released into the supernatant at high pH. Adams and Rose (1985b) reduced the anchor domain of VSV G by precisely deleting amino acids from within the hydrophobic core. When the length of the anchor was reduced from 20 amino acids to as few as 14, the protein was normally membrane associated and expressed in a bitopic fashion on the cell surface. On the other hand, proteins with an anchor domain of 12 or 8 amino acids, while spanning the membrane, appeared to be transported only as far as the Golgi where they accumulated; the surface expression of these proteins was greatly reduced (12 amino acid anchor) or undetectable (8 amino acids).

Doyle *et al.* (1986) have characterized a series of carboxy-terminal deletion mutants of the HA polypeptide in which the 27 amino acid anchor domain was truncated to 17, 14, and 9 amino acids, respectively. In this case, molecules with a 17 amino acid long transmembrane domain were stably anchored but were transported less efficiently to the plasma membrane. Truncation of the hydrophobic anchor to 9 or 14 residues resulted in HA proteins that were unstable and whose transport appeared to be blocked in the RER or in a pre-Golgi compartment—resembling the T11 mutant of the RSV *env* gene described above. From the results of these different systems and approaches it would seem that in strictly physical terms, anchors may be significantly reduced in their length without any consequence to the membrane association. The limits for this length appear to be about 8–12 amino acids, but it must be appreciated that the mere presence of a stretch of hydrophobic amino acids within a protein does not serve to constitute an anchor. Several mammalian virus envelope proteins contain in their external domain long hydrophobic amino acid regions that are equivalent to the truncated bitopic anchors have described here (Gething *et al.*, 1978; White *et al.*, 1981). Indeed, gp85 contains a strongly hydrophobic 11 amino acid region that clearly does not act as a stop-translocation sequence or play a role in membrane association (Hunter *et al.*, 1983). The work of Davis and Model (1985), on the other hand, implies that the length of a hydrophobic region is the major determinant as to whether or not it will confer membrane association properties to a protein, although they point out that the position of such sequences within the molecule may play a role.

Most eukaryotic membrane-spanning polypeptides have a complex tertiary structure that is stabilized by multiple disulfide linkages, and it is possible that the entropy of a correctly folded molecule is sufficient to pull potential stop-transfer regions through the membrane. A corollary of this hypothesis, therefore, is that, once folding is complete, short hydrophobic regions that can potentially span the membrane as an α helix would stop translocation. The length requirements for such a region could be shorter than the 20 residues predicted from α-helix dimensions if the region were flanked by arginines or lysines. Since the latter have long side chains, equivalent in length to a single turn of an α helix, a stretch of hydrophobic amino acids 13–14 amino acids long might be sufficient. Such a prediction fits well with the data we have obtained and with those of Adams and Rose (1985b) and Davis and Model (1985). It will be of interest to determine what effect inserting the truncated anchors of mutants T18 and T16 into the middle of the *env* precursor has on translocation; if the above speculations are correct they should be extruded into the external domain.

Finally, it should be reemphasized that merely providing a bitopic membrane anchor/stop-translocation is not sufficient to confer wild-type biological activity on a polypeptide. Mutant T24 of the RSV *env* gene has a hydrophobic domain which might be expected to be sufficiently long and hydrophobic in character to span the membrane stably; it is modified normally by palmitic acid and can clearly be transported to its targeted cellular location. Nevertheless, it is degraded rapidly by the cell. These results imply that hydrophobic transmembrane domains contain additional (and perhaps subtle) signals that remain to be deciphered, a conclusion that is supported by the finding that deletion of the anchor domain of the rotavirus VP7 protein abolishes its specific targeting and retention in the RER (Poruchynsky *et al.*, 1985).

4. Transport and Assembly in Polarized Cells

Polarized epithelial cells exhibit apical and basolateral membrane domains that are separated by well-defined tight junctions. Each membrane domain has a unique protein composition (Louvard, 1980; Reggio *et al.*, 1982), indicating that mechanisms must exist to specifically target membrane proteins to different surfaces. The sorting process occurs during or shortly after passage of the glycoproteins through the Golgi complex (Rodriguez-Boulan *et al.*, 1984; Pfeiffer *et al.*, 1985; Rindler *et al.*, 1984, 1985). However, the mechanisms determining this directed transport to either the apical or basolateral membranes are not understood. Their study has been facilitated by the use of cultured epithelial cell lines, such as the MDCK cell line, and by the observation that certain RNA viruses bud exclusively from apical or basolateral domains of these polarized cells in culture (Rodriguez-Boulan and Sabatini, 1978; Herrler *et al.*, 1981; Roth *et al.*, 1983a; Rindler *et al.*, 1985). Avian and mammalian retroviruses together with rhabdoviruses such as VSV mature from the basolateral surface, while ortho- and paramyxoviruses bud from the apical surface. The carbohydrate residues present on the different proteins do not appear to play a role in this sorting process, since tunicamycin does not interfere with the polarized release of the viruses (Roth *et al.*, 1979; Green *et al.*, 1981b).

As with the maturation of viruses that assemble at intracellular locations within the secretory pathway, polarized budding of enveloped viruses is dependent on the site to which viral glycoproteins are transported. Expression of cloned viral glycoprotein genes from both SV40-based and vaccinia expression vectors in polarized cells has demonstrated that the HA and neuraminidase polypeptides of influenza virus are targeted to the apical surface (Roth *et al.*, 1983b; Jones *et al.*, 1985; Gottlieb *et al.*, 1986) while the G protein of VSV and the gp70/p15E complex of murine leukemia virus (MuLV) are transported exclusively to the basolateral membranes (Stephens et al., 1986). In an attempt to locate the signals which direct these glycoproteins to the apical or basolateral domains, recombinant DNA techniques have been employed to construct chimeric proteins and express these in polarized cells in culture. In experiments where sequences encoding the external domain of HA were fused to those encoding the transmembrane and cytoplasmic domains of the VSV G protein, the hybrid glycoprotein behaved in the same manner as wild-type HA and was transported to the apical domain of polarized cells (McQueen *et al.*, 1986; Roth *et al.*, 1986). Conversely, fusing the external domain of G protein to the anchor/cytoplasmic domain of HA results in basolateral transport (McQueen *et al.*, 1987). These experiments thus suggest that the ectodomains of HA and G protein contain signals for apical and basolateral transport, respectively. While expression of a secreted form of the HA glycoprotein in an apical polarized manner supports this conclusion (Roth *et al.*, 1986), the unanchored ectodomain of the MuLV gp70/p15E complex, which is normally targeted to the basolateral membrane, is secreted in a nonpolarized fashion (Stephens and Compans, 1986). It is possible that this soluble protein is improperly folded and thus is unable to interact with the sorting machinery, alternatively it also raises the possibility that targeting signals may be located in more than one domain of these molecules. Further analyses should shed light on this problem.

III. CONCLUSIONS

The studies described in this chapter demonstrate the breadth of information that has been and can be obtained from studies on enveloped virus glycoprotein biosynthesis. Many of the studies were performed at a time when the cloned genes and molecular probes for cellular glycoproteins were unavailable and thus provided valuable insights into the manner in which cells compartmentalized and transported membrane proteins. The exciting possibility of utilizing viral glycoprotein genes for genetic analyses of the transport pathway, in a way analogous to that pursued in prokaryotic systems (Michaelis and Beckwith, 1982; Silhavy *et al.*, 1983; Oliver, 1985), has led to a plethora of studies that utilized both classic and recombinant DNA genetic approaches. These investigations have resulted in great progress in our understanding of the general processes involved in intracellular transport of proteins through the secretory pathway but at the same time have raised difficult questions about the molecular interactions required for protein sorting.

The initial observation that proteins destined for secretion contain an amino-terminal sorting sequence provided a precedent on which to build models for protein targeting based on topogenic sequences (Blobel, 1980). To a large extent the identification of signal sequences was facilitated by their transient nature, not by a conserved primary sequence. Indeed, while some common characteristics of signal peptides can be recognized (von Heijne, 1985), the identification of signal peptides in proteins where they are not removed has proved difficult and has required the use of sophisticated recombinant DNA technology [e.g., ovalbumin (Tabe *et al.*, 1984)]. The requirement for a signal sequence to initiate translocation across the ER membrane has been clearly confirmed through the isolation and construction of mutants which lack this functional region (as discussed above) and by fusion of this sequence to proteins that are not normally translocated (Lingappa *et al.*, 1984). These approaches have also been facilitated by the transient, nonstructural nature of many amino-terminal, translocation signals. Recent experiments by Friedlander and Blobel (1985) and Kaiser *et al.* (1987), however, raise questions about the informational content of signal sequences. In particular Kaiser and colleagues showed that several random amino acid sequences derived from human genomic DNA fragments could act as signal sequences for translocation of the yeast invertase enzyme. Thus even in this well-defined situation, where an amino acid sequence is known to play a functional role in the sorting process, it can be impossible to predict with confidence its location in the protein; how then might we expect to identify additional sorting sequences that may or may not exist within the structural domain of a transported protein?

The possibility that additional sorting sequences might be involved in steering the transport of a membrane-spanning or secreted protein through the vesicular maze of the secretory pathway remains open. At the present time the necessity for a native tertiary structure cannot be separated from the possibility of such additional sorting sequences. It is clear that disruption of a polypeptide's normal folding can completely prevent its transport from the ER (Gething *et al.*, 1986; Kreis and Lodish, 1986), and the simplest explanation for the phenotypes of a variety of conditional and nonconditional transport mutants would be that they alter the tertiary structure of the mature protein (see Section II,A, above). A 3-dimensional structure has been determined for only a few molecules that traverse the secretory pathway, and even with these proteins the current, predictive algorithms are insufficiently accurate to model potential changes in molecular shape in response to mutations. The question of a direct role for tertiary structure in protein transport thus represents a major challenge to molecular biologists. Furthermore, one might argue that a change in protein shape could also mask or distort a necessary (peptide) sorting sequence. This possibility is supported by the observation that for a majority of eukaryotic proteins the amino-terminal signal peptide is unable to initiate translocation across the ER if translation is allowed to proceed to completion, presumably because the tertiary structure of the nascent polypeptide precludes the interaction of the signal peptide with the translocation machinery.

It is quite feasible that sorting signals and targeting signals could be represented by different entities within a single polypeptide, particularly if the latter were required to fix the intracellular location of a protein. For example, the rotavirus VP7 polypeptide accumulates within the ER unless its amino-terminal hydrophobic anchor region is deleted, whereupon it is transported to the cell surface and secreted (Poruchynsky *et al.,* 1985); in this instance the deleted region presumably contains a sequence that can fix the intracellular location of the protein despite the fact that the molecule has the potential to be exported from the cell.

Since most translocated polypeptides appear to follow a common pathway to a late compartment of the Golgi (Kelly, 1985), it might be argued that a native conformation is the sole requirement for transport to this organelle and that the observed differences in the transport rates of proteins to the Golgi merely reflect the time necessary for completion of the folding process. Nevertheless, proteins leaving the Golgi appear to be sorted into specific pathways; for example, in secretory cells proteins may follow either the constitutive pathway or be sequestered in secretory granules (Moore and Kelly, 1985; reviewed by Kelly, 1985), and in epithelial cells specific proteins appear to be transported directly to either the apical or basolateral membranes (see above). Thus, it would seem likely that some form of sorting signal must be present in the polypeptide at this point in the secretory pathway in order to correctly direct its transport; initial results from viral glycoprotein expression studies indicate that at least in polarized cells the ectodomain of the sorted protein plays a dominant role (McQueen *et al.,* 1986; Roth *et al.,* 1987). Additional studies should provide a clearer picture of this complex process.

In summary, studies on the biosynthesis and transport of enveloped virus glycoproteins have provided important insights into the general processes involved in the intracellular movement of these membrane-associated molecules. The specific questions that remain to be answered are many and difficult, but it is likely that these viral systems will continue to play a vital role by providing clues and direction in this important area of cell biology.

ACKNOWLEDGMENTS

I would like to express my sincere thanks to Dr. Gary Davis, Dr. Marie Hardwick, Dr. Lautaro Perez, Ms. Karen Shaw, and Dr. John Wills, who performed much of the research on the RSV *env* gene described in this chapter. Without their dedicated and enthusiastic help it would not have been possible. The research described here was supported by Grant CA-29884 from the National Institutes of Health. L. Perez was a predoctoral trainee supported by an ODEPLAN–Chile fellowship. G. Davis was a predoctoral fellow supported by NIH Grant CA 09202. E. Hunter was the recipient of Research Career Development Award CA-00685 from the National Cancer Institute. For helpful, critical comments on the manuscript, I thank Drs. G. Air, D. Miller, and C. Morrow. I am most grateful to Ms. Melanie Pollard for her assistance in preparing the manuscript for publication.

REFERENCES

Adams, G. A., and Rose, J. K. (1985a). Incorporation of a charged amino acid into the membrane spanning domain blocks cell surface transport but not membrane anchoring of a viral protein. *Mol. Cell. Biol.* **5**, 1442–1448.

Adams, G. A., and Rose, J. K. (1985b). Structural requirements of a membrane-spanning domain for protein anchoring and cell surface transport. *Cell* **41**, 1007–1015.

Arias, C., Bell, J. R., Lenches, E. M., Strauss, E. G., and Strauss, J. H. (1983). Sequence analysis of two mutants of Sindbis virus defective in the intracellular transport of their glycoproteins. *J. Mol. Biol.* **168**, 87–102.

Armstrong, J., Perham, R. N., and Walker, J. E. (1981). Domain structure of bacteriophage fed adsorption protein. *FEBS Lett.* **135**, 167–172.

Bankaitis, V. A., Rasmussen, B. A., and Bassford, P. J., Jr. (1984). Intragenic suppressor mutations that restore export of maltose binding protein with a truncated signal peptide. *Cell* **37**, 243–252.

Benson, S. A., and Silhavy, T. J. (1983). Information within the mature LamB protein necessary for localization to the outer membrane of *Escherichia coli* K-12. *Cell* **32**, 1325–1335.

Benson, S. A., Hall, M. N., and Silhavy, T. J. (1985). Genetic analysis of protein export in *Escherichia coli* K-12. *Annu. Rev. Biochem.* **54**, 101–134.

Bergmann, J. E., and Singer, S. J. (1983). Immunoelectron microscopic studies of the intracellular transport of the membrane glycoprotein (G) of vesicular stomatitis virus in infected Chinese hamster ovary cells. *J. Cell Biol.* **97**, 1777–1787.

Bergmann, J. E., Tokuyasu, K. T., and Singer, S. J. (1981). Passage of an integral membrane protein, the vesicular stomatitis virus glycoprotein, through the Golgi apparatus en route to the plasma membrane. *Proc. Natl. Acad. Sci. U.S.A.* **78**, 1746–1750.

Bishop, D. H. L., and Shope, R. E. (1979). Bunyaviridae. *In* "Comprehensive Virology" (H. Fraenkel-Conrat and R. R. Wagner, eds.), Vol. 14, pp. 301–385. Plenum, New York.

Blobel, G. (1980). Intracellular protein topogenesis. *Proc. Natl. Acad. Sci. U.S.A.* **77**, 1496–1500.

Blobel, G., and Dobberstein, B. (1975). Transfer of proteins across membranes. I. Presence of proteolytically processed and unprocessed nascent immunoglobulin light chains on membrane-bound ribosomes of murine myeloma. *J. Cell Biol.* **67**, 835–851.

Blok, J., Air, G. M., Laver, W. G., Ward, C. W., Lilley, G. G., Woods, E. F., Roxburgh, C. M., and Inglis, A. S. (1982). Studies on the size, chemical composition and partial sequence of the neuraminidase (NA) from type A influenza viruses show the N-terminal region of NA is not processed and serves to anchor NA in the viral membrane. *Virology* **199**, 109–121.

Boeke, J. D., and Model, P. (1982). A prokaryotic membrane anchor sequence: Carboxyl terminus of bacteriophage f1 gene III protein retains it in the membrane. *Proc. Natl. Acad. Sci. U.S.A.* **79**, 5200–5204.

Boeke, J. D., Russel, M., and Model, P. (1980). Processing of filamentous phage precoat protein: Effect of sequence variations near the signal peptidase cleavage site. *J. Mol. Biol.* **144**, 103–116.

Bos, T. J., Davis, A. R., and Nayak, D. P. (1984). NH_2-terminal hydrophobic region of influenza virus neuraminidase provides the signal function in translocation. *Proc. Natl. Acad. Sci. U.S.A.* **81**, 2327–2331.

Bosch, J. V., and Schwarz, R. T. (1984). Processing of gPr92env, the precursor to the glycoproteins of Rous sarcoma virus: Use of inhibitors of oligosaccharide trimming and glycoprotein transport. *Virology* **132**, 95–109.

Boursnell, M. E. G., Brown, T. D. K., and Binns, M. M. (1984). Sequence of the membrane protein gene from avian coronavirus IBV. *Virus Res.* **1**, 303–313.

Compton, T., and Courtney, R. J. (1984). Virus-specific glycoproteins associated with the nuclear fraction of herpes simplex virus type 1-infected cells. *J. Virol.* **49**, 594–597.

Creek, K. E., and Sly, W. S. (1984). *In* "Lysosomes in Pathology and Biology" (J. T. Dingle, R. T. Dean, W. Sly, eds.), pp. 63–82. Elsevier/North Holland, New York.

Cutler, D. F., and Garoff, H. (1986). Mutants of the membrane-binding region of Semliki Forest virus E2 protein. I. Cell surface transport and fusgenic activity. *J. Cell Biol.* **102**, 889–901.

Cutler, D. F., Melancon, P., and Garoff, H. (1986). Mutants of the membrane-binding region of Semliki Forest virus E2 protein. II. Topology and membrane binding. *J. Cell Biol.* **102**, 902–910.

Daniels, R. S., Downie, J. C., Hay, A. J., Knossow, M., Skehel, J. J., Wang, M. L., and Wiley, D. C. (1985). Fusion mutants of the influenza virus hemagglutinin glycoprotein. *Cell* **40**, 431–439.

Date, T., and Wickner, W. (1981). Isolation of the *Escherichia coli* leader peptidase gene and effects of leader peptidase overproduction *in vivo*. *Proc. Natl. Acad. Sci. U.S.A.* **78**, 6106–6110.

Davis, G. L., and Hunter, E. (1987). A charged amino acid substitution within the transmembrane anchor of the Rous sarcoma virus envelope glycoprotein affects surface expression but not intracellular transport. *J. Cell Biol.* (in press).

Davis, G. L., Shaw, K. E. S., and Hunter, E. (1987). Altered surface expression, membrane association and intracellular transport result from deletions within the transmembrane anchor of the Rous sarcoma virus envelope glycoprotein. *J. Cell Biol.* (submitted).

Davis, N. G., and Model, P. (1985). An artificial anchor domain: Hydrophobicity suffices to stop transfer. *Cell* **41**, 607–614.

Davis, N. G., Boeke, J. D., and Model, P. (1984). Fine structure of a membrane anchor domain. *J. Mol. Biol.* **181**, 111–121.

Doms, R. W., Gething, M. J., Henneberry, J., White, J., and Helenius, A. (1986). Variant influenza virus hemagglutinin that induces fusion at elevated pH. *J. Virol.* **57**, 603–613.

Doyle, C., Roth, M. G., Sambrook, J., and Gething, M.-J. (1985). Mutations in the cytoplasmic domain of influenza virus hemagglutinin affect different stages of intracellular transport. *J. Cell Biol.* **100**, 704–714.

Doyle, C., Sambrook, J., and Gething, M.-J. (1986). Analysis of progressive deletions of the transmembrane and cytoplasmic domains of influenza hemagglutinin. *J. Cell Biol.* **103**, 1193–1204.

Dubois-Dalcq, M., Holmes, K. V., and Rentier, B. (1984). "Assembly of Enveloped RNA Viruses," pp. 1–182. D. Kingsbury, ed. Springer-Verlag, Wien, New York.

Dunphy, W. G., Fries, E., Urbani, L. J., and Rothman, J. E. (1981). Early and late functions associated with the Golgi apparatus reside in distinct compartments. *Proc. Natl. Acad. Sci. U.S.A.* **78**, 7453–7457.

Emr, S. D., and Bassford, P. J. (1982). Localization and processing of outer membrane and periplasmic proteins in *Escherichia coli* strains harboring export-specific suppressor mutations. *J. Biol. Chem.* **257**, 5852–5860.

Emr, S. D., Hanley-Way, S., and Silhavy, T. (1981). Suppressor mutations that restore export of a protein with a defective signal sequence. *Cell* **23**, 79–88.

Emr, S. D., and Silhavy, T. J. (1983). Importance of secondary structure in the signal sequence for protein secretion. *Proc. Natl. Acad. Sci. U.S.A.* **80**, 4599–4603.

Engelman, D. M., Henderson, R., McLachlan, A. D., and Wallace, B. A. (1980). Path of the polypeptide in bacteriohodopsin. *Proc. Natl. Acad. Sci. U.S.A.* **77**, 2023–2027.

Finer-Moore, J., and Stroud, R. M. (1984). Amphipathic analysis and possible formation of the ion channel in an acetylcholine receptor. *Proc. Natl. Acad. Sci. U.S.A.* **81**, 155–159.

Fitting, T., and Kabat, D. (1982). Evidence for a glycoprotein "signal" involved in transport between subcellular organelles. *J. Biol. Chem.* **257**, 14011–14017.

Friedlander, M., and Blobel, G. (1985). Bovine opsin has more than one signal sequence. *Nature (London)* **318**, 338–343.

Gahmberg, N. (1984). Characterization of two recombinant–complementation groups of Uukuniemi virus temperature-sensitive mutants. *J. Gen. Virol.* **65**, 1079–1090.

Gahmberg, N., Kuismanen, E., Keranen, S., and Pettersson, R. F. (1986). Uukuniemi virus glycoproteins accumulate in and cause morphological changes of the Golgi complex in the absence of virus maturation. *J. Virol.* **57**, 899–906.

Gallione, C. J., and Rose, J. K. (1983). Nucleotide sequence of a cDNA clone encoding the entire glycoprotein from the New Jersey serotype of vesicular stomatitis virus. *J. Virol.* **46**, 162–169.

Gallione, C. J., and Rose, J. K. (1985). A single amino acid substitution in a hydrophobic domain causes temperature-sensitive cell-surface transport of a mutant viral glycoprotein. *J. Virol.* **54**, 374–382.

Garoff, H. (1985). Using recombinant DNA techniques to study protein targeting in the eucaryotic cell. *Annu. Rev. Cell Biol.* **1**, 403–445.

Garoff, H., Kondor-Koch, C., Pettersson, R., and Burke, B. (1983). Expression of Semliki Forest virus proteins from cloned complementary DNA. II. The membrane-spanning glycoprotein E2 is transported to the cell surface without its normal cytoplasmic domain. *J. Cell Biol.* **97**, 652–658.

Gebhardt, A., Bosch, J. V., Ziemiecki, A., and Friis, R. R. (1984). Rous sarcoma virus p19 and gp35 can be chemically crosslinked to high molecular weight complexes. An insight into viral association. *J. Mol. Biol.* **174**, 297–317.

Gething, M.-J. (1985). "Protein Transport and Secretion." Cold Spring Harbor Laboratory, Cold Spring Habor, New York.

Gething, M.-J., and Sambrook, J. (1982). Construction of influenza haemagglutinin genes that code for intracellular and secreted forms of the protein. *Nature (London)* **300**, 598–603.

Gething, M.-J., White, J. M., and Waterfield, M. D. (1978). Purification of the fusion protein of Sendai virus: Analysis of the NH_2-terminal sequence generated during precursor activation. *Proc. Natl. Acad. Sci. U.S.A.* **75**, 2737–2740.

Gething, M.-J., Bye, J., Skehel, J., and Waterfield, M. (1980). Cloning and DNA sequence of double-stranded copies of haemagglutinin genes from H2 and H3 strains elucidates antigenic shift and drift in human influenza virus. *Nature (London)* **287**, 301–306.

Gething, M.-J., McCammon, K., and Sambrook, J. (1986). Expression of wild-type and mutant forms of influenza hamagglutinin: The role of folding in intracellular transport. *Cell* **46**, 939–950.

Gibson, R., Leavitt, R., Kornfeld, S., and Schlesinger, S. (1978). Synthesis and infectivity of vesicular stomatitis viruses containing nonglycosylated G protein. *Cell,* **13**, 671–679.

Gibson, R., Schlesinger, S., and Kornfeld, S. (1979). The nonglycosylated glycoprotein of vesicular stomatitis virus is temperature-sensitive and undergoes intracellular aggregation at elevated temperatures. *J. Biol. Chem.* **254**, 3600–3607.

Gilmore, R., Blobel, G., and Walter, P. (1982a). Protein translocation across the endoplasmic reticulum. I. Detection in the microsomal membrane of a receptor for the signal recognition particle. *J. Cell Biol.* **95**, 463–469.

Gilmore, R., Walter, P., and Blobel, G. (1982b). Protein translocation across the endoplasmic reticulum. II. Isolation and characterization of the signal recognition particle receptor. *J. Cell Biol.* **95**, 470–477.

Gilmore, R., Walter, P., Anderson, D., Erickson, A., Muller, M., and Blobel, G. (1984). The mechanism of protein translocation across the endoplasmic reticulum membrane. *In* "Protein Transport and Secretion," pp. 291–296. D. L. Oxendev, ed. Alan R. Liss, New York.

Gottlieb, T. A., Gonzalez, A., Rizzolo, L., Rindler, M. J., Adesnik, M., and Sabatini, D. D. (1986). Sorting and endocytosis of viral glycoproteins in transfected polarized epithelial cells. *J. Cell Biol.* **102**, 1242–1255.

Green, J., Griffiths, G., Louvard, D., Quinn, P., and Warren, G. (1981a). Passage of viral membrane proteins through the Golgi complex. *J. Mol. Biol.* **152**, 663–698.

Green, R. F., Meiss, H. K., and Rodriguez-Boulan, E. (1981b). Glycosylation does not determine segregation of viral envelope proteins in the plasma membrane of epithelial cells. *J. Cell Biol.* **89**, 230–239.

Guan, J.-L., and Rose, J. K. (1984). Conversion of a secretory protein into a transmembrane protein results in its transport to the Golgi complex but not to the cell surface. *Cell* **37**, 779–787.

Guan, J.-L., Machamer, C. E., and Rose, J. K. (1985). Glycosylation allows cell-surface transport of an anchored secretory protein. *Cell* **42**, 489–496.

Gumbiner, B., and Kelly, R. B. (1982). Two distinct intracellular pathways transport secretory and membrane glycoproteins to the surface of pituitary tumor cells. *Cell* **28**, 51–59.

Hardwick, J. M., Shaw, K. E. S., Wills, J. W., and Hunter, E. (1986). Amino-terminal deletion mutants of the Rous sarcoma virus glycoprotein do not block signal peptide cleavage but block intracellular transport. *J. Cell Biol.* **103**, 829–838.

Hasilik, A., and Neufeld, E. F. (1980). Biosynthesis of lysosomal enzymes in fibroblasts: Phosphorylation of mannose residues. *J. Biol. Chem.* **255**, 4946–4950.

Hayday, A., Saito, H., Gillies, S. D., Kranz, D. M., Tanigawa, G., Eisen, H. N., and Tonegar, S. (1985). Structure, organization, and somatic rearrangement of T cell α genes. *Cell* **40**, 259.

Hercz, A., Katona, E., Cutz, E., Wilson, J. R., and Barton, M. (1978). α_1-Antitrypsin: The presence of excess mannose in the Z variant isolated from liver. *Science* **201**, 1229–1232.

Herrler, G., Nagele, A., Meier-Ewert, H., Bhown, A. S., and Compans, R. W. (1981). Isolation and structural analysis of influenza virus C virion glycoproteins. *Virology* **113**, 439–451.

Hickman, S., Kulczycki, A., Lynch, R. G., and Kornfeld, S. (1977). Studies on the mechanisms of tunicamycin inhibition of IgA and IgE secretion by plasma cells. *J. Biol. Chem.* **252**, 4402–4408.

Hopkins, C. R. (1983). The importance of the endosome in intracellular traffic. *Nature (London)* **304**, 684–685.

Hughes, S. H. (1982). Sequence of the long terminal repeat and adjacent segments of the endogenous avian virus Rous-associated virus. *J. Virol.* **43**, 191–200.

Hunter, E., Hill, E., Hardwick, M., Bhown, A., Schwartz, D., and Tizard, R. (1983). Complete sequence of the Rous sarcoma virus *env* gene: Identification of structural and functional regions of its product. *J. Virol.* **46**, 920–936.

Hussain, M., Ichihara, S., and Mizushima, S. (1982). Mechanism of signal peptide cleavage in the biosynthesis of the major lipoprotein of the *Escherichia coli* outer membrane. *J. Biol. Chem.* **257**, 5177–5182.

Jackson, R. C., and White, W. R. (1981). Phospholipid is required for the processing of presecretory proteins by detergent-solubilized canine pancreatic signal peptidase. *J. Biol. Chem.* **256**, 2545–2550.

Johnson, D. A., and Smiley, J. R. (1985). Intracellular transport of herpes simplex virus gD occurs more rapidly in uninfected cells than in infected cells. *J. Virol.* **54**, 682–689.

Jones, L. V., Compans, R. W., Davis, A. R., Bos, T. J., and Nayak, D. P. (1985). Surface expression of influenza virus neuraminidase an amino-terminally anchored viral membrane glycoprotein, in polarized epithelial cells. *Mol. Cell. Biol.* **5**, 2181–2189.

Kabcenell, A. K., and Atkinson, P. H. (1985). Processing of the rough endoplasmic reticulum membrane glycoproteins of rotavirus SA11. *J. Cell Biol.* **101**, 1270–1280.

Kaiser, C. A., Preuss, D., Grisafi, P., and Botstein, D. (1987). Many random sequences functionally replace the secretion signal sequence of yeast invertase. *Science* **235**, 312–317.

Kaufman, J. F., Krangel, M. S., and Strominger, J. L. (1984). Cysteines in the transmembrane region of major histocompatability complex antigens are fatty acylated via thioester bonds. *J. Biol. Chem.* **259**, 7230–7238.

Kelly, R. B. (1985). Pathway of protein secretion in eukaryotes. *Science* **230**, 25–32.

Kielian, M. C., and Helenius, A. (1985). pH-induced alterations in the fusogenic spike protein of Semlilci Forest virus. *J. Cell Biol.* **101**, 2284–2291.

Kielian, M. C., Keranen, S., Kaarianen, L., and Helenius, A. (1984). Membrane fusion mutants of Semliki Forest virus. *J. Cell Biol.* **98**, 139–145.

Knipe, D. M., Baltimore, D., and Lodish, H. F. (1977a). Separate pathways of maturation of the major structural proteins of vesicular stomatitis virus. *J. Virol.* **21**, 1128–1139.

Knipe, D. M., Baltimore, D., and Lodish, H. F. (1977b). Maturation of viral proteins in cells infected with temperature-sensitive mutants of vesicular stomatitis virus. *J. Virol.* **21**, 1149–1158.

Kopito, R. R., and Lodish, H. F. (1985). Primary structure and transmembrane orientation of the murine anion exchange protein. *Nature (London)* **316**, 234–238.

Kreis, T. E., and Lodish, H. F. (1986). Oligomerization is essential for transport of the vesicular stomatitis virus glycoprotein to the cell surface. *Cell* **46**, 927–937.

Kuismanen, E., Bang, B., Hurme, M., and Pettersson, R. F. (1984). Uukuniemi virus maturation: An immune fluorescence microscopy study using monoclonal glycoprotein-specific antibodies. *J. Virol.* **51**, 137–146.

Kyte, J., and Doolittle, R. F. (1982). A simple method for displaying the hydropathic character of a protein. *J. Mol. Biol.* **157**, 105–132.

Lamb, R. A. (1985). Influenza M2 protein is an integral membrane protein expressed on the infected-cell surface. *Cell* **40**, 627–632.

Leavitt, R., Schlesinger, S., and Kornfeld, S. (1977). Impaired intracellular migration and altered solubility of nonglycosylated glycoproteins of vesicular stomatitis virus and Sindbis virus. *J. Biol. Chem.* **252**, 9018–9023.

Ledford, B. E., and Davis, D. F. (1983). Kinetics of serum protein secretion by cultured hepatoma cells: Evidence for multiple secretory pathways. *J. Biol. Chem.* **258**, 3304–3308.

Lingappa, V. R., Chaidez, J., Yost, C. S., and Hedgpeth, J. (1984). Determinants for protein localization: β-Lactamase signal sequence directs globin across microsomal membranes. *Proc. Natl. Acad. Sci. U.S.A.* **81**, 456–460.

Lively, M. O., and Walsh, K. A. (1983). Hen oviduct signal peptidase is an integral membrane protein. *J. Biol. Chem.* **258**, 9488–9495.

Lodish, H. F., and Kong, N. (1983). Reversible block in intracellular transport and budding of mutant vesicular stomatitis virus glycoprotein. *Virology* **125**, 335–348.

Lodish, H. F., Kong, N., Snider, M., and Strous, G. J. A. M. (1983). Hepatoma secretory proteins migrate from rough endoplasmic reticulum to Golgi at characteristic rates. *Nature (London)* **304,** 80–83.

Louvard, D. (1980). Apical membrane aminopeptidase appears at site of cell–cell contact in cultured kidney epithelial cells. *Proc. Natl. Acad. Sci. U.S.A.* **77,** 4132–4136.

Machamer, C. E., Florkiewicz, R. Z., and Rose, J. K. (1985). A single N-linked oligosaccharide at either of the two normal sites is sufficient for transport of vesicular stomatitis virus G protein to the cell surface. *Mol. Cell. Biol.* **11,** 3074–3083.

McQueen, N., Nayak, D. P., Stephens, E. B., and Compans, R. W. (1986). Polarized expression of a chimeric protein in which the transmembrane and cytoplasmic domains of the influenza virus hemagglutinin have been replaced by those of the vesicular stomatitis virus G protein. *Proc. Natl. Acad. Sci. U.S.A.* **83,** 9318–9322.

McQueen, N. L., Nayak, D. P., Stephens, E. B., and Compans, R. W. (1987). Basolateral expression of a chimeric protein in which the transmembrane and cytoplasmic domains of vesicular stomatitis virus G protein have been replaced by those of the influenza virus hemagglutinin. *J. Biol. Chem.* (in press).

Markoff, L., Lin, B.-C., Sveda, M., and Lai, C.-J. (1984). Glycosylation and surface expression of the influenza virus neuraminidase requires the N-terminal hydrophobic region. *Mol. Cell Biol.* **4,** 8–16.

Marsh, M. (1984). The entry of enveloped viruses into cells by endocytosis. *Biochem. J.* **218,** 1–10.

Mellman, I., and Plutner, H. (1984). Internalization and degradation of macrophage Fc receptors bound to polyvalent immune complexes. *J. Cell Biol.* **98,** 1170–1183.

Mellman, I., Fuchs, R., and Helenius, A. (1986). Acidification of the endocytic and exocytic pathways. *Annu. Rev. Biochem.* **55,** 663–700.

Meyer, D. E., and Dobberstein, B. (1980). Identification and characterization of a membrane component essential for the translocation of nascent proteins across the membrane of the endoplasmic reticulum. *J. Cell Biol.* **87,** 503–508.

Meyer, D. I., Krause, E., and Dobberstein, B. (1982). Secretory protein translocation across membranes—the role of the "docking protein." *Nature (London)* **297,** 647–650.

Michaelis, S., and Beckwith, J. (1982). Mechanism of incorporation of cell envelope proteins in *Escherichia coli*. *Annu. Rev. Microbiol.* **36,** 435–465.

Moore, H. P., and Kelly, R. B. (1985). Secretory protein targeting in a pituitary cell line: Differential transport of foreign secretory proteins to distinct secretory pathways. *J. Cell Biol.* **101,** 1773–1781.

Mosmann, T. R., and Williamson, A. R. (1980). Structural mutations in a mouse immunoglobulin light chain resulting in failure to be secreted. *Cell* **20,** 283–292.

Murre, C., Reiss, C. S., Bernabeu, C., Chen, L. B., Burakoff, S. J., and Seidman, J. G. (1984). Construction, expression and recognition of an H-2 molecule lacking its carboxyl terminus. *Nature (London)* **307,** 432–436.

Nakajima, S., Brown, D. J., Ueda, M., Nakajima, K., Sugiura, A., Pattnaik, A. K., and Nayak, D. P. (1986). Identification of the defects in the hemagglutinin gene of two temperature-sensitive mutants of A/WSN/33 influenza virus. *Virology* **154,** 279–285.

Nielsen, J. B. K., and Lampen, J. O. (1982). Membrane-bound penicillinases in gram-positive bacteria. *J. Biol. Chem.* **257,** 4490–4495.

Olden, K., Parent, J. B., and White, S. L. (1982). Carbohydrate moieties of glycoproteins, a reevaluation of their function. *Biochem. Biophys. Acta* **650,** 209–232.

Oliver, D. (1985). Protein secretion in *Escherichia coli*. *Annu. Rev. Microbiol.* **39,** 615–648.

Palmiter, D. R., Gagnon, J., Walsh, K. A. (1978). Ovalbumin: A secreted protein without a transient hydrophobic leader sequence. *Proc. Natl. Acad. Sci. U.S.A.* **75,** 94–98.

Parsegian, A. (1969). Energy of an ion crossing a low dielectric membrane: Solutions to four relevant problems. *Nature (London)* **221,** 844–846.

Patarca, R., and Haseltine, W. A. (1984). Similarities among retrovirus proteins. *Nature (London)* **312,** 496.

Perez, L., and Hunter, E. (1987). Mutations within the proteolytic cleavage site of the Rous sarcoma virus glycoprotein precursor block processing to gp85 and gp37. *J. Virol.* **61,** 1609–1614.

Perez, L. G., Davis, G. L., and Hunter, E. (1987). Mutants of the Rous sarcoma virus envelope glycoprotein that lack the transmembrane anchor and/or cytoplasmic domains: Analysis of intracellular transport and assembly into virions. *J. Virol.* **62.**

Perlman, D., and Halvorson, H. O. (1983). A putative signal peptidase recognition site and sequence in eucaryotic and procaryotic signal peptides. *J. Mol. Biol.* **167,** 391–409.

Pesonen, M., Sarast, J., Hashimoto, K., and Kaariainen, L. (1981). Reversible defect in the glycosylation of the membrane proteins of Semliki Forest virus *ts1* mutant. *Virology* **109,** 165–173.

Petrie, B. L., Graham, D. Y., Hanssen, H., and Estes, M. K. (1982). Localization of rotavirus antigens in infected cells by ultrastructural immunocytochemistry. *J. Gen. Virol.* **63,** 457–467.

Pfeiffer, S., Fuller, S. D., and Simons, K. (1985). Intracellular sorting and basolateral appearance of the G protein of vesicular stomatitis virus in MDCK cells. *J. Cell Biol.* **101,** 470–476.

Porter, A. G., Barber, C., Carey, N. H., Hallewell, R. A., Threlfall, G., and Emtage, J. S. (1979). Complete nucleotide sequence of an influenza virus haemagglutinin gene from cloned DNA. *Nature (London)* **282,** 471–477.

Poruchynsky, M. S., Tyndall, C., Both, G. W., Sato, F., Bellamy, A. R., and Atkinson, P. H. (1985). Deletions into an NH$_2$-terminal hydrophobic domain result in secretion of rotavirus VP7, a resident endoplasmic reticulum glycoprotein. *J. Cell Biol.* **101,** 2199–2209.

Randall, L. L., and Hardy, S. J. S. (1984). Export of protein in bacteria. *Microb. Rev.* **48,** 290–298.

Rao, J. K. M., and Argos, P. (1986). A conformation preference parameter to predict helices in integral membrane proteins. *Biophys. Biochim. Acta* **869,** 197–214.

Reggio, H., Coudrier, E., and Louvard, D. (1982). Surface and cytoplasmic domains in polarized epithelial cells. *In* "Membranes in Growth and Development" (G. Biebisch, ed.), pp. 89–105. Alan R. Liss, New York.

Rettenmier, C. W., Roussel, M. F., Quinn, C. O., Kitchingman, G. R., Lokk, A. T., and Sherr, C. J. (1985). Transmembrane orientation of glycoproteins encoded by the v-*fms* oncogene. *Cell* **40,** 971–981.

Rindler, M. J., Ivanov, I. E., Plesken, H., Rodriguez-Boulan, E., and Sabatini, D. D. (1984). Viral glycoproteins destined for apical or basolateral plasma membrane domains traverse the same Golgi apparatus during their intracellular transport in doubly infected Madin–Darby canine kidney cells. *J. Cell Biol.* **98,** 1304–1319.

Rindler, M. J., Ivanov, I. E., Plesken, H., and Sabatini, D. D. (1985). Polarized delivery of viral glycoproteins to the apical and basolateral plasma membranes of Madin–Darby canine kidney cells infected with temperature-sensitive viruses. *J. Cell Biol.* **100,** 136–151.

Rodriguez-Boulan, E., and Sabatini, D. D. (1978). Asymmetric budding of viruses in epithelial monolayers: A model system for study of epithelial polarity. *Proc. Natl. Acad. Sci. U.S.A.* **75,** 5071–5075.

Rodriguez-Boulan, E., Paskiet, K. T., Salas, P. J. T., and Bard, E. (1984). Intracellular transport of influenza virus hemagglutinin to the apical surface of Madin–Darby canine kidney cells. *J. Cell Biol.* **98**, 308–319.

Rose, J. K., and Bergmann, J. E. (1982). Expression from cloned cDNA of cell-surface secreted forms of the glycoprotein of vesicular stomatitis virus in eukaryotic cells. *Cell* **30**, 753–762.

Rose, J. K., and Bergmann, J. E. (1983). Altered cytoplasmic domains affect intracellular transport of the vesicular stomatitis virus glycoprotein. *Cell* **30**, 513–524.

Rose, J. K., Welch, W. J., Sefton, B. M., Esch, F. S., and Ling, N. C. (1980). Vesicular stomatitis virus is anchored in the viral membrane by a hydrophobic domain near the COOH terminus. *Proc. Natl. Acad. Sci. U.S.A.* **77**, 3884–3888.

Rose, J. K., Adams, G. A., and Gallione, C. J. (1984). The presence of cysteine in the cytoplasmic domain of the vesicular stomatitis virus glycoprotein is required for palmitate addition. *Proc. Natl. Acad. Sci. U.S.A.* **81**, 2050–2054.

Roth, M. G., Fitzpatrick, J. P., and Compans, R. W. (1979). Polarity of influenza and vesicular stomatitis virus in MDCK cells; lack of a requirement for glycosylation of viral glycoproteins. *Proc. Natl. Acad. Sci. U.S.A.* **76**, 6430–6434.

Roth, M. G., Compans, R. W., Giusti, L., Damis, A. R., Nayak, D. P., Gething, M.-J., and Sambrook, J. (1983a). Influenza virus hemagglutinin expression is polarized in cells infected with recombinant SV40 viruses carrying cloned hemagglutinin DNA. *Cell* **33**, 435–443.

Roth, M. G., Srinivas, R. V., and Compans, R. W. (1983b). Basolateral maturation of retroviruses in polarized epithelial cells. *J. Virol.* **45**, 1065–1073.

Roth, M. G., Doyle, C., Sambrook, J., and Gething, M.-J. (1986). Heterologous transmembrane and cytoplasmic domains direct functional chimeric influenza virus hemagglutinins into the endocytic pathway. *J. Cell Biol.* **102**, 1271–1283.

Roth, M., Gething, M.-J., and Sambrook, J. (1987). Membrane insertion and intracellular transport of influenza virus glycoproteins. In "The Influenza Viruses" (R. Krug, ed.), Plenum, New York.

Roth, M. G., Gundersen, D., Patil, N., and Rodriguez-Boulan, E. (1987). The large external domain is sufficient for the correct sorting of secreted or chimeric influenza virus hemagglutinins in polarized monkey kidney cells. *J. Cell Biol.* **104**, 769–782.

Rott, R., Orlich, M., Klenk, H.-D., Wang, M. L., Skehel, J. J., and Wiley, D. C. (1984). Studies on the adaption of influenza viruses to MDCK cells. *EMBO J.* **3**, 3329–3332.

Russel, M., and Model, P. (1981). A mutation downstream from the signal peptidase cleavage site affects cleavage but not membrane insertion of phage coat protein. *Proc. Natl. Acad. Sci. U.S.A.* **28**, 1717–1721.

Sabatini, D. D., Kreibich, G., Morimoto, T., and Adesnik, M. (1982). Mechanisms for the incorporation of proteins in membranes and organelles. *J. Cell Biol.* **92**, 1–22.

Saito, H., Dranz, D. M., Takagaki, Y., Hayday, A. C., Eisen, H. N., and Tonegawa, S. (1984). Complete primary structure of a heterodimeric T-cell receptor deduced from cDNA sequences. *Nature (London)* **309**, 759–762.

Saraste, J., and Kuismanen, E. (1984). Pre- and post-Golgi vacuoles operate in the transport of Semliki Forest virsus membrane glycoproteins to the cell surface. *Cell* **38**, 535–549.

Schmidt, M. F. G., and Schlesinger, M. J. (1979). Fatty acid binding to vesicular stomatitis virus glycoprotein: A new type of posttranslational modification of the viral glycoprotein. *Cell* **17**, 813–819.

Schmidt, M. F. G., and Schlesinger, M. J. (1980). Relation of fatty acid attachment to the translation and maturation of vesicular stomatitis and Sindbis virus membrane glycoproteins. *J. Biol. Chem.* **255**, 3334–3339.

Schmidt, M. F. G., Bracha, M., and Schlesinger, M. J. (1979). Evidence for covalent attachment of fatty acids to Sindbis virus glycoproteins. *Proc. Natl. Acad. Sci. U.S.A.* **76**, 1687–1691.

Sekikawa, K., and Lai, C.-J. (1983). Defects in functional expression of an influenza virus hemagglutinin lacking the signal peptide sequences. *Proc. Natl. Acad. Sci. U.S.A.* **80**, 3563–3567.

Shida, H., and Matsumoto, S. (1983). Analysis of the hemagglutinin glycoprotein from mutants of vaccinia virus that accumulates on the nuclear envelope. *Cell* **33**, 423–434.

Silhavy, T. J., Benson, S. A., and Emr, S. D. (1983). Mechanisms of protein localization. *Microbiol. Rev.* **47**, 313–344.

Skehel, J. J., Bayley, P. M., Brown, E. B., Martin, S. R., Waterfield, M. D., Whitel, J. M., Wilson, I. A., and Wiley, D. C. (1982). Changes in the conformation of influenza virus hemagglutinin at the pH optimum of virus-mediated membrane fusion. *Proc. Natl. Acad. Sci. U.S.A.* **79**, 968–972.

Sly, W. S., and Fischer, H. D. (1982). The phosphomannosyl recognition system for intracellular and intercellular transport of lysosomal enzymes. *J. Cell. Biochem.* **18**, 67–85.

Spear, P. G. (1985). Glycoproteins specified by herpes simplex viruses. *In* "The Herpesviruses" (B. Roizman, ed.), Vol.3. pp. 315–356. Plenum, New York.

Spiess, M., and Lodish, H. F. (1986). An internal signal sequence: The asialoglycoprotein receptor membrane anchor. *Cell* **44**, 177–185.

Stephens, E. B., and Compans, R. W. (1986). Nonpolarized expression of a secreted murine leukemia virus glycoprotein in polarized epithelial cells. *Cell* **47**, 1053–1059.

Stephens, E. B., Compans, R. W., Earl, P., and Moss, B. (1986). Polarized transport of the VSV G surface expression of viral glycoproteins is polarized in epithelial cells infected with recombinant vaccinia viral vectors. *EMBO J.* **5**, 235–247.

Strous, G. J. A. M., and Lodish, H. F. (1980). Intracellular transport of secretory and membrane proteins in hepatoma cells infected by vesicular stomatitis virus. *Cell* **22**, 709–717.

Strous, G. J. A. M., Willemsen, R., Kerkhof, P. V., Slot, J. W., Geuze, H. J., and Lodish, H. F. (1983). Vesicular stomatis virus glycoprotein, albumin, and transferrin are transported to the cell surface via the same Golgi vesicles. *J. Cell Biol.* **97**, 1815–1822.

Struck, D. K., Siuta, P. B., Lane, M. D., and Lane, W. J. (1978). Effect of tunicamycin on the secretion of serum proteins by primary cultures of rat and chicken hepatocytes. *J. Biol. Chem.* **253**, 5332–5337.

Sturman, L. S., and Holmes, K. V. (1983). The molecular biology of coronaviruses. *Adv. Virus Res.* **28**, 35–112.

Sveda, M. M., Markoff, L. J., and Lai, C. J. (1982). Cell surface expression of the influenza virus hemagglutinin requires the hydrophobic carboxy-terminal sequences. *Cell* **30**, 649–656.

Sveda, M. M., Markoff, L. J., and Lai, C.-J. (1984). Influenza virus hemagglutinin containing an altered hydrophobic carboxy terminus accumulates intracellularly. *J. Virol.* **49**, 223–228.

Tabe, L., Krieg, P., Strachan, R., Jackson, D., Wallis, E., and Colman, A. (1984) Segregation of mutant ovalbumins and ovalbumin–globin fusion proteins in *Xenopus* oocytes: Identification of an ovalbumin signal sequence. *J. Mol. Biol.* **180**, 645–666.

Talmadge, K., Stahl, S., and Gilbert, W. (1980). Eukaryotic signal sequence transports insulin antigen in *Escherichia coli*. *Proc. Natl. Acad. Sci. U.S.A.* **77**, 3369–3373.

Tokunaga, M., Loranger, J. M., Wolfe, P. B., and Wu, H. C. (1982). Prolipoprotein signal peptidase in *Escherichia coli* is distinct from the M13 procoat protein signal peptidase. *J. Biol. Chem.* **257**, 9922–9925.

Ueda, M., and Kilbourne, E. D. (1976). Temperature-sensitive mutants of influenza virus: A mutation in the hemagglutinin gene. *Virology* **70**, 425–431.

Underdown, B. J., Simms, E. S., and Eisen, H. N. (1971). Structural studies of IgA myeloma proteins having anti-DNP antibody activity. *Biochemistry,* **10**, 4359–4364.

von Heijne, G. (1983). Patterns of amino acids near signal-sequence cleavage sites. *Eur. J. Biochem.* **133**, 17–21.

von Heijne, G. (1984). How signal sequences maintain cleavage specificity. *J. Mol. Biol.* **173**, 243–251.

von Heijne, G. (1985). Signal sequences, the limits of variation. *J. Mol. Biol.* **184**, 99–105.

Walter, P., and Blobel, G. (1980). Purification of a membrane-associated protein complex required for protein translocation across the endoplasmic reticulum. *Proc. Natl. Acad. Sci. U.S.A.* **77**, 7112–7116.

Walter, P., and Blobel, G. (1981a). Translocation of proteins across the endoplasmic reticulum II. Signal recognition protein (SRP) mediates the selective binding to microsomal membranes of *in-vitro*-assembled polysomes synthesizing secretory protein. *J. Cell Biol.* **91**, 551–556.

Walter, P., and Blobel, G. (1981b). Translocation of proteins across the endoplasmic reticulum III. Signal recognition protein (SRP) causes signal sequence-dependent and site-specific arrest of chain elongation that is released by microsomal membranes. *J. Cell Biol.* **91**, 557–561.

Walter, P., Ibrahimi, I., and Blobel, G. (1981). Translocation of proteins across the endoplasmic reticulum I. Signal recognition protein (SRP) binds to *in-vitro*-assembled polysomes synthesizing secretory protein. *J. Cell Biol.* **91**, 545–550.

Warren, G. (1981). Membrane proteins: Structure and assembly. *In* "Membrane Structure" (J. B. Finean and R. H. Michell, eds.), Vol. 1, pp. 215–257. Elsevier, New York.

Watson, M. E. (1984). Compilation of published signal sequences. *Nucleic Acids Res.* **12**, 5144–5164.

Watts, C., Wickner, W., and Zimmerman, R. (1983). M13 procoat and a preimmunoglobulin share processing specificity but use different membrane receptor mechanisms. *Proc. Natl. Acad. Sci. U.S.A.* **80**, 2809–2813.

White, J. K., Matlin, K., and Helenius, A. (1981). Cell fusion by Semliki Forest, influenza, and vesicular stomatitis viruses. *J. Cell Biol.* **89**, 674–679.

White, J., Kielian, M., and Helenius, A. (1983). Membrane fusion proteins of enveloped animal viruses. *Q. Rev. Biophys.* **16**, 151–195.

Wickner, W. T., and Lodish, H. F. (1985). Multiple mechanisms of protein insertion into and across membranes. *Science* **230**, 400–407.

Wills, J. W., Hardwick, J. M., Shaw, K., and Hunter, E. (1983). Alterations in the transport and processing of Rous sarcoma virus envelope glycoproteins mutuated in the signal and anchor regions. *J. Cell. Biochem.* **23**, 81–94.

Wills, J. W., Srinivas, R. V., and Hunter, E. (1984). Mutations of the Rous sarcoma virus *env* gene that affect the transport and subcellular location of the glycoprotein products. *J. Cell Biol.* **99**, 2011–2023.

Wold, W. S. M., Caldars, C., Deutscher, S. L., and Kapoor, Q. S. (1985). The 19-kDa glycoprotein precursor coded by region E3 of adenovirus. *J. Biol. Chem.* **260**, 2424–2431.

Wu, G. E., Hozumi, N., and Murialdo, H. (1983). Secretion of a λ2 immunoglobulin chain is prevented by a single amino acid substitution in its variable region. *Cell* **33**, 77–83.

Yoshida, A., Lieberman, J., Laima, G., and Ewing, C. (1976). Molecular abnormality of human α_1-antitrypsin variant (Pi-ZZ) associated with plasma activity deficiency. *Proc. Natl. Acad. Sci. U.S.A.* **73**, 1324–1328.

158 Eric Hunter

Yoshimura, A., and Ohnishi, S.-I. (1984). Uncoating of influenza virus in endosomes. *J. Virol.* **51,** 497–504.

Yost, C., Hedgpeth, J., and Lingappa, V. R. (1983). A stop transfer sequence confers predictable transmembrane orientation to a previously secreted protein in cell-free systems. *Cell* **34,** 759–766.

Zerial, M., Melancon, P., Schneider, C., and Garoff, H. (1986). The transmembrane segment of the human transferrin receptor functions as a signal peptide. *EMBO J.* **5,** 1543–1550.

Zilberstein, A., Snider, M. D., Porter, M., and Lodish, H. F. (1980). Mutants of vesicular stomatitis virus blocked at different stages in maturation of the viral glycoprotein. *Cell* **21,** 417–427.

Zuniga, M. C., Malissen, B., McMillan, M., Brayton, P. R., Clark, S. S., Forman, J., and Hood, L. (1983). Expression and function of transplantation antigens with altered or deleted cytoplasmic domains. *Cell* **34,** 535–544.

4

Posttranslational Modification during Protein Secretion

KURT W. RUNGE

I. INTRODUCTION

In Canto V of *Inferno* in the *Divine Comedy,* Dante and Virgil come upon Minos, the judge of the dead. Minos' task is to consider the various sins of the damned and then hurl them to their deserved fates in lower

159

Hell (Alighieri, 1321). This action represents mammoth sorting problems given both the number of sinners and the number of compartments in hell. How Minos accomplishes this sorting process is only briefly described and not explained in detail.

The extent of our understanding of the protein sorting in eukaryotes was not much better than this as little as a decade ago, when Palade and co-workers described the pathway of labeled proteins in mammalian cells from the cytoplasm to the extracellular medium (Palade, 1975). Since then, a number of different approaches have allowed the dissection of the mechanism of protein targeting in eukaryotes to the nucleus, the mitochondrion, and those compartments topologically equivalent to the outside of the cell [the lysosome, the lumen of the endoplasmic reticulum (ER) and Golgi, and the cell surface]. The path of proteins through the cell can be followed by means of the posttranslational modifications they acquire in transit. In the case of secreted proteins in yeast, these modifications include glycosylation, acylation, phosphorylation, and proteolytic cleavage, which occur in different subcellular compartments to various extents (summarized in Table I). An understanding of these reactions will be useful in determining the role of posttranslational modification in protein function and targeting and in predicting the type of processing different foreign proteins will experience when they are expressed in yeast. The purpose of this chapter is to discuss those modifications that secretory proteins experience en route from the cytoplasm to the cell surface or lysosome, the localization of these reactions, and the enzymatic machinery that accomplishes them.

II. REACTIONS IN THE CYTOPLASM

A. Synthesis of the Peptide Backbone and Targeting to the Endoplasmic Reticulum

All secreted proteins contain a "signal sequence" which allows protein insertion and translocation across the ER membrane and which is likely to be involved in targeting the nascent peptide to the ER. This sequence can be at the amino terminus, and may be subsequently removed by the signal peptidase, or it may be internal and remain uncleaved. Proteins containing the amino-terminal extensions are termed preproteins, and the signal sequence is occasionally referred to as the presequence or leader peptide. This type of preprotein is also found in mitochondrial precursors, the synthesis and processing of which is discussed elsewhere in this volume (Brandriss, Chapter 16; Nicholson and Neupert, Chapter 17).

TABLE I

Protein Modification during Secretion[a]

Organelle	Modifications
Endoplasmic reticulum	Signal sequence cleavage
	Lipid-linked oligosaccharide biosynthesis
	Attachment of N-linked oligosaccharides
	Trimming of N-linked oligosaccharides
	Attachment of first O-linked mannose
	Fatty acylation of proteins
	Phosphorylation of carboxypeptidase Y
Golgi apparatus	Synthesis of the outer chain of N-linked oligosaccharides including the addition of outer chain phosphate residues
	Elongation of O-linked mannose by the addition of 0–3 mannose residues
	Proteolytic processing of the α-factor and killer toxin precursors

[a] The location of posttranslational processing events within the yeast cell is reviewed above. The data leading to these conclusions and the biochemical reactions themselves are discussed in the text.

Current theories on the initiation of translation of secreted proteins (and the localization to the ER) are derived from plant and animal cell-free systems. The present model is based on a large body of work from a number of different laboratories. When secreted proteins are translated in a wheat germ extract, they can insert into the lumen of added dog pancreas microsomes (Blobel and Dobberstein, 1975). In many cases, this insertion occurs only if the microsomes are present early in translation (Rothman and Lodish, 1977). Dissection of the events between translation initiation and membrane insertion has led to the following picture: As the secretory preprotein is being translated, the signal recognition particle (SRP), a complex of six polypeptides and a 7 S RNA (Walter and Blobel, 1982), binds to the ribosome after approximately 80 residues have been synthesized and arrests translation. The SRP–ribosome complex then binds to its receptor, a 72 kDa integral membrane protein, (the docking protein or DP) in the rough ER (RER) (Meyer et al., 1982). On binding, SRP releases the polysome, and translation continues with protein insertion into the ER lumen. This process has been reviewed in greater detail by others (Walter et al., 1984; Wickner and Lodish, 1985).

The above summary describes the synthesis of proteins which insert into the ER membrane cotranslationally. This may not be the case for all proteins. The membrane trigger hypothesis deals with the mechanims of membrane insertion of a complete polypeptide (reviewed in Wickner, 1979). In this case, the protein is thought to fold into a confirmation which allows membrane insertion of different segments. This mechanism has broader applications to eukaryotes, as proteins such as the human glucose transporter can also insert into ER membranes posttranslationally *in vitro*. Interestingly, this posttranslational insertion is SRP dependent (Mueckler and Lodish, 1986).

None of the above mechanisms has been demonstrated directly in yeast. One reason for this is that the methods of subcellular fractionation available for the analysis of animal cell systems have not been reproduced in yeast with the same precision. Another reason was the lack of a cell-free yeast *in vitro* translation–translocation system. This latter problem has been recently overcome with the isolation of an extract which will synthesize yeast prepro-α-factor (prepro-αF) and an amino portion of invertase and which will insert some of these molecules into exogenous yeast microsomes, as determined by the acquisition of protein-linked carbohydrate and signal sequence processing, two ER luminal reactions (discussed below) (Rothblatt and Meyer, 1986; Waters and Blobel, 1986). One system derived from yeast cytoplasm and membranes would not translocate and process immunoglobin G (IgG) light chain nor would yeast membranes function in a wheat germ extract (Rothblatt and Meyer, 1986). The biochemical dissection of this system should allow both a comparison of the yeast system with that of larger eukaryotes and identification of the biochemical functions of proteins whose activities are affected by various secretion (*sec*) mutations.

B. Yeast Mutations Affecting the Early Steps of Protein Secretion

Two yeast mutations have been isolated that affect protein insertion into the ER membrane. The *sec53* and *sec59* complementation groups are both class B *sec* mutations; cells carrying these mutations do not accumulate active intracellular precursors of secreted enzymes (Ferro-Novick *et al.*, 1984a). The invertase that accumulates in these mutants becomes membrane associated. One interpretation of this result is that the protein becomes stuck in the membrane as it is being translocated, and is thus prevented from participating in certain reactions in the RER (see Table I). Invertase and the α-factor precursor (prepro-αF) from wild-type cells contain 9–10 and 3 asparagine-linked (Asn-linked) oligosaccharide chains, respectively. However, both invertase and prepro-αF acquire from 0 to 3 Asn-linked oligosaccharide chains in *sec59-1* mutant cells, while invertase from *sec53* receives a more heterogeneous modification (Ferro-Novick *et al.*, 1984a,b; Julius *et al.*, 1984a). The nature of these modifications and their significance will be discussed below.

Mutants in the *sec53* complementation group have also been isolated using a protocol very different from that used to isolate the *sec* mutants. In order to study the process of lipid-linked oligosaccharide biosynthesis (described below), Huffaker and Robbins (1982, 1983) isolated a number of mutants blocked in this pathway. These *alg* mutants (asparagine-linked glycosylation) underglycosylate secreted and membrane glycoproteins. The largest complementation group isolated by these workers, *alg4*, proves to be the same as *sec53*. *alg4* and *sec53* mutants fail to complement one another, and a cloned *ALG4* gene will complement the temperature-sensitive growth defect of a *sec53-6* mutant (see Notes Added in Proof, Huffaker and Robbins, 1983; Ferro-Novick *et al.*, 1984a; Couto, 1984). Both *alg4* and *sec53* cells are defective in lipid-linked oligosaccharide biosynthesis when examined *in vivo* and *in vitro* (Runge and Robbins, 1986). Interestingly, *sec59-1* cells also show aberrant lipid-linked oligosaccharide synthesis at the nonpermissive temperature *in vivo* (K. Runge, unpublished data). (Lipid-linked oligosaccharide biosynthesis is discussed in Sections III,C and III,D).

The phenotype of *sec53* mutants suggests that the *SEC53* gene encodes an RER membrane protein. However, the sequence of the *SEC53* gene shows no hydrophobic stretches one might expect for a membrane-bound protein. In addition, the *SEC53* protein also fractionates with the cytoplasm and not with the endoplasmic reticulum (Bernstein *et al.*, 1985). These observations indicate that the *ALG4–SEC53* gene product is a cytoplasmic or peripheral membrane protein that affects both protein secretion and glycosylation. As both *sec53* and *sec59* do not have the phenotypes expected for SRP or DP mutations but do affect secretion and lipid-linked oligosaccharide biosynthesis, these two genes may define products important in the overall organization of the translocation and glycosylation apparatus.

III. REACTIONS IN THE ENDOPLASMIC RETICULUM

A. Signal Sequence Cleavage

One of the first modifications performed on a polypeptide entering the ER is cleavage of the signal sequence. While a number of proteins possess either an internal signal sequence or one that is not normally cleaved [e.g., yeast prepro-αF (Julius *et al.,* 1984a; Rothblatt and Meyer, 1986; Waters and Blobel, 1986) and possibly the arginine permease (Hoffman, 1985)], secreted yeast proteins such as invertase (β-fructofuranosidase or sucrase) and acid phosphatase do lose their amino-terminal hydrophobic segments (Arima *et al.,* 1983; Perlman and Halvorson, 1981; Perlman *et al.,* 1982). The canonical signal sequence consists of 0–2 positively charged amino acids followed by about 20 hydrophobic residues and the 4–6 amino acids that make up the cleavage site. The signal sequence is important for insertion into the ER membrane because its deletion from the invertase peptide results in a cytoplasmic protein (Williams *et al.,* 1985; Kaiser and Botstein, 1986). The process of signal sequence cleavage is affected by the insertion of the protein into the membrane and the amino acids at the cleavage site (see below).

Two types of models account for the entry of the signal sequence into the ER membrane. The signal hypothesis proposes that the amino terminus precedes the rest of the polypeptide through a proteinaceous membrane channel. As the signal sequence cleavage site enters the ER lumen, it is recognized by the signal sequence peptidase (or signal peptidase, Fig. 1) and cleaved (Blobel and Doberstein, 1975; Blobel, 1980). Alternatively, the helical hairpin hypothesis (Engelmann and Steitz, 1981), the direct transfer model, and loop models (von Heijne and Blomberg, 1979; Halegoua and Inouye, 1980) propose that pairs of helices insert into the lipid bilayer, not necessarily requiring a protein channel. Each model also allows for the insertion of multiple segments of the polypeptide to account for integral membrane proteins that span the bilayer many times.

A major difference in these models in terms of signal sequence cleavage is summarized in Fig. 1. In the case of the signal hypothesis, the signal peptidase would cleave a linear molecule and release the signal peptide into the ER lumen. Insertion of the amino terminus as a pair of helices would present the peptidase with its substrate as a bend in the protein with the signal sequence now embedded in the membrane. Subsequent cleavage results in a new amino terminus that is located in the ER. Models invoking this method of insertion more readily explain the synthesis of integral membrane proteins with cytoplasmically localized amino termini such as influenza neuraminidase (Fields *et al.,* 1981) (see also Wickner and Lodish, 1985; Mueckler and Lodish, 1986). In the case of mutations that block signal peptide cleavage, the signal hypothesis would predict a protein in the ER lumen that retains this peptide. Insertion via helical hairpins without signal peptidase action could lead to a protein now anchored in the membrane via its amino terminus or floating in the ER lumen, depending on the affinity of the particular signal peptide for the lipid bilayer.

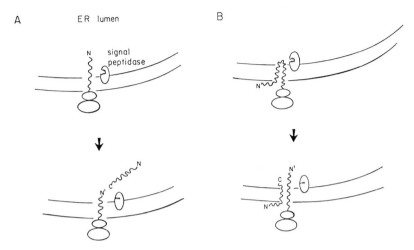

Fig. 1. Models of signal peptide insertion and cleavage. The proposed mechanisms of the insertion of a secreted protein into and through the endoplasmic reticular membrane in the signal hypothesis (A) and the helical hairpin hypothesis (B) are illustrated. (A) In the signal hypothesis, the amino terminus of the signal peptide precedes the protein through the membrane. As the signal peptidase cleavage site enters the ER lumen, signal peptidase removes the leader sequence. (B) In the helical hairpin hypothesis, the signal sequence inserts as a loop with the amino terminus on the cytoplasmic face of the ER. The leader peptidase cleavage site is presented as a bend in the protein. This proposed geometry of the cleavage site may invoke and an additional restraint on signal sequence processing and can be used to explain the behavior of certain mutant proteins that fail to be processed by signal peptidase, or are processed only slowly (see text).

Two extensive surveys of signal peptide cleavage sites have been conducted by von Heijne (1983) and Perlman and Halvorson (1983). By analyzing the mature termini of a variety of secreted proteins, and comparing them to the sequence of the primary translation product, both groups were able to formulate a number of predictive rules for the signal peptidase site. The first amino acid of a mature protein was numbered +1 and the residue on the immediate amino-terminal side numbered −1 (see Fig. 2). These workers found that the −1 and −3 sites were occupied by amino acids with small side chains: Ala, Ser, Thr, Cys, or Gly, with Ala being the most prevalent. The −2 position was less conserved; a large hydrophobic residue was often present, but charged residues could occur at this position. No Pro residues were found from positions −3 to +1. However, helix-breaking residues such as Gly and Pro are often found at positions −5 and −6. In addition, Perlman and Halvorson noted that, using the secondary structure predictive rules of Chou and Fasman (1978), β-turn nucleating amino acids were often found just before or after the signal

```
A
SUC2  Met-met-leu-leu-gln-ala-phe-leu-phe-leu-leu-ala-gly-phe-

#211  met-met-leu-leu-arg-                -leu-leu-ala-gly-...

                          -1 ↓ +1
SUC2  ala-ala-lys-ile-ser-ala-ser-met-thr-asn-glu-thr...

                     S1    val
                     S11   ile
                     S13        pro
                     S12   val-pro

B    PHO5
                                  -1 ↓ +1
     wild type  ...-ser-leu-ala-asn-ala-gly-thr-ile-pro...
                                  -1 ↓ +1
     ΔK         ...-ser-leu-ala-asn-ala-gly-thr-gln-lys...
     ΔBS        ...-ser-leu-val-asp-lys-ile-gly-thr-gln...

C    Glucoamylase
                                   -1 ↓ +1                              ↓
     wild type  ...-leu-val-cys-thr-gly-leu-ala-asn-val-ile-ser-lys-arg-

                                          units/liter
     wild type                                112
     mutations                                132        lys-pro
                                 pro           38        lys-pro
                                 asp           32        lys-pro
                                 trp           14        lys-pro
                delete [ala-asn-val-ile-ser]   12        lys-pro
```

Fig. 2. Mutations affecting signal sequence cleavage in invertase, acid phosphatase, and glucoamylase. (A) Amino acid substitutions in the invertase signal sequence. *suc2-211* is a four amino acid deletion and Gln to Arg amino acid change that inhibits signal sequence cleavage downstream (C. Kaiser and D. Botstein, personal communication). The *S1*, *S11*, *S13*, and *S12* mutations are single or double amino acid changes at the signal peptidase cleavage site that inhibit processing (Schauer *et al.*, 1985; Bohni *et al.*, in press). The arrow represents the point of signal peptide cleavage. The *S1* mutant is slowly cleaved between Ser and Met. (B) Deletion mutations in the acid phosphatase signal sequence. The amino acid sequence around the wild-type acid phosphatase molecule and two deletion mutants are shown. The ΔK mutation deletes 14 amino acids carboxy terminal to the cleavage site while the ΔBS mutation removes 13 amino acids including the leader peptidase processing site. The arrows indicate where the signal sequence is cleaved in the wild type and the ΔK mutant. The ΔBS mutant retains its signal sequence (Haguenauer-Tsapis and Hinnen, 1984, Haguenauer-Tsapis *et al.*, 1986). (C) Mutations in the glucoamylase signal sequence and their affect on active enzyme production. Amino acid substitutions of Lys-Arg to Lys-Pro eliminate a Golgi-localized *KEX2* enzyme cleavage (Innis *et al.*, 1985) while those substitutions for Ala at the −1 position (shown below that amino acid) affect signal sequence processing. The amount of active glucoamylase, expressed as units per liter, for each type of protein is shown. All of the signal sequence mutations are in a protein lacking the *KEX2* cleavage site (J. Meade, personal communication).

sequence cleavage site (Perlman and Halvorson, 1983). Both analyses involved aligning proteins by either their signal sequence cleavage sites or the start of the hydrophobic core, and not by the amino terminus of the protein. Thus, the signal peptidase recognizes specific determinants near the carboxy end of the signal sequence and not the number of residues in the signal sequence. Perlman and Halvorson argued that some of their observations were more consistent with the signal sequence inserting into the membrane as in a helical hairpin model.

Two proteins used extensively in studying signal sequence cleavage in yeast are invertase and acid phosphatase. Invertase is a sucrose-hydrolyzing enzyme encoded by a family of *SUC* genes (Carlson *et al.*, 1980, 1981; Grossmann and Zimmerman, 1979). Invertase is made in two forms, a constitutively synthesized cytoplasmic form and a derepressible secreted form. The derepressed form is made in cells grown on low-glucose medium. The mRNAs for both of these proteins are transcribed from the same gene, and the secreted form contains a 5' extension which encodes the signal sequence (Carlson *et al.*, 1981; Perlman *et al.*, 1982). Acid phosphatase is made in two secreted forms, the *PHO3* gene product, which is expressed in yeast grown in high-phosphate media, and an inducible enzyme encoded by the *PHO5* gene (Toh-e *et al.*, 1973; Rogers *et al.*, 1982). *PHO5* gene expression occurs during growth on low-phosphate media and was the gene employed in acid phosphatase signal sequence studies (discussed below).

A number of experiments in yeast have begun to test the models of von Heijne and Perlman and Halvorson. Kaiser and Botstein (1986) have made a number of deletions in the hydrophobic core of the invertase leader peptide. Invertase molecules with short deletions or substitutions had little effect and are secreted normally while larger deletions resulted in cytoplasmic localization of the normally secreted invertase. This cytoplasmically localized enzyme was completely active and capable of forming heterodimers with the normally made cytoplasmic enzyme. These results indicate that the invertase signal sequence is plastic enough to allow small changes in the length and structure of its hydrophobic core and still allow secretion.

One of the deletions possessed an unexpected phenotype different from those described above. Mutant *suc2-211* contains a four amino acid deletion near the beginning of the signal sequence which results in a core glycosylated enzyme that is inefficiently transported to the cell surface (Fig. 2A). Interestingly, the signal sequence in not efficiently removed from this protein (C. Kaiser and B. Botstein, personal communication). These results suggest that the amino terminus of the signal peptide can somehow affect its processing. One possibility is that the *SUC2* signal sequence inserts as a helical hairpin and that the four residue deletion in the *suc2-211* mutation allows membrane insertion but the shortened hydrophobic core keeps the signal peptidase cleavage site sequestered within the lipid bilayer. The resulting protein would retain its signal sequence, which may cause protein aggregation in the ER lumen or cause the invertase to remain membrane associated, which in turn prevents exit from this organelle (see below). If this hypothesis is true, one should be able to suppress the *suc2-211* mutation by introducing a new signal sequence cleavage site four or five residues carboxy terminal to the present site.

Experiments specifically directed at the invertase signal sequence cleavage site have been conducted by Schekman and co-workers. Schauer *et al.* (1985) searched for cis-acting point mutants in the *SUC2* gene that blocked invertase secretion. One of the mutants they isolated changed the cleavage site from Ala/Ser to Val/Ser (mutant S1, where "/" indicates the cleavage site; see Fig. 2A). This mutation greatly slowed down the rate of secretion. The enzyme that was secreted was found to have a different amino terminus, indicating signal peptidase cleavage at a new site (Schauer *et al.,* 1985). This new site has a high probility of signal sequence cleavage according to the algorithm of von Hiejne (1983). These investigators concluded that cleavage at the new site took place slowly, and it was this cleavage that allowed invertase to be secreted. The *suc2-s1* mutant provides a unique example of how substitution of the Ala residue at −1 with a similar but slightly larger residue can greatly affect signal peptidase action. The slower rate of processing of a presumably good signal sequence cleavage site may indicate that a specific leader peptide geometry is required for signal peptidase recognition.

Bohni *et al.* (in press) have followed up this work by making new amino acid substitutions at this site. These include substitution for Ser at +1 (mutant *suc2-s13*) and changing both amino acids at the cleavage site to Val (−1) -Pro (+1) (summarized in Fig. 2A). Cells carrying these mutant *suc2* alleles all secrete invertase at a rate 40-fold slower than the wild-type enzyme. All of these mutations reduced the amount of modification that oligosaccharide chains received in the Golgi. Preinvertase accumulates in these cells, and the small portion that is secreted to the periplasm is underglycosylated and appears to retain the signal peptide (P. C. Bohni and R. Schekman, personal communication). Thus, failure to remove the signal peptide from invertase appears to affect both exit from the ER and subsequent processing reactions such as glycosylation (discussed below). One mechanism for this effect might be that the uncleaved signal sequence mediates either membrane attachment or protein aggregation which sequesters the enzyme from those processes involved in both secretion and subsequent modification (discussed below). A second formal possibility is that retention of the 20 amino acid leader peptide drastically alters the folding of the 550 amino acid protein so that none of its normal export signals are accessible.

The role of signal sequence cleavage in the secretion of acid phosphatase has been studied using two deletions of either 13 or 14 amino acids, which span or immediately follow the signal sequence cleavage site (Fig. 2B). Proteolytic processing occurred in the protein that still contained the cleavage site, and this protein was secreted normally. However, the truncated protein lacking the signal sequence cleavage site was not cleaved and was secreted only slowly (Haguenauer-Tsapis and Hinnen, 1984). Further analysis showed that this protein accumulated to a high degree in the yeast ER, and those molecules that were secreted were not as extensively glycosylated as the wild-type enzyme (Haguenauer-Tsapis *et al.*, 1986). (i.e., these molecules appeared to possess only the oligosaccharides added in the ER and did not contain those modifications which occur in the Golgi body). The decreased glycosylation and slower transport parallel the results with the invertase signal sequence mutants and further illustrate that signal sequence removal is required in order for normal protein processing to occur.

A study of signal peptide cleavage sites has also been carried out on a foreign protein expressed in yeast, glucoamylase (Fig. 2C). This enzyme is capable of hydrolyzing starch and maltooligosaccharides to glucose and was isolated from the filamentous fungus, *Aspergillus awamori* (Innis *et al.*, 1985). After cloning the structural gene, deleting the intervening sequences, and attaching the gene to the yeast *ENO1* promoter, Innis *et al.* were able to obtain expression and secretion of functional enzyme. The enzyme was processed in yeast nearly identically as it was in *Aspergillus* in that the presequence was cleaved, a Lys-Arg/Ala cleavage that occurs in *Aspergillus* also took place in yeast, and O-linked and N-linked glycosylation occurred to approximately the same extent in both organisms. Meade and co-workers have used this protein as a model system to investigate the requirements for the signal peptidase. As summarized in Fig. 2C, these workers changed the amino acids at the Lys-Arg cleavage site (i.e., the *KEX2* cleavage site, discussed below) and then substituted various amino acids at the −1 position. The replacement of Ala with either Pro, Asp, or Trp greatly reduced the amount of active enzyme found in the supernatant. This result parallels those in the acid phosphatase and invertase systems in that lack of or slow signal sequence cleavage also appeared to retard secretion. Sequencing of the Pro substitution indicates that signal peptidase cleavage occurs at the Pro—Asn bond (J. Meade, personal communication), and not another site later in the protein.

The results of the invertase, acid phosphatase, and glucoamylase studies agree with the conclusions based on surveys of a number of signal sequences. The variable length of the hydrophobic core in the different proteins is consistent with the result that small deletions in the invertase presequence allow secretion while large deletions do not. Certain amino acids are rarely or never observed near the signal peptidase cleavage site, and when these amino acids are placed there processing is greatly inhibited. The rules established by consensus are not absolute, however, since the presence of "forbidden" amino acids at the cleavage site does not block all signal peptidase processing. Therefore, this aspect of processing in the yeast secretory pathway, while fine tuned for the endogenous proteins, still has the capacity for dealing with less optimal substrates. Future experiments should more precisely define the limitations of the system.

B. *sec* Mutants Commonly Used in Transport Studies

The isolation of mutants defective in protein secretion, the *sec* mutants, has been instrumental in elucidating the subcellular localization of posttranslational processing reactions. Novick *et al.* (1980) isolated 23 complementation groups that were blocked in protein secretion. These class A *sec* mutants are distinguished from class B *sec* mutants, such as *sec53* and *sec59*, in that class A mutants contain active intracellular precursors. Thus, mutants such as *sec16-1* and *sec18-1*, which accumulate ER-like structures when grown at the nonpermissive temperature, acquire large pools of active invertase (Novick *et al.*, 1980, 1981) and partially processed carboxypeptidase Y (CPY) (Stevens *et al.*, 1982) and α-factor (αF) (Julius *et al.*, 1984a). The existing *sec* mutants have been ordered with respect to which stage of protein secretion they inhibit. The mutants most commonly used in protein secretion studies are *sec1, sec7, sec14, sec16,* and *sec18,* and their order of action is as follows:

$$\text{Er} \xrightarrow{16,\ 18} \text{Golgi} \xrightarrow{7,\ 14} \text{vesicles} \xrightarrow{1} \text{cells surface}$$
$$\downarrow \ 7,\ 14$$
$$\text{vacuole}$$

In other words, the *sec16* and *sec18* mutations block protein exit from the ER, the *sec7* and *sec14* mutations block exit from the Golgi, and cells carrying the *sec1* mutation accumulate vesicles, apparently inhibiting vesicle fusion with the plasma membrane. As noted in the diagram, *sec1* does not block protein secretion to the vacuole (Stevens *et al.*, 1982). Results from higher eukaryotes suggest that this pathway is unidirectional in that resident ER proteins do not show protein modifications seen in the Golgi (Brands *et al.*, 1985; Yamamoto *et al.*, 1985).

There are two important caveats in interpreting experiments using these mutants. One possibility is that these mutations, the exact nature of which is unknown, affect protein processing in a manner other than denying a protein its subsequent modifications by blocking access to another subcellular compartment. For example, the possibility exists that enzymes which act in the Golgi may accumulate in *sec16* and *sec18* mutants and act in the ER. The other caveat is that these mutations may induce "piling up" of proteins in the secretory apparatus. Thus, a *sec7* cell grown at 37°C (the nonpermissive temperature) may accumulate both Golgi-processed proteins and proteins that only show ER processing, the latter possibly originating from proteins which did not enter the Golgi (e.g., the forms of αF found in a *sec7* mutant, discussed in Section IV,C).

C. Addition of Asparagine-Linked Carbohydrates

Asparagine-linked, or N-linked, carbohydrates are a posttranslational modification common to all eukaryotes. These structures function in maintenance of protein conformation, as important developmental antigens, and as useful markers of intracellular transport (Hubbard and Ivatt, 1981; Kornfeld and Kornfeld, 1985; Olden *et al.,* 1985). In all eukaryotes, the attachment of carbohydrate occurs by the transfer of an oligosaccharide precursor from a polyisoprenoid lipid (dolichol pyrophosphate, dol-PP) to asparagine residues in the acceptor sequence Asn-X-Ser/Thr where X can be any amino acid, with the possible exception of Pro, and the last two amino acids can be either Ser or Thr. The oligosaccharide precursor is synthesized by the stepwise addition of monosaccharide units to form the lipid-linked oligosaccharide precursor. This precursor has the composition glucose$_3$mannose$_9$N-acetylglucosamine$_2$-PP-dol (Glc-$_3$Man$_9$GlcNAc$_2$-PP-dol) in all eukaryotes so far examined with the exception of a few ciliated protozoa. In these few cases a truncated precursor identical to the intermediates formed in mammalian cells is made and transferred to protein, as opposed to the formation of a lipid-linked oligosaccharide with a different structure (see Kornfeld and Kornfeld, 1985). An outline of the yeast biosynthetic pathway leading to the production of the lipid-linked oligosaccharide precursor and its subsequent processing in yeast will be reviewed here. Those readers desiring a more detailed description are directed to the excellent reviews of Kornfeld and Kornfeld (1985) and Hubbard and Ivatt (1981).

The first step in lipid-linked oligosaccharide biosynthesis is the transfer of GlcNAc-P from UDPGlcNAc to dolichol phosphate (Molnar *et al.,* 1971). UDPGlcNAc provides another GlcNAc residue to form GlcNAc$_2$-PP-dol (Leloir *et al.,* 1973; Villemez and Carlo, 1980; Waechter and Hartford, 1979), followed by the addition of five Man residues, using the

translocation of the Man residue did not occur in vesicles made without the dol-P-Man synthase (Haselbeck and Tanner, 1982), and it had been shown previously that dolichol derivatives do not translocate spontaneously across the lipid bilayer (McCloskey and Troy, 1980). Presumably, the GDPMan on the outside of the vesicle donates a mannose residue to dol-P-Man which then actively or passively crosses the membrane. On the luminal face, the dol-P-Man synthase then catalyzes the back reaction to form GDPMan. One possible interpretation is that a translocase activity in this partially purified protein fraction is responsible for actively flipping dol-P-Man from one membrane side to the other. Further work is required in order to determine whether dol-P-Man translocation is mediated by the same enzymes that synthesize dol-P-Man from GDPMan and dol-P.

The addition of the first GlcNAc-P to dol-P is inhibitable by the antibiotic tunicamycin (Kuo and Lampen, 1974; Lehle and Tanner, 1976). Yeast and other cells treated continuously with this drug stop transferring carbohydrate to asparagine residues and eventually die. Barnes *et al.* (1984) isolated both dominant and recessive tunicamycin-resistant mutants, many of which map to the gene for this first transferase. Rine *et al.* (1983) have cloned the gene for this enzyme by transforming yeast with a genomic library in a multicopy vector and plating the transformants on media containing tunicamycin. Only cells that contained multiple copies of the gene encoding the first enzyme in the pathway were able to grow because an increased amount of gene product was able to overcome the competitive inhibitor. The gene for this enzyme has been named *ALG7*.

Several other *alg* mutants have been isolated which affect enzymes in this pathway. Cells carrying the *alg1-1* mutation have a temperature-sensitive (*ts*) defect in transferring the first Man residue to GlcNAc$_2$-PP-dol (Huffaker and Robbins, 1982). These cells show a *cdc* (cell division cycle) -like arrest at the nonpermissive temperature in G$_1$ (Klebl *et al.*, 1984; see Pringle and Hartwell, 1981, for a description of *cdc* mutants). Interestingly, tunicamycin causes a similar effect (Arnold and Tanner, 1982). The implications of these observations will be discussed below.

Other *alg* complementation groups include the following: *alg2* which causes cells to accumulate both Man$_2$GlcNAc$_2$-PP-dol and Man$_1$GlcNAc$_2$-PP-dol; *alg3* which leads to accumulation of Man$_5$GlcNAc$_2$-PP-dol; *alg5* and *alg6* which result in buildup of Man$_9$GlcNAc$_2$-PP-dol; and *alg8* which causes cells to accumulate Glc$_1$Man$_9$GlcNAc$_2$-lipid (see Huffaker and Robbins, 1983; Runge *et al.*, 1984; Runge and Robbins, 1986a). The *alg4* complementation group is the same as *sec53* (see above). The *alg4–sec53* mutations and the *sec59-1* mutation cause cells to accumulate a variety of lipid-linked oligosaccharide biosynthetic intermediates (Runge and Robbins, 1986b). The nature of the *alg2* defect is unclear as these mutant cells accumulate two different precursors. The *alg3* cells can synthesize dol-P-Man *in vitro* (K. Runge, unpublished), and so appear to be blocked in addition of the sixth Man residue. *alg5-1* cells are unable to synthesize dol-P-Glc while *alg6-1* cells cannot transfer the first glucose residue from dol-P-Glc to $Man_9GlcNAc_2$-PP-dol (Runge *et al.*, 1984), and cells carrying the *alg8-1* mutation cannot transfer the second glucose residue to Glc_1-$Man_9GlcNAc_2$-lipid (Runge and Robbins, 1986a).

A new mutation that blocks dol-P-Glc synthesis, *dpg1*, has also been isolated. This mutation is not allelic to *alg5-1* (Ballou *et al.*, 1986). Of these complementation groups, only *alg1*, *alg2*, and *alg7* affect genes that appear to be essential for growth (Barnes *et al.*, 1984; Huffaker and Robbins, 1983). Cells carrying the *alg3*, *alg5*, *alg6*, or *alg8* mutations grow normally at either 26° or 36°C and are defective in their various enzymatic steps at both temperatures (Huffaker and Robbins, 1983; Runge *et al.*, 1984; Runge and Robbins, 1986a). The last four types of mutations cause yeast cells to transfer truncated lipid-linked oligosaccharides to protein (Runge and Robbins, 1986a). All of these lack the three Glc residues thought to be a signal for oligosaccharide transfer. It has also been determined that all *alg* mutants underglycosylate invertase, proteinase A, and carboxypeptidase Y, but this underglycosylation does not significantly affect the growth rate of *alg3*, *alg5*, *alg6*, or *alg8* cells (Runge and Robbins, 1986a; Jones *et al.*, 1986; S. Garlow and E. Jones, personal communication). These results are in agreement with the hypothesis that the presence of the Glc residues increases oligosaccharide transfer efficiency (as those cells which cannot add all of these glucose residues do show the expected phenotype for slower transfer), but they are surprising in the large degree of tolerance yeast shows for underglycosylation of protein and a wide range of substrates for the oligosaccharide transferase.

D. Processing of Asn-Linked Oligosaccharides

Before discussing the modifications N-linked oligosaccharides undergo in the ER and Golgi, a short discussion of the enzymes used in their characterization is required. Until recently, the primary tool for removing N-linked carbohydrates was the enzyme endo-β-N-acetylglycosaminidase H (endo H). This enzyme hydrolyzes the GlcNAc—GlcNAc bond of certain Asn-linked oligosaccharides (see Fig. 3). The major structural requirement for efficient endo H cleavage is the α-1,3-linked Man present in the $Man_6GlcNAc_2$ oligosaccharide but not present in the $Man_5GlcNAc_2$ oligosaccharide shown in Fig. 3 (e.g., see Tai *et al.*, 1977; Prakash and Vijay, 1982). This makes endo H an excellent tool for removing any N-linked carbohydrates from yeast, most of which contain this determinant (discussed below). Some of the truncated oligosaccharides transferred in certain *alg* mutants are an important exception to this rule. The lipid-linked oligosaccharide accumulated in *alg2* and *alg3* cells are not cleavable by endo H, and, indeed, some endo H-resistant oligosaccharides are transferred in these mutants (Huffaker and Robbins, 1983).

These observations bear some relevance on the type of invertase observed in the *sec53* mutants, discussed earlier. Ferro-Novick *et al.* (1984a) reported that a heterogeneous form of invertase, with a slightly slower mobility on denaturing polyacrylamide gels than the nonglycosylated form, accumulated in *sec53-6* mutants at the nonpermissive temperature. The mobility of this protein was not changed by endo H treatment. A similar result was obtained with *alg4-9* cells (the *alg4* and *sec53* complementation groups are the same) (Huffaker and Robbins, 1983). Given that the *sec53-6* mutation also affects lipid-linked oligosaccharide biosynthesis, one possibility is that the invertase from these cells contains short oligosaccharides ($Man_5GlcNAc_2$ or smaller) on only a few of the 12 potential glycosylation sites. Ferro-Novick *et al.* (1984a) also reported that this form of invertase did not bind the mannose lectin concanavalin A. However, small mannose-containing oligosaccharides might not have been bound in their assay (see Narasimhan, 1982), so the possibility still exists that the heterogeneous form of invertase in *sec53-6* cells results from transfer of small lipid-linked oligosaccharides. Such caveats may be important when other proteins are examined in *sec53-6* cells.

A new glycosidase has recently become available that has certain advantages over endo H. Peptide *N*-glycosidase F (glycopeptide *N*-glycosidase) has the ability to cleave the GlcNAc—Asn bond of a wide variety of oligosaccharides, including those with the structure $ManGlcNAc_2Asn$, to GlcNAcAsn (Chu, 1986; Tarentino *et al.*, 1985). This enzyme should also be able to remove any of the truncated lipid-linked oligosaccharides transferred to protein in the *alg* mutants and, possibly, *alg4* and *sec53* mutants, and it provides a useful tool in analyzing endo H-resistant carbohydrates that are transferred to protein.

After the $Glc_3Man_9GlcNAc_2$ oligosaccharide is transferred to protein, the three Glc residues are removed, followed by removal of the central Man residue (Byrd *et al.*, 1982) to form the "core" oligosaccharide $Man_8GlcNAc_2$. Glucosidase I is the enzyme responsible for removing the first α-1,2-linked Glc residue while glucosidase II removes the two inner Glc residues (see Hubbard and Ivatt, 1981). The central Man residue is then removed by a specific mannosidase (Jelinek-Kelly *et al.*, 1985) (see Fig. 4). All of these processing enzymes have been partially purified from yeast (Jelinek-Kelly *et al.*, 1985; Kilker *et al.*, 1981; Saunier *et al.*, 1982).

The structure of the $Man_8GlcNAc_2$ oligosaccharide was originally determined in wild-type yeast (Byrd *et al.*, 1982) and more recently con-

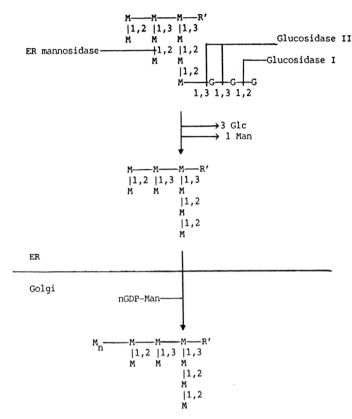

Fig. 4. Asn-Linked oligosaccharide processing. The N-linked $Glc_3Man_9GlcNAc_2$ oligosaccharide is trimmed to $Man_8GlcNAc_2$ by the sequential action of glucosidase I, which removes the $\alpha 1,2$-linked glucose, glucosidase II, which removes the two $\alpha 1,3$-linked glucose residues, and an ER α-mannosidase, which removes the terminal residue from the central branch. Glucose removal is not required for mannosidase action. After proteins enter the Golgi body, subsequent Man residues are added from GDPMan. It is unclear if Man residues are added to the inner core (however, see Trimble and Atkinson, 1986). The structure of one such Golgi-modified oligosaccharide is shown in Fig. 5. R' is GlcNAcGlcNAcAsn, M is mannose, G is glucose, and all saccharide linkages are in the α configuration.

firmed using yeast mutants in this stage of protein modification. *gls1-1* is a mutation that greatly reduces glucosidase I activity (Esmon *et al.*, 1984) but does not affect glucosidase II activity (Runge and Robbins, 1986a). Using a yeast strain containing this mutation, Tsai *et al.* (1984a) found a Glc_3-containing oligosaccharide that had the appropriate Man residue removed. Other oligosaccharides processed in a similar manner have been found in yeast containing the *mnn1, mnn2,* and *mnn9* mutations (Ballou *et al.*, 1980; Cohen *et al.*, 1982) (these mutations affect processing in the yeast Golgi and will be described below). It should be noted that the present structure of the "core" oligosaccharide is not the same as the one originally reported by Nakajima and Ballou (1974b). The original structure was for yeast mannan, a component of the yeast cell wall, and is likely to be different for secreted glycoproteins such as invertase (see Trimble and Atkinson, 1986). All of the present data suggest that $Man_8GlcNAc_2$ is the actual yeast core oligosaccharide. Thus, removal of the central Man residue from yeast N-linked oligosaccharides appear to be a ubiquitous step, and this processing does not require Glc removal in order to occur.

The localization of these reactions to the endoplasmic reticulum was first indicated by subcellular fractionation and by analogy with higher eukaryotes. The enzymatic activities responsible for both lipid-linked oligosaccharide biosyntheses cofractionate with membranes at the density expected for the ER (Marriott and Tanner, 1979). All of these reactions have been shown to take place in the ER of animals cells by means of subcellular fraction, pulse–chase experiments, and electron microscopy (reviewed in Hubbard and Ivatt, 1981).

The above observations have been supported using the *sec* mutants. Esmon *et al.* (1981) showed that those mutants which accumulated in the ER also accumulated invertase with short oligosaccharides while those blocked at the Golgi accumulated invertase with long oligosaccharide side chains. When *sec18-1* cells are labeled with radioactive Man at the nonpermissive temperature, the major protein-linked oligosaccharide seen is $Man_8GlcNAc_2$ with larger high-mannose oligosaccharides conspicuously absent, in contrast to the wild type where a background of larger oligosaccharides is always present (Byrd *et al.*, 1982; Esmon *et al.*, 1984; Runge *et al.*, 1984). The sum of these observations suggest that all of the above reactions occur in the ER.

E. Addition of O-Linked Oligosaccharides

A second class of carbohydrate–protein linkage is the attachment of sugar residues to the hydroxyl side chain of serine and threonine, forming O-linked oligosaccharides. Unlike N-linked oligosaccharides, there is no consensus sequence for transfer of monosaccharides to Ser or Thr, but placement of a Pro residue near the accepting amino acid increases the rate of sugar transfer to Ser and Thr residues *in vitro* (Lehle and Bause, 1984). These oligosaccharides are chemically distinct from their N-linked counterparts in that only the O-linked carbohydrates are released by base treatment. The alkali-labile oligosaccharides of *Saccharomyces cerevisiae* are much more simple than the complex structures seen in Asn-linked oligosaccharides. They consist of short Man oligomers containing one to four residues (Nakajima and Ballou, 1974a) (Fig. 4). The synthesis of these chains differs from similar reactions in higher cells in one important respect. O-Linked glycans in animal cells are synthesized exclusively in the Golgi apparatus while in yeast they are made in both the ER and the Golgi (Dunphy and Rothman, 1985).

Early work with crude yeast membrane preparations established that the first mannosyl residue attached to Ser or Thr was donated by dol-P-Man, while subsequent monosaccharides were added from GDPMan (Sharma *et al.*, 1974). As later work on the synthesis of N-linked carbohydrates proceeded, it became clear that some ER-localized reactions utilized oligosaccharide–lipid as a donor while subsequent elongation used the nucleotide sugar GDPMan (Parodi, 1979; Lehle, 1980). These observations led to the hypothesis that the lipid-linked monosaccharide was transferred to acceptor Ser and Thr residues in the ER and subsequent elongation might occur in the Golgi.

Support for this idea came from subcellular fraction studies which indicated that O-linked glycosylation begins in the ER (Marriott and Tanner, 1979). More definitive proof was derived from studies involving the *sec18* and *sec1* mutants. Cells bearing the *sec18-1* mutation that are labeled with [2-^{3}H]mannose accumulate protein containing primarily one O-linked sugar while *sec1* cells contain mannose to mannotetrose (Haselbeck and Tanner, 1983). A small amount of mannobiose was observed in the *sec18* strain, raising the possibility that more than one O-linked Man residue is added in the ER. This may not necessarily be the case as other mannosyltransferases in the ER may recognize the single O-linked Man as poor substrate and add a second residue. This type of addition would probably not occur in the normal *in vivo* situation where proteins are rapidly transported from the ER to the Golgi, but in the *sec18* cell proteins accumulate in the ER for longer periods of time at higher intracellular concentrations. These conditions may allow the ER mannosyltransferases involved in lipid-linked oligosaccharide biosynthesis (see Section III,D) to utilize the O-linked Man as poor substrate and thus add a second residue. A similar activity has been demonstrated for the lipid-linked oligosaccharide biosynthetic enzymes of rabbit liver which can transfer Man residues from endogenous donor to free Man (Verma *et al.*, 1977). Thus, the sum of the evidence strongly indicates that only the first O-linked Man is transferred in the ER and that subsequent residues are added in the Golgi; the existence of mannobiose in *sec18* cells may indicate an interesting side reaction that can take place *in vivo*.

F. Function of Core Oligosaccharide Addition

Given that CPY and alkaline phosphatase are correctly localized to the vacuole in the presence of tunicamycin (Hasilik and Tanner, 1978; Onishi *et al.*, 1979; Schwaiger *et al.*, 1982) and that a number of *alg* mutants which partially disrupt normal protein glycosylation show no obvious growth defects, one might wonder what purpose protein glycosylation serves in yeast. The activity of the secreted proteins invertase, CPY, and acid phosphatase are all greatly reduced when synthesized in the presence of tunicamycin (Kuo and Lampen, 1974). This drug also induces a temperature-sensitive defect in cell surface localization for invertase and αF; nonglycosylated precursors are correctly localized at 25° but not (or only poorly) at 37°C (Ferro-Novick *et al.*, 1984a; Julius *et al.*, 1984a). Acid phosphatase is slowly secreted from tunicamycin-treated cells (Schönholzer *et al.*, 1985). Invertase represents a particularly interesting case because a cytoplasmic form of this enzyme is synthesized from the same gene. This enzyme lacks only the signal peptide when compared to secreted invertase and yet is fully active in nonglycosylated form. Interestingly, the signal sequence deletions constructed by Kaiser and Botstein that block invertase secretion result in cytoplasmic localization of an active enzyme (Kaiser and Botstein, 1986).

A partial explanation of these observations comes from the work of Gibson *et al.* (1979) and Leavitt *et al.* (1977). These workers determined that the nonglycosylated spike glycoproteins of both vesicular stomatitis virus (VSV) and Sindbis virus possessed altered solubilities and tended to aggregate both in *in vivo* and *in vitro*. The nonglycosylated VSV G protein shows temperature-sensitive secretion *in vivo* in parallel with altered solubility *in vitro* (Gibson *et al.*, 1979); the protein appeared at the cell surface at 30° but not at 37°C and formed aggregates only at 37°C. The authors suggested that G protein was probably forming aggregates in the ER lumen at 37°C which prevented its secretion. A similar supposition could be applied to invertase and αF. In order to explain the differences between secreted invertase and the cytoplasmic form discussed above, one must suppose a protein extruded through the ER membrane has different folding constraints than a cytoplasmic protein synthesized on free ribosomes. In this model, the transfer of N-linked oligosaccharides determines the appropriate folding pathway of the secreted protein.

The importance of glycosylation for protein activity may have specific effects in the cell cycle. *alg1-1* cells do not grow at 37°C but arrest at a specific stage of the cell cycle (Klebl *et al.*, 1984). These cells arrest at G_1 before the onset of DNA synthesis. This type of cell cycle arrest is similar to that caused by αF or tunicamycin treatment (Arnold and Tanner, 1982;

Pringle and Hartwell, 1981; Thorner, 1981). Indeed, *alg1-1* cells grown and arrested with αF at 26° and then subsequently transferred to 37°C were unable to leave G_1 after αF was removed (Klebl *et al.*, 1984). The sum of these data suggests that protein glycosylation is involved in the transition from G_1 to S phase. Support for this idea comes from recent experiments with the cell division cycle mutant *cdc25*. This mutant shows growth arrest at 37°C prior to DNA synthesis in G_1. By means of pulse–chase experiments with [^{35}S]Met, Popolo *et al.* (1986) demonstrated that one specific protein which was labeled only in *cdc25* cells after release from G_1 arrest could be chased into a glycoprotein. While the biochemical function of this stage-specific glycoprotein is unknown, it is interesting to speculate that a glycoprotein is required for commitment to DNA synthesis, and that proper glycosylation is required for its activity. However, any direct connection between protein glycosylation and the cell cycle will have to await the isolation and characterization of these glycoproteins and the genes that encode them.

G. Addition of Phosphoryl Groups

The phosphorylation of N-linked oligosaccharides is a common mechanism of designating a protein for transport to the lysosomes in higher cells. In animal cells this reaction occurs by the addition of a GlcNAc-P to the sixth carbon of a Man residue. The GlcNAc residue is then removed to yield Man-6-P, which is thought to serve as a signal for lysosomal transport (Tabas and Kornfeld, 1980; reviewed by Sly and Fischer, 1982). In animal cells, this reaction occurs in the Golgi apparatus. In yeast, phosphorylation of the oligosaccharides occurs in both the ER and the Golgi compartments.

The yeast vacuole is the structural equivalent of the mammalian lysosome (see Jones, 1984). Mature carboxypeptidase Y, a vacuolar protein, contains relatively short oligosaccharides containing 12–15 Man residues. These oligosaccharide chains are phosphorylated with zero, one, or two phosphoryl groups (Hashimoto *et al.*, 1981; Trimble *et al.*, 1983). These phosphates can occur in a diester linkage with the structure Man-6-P-Man or Man-6-P-Man$_2$ (Hashimoto *et al.*, 1981). CPY acquires at least the phosphoryl residues in the ER (see below).

Stevens *et al.* (1982) used the *sec* mutants to analyze the ratio of ^{32}P to ^{35}S in CPY and invertase. The ratio for CPY was the same in ER- and Golgi-blocked mutants while barely any ^{32}P was detected in ER-accumulated invertase. Phosphorylation occurs during the addition of outer chain oligosaccharides in the Golgi (described below), and invertase from the Golgi-blocked mutants had a high ^{32}P to ^{35}S ratio. The nature of the phosphoryl group added in the *sec18* and *sec16* (ER-accumulating) mutants was not chemically determined. Thus, it is unclear whether the addition of phosphate in the ER occurs by the same mechanism as the addition of outer chain residues in the Golgi.

With regard to this question, it is necessary to consider the processing pathway of CPY. CPY is first synthesized as a proenzyme. The core glycosylated form found in the ER is designated P1. A more heavily glycosylated protein, P2, is found in Golgi-blocked mutants, while mature CPY is formed in the vacuole, a step involving cleavage of an amino-terminal segment (Stevens *et al.*, 1982). Based on our present understanding of yeast glycoprotein biosynthesis, it seems likely that the P1 form of CPY contains $Man_8GlcNAc_2$ oligosaccharides, some of which are phosphorylated. It is not known if these phosphate residues exist as Man-6-P monoesters or Man-6-P-Man diesters. The formation of P2 occurs by means of outer chain addition in the Golgi apparatus prior to transport to the yeast vacuole (discussed below). Further work is needed to resolve whether the mechanism of phosphate addition in the ER is distinct from that used in the Golgi, or if both represent the action of one enzyme in two compartments.

Another significant difference from mammalian cells is that yeast proteases do not require carbohydrate to be localized correctly to the vacuole. Both CPY and alkaline phosphatase are apparently localized to the yeast vacuole in the presence of tunicamycin (Hasilick and Tanner, 1978; Onishi *et al.*, 1979; Schwaiger *et al.*, 1982). Mammalian cells lacking the GlcNAc-P transferase secrete their lysosomal enzymes into the medium (Sly and Fischer, 1982). Thus, the role of oligosaccharide phosphorylation is different in yeast and larger eukaryotes.

H. Addition of Fatty Acid

Protein acylation is a common modification of membrane proteins in many higher eukaryotes. Fatty acids are attached to a number of viral envelope glycoproteins (Schmidt and Schlesinger, 1979; Schmidt *et al.*, 1979; Schmidt, 1982) and to certain cellular integral membrane proteins (Agrawal *et al.*, 1981; Schlesinger *et al.*, 1980) including the HLA-B and HLA-DR heavy chains of the major histocompatability complex (Kaufman *et al.*, 1984). In the case of the HLA antigens, the C_{16} fatty acid palmitate was shown to be attached via a thioester linkage to a cysteine residue, and Kaufman *et al.* also stated that several viral spike glycoproteins were acylated via a cysteine residue.

This type of posttranslational processing has also been recently demonstrated in yeast. Wen and Schlesinger (1984) labeled whole yeast cells with [³H]palmitate and analyzed the different membrane proteins labeled in wild-type yeast and various *sec* mutants. While no palmitate-labeled proteins were seen in wild-type or *sec53-6* cells, four proteins were detected in *sec18-1* cells and one protein in both *sec7* and *sec1* cells. All of these proteins were labeled with [³H]mannose, and the electrophoretic mobilities of most of them were changed by endo H treatment, indicating that they were glycoproteins. The reason that four proteins were not observed in the *sec7* and *sec1* cells was not necessarily loss of the labeled palmitic acid but could also be explained by elongation of N-linked oligosaccharides. Such processing gives rise to glycoproteins of heterogeneous molecular weights (discussed below) which would migrate as diffuse bands on polyacrylamide gels and may render these proteins undetectable above background. These results indicate that fatty acid attachment to secreted proteins first occurs in the endoplasmic reticulum. Not all acylated yeast proteins are secreted, however; the *RAS1* and *RAS2* proteins acquire lauric, myristic, or palmitic acid and fractionate with the plasma membrane (Fujiyama and Tamanoi, 1986). These proteins did not accumulate in *sec53, sec7,* or *sec18* cells, suggesting that they are localized to the cytoplasmic surface of the plasma membrane, as are their mammalian counterparts (Shih *et al.,* 1982; Weeks *et al.,* 1985; Willumsen *et al.,* 1984).

IV. PROTEIN MODIFICATION IN THE GOLGI APPARATUS

With the exception of proteins that reside in the ER, the next step in protein secretion is transport to the Golgi apparatus. Evidence from higher eukaryotes suggests that proteins that are normally found in the ER membrane do not travel to the Golgi body (Brands *et al.,* 1985; Yamamoto *et al.,* 1985). The cloning of ER proteins such as those encoded by *ALG* and *GLS1* genes should provide valuable tools for similar analyses in yeast.

Transport from the ER to the Golgi is an energy-requiring process (Novick *et al.,* 1980) and in higher eukaryotes involves vesicular transport (reviewed in Dunphy and Rothman, 1985). The primary means of measuring the rate of transport involved monitoring the carbohydrate modifications that proteins experience in Golgi body. The time course of this transition in higher cells is of the order of 30–40 min (Hubbard and Ivatt, 1981). This process occurs much more rapidly in yeast; yeast invertase is synthesized and reaches the Golgi in 5–10 min (Novick *et al.,* 1981). It is interesting to note that yeast mutants in the cytoskeletal protein action show a slower transition from ER to Golgi than do wild-type cells (Novick and Botstein, 1985).

A. Oligosaccharide Processing in the Golgi Apparatus

Carbohydrate addition in the Golgi involves the decoration of N-linked oligosaccharides with large numbers of mannose and phosphate residues (in the synthesis of a structure known as the "outer chain") and the addition of one to three mannose residues to O-linked chains. The structures of the O-linked oligosaccharides are described above and are shown in Fig. 5. Both N-linked and O-linked oligosaccharide biosynthesis in the Golgi use GDPMan as substrate (Parodi 1979; Lehle, 1980; Sharma *et al.*, 1974). Analyses with *sec7* and *sec14* cells have shown that only mutants blocked in the Golgi or later stages accumulate proteins with these large oligosaccharides (Esmon *et al.*, 1981).

The structure of these elongated N-linked oligosaccharides has been determined for only a few yeast proteins. Nakajima and Ballou (1974b) analyzed bulk mannoprotein from yeast cells and elucidated the structure shown in Fig. 5. Even though this study characterized the oligosac-

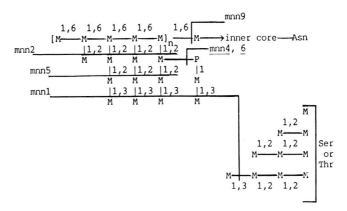

Fig. 5. The structure of N- and O-linked oligosaccharides on mannoprotein and mutants that affect their synthesis. The outer chain of mannoprotein, a constituent of the yeast cell wall, is formed by the elongation of an α-1,6-linked mannose backbone (see Fig. 4) to which short oligosaccharides are added. All linkages shown are in the α configuration. The *mnn9* mutation prevents elongation of the mannose backbone while the *mnn2* mutation blocks addition of the first α-1,2-linked Man. The *mnn5* mutation prevents addition of the second α-1,2-linked Man. The *mnn3* mutation has the same phenotype as the *mnn5* mutation. The *mnn1* mutation affects both N- and O-linked polysaccharides by interfering with the addition of the terminal α-1,3-linked Man. The *mnn4* and *mnn6* mutations both block the addition of mannose phosphate groups. M is mannose, and the inner core consists of the $Man_8GlcNAc_2$ oligosaccharide shown in Fig. 4.

charides from a heterogeneous mixture of proteins, these workers were able to determine the repeating unit added to the majority of oligosaccharides.

By analogy with the terminology in higher eukaryotes, I will refer to the "inner core" of the processed oligosaccharide as the portion that is the same in all such macromolecules. In yeast, the inner core is the $Man_8GlcNAc_2$ structure formed in the ER (Fig. 4). Additional mannose residues can also be added to this structure, and oligosaccharides with similar processed "core" structures have been found in both cell wall mannan and on carboxypeptidase Y (Cohen *et al.,* 1982; Tsai *et al.,* 1984a,b). However, these modifications are not the end result of all processing pathways, as Trimble and Atkinson (1986) have found a variety of other structural isomers on invertase. The structure of the inner core defined in this way is not modified by mutations that affect oligosaccharide processing in the Golgi but can be changed by *alg* mutations that act in ER.

The length of some of the N-linked chains synthesized in the Golgi can number up to 100 Man residues or more. These extremely long N-linked chains are not peculiar to mannoprotein as they have also been found on invertase (Lehle *et al.,* 1979; Trimble and Maley, 1977). The structure of the repeating unit of outer chain mannan (the oligosaccharide on mannoprotein) is shown in Fig. 5, and mutants in its synthesis are described below.

1. Mutants in Outer Chain Biosynthesis

Ballou and co-workers have isolated a number of mutants in outer chain addition in both *Saccharomyces cerevisiae* and other yeast species (reviewed in Ballou, 1982). Their primary method of mutant isolation involved antibodies specific for outer chain determinants. The mutants *mnn1–mnn3, mnn5,* and *mnn6* were selected by failure to be agglutinated with these reagents (Raschke *et al.,* 1973). *mnn4* was isolated by mutagenesis of *mnn1* cells followed by screening cells for failure to bind the basic dye, Alcian blue. Alcian blue binds to acidic residues and recognizes the phosphates in yeast outer chains (Ballou, 1982).

The effect of the various *mnn* mutations on outer chain biosynthesis is summarized in Fig. 5. The *mnn1* mutation lacks an α-1,3-mannosyltransferase which adds terminal Man residues on both N-linked and O-linked chains. The *mnn2* mutation eliminates almost all of the α-1,2 mannosyltransferase I activity in the cell, which results in a linear α-1,6 Man chain attached to the inner core. This mutant fails to bind Alcian blue, presumably because the substrate for phosphate addition, the first α-1,2-linked Man, is no longer present (Nakajima and Ballou, 1974b). Cells carrying either the *mnn3* or *mnn5* mutation fail to add the second α-1,2-

linked Man residue, though the *mnn3* mutation also affects the length of O-linked chains, leading to the hypothesis that *mnn3* cells may have a lower level of GDPMan in the Golgi lumen. *mnn5* has been designated as a mutation that affects α-1,2-mannosyltransferase II.

Cells bearing the *mnn4* or *mnn6* mutation fail to bind Alcian blue. *mnn4* is a dominant mutation, while the rest of the *mnn* mutations are recessive. It has been suggested that *mnn4* is a regulatory mutation while *mnn6* may affect the structural gene involved in phosphomannose addition (Ballou, 1982). Karson and Ballou (1978) have demonstrated that the phosphomannose is donated by GDPMan, in marked contrast to the mechanism used in higher eukaryotes to attach phosphoryl residues to the sixth carbon of mannose (see above). It is not yet known if oligosaccharide phosphorylation in the yeast Golgi occurs by the same mechanism that adds phosphoryl groups to CPY oligosaccharides in the ER.

The original search for *mnn* mutants found no mutants in the synthesis of the α-1,6 Man backbone of the outer chain. One possible reason for this was that the original selection required that mutated cells be able to outgrow the agglutinated wild-type cells. Yeast strains lacking outer chain Man structures might possess a growth disadvantage such that they are unable to outgrow the agglutinated wild-type cells. In order to overcome this limitation and isolate mutants in α-1,6 chain synthesis, Ballou *et al.* (1980) used *Fab* fragments to the *mnn2* outer chain and fluorescence-activated cell sorting to screen mutagenized *mnn2* cells. Four complementation groups, *mnn7–mnn10,* were recovered. These four mutants grow poorly and show severe truncations in their outer chain mannan. Cells carrying the *mnn9* mutation only add one α1,6 Man to the Man$_8$GlcNAc$_2$ core. In the absence of other *mnn* mutations, this short oligosaccharide can be modified by the addition of other Man residues. These residues are added by outer chain mannosyl transferases that now act on the inner core to produce a Man$_{13}$GlcNAc$_2$ structure (see Fig. 5 and the oligosaccharide structures in Cohen *et al.,* 1982 and Tsai *et al.,* 1984b) without an outer chain (see Tsai *et al.,* 1984b). Interestingly, *mnn7–mnn10* all bind Alcian blue and react with anti-α1,3-mannose antibody (while the *mnn2* parent did not), presumably because the shorter outer chain allows the Golgi processing enzymes and antibodies access to the inner core oligosaccharide. *mnn7–mnn10* all show enhanced growth with osmotic support.

The isolation of four complementation groups affecting the length of the mannan outer chain presents a conundrum. While one obvious defect leading to this phenotype would be a mutation in the α-1,6-mannosyltransferase, the other types are by no means clear. Wild-type levels of invertase are synthesized and secreted to the yeast periplasm in these cells (Ballou *et al.,* 1980), indicating that these mutations do not block secretion. One formal possibility is that some of the *mnn* mutations affect subcellular routing from the ER to the cell surface. In this way, some proteins may bypass the Golgi or subcompartments of the Golgi and proceed to the cell surface. The appropriate processing enzymes would then be sequestered from secreted proteins and processing would never occur. A number of mutations affecting protein localization to the yeast vacuole may alter protein processing in just this way (discussed below). Further work is required before any secretion defects can be linked to the *mnn7–mnn10* complementation groups.

2. Factors Affecting Oligosaccharide Processing

The control of oligosaccharide processing in wild-type cells involves other factors beside the *mnn* gene products. These include conformation of the secreted protein and sequestering of processing enzymes in subcompartments of the Golgi (the latter aspect will be discussed below). The effect of protein conformation on oligosaccharide processing has been approached by several workers. The usual method is to use endoglycosidases, such as endo H, to determine the accessibility of an oligosaccharide on both native and denatured protein. When the oligosaccharide is on the surface of the native protein, it can be removed. If the carbohydrate chain is rendered cryptic by protein folding, the native protein will sterically exclude endo H from the oligosaccharide which will remain attached to the protein. All oligosaccharides should be removed from the denatured protein, thereby controlling for lack of cleavage in the native protein.

Such studies have been carried out in both yeast (Trimble *et al.*, 1983) and higher eukaryotes (Hsieh *et al.*, 1983). Trimble *et al.* (1983) used this approach in examining the oligosaccharides on invertase and CPY. CPY has four sites for N-linked carbohydrate addition, and three of these are readily released from native protein. These oligosaccharides had 11–18 Man residues and all of the phosphate. The oligosaccharides at the inaccessible site could only be removed from denatured protein, contained no phosphoryl residues, and ranged in size from $Man_8GlcNAc_2$ to $Man_{12}GlcNAc_2$. These oligosaccharides could be mapped to a specific tryptic peptide fragment separate from the other glycosylation sites (Trimble *et al.*, 1983). A similar analysis of invertase oligosaccharides by these workers showed that all of the material released from the native enzyme (seven of nine chains) contained greater the 20 Man residues while the carbohydrate released from the two protected sites was of the size $Man_{8-12}GlcNAc_2$. These data suggest that the native protein is folded around certain oligosaccharides, preventing the addition of outer chain Man residues and phosphate. These observations also suggest that the

enzymes that add the first three or four Man residues may be less susceptible to steric hindrance than those that add the later residues. The addition of phosphate to CPY in the ER must occur after protein folding is
complete since the "protected" oligosaccharide on this protein was phosphate free. The fact that the easily released CPY oligosaccharides were
much smaller than the analogous structures on invertase does not necessarily mean that native CPY sterically prevents further elongation of its
N-linked oligosaccharides, but may reflect some other control mechanism, such as the transport of CPY to the vacuole before the glycoprotein
encounters the cellular machinery responsible for outer chain synthesis.

B. Protein Sorting in the Golgi Apparatus and Its Effects on Protein Modification

Much of the current interest in the Golgi body stems from its function
as a decision point in protein sorting. Secretory proteins are directed to
either the yeast periplasm or vacuole in this organelle. In more complex
eukaryotes, the Golgi apparatus consists of stacked lamelli and is the site
of addition of phosphate, N-acetylglucosamine, galactose, sialic acid, and
other sugars. These reactions occur in different subcompartments designated cis, for the "stacks" closest to the ER, and trans, for those that are
distal (reviewed in Hubbard and Ivatt, 1981; Dunphy and Rothman, 1985).
A variety of different reactions have been localized to different regions of
this organelle, and an *in vitro* translocation system that allows transport
between different lamelli has been described (Fries and Rothman, 1980).
The important points relevant to discussion of the yeast system is that in
animals cells oligosaccharide trimming is thought to occur in the cis Golgi,
addition of new monosaccharides begins in the medial Golgi, and completion of oligosaccharide processing appears to take place in the trans compartment. A similar subdivision of processing reactions could exist in
yeast.

All secretory proteins so far investigated that traverse the entire pathway from ER to Golgi to cell surface, such as acid phosphatase and
invertase, are highly glycosylated and about 50% carbohydrate by weight
(Lehle *et al.*, 1979; Trimble and Maley, 1977). The αF precursor, when
not correctly processed in *kex2* cells (see below), also shows similar glycosylation (Julius *et al.*, 1984b). Those proteins that are targeted to the
yeast vacuole, such as CPY, proteinase A, proteinase B, and alkaline
phosphatase, have a much smaller carbohydrate content of about 10–20%
by weight (see Onishi *et al.*, 1979). It has been shown for CPY that its
transport to the vacuole is blocked in *sec16, sec18, sec7,* and *sec14* but
not in *sec1* mutants (interestingly, the block in *sec14* cells is not com-

pletely restrictive) (Stevens *et al.*, 1982). Stevens *et al.* (1982) concluded that CPY followed the secretory pathway through the ER (thus the *sec16* and *sec18* blocks) and part of the Golgi (the *sec7* and *sec14* blocks) but did not become packaged in secretory vesicles of the type that accumulate in *sec1* cells. Thus there is a branch point in the yeast Golgi where some proteins are routed to the cell surface while others go to the vacuole.

If the yeast Golgi is subdivided laterally as the mammalian organelle is, one possible explanation for the lower amount of outer chain mannose decoration of the N-linked chains of vacuolar enzymes is that these proteins are shunted out of the Golgi before extensive carbohydrate addition can occur. Only after the decision to target a protein to the periplasm or the vacuole does much of the outer chain glycosylation occur, and enzymes such as CPY and proteinase A simply miss out because they have left the Golgi. A second possibility is that these proteins contain some unknown cis-acting signal which affects protein glycosylation.

A number of recently isolated mutants in protein transport to the yeast vacuole provide data which relate to this question. Rothman and Stevens have isolated several vacuolar protein localization (*vpl*) mutants that mislocalize CPY and other vacuolar proteases to the cell surface. These mutants fall into 14 complementation groups (J. H. Rothman and T. Stevens, personal communication). Vacuolar proteases secreted to the cell surface in these mutants possess short oligosaccharides, characteristic of those proteins found in the vacuoles of wild-type cells. Bankaitis *et al.* (1986) have also isolated a number of mutants defective in vacuolar protein targeting. The *vpt* mutants were isolated using a CPY–invertase protein fusion that is localized to the vacuole, selecting for cells that now transported this fusion protein to the cell surface. These workers obtained mutants that define eight complementation groups. Cells bearing the *vpt* mutations also secrete other vacuolar proteins, and their secretion is blocked by the *sec1* mutation. A similar finding has been made for the *vpl1–vpl8* mutations (T. Stevens and J. Rothman, personal communication). The *vpt* complementation groups show only small overlap with the *vpl* groups (Johnson *et al.*, in press; S. Emr and T. Stevens, personal communication).

All of the invertase fusions constructed by this group possess invertase activity, indicating that the invertase portion of the protein retains its native conformation. It is therefore interesting to note that CPY and the CPY–invertase protein fusion secreted to the cell surface in these mutants show short chain, CPY-like glycosylation. Thus, even though the invertase portion of this fusion is in a native or near-native conformation, it is not glycosylated in the way native invertase is (L. Johnson and S. Emr, personal communication). (Further discussion of protein targeting to the vacuole and protein modification in that organelle can be found in Chapter 8 by Rothman and Stevens, this volume.)

Stevens *et al.* (1986) have shown that when CPY is overproduced on a multicopy plasmid, a portion of it is mislocalized to the cell surface where some of it is activated (activation of proCPY to CPY normally occurs in the vacuole). Rothman *et al.* (1986) have shown a similar phenomenon for proteinase A. In either case, only the overproduced enzyme was mislocalized to the cell surface. The vacuolar proteases that were secreted to the cell surface did not possess the large oligosaccharide chains commonly seen on invertase and acid phosphatase, and both CPY and proteinase A mislocalizations are blocked by *sec1*.

The sum of the results from the groups of Stevens and Emr show that when vacuolar enzymes are secreted, or are "directing" secretion as with the fusion protein, they do not acquire the large oligosaccharides that invertase, acid phosphatase, and the αF precursor do. One explanation for this might be that unknown protein determinants specify the extent to which a protein may be glycosylated. In this case, the CPY determinants directing the addition of the short chain oligosaccharides must be present in the fusions made by Bankaitis *et al.* (1986), and these signals must supercede any in the invertase portion of the molecule. This type of control over oligosaccharide processing would be different from the type described by Trimble *et al.* (1983) (see above) because even the freely accessible oligosaccharide chains of CPY are not extensively glycosylated.

A second explanation is that all of these proteins leave the Golgi en route to the vacuole but cannot reach the vacuole as a consequence of titration of some receptor for either a specific protein or transport vesicle (as in the overproduction of CPY and proteinase A) or because of a *vpl* or *vpt* mutation. These proteins, presumably in a transport vesicle of some sort, are now rerouted to the cell surface to fuse with the plasma membrane. These vesicles have left the Golgi and do not refuse with it, and so bypass the glycosylation machinery there. In addition to the many assumptions made above, this argument also requires that the *sec1* mutation block many types of vesicles from fusing with the cell membrane, including the putative transport vesicles mentioned above. One advantage of this hypothesis is that the normally short carbohydrate chains of CPY and proteinase A can be explained by simple withdrawal of these proteins from the glycosylation apparatus. One can approach the problem of whether the CPY molecule can acquire large oligosaccharide chains by constructing an αF–or invertase–CPY fusion that localizes CPY to the cell surface. The second explanation would demand that CPY contain long oligosaccharides while the first explanation would be compatible with either long or short chains. Through experiments such as these, one should be able to investigate both protein targeting and the control of oligosaccharide processing.

Another mutation that merits discussion in this section is *pep7*. This complementation group is one of many isolated that showed reduced levels of vacuolar protease activity (reviewed by Jones, 1984). When CPY is isolated from this mutant, it runs as a diffuse band on denaturing polyacrylamide gels with a mobility similar to that of CPY from wild-type cells. If the protein is first treated with endo H, it comigrates with wild-type endo H-treated CPY. CPY isolated from *alg1–alg3*, *alg5*, *alg6*, *gls1-1*, *mnn1*, *mnn2*, and *mnn4* cells all migrate as sharp bands on gels and differently from CPY from wild-type cells or *pep7* cells (*mnn2* and *mnn4* have no effect on CPY mobility), and the *pep7* mutations complement the *alg1–alg6*, *gls1*, *mnn1*, *mnn2*, *mnn4*, and all previously described *sec* mutations (S. Garlow and E. Jones, personal communication). The source of this heterogeneity could represent increased phosphorylation, outer chain addition or trimming, or a combination of these. It is unclear what the cause of this aberrant glycosylation is; it could represent a bona fide defect in a mannosyltransferase or mannosidase, or it could stem from slow or incorrect routing of CPY to the vacuole. Along these lines, Garlow and Jones have noted that proCPY requires up to 3 hr to be processed to the mature form in pep7 cells and that other vacuolar protease activities are depressed in this mutant (S. Garlow and E. Jones, personal communication). Further analysis of protein modifications in this mutant and various *sec–pep7* double mutants should lend insight into whether this mutant is defective in glycosylation or subcellular trafficking.

C. Proteolytic Processing in the Golgi Apparatus and Periplasmic Space

The maturation of both αF and killer toxin involve proteolytic cleavage of a larger precursor to liberate either the mature mating pheromone (Emter *et al.,* 1983; Julius *et al.,* 1983; Kurjan and Herskowitz, 1982; Singh *et al.,* 1983) or the two toxin subunits (Bostian *et al.,* 1980, 1983; Bussey *et al.,* 1982, 1983a,b). The processing of both of these precursors involves at least one common step (Leibowitz and Wickner, 1976), and many of these reactions strongly resemble those used by higher eukaryotes in the synthesis and posttranslational modifications of peptide hormones (reviewed in Julius *et al.,* 1983). Thus, the study of the procesing pathway of αF and killer toxin has not only revealed the presence of certain proteases in the Golgi body and, possibly, the periplasm, but may also have wider implications for hormone maturation in multicellular organisms.

The mating pheromone αF is a 13 amino acid peptide involved in conjugation of MATa and MATα cells (reviewed in Thorner, 1981; Herskowitz and Oshima, 1981). The αF precursor comes in two forms encoded by two genes. The *MFα1* gene encodes a 165 amino acid precursor that contains four repeats of the mature αF peptide while the *MFα2* locus encodes a 120 amino acid protein that contains two αF-like peptides (see Fig. 6). When Julius *et al.* (1984a) followed the maturation of αF through the secretory

```
MFα1              KEX2
...Asp-lys-arg-        -glu-ala-glu-ala-   alpha factor  -
          lys-arg-+glu-ala-glu-ala-glu-ala-  alpha factor  -
          lys-arg-+glu-ala-asp-ala-glu-ala-  alpha factor  -
          lys-arg-+glu-ala-asp-ala-glu-ala-  alpha factor  -end

MFα2

...Ala-lys-arg-+glu-ala-val-ala-asp-ala-  'alpha factor' -
          lys-arg-+glu-ala-asn-ala-asp-ala-   alpha factor - end

                          ------> diamino peptidase A

          <--|
 carboxypeptidase
        or
 chymotrypsin-like
     activity


...Leu-pro-arg-+glu-ala-pro...     alpha subunit of killer toxin

...ala-lys-arg-+tyr-val-tyr...     beta subunit of killer toxin

           KEX2 sites?

 alpha factor is
     trp-his-trp-leu-gln-leu-lys-pro-gly-gln-pro-met-tyr-

 'alpha factor' is
     trp-his-trp-leu-ASN-leu-ARG-pro-gly-gln-pro-met-tyr-
```

Fig. 6. α-Factor and killer toxin precursor processing. The vertical line represents the cleavage sites of the *KEX2* protease in the αF precursor and the presumed sites in the killer toxin precursor. In the case of αF, this cleavage generates separate fragments that contain the pheromone with additional amino and carboxy-terminal amino acids. The horizontal arrows indicate the trimming reactions which remove these residues to release mature αF. Diaminopeptidase A removes the X-Ala residues that are amino terminal to the peptide. The Lys-Arg carboxy-terminal residues are removed by either a chymotrypsin-like or carboxy-peptidase activity. The sequence of αF and the homologous 'α-factor' peptides are shown below. Only a portion of the αF and killer toxin precursors are shown, and other proteolytic activities are probably required for killer toxin maturation.

pathway using an antibody against the mature, 13 amino acid peptide, only the 165 amino acid precursor was detected. These results strongly suggest that the *MFα1* gene is responsible for the majority of αF produced in α cells, but they are complicated by the fact that the *MFα1* gene product contains more antigenic determinants than the *MFα2*-encoded protein and that one of these αF peptides in the *MFα2* protein contains two conservative amino acid changes (Gln to Asn and Arg to Lys) which may affect its ability to be recognized by anti-αF antibody. When either gene is present on a multicopy plasmid in α cells, however, functional αF is overproduced (Singh *et al.*, 1983). It has not yet been determined whether or not the altered αF peptide encoded in the *MFα2* gene is functionally equivalent to the peptides encoded in the *MFα1* gene.

The processing of both αF precursors requires three proteolytic activities: *KEX2* protein, dipeptidyl aminopeptidase A (DPAP A), and a carboxypeptidase or chymotrypsin-like activity. The specificities of these three enzymes and their roles in αF maturation are summarized in Fig. 6. The *KEX2* gene encodes a protease that cleaves after Lys and Arg residues. The action of this enzyme liberates the αF pheromone repeats from the precursor with two or three Glu/Asp-Ala dipeptides at the amino terminus and a Lys-Arg dibasic pair at the carboxy terminus (Julius *et al.*, 1984b). The amino-terminal residues are removed by the action of DPAP A, which is probably encoded by the *STE13* gene (Julius *et al.*, 1983). This protease should also be capable of removing Val-Ala and Asn-Ala residues which are present in the *MFα2* precursor (Singh *et al.*, 1983). A second enzyme, DPAP B, has activity similar to DPAPase A (see below). A carboxypeptidase or chymotryptic enzyme activity is required to remove the Lys-Arg residues and generate a free Tyr at the end of αF.

The processing of killer toxin is similar to that of αF in many respects. The recent cloning and sequencing of a cDNA complementary to the piece of double-stranded RNA that codes for the killer and immunity functions predicts an mRNA which encodes one 316 amino acid protein (Bostian *et al.*, 1984; Skipper *et al.*, 1984). This protein contains an amino-terminal region with aspects of a signal sequence [for which the *PHO5* signal sequence can substitute (Hanes *et al.*, 1986)], a central domain that includes three Asn-linked glycosylation sites (the γ domain), and two regions which encode the α and β subunits of killer toxin (Bostian *et al.*, 1984). The identity of the α and β peptides was determined by comparison of the predicted amino acid sequence with that of the amino-terminal sequence of the purified toxin subunits (Bostian *et al.*, 1984). Expression of this cDNA is sufficient to allow wild-type cells both to kill sensitive cells and to be immune to the toxin they secrete (Hanes *et al.*, 1986; Lolle *et al.*, 1984). Interestingly, a *KEX2*-like activity would be required to release these peptides from the precursor protein, and it is hypothesized that this step is the actual block in *kex2* mutants. Thus, both αF and killer toxin are synthesized as preproproteins and both require the *KEX2*-encoded protease for correct processing.

Killer toxin processing differs from that of αF in the effect of the *kex1* mutations. *kex1* mutants block toxin maturation but have no affect on αF production (Leibowitz and Wickner, 1976). The α and β subunits of the toxin are separated by the γ domain in the precurser, which is thought to play some role in immunity (Bostian *et al.*, 1984). This juxtaposition of subunits suggests the existence of another protease (or proteases) necessary for the formation of processed killer toxin. An additional difference is the effect of tosyl-L-phenylalanylchloromethyl ketone (TPCK) on toxin secretion. TPCK is an inhibitor of chymotrypsin. When yeast cells are grown in the presence of this compound, toxin secretion is blocked, and immunoreactive forms of the toxin with M_r 13K, 36K, and 43K are found intracellularly (Bussey *et al.*, 1983a). TPCK is also known to slow the secretion of acid phosphatase (Hanes *et al.*, 1986), and so the effect on toxin secretion may represent a more general affect on protein processing and transport.

1. KEX2 Enzyme

Mutations in the *kex2* gene (as well as the *kex1* gene) were first isolated in cells which were immune to killer toxin but were unable to secrete it (Wickner and Leibowitz, 1976). MATα cells bearing the *kex2* mutation are unable to secrete active αF, and *kex2/kex2* homozygotes cannot sporulate (Leibowitz and Wickner, 1976). Cells bearing the *kex2* mutation also show alterations in their cell surface proteins (Rogers *et al.*, 1979). The form of αF found in *kex2* mutants has an M_r of ~120,000, more than 90,000 of which is due to carbohydrate addition (Julius *et al.*, 1984b). Using *kex2–sec* double mutants, Julius *et al.* (1984b) showed that this form of αF was only seen in *kex2–sec1* or *Kex2–sec7* double mutants but not in *kex2–sec18* cells. The forms of αF found in *sec18* and *kex2–sec18* cells have identical mobilities on denaturing polyacrylamide gels, suggesting that the normal *KEX2* processing event occurs after the endoplasmic reticulum and, based on the *kex2–sec7* results, probably in the Golgi. [Interestingly, the hyperglycosylated form of the αF precursor, as well as mature α-F peptide, are also seen in *sec7* cells; however, *kex2–sec7* mutants do not accumulate any mature αF peptide (Julius *et al.*, 1984a).]

Julius *et al.* (1984b) have identified the proteolytic activity of the *KEX2* gene product and cloned the *KEX2* gene. The *KEX2* enzyme is a Zn^{2+}-sensitive membrane-associated endopeptidase that can cleave on the carboxy side basic residues. When presented with a pair of basic residues such as Y-Arg-Arg-X, this protease will release Y-Arg-Arg and X and does not appear to cleave between the two basic residues. This gene is expressed in both MATa and MATα cells as indicated by the effect of the *kex2* mutation on both killer toxin secretion (Wickner and Leibowitz, 1976) and *KEX2* enzyme activity (Julius *et al.*, 1984b) in cells of either mating type. The *KEX2* enzyme will also cleave the *Aspergillus* glucoamylase enzyme after a Lys-Arg pair. This processing site is also used in *Aspergillus* (Innis *et al.*, 1985). No pairs of Glu-Ala residues follow the Lys-Arg residues in this protein (see Fig. 2), indicating that the *KEX2* enzyme does not require them for activity. This has also been demonstrated by fusing the αF signal sequence to the Lys-Arg residues of heterologous proteins. These fusions produce proteins that are processed by *KEX2* even though no Glu-Ala residues follow the cleavage site (Brake *et al.*, 1984; Singh *et al.*, 1983; Vlasuk *et al.*, 1986).

The phenotypes of *kex2* cells have important implications about protein processing in the Golgi. First, since *kex2* cells containing genes for the killer toxin precursor are immune to exogenous toxin, the *KEX2* cleavages hypothesized by Bostian *et al.* (1984) are not required for the immunity phenotype. Second, Bussey *et al.* (1983a) have reported that killer toxin is unstable in *kex2* cells, based on immune precipitations. The type of hyperglycosylation seen for the αF precursor, however, may also happen to the killer toxin precursor, and the large Asn-linked polymannose chains may interfere with antibody binding. This interpretation suggests that killer toxin is not degraded in the Golgi but is simply undetectable. This hypothesis could be easily tested by examining protoxin stability in a *kex2–mnn9* strain where outer chain addition is inhibited and antibody binding would be unhindered. Third, the hyperglycosylation of the αF precursor, and possibly the killer toxin precursor, in *kex2* cells suggests that the cleavage reaction occurs before the majority of the N-linked outer chain addition reactions. This leads one to speculate that the yeast Golgi may contain a spatial organization of enzymatic activities similar to that in mammalian cells (reviewed in Dunphy and Rothman, 1985).

2. Dipeptidyl Aminopeptidases

After the *KEX2* enzyme liberates the αF peptides from the precursor, the remaining amino-terminal extension of four to six amino acids is removed by DPAP A, which is thought to be the product of the *STE13* gene (Julius *et al.*, 1983). Yeast contains two membrane-bound enzymes with this activity, a heat-stable enzyme (DPAP A) and a heat-labile enzyme (DPAP B) (Julius *et al.*, 1983). Two genes have been cloned that can complement the *ste13* defect when present in a *ste13* cell on a multicopy plasmid; however, one of these complements only poorly. This poorly complementing gene probably encodes the vacuolar membrane protein DPAP B (Bordallo *et al.*, 1984). Stevens has suggested that DPAP B may represent a vacuolar protease that is mislocalized when overproduced on a multicopy plasmid (T. Stevens, personal communication), similar to the cases of carboxypeptidase Y and proteinase A (see above). If this proves to be the case and DPAP B is being mislocalized to the cell surface, it would be a unique example of an enzyme in transit that can act on other secreted proteins.

In a wild-type cell, the amount of DPAP A is limited. Thus, when the αF precursor is overproduced, a large amount of αF still contains some amino-terminal Glu-Ala or Asp-Ala residues (Julius *et al.*, 1983). A similar phenomenon has been noted in the expression of heterologous genes fused to the αF signal sequence and Glu-Ala residues; that is, a portion of the fusion protein which was secreted contained amino-terminal Glu-Ala residues (Brake *et al.*, 1984; Bitter *et al.*, 1984). It remains to be determined whether or not this limitation of DPAP A activity is involved in regulating the amount of pheromone secreted by MATα cells.

3. Other Processing Enzymes in the Golgi Apparatus

As mentioned above, a chymotrypsin-like or carboxypeptidase activity is required to remove the Lys-Arg residues at the carboxy terminus of the αF peptide after *KEX2* enzyme processing. Achstetter and Wolf (1985) have detected two membrane-bound peptidases that can cleave Cbz-Tyr-Lys-NA and Cbz-Tyr-Lys-Arg-NA (where Cbz is benzyloxycarbonyl and NA is 4-nitroanilide), two putative intermediates in αF processing after *KEX2* enzyme cleavage. Interestingly, an αF fusion of the precursor leader to human atrial natriuretic peptide (hANP) at the Lys-Arg site yielded hANP with a correct amino terminus and two carboxy termini. The hANP carboxy terminus of R-Phe-Arg-Tyr-COOH was cleaved to give R-Phe as the major product (Vlasuk *et al.*, 1986). The specific processing of the bond between a hydrophobic amino acid and a basic amino acid may reflect a normal aspect of αF maturation *in vivo*.

Additional enzymes that function in killer toxin secretion also remain to be isolated. The *kex1* mutation identifies a gene for one of these activities. Another processing protease has been suggested by experiments with cells treated with TPCK. Cells grown in the presence of this chymotrypsin inhibitor do not secrete active toxin (Bussey *et al.*, 1983a) and show delays in acid phosphatase secretion (Hanes *et al.*, 1986). While these results allow many interpretations, they are suggestive of a chymotrypsin-like activity in the Golgi that may be involved in posttranslational modification of secretory proteins.

D. Processing in the Periplasmic Space and the Extracellular Medium

Proteolytic processing may also occur beyond the plasma membrane. Certain aminopeptidases are thought to be localized in the yeast periplasm. These enzymes are involved in the exoproteolytic degradation of oligopeptides (reviewed in Jones, 1984). Another protease is encoded by the *SST1* (or *BAR1*) locus, which inactivates αF. This enzyme is made only in MATa cells and allows them to recover from αF-mediated arrest (Chan and Otte, 1982a,b; Finkelstein and Strausberg, 1979; Maness and Edelman, 1978; Sprague and Herskowitz, 1981). The final enzyme to be listed here is involved wtih the *SKI5* gene. The *ski5* mutation causes a 10-fold oversecretion of killer toxin and an oversecretion of αF (Bussey *et al.*, 1983b). In the case of killer toxin, treatment of wild-type cells with the serine protease inhibitor phenylmethylsuffonyl (PMSF) produces effects similar to the *ski5* mutation (Bussey *et al.*, 1983b). Thus the *ski5* gene may affect the expression of a serine protease at the cell surface. The cellular role of the *SKI5* enzyme is unknown at this time, but *ski5* cells are not supersensitive to either aF, αF, or other forms of killer toxin.

All of the enzymes described above are primarily degradative and do not appear to be involved in a specific maturation event as the *KEX2* and DPAPase A enzymes are. However, overproduction of the vacuolar enzyme CPY results in its secretion to the cell surface and its processing from proCPY to CPY, and this occurs even in mutant cells where this conversion is blocked in the vacuole (Stevens *et al.*, 1986). Pulse–chase investigations of this processing event indicate that the time required for production of substantial amounts of CPY greatly exceeds the time required for known proteins to transit the secretory pathway, suggesting that this event probably occurs in the periplasm. This reaction might be the result of one of the degradative enzymes described above fortuitously activating proCPY via proteolytic cleavage near the appropriate *in vivo* site. It is also possible that some periplasmic enzymes are converted from zymogens to their active forms in a similar manner, or that specific processing enzymes may exist in the periplasm space. Insight into these possibilities may come from analysis of the synthesis and maturation of the aminopeptidases, *SKI5* protein, and *BAR1* protein which populate the extracellular medium.

V. CONCLUSION

The current understanding of protein modification in the yeast secretory pathway consists primarily of the classification of various reactions and their subcellular location. Almost nothing is known about the control of these processes. For example, what is to prevent any of the various protease activities in the ER and Golgi, such as signal peptidase or the *KEX2* protease, from cleaving inappropriate substrates? What factors control the extent of glycosylation of vacuolar versus secreted enzymes? Some of these controls may be elucidated by work on the specific proteins that perform these modifications, while other forms of regulation may involve the sequestering of certain processing enzymes to different subcompartments of the Golgi body. For this purpose it will be interesting to determine if the yeast Golgi is organized in a fashion similar to higher eukaryotes. Along these lines, the recent development of an *in vitro* system that mimics ER to Golgi transport in yeast (Haselbeck and Schekman, 1986) and a higher eukaryotic *in vitro* intra-Golgi transport system that can use yeast components (Dunphy *et al.*, 1986) should be useful in approaching these problems.

The processing enzymes themselves raise interesting questions about the secretory apparatus. In addition to the various reactions they perform, these enzymes are most likely residents of specific organelles. Do the proteins involved in ER- and Golgi-associated reactions always remain in their respective organelles, and, if so, what prevents these enzymes from being secreted? One possibility is that these enzymes are somehow anchored via attachment to the yeast cytoskeleton and it is the internal architecture of the yeast cell that maintains the proper juxtaposition of secretory organelles and their enzymatic constitutents. In this context it is interesting to note that temperature-sensitive actin mutants show delays in invertase secretion under nonpermissive conditions (Novick and Botstein, 1985). Thus, the study of the secretion and processing of enzymes involved in protein modification should allow one to investigate new aspects of the secretory process.

ACKNOWLEDGMENTS

I would like to thank C. Buluwa and M. Kukuruzinska for their comments on the manuscript. I would also like to thank those investigators who communicated their results prior to publication.

REFERENCES

Achstetter, T., and Wolf, D. H. (1985). Hormone processing and membrane-bound protein-ases in yeast. *EMBO J.* **4,** 173–177.

Agrawal, H. C., Randle, C. L., and Agrawal, D. (1981). *In vivo* acylation of rat brain myelin proteolipid protein. *J. Biol. Chem.* **257,** 4588–4592.

Alighieri, D. (1321). "The Divine Comedy" (H. R. Huse, transl.). Rinehart, New York (1954).

Arima, K., Oshima, T., Kubota, I., Nakamura, N., Mizunaga, T., and Toh-e, A. (1983). The nucleotide sequence of the yeast *PHO5* gene: A putative precursor of repressible acid phosphatase contains a signal peptide. *Nucleic Acids Res.* **11,** 1657–1672.

Arnold, E., and Tanner, W. (1982). An obligatory role of protein glycosylation in the life cycle of yeast cells. *FEBS Lett.* **148,** 49–53.

Ballou, C. E. (1982). Yeast cell wall and cell surface. *In* "The Molecular Biology of the Yeast *Saccharomyces:* Metabolism and Gene Expression" (J. N. Strathern, E. W. Jones, and J. R. Broach, eds.), pp. 335–360. Cold Spring Harbor Laboratory, Cold Spring Harbor, New York.

Ballou, L., Cohen, R. E., and Ballou, C. E. (1980). *Saccharomyces cerevisiae* mutants that make mannoproteins with a truncated carbohydrate outerchain. *J. Biol. Chem.* **255,** 5985–5991.

Ballou, L., Gopal, P., Krummel, B., Tammi, M., and Ballou, C. (1986). A mutation that prevents glucosylation of the lipid-linked oligosaccharide precursor leads to undergly-cosylation of secreted yeast invertase. *Proc. Natl. Acad. Sci. U.S.A.* **83,** 3081–3085.

Bankaitis, V. A., Johnson, L. M., and Emr, S. (1986). Isolation of yeast mutants defective in protein targetting to the vacuole. *Proc. Natl. Acad. Sci. U.S.A.* **83,** 9075–9079.

Barnes, G., Hansen, W. J., Holcomb, C. L., and Rine, J. (1984). Asparagine-linked glyco-sylation in *Saccharomyces cerevisiae:* Genetic analysis of an early step. *Mol. Cell. Biol.* **4,** 2381–2388.

Bernstein, M., Hoffman, W., Ammerer, G., and Schekman, R. (1985). Characterization of a gene product (sec53p) required for protein assembly in the yeast endoplasmic reticu-lum. *J. Cell Biol.* **101,** 2374–2382.

Bitter, G. A., Chen, K. K., Banks, A. R., and Lai, P.-H. (1984). Secretion of foreign proteins from *Saccharomyces cerevisiae* directed by α-factor gene fusions. *Proc. Natl. Acad. Sci. U.S.A.* **81,** 5330–5334.

Blobel, G. (1980). Intracellular protein topogenesis. *Proc. Natl. Acad. Sci. U.S.A.* **77,** 1496–1500.

Blobel, G., and Dobberstein, B. (1975). Transfer of proteins across membranes: II. Recon-stitution of functional rough microsomes from heterologous components. *J. Cell Biol.* **67,** 852–862.

Bohni, P. C., Schauer, I., Tekamp-Olson, P., and Schekman, R. (in press). Signal peptide cleavage mutants of yeast invertase. *In* "Proteases in Biological Control and Biotech-nology" (UCLA–Cetus Symposium).

Bordallo, C., Schwencke, J., and Rendueles, M. S. (1984). Localization of the thermosensi-tive X-prolyl dipeptidyl aminopeptidase in the vacuolar membrane of *Saccharomyces cerevisiae. FEBS Lett.* **173,** 199–203.

Bostian, K. A., Hopper, J. E., Rogers, D. T., and Tipper, D. J. (1980). Translational analysis of the killer-associated virus-like particle dsRNA genome of *S. cerevisiae:* M dsRNA encodes toxin. *Cell* **19,** 403–414.

Bostian, K. A., Jayachandran, S., and Tipper, D. J. (1983). A glycosylated protoxin in killer yeast: Models for its structure and maturation. *Cell* **32,** 169–180.

Bostian, K. A., Elliott, Q., Bussey, H., Burn, V., Smith, A., and Tipper, D. J. (1984). Sequence of the preprotoxin dsRNA gene of type 1 killer yeast: Multiple processing events produce a two-component toxin. *Cell* **36,** 741–751.

Brake, A. J., Merryweather, J. P., Coit, D. G., Heberlein, U. A., Masiarz, F. R., Mullenbach, G. T., Urdea, M. S., Valenzuela, P., and Barr, P. J. (1984). α-Factor-directed synthesis and secretion of mature foreign proteins in *Saccharomyces cerevisiae*. *Proc. Natl. Acad. Sci. U.S.A.* **81,** 4642–4646.

Brands, R., Snider, M. D., Hiro, Y., Park, S. S., Gelboin, H. V., and Rothman, J. E. (1985). Retention of membrane proteins by the endoplasmic reticulum. *J. Cell Biol.* **101,** 1724–1732.

Bussey, H., Sacks, W., Galley, D., and Saville, D. (1982). Yeast killer plasmid mutations affecting toxin secretion and activity and toxin immunity function. *Mol. Cell. Biol.* **2,** 346–354.

Bussey, H., Saville, D., Greene, D., Tipper, D. J., and Bostian, K. A. (1983a). Secretion of yeast killer toxin: Processing of the glycosylated precursor. *Mol. Cell. Biol.* **3,** 1362–1370.

Bussey, H., Steinmetz, O., and Saville, D. (1983b). Protein secretion in yeast: Two chromosomal mutants that oversecrete killer toxin in *Saccharomyces cerevisiae*. *Curr. Genet.* **7,** 449–456.

Byrd, J. C., Tarentino, A. L., Maley, G., Atkinson, P. H., and Trimble, R. B. (1982). Glycoprotein synthesis in yeast: Identification of $Man_8GlcNAc_2$ as an essential intermediate in oligosaccharide processing. *J. Biol. Chem.* **257,** 14657–14666.

Carlson, M., Osmond, B. C., and Botstein, D. (1980). *suc* genes of yeast: A dispersed gene family. *Cold Spring Harbor Symp. Quant. Biol.* **49,** 799–803.

Carlson, M., Osmond, B. C., and Botstein, D. (1981). Mutants of yeast defective in sucrose utilization. *Genetics* **98,** 25–40.

Chambers, J., Forsee, W. T., and Elbein, A. D. (1977). Enzymatic transfer of mannose from mannosyl-phosphoryl-polyprenol to lipid-linked oligosaccharides by pig aorta. *J. Biol. Chem.* **252,** 2498–2506.

Chan, R., and Otte, C. (1982a). Isolation and genetic analysis of *Saccharomyces cerevisiae* mutants supersensitive to G_1 arrest by a-factor and α-factor pheromone. *Mol. Cell. Biol.* **2,** 11–20.

Chan, R., and Otte, C. (1982b). Physiological characterization of *Saccharomyces cerevisiae* mutants supersensitive to G_1 arrest by a-factor and α-factor pheromone. *Mol. Cell. Biol.* **2,** 21–29.

Chapman, A., Li, E., and Kornfeld, S. (1979). The biosynthesis of the major lipid-linked oligosaccharide of Chinese hamster ovary cells occurs by the ordered addition of mannose residues. *J. Biol. Chem.* **254,** 10243–10249.

Chou, P. Y., and Fasman, G. D. (1978). Prediction of the secondary structure of proteins from their amino acid sequence. *Adv. Enzymol.* **47,** 45–148.

Chu, F. K. (1986). Requirements of cleavage of high mannose oligosaccharides in glycoproteins by peptide *N*-glycosidase F. *J. Biol. Chem.* **261,** 172–177.

Cohen, R. E., Zhang, W., and Ballou, C. E. (1982). Effects of mannoprotein mutations on *Saccharomyces cerevisiae* core oligosaccharide structure. *J. Biol. Chem.* **255,** 7700–7707.

Couto, J. (1984). Ph.D. thesis, Massachusetts Institute of Technology, Cambridge.

Dunphy, W. G., and Rothman, J. E. (1985). Compartmental organization of the Golgi stack. *Cell* **42,** 13–21.

Dunphy, W. G., Ofeffer, S. R., Douglas, O. C., Wattenberg, B. W., Glick, B. S., and Rothman, J. E. (1986). Yeast and mammals utilize similar cytosolic components to drive protein transport through the Golgi complex. *Proc. Natl. Acad. Sci. U.S.A.* **83**, 1622–1626.

Emter, O., Mechler, B., Achstetter, T., Muller, H., and Wolf, D. H. (1983). Yeast pheromone α-factor is synthesized as a high molecular weight precursor. *Biochem. Biophys. Res. Commun.* **116**, 822–829.

Engelman, D. M., and Steitz, T. A. (1981). The spontaneous insertion of proteins into and across membranes: The helical hairpin hypothesis. *Cell* **23**, 411–422.

Esmon, B., Novick, P., and Schekman, R. (1981). Compartmentalized assembly of oligosaccharides on exported glycoproteins in yeast. *Cell* **25**, 451–460.

Esmon, B., Esmon, P. C., and Schekman, R. (1984). Early steps in processing of yeast glycoproteins. *J. Biol. Chem.* **259**, 10322–10327.

Ferro-Novick, S. Novick, P., Field, C., and Schekman, R. (1984a). Yeast secretory mutants that block the formation of active cell surface enzymes. *J. Cell Biol.* **98**, 35–43.

Ferro-Novick, S., Hansen, W., Schauer, I., and Schekman, R. (1984b). Genes required for completion of import of proteins into the endoplasmic reticulum in yeast. *J. Cell Biol.* **98**, 44–53.

Fields, S., Winter, G., and Brownlee, G. C. (1986). Structure of the neuramindase gene in human influenza virus A/PR/8/34. *Nature (London)* **290**, 213–217.

Finkelstein, D., and Strausberg, S. (1979). Metabolism of α-factor by *a* mating type cells of *Saccharomyces cerevisiae*. *J. Biol. Chem.* **254**, 796–803.

Forsee, W. T., Griffin, J. A., and Schutzback, J. S. (1977). Mannosyltransfer from GDP-mannose to oligosaccharide lipids. *Biochem. Biophys. Res. Commun.* **75**, 799–805.

Fries, E., and Rothman, J. (1980). Transport of vesicular stomatitis virus glycoprotein in a cell-free extract. *Proc. Natl. Acad. Sci. U.S.A.* **77**, 3870–3874.

Fujiyama, A., and Tamanoi, F. (1986). Processing and fatty acid acylation of *RAS1* and *RAS2* proteins in *Saccharomyces cerevisiae*. *Proc. Natl. Acad. Sci. U.S.A.* **83**, 1266–1270.

Gibson, R., Schlesinger, S., and Kornfeld, S. (1979). The nonglycosylated glycoproteins of vesicular stomatitis virus is temperature-sensitive and undergoes intracellular aggregation at elevated temperatures. *J. Biol. Chem.* **254**, 3600–3607.

Grossman, M. K., and Zimmermann, F. K. (1979). The structural genes of internal invertases in *Saccharomyces cerevisiae*. *Mol. Gen. Genet.* **175**, 223–229.

Haguenauer-Tsapis, R., and Hinnen, A. (1984). A deletion that includes the signal peptidase cleavage site impairs processing, glycosylation, and secretion of cell surface yeast acid phosphatase. *Mol. Cell. Biol.* **4**, 2668–2675.

Haguenauer-Tsapis, R., Nagy, M., and Ryter, A. (1986). A deletion that includes the segment coding for the signal peptidase cleavage site delays release of *Saccharomyces cerevisiae* acid phosphatase from the endoplasmic reticulum. *Mol. Cell. Biol.* **6**, 723–729.

Halegoua, S., and Inouye, M. (1980). Biosynthesis and assembly of the outer membrane proteins. *In* "Bacterial Outer Membranes" (M. Inouye, ed.), pp. 67–114. Wiley, New York.

Hanes, S. D., Burn, V. E., Sturley, S. L., Tipper, D. J., and Bostian, K. A. (1986). Expression of a cDNA derived from the yeast killer preprotoxin gene: Implications for processing and immunity. *Proc. Natl. Acad. Sci. U.S.A.* **83**, 1675–1679.

Haselbeck, A., and Schekman, R. (1986). Interorganelle transfer and glycosylation of yeast invertase *in vitro*. *Proc. Natl. Acad. Sci. U.S.A.* **83**, 2017–2021.

Haselbeck, A., and Tanner, W. (1982). Dolichol phosphate-mediated mannosyl transfer through liposomal membranes. *Proc. Natl. Acad. Sci. U.S.A.* **79,** 1520–1524.

Haselbeck, A., and Tanner, W. (1983). O-Glycosylation in *Saccharomyces cerevisiae* is initiated at the endoplasmic reticulum. *FEBS Lett.* **158,** 335–338.

Hashimoto, C., Cohen, R. E., Zhang, W., and Ballou, C. E. (1981). Carbohydrate chains on yeast carboxypeptidase Y are phosphorylated. *Proc. Natl. Acad. Sci. U.S.A.* **78,** 2244–2248.

Hasilik, A., and Tanner, W. (1978). Carbohydrate moiety of carboxypeptidase Y and perturbation of its synthesis. *Eur. J. Biochem.* **91,** 567–575.

Herskowitz, I., and Oshima, Y. (1981). Control of cell type in *Saccharomyces cerevisiae:* Mating type and mating type interconversion. *In* "The Molecular Biology of the Yeast *Saccharomyces:* Life Cycle and Inheritance" (J. N. Strathern, E. W. Jones, and J. R. Broach, eds.), pp. 181–210. Cold Spring Harbor Laboratory, Cold Spring Harbor, New York.

Hoffman, W. (1985). Molecular characterization of the *CAN1* locus in *Saccharomyces cerevisiae:* A transmembrane protein without N-terminal hydrophobic signal sequence. *J. Biol. Chem.* **260,** 11831–11837.

Hsieh, P., Rosner, M. R., and Robbins, P. W. (1983). Selective cleavage of endo-β-*N*-acetylglucosaminidase H at individual sites of Sindbis virion envelope glycoproteins. *J. Biol. Chem.* **258,** 2555–2561.

Hubbard, S. C., and Ivatt, R. J. (1981). Synthesis and processing of asparagine-linked oligosaccharides. *Annu. Rev. Biochem.* **50,** 555–583.

Huffaker, T. C., and Robbins, P. W. (1982). Temperature-sensitive yeast mutants deficient in asparagine-linked glycosylation. *J. Biol. Chem.* **257,** 3203–3210.

Huffaker, T. C., and Robbins, P. W. (1983). Yeast mutants deficient in protein glycosylation. *Proc. Natl. Acad. Sci. U.S.A.* **80,** 7466–7470.

Innis, M. A., Holland, M. J., McCabe, P. C., Cole, G. E., Wittman, V. P., Tal, R., Watt, K. W. K., Gelfand, D. H., Holland, J. P., and Meade, J. H. (1985). Expression, glycosylation, and secretion of an *Aspergillus* glucoamylase by *Saccharomyces cerevisiae*. *Science* **228,** 21–26.

Jelinek-Kelly, S., Akiyoma, T., Saunier, B., Tkacz, J. S., and Herscovics, A. (1985). Characterization of a specific α-mannosidase involved in oligosaccharide processing in *Saccharomyces cerevisiae*. J. Biol. Chem. **260,** 2253–2257.

Johnson, L. M., Bankaitis, V. A., and Emr, S. A genetic approach to study protein sorting and organelle assembly in yeast. *CRC Crit. Rev. J.* (in press).

Jones, E. W. (1984). The synthesis and function of proteases in *Saccharomyces:* Genetic approaches. *Annu. Rev. Genet.* **18,** 233–270.

Jones, E. W., Moehler, C., Kolodry, M., Aynardi, M., Park, F., Daniels, L., and Garlow, S. (1986). Genetics of vacuolar proteases. *In* "Yeast Cell Biology" (James Hicks, ed.), pp. 505–518. Alan R. Liss, New York.

Julius, D., Blair, L., Brake, A., Sprague, G., and Thorner, J. (1983). Yeast α-factor is processed from a larger precursor polypeptide: The essential role of a membrane-bound dipeptidyl aminopeptidase. *Cell* **32,** 839–852.

Julius, D., Schekman, R., and Thorner, J. (1984a). Glycosylation and processing of prepro-α-factor through the yeast secretory pathway. *Cell* **36,** 309–318.

Julius, D., Brake, A., Blair, L., Kunisawa, R., and Thorner, J. (1984b). Isolation of the putative structural gene for the lysine–arginine cleaving endopeptidase required for processing of yeast prepro-α-factor. *Cell* **37,** 1075–1089.

Kaiser, C. A., and Botstein, D. (1986). Secretion-defective mutations in the signal sequence for *Saccharomyces cerevisiae* invertase. *Mol. Cell. Biol.* **6,** 2382–2391.

Karson, E. M., and Ballou, C. E. (1978). Biosynthesis of yeast mannan: Properties of a mannosylphosphate transferase in *Saccharomyces cerevisiae*. *J. Biol. Chem.* **253**, 6484–6492.

Kaufman, J. F., Krangel, M. S., and Streminger, J. L. (1984). Cysteines in the transmembrane region of major histocompatibility complex antigens are fatty acylated via thioester bonds. *J. Biol. Chem.* **259**, 7230–7238.

Kilker, R. D., Saunier, B., Tkacz, J. S., and Herskovics, A. (1981). Partial purification from *Saccharomyces cerevisiae* of a soluble glucosidase which removes the terminal glucose from the oligosaccharide $Glc_3Man_9GlcNAc$. *J. Biol. Chem.* **256**, 5299–5303.

Klebl, F., Huffaker, T., and Tanner, W. (1984). A temperature-sensitive N-glycosylation mutant of *S. cerevisiae* that behaves like a cell-cycle mutant. *Exp. Cell Res.* **150**, 309–313.

Kornfeld, R., and Kornfeld, S. (1985). Assembly of asparagine-linked oligosaccharides. *Annu. Rev. Biochem.* **54**, 631–664.

Kuo, S., and Lampen, J. (1974). Tunicamycin, an inhibitor of yeast glycoprotein synthesis. *Biochem. Biophys, Res. Commun.* **58**, 287–295.

Kurjan, J., and Herskowitz, I. (1982). Structure of a yeast pheromone gene (MFα): A putative α-factor precursor contains four tandem copies of mature α-factor. *Cell* **30**, 933–943.

Leavitt, R., Schlessinger, S., and Kornfeld, S. (1977). Impaired intracellular migration and altered solubility of nonglycosylated glycoproteins of vesicular stomatitis virus and Sindbis virus. *J. Biol. Chem.* **252**, 9018–9023.

Lehle, L. (1980). Biosynthesis of the core region of yeast mannoproteins: Formation of a glycosylated dolichol-bound oligosaccharide precursor, its transfer to protein and subsequent modification. *Eur. J. Biochem.* **109**, 589–601.

Lehle, L., and Bause, E. (1984). Primary structural requirements for N- and O-linked glycosylation of yeast mannoproteins. *Biochim. Biophys. Acta* **799**, 246–251.

Lehle, L., and Tanner, W. (1976). The specific site of tunicamycin inhibition in the formation of dolichol-bound *N*-acetylglucosamine derivatives. *FEBS Lett.* **71**, 167–170.

Lehle, L., Cohen, R. E., and Ballou, C. (1979). Carbohydrate structure of yeast invertase. *J. Biol. Chem.* **254**, 12209–12218.

Leibowitz, M. J., and Wickner, R. B. (1976). A chromosomal gene required for killerplasmid expression, mating and sporulation in *Saccharomyces cerevisiae*. *Proc. Natl. Acad. Sci. U.S.A.* **73**, 2061–2065.

Leloir, L. F., Staneloni, R. J., Carminatti, H., and Behrens, N. H. (1973). The biosynthesis of a *N,N'*-diacetylchitobiose containing lipid by liver microsomes: A probable dolichol pyrophosphate derivative. *Biochem. Biophys. Res. Commun.* **52**, 1285–1292.

Liu, T., Stetson, B., Turco, S. J., Hubbard, S. C., and Robbins, P. W. (1979). Arrangement of glucose residues in the lipid-linked oligosaccharide precursor of asparaginyl oligosaccharides. *J. Biol. Chem.* **254**, 4554–4559.

Lolle, S., Skipper, N., Bussey, H., and Thomas, D. Y. (1984). The expression of cDNA clones of yeast M1 double-stranded RNA in yeast confers both killer and immunity phenotypes. *EMBO J.* **3**, 1383–1387.

McCloskey, M. A., and Troy, F. A. (1980). Paramagnetic isoprenoid carrier lipids. 2. Dispersion and dynamics in lipid membranes. *Biochemistry* **19**, 2061–2066.

Maness, P., and Edelman, G. (1978). Inactivation and chemical alteration of mating factor α by cells and spheroplasts of yeast. *Proc. Natl. Acad. Sci. U.S.A.* **75**, 1304–1308.

Marriot, M., and Tanner, W. (1979). Localization of dolichyl-phosphate and -pyrophosphate dependent glycosyl transfer reactions in *Saccharomyces cerevisiae*. *J. Bacteriol.* **139**, 565–572.

Meyer, D. I., Krause, E., and Dobberstein, B. (1982). Secretory protein translocation across membranes—the role of the "docking protein." *Nature (London)* **297,** 647–650.

Molnar, J., Chao, H., and Hartford, J. B. (1971). Phosphoryl-*N*-acetylglucosamine transfer to a lipid acceptor of liver microsomal preparations. *Biochim. Biophys. Acta* **239,** 401–410.

Mueckler, M., and Lodish, H. F. (1986). The human glucose transporter can insert posttranslationally into microsomes. *Cell* **44,** 629–637.

Nakajima, T., and Ballou, C. E. (1974a). Characterization of the carbohydrate fragments obtained from *Saccharomyces cerevisiae* mannan by alkaline degradation. *J. Biol. Chem.* **249,** 7679–7684.

Nakajima, T., and Ballou, C. E. (1974b). Structure of the linkage region between the polysaccharide and protein parts of *Saccharomyces cerevisiae* mannan. *J. Biol. Chem.* **249,** 7685–7694.

Narasimhan, S. (1982). Control of glycoprotein synthesis: UDP–GlcNAc:glycopeptide β4-*N*-acetylglucosaminlytransferase III, an enzyme in hen oviduct which adds GlcNAc in β1,4 linkage to the β-linked mannose of the trimannosyl core of N-glycosyl oligosaccharides. *J. Biol. Chem.* **257,** 10235–10242.

Novick, P., and Botstein, D. (1985). Phenotypic analysis of temperature-sensitive yeast actin mutants. *Cell* **40,** 405–416.

Novick, P., Field, C., and Schekman, R. (1980). Identification of 23 complementation groups required for posttranslational events in the yeast secretory pathway. *Cell* **21,** 205–215.

Novick, P., Ferro, S., and Schekman, R. (1981). Order of events in the yeast secretory pathway. *Cell* **25,** 461–469.

Olden, K., Bernard, B. A., Humphries, M. J., Yeo, T.-K., Yeo, K.-T., White, S. L., Newton, S. A., Bauer, H. C., and Parent, J. B. (1985). Function of glycoprotein glycans. *Trends Biochem. Sci.* **10,** 78–82.

Onishi, H. R., Tkacz, J. S., and Lampen, J. O. (1979). Glycoprotein nature of yeast alkaline phosphatase: Formation of active enzyme in the presence of tunicamycin. *J. Biol. Chem.* **254,** 11943–11952.

Palade, G. (1975). Intracellular aspects of the process of protein synthesis. *Science* **189,** 347–358.

Parodi, A. J. (1979). Biosynthesis of yeast mannoproteins: Synthesis of mannose outerchain and of dolichol dervitives. *J. Biol. Chem.* **254,** 8343–8352.

Perlman, D., and Halvorson, H. O. (1981). Distinct repressible mRNAs for cytoplasmic and secreted yeast invertase are encoded a single gene. *Cell* **25,** 525–536.

Perlman, D., and Holvorson, H. O. (1983). A putative signal pepidase recognition site and sequence in eukaryotic and prokaryotic signal peptides. *J. Mol. Biol.* **167,** 391–409.

Perlman, D., Holvorson, H. O., and Cannon, C. E. (1982). Presecretory and cytoplasmic invertase polypeptides encoded by distinct mRNAs derived from the same structural gene differ by a signal sequence. *Proc. Natl. Acad. Sci. U.S.A.* **79,** 781–785.

Popolo, L., Vai, M., and Alberghina, L. (1986). Identification of a glycoprotein involved in cell cycle progression in yeast. *J. Biol. Chem.* **261,** 3470–3482.

Prakash, C., and Vijay, I. K. (1982). Characterization of intermediates up to lipid-linked heptasaccharide implicated in the biosynthesis of *Saccharomyces cerevisiae* mannoproteins. *Biochemistry* **21,** 4810–4818.

Pringle, J. R., and Hartwell, L. H. (1981). The *Saccharomyces cerevisiae* cell cycle. *In* "The Molecular Biology of the Yeast *Saccharomyces:* Life Cycle and Inheritance" (J. N. Strathern, E. W. Jones, and J. R. Broach, eds.), pp. 97–142. Cold Spring Harbor Laboratory, Cold Spring Harbor, New York.

Raschke, W. C., Kern, K. A., Antalise, C., and Ballou, C. E. (1973). Genetic control of yeast mannan structure: Isolation and characterization of mutants. *J. Biol. Chem.* **248,** 4660–4666.

Rearick, J. I., Fujimoto, K., and Kornfeld, S. (1981). Identification of the mannosyl donors involved in the synthesis of lipid-linked oligosaccharides. *J. Biol. Chem.* **256,** 3762–3769.

Rine, J., Hansen, W., Hardeman, E., and Davis, R. (1983). Targeted selection of recombinant clones through gene dosage effects. *Proc. Natl. Acad. Sci. USA* **80,** 6750–6754.

Rogers, D. T., Saville, D., and Bussey, H. (1979). *Saccharomyces cerevisiae* killer expression mutant *kex2* has altered secretory proteins and glycoproteins. *Biochem. Biophys. Res. Commun.* **90,** 187–193.

Rogers, D. T., Lemire, J. M., and Bostian, K. A. (1982). Acid phosphatase polypeptides in *Saccharomyces cerevisiae* are encoded by a differentially regulated multigene family. *Proc. Natl. Acad. Sci. U.S.A.* **79,** 2157–2161.

Rothblatt, J. A., and Meyer, D. I. (1986). Secretion in yeast: Reconstitution of the translocation and glycosylation of α-factor and invertase in a homologous cell-free system. *Cell* **44,** 619–628.

Rothman, J. E., and Lodish, H. F. (1977). Synchronized transmembrane insertion and glycosylation of a nascent membrane protein. *Nature (London)* **269,** 775–780.

Rothman, J. H., Hunter, C. P., Valls, L. A., and Stevens, T. H. (1986). Overproduction-induced mislocation of a yeast vacuolar protein allows isolation of its structural gene. *Proc. Natl. Acad. Sci. U.S.A.* **83,** 3248–3252.

Runge, K. W., and Robbins, P. W. (1986b). *Saccharomyces cerevisiae* mutants in the early stages of protein glycosylation. *Microbiology* **1986,** 312–316.

Runge, K. W., and Robbins, P. W. (1986a). *alg8,* A new yeast mutation in the glycosylation steps of the asparagine glycosylation pathway: Formation of a novel Glc_2 oligosaccharide *J. Biol. Chem.* **261,** 15582–15590.

Runge, K. W., Huffaker, T. C., and Robbins, P. W. (1984). Two yeast mutations in the glycosylation steps of the asparagine glycosylation pathway. *J. Biol. Chem.* **259,** 412–417.

Saunier, B., Kilker, R. D., Tkacz, J. S., Quaroni, A., and Herscovics, A. (1982). Inhibition of N-linked complex oligosaccharide formation by 1-deoxynorjirimycin, an inhibitor of processing glucosidases. *J. Biol. Chem.* **257,** 14155–14161.

Schauer, I., Emr, S., Gross, C., and Schekman, R. (1985). Invertase signal and mature sequence substitutions that delay intercompartmental transport of active enzyme. *J. Cell. Biol.* **100,** 1664–1675.

Schlesinger, M. J., Magee, A. I., and Schmidt, M. F. G. (1980). Fatty acid acylation of proteins in cultured cells. *J. Biol. Chem.* **255,** 10021–10024.

Schmidt, M. F. G. (1982). Acylation of viral spike glycoproteins: A feature of enveloped RNA viruses. *Virology* **116,** 327–338.

Schmidt, M. F. G., and Schlesinger, M. J. (1979). Fatty acid binding to vesicular stomatitis virus glycoprotein: A new type of posttranslational modification of the viral glycoprotein. *Cell* **17,** 813–819.

Schmidt, M. F. G., Bracha, M., and Schlesinger, M. J. (1979). Evidence for covalent attachment of fatty acids to Sindbis virus glycoproteins. *Proc. Natl. Acad. Sci. U.S.A.* **76,** 1687–1691.

Schönholzer, F., Schweingruber, A.-M. Trachsel, H., and Schweingruber, M. E. (1985). Intracellular maturation and secretion of acid phosphatase of *Saccharomyces cerevisiae. Eur. J. Biochem.* **147,** 273–279.

Schwaiger, H., Hasilik, A., von Figura, K., Wiemken, A., and Tanner, W. (1982). Carbohydrate-free carboxypeptidase Y is transferred into the lysosome-like yeast vacuole. *Biochem. Biophys. Res. Commun.* **104**, 950–956.

Sharma, C. B., Babczinski, P., Lehle, L., and Tanner, W. (1974). The role of dolichol monophosphate in glycoprotein biosynthesis in *Saccharomyces cerevisiae*. *Eur. J. Biochem.* **46**, 35–41.

Shih, T. Y., Weeks, M. O., Gruss, P., Dhar, R., Oroszlan, S., and Scholnick, E. M. (1982). Identification of a precursor in the biosynthesis of the p21 transforming protein of Harvey murine sarcoma virus. *J. Virol.* **42**, 253–261.

Singh, A., Chen, E. Y., Lugovoy, J. M., Chang, C. N., Hitzeman, R. A., and Seeburg, P. H. (1983). *Saccharomyces cerevisiae* contains two discrete genes coding for the α-factor pheromone. *Nucleic Acids Res.* **11**, 4049–4063.

Skipper, N., Thomas, D. Y., and Lau, P. C. K. (1984). Cloning and sequencing of the preprotoxin-coding region of the yeast M1 double-stranded RNA. *EMBO J.* **3**, 107–111.

Sly, W. S., and Fischer, H. D. (1982). The phosphomannosyl recognition system for intercellular and intracellular transport of lysosomal enzymes. *J. Cell. Biochem.* **18**, 67–85.

Snider, M., and Rogers, O. (1984). Transmembrane movement of oligosaccharide–lipids during glycoprotein synthesis. *Cell* **36**, 753–761.

Sprague, G., Jr., and Herskowitz, I. (1981). Control of yeast cell type by the mating type locus. I. Identification and control of expression of the *a*-specific gene, *BAR1. J. Mol. Biol.* **153**, 305–321.

Staneloni, R. J., Ugalde, R. A., and Leloir, L. F. (1980). Addition of glucose to dolichyl diphosphate oligosaccharide and transfer to protein. *Eur. J. Biochem.* **105**, 275–278.

Stevens, T., Esmon, B., and Schekman, R. (1982). Early stages in the yeast secretory pathway are required for transport of carboxypeptidase Y to the vacuole. *Cell* **30**, 439–448.

Stevens, T. H., Rothman, J. H., Payne, G. S., and Schekman, R. (1986). Gene dosage-dependent secretion of yeast vacuolar carboxypeptidase Y. *J. Cell. Biol.* **102**, 1551–1557.

Tabas, I., and Kornfeld, S. (1980). Biosynthetic intermediates of β-glucouronidase contain high mannose oligosaccharides with blocked phosphate residues. *J. Biol. Chem.* **255**, 6633–639.

Tai, T., Yamashita, K., and Kobata, A. (1977). The substrate specificities of endo-β-N-acetylglucosaminidases C_{II} and H. *Biochem. Biophys. Res. Commun.* **78**, 434–441.

Tarentino, A. L., Gomez, C. M., and Plumner, T. H. (1985). Deglycosylation of asparagine-linked glycans by peptide:N-glycosidase F. *Biochemistry* **24**, 4665–4671.

Thorner, J. (1981). Pheromonal regulation of development in *Saccharomyces ceresiviae*. *In* "The Molecular Biology of the Yeast *Saccharomyces:* Life Cycle and Inheritance" (J. N. Strathern, E. W. Jones, and J. R. Broach, eds.), pp. 143–180. Cold Spring Harbor Laboratory, Cold Spring Harbor, New York.

Toh-e, A., Ueda, Y., Kakimoto, S., and Oshima, Y. (1973). Isolation and characterization of acid phosphatase mutants in *Saccharomyces cerevisiae*. *J. Bacteriol.* **113**, 727–738.

Trimble, R. B., and Atkinson, P. H. (1986). Structure of yeast external invertase $Man_{8-14}GlcNAc$ processing intermediates by 500–megahertz ^1H spectroscopy. *J. Biol. Chem.* **261**, 9815–9824.

Trimble, R. B., and Maley, F. (1977). Subunit structure of external invertase from *Saccharomyces cerevisiae*. *J. Biol. Chem.* **252**, 4409–4412.

Trimble, R. B., Maley, F., and Tarentino, A. L. (1980a). Characterization of large oligosaccharide–lipids synthesized *in vitro* by microsomes from *Saccharomyces cerevisiae*. *J. Biol. Chem.* **255**, 10232–10238.

Trimble, R. B., Byrd, J. C., and Maley, F. (1980b). Effect of glucosylation of lipid intermediates on oligosaccharide transfer in solubilized microsomes from *Saccharomyces cerevisiae*. *J. Biol. Chem.* **255,** 11892–11895.

Trimble, R. B., Maley, F., and Chu, F. K. (1983). Glycoprotein biosynthesis in yeast: Protein conformation affects processing of high mannose oligosaccharides on carboxypeptidase Y and invertase. *J. Biol. Chem.* **258,** 2562–2567.

Tsai, P.-K., Ballou, L., Esmon, B., Schekman, R., and Ballou, C. (1984a). Isolation of glucose-containing high-mannose glycoprotein core oligosaccharides. *Proc. Natl. Acad. Sci. U.S.A.* **81,** 6340–6343

Tsai, P.-K., Frevert, J., and Ballou, C. E. (1984b). Carbohydrate structure of *Saccharomyces cerevisiae mnn9* mannoprotein. *J. Biol. Chem.* **259,** 3805–3811.

Turco, S. J., Stetson, B., and Robbins, P. W. (1977). Comparative rates of transfer of lipid-linked oligosaccharides to endogenous glycoprotein acceptors *in vitro*. *Proc. Natl. Acad. Sci. U.S.A.* **74,** 4411–4414.

Verma, A. K., Raizada, M. K., and Schutzback, J. S. (1977). Formation of α1,2-mannosyl-mannose by an enzyme preparation from rabbit liver. *J. Biol. Chem.* **252,** 7235–7242.

Villemez, C., and Carlo, P. (1980). Properties of a soluble polyprenyl phosphate: UDP–D-*N*-acetylglucosamine *N*-acetylglucosamine-1-phosphate transferase. *J. Biol. Chem.* **255,** 8174–8178.

Vlasuk, G. P., Bencen, G. H., Scarborough, R. M., Tsai, P.-K., Whang, J. L., Maack, T., Carmargo, J. F., Kirsher, S. W., and Abraham, J. A. (1986). Expression and secretion of biologically active human atrial natriuretic peptide in *Saccharomyces cerevisiae*. *J. Biol. Chem.* **261,** 4789–4796.

von Heijne, G. (1983). Patterns of amino acids near signal-sequence cleavage sites. *Eur. J. Biochem.* **133,** 17–21.

von Heijne, G., and Blomberg, C. (1979). Transmembrane translocation of proteins: The direct transfer model. *Eur. J. Biochem.* **97,** 175–181.

Waechter, C. J., and Hartford, J. B. (1979). A dolichol-linked trisaccharide from central nervous tissue: Structure and biosynthesis. *Arch. Biochem. Biophys.* **192,** 380–390.

Walter, P., and Blobel, G. (1982). Signal recognition particle contains a 7S RNA essential for protein translocation across the endoplasmic reticulum. *Nature (London)* **299,** 691–698.

Walter, P., Gilmore, R., and Blobel, G. (1984). Protein translocation across the endoplasmic reticulum. *Cell* **38,** 5–8.

Waters, M. G., and Blobel, G. (1986). Secretory protein translocation in a yeast cell-free system can occur posttranslationally and requires ATP hydrolysis. *J. Cell Biol.* **102,** 1543–1550.

Weeks, M. O., Hager, G. L., Loe, R., and Scolnick, E. M. (1985). Development and analysis of a transformation-defective mutant of Harvey murine sarcoma tk virus and its gene product. *J. Virol.* **54,** 586–597.

Wen, D., and Schlesinger, M. J. (1984). Fatty acid-acylated protiens in secretory mutants of *Saccharomyces cerevisiae*. *Mol. Cell. Biol.* **4,** 688–694.

Wickner, R. B., and Leibowitz, M. J. (1976). Two chromosomal genes required for killing expression in killer strains of *Saccharomyces cerevisiae*. *Genetics* **82,** 429–442.

Wickner, W. (1979). The assembly of proteins into biological membranes: The membrane trigger hypothesis. *Annu. Rev. Biochem.* **48,** 23–45.

Wickner, W., and Lodish, H. F. (1985). Multiple mechanisms of protein insertion into and across membranes. *Science* **230,** 400–407.

Williams, R. S., Trumbly, R. J. MacColl, R., Trimble, R. B., and Maley, F. (1985). Comparative properties of amplified external and internal invertase from the yeast *SUC2* gene. *J. Biol. Chem.* **260,** 13334–13341.

Willumsen, B., Christensen, A., Hubbert, N. L., Papageorge, A. G., and Lowy, D. (1984). The p21 *ras* C-terminus is required for transformation and membrane association. *Nature (London)* **310,** 583–586.

Yamamoto, A., Masaki, R., and Tashiro, Y. (1985). Is cytochrome P-450 transported from the endoplasmic reticulum to the Golgi apparatus in rat hepatocytes? *J. Cell Biol.* **101,** 1733–1740.

5

Mannosidases in Mammalian Glycoprotein Processing

KELLEY W. MOREMEN AND OSCAR TOUSTER

I. INTRODUCTION

Processing of Asparagine-Linked Oligosaccharides: Involvement of α-Mannosidases

Most secretory and membrane proteins contain carbohydrate moieties covalently attached through an amide linkage to asparagine residues. This has been shown to occur through the en bloc trasfer of a preformed oligosaccharide from a lipid-linked donor to specific glycosylation sites containing the required Asn-X-Ser(Thr) acceptor consensus sequence (for reviews, see Hubbard and Ivatt, 1981; Kornfeld and Kornfeld, 1985; Struck and Lennarz, 1980). The mature lipid-linked oligosaccharide,[1] containing a $Glc_3Man_9GlcNAc_2$ structure (Fig. 1), is transferred cotranslationally to the nascent polypeptide chain on the luminal face of the rough

---

[1] Abbreviations: Glc, glucose; Man, mannose; GlcNAc, N-acetylglucosamine; Gal, galactose.

PROTEIN TRANSFER AND ORGANELLE BIOGENESIS

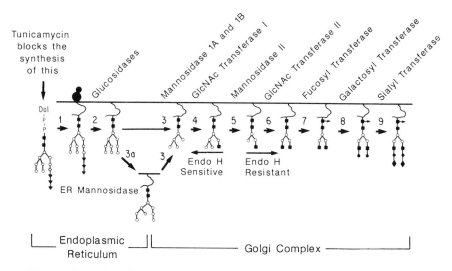

Fig. 1. Processing of Asn-linked oligosaccharides to a representative biantennary complex structure. The scheme depicts the processing from the transfer of $Glc_3Man_9GlcNAc_2$ from its dolichol pyrophosphoryl derivative to the nascent polypeptide chain still bound to the ribosome, followed by processing reactions in the ER and Golgi complex. Oligosaccharide processing enzymes are listed above the line; the reaction they catalyze is diagramed below the line (except for the alternate processing reaction, 3a). The subcellular localization of processing events are depicted by the brackets. Structures susceptible or resistant to digestion by endo H, an enzyme frequently used as a diagnostic test for processing to complex structures, are indicated. Symbols: ▼, glucose; ○, mannose; ■, *N*-acetylglucosamine; ●, galactose; ▶, fucose; ◆, sialic acid.

endoplasmic reticulum (RER) (Struck and Lennarz, 1980). The luminal orientation of these oligosaccharides results in a defined luminal polarity for all subsequent oligosaccharide processing events. Oligosaccharides of the nascent glycoproteins can then be modified from this common precursor to a variety of mature structures. Minimal modification will result in oligosaccharides containing mannose-terminal branches (high-mannose-type), while more extensive modification may result in structures containing GlcNAc, Gal, fucose, and sialic acid characteristic of complex-type oligosaccharides (Fig. 1). Partial processing of one or more oligosaccharide branches may also result in both mannose-terminal and complex-type modifications on different branches of the same oligosaccharide (hybrid type) (for review, see Kornfeld and Kornfeld, 1985).

Modification of the common precursor to form these mature structures begins almost immediately after the cotranslational oligosaccharide transfer. The outer α-1,2-glucose and the inner two α-1,3-glucosyl residues are removed by glucosidases I and II, respectively. Each of these enzymes has been purified to homogeneity and has been localized to the mem-

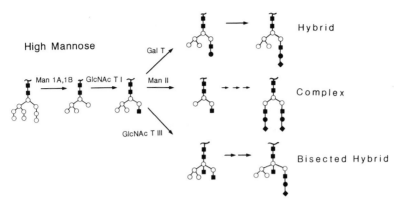

Fig. 2. The processing reactions of high-mannose oligosaccharides to hybrid, complex, or bisected hybrid structures. Enzymes catalyzing the above reactions are mannosidases IA and IB (Man 1A, 1B), GlcNAc transferase I (GlcNAc T I), galactosyltransferase (Gal T), mannosidase II (Man II), and GlcNAc transferase III (GlcNAc T III).

branes of the RER (Burns and Touster, 1982; Brada and Dubach, 1984; Hettkamp *et al.*, 1984; Hino and Rothman, 1985). The removal of a single α-1,2-mannosyl residue on some glycoproteins by the action of an ER mannosidase apparently occurs prior to the transfer of these molecules to the Golgi complex (see Section II,A). Further trimming of α-1,2-mannosyl residues proceeds in the Golgi complex by the action of mannosidase I (see Section II,B). This enzyme is responsible for the complete removal of α-1,2-mannosyl residues to form $Man_5GlcNAc_2$. Removal of the final two mannosyl residues to yield the $Man_3GlcNAc_2$ core structure present on complex-type oligosaccharides requires the prior addition of a GlcNAc residue to the 2 position of the mannose on the α-1,3 branch of $Man_5GlcNAc_2$ (see Fig. 1). This transfer is accomplished by GlcNAc transferase I, an enzyme which has been purified and characterized from bovine colostrum (Harpaz and Schachter, 1980a) and rabbit liver (Oppenheimer and Hill, 1981). After the formation of $GlcNAcMan_5GlcNAc_2$, a second Golgi mannosidase, mannosidase II (see Section II,C) catalyzes the removal of the two remaining terminal α-1,3- and α-1,6-linked mannosyl residues. The rigid substrate specificity of this enzyme therefore results in an overlap of the trimming phase of complex-type oligosaccharide synthesis with the subsequent elongation phase.

As shown in Fig. 2, the formation of $GlcNAcMan_5GlcNAc_2$ by GlcNAc transferase I is the committed step in the synthesis of either hybrid or complex structures while mannosidase II catalyzes the committed step toward exclusively complex-type structures. Further elaboration of hybrid and complex branches by the addition of GlcNAc, Gal, fucose, and sialic acid results in the varied mature structures found in mammalian glycoproteins. The order of addition and substrate requirements for many of these terminal transferase reactions have been established (Schachter *et al.*, 1983), and several of the enzymes have been purified to homogeneity (for review, see Kornfeld and Kornfeld, 1985).

With the exception of the two glucosidases and the ER mannosidase, all of the enzymes involved in the later stages of oligosaccharide processing were found to be enriched in the membranes of the Golgi complex (Dunphy and Rothman, 1985). The recent use of immunolocalization techniques has also revealed that several of these activities appear to be enriched in discrete subregions of the Golgi elements (for review, see Dunphy and Rothman, 1985).

The α-mannosidase activities involved in processing of nascent high-mannose oligosaccharides to complex-type structures can therefore be reduced to three steps. Evidence will be presented that the removal of one or more α-1,2-mannosyl residues from some nascent oligosaccharides may occur in the RER. This cleavage is followed by a complete removal of α-1,2-mannosyl residues by mannosidase I in the Golgi complex. Finally, α-1,3- and α-1,6-mannosyl residues are removed by mannosidase II following the prior addition of a GlcNAc residue to a specific position by GlcNAc transferase I. A more detailed examination of these three enzyme systems will be presented in this chapter along with a brief discussion of other alternate enzyme systems.

II. PROCESSING MANNOSIDASES AND MULTIPLE ROUTES

A. Endoplasmic Reticulum and Cytosolic α-Mannosidases

1. General Aspects

The first evidence for the presence of a neutral, nonlysosomal mannosidase which could be assayed with the synthetic substrate *p*-nitrophenyl-α-D-mannoside was presented by March and Gourlay (1971). This enzyme was found to differ significantly from the lysosomal mannosidase and subsequently from an α-mannosidase activity of Golgi membranes (see Table I). It was characterized by an extreme instability during purification, a pH optimum significantly higher than the previously characterized mannosidase activities and a requirement for divalent cations for activity and/or stability (Shoup and Touster, 1976). Utilizing the stabilizing effect of the competitive inhibitor mannosylamine, Shoup and Touster (1976) purified the enzyme from a rat liver high-speed supernatant fraction 12,000-fold to 90% purity and characterized it as an oligomeric enzyme with a subunit molecular weight of 110,000. Sucrose gradient centrifugation and gel filtration suggested that the enzyme was a tetramer of four

TABLE I

Properties of Rat α-Mannosidases

Parameter	ER/cytosolic mannosidase[a]	Mannosidase IA and IB[b]	Mannosidase II[c]	Brain microsomal mannosidase[d]	Lysosomal mannosidase[e]
Subunit M_r	110,000[f]	56,000–58,000[g]	124,000	?	80,000, 55,000, 40,000
Oligomeric M_r	420,000[f]	230,000[j]	~300,000	?	200,000–335,000
Natural substrates	M_9N[h,i]	$M_{9-6}N$	NM_5N	$M_{9-4}N$	Linear $M_{4-2}N$[k]
Inhibitors	None	Deoxymannojirimycin	Swainsonine	None	Swainsonine
pH optimum	6.0	6.0	5.6	6.0	4.6
K_m	0.14mM (pNPM substrate)	0.1 mM (M_9N substrate)	30 mM (pNPM substrate)	?	10 mM (pNPM substrate)
Activity toward pNPM	+	–	+	–	+
Purified to homogeneity	+	+[g,j]	+	–	+
Immunologically distinct	–[l]	+	+	–[l]	+
Glycoprotein	–	+	+	–	+

[a] Shoup and Touster (1976), Bischoff and Kornfeld (1983).

[b] Tulsiani and Touster (1986), Tulsiani et al. (1982b), Tabas and Kornfeld (1979).

[c] Tulsiani et al. (1977, 1982a,b), Moremen and Touster (1985).

[d] Tulsiani and Touster (1985).

[e] Opheim and Touster (1978).

[f] These values refer to the soluble/cytosolic α-mannosidase (Shoup and Touster, 1976).

[g] These values refer to mannosidase IA (Tulsiani and Touster, 1986).

[h] abbreviations; M, mannose; N, N-acetylglucosamine; pNPM, p-nitrophenylmannoside.

[i] Based on in vivo studies.

[j] D. R. P. Tulsiani and O. Touster (unpublished observations).

[k] Tulsiani and Touster (1987).

[l] Antibody to the rat liver cytosolic mannosidase (Shoup and Touster, 1976) reacts with the ER mannosidase and brain microsomal mannosidase.

identical subunits. The enzyme had an unusually low K_m for the synthetic substrate p-nitrophenyl mannoside (0.14 mM) relative to the lysosomal and Golgi enzymes (10 and 30 mM, respectively) and differed also in its inability to bind to concanavalin A (Con A) and cross-react with antiserum raised against the other two enzymes (see Table I).

The physiological role of this mannosidase activity in the cytosol is, however, unclear in view of the fact that the known metabolic processes requiring mannosidase activities are restricted to the membrane systems of the Golgi complex, lysosomes, and possibly the endoplasmic reticulum. Cell fractionation of rat liver by the method of DeDuve *et al.* (1955) demonstrated that less than 5% of the pH 6.5 mannosidase activity could be found in sedimentable, membrane-bound form (Dewald and Touster, 1973). These results obviously suggested that the enzyme was cytosolic in origin. The role of the enzyme *in vivo* therefore remained in question.

In 1983, with a growing body of evidence for the existence of a previously undetected mannosidase activity in the membranes of the ER (see Section III,A,2), Bischoff and Kornfeld (1983), using the rat liver cell fractionation procedure of Kreibich *et al.* (1978) to isolate membranes of the endoplasmic reticulum, found a mannosidase activity similar to that of the cytoplasmic enzyme in a washed ER membrane preparation. This residual activity in the membranes, however, required 0.2% deoxycholate for solubilization. Although the membrane-bound activity could have been at most 25% of the total cytosolic-like mannosidase activity (i.e., ER plus cytosol) even when protease inhibitors were added to the homogenization medium, Bischoff and Kornfeld hypothesized, by analogy to other ER proteins which are readily released from membranes during homogenization, that the soluble cytosolic form was generated by proteolytic degradation of the membrane-bound enzyme.

Glucosidase II in rat liver is also found in both soluble and membrane-bound forms when the tissue is homogenized in the absence of protease inhibitors (Burns and Touster, 1982) but is found almost exclusively in the membrane-bound form in the presence of protease inhibitors (Brada and Dubach, 1984). Hydroxymethylglutaryl-coenzyme A reductase (HMG-CoA reductase), an integral membrane protein of the ER, is also readily released in soluble form from cell extracts by endogeneous proteases which cleave the intact 97 kDa polypeptide to a soluble 50–55 kDa species (Chin *et al.*, 1982). Addition of protease inhibitors during cell lysis can also protect the enzyme from proteolytic release in soluble form. A similar bimodal distribution of enzyme activities has been found for yeast α-glucosidase (Kilker *et al.*, 1981) and α-mannosidase (Jelinek-Kelly *et al.*, 1985). A mannosidase activity responsible for the hydrolysis of the single mannose removed during yeast oligosaccharide processing (Byrd *et al.*, 1982) was demonstrated to be predominantly in the soluble fraction of yeast homogenates (Jelinek-Kelly *et al.*, 1985) even though *in vivo* studies on yeast transport mutants, which block the exit of glycoproteins from the ER, suggested that this activity resides in the ER. Glycoproteins accumulating in the ER of these cells contained the structure $Man_8GlcNAc_2$ (Esmon *et al.*, 1984).

By analogy to glucosidase I and HMG-CoA reductase one would therefore expect to enhance the ER mannosidase activity in a membrane fraction when cells were disrupted in the presence of protease inhibitors. There was, however, no change in the distribution of the mannosidase activity in the presence of a wide variety of protease inhibitors (Bischoff and Kornfeld, 1983). In fact, in Chinese hamster ovary (CHO) cells and rat hepatoma H35 cells, the microsome-bound mannosidase activity represented as little as 3–5% of the total "cytosolic/ER-like" activity (Bischoff and Kornfeld, 1986).

More recently Bischoff and Kornfeld (1986) have presented additional evidence that the soluble and membrane-bound forms of the enzyme are two forms of the same enzyme, exhibiting similar substrate specificity, K_m for *p*-nitrophenyl mannoside, and pH optima. Neither enzyme is inhibited by swainsonine or deoxymannojirimycin (dMM) (Bischoff and Kornfeld, 1984), and both are quite unstable. In addition, antibody raised against the soluble form of the enzyme purified essentially by the method of Shoup and Touster (1976) could cross-react and immunoprecipitate the enzyme solubilized from ER membranes. The only significant differences between the two forms of the enzyme were (1) the membrane-bound form requires deoxycholate for solubilization and (2) the two enzymes differ in their specificity toward $Man_9GlcNAc_2$. The latter difference will be discussed in greater detail below.

The presence of a small portion of an enzyme activity associated with the membranes of the ER would not, by itself, be very convincing evidence for the assignment to that subcellular compartment. Additional support for the hypothesis that this enzyme is the ER processing mannosidase comes from an examination of the relative activity of the enzyme toward individual high-mannose oligosaccharides. The crude enzyme solubilized from ER membranes has a higher initial rate of hydrolysis for the $Man_9GlcNAc$ species when compared to the $Man_{8-6}GlcNAc$ structures (Bischoff and Kornfeld, 1983). As discussed below (Section II,A,2), evidence is accumulating for a dMM-insensitive mannosidase activity in ER membranes capable of cleaving $Man_9GlcNAc_2$ to a specific $Man_8GlcNAc_2$ isomer; thus the specificity of the ER mannosidase activity was consistent with such a role. Unfortunately, when a more detailed examination of the substrate specificity of the purified soluble enzyme was performed, no preference for $Man_9GlcNAc_2$ was found (Bischoff and Kornfeld, 1986). Analysis of the intermediate structures generated by the digestion of $Man_9GlcNAc$ revealed that there is no apparent specificity in the removal of mannosyl residues by either the soluble or the membrane-associated enzyme.

Bischoff and Kornfeld (1986) suggest that the difference between the free oligosaccharides and the protein-bound structures may confer steric or conformational features to the latter substrates allowing a greater specificity of hydrolysis. There may also be differences in specificity between the enzyme in association with the ER lipid bilayer in comparison to a soluble or detergent-solubilized form of the enzyme. Finally, in a temporal respect, nascent glycoproteins would presumably encounter an ER mannosidase for only a brief period prior to routing elsewhere in the cell. This would result in a limited availability for mannose removal.

The view that there is a specific $Man_9GlcNAc$ to $Man_8GlcNAc$ cleaving activity in mammalian cells is supported by consideration of oligosaccharide processing in yeast. Precursors of yeast glycoproteins contain the same $Glc_3Man_9GlcNAc_2$ structure as mammalian cells and convert it to an extended phosphomannan structure by the sequential removal of the three glucose residues and a single central mannose residue prior to extensive mannose and phosphate addition (Byrd et al., 1982). The trimming of the mannose residue to form a single $Man_8GlcNAc_2$ isomer is accomplished by a specific ER processing enzyme (see discussion above). There are significant differences between this enzyme and the ER processing mannosidase of rat liver, however. The yeast enzyme shows the appropriate substrate specificity in vitro, has no activity toward the synthetic substrate p-nitrophenylmannoside, and is inhibited by the mammalian mannosidase I inhibitor, dMM (Jelinek-Kelly et al., 1985). Although there are differences between the yeast and mammalian mannosidase processing systems it appears that at least the preferential processing of the central terminal α-1,2-mannose residue has been retained through evolution (Bischoff, 1985).

In summary, there is a considerable amount of evidence that the ER mannosidase is closely related to the cytosolic enzyme. The relationship between the two enzymes remains to be established. Other questions are the following: What is the nature of the association of this mannosidase with the ER membrane? What is the basis for the significant differences in the in vivo and in vitro substrate specificity of this enzyme? Finally, what is the physiological role of redundant α-1,2-mannosidase activities in the cell? The recent report on the substrate specificities of cytosolic and lysosomal mannosidases should be considered in relation to possible roles of the enzymes (Tulsiani and Touster, 1987).

2. In Vivo Evidence for an Endoplasmic Reticulum α-Mannosidase Involved in Glycoprotein Processing

The presence of an α-1,2-mannosidase activity in ER membranes has been supported by studies in several laboratories. The earliest studies employed inhibitors of protein translocation. Tartakoff and Vassali (1979) presented suggestive evidence of mannosidase cleavage of immunoglobulin M (IgM) μ-chain oligosaccharides in the presence of carbonyl cyanide N-chlorophenylhydrazone (CCCP), an oxidative phosphorylation uncoupler which inhibits the transport of nascent glycoproteins from the ER to the Golgi complex. The cleavage of thyroglobulin oligosaccharides to $Man_8GlcNAc_2$ in the presence of CCCP in calf thyroid slices (Godelaine et al., 1981) also supported the view that a single mannose residue could be removed in the ER. Processing of IgA oligosaccharides in the presence of CCCP, in contrast, could proceed well beyond $Man_8GlcNAc_2$ to a $Man_6GlcNAc_2$ structure (Hickman et al., 1984).

Evidence for the ER mannosidase cleavage of viral and cellular proteins in the absence of translocation inhibitors has also been presented. The precursor for Sindbis virus glycoproteins E2 and E3 is a single polypeptide (PE2) in the ER containing $Man_8GlcNAc_2$ oligosaccharides. Following proteolytic cleavage in the Golgi complex these glycoproteins are rapidly converted to complex oligosaccharides (Hakimi and Atkinson, 1982). In fact, nascent vesicular stomatitis virus (VSV) G glycoprotein still bound to polysomes has been found to contain a variety of intermediates of glucose trimming ($Glc_3Man_9GlcNAc$ to $Man_9GlcNAc$) as well as a significant amount of $Man_8GlcNAc$ (Atkinson and Lee, 1984). Similar results were obtained with a mutant of VSV G protein which accumulates in the ER, namely, the glycoprotein accumulates with $Man_8GlcNAc$ (Gabel and Bergmann, 1985). A note of caution is appropriate in interpreting these results and some of those discussed below because the transfer of nonglucosylated oligosaccharide from lipid to protein has been detected in mammalian cells (Romero and Herscovics, 1986).

Rotavirus SA11, a lytic virus which assembles and buds into the lumen of the ER, contains a coat glycoprotein, VP7, with high-mannose oligosaccharides. Since viral particles have been localized by electron microscopy exclusively to the ER, these membrane proteins should be analogous to endogenous ER membrane glycoproteins. VP7 in cellular extracts pulse-labeled with [^{3}H]mannose for 5 min followed by a 90 min chase accumulated $Man_{8-6}GlcNAc_2$ while VP7 of the mature virus contained

$Man_{8-5}GlcNAc_2$ (Kabcenell and Atkinson, 1985). Treatment with CCCP resulted in the production of predominantly $Man_8GlcNAc_2$ structures. These results may therefore suggest that VP7 must be transported across an energy barrier in order to be processed beyond $Man_8GlcNAc_2$.

Similar structures for endogenous ER proteins have also been obtained. HMG-CoA reductase (Liscum *et al.*, 1983), ribophorin I (Rosenfeld *et al.*, 1984), and prolyl hydroxylase (Kedersha *et al.*, (1985), all integral ER membrane proteins, were found to contain a collection of high-mannose oligosaccharides ranging from $Man_8GlcNAc_2$ to as low as $Man_5GlcNAc_2$. The human liver z variant of α_1-antitrypsin, which accumulates in the ER as a result of a genetic defect, was also found to contain oligosaccharides ranging from $Man_7GlcNAc_2$ to $Man_5GlcNAc_2$ (Hercz and Harpaz, 1980).

Additional support for the presence of one or more ER mannosidase activities was presented when the ER membrane protein HMG-CoA reductase was synthesized in the presence of the mannosidase I inhibitor, dMM (Bischoff *et al.*, 1986). The glycoprotein synthesized in the absence of dMM contains $Man_{8-6}GlcNAc_2$ structures. In the presence of the inhibitor, however, the HMG-CoA reductase-bound oligosaccharides were predominantly $Man_8GlcNAc_2$. These results suggest that there are two α-1,2-mannosidase activities involved in the processing of this ER membrane glycoprotein. The first activity is very close to the site of protein synthesis (RER) and is insensitive to inhibition by dMM, while the second activity is sensitive to dMM and, based on the rotavirus CCCP results discussed above, may be spatially separated from the first activity. Preliminary evidence for a dMM-sensitive ER mannosidase involved in the processing of the oligosaccharides on a rat growth hormone–influenza hemagglutinin hybrid protein has recently been presented (Rizzolo and Kornfeld, 1987). In addition, Schweden *et al.* (1986) have isolated a dMM-sensitive α-1,2-mannosidase from calf liver microsomes which recognizes $Man_{9-7}GlcNAc_2$ but not $Man_6GlcNAc_2$ and has a requirement for divalent cations (see section B.2). Whether this second dMM-sensitive activity represents an alternate form of the previously characterized Golgi mannosidase I which may reside in low levels in ER membranes and slowly act on resident ER membrane proteins or is an additional previously uncharacterized ER mannosidase is presently under investigation.

It is interesting to note that when analogous experiments with dMM treatment were performed on the total cell and secreted glycoproteins of rat hepatocytes (Bischoff *et al.*, 1986) the predominant structure on the oligosaccharides on these glycoproteins was $Man_9GlcNAc_2$ with smaller amounts of $Man_8GlcNAc_2$. The dMM-insensitive α-1,2-mannosidase activity in this cell line is apparently acting on less than half of the total cellular glycoproteins. The mannosidase processing of cellular and secreted glycoprotiens may therefore have more than one route depending on the polypeptide structure of the substrate.

B. Multiple Routes of Mannose Processing and Factors Influencing the Extent of Oligosaccharide Processing

The extent of processing of cellular glycoproteins would be expected to reflect the cellular processing machinery that they have encountered. Thus, ER membrane glycoproteins contain predominantly high-mannose oligiosaccharides while cell surface glycoproteins contain predominantly complex-type structures, reflecting the differences in transport of the respective macromolecules through the Golgi processing machinery. An alternate mechanism for the regulation in processing of the bulk cell surface or secreted glycoproteins is to regulate the levels of individual processing enzymes in the cell. Many cell surface or secreted proteins, however, contain incompletely processed oligosaccharides at individual sites. In general, this does not correlate with a lack of the appropriate processing enzyme in the host cell. The factors influencing the extent of oligosaccharide processing at individual sites has, therefore, been the subject of study by several investigators.

The maturation of Sindbis virus oligosaccharides was examined by determining the accessibility of individual glycosylation sites to digestion by endo-β-N-acetylglucosaminidase H (endo H) (Hsieh *et al.*, 1983a). In this case, glycosylation sites on mature viral glycoproteins were examined in two host cell lines. Glycosylation sites which were readily accessible to digestion by endo H in a mutant cell line lacking GlcNAc transferase I were the same sites which were destined to become complex structures in the corresponding control cell line. Sites which were destined to remain high-mannose structures in the intact native viral glycoproteins were inaccessible to endo H cleavage in the absence of prior denaturation. Similar results were obtained in the processing of the high-mannose oligosaccharides of yeast carboxypeptidase Y and invertase (Trimble *et al.*, 1983). Sites which contained large phosphorylated mannan side chains were readily accessible to endo H digestion while the sites which could be cleaved by endo H only after protein denaturation were smaller and contained no phosphate.

The processing of oligosaccharides at a particular glycosylation site can depend not only on the accessibility of the entire structure to processing enzymes but also on the ability of a given processing enzyme in different host systems to gain conformational access to the oligosaccharide. The oligosaccharides of Sindbis virus glycoproteins, when grown on three different host cell lines, were found to vary in the extent of host-dependent processing at the individual glycosylation sites. Oligosaccharides at one site on glycoprotein E1 were predominantly complex-type structures in baby hamster kidney cells but were high-mannose structures at the same site in chick embryo fibroblasts. Other glycosylation sites in the same molecule were maintained as exclusively complex-type structures (Hsieh *et al.*, 1983b). The observation that a second glycosylation site on the same molecule can be processed to a complex structure in all three host cell lines suggests that the host-dependent variation in glycosylation did not result from alternate transport routes containing differing processing machinery. Since the polypeptide structure and presumably the conformation of the viral glycoprotein substrate remained constant, the only differences between the cell lines were probably the limits in the steric constraints required by one host cell processing enzyme to gain access to the viral glycosylation site when compared to the equivalent processing enzyme of the other cell lines.

Additional evidence for site-specific rates of oligosaccharide processing has recently been presented in the processing of the μ chain of IgM in a murine plasmacytoma cell line (Brown and Hickman, 1986). Individual glycosylation sites on this glycoprotein were isolated, and the time course of oligosaccharide processing at each site was assessed. Surprisingly, sites which were destined to be high-mannose oligosaccharides with $Man_{6-5}GlcNAc_2$ structures were processed to these structures after only a 30-min chase, while those destined to eventually become complex structures remained $Man_{8-7}GlcNAc$ even after a 180-min chase. Since the isomeric forms of $Man_8GlcNAc_2$ at each of the sites were identical, the differences in processing time could not result from a difference in the order of mannose removal, but merely the rate of removal. These authors hypothesized that the rate of processing of the oligosaccharides at these individual sites depends on the extent of substrate accessibility of each processing intermediate to the processing enzymes at each site. Rate-limiting steps in processing might therefore reflect conditions where the constraints on substrate accessibility were most severe. Accumulation of specific intermediates of processing would not need to reflect a delay in the transport of a glycoprotein substrate to the site of a new processing mannosidase, as suggested previously by these authors (Hickman *et al.*, 1984), but merely reflects a rate-limiting processing step.

Host-dependent processing of glycoproteins has also been studied *in vitro*. Bovine pancreatic ribonuclease B, a glycoprotein containing predominantly $Man_5GlcNAc_2$ oligosaccharides, was added to Golgi membrane preparations from either rat liver or bovine pancreas in the presence of the GlcNAc transferase donor substrate UDP-GlcNAc (Williams and Lennarz, 1984). The rat liver Golgi membranes proceeded to process this native substrate through the GlcNAc transferase I and mannosidase

II steps to an endo H-resistant form in contrast to the bovine membranes which processed the glycoprotein only after prior denaturation.

Further evidence for the influence of the combined 3-dimensional conformation of a polypeptide and oligosaccharide at a glycosylation site has been presented by several investigators. The oligosaccharide structures of myeloma IgG heavy (H) and light (L) chains were analyzed by Savvidou et al. (1984) using proton nuclear magnetic resonance spectroscopy (^{1}H NMR). In this case the majority of the L-chain oligosaccharides contained fully bisected biantennary complex oligosaccharides while the H chain contained predominantly nonbisected biantennary complex structures. Since the oligosaccharides at both sites contained complex structures, and since the H and L chains had already become covalently associated prior to the addition of the bisecting GlcNAc, the differences in structure could not be due to host-dependent variation or to a general lack of "accessibility" of the oligosaccharide. These authors present evidence, however, that the conformation of the oligosaccharide on the H chain is constrained through an interaction with the polypeptide chain in such a way that is renders the oligosaccharide unable to be recognized by GlcNAc transferase III, the enzyme which adds the bisecting GlcNAc.

An analogous situation was revealed when the ovalbumin gene was expressed in a heterologous mouse L cell line (Sheares and Robbins, 1986). Mature chicken ovalbumin contains a variety of high-mannose and bisected hybrid oligosaccharides at the single glycosylation site (Yamashita et al., 1978). When the protein was expressed in mouse L cells the polypeptide was glycosylated at the same single glycosylation site, but the oligosaccharides at this site contained 27% complex, 23% high-mannose, and 50% nonbisected hybrid-type oligosaccharides. Although the addition by GlcNAc transferase III of a bisecting GlcNAc on chicken ovalbumin would convert these oligosaccharides to a structure which is no longer a substrate for mannosidase II, an explanation which has been used to account for bisected hybrid structures on this protein (Schachter et al., 1982), in the mouse L cells ovalbumin oligosaccharides contain hybrid structures even in the absence of a bisecting GlcNAc. The negligible amount of hybrid oligosaccharides in the total cell glycoproteins suggests that ovalbumin hybrid structures do not result merely from a stoichiometric competition between mannosidase II and galactosyltransferase, but that an interaction between the oligosaccharide and the polypeptide on ovalbumin hinders the action of mannosidase II even in the absence of a bisecting GlcNAc. Whether this is caused by the mere proximity of the polypeptide to essential oligosaccharide structural features or an interaction between the oligosaccharide and the polypeptide

Mannosidase IA had approximately 4 times greater activity toward this substrate than mannosidase IB. Both activities could also be fractionated on DEAE–cellulose into binding and nonbinding fractions, but no catalytic differences between these forms could be detected. This behavior had previously been observed for mannosidase II (Tulsiani et al., 1977) and mannosidase IA (Tabas and Kornfeld, 1979) and suggests that all of these enzymes exhibit charge heterogeneity due to posttranslational modifications or protein sequence variations producing structural isozymes.

The substrate specificities of the enzymes have been studied by two groups with conflicting results. Tabas and Kornfeld (1979) assayed the activity of mannosidase IA against high-mannose substrates and found the greatest activity toward $Man_7GlcNAc$. Tulsiani et al. (1982b), however, found that the highest activities for both mannosidase IA and IB were toward the largest substrates, $Man_9GlcNAc$ and $Man_8GlcNAc$. The K_m of mannosidase IA toward $Man_8GlcNAc$ was 100 μM (Tabas and Kornfeld, 1979). The activity toward $Man_5GlcNAc$, which contains no α-1,2-linked mannosyl residues, was quite low for both enzymes, confirming a specificity for α-1,2-linkages (Tulsiani et al., 1982b).

When the hydrolysis products of a mannosidase IA digestion were examined by acetolysis, a very specific sequence of mannose removal was detected (Tabas and Kornfeld, 1979). Surprisingly, the enzyme readily cleaved the upper and lower mannosyl branches of the $Man_9GlcNAc$ structure but cleaved the central α-1,2-mannose only after prolonged incubation. This is in contrast to recent work by D. R. P. Tulsiani and O. Touster (unpublished) on purified mannosidase IA (see below) and also in contrast to several other studies, all of which indicate that the central α-1,2-mannosyl residue is the first to be removed from mammalian glycoproteins in vivo (Kornfeld et al., 1978; Cohen and Ballou, 1980; Brown and Hickman, 1986; Bischoff and Kornfeld, 1986). Since the processing of approximately half of the total cellular glycoproteins in rat hepatocytes is blocked by dMM at $Man_9GlcNAc$ rather than at the $Man_8GlcNAc$ product (Bischoff et al., 1986), and since mannosidase IA appears to produce the same $Man_8GlcNAc$ isomer found in vivo, the initial processing of these glycoproteins is apparently being carried out by Golgi mannosidase I, or by another dMM-sensitive enzyme, rather than by ER mannosidase.

Mannosidase IA purified to homogeneity from rat liver Golgi membranes by several chromatographic steps (Tulsiani and Touster, 1986) has a subunit molecular weight of approximately 57,000 and an oligomeric molecular weight of approximately 230,000. It therefore appears to be a tetramer composed of identical or very similar subunits. Since several proteins including mannosidase II (Moremen and Touster, 1985) and glucosidase II (Brada and Dubach, 1984) have been shown to be cleaved by proteases during purification, further studies are presently underway to confirm that the isolated enzyme represents the intact macromolecule present *in vivo*. Recent work (D.R.P. Tulsiani and O. Touster, unpublished) indicates that mannosidase IA is a glycoprotein since it binds to Con A. However, it appears not to have the terminal sugars characteristic of complex oligosaccharides, since it does not bind to immobilized ricin, lentil lectin, or serotonin. The enzyme is inhibited by dMM, with a 50% inhibition at 1 μM, but as much as 5% of the activity is resistant to concentrations as high as 2 mM.

Recent substrate specificity studies on purified mannosidase IA confirm the earlier finding (Tulsiani *et al.*, 1982b) that the enzyme has the highest activity toward Man$_9$GlcNAc, with activity descending through the series from Man$_8$GlcNAc to Man$_5$GlcNAc. Analysis by acetolysis of the intermediates of Man$_9$GlcNAc degradation by homogeneous mannosidase IA as stated above is in contradiction to the previous studies by Tabas and Kornfeld (1979), which employed a partially purified preparation of mannosidase IA. The homogeneous enzyme preferentially removes the central mannose on Man$_9$GlcNAc yielding a single Man$_8$GlcNAc isomer and further cleaves through two Man$_7$GlcNAc isomers, and a single Man$_6$GlcNAc isomer, to the final Man$_5$GlcNAc structure which is a substrate for GlcNAc transferase I.

Immunoprecipitation studies with antibody raised to homogeneous mannosidase IA cross-reacted with mannosidase IB, a result suggesting that these two forms of the enzyme share antigenic determinants (D. R. P. Tulsiani and O. Touster, unpublished). It is likely, therefore, that mannosidases IA and IB share a common subunit or that they represent two different isoforms of the same molecule by analogy to the A and B forms of human liver lysosomal α-mannosidase (Cheng *et al.*, 1986). This acidic mannosidase is composed of two immunologically identical charge isomers generated from the differential processing of the large subunit of one form.

A partial purification of a neutral α-1,2-mannosidase activity from rabbit liver microsomes has been accomplished (Forsee and Schutzbach, 1981), and its interaction with phospholipids has been examined (Forsee and Schutzbach, 1981, 1983; Forsee *et al.*, 1982). Rat liver mannosidase IA differs from this enzyme in not being activated by phospholipid (D. R. P. Tulsiani and O. Touster, unpublished).

3. Inhibitor Studies

The use of a mannosidase I inhibitor to study the role of mannosidase processing on subsequent transport was first described by Fuhrmann *et al.* (1984). These authors, in examining the effect of dMM on the processing of IgM and IgD, found that while the secretion of these glycoproteins was not affected by the inhibitor, the oligosaccharides of the secreted material contained mostly $Man_9GlcNAc$ (75%) along with $Man_8GlcNAc$ (25%) structures. Subsequent studies have shown that this inhibitor will cause the accumulation of $Man_9GlcNAc_2$ on the glycoproteins of Rous sarcoma virus (Bosch *et al.*, 1985), influenza virus (Elbein *et al.*, 1984; Burke *et al.*, 1984), and vesicular stomatitis virus (Burke *et al.*, 1984) while not interfering with their cell surface expression or viral infectivity. The cell surface expression of the human class I histocompatibility antigens was also not altered by this inhibitor (Burke *et al.*, 1984). In HepG2 cells dMM caused a delay in the proteolytic processing of cathepsin D while increasing the fraction of the enzyme which was secreted (Nauerth *et al.*, 1985), but in parallel studies in fibroblasts the inhibitor did not alter either the segregation or maturation of the enzyme. 1-Deoxymannojirimycin in combination with other analogous glycosidase inhibitors such as deoxynorjirimycin, castanospermine, and swainsonine should therefore by a useful tool in determining the precise role of the sequential steps of oligosaccharide maturation on localization and biological activity of cellular and secreted glycoproteins.

A second type of mannosidase inhibitor which will be a valuable tool for the analysis of the role of oligosaccharide processing on glycoprotein metabolism is the methyl-*p*-nitrophenyltriazene (MNT) derivative of mannose (Docherty *et al.*, 1986). Similar MNT derivatives of galactose and glucose have been shown to irreversibly inhibit the corresponding galactosidases and glucosidases (Van Diggelen *et al.*, 1980) and have been instrumental in both examining the expression of these enzymes in human cells (Van Diggelen *et al.*, 1981) and uncovering the metabolic defect in combined galactosidase and neuraminidase deficiency (Van Diggelen *et al.*, 1981). Treatment of perfused rat liver by Docherty *et al.* (1986) with mannose-MNT resulted in the secretion of α_1-acid glycoprotein containing $Man_{9-7}GlcNAc$. Recovery of the secretion of normal complex-type structures following the removal of the inhibitor took over 24 hr, probably reflecting the synthesis of new enzyme. Since this compound results in the irreversible inhibition of mannosidases it could also theoretically be used as a covalent active site probe. Further studies on the inhibition of isolated processing mannosidases would be helpful, however, to further assess the inhibitor studies *in vivo*. Unfortunately, the restricted availability of the mannose-MNT derivative and the difficulty of synthesis of mannosylmethylamine, the precursor of mannose-MNT, may also prove to be factors affecting the general use of the inhibitor in processing studies.

D. Mannosidase II

1. General Aspects

Early studies on the processing of N-linked oligosaccharides suggested that the maturation of newly synthesized oligosaccharides to mature complex structures involved cleavage of all but the final three core mannose residues prior to addition of complex terminal sugars (Tabas *et al.*, 1978). These authors also mentioned concurrent studies on a ricin-resistant CHO cell line, clone 15B, which revealed that a deficiency in a GlcNAc transferase caused the accumulation of oligosaccharides with a $Man_5GlcNAc_2$ structure. This observation was, in fact, the first evidence that the full cleavage of mannosyl residues to the $Man_3GlcNAc_2$ core found in complex oligosaccharides required the prior addition of a GlcNAc. Several investigators (Tabas and Kornfeld, 1978; Tulsiani *et al.*, 1982b; Harpaz and Schachter, 1980b) have since shown that rat liver contains an enzyme which is dependent on the prior addition by GlcNAc transferase I of a single GlcNAc in an α-1,2 linkage to the lower α-1,3 branch of $Man_5GlcNAc_2$. This mannosidase cleaves both the α-1,3- and α-1,6-mannose residues of $GlcNAcMan_5GlcNAc_2$ but does not recognize as a substrate $Man_5GlcNAc_2$, $GlcNAc_2Man_5GlcNAc_2$ (with one bisecting β-1,4-GlcNAc on the core β-1,4-mannose), or $GalGlcNAcMan_5GlcNAc_2$ (Tulsiani *et al.*, 1982b; Schachter *et al.*, 1982). Thus this enzyme, with a very restricted substrate specificity toward two specific α-1,3- and α-1,6-mannosyl residues on $GlcNAcMan_5GlcNAc_2$, was named mannosidase II for its position as the second Golgi processing mannosidase in the maturation of complex oligosaccharides (Tulsiani *et al.*, 1982b).

2. Purification and Characterization

Several years before the discovery of a GlcNAc transferase I-dependent α-1,3/α-1,6-mannosidase, Dewald and Touster (1973) observed a mannosidase in Golgi membranes that was active toward the synthetic substrate *p*-nitrophenylmannose. This enzyme was an integral membrane protein enriched in Golgi membranes and was clearly distinct from the lysosomal and soluble mannosidases on the basis of pH optimum, immunological cross-reactivity, molecular weight, and kinetic properties. The enzyme was purified from rat liver Golgi membranes (Tulsiani *et al.*, 1977) and found to have a subunit molecular weight of 124,000 (Moremen and Touster, 1985). Early work indicated that the enzyme is a glycoprotein inasmuch as it stained for carbohydrate and bound to Con A.

Substrate specificity studies (Tulsiani *et al.*, 1982b) have shown that this enzyme is the $GlcNAcMan_5GlcNAc_2$ cleaving enzyme, since the activities toward p-nitrophenylmannose and $GlcNAcMan_5NAc_2$ copurify, respond similarly to effectors, and show similar rates of inactivation. In addition, both activities are inhibited by swainsonine, an indolizidine alkaloid which also inhibits the lysosomal enzyme (Tulsiani *et al.*, 1982a). Activity of the enzyme toward $GlcNAcMan_5GlcNAc$ is at least 10-fold higher than the corresponding activity toward the high-mannose oligosaccharides $Man_{9-5}GlcNAc$. Lack of activity toward oligosaccharides containing a bisecting GlcNAc or a galactose added to the nonreducing GlcNAc of $GlcNAcMan_5GlcNAc_2$ implies that the relative levels of any of three key processing enzymes, namely, mannosidase II, GlcNAc transferase III, and galactosyltransferase, will determine the fate of oligosaccharide processing toward complex, bisected hybrid, or hybrid structures, respectively. Thus, a relative excess of mannosidase II will result in the cleavage to $GlcNAcMan_3GlcNAc_2$ prior to the substrate being "frozen" in a hybrid structure by GlcNAc transferase III or galactosyltransferase activities. This cleavage will route the resultant oligosaccharide toward the formation of complex structures.

Suggestive evidence for the regulation of the route of glycoprotein maturation by adjusting the processing enzyme levels was presented in studies on the chicken oviduct. The relative abundance of GlcNAc transferase III over mannosidase II was used as an explanation for the abundance of bisected hybrid structures on glycoproteins from these tissues (Schachter *et al.*, 1982), but as discussed earlier (see Section II,A,3) other factors can also contribute to the abundance of hybrid structures on ovalbumin.

The relative abundance of complex oligosaccharides on cellular and secreted glycoproteins can also be accounted for by the spatial separation of mannosidase II and galactosyltransferase activities. Mannosidase II has been shown by immunocytochemistry to have a broad distribution throughout Golgi membranes (Novikoff *et al.*, 1983), with a relative abundance of activity in the medial Golgi cisternae (Farquhar, 1985). Galactosyltransferase, in contrast, is found localized in the last few cisternae on the trans side of the Golgi complex (Roth and Berger, 1982). This separation into subregions of Golgi localization for competing processing reactions would, therefore, act to assure the appropriate routing of modification for cellular glycoproteins toward complex structures.

3. Inhibitor Studies

Swainsonine was originally found to be an inhibitor of glycoprotein processing, with the authors suggesting that the accumulating oligosaccharides were $Man_{9-8}GlcNAc_2$ (Elbein *et al.*, 1981, 1982). Inhibition studies on isolated mannosidases suggested, however, that the compound inhibited mannosidase II, not mannosidase I (Tulsiani *et al.*, 1982a).

TABLE II

Effect of Swainsonine on Glycoprotein Transport and Expression

Protein	Cell line	Hybrid oligosaccharides	Effect on Transport	Effect on Expression	Reference
VSV G protein	BHK cells	+	None	None	Kang and Elbein (1983)
Aminopeptidase M	Pig intestinal explant	+	Minor decrease	None	Danielsen et al. (1983)
erbB oncogene	Avian erythroblasts	+	None	None	Schmidt et al. (1985)
Fibronectin	Human fibroblasts	+	None	None	Arumugham and Tanzer (1983b)
α_1-Antitrypsin	Rat hepatocytes	+	None	None	Gross et al. (1983)
α_1-Antitrypsin	Human hepatoma	+	None	None	Lodish and Kong (1984)
α_1-antichymotrypsin	Human hepatoma	+	None	None	Lodish and Kong (1984)
C3	Human hepatoma	+	None	None	Lodish and Kong (1984)
Transferrin	Human hepatoma	+	None	None	Lodish and Kong (1984)
Albumin	Human hepatoma	No carbohydrate	None	None	Lodish and Kong (1984)
Insulin receptor	Human lymphocytes	+	None	None	Duronio et al. (1986)
Insulin-like growth factor I receptor	Human lymphocytes	+	None	None	Duronio et al. (1986)
von Willebrand protein	Human endothelial cells	+	None	None	Wagner et al. (1985)
Asialoglycoprotein receptor	Human hepatoma	+	None	None	Breitfeld et al. (1984)
Mouse hepatitis virus EZ protein	Mouse fibroblasts	+	None	None	Repp et al. (1985)
Surfactant-associated glycoprotein A	Rat lung epithelial cells	+	None	None	Whitsett et al. (1985)
Epidermal growth factor receptor	Epidermal carcinoma	+	None	None	Soderquist and Carpenter (1984)
Rous sarcoma virus glycoproteins	Chick embryo fibroblasts	+	None	None	Bosch et al. (1985)
IgM, IgD	Mouse hybridoma	+	None	None	Peyrieras et al. (1983)
HLA A, B, C	Human lymphoid	+	None	None	Peyrieras et al. (1983)
Transferrin	Human hepatoma	+	Accelerated	None	Yeo et al. (1985)
Ceruloplasmin	Human hepatoma	+	Accelerated	None	Yeo et al. (1985)
α_2-Macroglobulin	Human hepatoma	+	Accelerated	None	Yeo et al. (1985)
α_1-Antitrypsin	Human hepatoma	+	Accelerated	None	Yeo et al. (1985)

The inhibition was subsequently shown to cause the synthesis of hybrid structures both *in vitro* and *in vivo* (Tulsiani and Touster, 1983) supporting the conclusions that, in the absence of mannosidase III, galactosyltransferase could route oligosaccharide processing toward hybrid structures.

The effect of swainsonine on glycoprotein transport and expression are quite uniform (Table II). In general, though hybrid structures are induced, there are few cases of swainsonine causing an altered function of cellular or secreted glycoproteins. Yeo *et al.* (1985) detected a slight acceleration in intracellular transport of the secretory glycoproteins transferrin, ceruloplasmin, α_2-macroglobulin, and α_1-antitrypsin while no alteration in the rate of transport of the nonglycoprotein albumin was observed. The inhibitor was also shown to inhibit receptor-mediated uptake by the mannose receptor (Arumugham and Tanzer, 1983a; Chung *et al.*, 1984) but this probably results from the production of cell surface glycoproteins containing hybrid oligosaccharides. These structures, which are effective ligands for the mannose receptor, probably compete with exogenously added ligand for receptor binding. Swainsonine was also found to be a potent immunomodulator both *in vitro* (Hino *et al.*, 1985) and *in vivo* (Kino *et al.*, 1985), while inhibitor treatment of tumor cells has been shown to alter the recognition of these cells by allogeneic lymphocytes (Powell *et al.*, 1985). Therefore, although mannosidase II and the resulting complex oligosaccharides it produces may not be essential for the expression or localization of specific glycoproteins in homogeneous cell populations in culture, the interaction of heterogeneous cell populations in mammalian tissues may be more strongly effected by the shift in oligosaccharide structure.

Swainsonine was originally isolated from toxic plants of the genus *Swainsona*. Since the symptoms of *Swainsona* toxicity and hereditary lysosomal mannosidosis are similar and since this compound also inhibits the lysosomal enzyme *in vitro*, it was originally proposed (Colegate *et al.*, 1979; Dorling *et al.*, 1978, 1980) that this compound induces a phenocopy of the hereditary mannosidosis. Subsequent studies have shown, however, that swainsonine induced kidney and excreted oligosaccharides have branched $Man_{3-5}GlcNAc$ structures (Daniel *et al.*, 1984) while in human hereditary mannosidosis urinary oligosaccharides contain predominantly linear $Man_{5-2}GlcNAc$ structures (Matsuura *et al.*, 1981). At the present time it appears that the toxic effect of swainsonine may result from a combination of the inhibition of the lysosomal enzyme and an alteration of glycoprotein processing. Swainsonine has been found to be the active agent in locoweed toxicity (Tulsiani *et al.*, 1984).

4. Biosynthesis and Topology

Studies examining the synthesis, topology, posttranslational modification, and turnover of cellular and secreted glycoproteins have been useful in determining the routing and life history of these macromolecules. The availability of a highly specific antibody against mannosidase II has allowed these types of studies also to be performed on this Golgi enzyme (Moremen and Touster, 1985). The enzyme is synthesized in both 3T3 and HeLa cells as a 124 kDa polypeptide containing endo H-sensitive oligosaccharides which contribute ~7 kDa to the apparent molecular mass of the molecule. In both cell lines the oligosaccharides are eventually converted to mostly (HeLa) or entirely (3T3) endo H-resistant structures. The HeLa cell enzyme is further processed by the addition of O-linked oligosaccharides which cause a shift in apparent molecular mass to 134–136 kDa. These studies demonstrated the addition of sulfate to the N-linked oligosaccharides of mannosidase II in both cell lines as well as covalently bound phosphate and palmitate. The phosphate on the HeLa cell enzyme was found as serine phosphate moieties. The half-life of the enzyme in both cell lines was about 20 hr, which was slightly shorter than the rate of turnover of the total cellular protein.

Topology studies (Moremen and Touster, 1986a) have revealed that the catalytic domain, as expected, was oriented toward the lumen of the Golgi membrane. An unexpected result, however, was that the catalytic domain was quite resistant to digestion by a wide variety of the proteases used in the topology studies. Chymotrypsin, pronase, and proteinase K could all cleave the intact 124 kDa polypeptide to a 110 kDa form which, at least after chymotrypsin digestion, retained all of the original catalytic activity. A similar cleavage could be accomplished by trypsin or V-8 protease to a 115 kDa form. This cleavage could not be demonstrated in nonpermeabilized intact Golgi membranes. In saponin-permeabilized membranes chymotrypsin cleavage resulted in the production of the 110 kDa cleaved form, which, if the membranes were sedimented, was found in soluble form in the supernatant. Both the cleavage and release were dependent on the presence of both detergent and protease. These results suggest that the 110 kDa form represents a soluble form of the enzyme generated by protease digestion. Triton X-114 phase separation studies confirmed that the 110 kDa polypeptide represents a hydrophilic catalytic domain while the 124 kDa intact enzyme is a hydrophobic integral membrane protein.

The unusual nature of conversion in phase separation behavior following chymotrypsin treatment and the unusual stability of the catalytic domain to further digestion have allowed the development of a novel, high-yield purification procedure for the 110 kDa catalytic domain. Comparison of this species with the intact enzyme has revealed that the 110 kDa form exhibits an identical K_m and inhibition by swainsonine as the intact enzyme. This cleavage product also contains most, if not all, of the carbohydrate of the intact glycoprotein. Sequencing and amino acid analysis of both forms of the enzyme have revealed that the 110 kDa form differs from the intact enzyme by an amino-terminal fragment highly enriched in hydrophobic amino acids (Moremen and Touster, 1986b).

These results suggest that mannosidase II is anchored to the Golgi membranes through an amino-terminal membrane-anchoring domain, an orientation characteristic of an increasing number of integral membrane enzymes and receptor proteins (for review, see Wickner and Lodish, 1985). The fact that the catalytic domain does not appear to be independently associated with the membrane suggests that the proteolytically sensitive ~14 kDa region may function to anchor the catalytic domain to the membrane and also may contain the structural information necessary for Golgi localization. This hypothesis is presently being examined by cloning the enzyme and testing the expression and localization of the intact and modified versions of the gene in cultured cells.

E. Brain Microsomal α-Mannosidase

1. General Aspects and Comparison with Liver Processing Mannosidases

Tulsiani and Touster (1985) presented evidence for a novel α-mannosidase in rat brain. This enzyme, which was partially purified from a crude rat brain microsomal fraction, has a pH optimum of 6.0 and was found to have specificity toward the α-1,2, α-1,3, and α-1,6 linkages present in high-mannose oligosaccharides. The enzyme can cleave $Man_{9-5}GlcNAc$ down to $Man_3GlcNAc$ and exhibits similar rates of reaction toward $Man_8GlcNAc$, $Man_5GlcNAc$, and $GlcNAcMan_5GlcNAc$. p-Nitrophenyl mannose is a very poor substrate for the enzyme, and the enzyme is not inhibited by swainsonine or dMM. The fact that it does not bind to Con A suggests that it is not a glycoprotein.

These results indicate that this brain membrane-bound enzyme is distinct from any of the previously characterized processing mannosidases, the cytosolic mannosidase, or the lysosomal mannosidase (see Table I). The fact that no detectable mannosidase II was present in rat brain was also used as an explanation for the absence of neurological effects or oligosaccharide accumulation in rat brain following the feeding of swainsonine. A surprising finding was that this novel brain mannosidase cross-reacted with an antibody raised to the rat liver cytosolic mannosidase.

The two enzymes differ in several respects, however. The brain membrane enzyme does not act on p-nitrophenyl mannoside and is not inhibited by zinc ions. The two enzymes are similar in their pH optima, lack of binding to Con A, and lack of sensitivity to swainsonine or dMM. Immunohistochemical staining using an antibody raised against the brain cytosolic enzyme revealed that, while most of the reactivity was in the cytosol, postsynaptic densities were also immunoreactive (Zannetta *et al.*, 1983). The possibility was presented that the cytosolic and membrane-bound forms both possess a common subunit or are derived from a common precursor (Tulsiani and Touster, 1985).

2. Implications for Brain Glycoprotein Processing

The presence of a single enzyme in brain microsomes of the rat, pig, and sheep capable of fully processing $Man_9GlcNAc$ to a $Man_3GlcNAc$ structure without the intermediate synthesis of the $GlcNAcMan_5GlcNAc$ structure necessary for mannosidase II activity suggests that this tissue may have a novel method for oligosaccharide processing. Glycosyltransferases might extend the trimannosyl core structure of this product into the complex oligosaccharides of brain tissue. However, in spite of the apparent absence of mannosidase II in brain, swainsonine administration to the pig and sheep produces the accumulation of brain glycoproteins containing oligosaccharides that appear to be of the hybrid type (D. R. P. Tulsiani and O. Touster, unpublished). This result suggests that the brain of at least two species contains a swainsonine-sensitive α-mannosidase other than mannosidase II.

III. CONCLUSIONS

This chapter has summarized the present state of research on mannosidases involved in glycoprotein processing. In spite of the advances made in understanding the roles of these enzymes, many intriguing questions still remain. The apparent multiplicity of mannosidases cleaving early high-mannose intermediates suggests that there may be a distinct physiological function for each of the α-1,2-mannosidase activities. The finding of a broad-specificity membrane-bound mannosidase in brain also raises questions about alternate routes of biosynthesis of glycoproteins. Finally, fundamental questions in cell biology, such as determinants for the localization of organellar glycoproteins and regulation of the levels of glycoprotein processing enzymes in the cell, are being pursued through the cloning and expression of processing mannosidases. The solutions to these and other problems will, we hope, lead to a further understanding of the role of glycoprotein structure in the physiology of a cell.

ACKNOWLEDGMENTS

The continued advice and major contributions of Dr. D. R. P. Tulsiani are gratefully acknowledged, as is the dedicated assistance of Mrs. Vera Coleman. Unpublished work in this chapter was supported in part by Grant GM 26430 and BRSG Grant S07-RR07201 from the National Institutes of Health.

REFERENCES

Arumugham, R. G., and Tanzer, M. L. (1983a). Swainsonine inhibits macrophage receptor-mediated uptake and degradation of a mannosyl-oligosaccharide. *Biochem. Biophys. Res. Commun.* **116,** 922–930.

Arumugham, R. G., and Tanzer, M. L. (1983b). Abnormal glycosylation of human cellular fibronectin in the presence of swainsonine. *J. Biol. Chem.* **258,** 11883–11889.

Atkinson, P. H., and Lee, J. T. (1984). Cotranslational excision of α-glucose and α-mannose in nascent vesicular stomatitis virus G protein. *J. Cell Biol.* **98,** 2245–2249.

Bischoff, J. (1985). The identification and characterization of an α-mannosidase in the endoplasmic reticulum of rat liver. Ph.D. dissertation, Washington University, St. Louis, Missouri.

Bischoff, J., and Kornfeld, R. (1983). Evidence for an α-mannosidase in endoplasmic reticulum of rat liver. *J. Biol. Chem.* **258,** 7907–7910.

Bischoff, J., and Kornfeld, R. (1984). The effect of 1-deoxymannojirimycin on rat liver α-mannosidases. *Biochem. Biophys. Res. Commun.* **125,** 324–331.

Bischoff, J., and Kornfeld, R. (1986). The soluble form of rat liver α-mannosidase is immunologically related to the endoplasmic reticulum membrane α-mannosidase. *J. Biol. Chem.* **261,** 4758–4765.

Bischoff, J., Liscum, L., and Kornfeld, R. (1986). The use of 1-deoxymannojirimycin to evaluate the role of various α-mannosidases in oligosaccharide processing in intact cells. *J. Biol. Chem.* **261,** 4766–4774.

Bosch, J. V., Tlusty, A., McDowell, W., Legler, G., and Schwarz, R. T. (1985). The mannosidase inhibitors 1-deoxymannojirimycin and swainsonine have no effect on the biosynthesis and infectivity of Rous sarcoma virus. *Virology* **143,** 342–346.

Brada, D., and Dubach, U. (1984). Isolation of a homogeneous glucosidase II from pig kidney microsomes. *Eur. J. Biochem.* **141,** 149–156.

Breitfeld, P. P., Rup, D., and Schwartz, A. L. (1984). Influence of the N-linked oligosaccharides on the biosynthesis, intracellular routing, and function of the human asialoglycoprotein receptor. *J. Biol. Chem.* **259,** 10414–10421.

Brown, P. H., and Hickman, S. (1986). Oligosaccharide processing at individual glycosylation sites on MOPC 104E immunoglobulin M. *J. Biol. Chem.* **261,** 2575–2582.

Burke, B., Matlin, K., Bause, E., Legler, G., Peyrieras, N., and Ploegh, H. (1984). Inhibition of N-linked oligosaccharide trimming does not interfere with surface expression of certain integral membrane proteins. *EMBO J.* **3,** 551–556.

Burns, D. M., and Touster, O. (1982). Purification and characterization of glucosidase II, an endoplasmic reticulum hydrolase involved in glycoprotein biosynthesis. *J. Biol. Chem.* **257,** 9991–10000.

Byrd, J. C., Tarentino, A. L., Maley, F., Atkinson, P. H., and Trimble, R. B. (1982). Glycoprotein synthesis in yeast. Identification of $Man_8GlcNAc_2$ as an essential intermediate in oligosaccharide processing. *J. Biol. Chem.* **257,** 14657–14666.

Cheng, S. H., Malcolm, S., Pemble, S., and Winchester, B. (1986). Purification and comparison of the structure of human liver acidic α-D-mannosidases A and B. *Biochem. J.* **233**, 65–72.

Chin, D. L., Luskey, K. L., Anderson, R. G., Faust, J. R., Goldstein, J. L., and Brown, M. S. (1982). Appearance of crystalloid endoplasmic reticulum in compactin-resistant Chinese hamster cells with a 500-fold increase in 3-hydroxy-3-methylglutaryl-coenzyme A reductase. *Proc. Natl. Acad. Sci. U.S.A.* **79**, 1185–1189.

Chung, K., Shepherd, V. L., and Stahl, P. (1984). Swainsonine and castanospermine blockade of mannose glycoprotein uptake by macrophages. Apparent inhibition of receptor-mediated endocytosis by endogenous ligands. *J. Biol. Chem.* **259**, 14637–14641.

Cohen, R. E., and Ballou, C. E. (1980). Linkage and sequence analysis of mannose-rich glycoprotein core oligosaccharides by proton nuclear magnetic resonance spectroscopy. *Biochemistry* **19**, 4345–4358.

Colegate, S. M., Dorling, P. R., and Huxtable, C. R. (1979). A spectroscopic investigation of swainsonine: An α-mannosidase inhibitor isolated from *Swainsona canescens*. *Aust. J. Chem.* **32**, 2257–2264.

Daniel, P. F., Warren, C. D., and James, L. F. (1984). Swainsonine-induced oligosaccharide excretion in sheep. Time dependent changes in oligosaccharide profile. *Biochem. J.* **221**, 601–607.

Danielsen, E. M., Cowell, G. M., Noren, O., and Sjostrom, H. (1983). Biosynthesis of intestinal microvillar proteins. The effect of swainsonine on posttranslational processing of aminopeptidase N. *Biochem. J.* **216**, 325–331.

DeDuve, C., Pressman, B. C., Gianetto, R., Wattiaux, R., and Appelmans, F. (1955). Tissue fractionation studies. 6. Intracellular distribution patterns of enzymes in rat-liver tissue. *Biochem. J.* **60**, 604–621.

Dewald, B., and Touster, O. (1973). A new α-D-mannosidase occurring in Golgi membranes. *J. Biol. Chem.* **248**, 7223–7233.

Docherty, P. A., Kuranda, M. J., Aronson, N. N., BeMiller, J. N., Myers, R. W., and Bohn, J. A. (1986). Effect of α-D-mannopyranosylmethyl-*p*-nitrophenyltriazene on hepatic degradation and processing of the N-linked oligosaccharide chains of α_1-acid glycoprotein. *J. Biol. Chem.* **261**, 3457–3463.

Dorling, P. R., Huxtable, C. R., and Vogel, P. (1978). Lysosomal storage in *Swainsona* spp. toxicosis: An induced mannosidosis. *Neuropathol. Appl. Neurobiol.* **4**, 285–295.

Dorling, P. R., Huxtable, C. R., and Colegate, S. M. (1980). Inhibition of lysosomal α-mannosidase by swainsonine, an indolizidine alkaloid isolated from *Swainsona canescens*. *Biochem. J.* **191**, 649–651.

Dunphy, W. G., and Rothman, J. E. (1985). Compartmental organization of the Golgi stack. *Cell* **42**, 13–21.

Dunphy, W. G., Fries, E., Urbani, L. J., and Rothman, J. E. (1981). Early and late functions associated with the Golgi apparatus reside in distinct compartments. *Proc. Natl. Acad. Sci. U.S.A.* **78**, 7453–7457.

Duronio, V., Jacobs, S., and Cuatrecasas, P. (1986). Complete glycosylation of the insulin and insulin-like growth factor I receptors is not necessary for their biosynthesis and function. *J. Biol. Chem.* **261**, 970–975.

Elbein, A. D., Solf, R., Dorling, P. R., and Vosbeck, K. (1981). Swainsonine: An inhibitor of glycoprotein processing. *Proc. Natl. Acad. Sci. U.S.A.* **78**, 7393–7397.

Elbein, A. D., Dorling, P. R., Vosbeck, K., and Horisberger, M. (1982). Swainsonine prevents the processing of the oligosaccharide chains of influenza virus hemagglutinin. *J. Biol. Chem.* **257**, 1573–1576.

Elbein, A. D., Legler, G., Tlusty, A., McDowell, W., and Schwarz, R. (1984). The effect of deoxymannojirimycin on the processing of the influenza viral glycoproteins. *Arch. Biochem. Biophys.* **235**, 579–588.

Esmon, B., Esmon, P. C., and Schekman, R. (1984). Early steps in processing of yeast glycoproteins. *J. Biol. Chem.* **259**, 10322–10327.

Farquhar, M. G. (1985). Progress in unraveling pathways of Golgi traffic. *Annu. Rev. Cell Biol.* **1**, 447–488.

Forsee, W. T., and Schutzbach, J. S. (1981). Purification and characterization of a phospholipid-dependent α-mannosidase from rabbit liver. *J. Biol. Chem.* **256**, 6577–6582.

Forsee, W. T., and Schutzbach, J. S. (1983). Interaction of α-1,2-mannosidase with anionic phospholipids. *Eur. J. Biochem.* **136**, 577–582.

Forsee, W. T., Springfield, J. D., and Schutzbach, J. S. (1982). Effect of phospholipids on α-1,2-mannosidase activity. *J. Biol. Chem.* **257**, 9963–9967.

Fuhrmann, U., Bause, E., Legler, G., and Ploegh, H. (1984). Novel mannosidase inhibitor blocking conversion of high mannose to complex oligosaccharides. *Nature (London)* **307**, 755–758.

Gabel, C. A., and Bergmann, J. E. (1985). Processing of the asparagine-linked oligosaccharides of secreted and intracellular forms of the vesicular stomatitis virus G protein: *In vivo* evidence of Golgi apparatus compartmentalization. *J. Cell Biol.* **101**, 460–469.

Godelaine, D., Spiro, M. J., and Spiro, R. G. (1981). Processing of the carbohydrate units of thyroglobulin. *J. Biol. Chem.* **256**, 10161–10168.

Goldberg, D. E., and Kornfeld, S. (1983). Evidence for extensive subcellular organization of asparagine-linked oligosaccharide processing and lysosomal enzyme phosphorylation. *J. Biol. Chem.* **258**, 3159–3165.

Gross, V., Tran-Thi, T., Vosbeck, K., and Heinrich, P. C. (1983). Effect of swainsonine on the processing of the asparagine-linked carbohydrate chains of α_1-antitrypsin in rat hepatocytes. Evidence for the formation of hybrid oligosaccharides. *J. Biol. Chem.* **258**, 4032–4036.

Hakimi, J., and Atkinson, P. H. (1982). Glycosylation of intracellular Sindbis virus glycoproteins. *Biochemistry* **21**, 2140–2145.

Harpaz, N., and Schachter, H. (1980a). Control of glycoprotein synthesis. Bovine colostrum UDP–*N*-acetylglucosamine:α-D-mannoside β2-*N*-acetylglucosaminlytransferase I. Separation from UDP–*N*-acetylglucosamine:α-D-mannoside β2-*N*-acetylglucosaminlytransferase II, partial purification, and substrate specificity. *J. Biol. Chem.* **255**, 4885–4893.

Harpaz, N., and Schachter, H. (1980b). Control of glycoprotein synthesis. Processing of asparagine-linked oligosaccharides by one or more rat liver Golgi α-D-mannosidases dependent on the prior action of UDP–*N*-acetylglucosamine:α-D-mannoside β2-*N*-acetylglucosaminyltransferase I. *J. Biol. Chem.* **255**, 4894–4902.

Hercz, A., and Harpaz, N. (1980). Characterization of the oligosaccharides of the liver Z variant α_1-antitrypsin. *Can. J. Biochem.* **58**, 644–648.

Hettkamp, H., Legler, G., and Bause, E. (1984). Purification by affinity chromotography of glucosidase I, an endoplasmic reticulum hydrolase involved in the procession of asparagine-linked oligosaccharides. *Eur. J. Biochem.* **142**, 85–90.

Hickman, S., Theodorakis, J. L., Greco, J. M., and Brown, P. H. (1984). Processing of MOPC 315 immunoglobulin A oligosaccharides: Evidence for endoplasmic reticulum and *trans* Golgi α-1,2-mannosidase activity. *J. Cell Biol.* **98**, 407–416.

Hino, M., Nakayama, O., Tsurumi, Y., Adachi, K., Shibata, T., Terano, H., Kohsaka, M., Aoki, H., and Imanaki, H. (1985). Studies of an immunomodulator, swainsonine. I. Enhancement of immune response by swainsonine *in vitro*. *J. Antibiot.* **38**, 926–935.

Hino, Y., and Rothman, J. E. (1985). Glucosidase II, a glycoprotein of the endoplasmic reticulum membrane. Proteolytic cleavage into enzymatically active fragments. *Biochemistry* **24,** 800–805.

Hsieh, P., Rosner, M. R., and Robbins, P. W. (1983a). Selective cleavage by endo-β-*N*-acetylglucosaminidase H at individual glycosylation sites of Sindbis virion envelope glycoproteins. *J. Biol. Chem.* **258,** 2555–2561.

Hsieh, P., Rosner, M. R., and Robbins, P. W. (1983b). Host-dependent variation of asparagine-linked oligosaccharides at individual glycosylation sites of Sindbis virus glycoproteins. *J. Biol. Chem.* **258,** 2548–2554.

Hubbard, S. C., and Ivatt, R. J. (1981). Synthesis and processing of asparagine-linked oligosaccharides. *Annu. Rev. Biochem.* **50,** 555–583.

Jelinek-Kelly, S., Akiyama, T., Saunier, B., Tkacz, J. S., and Herscovics, A. (1985). Characterization of a specific α-mannosidase involved in oligosaccharide processing in *Saccharomyces cerevisiae. J. Biol. Chem.* **260,** 2253–2257.

Kabcenell, A. K., and Atkinson, P. H. (1985). Processing of the rough endoplasmic reticulum membrane glycoproteins of rotavirus SA11. *J. Cell Biol.* **101,** 1270–1280.

Kang, M. S., and Elbein, A. D. (1983). Alterations in the structure of the oligosaccharide of vesicular stomatitis virus G protein by swainsonine. *J. Virol.* **46,** 60–69.

Kedersha, N. L., Tkacz, J. S., and Berg, R. A. (1985). Characterization of the oligosaccharides of prolyl hydroxylase, a microsomal glycoprotein. *Biochemistry* **24,** 5952–5960.

Kilker, R. D., Saunier, B., Tkacz, J. S., and Herscovics, A. (1981). Partial purification from *Saccharomyces cerevisiae* of a soluble glucosidase which removes the terminal glucose from the oligosaccharide $Glc_3Man_9GlcNAc_2$. *J. Biol. Chem.* **256,** 5299–5303.

Kino, T., Inamura, N., Nakahara, K., Kiyoto, S., Goto, T., Terano, H., Kohsaka, M., Aoki, H., and Imanaka, H. (1985). Studies on an immunomodulator, swainsonine. *J. Antibiot.* **38,** 936–940.

Kornfeld, R., and Kornfeld, S. (1985). Assembly of asparagine-linked oligosaccharides. *Annu. Rev. Biochem.* **54,** 631–664.

Kornfeld, S., Li, E., and Tabas, I. (1978). The synthesis of complex-type oligosaccharides. *J. Biol. Chem.* **253,** 7771–7778.

Kreibich, G., Ulrich, B. L., and Sabatini, D. D. (1978). Proteins of rough microsomal membranes related to ribosome binding. *J. Cell Biol.* **77,** 464–487.

Liscum, L., Cummings, R. D., Anderson, R. G. W., De Martino, G. N., Goldstein, J. L., and Brown, M. S. (1983). 3-Hydroxy-3-methylglutaryl-CoA reductase: A transmembrane glycoprotein of the endoplasmic reticulum with N-linked "high-mannose" oligosaccharides. *Proc. Natl. Acad. Sci. U.S.A.* **80,** 7165–7169.

Lodish, H. F., and Kong, N. (1984). Glucose removal from N-linked oligosaccharides is required for efficient maturation of certain secretory glycoproteins from the rough endoplasmic reticulum to the Golgi complex. *J. Cell Biol.* **98,** 1720–1729.

Marsh, C. A., and Gourlay, G. C. (1971). Evidence for a non-lysosomal α-mannosidase in rat liver homogenates. *Biochim. Biophys, Acta* **235,** 142–148.

Matsuura, F. H., Nunez, A., Grabowski, G. A., and Sweeley, C. C. (1981). Structure studies of urinary oligosaccharides for patients with mannosidosis. *Arch. Biochem. Biophys.* **207,** 337–353.

Moremen, K. W., and Touster, O. (1985). Biosynthesis and modification of Golgi mannosidase II in HeLa and 3T3 cells. *J. Biol. Chem.* **260,** 6654–6662.

Moremen, K. W., and Touster, O. (1986a). Topology of mannosidase II in rat liver Golgi membranes and release of the catalytic domain by selective proteolysis. *J. Biol. Chem.* **261,** 10945–10951.

Moremen, K. W., and Touster, O. (1986b). A novel purification of the catalytic domain of

Golgi mannosidase II: comparison with the intact enzyme. *Fed. Proc., Fed. Am. Soc. Exp. Biol.* **45,** 1680.

Nauerth, A., Lemansky, P., Hasilik, A., von Figura, K., Bause, E., and Legler, G. (1985). Cell type dependent inhibitor of transport of cathepsin D in HepG2 cells and fibroblasts exposed to deoxymannonorjirimycin and deoxynorjirimycin. *Biol. Chem. Hoppe-Seyler* **366,** 1009–1016.

Novikoff, P. M., Tulsiani, D. R. P., Touster, O., Yam, A., and Novikoff, A. B. (1983). Immunocytochemical localization of α-D-mannosidase II in the Golgi apparatus of rat liver. *Proc. Natl. Acad. Sci. U.S.A.* **80,** 4364–4368.

Opheim, D. J., and Touster, O. (1978). Lysosomal α-D-mannosidase of rat liver. *J. Biol. Chem.* **253,** 1017–1023.

Oppenheimer, C. L., and Hill, R. L. (1981). Purification and characterization of a rabbit liver α1-3 mannoside β1-2 *N*-acetylglucosaminyltransferase. *J. Biol. Chem.* **256,** 799–804.

Peyrieras, N., Bause, E., Legler, G., Vasilov, R., Claeson, L., Peterson, P., and Ploegh, H. (1983). Effects of the glucosidase inhibitors norjirimycin and deoxynorjirimycin on the biosynthesis of membrane and secretory glycoproteins. *EMBO J.* **2,** 823–832.

Powell, L. D., Bause, E., Legler, G., Molyneux, R. J., and Hart, G. W. (1985). Influence of asparagine-linked oligosaccharides on tumor cell recognition in the mixed lymphocyte reaction. *J. Immunol.* **135,** 714–724.

Repp, R., Tamura, T., Boschek, C. B., Wege, H., Schwarz, R., and Niemann, H. (1985). The effects of processing inhibitors of N-linked oligosaccharides on the intracellular migration of glycoprotein E2 of mouse hepatitis virus and the maturation of corona virus particles. *J. Biol. Chem.* **260,** 15873–15879.

Rizzolo, L. J., and Kornfeld, R. (1987). Posttranslational modifications of a growth hormone–HA hybrid protein that accumulates in a pre-Golgi region of the endoplasmic reticulum *J. Cell. Biol.* **105,** 2672.

Romero, P. A., and Herscovics, A. (1986). Transfer of nonglucosylated oligosaccharide from lipid to protein in a mammalian cell. *J. Biol. Chem.* **261,** 15936–15940.

Rosenfeld, M. G., Marcantonio, E. E., Hakimi, J., Ort, V. M., Atkinson, D. H., Sabatini, D., and Kreibich, G. (1984). Biosynthesis and processing of ribophorins in the endoplasmic reticulum. *J. Cell Biol.* **99,** 1076–1082.

Roth, J., and Berger, E. G. (1982). Immunocytochemical localization of galactosyltransferase in HeLa: Codistribution with thiamine pyrophosphatase in trans-Golgi cisternae. *J. Cell Biol.* **93,** 223–229.

Rothman, J. E. (1981). The Golgi apparatus: Two organelles in tandem. *Science* **213,** 1212–1220.

Savvidou, G., Klein, M., Grey, A. A., Dorrington, K. J., and Carver, J. D. (1984). Possible role for peptide–oligosaccharide interactions in differential oligosaccharide processing at asparagine-107 of the light chain and asparagine-297 of the heavy chain in a monoclonal IgG$_{1K}$. *Biochemistry* **23,** 3736–3740.

Schachter, H., Narasimhan, S., Gleeson, P., Vella, G. J., and Brockhausen, I. (1982). Oligosaccharide branching of glycoproteins: Biosynthetic mechanisms and possible biological functions. *Philos. Trans. R. Soc. London, Ser. B* **300,** 145–159.

Schachter, H., Narasimhan, S., Gleeson, P., and Vella, G. (1983). Control of branching during the biosynthesis of asparagine-linked oligosaccharides. *Can. J. Biochem. Cell Biol.* **61,** 1049–1066.

Schmidt, J. A., Beug, H., and Hayman, M. J. (1985). Effects of inhibitors of glycoprotein processing on the synthesis and biological activity of the *erbB* oncogene. *EMBO J.* **4,** 105–112.

Schweden, J., Legler, G., and Bause, E. (1986). Purification and characterization of a neutral processing mannosidase from calf liver acting on (Man)₉(GlcNAc)₂ oligosaccharides. *Eur. J. Biochem.* **157**, 563–570.

Sheares, B. T., and Robbins, P. W. (1986). Glycosylation of ovalbumin in a heterologous cell: Analysis of oligosaccharide chains of the cloned glycoprotein in mouse L cells. *Proc. Natl. Acad. Sci. U.S.A.* **83**, 1993–1997.

Shoup, V. A., and Touster, O. (1976). Purification and characterization of the α-D-mannosidase of rat liver cytosol. *J. Biol. Chem.* **251**, 3845–3852.

Soderquist, A. M., and Carpenter, G. (1984). Glycosylation of the epidermal growth factor receptor in A-431 cells. The contribution of carbohydrate to receptor function. *J. Biol. Chem.* **259**, 12586–12594.

Struck, D. K., and Lennarz, W. J. (1980). The function of saccharide–lipids in synthesis of glycoproteins. *In* "The Biochemistry of Glycoproteins and Proteoglycans" (W. J. Lennarz, ed.), pp. 35–83. Plenum, New York.

Tabas, I., and Kornfeld, S. (1978). The synthesis of complex-type oligosaccharides. III. Identification of an α-D-mannosidase activity involved in a late stage of processing of complex-type oligosaccharides. *J. Biol. Chem.* **253**, 7779–7786.

Tabas, I., and Kornfeld, S. (1979). Purification and characterization of a rat liver Golgi α-mannosidase capable of processing asparagine-linked oligosaccharides. *J. Biol. Chem.* **254**, 11655–11663.

Tabas, I., Schlesinger, S., and Kornfeld, S. (1978). Processing of high mannose oligosaccharides to form complex type oligosaccharides on the newly synthesized polypeptides of vesicular stomatitis virus G protein and the IgG heavy chain. *J. Biol. Chem.* **253**, 716–722.

Tartakoff, A., and Vassali, P. (1979). Plasma cell immunoglobulin M molecules. Their biosynthesis, assembly, and intracellular transport. *J. Cell Biol.* **83**, 284–299.

Trimble, R. B., Maley, F., and Chu, F. K. (1983). Glycoprotein biosynthesis in yeast. Protein conformation affects processing of high mannose oligosaccharides on carboxypeptidase Y and invertase. *J. Biol. Chem.* **258**, 2562–2567.

Tulsiani, D. R. P., and Touster, O. (1983). Swainsonine causes the production of hybrid glycoproteins by human skin fibroblasts and rat liver Golgi preparations. *J. Biol. chem.* **258**, 7578–7585.

Tulsiani, D. R. P., and Touster, O. (1985). Characterization of a novel α-D-mannosidase from rat brain microsomes. *J. Biol. Chem.* **260**, 13081–13087.

Tulsiani, D. R. P., and Touster, O. (1986). Purification of α1,2-specific mannosidase from rat liver Golgi membranes. *Fed. Proc., Fed. Am. Soc. Exp. Biol.* **45**, 1680.

Tulsiani, D. R. P., and Touster, O. (1987). Substrate specificities of rat kidney lysosomal and cytosolic α-D-mannosidases and effects of swainsonine suggest a role of the cytosolic enzyme in glycoprotein catabolism. *J. Biol. Chem.* **262**, 6506–6514.

Tulsiani, D. R. P., Opheim, D. J., and Touster, O. (1977). Purification and characterization of α-D-mannosidase from rat liver Golgi membranes. *J. Biol. Chem.* **252**, 3227–3233.

Tulsiani, D. R. P., Harris, T. M., and Touster, O. (1982a). Swainsonine inhibits the biosynthesis of complex glycoproteins by inhibition of Golgi mannosidase II. *J. Biol. Chem.* **257**, 7936–7939.

Tulsiani, D. R. P., Hubbard, S. C., Robbins, D. W., and Touster, O. (1982b). α-D-Mannosidases of rat liver Golgi membranes. Mannosidase II is the GlcNAcMan₅-cleaving enzyme in glycoprotein biosynthesis and mannosidases IA and IB are the enzymes converting Man₉ precursors to Man₅ intermediates. *J. Biol. Chem.* **257**, 3660–3668.

Tulsiani, D. R. P., Broquist, H. P., James, L. F., and Touster, O. (1984). The similar effects of swainsonine and locoweed on tissue glycosidases and oligosaccharides of the pig indicate that the alkaloid is the principal toxin responsible for the indication of loco-ism. *Arch. Biochem. Biophys.* **232,** 76–85.

Van Diggelen, O. P., Galjaard, H., Sinnott, M. L., and Smith, P. J. (1980). Specific inactivation of lysosomal glycosidases in living fibroblasts by the corresponding glycosylmethyl-*p*-nitrophenyltriazines. *Biochem. J.* **188,** 337–343.

Van Diggelen, O. P., Schram, A. W., Sinnott, M. L., Smith, P. J., Robinson, D., and Galjaard, H. (1981). Turnover of β-galactosidase in fibroblasts from patients with genetically different types of β-galactosidase deficiency. *Biochem. J.* **200,** 143–151.

Wagner, D. D., Mayadas, T., Urban-Pickering, M., Lewis, B. H., and Marder, V. J. (1985). Inhibition of disulfide bonding of von Willebrand protein by monensin results in small, functionally defective multimers. *J. Cell Biol.* **101,** 112–120.

Whitsett, J. A., Ross, G., Weaver, T., Rice, W., Dion, C., and Hull, W. (1985). Glycosylation and secretion of surfactant-associated glycoprotein A. *J. Biol. Chem.* **260,** 15273–15279.

Wickner, W. T., and Lodish, H. F. (1985). Multiple mechanisms of protein insertion into and across membranes. *Science* **230,** 400–407.

Williams, D. B., and Lennarz, W. J. (1984). Control of asparagine-linked oligosaccharide chain processing studies on bovine ribonuclease B. An *in vitro* system for the processing of exogenous glycoproteins. *J. Biol. Chem.* **259,** 5104–5114.

Yamashita, K., Tachibana, Y., and Kobata, A. (1978). The structures of the galactose-containing sugar chains of ovalbumin. *J. Biol. Chem.* **253,** 3862–3869.

Yeo, T., Yeo, K., Parent, J. B., and Olden, K. (1985). Swainsonine treatment accelerates intracellular transport and secretion of glycoproteins in human hepatoma cells. *J. Biol. Chem.* **260,** 2565–2569.

Zanetta, J. P., Roussel, G., Dontenwill, M., and Vincendon, G. (1983). Immunohistochemical localization of α-mannosidase during postnatal development of the rat cerebellum. *J. Neurochem.* **40,** 202–208.

III

Sorting and Organelle Assembly

6

Biosynthesis and Sorting of Proteins of the Endoplasmic Reticulum

MICHAEL GREEN AND RICHARD A. MAZZARELLA

I. INTRODUCTION

A. Protein Sorting in General

How proteins are sorted to their proper location within the cell is one of the most intriguing questions in cell biology. The answers to this question will enable us to begin to understand how a cell maintains and regulates the structural and functional distinctiveness of its intracellular membrane compartments. A significant amount of progress has been made in understanding many aspects of the general problem of protein sorting. The mechanisms which discriminate between proteins that remain in the cytoplasm and those that are either to be secreted or to become integral membrane proteins continue to be elucidated in great detail (for review, see Lodish *et al.*, 1981; Sabatini *et al.*, 1982; Walter *et al.*, 1984; Wickner

PROTEIN TRANSFER AND ORGANELLE BIOGENESIS

and Lodish, 1985). Recent studies have shown that many nuclear-encoded proteins destined for the mitochondria (for review, see Hay *et al.*, 1984) or the chloroplast (for review, see Schmidt *et al.*, 1980) are synthesized as preproteins with an amino-terminal extension which acts as a signal for the posttranslational uptake of the precursors into these organelles. In the case of lysosomal hydrolases in fibroblasts, it has been shown that appropriately oriented mannose 6-phosphate moieties on the enzymes act as one of the signals which results in the sorting of these proteins into lysosomes (for review, see Sly and Fischer, 1982; Sly, 1985).

It has been shown that the rate-limiting and distinctive step in the intracellular maturation of different secretory proteins is the movement of the proteins from their site of synthesis, the endoplasmic reticulum (ER),[1] to the Golgi (Lodish *et al.*, 1983). These results are incompatible with the hypothesis that vesicles bud off from a specialized region of the ER and merely contain a sample of the luminal content proteins and the ER membrane proteins. Therefore, it has been proposed that this type of sorting may also be a receptor-mediated process. In addition, Kalderon *et al.* (1984) have described a short amino acid sequence of SV40 T antigen, Pro-Lys-Lys-Lys-Arg-Lys-Val, which appears to specify the nuclear localization of that protein and which can mediate the transport to the nucleus of chimeric proteins which have been constructed to contain this sequence. This chapter describes experiments which have as their ultimate goal to test the hypothesis that the proteins of the internal membranes of the cell have analogous sequence or structural features which enable the cell to sort them to their correct subcellular compartments (Nelson and Robinson, 1983).

The work of many laboratories has made it clear that the biochemical tools already exist to begin a detailed investigation of the sorting problem. Experiments involving the ability to express the coding sequences of membrane proteins in suitable vector–host systems, coupled with the ability to modify these sequences through the use of *in vitro* mutagenesis techniques, provide a powerful way to approach the answers to these questions (for review, see Garoff, 1985; Schekman, 1985). This type of

---

[1] Abbreviations: ER, endoplasmic reticulum; VSV, vesicular stomatitis virus: ERp, endoplasmic reticulum protein; endo H, endo-β-N-acetylglucosaminidase H; HMG-CoA reductase, 3-hydroxy-3-methylglutaryl-coenzyme A reductase; MOPC-315, mineral oil-induced plasmacytoma 315; RER, rough endoplasmic reticulum; EDTA, ethylenediaminetetraacetic acid; SDS–PAGE, sodium dodecyl sulfate–polyacrylamide gel electrophoresis; LPS, bacterial lipopolysaccharide (*Escherichia coli* serotype B5:055); H^{M315}, heavy chain of MOPC-315 immunoglobulin A (IgA) molecule; MDBK, Madin–Darby bovine kidney; HPLC, high-performance liquid chromatography; CHO, carbohydrate; bp, base pairs.

approach has provided a valuable way to identify possible protein sorting signals in plasma membrane proteins, such as VSV glycoprotein (Rose and Bergmann, 1982, 1983; Guan and Rose, 1984, Guan *et al.*, 1985; Adams and Rose, 1985a,b), the influenza virus hemagglutinin (Gething and Sambrook, 1982; Doyle *et al.*, 1985; Gething *et al.*, 1986), and the polymeric immunoglobulin receptor (Mostov *et al.*, 1986).

The work of Rose and his co-workers provides an excellent example of such studies as applied to the intracellular transport of plasma membrane proteins. Rose and colleagues were able to examine the effects of alterations in various regions of VSV G protein on G protein sorting. For example, it was demonstrated that alteration of the cytoplasmic domain produced two classes of mutations: Class I, in which G protein transport was arrested and G protein accumulated in an early subcellular compartment, and Class II, in which a reduced rate of transport was observed with G protein eventually reaching the cell surface. The existence of the Class I alterations was interpreted to provide preliminary evidence for the existence of an "appropriate" cytoplasmic domain leading to ER localization.

Additional studies (Puddington *et al.*, 1986), utilizing chimeric cDNAs that encode the extracellular and transmembrane domains of VSV G protein linked to the cytoplasmic domain of other plasma membrane proteins, further emphasized the importance of the cytoplasmic domain for efficient transport to the cell surface. Furthermore, in studies of the transmembrane region, it was demonstrated the deletion of 2,4, or 6 amino acids in the 20 amino acid transmembrane region had no significant effect on transport, while deletions of 8 or 12 amino acids resulted in a G protein whose transport was blocked in the Golgi. Deletion of the entire transmembrane region resulted in a G protein which accumulated in the ER and was secreted only slowly. In other studies, it was found that the incorporation of a charged amino acid in the middle of the transmembrane region blocked transport of G protein to the surface, but not membrane insertion. Finally, studies in which a chimeric protein was created, containing the coding sequences for growth hormone fused to the transmembrane and cytoplasmic domains of G protein, demonstrated that this construction was only transported to the Golgi and that the signal for secretion of growth hormone was not operative when it was membrane bound.

Rizzolo *et al.* (1985) and Guan *et al.* (1985) have studied growth hormone–viral glycoprotein chimeric molecules to gain insights into the role of both luminal and cytoplasmic protein sequences in protein sorting. Although some of the results from these studies are not easily interpretable, it is clear that both luminal and cytoplasmic sequences influence protein sorting either directly as signals or indirectly through altered conformations of the chimeric molecules. In general, a major difficulty encountered in carrying this work further has been that alterations in protein sequences of membrane proteins have produced negative effects, i.e., the transport of the protein is blocked at an intermediate step in the transport pathway. It is difficult, therefore, to determine the direct cause of this transport lesion. It could be that sorting signals have been specifically altered and that the altered protein is no longer recognized by the cellular sorting machinery. It also could be the case, however, that the altered protein merely precipitates more easily in the cell and thus becomes physically unable to be transported. In this regard, it now appears that the retention in the ER of the altered forms of many normally exported proteins is due to their inability to fold properly. In this state, they are retained in the ER by their association with the immunoglobulin heavy chain binding protein (BiP) (Haas and Wabl, 1983; Bole et al., 1986; Gething et al., 1986).

Guan et al. (1985) have observed that the addition of functional carbohydrate attachment sites to the growth hormone sequence of a growth hormone–VSV G protein chimera, which accumulated in the Golgi apparatus, provided this molecule with a signal for protein transport to the plasma membrane. Thus, these workers have produced positive effects on transport by altering the luminal portion of a hybrid molecule.

While these data indicate that all domains of a membrane protein contribute to its sorting properties, they also underscore the need for more experiments to determine the sorting behavior of different types of membrane proteins before a truly coherent picture of protein transport signals can be developed. The design of systems in which positive effects on transport are created should be of particular value. The application of recombinant DNA technology to the study of ERp sorting should provide such model systems in that the deletion or mutation of signals that lead to ER localization could allow the ERps to traverse the complete protein transport pathway.

B. Sorting of ER Proteins in Particular

1. Site of Sorting of ER Proteins in the ER

One of the early questions asked about ERp sorting was where the sorting took place. Were the ERps preferentially retained in the ER or did they leave the ER with the rest of the proteins being transported from the ER only to be returned by a recycling mechanism? Based on the composition and structural organization of the Golgi apparatus and its well-established role in the sorting of other membrane proteins, it had been postulated (Rothman, 1980) that ERps are sorted in a process analogous to fractional distillation. This idea has been invoked recently by Munro and Pelham (1987) to explain their observations concerning a retention signal for a luminal ER protein (see below). According to this model, the ER is not a site for protein sorting. Instead, all proteins are first transported to the Golgi, and those proteins which are destined for the ER are recovered from the cis-most stacks of this organelle by recycling. Our studies investigating the sorting of six major ERps from murine plasmactyoma cells, ERp49, ERp59, ERp60, ERp61, ERp72, and ERp99, have produced no evidence to indicate that these proteins ever leave the ER (Lewis *et al.,* 1985b, 1986). Similar conclusions have been reached in studies of hexose-6-phosphate dehydrogenase (Brands *et al.,* 1985), glucosidase II (Brands *et al.,* 1985), and ribophorin I (Rosenfeld *et al.,* 1984) and in studies utilizing tripeptides containing the acceptor sequence for Asn-linked glycosylation to examine the rate of bulk flow from the endoplasmic reticulum to the cell surface (Wieland *et al.,* 1987).

The evidence against a recycling mechanism for ER localization can be summarized by three general observations. For those ERps with N-linked oligosaccharides, it has been determined that the oligosaccharides are of the high-mannose type and are sensitive to endo H. This is consistent with the hypothesis that the ERps never leave the ER for the Golgi where the oligosaccharides can be processed into complex, endo H-resistant moieties. In addition, in experiments where protein transport was assayed directly by sucrose gradient fractionation of pulse-labeled cell extracts, no movement of the proteins from the ER to the Golgi could be detected. Finally, the secretion of glycopeptides is faster than that of any known secretory proteins (Wieland, 1987). Since much evidence suggests that oligosaccharides are not signals for transport (Hickman and Kornfeld, 1978; Keller and Swank, 1978; Olden *et al.,* 1978; Struck *et al.,* 1978; Bell-Quint *et al.,* 1981; Sidman *et al.,* 1981; Gotlieb and Wallace, 1982), it appears that no signal is required for rapid and efficient transport from the ER to the cell surface. Thus, the unique protein composition of the ER appears to be established and maintained by the selective export of non-ER proteins. This suggests that the ER is the initial site for the sorting of ER membrane proteins and that ERps must contain specific retention signals.

2. Signals Directing the Sorting of ER Proteins

Given the conclusion that ERps are specifically retained in the ER, the problem becomes the elucidation of the structural features which identify newly synthesized membrane proteins as ERps. Many ER proteins and enzymes have been studied in a large number of laboratories. ER proteins are involved in the cotranslational insertion of membrane proteins into the ER membrane (for review, see Walter *et al.*, 1984) as well as the cotranslational translocation of lysosomal and secretory proteins across the ER membrane (Gilmore and Blobel, 1985). In the majority of cases, the cotranslational insertion or translocation of a protein in or through the membrane is accompanied by the removal of an amino-terminal signal peptide by the ER signal peptidase. Evans *et al.* (1986) have recently described the purification of a complex of microsomal proteins which contain signal peptidase activity. Other ER-associated activites include the enzymes of glycerolipid synthesis (for review, see Bell and Coleman, 1980; Bell *et al.*, 1981), the enzymes responsible for the synthesis and transfer of the lipid-linked oligosaccharide involved in N-linked glycosylation of nascent proteins (Snider and Robbins, 1982; Snider and Rogers, 1984), and the enzymes which process the newly transferred oligosaccharide, such as glucosidase I, glucosidase II, and the ER α-mannosidase (for review, see Kornfeld and Kornfeld, 1985). In addition, protein disulfide isomerase (for review, see Freedman and Hillson, 1980) and prolyl hydroxylase (for review, see Kivirikko and Myllyla, 1980) are confined to the ER.

For the purpose of this chapter, we confine our attention to those ERps whose structure and/or sorting are the best characterized. The structural features of these ERps are given in Table I. Cytochrome b_5 and NADH–cytochrome b_5 reductase have been included in Table I because both components have been shown to be enriched in the ER. They are, however, present in other cell fractions including the Golgi apparatus, the plasma membrane, mitochondria, and peroxisomes. It is possible that the reason for this wide distribution can be found in the way in which these proteins are made and the way in which they subsequently interact with the intracellular membranes. Cytochrome b_5 and NADH–cytochrome b_5 reductase are synthesized on free polysomes, released into the cytoplasm,a nd the completed polypeptides are then incorporated into membranes posttranslationally (Okada *et al.*, 1982; Anderson *et al.*, 1983). Analysis of the amino acid sequence of cytochrome b_5 has led to the idea that this protein interacts with the membrane bilayer by virtue of its hydrophobic carboxy-terminal region (Ozols and Gerard, 1977). What is known about the structure of NADH–cytochrome b_5 reductase is consistent with a similar conclusion for this membrane component (Tajima *et al.*, 1979). Thus, the finding that cytochrome b_5 and NADH–cytochrome b_5 reductase are enriched in the ER may be the result of the interaction of these proteins with proteins of the ER membrane before they themselves are incorporated into the membrane.

TABLE I

Properties of Endoplasmic Reticulum Membrane Proteins[a]

Protein	Molecular weight (Ref.)	Number of N-linked oligosaccharides	Cleaved signal sequence	Orientation[b]
Cellular proteins				
Cytochrome b_5	17,000 (ref. 1)	—	No	Cytoplasmic
Cytochrome P-450 (induced by phenobarbital)	56,000 (ref. 2)	—	No	Cytoplasmic
Egasyn (monomer → tetramer)	64,000 (ref. 3) (monomer)	+[c]	?	Luminal
Epoxide hydrolase	53,000 (ref. 4)	—	No	Cytoplasmic
ERp49	49,000	—	No	Transmembrane
ERp60	60,000	—	Yes	Luminal
ERp61	61,000	—	Yes	Luminal
ERp72	72,000	—	Yes	Luminal
ERp99 (also GRP94) (monomer → dimer)	92,500 (ref. 5–9) (monomer)	1	Yes	Transmembrane
Glucosidase II	123,000	+[c]	?	Luminal
Hexose-6-phosphate dehydrogenase	108,000	+[c]	?	Luminal
Heavy chain binding protein (BiP; also GRP78)	70,000 (ref. 10)	–	Yes	Luminal
3-Hydroxy-3-methyl-glutaryl-coenzyme A reductase	97,000 (ref. 11)	1	No	Transmembrane
NADH–cytochrome b_5 oxidoreductase	31,000 (ref. 12)	—	No	Cytoplasmic
NADPH–cytochrome P-450 oxidoreductase	77,000 (ref. 13)	—	No	Cytoplasmic
Prolyl hydroxylase (tetramer)	230,000 (ref. 14)	—	—	Luminal
α subunit	64,000 (refs. 15,16)	1	?	
α' subunit	64,000 (refs. 15,16)	2	?	
β subunit	60,000 (ref. 15)	—	Yes	
(Note: β subunit is PDI) (refs. 16–18)				
Protein Disulfide Isomerase (PDI; also ERp59)	55,000 (ref. 19)	—	Yes	Luminal
Ribophorin I	65,000	1	Yes	Transmembrane

continued

TABLE I (*Continued*)

Protein	Molecular weight (Ref.)	Number of N-linked oligosac- charides	Cleaved signal sequence	Orientation[b]
Ribophorin II	63,000	1	Yes	Transmembrane
Signal recognition particle receptor ("docking protein")	70,000 (ref. 20)	—	No	Cytoplasmic
Viral Proteins				
Adenovirus E19	19,000 (ref. 21)	2	Yes	Transmembrane
Coronavirus E1	26,000 (ref. 22)	—[d]	No	Transmembrane
Rotavirus NCVP5	29,000	2	No	Transmembrane
Rotavirus VP7	38,000	1	No	Transmembrane

[a] The information was compiled from the references given here and those cited in the text. Key to references: (1) Ozols and Heinemann (1982), (2) Fujii-Kuriyama et al. (1982), (3) Swank and Paigen (1973), (4) Heinemann and Ozols (1984), (5) Sorger and Pelham, 1987, (6) Mazzarella and Green (1987), (7) Kulomaa et al. (1986), (8) Sargan et al. (1986), (9) Kleinsek et al. (1986), (10) Munro and Pelham (1986), (11) Chin et al., 1984, (12) Yubisui et al. (1984), (13) Porter and Kasper (1985), (14) Berg and Prockop (1973), (15) Berg et al. (1980), (16) Koivu and Myllyla (1986), (17) Koivu and Myllyla (1987), (18) Koivu et al. (1987), (19) Edman et al. (1985), (20) Lauffer et al. (1985), (21) Herrisse' et al. (1980), and (22) Armstrong et al. (1984).

[b] The classification of a protein as either cytoplasmically or luminally oriented is based on the protein's activity being either sensitive or insensitive to proteases in the absence of detergent in intact microsomes. A protein is classified as transmembrane if the protein is shown to be available to protease in intact microsomes and the protein possesses an oligosaccharide or if a substantial proteolytic fragment which is detected in the absence of detergent is also degraded in the presence of detergent.

[c] The number of endo H-sensitive oligosaccharides is not known.

[d] Glycosylation of E1 is a posttranslational event. It is not inhibited by tunicamycin, 2-deoxyglucose, or 2-deoxy-2-fluoroglucose, all of which are inhibitors of asparagine-linked glycosylation. Analysis of the E1 carbohydrate and its sensitivity to mild alkaline treatment suggest that there is an O-glycosidic linkage between the oligosaccharide and the polypeptide. O-Linked glycosylation of E1 presumably occurs after the virus assembles and subsequently migrates to the Golgi apparatus. Glycosylation of E1 increases its molecular weight by about 3,000 (Niemann and Klenk, 1981).

While this is a type of protein sorting, it is not the kind that is the main focus of this chapter. We are interested in those ER proteins that are cotranslationally incorporated into the ER membrane and remain there in spite of the continuous export of proteins from the ER compartment. These proteins begin the transport pathway like all other membrane proteins and secretory proteins, but for some reason they are normally arrested at an early stage of the process. We are interested in defining those features of ER proteins which are responsible for this arrest and determining how these signals accomplish the ER localization of these proteins.

Drawing on the work of other laboratories investigating the transport of plasma membrane proteins, it is reasonable to assume that sorting information, or at least structural features which could influence sorting, could be located in all three domains of a typical ER membrane protein: the luminal domain, the transmembrane domain(s), and the cytoplasmic domain. Two common features of the luminal domain which could influence sorting are the presence of an uncleaved signal peptide or the presence of an N-linked oligossaccharide moiety. It has been reported that HMG-CoA reductase (Brown and Simoni, 1984), cytochrome *P*-450 (Bar-Nun *et al.*, 1980), NADPH–cytochrome *P*-450 reductase (Okada *et al.*, 1982), epoxide hydrolase (Okada *et al.*, 1982), rotavirus NCVP5 (Both *et al.*, 1983), and coronavirus E1 (Rottier *et al.*, 1984) lack cleavable signal peptides, while the signal peptides of rotavirus VP7 (Ericson *et al.*, 1983), adenovirus E19 (Persson *et al.*, 1980; Wold *et al.*, 1985), and ribophorin I and II are cleaved (Rosenfeld *et al.*, 1984). Our own experiments have shown that ERp99, ERp72, ERp61, ERp60, and protein disulfide isomerase (ERp59) have cleavable signal peptides (R. Mazzarella, unpublished results). Thus, it appears that the mere possession of an uncleaved signal peptide is not sufficient to determine the sorting properties of ER proteins. In addition, our finding that, of the six ER proteins we have studied, only ERp99 is a glycoprotein (Lewis *et al.*, 1985a,b, 1986) taken together with analogous studies of ER glycoproteins such as HMG-CoA reductase (Liscum *et al.*, 1983), the ribophorins (Kreibich *et al.*, 1983a; Rosenfeld *et al.*, 1984), hexose-6-phosphate dehydrogenase (Hino and Minakami, 1982; Brands *et al.*, 1985), glucosidase II (Hino and Rothman, 1985), prolyl hydroxylase (Kedersha *et al.*, 1985a,b), egasyn (Swank and Paigen, 1973), adenovirus E19 (Kornfeld and Wold, 1981), and rotavirus VP7 (Kabcenell and Atkinson, 1985) and NCVP5 (Both *et al.*, 1983) indicates that the mere presence or absence of an N-linked oligosaccharide is not sufficient to define an ER protein.

As the amino acid sequences of ER proteins have become available from the sequencing of cDNA clones encoding them, it has become possible to examine the transmembrane regions of these proteins with regard to their possible role in ER localization. Liscum *et al.* (1985) have proposed that the seven membrane spanning domains of HMG-CoA reductase may be involved in the localization of this protein to the cholesterol-poor membranes of the ER. Recently, Kabcenell and Atkinson (1985) and Poruchynsky *et al.* (1985) have described studies investigating the subcellular localization and intracellular sorting of the VP7 glycoprotein of rotavirus SA11. These workers have shown that while wild-type VP7 is activity, a Golgi marker, as well as an analysis of the intracellular transport of H^{M315}, demonstrated that the procedure was suitable for separating these two intracellular membrane compartments. The amount of ERp99 in the gradient fractions was assayed by immunoprecipitation. Nearly all of ERp99 was always recovered from the densest of the ER fractions, well separated from the Golgi region. This restricted localization of ERp99 was entirely consistent with the results obtained from subcellular fractionation studies. These results suggest that ERp99 may be exclusively localized in the RER. There was no evidence from this experiment that ERp99 ever leaves the ER, in that the gradient distribution of ERp99 was consistent throughout the 180-min chase. We feel, therefore, that it is unlikely that transport to and recycling from the Golgi apparatus plays a major role in the sorting of ER proteins. It is, of course, impossible to exclude this mechanism for all ER proteins, and it would be useful to apply these tests to newly identified ERps as they become available.

C. Tissue Distribution and Antigenic Conservation of ERp99

Using our standard immunoprecipitation methods, we have demonstrated that ERp99 is present in normal mouse tissues (Table III). These results indicate that ERp99 is a significant component of the microsomes

TABLE III

Tissue Distribution of ERp99[a]

Tissue	Quantity of ERp99 per 500 ng of total microsomal protein	
	Total amount	Relative amount
Plasmacytoma	1.92	1
Spleen	0.35	0.18
Liver	4.44	2.3
Kidney	0.38	0.20
Muscle	Not detected	—

[a] ERp99 was immunoprecipitated from ^{125}I-labeled preparations of total microsomes of various mouse tissues. ERp99 was immunoprecipitated from 500-ng aliquots of protein and analyzed by SDS–PAGE and autoradiography. The quantity of ERp99 was determined from a densitometric scan of the autoradiograph and is expressed in arbitrary density units.

of several tissues, especially liver, and serve to emphasize that the results we obtain by studying the sorting of ERp99 in plasmacytoma cells will be relevant to the assembly of intracellular membranes in other tissues. Proteins which are immunologically related to and of similar size to the ERp99 were present in detergent extracts of various cell types from other eukaryotic species (Fig. 3). Proteins were metabolically labeled for 4 hr with [^{35}S]methionine, and aliquots of a detergent-soluble extract were treated with anti-ERp99 antibody. The immunoabsorbed proteins were treated overnight with endo H and analyzed by SDS–PAGE. A prominent band which comigrated with murine ERp99 was detected in rat, hamster,

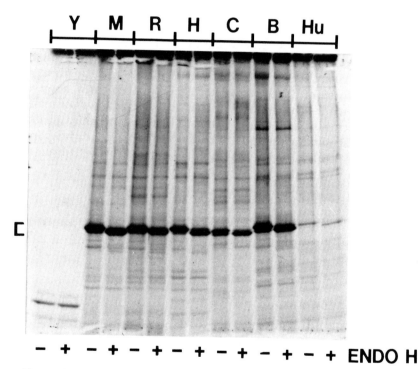

Fig. 3. Proteins immunologically related to ERp99 are found in several eukaryotic species. Seven eukaryotic cell lines were labeled with [^{35}S]methionine. Detergent-soluble lysates were prepared from the labeled cells and treated with anti-ERp99 antibody. Immunoprecipitated proteins were incubated in the absence and presence of endo H and analyzed by SDS–PAGE on a 7.5% polyacrylamide gel. The cell lines used were as follows: yeast cells (Y), MOPC-315 cells (M), rat Y3 myeloma cells (R), Chinese hamster ovary cells (H), chick embryo fibroblasts (C), MDBK (bovine) cells (B), and HeLa cells (human) (Hu). The position of the authentic ERp is indicated by the bracket. Adapted from Lewis et al. (1985a).

consistent with the idea that these proteins are not transported from the ER to the Golgi. In addition, these data taken together with the data obtained from subcellular fractionation and biosynthetic sorting experiments confirm the localization of ERp99 to the RER membrane.

E. Membrane Topography of ERp99

The results described above provide a framework for a study of the structural features of ERp99 which result in its sorting to the ER membrane. One of the initial questions which must be answered is the manner in which ERp99 is oriented in the membrane. We were able to obtain a general picture of the membrane topography of ERp99 by the use of proteolysis experiments (Fig. 6). ERp99 was completely sensitive to both papain and trypsin digestion in the presence of detergent. Interestingly, ERp99 was completely degraded by papain and significantly degraded by trypsin even in the absence of detergent. These results indicate that the bulk of ERp99 is exposed on the cytoplasmic side of the ER membrane.

An additional piece of evidence which helped us to create a picture of the way in which ERp99 interacts with the membrane came from the results of experiments performed to investigate the synthesis and processing of ERp99 in a cell-free system (Fig. 7). The 15–19 S fraction of MOPC-315 poly(A)$^+$ mRNA, which had been isolated by standard procedures (McCandliss *et al.,* 1981), was incubated in reticulocyte extracts in the absence or presence of intracellular membranes isolated from Krebs II ascites tumor cells (Szczesna and Boime, 1976; Green, 1979; Green and Gleiber, 1980). The membranes were added either cotranslationally or posttranslationally. ERp99-related cell-free products were isolated by immunoprecipitation, incubated in the presence or absence of endo H to detect N-linked glycosylation, and analyzed by SDS–PAGE. ERp99 was proteolytically cleaved by the membranes when they were added cotranslationally but not posttranslationally. In addition, ERp99 was cotranslationally, but not posttranslationally, N-glycosylated by the oliogsaccharyltransferase of the ascites membranes.

From these data, we have formulated a general model of the interaction of ERp99 with the ER membrane. Since ERp99 has a cleavable signal peptide, we propose that the mature amino terminus is inside the lumen of the ER. Since ERp99 appears to be completely sensitive to papain and trypsin in intact microsomes, showing no evidence of a major membrane-protected fragment, we propose that the bulk of the protein is exposed on the cytoplasmic face of the ER. Since, however, ERp99 possesses an N-linked oligosaccharide moiety, we propose that this moiety is attached at a site in the amino-terminal end of the molecule and that enough of this

A. Papain   B. Trypsin

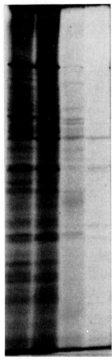
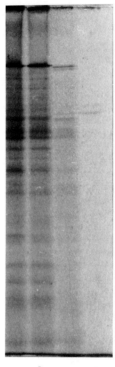

TX100   +   +   +   +
ENZ   + +   + +

Fig. 6.   Orientation of the ERp99 in the ER membrane. MOPC-315 cells were labeled for 60 min with [^{35}S]methionine, and cell homogenates were prepared as described by Goldberg and Kornfeld (1983). (A) Papain digestion. Where indicated, papain was added to a final concentration of 40 μg/ml and Triton X-100 added to a final concentration of 0.067% (w/v). All samples were incubated for 30 min at 37°C. (B) Trypsin digestion. Where indicated, trypsin was added to a final concentration of 250 μg/ml and Triton X-100 added to a final concentration of 0.067% (w/v). All samples were incubated for 30 min at room temperature. After inactivation of the enzymes, ERp99 was isolated by immunoprecipitation and analyzed on 10% polyacrylamide gels by SDS–PAGE and autofluorography. The arrows indicate the position of ERp99. Adapted from Lewis *et al.* (1985b).

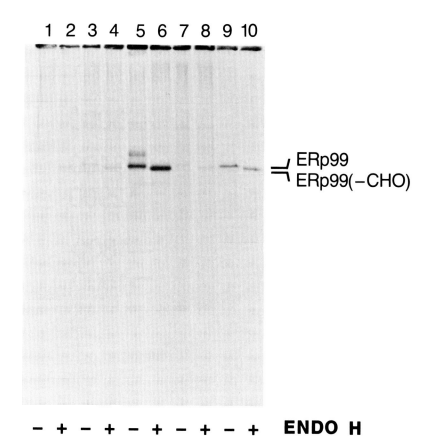

Fig. 10. Expression of cloned murine ERp99 cDNA in COS cells. Two clones containing ERp99 cDNA, pcD99-1 and pcD99-2, were obtained from a transformed mouse fibroblast cDNA library prepared in the pcD shuttle vector. Plasmid DNA was purified from the clones and used to transfect COS cells according to standard procedures. After 72 hr of culture, the cells were labeled with [^{35}S]methionine for 4 hr, washed, and lysed with detergent. Aliquots (2×10^6 TCA precipitable cpm) of the cell lysates were assayed for the presence of murine ERp99 by immunoprecipitation with anti-ERp99 antibody. The immunoprecipitated ERp99 was divided into two aliquots and incubated in the presence (+) or absence (−) of endo H. After this incubation, the products were analyzed by SDS–PAGE. The samples are as follows: Lanes 1,2—mock transfected COS cells; Lanes 3,4—COS cells transfected with pcD99-1 DNA; Lanes 5,6—COS cells transfected with pcD99-2 DNA; Lanes 7,8—untransfected COS cells; Lanes 9,10—authentic ERp99 immunoprecipitated from murine plasmacytoma cells labeled for 4 hr with [^{35}S]methionine. The migration of endoH treated ERp99 is indicated by ERp99-CHO. Taken from Mazzarella and Green (1987).

ERp99 cDNA clone pcD99-2, is depicted in Fig. 11. The entire sequence of both DNA strands of the insert was completely determined by isolating a nested set of subclones for each strand (Dale *et al.*, 1985). The ERp99 amino acid sequence derived from the nucleotide sequence is presented in Fig. 12. The ERp99 precursor encoded by the cDNA consists of 802 amino acids and has a calculated molecular weight of 92,475. The first AUG codon in the cDNA sequence (position −21 of the protein) is preceded by a purine nucleotide three nucleotides upstream of the coding sequence. This is typical of the large majority of functional initiation sites in eukaryotic mRNA (Kozak, 1983). The consensus poly(A) addition signal (Moldave, 1985) is present 14 nucleotides from the poly(A) tail. A comparison of the derived amino acid sequence to the N-terminal sequence obtained directly from the mature protein indicated that the site of

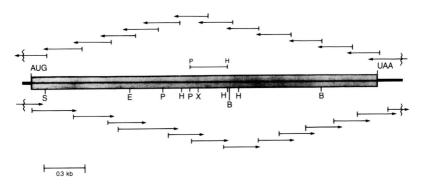

Fig. 11. Sequencing strategy for full-length ERp99 cDNA clone (pcD99-2 insert). The arrows depict the sequences determined from the overlapping set of clones obtained from each strand by the method described in the text. The sequence of each strand was completely determined. The thick line in the middle depicts the relative position of the ERp99 coding region within the insert. The *Pst*I–*Hin*dIII fragment used as a probe in these studies is indicated above the coding region (P—H). Some six-base restriction enzyme sites are indicated: B, *Bcl*I; E, *Eco*RI; H, *Hin*dIII; P, *Pst*I; S, *Sal*I; X, *Xho*I. Taken from Mazzarella and Green (1987).

Fig. 12. Nucleotide sequence and deduced amino acid sequence of murine ERp99 (pcD99-2 cDNA insert). Nucleotides are numbered on the right of each row. The deduced amino acid sequence is shown below the nucleotide sequence. Amino acids are numbered positively beginning with the N-terminus of mature ERp99 and negatively from the site of signal peptide cleavage (▲). The underlined amino acid sequence indicates the sequence confirmed by the direct amino acid sequencing of the N-terminus of mature ERp99. Potential N-linked glycosylation sites are identified by the asterisks. The putative polyadenylation signal is indicated by the box (end of sequence). A poly(A) tail of about 100 nucleotides was found following the T at position 2759. Taken from Mazzarella and Green (1987).

```
                                        TGGAAGTGTGAGGAGCTTAGACTCGGGGTGGGGGAGCTGGAGCGGCTCCTGAG   54

ACCGAAAAGGACTTGCGACTCGCCGGCCACGCACC
            -21                                                 -10                                      -1
            ATG AAG GTC GTC TGC TGT GTC CTG CTG ACC TTC GGG TTC GTC AGA GCT  ...  GTG CAG AGA        152
            Met Arg Val Val Cys Cys Val Leu Leu Thr Phe Gly Phe Val Arg Ala       Val Gln Arg
                                                                   30
                                          *****
  1                                          10                         20                         30
  GAT GAG GTC ATT CAG GAT GGC ACA  ...  GAA  ...  AGA  ...  AAT  ...  GAA  ...  GTT  ...  AGA        242
  Asp Glu Val Ile Gln Asp Gly Thr       Glu       Arg       Asn       Glu       Val       Arg
                                                                                                 60
 40                                      50                          60
 GTG AAC ATA AGG CAT GTC AAA ACG  ...  TTA  ...  GCA  ...  GAG  ...  AAT  ...  CTC  ...  GAA        332
 Val Asn Ile Arg His Val Lys Thr       Leu       Ala       Glu       Asn       Leu       Glu
                                                                                                 90
 70                                      80                          90
 TTA GAC AAG AAA AGG CTC ATC TCC  ...  CTG  ...  TTC  ...  ATT  ...  TCA  ...  GCT  ...  GCT        422
 Leu Asp Lys Lys Arg Leu Ile Ser       Leu       Phe       Ile       Ser       Ala       Ala
                          *****                                      *****
                                                                                                120
100                                     110                        120
TTA GAC AAG GTC CAT GTC ACA ACG  ...  GGA  ...  GAG  ...  GTT  ...  GGT  ...  TCT  ...  ACA        512
Leu Asp Lys Val His Val Thr Thr       Gly       Glu       Val       Gly       Ser       Thr
                                                                                                150
130                                     140                        150
AAA AAC CTG TTA AAC ACG ACA ACG  ...  TCA  ...  CTC  ...  CGC  ...  CAG  ...  GGT  ...  GAG        602
Lys Asn Leu Leu Asn Thr Thr Thr       Ser       Leu       Arg       Gln       Gly       Glu
                                                                                                180
160                                     170                        180
AGC GAG TTT TTA AAC AAA ATG GAA  ...  AAT  ...  ACC  ...  CTC  ...  ATT  ...  TTT  ...  TCT        692
Ser Glu Phe Leu Asn Lys Met Glu       Asn       Thr       Leu       Ile       Phe       Ser
                          *****
                                                                                                210
190                                     200                        210
GCC TTC CTT GTA GCA GAT GTC ATT  ...  CAC  ...  ACC  ...  ATC  ...  GAA  ...  GAA  ...  TCT        782
Ala Phe Leu Val Ala Asp Val Ile       His       Thr       Ile       Glu       Glu       Ser
                 *****
                                                                                                240
220                                     230                        240
GTA ATT GCT CCA GAC AAC GGA ACA  ...  TTA  ...  CTT  ...  GCA  ...  GAA  ...  TAC  ...  TTG        872
Val Ile Ala Pro Asp Asn Gly Thr       Leu       Leu       Ala       Glu       Tyr       Leu
                                                                                                270
250                                     260                        270
GAC ACA ATT AAT CTT TAC ATC ATC  ...  GTG  ...  TTC  ...  AGT  ...  AAG  ...  GTT  ...  GAG        962
Asp Thr Ile Asn Leu Tyr Ile Ile       Val       Phe       Ser       Lys       Val       Glu
                                                                                                300
280                                     290                        300
CCC TTG GAA GAT GAT GCA AAA AAA  ...  GTA  ...  GTA  ...  GAG  ...  GAA  ...  GAA  ...  CCA       1052
Pro Leu Glu Asp Asp Ala Lys Lys       Val       Val       Glu       Glu       Glu       Pro
                                                                                                330
310                                     320                        330
AAA ACT AAG GAA GAT ACT GTG TGG  ...  ATA  ...  ATC  ...  TGG  ...  CCA  ...  AAA  ...  GAA       1142
Lys Thr Lys Glu Asp Thr Val Trp       Ile       Ile       Trp       Pro       Lys       Glu
                                                                                                360
340                                     350                        360
GAA GAC TAC GCT AAA TCA TTT TCA  ...  GAT  ...  CCC  ...  ATG  ...  GCT  ...  CAC  ...  GTC       1232
Glu Asp Tyr Ala Lys Ser Phe Ser       Asp       Pro       Met       Ala       His       Val
                                                                                                390
370                                     380                        390
ACC TTC AAG TCG ATT TTG GTA CCC  ...  GGT  ...  GAT  ...  GAA  ...  GAT  ...  AAG  ...  CTG       1322
Thr Phe Lys Ser Ile Leu Val Pro       Gly       Asp       Glu       Asp       Lys       Leu
                                                                                                420
400                                     410                        420
TAT GTA CGC GTA TTC ATC ACA ACA  ...  AAT  ...  CTT  ...  TAC  ...  GTT  ...  TCC  ...  GAT       1412
Tyr Val Arg Val Phe Ile Thr Thr       Asn       Leu       Tyr       Val       Ser       Asp
```

```
          430                                              440                                      450
CTC CCC CTC AAT GTT TTC CGT GAG ACT CTT CAG CAA CAT AAA TTG CTC AAG AAG CTT GTC CGA AAA ACT CTG GAC ATG  1502
Leu Pro Leu Asn Val Phe Arg Glu Thr Leu Gln Gln His Lys Leu Leu Lys Lys Leu Val Arg Lys Thr Leu Asp Met
*********

          460                                              470                                      480
ATC AAG ATT GCT GAT GAG AAG TAT AAC GAC ACT TTC TGG AAG GAG TTC ACG AAT ATC AAG CTT GGT GTG ATT GAC CAC TCA  1592
Ile Lys Ile Ala Asp Glu Lys Tyr Asn Asp Thr Phe Trp Lys Glu Phe Thr Asn Ile Lys Leu Gly Val Ile Asp His Ser
                        *****

          490                                              500                                      510
AAT CGA ACA CGG CTT AAA CTT AGG TTC CAG TCT TCT CAC CAT TCA ACT GAC ATT ACT AGT TTA GAC CAG TAT GAA AGA ATG  1682
Asn Arg Thr Arg Leu Lys Leu Arg Phe Gln Ser Ser His His Ser Thr Asp Ile Thr Ser Leu Asp Gln Tyr Glu Arg Met
*******

          520                                              530                                      540
AAG GAA AAA CAG GAC AAA ATC TAC TTC GCT ATG GGG TCA GCG GAG AGC CCA TTT GTT GAT CTT AGA CTT CTG AAG AAG  1772
Lys Glu Lys Gln Asp Lys Ile Tyr Phe Ala Met Gly Ser Ala Gln Ser Pro Phe Val Asp Leu Arg Leu Leu Lys Lys

          550                                              560                                      570
GGC TAT GAA GTC ATT CTC ACA GAG GTG GAT TTT GAT GAG CAG GCT GCA CTT CCC AAG TTT GAT AGG CAG AAT GTT  1862
Gly Tyr Glu Val Ile Thr Leu Tyr Phe Asp Val Pro Asp Glu Gln Ala Leu Pro Lys Phe Asp Arg Gln Asn Val

          580                                              590                                      600
GCC AAA GGG GTG AGT GAG TTT GAG GAT GAG AGT ACA AAG AGT CGG GAG ACA TTT GAA CTG CTC AAC TGG  1952
Ala Lys Gly Gly Val Lys Phe Glu Asp Glu Ser Thr Lys Ser Arg Glu Thr Phe Glu Leu Leu Asn Trp

          610                                              620                                      630
ATG AAG GAT AAG CTC ATA GAA GCT GTG TCG CAG GCA GCA TAC CAG ATG CGC CTC TGT GCT GTG GCC AGT CAG  2042
Met Lys Asp Lys Leu Ala Lys Ile Glu Ala Val Ser Ala Gln Ile Arg Ser Pro Cys Ala Val Ser Gln

          640                                              650                                      660
TAT GGA TGG TCT GGC AAC ATG AGG ATA ATG AAG CAA GCA TAC CAG ACG GAC ATC TCT ACA AAT TAC TAT GCC AGT CAA  2132
Tyr Gly Trp Ser Gly Asn Met Arg Ile Met Lys Gln Ala Tyr Gln Thr Asp Ile Ser Thr Asn Tyr Tyr Ala Ser Gln

          670                                              680                                      690
AAG AAA ACG TTC GAA ATC AAT CCA CGG CAC CCA CTG ATC AGA GAC ATG AAG AGG ATT AAG GAC GAT GAA AAG ACA GTC ATG  2222
Lys Lys Thr Phe Glu Ile Asn Pro Arg His Pro Leu Ile Arg Asp Met Lys Arg Ile Lys Asp Asp Glu Lys Thr Val Met

          700                                              710                                      720
GAT CTT GCT GTA GTT TTT GAA ACG AAC ACA GCG ACC CTT CGG TCA GGA TAT GCG TAT GAT GGA AGA ATA GAA AGA  2312
Asp Leu Ala Val Val Phe Glu Thr Asn Thr Ala Thr Leu Arg Ser Gly Tyr Ala Tyr Asp Gly Arg Ile Glu Arg

          730                                              740                                      750
ATG CTT CGC CTC AGT ATT GAC CTC AAC ATT GAC CCT GAA GAA GTG CAG GTG CAG GAA GAA CCA ACC TCA GAA GCA GAA  2402
Met Leu Arg Leu Ser Ile Asp Leu Asn Ile Asp Pro Glu Glu Val Gln Val Gln Glu Glu Pro Thr Ser Glu Ala Glu

          760                                              770                                      780
GAC TCA GAG CAG GAA GAT GGA GGG GGA GGG ACA GAT ATG GAT GCA GGA ACA GAA GAG GAG ACA TCT GAA ACA GAG AAG GAT GAA  2492
Asp Ser Glu Gln Glu Asp Gly Gly Gly Gly Thr Asp Met Asp Ala Gly Thr Glu Glu Glu Thr Ser Glu Thr Glu Lys Asp Glu

  781
TTG TAA  ATTATACTCCTCGCTATGAATCCCGTGTGGAGAGGGAATGTGAAGTTTGAAGTCATTCTTTGGATGCTTCCCCAAGCCTCCTTCTCCCCTGCGAC  2608
Leu Stop

TGTAAAATGTTGGATTATGGGTCACAGGAAGAAGTGGTTTTTTTTAGTTGAATTTTTTTTAACCATTGGTAATGTAAATTGTACTATTTAACTGACTATTGGTGTAAAATCTTGT  2727

CATGTGTATAAA ATAAA AAAGAATCCCAAAT  2759
```

271

signal peptide cleavage was between the Ala–Asp residues (positions −1 and 1, respectively), removing the first 21 amino acids. The ERp99 signal peptide has the typical features of a leader peptide. It possesses a hydrophobic central section preceded by a positively charged N-terminus and followed by two small hydrophobic residues separated by one amino acid (Watson, 1984). The calculated molecular weight of mature ERp99, 90,096, is in good agreement with the size of the polypeptide moiety of ERp99 estimated by SDS–PAGE. There are six potential N-glycosylation sites in the ERp99 sequence. Our earlier results have shown that only one of these sites is used under normal circumstances (Lewis *et al.*, 1985b). We have used this information to develop a possible model for the orientation of ERp99 in the membrane.

Hydropathy analysis of the protein sequence (Fig. 13) reveals, in addition to the expected hydrophobic signal peptide, a strongly hydrophobic 21–23 residue region, beginning at Leu 170, which has the characteristics of a membrane-spanning region. There are other regions of the polypeptide that have moderate hydrophobicity and that might also be membrane-spanning segments, but analysis of these regions using the algorithm of Eisenberg *et al.* (1984) showed that only the segment beginning at position 170 possessed the necessary characteristics of a membrane-spanning region. ERp99 also possesses two long, highly charged regions from positions 265 to 347 and from 731 to the C terminus.

A large data base (Barker *et al.*, 1987) was used to search for homology between ERp99 and other protein sequences. Extensive homology was observed between ERp99 and hsp90 of *Saccharomyces cerevisiae* and hsp83 of *Drosophila melanogaster*. A comparison of the sequences of ERp99 and hsp90 is shown in Fig. 14. The two sequences show 46.5% identity over a 719 amino acid overlap with the indicated gaps introduced for optimal alignment. Similarly, ERp99 showed 47.4% identity over a 382 amino acid overlap with hsp83. A search of the literature revealed that the

Fig. 13. Hydropathy plot of the predicted amino acid sequence of murine ERp99. The hydropathy plot was obtained using the algorithm and hydropathy values of Kyte and Doolittle (1982) for an amino acid segment length of 21. Hydrophobic regions are above the line and drophilic regions are below. The position of the putative membrane-spanning region is indicated by the dark box. Taken from Mazzarella and Green (1987).

```
        10         20         30         40         50         60         70         80         90        100        110        120
ERp99 MRVLWVLGLCCVLLTFGFVRADDEVDVDGTVEEDLGKSREGSRTDDEVVQREEEAIQLDGLNASQIRELREKSEKFAFQAEVNRMKLIINSLYKNKEIFLRELISNASDALDKIRILSL
                                                                        : :::   : :::::::: :::::::::::::::::::::
hsp90                                                MASETFEFQAEITQLMSLIINTVYSNKEIFLRELISNASDALDKIRYKSL
                                                                10         20         30         40         50

       130        140        150        160        170        180        190        200        210        220        230
TDENALAGNEELTVKIKCDKEKNLLHVTDTGVGMTREELVKNLGTIAKSGTSEFLNKMTEAQEDGQSTSELIGQFGVGFYSAFLVADKVIVTSKHNNDTQHIWESD-SNEFSVIADPRGN
       :  ::  :::   :::    :  : ::::::::::::::::::::::     ::::::   :::::::::::::: ::::: : : ::::  : :::::  :  :::: :: :
SDPKQLETEPDLFIRITPKPEQKVLEIRDSGIGMTKAELINNLGTIAKSGTKAFMEAL-----SAGADVSMIGQFGVGFYSLFLVADRVQVISKSNDDEQYIWESNAGGSFTVTLDEVNE
        60         70         80         90        100        110        120        130        140        150        160

       240        250        260        270        280        290        300        310        320        330        340        350
TLGRGTTITLVLKEEASDYLELDTIKNLVRKYSQFINFPIYVWSSKTETVEEPL------EEDEAAKEEESDDEAAVEE--EEEEKKPKTKKVEKTVWDWELMNDIKPIWQRPSKEVE
 ::::::  ::  :    :::  :: ::::::: :: :: :: :  ::::: :::       :: : ::::::::   : :   :::::::::::::: ::  : : :::
RIGRGTILRLFLKDDQLEYLEEKRIKEVIKRHSEFVAYPIQLVVTKEVEKEVPIPEEEKKDEKKDEKKDDKPKLEVDEEEKPKTKKVKEEVQEIELNKTKPLWTRNPSDIT
       170        180        190        200        210        220        230        240        250        260        270        280

       360        370        380        390        400        410        420        430        440        450        460        470
EDEYKAFYKSFSKESDDPMAYIHFTAEGEVTFKSILFVPTSAPRGLFDEYGSKKSDYIKLVRRVFITDDFHDMMPKYLNFVKGVVDSDDLPLNVSRETLQQHKLLKVIRKKLVRKTLDM
 :::::::: ::::::::: :: :  :: :: :: ::: :: ::  :::  :  : ::::::::::::::  :: :::::::::::::: :::::::: :: :::  ::::: :::  : :
QEEYNAFYKSISNDWEDPLVVKHFSVEGQLEFRAILFIPKRAPFDLFES--KKKKNNIKLYVRRVFITDEAEDLIPEWLSFVKGVVDSEDLPLNLSREMLQQNKIMKVIRKNIVKKLIEA
       290        300        310        320        330        340        350        360        370        380        390        400

       480        490        500        510        520        530        540        550        560        570        580        590
IKKIADEKYN-DTFWKFGTNIKLGVIEDHSNRTRLAKLLRFQSSHSSTDITSLDQYVERMEKQDKIYFMAGSSRKEAESSPFVERLLKKGYEVIYLTEPVDEYCIQALPEFDGKRFQN
 :::: :::   ::: ::  :::::::: :  ::  ::: :::: : ::: :::: ::: ::: ::::: :::: :::: :::::::::: :::: ::: :::::: : :: ::::
FNEIAEDSEQFEKFYSAFSKNIKLGVHEDTQNRAALAKLRYNSTKSVDELTSLTDYVTRMPEHQKNIYIITGESLKAVEKSPFLDALKAKNFEVLFLTDPIDEYAFTQLKEFEGKTLVD
       410        420        430        440        450        460        470        480        490        500        510        520

       600        610        620        630        640        650        660        670        680        690        700
VAKEGVKFDESEKTKESREATEKEFEPLLNWMKDKALKDKIEKAVVSQRLTESPCALVASQYGWSGNMERIMKAQAYQTGKDISTNYSASQKKTFEINPRHPLIRDMLRRIKE-DEDDKT
 :::   : :  ::::: ::::::::::: ::::::: :: ::: ::::: : ::   :  ::::: ::::::::::::::  ::    :: ::: ::::::   :::::::::  : ::::
ITKD-FELEETDEEKAEREKEIKEYEPLTKALKE-ILGDQVEKVVVSVKLLDAPAAIRTGQFGWSANMERIMKAQAL--RDSSMSSYMSSKKTFEISPKSPIIKELKKRVDEGGAQDKT
       530        540        550        560        570        580        590        600        610        620        630

       710        720        730        740        750        760        770        780        790        800
VMDLAVVLFETATLRSGYLLPDTKAYGDRIERMLRLSLNIDPEAQVEEPEEPEDTSED-AEDSEQDEGEEMDAGTEEEETEKESTEKDEL
 ::::  :::::: : ::  :: :::  : ::: :: ::::::::    :   :  :::  ::    :  :: :::  :  :
VKDLTKLLYETALLTSGFSLDEPTSFASRINRLISLGLNIDEDEETETAPEASTAAPVEEVPADTEMEEVD
       640        650        660        670        680        690        700
```

Fig. 14. Comparison of the sequence of murine ERp99 and hsp90 of *Saccharomyces cerevisiae*. The sequences have been aligned for maximum homology. The identical amino acids are indicated by the dots. Taken from Mazzarella and Green (1987).

first 14 amino acids of mature ERp99 are identical to the N-terminal sequence reported for GRP94 (Lee *et al.*, 1984) and that the C-terminal half of ERp99 is 96% identical to the sequence derived from a cDNA clone encoding the C-terminal half of GRP94 (Lee *et al.*, 1983; Munro and Pelham, 1986). Other similarities which strengthen this finding are that, like ERp99, GRP94 is a dimeric glycoprotein (Lee *et al.*, 1984) and is encoded by an abundant mRNA of approximately 3000 nucleotides (Lee *et al.*, 1983). Furthermore, a comparison of the ERp99 cDNA sequence with the cDNA sequence determined for "hsp108" by Kulomaa (1986) indicates that these proteins are 91% identical. An alignment of these sequences is given in Fig. 15. Thus, the relationship between the ER protein ERp99, and hsp90, hsp83, and GRP94 is analogous to that reported for the rat liver hsp70-related protein, p72 and *D. melanogaster* hsp70, GRP78 and heavy chain binding protein (BiP) (Munro and Pelham, 1986).

III. CONCLUSIONS AND FUTURE DIRECTIONS

The elucidation of the specific sorting signal(s) that define a membrane protein as an ER protein will require further investigation. The studies that have been reported so far help to focus our attention on protein domains which could contain sorting signals dictating ER localization. The results described by Kabcenell and Atkinson (1985) in their investigation of the sorting of the rotavirus VP7 protein emphasized the role of an amino-terminal transmembrane domain as a positive signal for ER localization. In addition, the observations that the incorporation of a charged amino acid in the middle of the transmembrane region (Adams and Rose, 1985a), or the deletion of a substantial part of the transmembrane region (Adams and Rose, 1985b) blocked transport of VSV G protein to the cell surface, but not membrane insertion, are consistent with the view that the transmembrane domains of membrane proteins can contain important information for protein sorting as the proteins traverse the transport pathway. On the other hand, the observation that many ER proteins have a significant portion of their protein chain exposed on the cytoplasmic side of the ER membrane has been interpreted to support the possibility that the cytoplasmic domain of ER proteins may contain positive or negative sorting signals which result in ER localization of the particular protein (Kreibich *et al.*, 1983b; Paabo *et al.*, 1987). Recent observations by Munro and Pelham (1987) and Medda *et al.* (1987) implicate the importance of luminal domains resulting in the ER localization of ER proteins. As more ER proteins are purified, cloned, sequenced, and analyzed in

```
         10         20         30         40         50         60         70         80         90        100        110        120
MOUSE    MRVLWLGLCCVLLTFGFVRADDEVDVDGTVEEDLGKSREGSRTDDEVVQREEAIQLDGLNASQIRELREKSEKFAFQAEVNRMMKLIINSLYKNKEIFLRELISNASDALDKIRLISL
CHICKEN  .KSA.A.A.T.LAAS.T-.E-...................................................................K.I.......................

        130        140        150        160        170        180        190        200        210        220        230        240
MOUSE    TDENALAGNEELTVKIKCDKEKNLLHVTDTGVGMTREELVKNLGTIAKSGTSEFLNKMTEAQEDGQSTSELIGQFGVGFYSAFLVADKVIVTSKHNNDTQHIWESDSNEFSVIADPRGNT
CHICKEN  .............G....M....I..K..I...............................M.D.S...................R......................D........

        250        260        270        280        290        300        310        320        330        340        350        360
MOUSE    LGRGTTITLVLKEEASDYLELDTIKNLVRKYSQFINFPIYVWSSKTETVEEPLEDEAAKEEKKESDD-EAAVEEEEEEKKPKTKKVEKTVWDWELMNDIKPIWQRPSKEVEEDEYKAFYK
CHICKEN  ...............V....K...........V..E..................T..N.............................................................

        370        380        390        400        410        420        430        440        450        460        470        480
MOUSE    SFSKESDDPMAYIHFTAEGEVTFKSILFVPRSAPRGLFDEYGSKKSDYIKLYVRRVFITDDFHDMMPKYLNFVKGVVDSDDLPLNVSRETLQQHKLKLVIRKKLVRKTLDMIKKIADEKY
CHICKEN  T.......H..........N...............F................................................E................................
HAMSTER  T..................................................................................................................

        490        500        510        520        530        540        550        560        570        580        590        600
MOUSE    NDTFWKEFGTNIKLGVIEDHSNRTRLAKLLRFQSSHHSTDITSLDQYVERMKEKQDKIYFMAGSSRKEAESSPFVERLLKKGYEVIYLTEPVDEYCIQALFEFDGKRFQNVAKEGVKFDE
CHICKEN  ........V..............ESNL...........................A.............................................................E.
HAMSTER  

        610        620        630        640        650        660        670        680        690        700        710        720
MOUSE    SEKTKESREATEKEFEPLLNWMKDKALKDKIEKAVVSQRLTESPCALVASQYGWSGNMERIMKAQAYQTGKDISTNYYASQKKTEINPRHPLIRDMIRRIKEDEDKTVMDLAVVLFET
CHICKEN  ...S.....L.......R...........Q................................K...V..N....S.........................................
HAMSTER  ...N.....................................................A.......V......L...........................................

        730        740        750        760        770        780        790        800
MOUSE    ATLRSGVLLPDTKAVGDRIERMLRLSLNIDPEAQVEEPEEPEDTSEDAEDSEQDEGEMDAGTEEEEE--TEKEST-EKDEL
CHICKEN  .......M....E.............LD.D..........AA.E.----..V.DA.DS.----.Q...DV...A...
HAMSTER  ...........A..........T.--T.....E..V.......EQ.A....A....
```

Fig. 15. Amino acid conservation of ERp99 (GRP94). A comparison of the amino acid sequences of ERp99 from mouse (Mazzarella and Green, 1987), chicken (Kulomaa *et al.*, 1986), and hamster (Sorger and Pelham, 1987) is shown. Identity to murine ERp99 is indicated by a dot. In cases where the sequences differ, the particular amino acid found at that position is listed. A dash is used to indicate where gaps were introduced to produce optimal alignment. A comparison of mouse ERp99 and chicken ERp99 reveals that they have 91% homology; mouse ERp99 and hamster ERp99 have 96% identity.

sorting studies, a description of which signal or mixture of signals is operating to localize a particular protein to the ER membrane should emerge. Our studies of ERp99 should provide significant contributions to our understanding of the ER protein sorting problem.

We have formulated a simple model for the way in which ERp99 is oriented in the membrane (Fig. 16). This model is based on data from our earlier work (Lewis *et al.*, 1985b) as well as from the results described in this paper. ERp99 is synthesized as a precursor with a cleavable signal peptide (Figure 7). We have placed the N-terminus within the lumen to account for this fact. ERp99 is a glycoprotein with, in all likelihood, only one N-linked oligosaccharide. Analysis of the sequence indicates that ERp99 has six potential sites for N-linked glycoslation. Our model places only two of these sites in the lumen where they have the potential to be acceptors for the core oligosaccharide. The presence of two sites in the lumen, one utilized and the other cryptic, could explain our observations concerning ERp99 expression in COS cells. An overabundance of ERp99 in the ER membrane may allow detection of an abnormal N-glycosylation reaction at the second site. Analysis of the hydrophobicity of the ERp99 sequence indicates that the best candidate for a transmembrane region is the 21–23 residue hydrophobic segment beginning at position 170. Our model uses this segment as the membrane-spanning, stop transfer region allowing ERp99 to pass through the membrane only once with 75% of the protein remaining on the cytoplasmic side of the membrane. Analysis of the hydrophobicity and hydrophobic moment of this segment using the algorithm of Eisenberg *et al.* (1984) indicated that the values for this region are characteristic of transmembrane regions of multimeric proteins

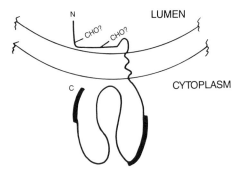

Fig. 16.   Model for the membrane orientation of murine ERp99. The model employs one membrane-spanning region (residues 170 to 192). Two potential glycosylation sites (CHO) are within the lumen. It is not known which is utilized as the actual acceptor. The positions of the highly hydrophilic regions are indicated by the thick lines. Taken from Mazzarella and Green (1987).

(e.g., membrane IgM). In this regard, it is interesting to note that when ERp99 is analyzed under nonreducing conditions, approximately 50% of the protein is found in a homodimer (R. Mazzarella, unpublished observation). The proposed orientation places all the hydrophilic regions of ERp99 on the cytoplasmic side of the membrane. This is similar to the model proposed for the orientation of the signal recognition particle (SRP) receptor (Lauffer *et al.*, 1985). It is also consistent with our earlier observation that ERp99 is completely susceptible to proteases in intact membrane preparations (Lewis *et al.*, 1985b). Under the conditions used in these experiments, known luminal proteins were protected from degradation. The newly synthesized IgA produced by the MOPC-315 plasmacytoma cell used in these studies was resistant to papain digestion, and another ERp, ERp59, which we have recently shown by immunological criteria (R. B. Freedman, personal communication) and DNA sequence analysis (R. Mazzarella and M. Green, unpublished results) to be the enzyme protein disulfide isomerase, a known luminal enzyme, was resistant to both papain and trypsin digestion (Lewis *et al.*, 1986). From our model, a peptide of approximately 20,000 Da would be expected to survive protease treatment. We have not observed such a fragment. It is possible that the anti-ERp99 polyclonal antibody used to analyze the protease protection experiments cannot recognize, or binds only weakly to, this fragment of ERp99. It should be possible to test our model. Using the deduced amino acid sequence, we can prepare antipeptide antibodies against different regions of ERp99. These antibodies can be assayed for their ability to react with ERp99 in intact microsomes. Antibodies against putative luminal regions should bind to the protein only after the disruption of microsomes or the solubilization of the protein. This approach was used successfully to analyze the orientation of HMG-CoA reductase, another ER protein (Liscum *et al.*, 1985).

A comparison of the ERp99 sequence with the sequences of other proteins has shown that ERp99 is homologous to hsp90 of *S. cerevisiae* and hsp83 of *D. melanogaster*. In addition, ERp99 is identical to GRP94. This observation is analogous to the finding by Munro and Pelham (1986) that the heat shock-related protein designated p72 is homologous to hsp70 from *D. melanogaster*, GRP78 and the ER protein BiP, the immunoglobulin heavy chain binding protein. It is interesting to note that, similar to the relationship of p72 and hsp70, the overlap of ERp99 with hsp90 is not observed until approximately 70 amino acids into the ERp99 sequence. This extra N-terminal sequence contains the signal peptide which is responsible for mediating the initial sorting of ERp99 to the ER membrane in addition to one of the proposed N-glycosylation sites. The other potential site (ERp99 position 86) is conserved between ERp99 and hsp90. In contrast to the overlap between p72 and hsp70, however, ERp99 also has extra amino acids at its C-terminal end. Our model of ERp99 places most of the protein on the cytoplasmic side of the ER membrane, unlike p72 which is in the lumen of the ER. Thus, our hypothesis differs from that of Pelham who has indicated that GRP94 may be a luminal ER protein by virtue of the fact that it, like GRP78 and protein disulfide isomerase (Pelham, 1986), possesses the common C-terminal sequence Lys-Asp-Glu-Leu-COOH (KDEL) (Edman *et al.*, 1985). It does not necessarily follow, however, that because ERp99 possess this sequence, it is a luminal protein. Indeed, if this sequence fulfills the same function for ERp99 that it does for GRP78, it would only be necessary that this part of the protein be in the lumen of the ER. In this case, our model (Fig. 16) would have to be modified by postulating that the ERp99 protein chain passes through the membrane twice, once at the indicated region and again near the C terminus. Analysis of the protein sequence, using the algorithm of Eisenberg *et al.* (1984), indicates that there is another potential membrane-spanning region, albeit a relatively poor one, from residues 687 to 707 that could serve this function. It is our aim to test our working model directly by using methods designed to produce specific truncations, internal deletions, and point mutations within the ERp99 coding sequence and to investigate the effect of these alterations on the sorting and membrane orientation of the mutated proteins expressed from these clones.

   In addition to using ERp99 as a model system to study ER protein sorting, it is interesting to determine the function of ERp99 in the ER membrane. No enzymatic activity has been assigned to hsp90, GRP94, or related proteins. Hsp90, however, is abundant in many cells and has been reported to have several activities. Hsp90 forms a complex with pp60[src] and a 50,000 Da phosphoprotein in cultured avian cells (Opperman *et al.*, 1981; Brugge *et al.*, 1981; Adkins *et al.*, 1982). Other viral transforming proteins containing tyrosine specific kinase activity, pp140[fps] and pp94[yes], have also been shown to be transiently associated with hsp90 (Lipsich *et al.*, 1982). Recently, hsp90 has been shown to bind to the steroid hormone receptor (Sanchez *et al.*, 1985; Schuh *et al.*, 1985; Catelli *et al.*, 1985) and to actin (Catelli *et al.*, 1985). It has been postulated that all of these findings point to the involvement of hsp90 as a carrier for the transportation of important biological molecules within the cell (Koyasu, 1986). In contrast to our finding that ERp99 is localized to the ER, other laboratories have reported that the larger glucose regulated protein is localized to the Golgi (Lin and Quelly, 1982; Welch *et al.*, 1983) or to the plasma membrane (Pouyssegur and Yamado, 1978; McCormick *et al.*, 1982). Given the fact that these proteins are identified merely by their molecular weight on SDS–PAGE and, in some cases, their isoelectric point and that there is a paucity of sequence information available about them, it is possible that these other, larger glucose-regulated proteins are not identical to ERp99. When more sequence data are obtained, it should be possible to reconcile these findings.

We have described what appears to be a novel relative in the hsp90 family of proteins. Future work will focus on trying to understand if and how the ER localization of ERp99 is involved in its function. It is possible, for example, that ERp99 has a function analogous to the one proposed for p72 (BiP) but performs this function on the cytoplasmic side of the ER membrane. It is also possible, however, that although there is extensive sequence homology among the hsp90 family, they fulfill different functions within the cell. Given the relatively strict localization of ERp99 to the RER and its transmembrane orientation in the ER membrane, we are currently investigating the possibility that ERp99 has a role in the translocation of nascent proteins through the ER membrane. Approaches being used include the expression of cloned ERp99 and derivatives in yeast to determine if ERp99 can restore viability to mutants in which the HSP90 genes have been inactivated. In addition, the use of antisense RNA strategies (Izant and Weintraub, 1984; Kim and Wold, 1985) could provide insights into the role of ERp99 in cells under normal and "stressful" growth conditions. Our study of ERp99 should provide valuable information concerning the mechanism of protein transport and sorting, as well as the possible functions of heat shock and "stress" proteins.

In this way, it should be possible to construct a sorting fate map of a protein, such as ERp99, in which alterations producing changes in the protein localization could be superimposed on the ERp99 amino acid sequence and structure. This map will be useful in several ways. By determining which alterations in the ERp99 sequence produce the positive effect of movement out of the ER compartment, we hope to gain insight into the manner in which ERp99 is initially localized in the ER membrane. In addition, depending upon how far along the normal transport pathway the altered ERp99 molecules travel, we may also gain insights into features of the protein that are necessary for the complete transport of the protein to the plasma membrane or into the medium as a secreted protein. Furthermore, if specific features of the ERp99 molecule (i.e., the transmembrane regions or the cytoplasmic domain) can be shown to be responsible for its ER localization, it should be possible to confirm this observation by constructing suitable chimeric molecules between ERp99 and either secretory protein sequences, or plasma membrane protein sequences. These types of experiments have the drawback that the construction of these chimeras result in "new" proteins that may not behave in simple, predictable ways. They should prove useful, however, to confirm the results of mutagenesis studies and to determine if any ERp99 domains contain ER localization information when excised from the molecule and fused to foreign proteins.

In the future, it should be possible to use the results obtained from sorting studies of proteins such as ERp99 to gain further insight into the mechanism of protein transport in eukaryotic cells. It would be worthwhile, for example, to determine if there are specific proteins which interact with ERp99 to either keep it in the ER normally or allow it to be transported out of the ER, if it is altered in a suitable way. In addition, it may be possible to select mutants of eukaryotic cells with altered components of the transport pathway by using the recombinant DNA constructions to create a situation in which such mutants can be selected. For example, if it was observed that a VSV-G–ERp99 chimera was sorted to the ER, it should be possible to construct a stable expression vector containing this chimera, transform a suitable host cell with it, and select a mutant cell in which this chimera was transported to the surface and could be detected by the expression of the VSV domain on the plasma membrane. This type of study should help us to begin to elucidate the cellular machinery involved in protein transport. Finally, it should be possible to accomplish the expression of ERp99 in yeast in order to take advantage of both the availability of secretion mutants in that system (for review, see Schekman, 1985) and the possibility of producing large amounts of altered ERp99 molecules for chemical and physical studies which could complement our studies of the effects of protein sequence alterations on ERp99 sorting. We believe that the study of the sorting of proteins such as ERp99 will provide a rich source of experimental results and scientific insights that will make a significant contribution to our knowledge of protein transport and sorting in general.

ACKNOWLEDGMENTS

The research described in this chapter was supported in part by U.S. Public Health Service Grant GM-33575, BRSG RR05388 No. 578, and CRIC 85007. The authors gratefully acknowledge Sharon McKenzie for her patience and secretarial skills.

REFERENCES[2]

Adams, G. A., and Rose, J. K. (1985a). Incorporation of a charged amino acid into the membrane-spanning domain blocks cell surface transport but not membrane anchoring of a viral glycoprotein. *Mol. Cell. Biol.* **5,** 1442–1448.

[2] This article reflects a literature search of August 1, 1987.

Adams, G. A., and Rose, J. K. (1985b). Structural requirements of a membrane-spanning domain for protein anchoring and cell surface transport. *Cell* **41,** 1007–1015.

Adkins, B., Hunter, T., and Sefton, B. M. (1982). The transforming proteins of PRCII virus and Rous sarcoma virus form a complex with the same two cellular phosphoproteins. *J. Virol.* **43,** 448–455.

Anderson, D. J., Mostov, K. E., and Bloble, G. (1983). Mechanisms of integration of *de novo*-synthesized polypeptides into membranes: Signal-recognition particle is required for integration into microsomal membranes of calcium ATPase and of lens MP26 but not of cytochrome b_5. *Proc. Natl. Acad. Sci. U.S.A.* **80,** 7249–7253.

Armstrong, J., Niemann, H., Smeekens, S., Rottier, P., and Warren, G. (1984). Sequence and topology of a model intracellular membrane protein, E1 glycoprotein from a coronavirus. *Nature (London)* **308,** 751–752.

Barker, W. C., Hunt, L. T., George, D. G., Yeh, L. S., Chen, H. R., Blomquist, M. C., Seibel-Ross, E. I., Elzanowski, A., Hong, M. K., Ferrick, D. A., Bair, J. K., Chen, S. L., and Ledley, R. S. (1986). (3800 sequences). Protein sequence database of the Protein Identification Resource, National Biomedical Research Foundation, Washington, D.C.

Bar-Nun, S., Kreibich, G., Adesnik, M., Alterman, L., Negishi, M., and Sabatini, D. D. (1980). Synthesis and insertion of cytochrome *P*-450 into endoplasmic reticulum membranes. *Proc. Natl. Acad. Sci. U.S.A.* **77,** 965–969.

Bell, R. M., and Coleman, R. A. (1980). Enzymes of glycerolipid synthesis n eukaryotes. *Annu. Rev. Biochem.* **49,** 459–487.

Bell, R. M., Ballas, L. M., and Coleman, R. A. (1981). Lipid topogenesis. *J. Lipid Res.* **22,** 391–403.

Bell-Quint, J., Forte, T., and Graham, P. (1981). Glycosylation of apolipoproteins by cultured rat hepatocytes. *Biochem. J.* **200,** 409–414.

Berg, R. A., and Prockop, D. J. (1973). Affinity column purification of protocollagen proline hydroxylase from chick embryos and further characterization of the enzyme. *J. Biol. Chem.* **248,** 1175–1182.

Berg, R. A., Kao, W. W.-Y., and Kedersha, N. L. (1980). The assembly of tetrameric prolyl hydroxylase in tendon fibroblasts from newly synthesized α-subunits and from preformed cross-reacting protein. *Biochem. J.* **189,** 491–499.

Black, S. D., French, J. S., Williams, C. H., Jr., and Coon, M. J. (1979). Role of a hydrophobic polypeptide in the N-terminal region of NADPH–cytochrome *P*-450 reductase in the complex formation with $P\text{-}450_{LM}$. *Biochem. Biophys. Res. Commun.* **91,** 1528–1535.

Bole, P. G., Hendershot, L. M., and Kearney, J. F. (1986). Postranslational association of immunoglobulin heavy chain binding protein with nascent heavy chains in nonsecreting and secreting hydridomas. *J. Cell Biol.* **102,** 1558–1566.

Both, G. W., Siegman, L. J., Bellamy, A. R., and Atkinson, P. H. (1983). Coding assignment and nucleotide sequence of simian rotavirus SA11 gene segment 10: Location of glycosylation sites suggests that the signal peptide is not cleaved. *J. Virol.* **48,** 335–339.

Brands, R., Snider, M. D., Hino, Y., Park, S. S., Gelboin, H. V., and Rothman, J. E. (1985). Retention of membrane proteins of the endoplasmic reticulum. *J. Cell. Biol.* **101,** 1724–1732.

Brown, D. A., and Simoni, R. D. (1984). Biogenesis of 3-hydroxy-3-methylglutaryl-coenzyme A reductase, an integral membrane protein of the endoplasmic reticulum. *Proc. Natl. Acad. Sci. U.S.A.* **81,** 1674–1678.

Brugge, J. S., Erikson, E., and Erikson, R. L. (1981). The specific interaction of the Rous sarcoma virus transforming protein, pp60[src], with two cellular proteins. *Cell* **25,** 363–372.

Catelli, M. G., Binart, N., Jung-Testas, I., Renoir, J. M., Baulieu, E. E., Feramisco, J. R., and Welch, W. J. (1985). The common 90-Kd protein component of non-transformed '8S' steriod receptors is a heat-shock protein. *EMBO J.* **4**, 3131–3135.

Chin, D. L., Gil, G., Russell, D. W., Liscum, L., Luskey, K. L., Basu, S. K., Okayama, H., Berg, B., Goldstein, J. L., and Brown, M. S. (1984). Nucleotide sequence of 3-hydroxy-3-methyl-glutaryl-coenzyme A reductase, a glycoprotein of the endoplasmic reticulum. *Nature (London)* **308**, 613–617.

Dale, R. M. K., McClure, B. A., and Houchins, J. P. (1985). A rapid single-stranded cloning strategy for producing a sequential series of overlapping clones for use in DNA sequencing: Application to sequencing the corn mitochondrial 18S rDNA. *Plasmid* **13**, 31–40.

Doyle, C., Roth, M. G., Sambrook, J., and Gething, M.-J. (1985). Mutations in the cytoplasmic domain of the influenza hemagglutinin affect different stages of intracellular transport. *J. Cell. Biol.* **100**, 704–714.

Edman, J. C., Ellis, L., Blacher, R. W., Roth, R. A., and Rutter, W. J. (1985). Sequence of protein disulfide isomerase and implications of its relationship to thionedoxin. *Nature (London)* **317**, 267–270.

Eisenberg, D., Schwarz, E., Komaromy, M., and Wall, R. (1984). Analysis of membrane and surface protein sequences with the hydrophobic moment plot. *J. Mol. Biol.* **179**, 125–142.

Ericson, B. L., Graham, D. Y., Mason, B. B., Hansson, H. H., and Estes, M. K. (1983). Two types of glycoprotein precursors are produced by the simian rotavirus SA11. *Virology* **127**, 320–332.

Evans, E. A., Gilmore, R., and Blobel, G. (1986). Purification of microsomal signal peptidase as a complex. *Proc. Natl. Acad. Sci. U.S.A.* **83**, 581–585.

Freedman, R. B., and Hillson, D. A. (1980). In "The Enzymology of Post-Translational Modification of Proteins (R. B. Freedman and H. C. Hawkins, eds.), Vol. I, pp. 167–212. Academic Press, London.

Fujii-Kariyama, Y., Mizukami, Y., Kawajiri, K., Sogawa, K., and Muramatsu, M. (1982). Primary structure of a cytochrome *P*-450: Coding nucleotide sequence of phenobarbital-inducible cytochrome *P*-450 cDNA from rat liver. *Proc. Natl. Acad. Sci. U.S.A.* **79**, 2793–2797.

Galteau, M.-M., Antoine, B., and Reggio, H. (1985). Epoxide hydrolase is a marker for the smooth endoplasmic reticulum. *EMBO J.* **11**, 2793–2800.

Garoff, H. (1985). Using recombinant DNA techniques to study protein targeting in the eukaryotic cell. *Annu. Rev. Cell Biol.* **1**, 403–445.

Gething, M.-J., and Sambrook, J. (1982). Construction of influenza haemagglutinin genes that code for intracellular and secreted forms of the protein. *Nature (London)* **300**, 598–603.

Gething, M.-J., McCammon, K., and Sambrook, S. (1986). Expression of wild-type and mutant forms of influenza hemagglutinin: The role of folding in intracellular transport. *Cell* **46**, 939–950.

Gilmore, R., and Blobel, G. (1985). Translocation of secretory proteins across the microsomal membrane occurs through an environment accessible to aqueous perturbants. *Cell* **42**, 497–505.

Gold, P., and Green, M. (1983). Partial purification and properties of a murine plasmacytoma glucosyltransferase. *J. Biol. Chem.* **258**, 12967–12975.

Goldberg, D. E., and Kornfeld, S. (1983). Evidence for extensive subcellular organization of asparagine-linked oligosaccharide processing and lysosomal enzyme phosphorylation. *J. Biol. Chem.* **258**, 3159–3165.

Gottlieb, T. A., and Wallace, R. A. (1982). Intracellular glycosylation of vitellogenin in the liver of estrogen-stimulated Xenopus laevis. *J. Biol. Chem.* **257**, 95–103.

Green, M. (1979). Cotranslational cleavage of immunoglobulin light chain precursors by plasmacytoma microsomes. *Arch. Biochem. Biophys.* **195**, 368–377.

Green, M., and Gleiber, W. E. (1980). Cotranslational cleavage and glycosylation of the mineral oil-induced plasmacytoma-46B κ chain precursor by plasmacytoma microsomes. *Arch. Biochem. Biophys.* **199**, 37–42.

Griffen, M. J., and Noda, K.-I. (1980). Quantitation of epoxide hydrolase released from hyperplastic nodule and hepatoma microsomes. *Cancer Res.* **40**, 2768–2773.

Guan, J.-L., and Rose, J. K. (1984). Conversion of a secretory protein into a transmembrane protein results in its transport to the Golgi complex but not to the cell surface. *Cell* **37**, 779–787.

Guan, J.-L., Machamer, C. E., and Rose, J. K. (1985). Glycosylation allows cell-surface transport of an anchored secretory protein. *Cell* **42**, 489–496.

Haas, I. G., and Wabl, M. (1983). Immunoglobulin heavy chain binding protein. *Nature (London)* **306**, 387–389.

Hay, R., Bohni, P., and Gasser, S. (1984). How mitochondria import proteins. *Biochim. Biophys. Acta* **779**, 65–87.

Heinemann, F. S., and Ozols, J. (1982). The covalent structure of rabbit phenobarbital-induced cytochrome *P*-450. Partial amino acid sequence and order of cyanogen bromide peptides. *J. Biol. Chem.* **257**, 14988–14999.

Heinemann, F. S., and Ozols, J. (1984). The covalent structure of hepatic microsomal epoxide hydrolase. II. The complete amino acid sequence. *J. Biol. Chem.* **259**, 797–804.

Hickman, S., and Kornfeld, S. (1978). Effect of tunicamycin of IgM, IgA, and IgG secretion by mouse plasmacytoma cells. *J. Immunol.* **121**, 990–996.

Hino, Y., and Minakami, S. (1982). Hexose-6-phosphate dehydrogenase of rat liver microsomes. Isolation by affinity chromatography and properties. *J. Biol. Chem.* **257**, 2563–2568.

Hino, Y., and Rothman, J. E. (1985). Glucosidase II, a glycoprotein of the endoplasmic reticulum membrane. Proteolytic cleavage into enzymatically active fragments. *Biochemistry* **24**, 800–805.

Hubbard, S. C., and Ivatt, R. J. (1981). Synthesis and processing of asparagine-linked oligosaccharides. *Annu. Rev. Biochem.* **50**, 555–583.

Izant, J. G., and Weintraub, H. (1984). Inhibition of thymidine kinase gene expression by anti-sense RNA: A molecular approach to genetic analysis. *Cell* **36**, 1007–1015.

Kabcenell, A. K., and Atkinson, P. H. (1985). Processing of the rough endoplasmic reticulum membrane glycoproteins of rotavirus SA11. *J. Cell. Biol.* **101**, 1270–1280.

Kalderon, D., Roberts, B. L., Richardson, W. D., and Smith, A. E. (1984). A short amino acid sequence able to specify nuclear location. *Cell* **39**, 499–509.

Kedersha, N. L., Tkacz, J. S., and Berg, R. A. (1985a). Characterization of the oligosaccharides of prolyl hydroxylase, a microsomal glycoprotein. *Biochemistry* **24**, 5952–5960.

Kedersha, N. L., Tkacz, J. S., and Berg, R. A. (1985b). Biosynthesis of prolyl hydroxylase: Evidence for two separate dolichol-mediated pathways of glycosylation. *Biochemistry* **24**, 5960–5967.

Keller, R. K., and Swank, G. P. (1978). Tunicamycin does ot block ovalbumin secretion in the oviduct. *Biochem. Biophys. Res. Commun.* **85**, 762–768.

Kim, S. K., and Wold, B. J. (1985). Stable reduction of thymidine kinase activity in cells expressing high levels of anti-sense RNA. *Cell* **42**, 129–138.

Kivirikko, K. I., and Myllyla, R. (1980). In "The Enzymology of Post-Translational Modification of Proteins" (R. B. Freedman and H. C. Hawkins, eds.), Vol. I, pp. 157–212. Academic Press, London.

Kleinsck, D. A., Beattie, W. G., Tsai, M.-J., and O'Malley, B. W. (1986). Molecular cloning for a steriod-regulated 108K heat shock protein gene from hen oviduct. *Nucl. Acids Res.* **14,** 10053–10069.

Koivu, J., and Myllyla, R. (1986). Protein disulfide-isomerase retains procollagen prolyl 4-hydroxylase structure in its native configuration. *Biochemistry* **25,** 5982–5986.

Koivu, J., and Myllyla, R. (1987). Interchain disulfide bond formation in types I and II procollagen. Evidence for a protein disulfide isomerase catalyzing bond formation. *J. Biol. Chem.* **262,** 6159–6164.

Koivu, J., Myllyla, K., Helaakoski, T., Pihlajanemi, T., Tasaheh, K., and Kivirikko, K. I. (1987). A single polypeptide acts both as the β subunit of prolyl 4-hydroxylase and as protein disulfide-isomerase. *J. Biol. Chem.* **262,** 6447–6449.

Kornfeld, R., and Kornfeld, S. (1985). Assembly of asparagine-linked oligosaccharides. *Annu. Rev. Biochem.* **54,** 631–664.

Kornfeld, R., and Wold, W. S. M. (1981). Structure of the oligosaccharides of the glycoprotein coded by early region E3 of adenovirus 2. *J. Virol.* **40,** 440–449.

Koyasu, S., Nishida, E., Kadowaki, T., Matsuzaki, F., Iida, K., Harada, F., Ksauga, M., Sakai, H., and Yahara, I. (1986). Two mammalian heat shock proteins, HSP90 and HSP100, are actin-binding proteins. *Proc. Natl. Acad. Sci. U.S.A.* **83,** 8054–8058.

Kozak, M. (1983). Comparison of initiation of protein synthesis in prokaryotes, eukaryotes and organelles. *Microbiol. Rev.* **47,** 1–45.

Kreibich, G., Ulrich, B. L., and Sabatini, D. D. (1978). Proteins of rough microsomal membranes related to ribosome binding. I. Identification of ribophorins I and II, membrane proteins characteristic of rough microsomes. *J. Cell. Biol.* **77,** 464–487.

Kreibich, G., Marcantonio, E. E., and Sabatini, D. D. (1983a). Ribophorins I and II: Membrane proteins characteristic of the rough endoplasmic reticulum. *Methods Enzymol.* **96,** 520–530.

Kreibich, G., Sabatini, D. D., and Adesnik, M. (1983b). Biosynthesis of hepatocyte endoplasmic reticulum proteins. *Methods Enzymol.* **96,** 530–542.

Kulomaa, M. S., Weigel, N. L., Kleinsek, D. A., Beattie, W. G., Connelly, O. M., March, C., Zarucki-Schulze, T., Schrader, W. T., and O'Malley, B. W. (1986). Amino acid sequence of a chicken heat shock protein derived from the complementary DNA nucleotide sequence. *Biochemistry* **25,** 6244–6251.

Kyte, J., and Doolittle, R. F. (1982). A simple method of displaying the hydropathic character of a protein. *J. Mol. Biol.* **157,** 105–132.

Lambert, N., and Freedman, R. B. (1985). The latency of rat liver microsomal protein disulphide-isomerase. *Biochem. J.* **228,** 635–645.

Lauffer, L., Garcia, P. D., Harkins, R. N., Coussens, L., Ullrich, A., and Walter, P. (1985). Topology of signal recognition particle receptor in endoplasmic reticulum membrane. *Nature (London)* **318,** 334–338.

Lee, A. S., Delegeane, A. M., Baker, V., and Chow, P. C. (1983). Transcriptional regulation of two genes specifically induced by glucose starvation in a hamster mutant fibroblast cell line. *J. Biol. Chem.* **258,** 597–603.

Lee, A. S., Bell, J., and Ting, J. (1984). Biochemical characterization of the 94- and 78-kilodalton glucose-regulated proteins in hamster fibroblasts. *J. Biol. Chem.* **259,** 4616–4621.

Lewis, M. J., Mazzarella, R. A., and Green, M. (1985a). Structure and assembly of the endoplasmic reticulum. The synthesis of three major endoplasmic reticulum proteins during lipopolysaccharide-induced differentiation of murine lymphocytes. *J. Biol. Chem.* **260,** 3050–3057.

Lewis, M. J., Turco, S. J., and Green, M. (1985b). Structure and assembly of the endoplasmic reticulum. Biosynthetic sorting of endoplasmic reticulum proteins. *J. Biol. Chem.* **260,** 6926–6931.

Lewis, M. J., Mazzarella, R. A., and Green, M. (1986). Structure and assembly of the endoplasmic reticulum. Biosynthesis and intracellular sorting of ERp61, ERp59 and ERp49, three protein components of murine endoplasmic reticulum. *Arch. Biochem. Biophys.* **245,** 389–403.

Lin, J. J. C., and Queally, S. A. (1982). A monoclonal antibody that recognizes golgi-associated protein of cultured fibroblast cells. *J. Cell Biol.* **92,** 108–112.

Lipsich, L. A., Cutt, J. R., and Brugge, J. S. (1982). Association of the transforming proteins of Rous, Fujinami, and Y73 avian sarcoma viruses with the same two cellular proteins. *Mol. Cell. Biol.* **2,** 875–880.

Liscum, L., Cummings, R. D., Anderson, R. G. W., DeMartino, G. N., Goldstein, J. L., and Brown, M. S. (1983). 3-Hydroxy-3-methylglutaryl-CoA reductase: A transmembrane glycoprotein of the endoplasmic reticulum with N-linked "high mannose" oligosaccharides. *Proc. Natl. Acad. Sci. U.S.A.* **80,** 7165–7169.

Liscum, L., Finer-Moore, J., Stroud, R. M., Luskey, K. L., Brown, M. S., and Goldstein, J. L. (1985). Domain structure of 3-Hydroxy-3-methylglutaryl coenzyme A reductase, a glycoprotein of the endoplasmic reticulum. *J. Biol. Chem.* **260,** 522–530.

Lodish, H. F., Braell, W. A., Schwartz, A. L., Strous, G. J. A. M., and Zilberstein, A. (1981). Synthesis and assembly of membrane and organelle proteins. *Int. Rev. Cytol.* **12,** (Suppl.), 247–307.

Lodish, H. F., Kong, N., Snider, M., and Strous, G. J. A. M. (1983). Hepatoma secretory proteins migrate from rough endoplasmic reticulum to Golgi at characteristic rates. *Nature (London)* **304,** 80–83.

Matsuura, S., Fujii-Kuriyama, Y., and Tashiro, Y. (1978). Immunoelectron microscope localization of cytochrome *P*-450 on microsomes and other membrane structures of rat hepatocytes. *J. Cell. Biol.* **78,** 503–519.

Mazzarella, R. A., and Green, M. (1987). ERp99, an abundant, conserved glycoprotein of the endoplasmic reticulum, is homologous to the 90kDa heat shock protein (hsp90) and the 94-kDa glucose regulated protein (GRP94). *J. Biol. Chem.* **262,** 8875–8883.

McCandliss, R., Sloma, A., and Pestka, S. (1981). Isolation and cell-free translation of human interferon mRNA from fibroblasts and leukocytes. *Methods Enzymol.* **79,** 51–59.

McCormick, P. J., Millis, A. J. T., and Babiarz, B. (1982). Distribution of a 100K dalton glucose regulated cell surface protein in mammalian cell cultures and sectioned tissues. *Exp. Cell. Res.* **138,** 63–72.

Medda, S., Stevens, A. M., and Swank, R. T. (1987). Involvement of the esterase active site of egasyn in compartmentalization of β-glucuronidase within the endoplasmic reticulum. *Cell* **50,** 301–310.

Meyer, D. I., and Dobberstein, B. (1980a). A membrane component essential for vectorial translocation of nascent proteins across the endoplasmic reticulum: Requirements for its extraction and reassociation with the membrane. *J. Cell. Biol.* **87,** 498–502.

Meyer, D. I., and Dobberstein, B. (1980b). Identificaiton and characterization of a membrane component essential for the translocation of nascent proteins across the membrane of the endoplasmic reticulum. *J. Cell. Biol.* **87,** 503–508.

Moldave, K. (1985). Eukaryotic protein synthesis. *Annu. Rev. Biochem.* **54,** 1109–1149.

Mostov, K. E., de Bruyn Kops, A., and Deitcher, D. L. (1986). Deletion of the cytoplasmic domain of the polymeric immunoglobulin receptor prevents basolateral localization and endocytosis. *Cell* **47,** 359–364.

Munro, S., and Pelham, H. R. B. (1986). An hsp70-like protein in the ER: Identify with the 78 kd glucose-regulated protein and immunoglobulin heavy chain binding protein. *Cell* **46,** 291–300.

Munro, S., and Pelham, H. R. B. (1987). A C-terminal signal prevents secretion of luminal ER proteins. *Cell* **48,** 899–907.

Nelson, D. R., and Robinson, N. C. (1983). Membrane proteins: A summary of known structural information. *Methods Enzymol.* **97,** 571–618.

Niemann, H., and Klenk, H.-D. (1981). Coronavirus glycoprotein E1, a new type of viral glycoprotein. *J. Mol. Biol.* **153,** 993–1010.

Nilsson, O. S., DePierre, J. W., and Dallner, G. (1978). Investigation of the transverse topology of the microsomal membrane using combinations of proteases and the non-penetrating reagent diazobenzene sulfonate. *Biochem. Biophys. Acta* **511,** 93–104.

Okada, Y., Frey, A. B., Guenthner, T. M., Oesch, F., Sabatini, D. D., and Kreibich, G. (1982). Studies on the biosynthesis of microsomal membrane proteins: Site of synthesis and mode of insertion of cytochrome b_5, cytochrome b_5 reductase, cytochrome P-450 reductase and epoxide hydrolase. *Eur. J. Biochem.* **122,** 393–402.

Okayama, H., and Berg, P. (1982). High-efficiency cloning of full-length cDNAs. *Mol. Cell. Biol.* **2,** 161–170.

Olden, K., Pratt, R. M., and Yamada, K. M. (1978). Role of carbohydrate in protein secretion and turnover: Effects of tunicamycin on the major cell surface glycoprotein of chick embryo fibroblasts. *Cell* **13,** 461–473.

Olsen, B. R., Berg, R. A., Kishida, Y., and Prockop, D. J. (1973). Collagen synthesis: Localization of prolyl hydroxylase in tendon cells detected with ferritin-labeled antibodies. *Science* **182,** 825–827.

Oppermann, H., Levinson, A. D., Levintow, L., Varmus, H. E., Bishop, J. M., and Kawai, S. (1981). Two cellular proteins that immunoprecipitate with the transforming protein of Rous sarcoma virus. *Virology* **113,** 736–751.

Ozols, J., and Gerard, C. (1977). Primary structure of the membranous segment of cytochrome b_5. *Proc. Natl. Acad. Sci. U.S.A.* **74,** 3725–3729.

Ozols, J., and Heinemann, F. S. (1982). Chemical structure of rat liver cytochrome b_5. Isolation of peptides by high-pressure liquid chromatography. *Biochim. Biophys. Acta* **704,** 163–173.

Paabo, S., Bhat, B. M., Wold, W. S. M., and Peterson, P. A. (1987). A short sequence in the COOH-terminus makes an adenovirus membrane glycoprotein a resident of the endoplasmic reticulum. *Cell* **50,** 311–317.

Pelham, H. R. B. (1986). Speculations on the functions of the major heat shock and glucose-regulated proteins. *Cell* **46,** 959–961.

Persson, H., Jornvall, H., and Jalielski, J. (1980). Multiple on RNA species for the precursor to an adenovirus-encoded glycoprotein: Identification and structure of the signal sequence. *Proc. Natl. Acad. Sci. U.S.A.* **77,** 6349–6353.

Pihlajaniemi, T., Helaakoski, T., Tasanen, K., Myllyla, R., Huhtala, M. L., Koivu, J., and Kivirikko, K. I. (1987). Molecular cloning of the β-subunit of human prolyl 4-hydroxylase. This subunit and protein disulfide isomerase are products of the same gene. *EMBO J.* **6,** 643–649.

Porter, T. D., and Kasper, C. B. (1985). Coding nucleotide sequence of rat NADPH–

cytochrome *P*-450 oxidoreductase cDNA and identification of flavin binding domains. *Proc. Natl. Acad. Sci. U.S.A.* **82,** 973–977.

Poruchynsky, M. S., Tyndall, C., Both, G. W., Sato, F., Bellamy, A. R., and Atkinson, P. H. (1985). Deletions into an NH$_2$-terminal hydrophobic domain result in secretion of rotavirus VP7, a resident endoplasmic reticulum membrane glycoprotein. *J. Cell Biol.* **101,** 2199–2209.

Pouyssegur, J., and Yamada, K. M. (1978). Isolation and immunological characterization of a glucose-regulated fibroblast cell surface glycoprotein and its nonglycosylated precursor. *Cell* **13,** 139–150.

Puddington, L., Machamer, C. E., and Rose, J. K. (1986). Cytoplasmic domains of cellular and viral integral membrane proteins substitute for the cytoplasmic domain of the vesicular stomatitis virus glycoprotein in transport to the plasma membrane. *J. Cell Biol.* **102,** 2147–2157.

Rizzolo, L. J., Finidori, J., Gonzalez, A., Arpin, M., Ivanov, I. E., Adesnik, M., and Sabatini, D. D. (1985). Biosynthesis and intracellular sorting of growth hormone–viral envelope glycoprotein hybrids. *J. Cell. Biol.* **101,** 1351–1362.

Rose, J. K., and Bergmann, J. E. (1982). Expression from cloned cDNA of cell-surface secreted forms of the glycoprotein of vesicular stomatitis virus in eukaryotic cells. *Cell* **30,** 753–762.

Rose, J. K., and Bergmann, J. E. (1983). Altered cytoplasmic domains affect intracellular transport of the vesicular stomatitis virus glycoprotein. *Cell* **34,** 513–524.

Rosenfeld, M. G., Marcantonio, E. E., Hakimi, J., Ort, V. M., Atkinson, P. H., Sabatini, D. D., and Kreibich, G. (1984). Biosynthesis and processing of ribophorins in the endoplasmic reticulum. *J. Cell Biol.* **99,** 1076–1082.

Rothman, J. E. (1981). The Golgi apparatus: Two organelles in tandem. *Science* **213,** 1212–1219.

Rottier, P., Brandenburg, D., Armstrong, J., van der Zeijst, B., and Warren, G. (1984). Assembly in vitro of a spanning membrane protein of the endoplasmic reticulum: The E1 glycoprotein of the coronavirus mouse hepatitis virus A59. *Proc. Natl. Acad. Sci. U.S.A.* **81,** 1421–1425.

Sabatini, D. D., Kreibich, G., Morimoto, T., and Adesnik, M. (1982). Mechanisms for the incorporation of proteins in membranes and organelles. *J. Cell Biol.* **92,** 1–22.

Sargan, D. R., Tsai, M.-J., and O'Malley, B. W. (1986). hsp108, a novel heat shock inducible protein of chicken. *Biochemistry* **25,** 6252–6258.

Schekman, R. (1985). Protein localization and membrane traffic in yeast. *Annu. Rev. Cell. Biol.* **1,** 115–143.

Schmidt, G. W., Bartlett, S. G., Grossman, A. R., Cashmore, A. R., and Chua, N.-H. (1980). *In* "Genome Organization and Expression in Plants" (C. J. Leaver, ed.), pp. 337–351. Plenum, New York.

Sidman, C., Potash, M. S., and Kohler, G. (1981). Roles of protein and carbohydrates in glycoprotein processing and secretion. *J. Biol. Chem.* **256,** 13180–13187.

Sly, W. S. (1985). Receptor-mediated transport of acid hydrolases to lysosomes. *Curr. Topics Cell. Recognition* **28,** 27–38.

Sly, W. S., and Fischer, H. D. (1982). The phosphorylation recognition system for intracellular transport of lysosomal enzymes. *J. Cell. Biochem.* **18,** 67–85.

Snider, M. D., and Robbins, P. W. (1982). Transmembrane organization of protein glycosylation. Mature oligosaccharide–lipid is located on the luminal side of microsomes from Chinese hamster ovary cells. *J. Biol. Chem.* **257,** 6796–6801.

Snider, M. D., and Rogers, O. C. (1984). Transmembrane movement of oligosaccharide–lipids during glycoprotein synthesis. *Cell* **36,** 753–761.

Sanchez, E. R., Toft, D. O., Schlesinger, M. J., and Pratt, W. B. (1985). Evidence that the 90-kDa phosphoprotein associated with the untransformed L-cell glucocorticoid receptor is a murine heat shock protein. *J. Biol. Chem.* **260,** 12398–12401.

Schuh, S., Yonemoto, W., Brugge, J., Bauer, V. J., Riehl, R. M., Sullivan, W. P., and Taft, D. O. (1985). A 90,000-dalton binding protein common to both steriod receptors and the Rous sarcoma virus transforming protein, pp60$^{v\text{-src}}$. *J. Biol. Chem.* **260,** 14292–14296.

Sorger, P. K., and Pelham, H. R. B. (1987). The glucose regulated protein grp94 is related to heat shock protein hsp90. *J. Mol. Biol.* **194,** 341–344.

Struck, D. K., Sinta, P. B., Lane, M. D., and Lennarz, W. S. (1978). Effect of tunicamycin on the secretion of serum proteins by primary cultures of rat and chick hepatocytes. *J. Biol. Chem.* **253,** 5332–5337.

Swank, R. T., and Paigen, K. (1973). Biochemical and genetic evidence for a macromolecular β-glucoronidase complex in microsomal membranes. *J. Mol. Biol.* **77,** 371–389.

Szczesna, E., and Biome, I. (1976). mRNA-dependent synthesis of authentic precursor to human placental lactogen: Conversion to its mature hormone form in ascites cell-free extracts. *Proc. Natl. Acad. Sci. U.S.A.* **73,** 1179–1183.

Tajimi, S., Mihara, K., and Sato, R. (1979). Two-domain structure of microsomal reduced nicotinamide adenine dinucleotide–cytochrome b_5 reductase. *Arch. Biochem. Biophys.* **198,** 137–144.

Walter, P., Jackson, R. C., Marcus, M. M., Lingappa, V. R., and Blobel, G. (1979). Tryptic dissection and reconstitution of translocation activity for nascent presecretory proteins across microsomal membrane. *Proc. Natl. Acad. Sci. U.S.A.* **76,** 1795–1799.

Walter, P., Gilmore, R., and Blobel, G. (1984). Protein translocation across the endoplasmic reticulum. *Cell* **38,** 5–8.

Watson, M. E. E. (1984). Compilation of published signal sequences. *Nucleic Acids Res.* **12,** 5145–5164.

Welch, W. J., Gamels, J. I., Thomas, G. P., Lin, J. J. C., and Feramisco, J. R. (1983). Biochemical characterization of the mammalian stress proteins and identification of two stress proteins as glucose- and Ca^{2+}-ionophase-regulated proteins. *J. Biol. Chem.* **258,** 7102–7111.

Wieland, F. T., Gleason, M. L., Serafini, T. A., and Rothman, J. E. (1987). The rate of bulk flow from the endoplasmic reticulum to the cell surface. *Cell* **50,** 289–300.

Wickner, W. T., and Lodish, H. F. (1985). Multiple mechanisms of protein insertion into and across membranes. *Science* **230,** 400–407.

Wold, W. S. M., Cladaras, C., Deutscher, S. L., and Kapoor, Q. S. (1985). The 19 kDa glycoprotein coded by region E3 of adenovirus. *J. Biol. Chem.* **260,** 2424–2431.

Yubisui, T., Miyata, T., Iwanaga, S., Tamura, M., Yoshida, S., Takeshita, M., and Nakajima, H. (1984). Amino acid sequence of NADH–cytochrome b_5 reductase of human erythrocytes. *J. Biochem.* **96,** 579–582.

7

Functional Topology of Golgi Membranes

BECCA FLEISCHER

I. INTRODUCTION

The overall objective of this chapter is to summarize our current understanding of the organization of the enzymatic functions of the Golgi apparatus with particular emphasis on their transmembrane orientation and its consequences. Several comprehensive reviews have appeared detailing the central role of the Golgi apparatus in protein secretion (Palade, 1975; Farquhar and Palade, 1981) and in glycoprotein biosynthesis (Kornfeld and Kornfeld, 1985; Farquhar, 1986). At present most of the well-characterized functions of the Golgi are glycosylating enzymes involved in the terminal glycosylation of asparagine-linked glycoproteins or the elaboration of the carbohydrate region of glycosphingolipids. Thus most of this discussion will deal with the topology of glycosylation reactions in the Golgi apparatus; however, existing information on the orientation of other Golgi constituents will be summarized as well.

II. GENERAL PROPERTIES OF THE GOLGI APPARATUS

A. Structure

The Golgi apparatus is a complex membranous organelle, present in most eukaryotic cells, whose characteristic structure has been defined by electron microscopy. It consists of a series of flattened, membrane-bound cisternae, numerous peripheral small vesicles, and large secretory vesicles often clustered near one face of the stacked cisternae. The complex often shows a marked polarity both in structure and in cytochemical staining (Goldfischer, 1982). This, in addition to its characteristic orientation in secretory cells between the endoplasmic reticulum (ER) and the lumen, has led to one side being designated the cis face and the opposite side the trans face. The cis face apposes the rough endoplasmic reticulum while the trans side and its associated secretory vesicles and/or granules face the luminal plasma membrane.

B. Function

The Golgi functions to both transport and direct proteins and lipids from the endoplasmic reticulum, where they are newly synthesized, to their final destination. The latter is the extracellular space in the case of secreted proteins and glycosaminoglycans (GAGs), the plasma membrane in the case of glycolipids or plasma membrane glycoproteins, or the lysosome in the case of the acid hydrolases which characterize that organelle. Secreted proteins are first segregated into the lumen of the endoplasmic reticulum during their synthesis on ribosomes attached to the rough endoplasmic reticulum (RER).[1] The Golgi maintains the segregation of these components within its lumen until they can be directed to their final extracellular or lysosomal destination. The lumen of the Golgi is therefore

[1] Abbreviations: UMP, uridine 5'-monophosphate; CMP, cytidine 5'-monophosphate; dTMP, deoxythymidine 5'-monosphosphate; AMP, adenosine 5'-monophosphate; GMP, guanosine 5'-monophosphate; UDP, uridine diphosphate; ADP, adenosine diphosphate; ATP, adenosine triphosphate; UDPGal, uridine diphosphogalactose; UDPGlc, uridine diphosphoglucose; UDPGlcNAc, uridine diphospho-N-acetylglucosamine; UDPGalNAc, uridine diphospho-N-acetylgalactosamine; CMPNeuAc, cytidine monophospho-N-acetylneuraminic acid; G_{M2}, GalNAcGal (NeuAc)Glc ceramide; G_{M3}, Gal (NeuAc)Glc ceramide.

topologically continuous with the lumen of the ER and of lysosomes as well as with the external face of the plasma membrane and the extracellular space.

A large variety of enzymatic modifications of the transported proteins and lipids occurs in the Golgi apparatus. The best characterized of these are the enzymes involved in the terminal glycosylation of N-asparatgine-linked glycoproteins (Schachter and Roseman, 1980; Schachter et al., 1983; Kornfeld and Kornfeld, 1985), a major class of secreted proteins. Two distinct stages of glycosylation occur in the formation of N-aspara-gine-linked glycoproteins. "Core" sugars, that is, sugars close to the polypeptide backbone of the protein, are assembled stepwise in the RER on a carrier lipid, dolichol pyrophosphate, and then transferred to an asparagine residue of the protien close to the time the protein is being synthesized on polysomes attached to the membrane and concomitantly being secreted into the lumen. The more terminal sugars of these glyco-proteins are added in the Golgi apparatus by a different mechanism. They are attached stepwise to the nonreducing end of the carbohydrate side chain of the incomplete glycoprotein by a series of specific glycosyltrans-ferases using nucleotide sugars as donors (Schachter et al., 1983). Before terminal glycosylation can occur, excess glucose and mannose residues are removed by specific membrane-bound glucosidases in the endoplas-mic reticulum (Grinna and Robbins, 1979; Tabas and Kornfeld, 1982; Lucocq et al., 1986) and mannosidases in the Golgi apparatus (Tabas and Kornfeld, 1979; Tulsiani et al., 1982).

A second major type of oligosaccharide linkage that occurs in secreted and membrane glycoproteins involves an O-glycosidic linkage of N-ace-tylgalactosamine to serine or threonine residues of the protein. These glycosylations appear to occur entirely in the Golgi apparatus by the stepwise addition of sugar units directly from sugar nucleotides (Schach-ter and Roseman, 1980; Lennarz, 1983; Abeijon and Hirschberg, 1987).

In addition to glycoproteins, the Golgi apparatus is the site of most of the glycosylations involved in the formation of glycosphingolipids and gangliosides (Keenan et al., 1974; Fleischer, 1977). The initial glycosyla-tions of ceramide to form cerebrosides, however, have not been clearly localized in the Golgi (Fleischer, 1977).

Other modifications known to occur in the Golgi include sulfation of cerebrosides to from sulfatides (Fleischer and Zambrano, 1974), sulfation of the sugar moieties of glycosaminoglycans (De Luca et al., 1973) and glycoproteins (Hoshina et al., 1982), and the sulfation of tyrosine residues of protiens (Lee and Huttner, 1985). Not all sulfation reactions are clearly localized in the Golgi, however. Formation of steroid sulfates, for exam-ple, may occur in the cytoplasm of hepatocytes (Lyon and Jakoby, 1980).

It is almost an article of faith that the large variety of modifications of glycoproteins that occur in the Golgi apparatus are all somehow related to the central role the Golgi plays in directing constituents to the plasma membrane or lysosomes during biogenesis of these organelles or in resorting membrane constituents from contents during the recycling of endocytic vesicles (Farquhar, 1986). Several clear examples of the targeting function of carbohydrate moieties of glycoproteins modified in the Golgi apparatus have been elucidated. These include recognition sites in secreted glycoproteins for receptors on cell surfaces (Ashwell and Harford, 1982) and the mannose 6-phosphate moiety which appears on enzymes destined for packaging in lysosomes (Sly and Fischer, 1982).

C. Isolation and Overall Characterization

Although large, highly organized, and membranous, the Golgi apparatus can be isolated from a number of mammalian tissues in a fairly intact and recognizable form provided gentle homogenization is employed and osmotic shock minimized. The latter is difficult to avoid, since sucrose gradients are generally used as final steps in the purification procedure. Dilution and recentrifugation of fractions in order to recover the membranes results in fragmentation of the Golgi into vesicles whose intactness may vary considerably depending on the conditions employed (Fleischer, 1983a). The orientation of the vesicles formed is predominantly that of the original intact Golgi (Fleischer, 1981b), and the contents of secretory products even in shocked preparations can constitute as much as 30% of the total protein of the Golgi vesicles prepared from rat liver (Fleischer, 1974).

Isolation of the Golgi apparatus from tissues such as liver has led to extensive analysis of its overall chemistry and enzymology in comparison to other well-characterized membranous organelles such as endoplasmic reticulum and plasma membranes with which it interacts (Fleischer and Fleischer, 1977). Its unique function discussed in section I,B above is reflected in its unique protein composition. It is also low or lacking in a number of enzymes characteristic of the endoplasmic reticulum such as glucose-6-phosphatase, cytochrome P-450 and a number involved in lecithin biosynthesis. It is distinct from plasma membranes in having low levels of 5'-nucleotidase and Mg^{2+}-ATPase.

The Golgi apparatus is also unique in its lipid composition. It resembles the plasma membrane in its high content of neutral lipid, mainly cholesterol, and in its content of sphingomyelin, while its content of other phospholipids is very similar to that of the endoplasmic reticulum. It may not be possible to generalize about these relationships, however, since they do not appear to be valid in kidney where the Golgi lipid composition resembles that of liver Golgi more than that of kidney endoplasmic reticulum (Zambrano *et al.*, 1975). Golgi membranes from both tissues show a high content of ubiquinone, a lipid oxidation–reduction component originally discovered in mitochondria. Its function in the Golgi is unknown.

D. Functional Polarity of the Complex

The elucidation of the steps involved in the formation of glycoproteins containing *N*-asparagine-linked oligosaccharides of the complex type together with the isolation and purification of some of the Golgi enzymes involved in terminal glycosylation has provided direct evidence for the localization of several of these enzymes in discrete positions or subcompartments of the Golgi apparatus (for reviews, see Kornfeld and Kornfeld, 1985; Farquhar, 1986). Thus, the more terminal glycosyltransferases such as galactosyltransferase (Roth and Berger, 1982) and sialyltransferases (Roth *et al.*, 1985) appear to be concentrated in the trans cisternae while GlcNAc transferase II appears predominantly in medial cisternae (Dunphy and Rothman, 1985). Fractionation studies have provided additional evidence that enzymes involved in earlier events in Golgi processing of these glycoproteins such as mannosidase I and GlcNAc transferase I occur in more dense fractions of Golgi vesicles, possibly derived from cis Golgi, while less dense fractions are enriched in galactosyl- and sialyltransferases, derived from trans Golgi (Dunphy and Rothman, 1983). In addition to the heterogeneity due to the distribution of membrane-bound enzymes among the Golgi cisternae, there are believed to be variations in cholesterol content across the stacks (Orci *et al.*, 1981) and possibly of other lipids as well. The presence of associated secretory vesicles, endocytic vesicles, and vesicles involved in intraorganelle traffic (Farquhar, 1986) each with distinctive membrane composition as well as contents should also be kept in mind when considering the data obtained using isolated Golgi vesicle fractions. Correlations with morphological and histochemical data are obviously necessary to fully interpret the significance of biochemical findings on such a heterogeneous population.

III. TOPOLOGY OF MEMBRANE-BOUND FUNCTIONS

A. Methods

1. Cytochemistry

In a number of instances, histochemistry at the ultrastructural level has given useful evidence not only as to the localization of certain activities in various subcellular organelles but also as to the orientation of some membrane-bound enzymes. The central role that phosphatases play in the energy metabolism of the cell coupled with the insolubility and electron opaqueness of lead phosphate salts used to visualize product accumulation has resulted in extensive investigations on the localization of a wide range of enzymes which release inorganic phosphate from organic phosphate esters. Thus it was clear quite early in the studies aimed at isolating purified Golgi fractions that glucose-6-phosphatase activity is characteristic of endoplasmic reticulum in liver and is not present in the Golgi apparatus (Goldfischer *et al.,* 1964; Farquhar *et al.,* 1974). Phosphate released by this membrane-bound enzyme clearly occurs on the luminal side of the ER membrane *in situ* and can be used to determine the orientation and leakiness of isolated ER vesicles provided mannose 6-phosphate, which is not transported across the ER membrane, is used as substrate. In contrast to ER, Golgi membranes were shown to contain thiamin pyrophosphatase (TPPase) activity and, in some tissues, acid phosphatase activity. Nucleoside diphosphatase activity is present in both organelles in liver, and all of these activities show phosphate accumulation in the lumen (Goldfischer *et al.,* 1964). An interesting exception is 5'-nucleotidase which, in liver, is predominantly present in plasma membranes where it is oriented extracellularly but is also present to a much lesser degree in ER, where it is cytoplasmically oriented (Widnell, 1972). It also occurs in the Golgi apparatus where it has a dual orientation, namely, on the cytoplasmic side in Golgi cisternae and on the luminal side in Golgi secretory vesicles (Farquhar *et al.,* 1974). *In vitro* experiments on isolated ER and Golgi vesicles, to be discussed in greater detail below, are in agreement with these observations.

Autoradiography was used extensively in early studies on the role of the Golgi in cell function to define the types of products accumulating within the organelles and the sequence of events involved in their ultimate secretion. Although the ultrastructural resolution of the early studies is not sufficient to distinguish between localization in the luminal contents of the Golgi or in its membranes, the fact that the labeled products were finally secreted indicates that the products of galactosylation (Neutra and Leblond, 1966), sulfation (Berg and Young, 1971), and sialylation (Bennett and O'Shaughnessy, 1981) are present in the lumen of the Golgi apparatus in the cells studied, so that the acceptors of the transferases probably interact with the enzymes on the luminal face of the Golgi.

2. Accessibility of Substrates to Catalytic Sites

Although the isolation of subcellular organelles often results in the vesiculation of the membranes involved, the orientation of the vesicles is often not random. Two well-known examples are the rough endoplasmic reticulum, which normally vesiculates with the ribosomes (the cytoplasmic face) on the outside, and the mitochondrial inner membrane, which is oriented inside out on disruption of the mitochondrion by sonication (i.e., matrix face outside). As will be discussed in greater detail below, the Golgi apparatus similarly is vesiculated during isolation but remains largely oriented with its cytoplasmic face outside.

One way to demonstrate the orientation of membrane-bound enzymes in isolated vesicles is to use impermeable substrates and to compare the rate of reaction in intact vesicles with that in disrupted vesicles. Similarly, accessibility to high molecular weight inhibitors such as lectins or specific antibodies can be assessed. Methods of disrupting the vesicles often involve the use of detergents, and it is prudent to use a number of detergents of differing properties to rule out activation effects of a particular detergent on the enzymes themselves. Sometimes membrane-bound enzymes can be activated simply by solubilization from the membrane or removal of some inhibitor present, and such effects can mimic the "opening" effect of low levels of detergents on membrane vesicles.

3. Accessibility to Enzymatic Digestion

Since enzymes are large molecules that usually do not penetrate membranes, accessibility of components to digestion by proteolytic enzymes can be used to determine their orientation in the membrane. In this case also appropriate controls must be carried out to preclude disruption of the vesicles by the treatment as well as to show susceptibility of the component to breakdown once the vesicle has been "opened." Similarly, accessibility to phospholipases (Op den Kamp, 1979), to impermeable chemical probes (Gordesky et al., 1975), and to lipid exchange proteins (Zilversmit, 1983) can be used to evaluate the accessibility of various lipids on the outside of vesicles provided there is no intrinsic mechanism for rapid transmembrane movement of lipids from the inner to the outer side of the membrane as there appears to be in liver ER (Zilversmit and Hughes, 1977). Good agreement among the various methods has been obtained for the lipids of the red cell membrane but not for those of ER, probably owing to the greater stability of the erythrocyte membrane (van Deenen, 1981).

B. Orientation of Some Golgi Components

1. Lipids

There have been several studies on the relative distribution of individual classes of phospholipids on different sides of the Golgi membrane, with mostly contradictory results. The results and the possible technical problems which could lead to such contradictions have been summarized by Op den Kamp (1979). Methods which deplete one side of the membrane of polar head groups or complete lipid molecules may cause compensatory changes in the orientation of the remaining phospholipids in some membranes, although this does not appear to be the case in the red blood cell membrane. In this regard the use of lipid-exchange proteins, which can catalyze the exchange of a radioactive lipid from an added pool with accessible membrane lipids causing minimal changes in membrane structure, should be applied more widely to this problem. This approach is feasible only if rapid exchange across the membrane is not present endogenously. In contrast to ER, Golgi vesicles do not appear to undergo such exchange. We have treated Golgi vesicles with phosphatidylcholine-specific lipid-exchange protein (PC-PLEP) from bovine liver in the presence of unilamellar phospholipid vesicles containing [^{14}C]phosphatidylcholine (Fig. 1). With intact Golgi vesicles 54% of the PC of the vesicles is

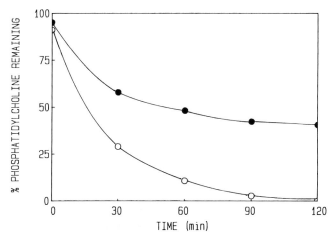

Fig. 1. Exchange of phosphatidylcholine (PC) of Golgi vesicles with added ^{14}C-labeled PC vesicles in the presence of PC-specific phospholipid-exchange protein (PC-PLEP) as a function of time. ^{14}C-Labeled PC vesicles were prepared by sonication and contained a trace level of [^{3}H]triolein in order to enable corrections to be made for nonspecific binding of the PC vesicles to Golgi. The sonicated PC was enriched for small unilamellar vesicles by chromatography on a Sepharose 4B column. For exchange measurements, Golgi vesicles (150 μg protein/ml) were incubated at 20°C with PC vesicles (40 μg phosphorus/ml) and partially purified PC-PLEP from bovine liver (8–10 exchange units/ml) in 0.25 M sucrose containing 10 mM potassium HEPES, pH 7.6. At various times, aliquots were taken, centrifuged in the cold to recover the Golgi, washed with cold buffer by centrifugation, and analyzed for the amount of PC exchanged. ●, Intact Golgi vesicles; ○, Golgi vesicles pretreated with filipin (0.2 mg/mg protein) at 37°C for 15 min.

readily exchangeable with the labeled PC of the added lipid vesicles while the remainder exchanges very slowly. If the Golgi is pretreated with filipin, however, under conditions which make the vesicles permeable to trypsin (a protein of about the same molecular weight as PC-PLEP), the Golgi PC is now 100% rapidly exchangeable.

We have also found that cholesterol present in Golgi membranes spontaneously and rapidly equilibrates with added unilamellar lipid vesicles containing [^{14}C]cholesterol as a single pool (E. Brandan and B. Fleischer, unpublished observations), as does the cholesterol present in most plasma membranes studied in other laboratories. *In vivo,* however, cholesterol obviously does not equilibrate freely among internal membranes. About 50% of the total cholesterol of the liver cell is in the plasma membrane, which does not equilibrate with internal cholesterol (Lange and Ramos, 1983). *In vivo* studies utilizing cultured Chinese hamster ovary (CHO) cells have shown that the transfer of newly synthesized PC from its site of synthesis in the ER to the plasma membrane is a very rapid process, with a $t_{1/2}$ at 25°C of 2 min, and is blocked by cooling to 2°C but not by energy poisons, monensin, or colchicine (Kaplan and Simoni, 1985a). The transfer of newly formed cholesterol, on the other hand, is about 5-fold slower, is energy dependent, and ceases at 15°C, whereupon the cholesterol accumulates in a light, smooth vesicle fraction. The transfer to the plasma membrane is not blocked by monensin or colchicine treatment of the cells (Kaplan and Simoni, 1985b). Thus cholesterol may be transferred intracellularly via a vesicular pathway which differs from the Golgi pathway of secretory proteins and plasma membrane proteins and glycolipids.

Some indirect evidence for the orientation of sphingomyelin and glycosphingolipids within the Golgi apparatus and their mode of transfer to the plasma membrane has been provided by the work of Lipsky and Pagano (1985). Using a fluorescent ceramide analog, these authors found that lung fibroblasts readily take up the lipid at 2°C into mitochondria, endoplasmic reticulum, and nuclear envelope membranes. After further incubation at 37°C the fluorescent lipids, now in the form of sphingomyelin and glucocerebrosides, are concentrated in the Golgi apparatus and later appear in the plasma membrane. Agents such as monensin and colchicine, which block the movement of secretory products through the Golgi or from the Golgi to the plasma membrane, respectively, block the transfer of fluorescent sphingolipid analogs from the Golgi to the plasma membrane. Fractionation of the cells on sucrose gradients showed considerable colocalization of the *in vitro* formation of both fluorescent sphingolipids with galactosyltransferase activity. These results lend support to the idea that the ceramide analog used readily equilibrated among the internal membranes and transversely across the membranes of the

Golgi. There, it is probably converted to sphingomyelin or glucocerebroside on the luminal side of the Golgi and thus becomes fixed, no longer able to equilibrate spontaneously back to the cytoplasmic side of the Golgi membrane or to other organelles via a cytoplasmic route. The lipids are transferred to the plasma membrane via secretory vesicles. It is not clear that this pathway of synthesis and transfer of these particular lipids holds for all cell types since sphingomyelin biosynthesis appears to be present predominantly in purified plasma membrane fractions from rat liver (Voelker and Kennedy, 1982), although purified Golgi fractions also showed significant activity.

2. Enzymes

Golgi fractions prepared from rat liver, using D_2O–sucrose gradients to minimize osmotic shock, have been employed in our laboratory to determine the orientation of several protein glycosylating enzymes and nucleotide phosphatases. Using accessibility to high molecular weight substrates in the presence and absence of detergents, we showed that the vesicles are greater than 90% oriented with protein galactosyltransferase and sialyltransferase on the luminal side (Fleischer, 1981a). This orientation was also confirmed using accessibility to degradation by trypsin following membrane disruption by filipin (Fleischer, 1981b). The action of the polyene antibiotic filipin differs in an important way from that of a detergent. It complexes specifically with cholesterol in the membrane to cause leakiness and eventually disruption rather than solubilizing all membrane components. Thus it is not likely to render an enzyme on a vesicle surface more accessible to a substrate or enzyme nor to selectively remove an inhibitor. Using this approach, we also demonstrated that a nucleoside diphosphatase unique to the Golgi, probably identical with the widely used cytochemical Golgi marker activity thiamin pyrophosphatase, was about 80% luminally oriented, while 5'-nucleotidase was equally distributed between the cytoplasmic and luminal sides of the Golgi vesicles (Brandan and Fleischer, 1982) in agreement with cytochemical evidence.

The orientation of glycolipid glycosyltransferases are more difficult to demonstrate since addition of exogenous lipid substrates usually is done in the form of detergent–lipid complexes in order to make the substrates dispersible in water and accessible to membrane-bound enzymes. Recently, however, it has been shown that codispersion of several glycolipid acceptors with phosphatidylglycerol allows lipid glycosyltransferase activities to be measured in the absence of detergents without opening rat liver Golgi vesicles to high molecular weight substrates (Yusuf *et al.*, 1984). Although the exact mechanism of the PG activation is not clear, it occurs only with intact Golgi vesicles and probably involves accessibility of the lipid substrates to the lipid glycosyltransferases. The effect was limited to UDPGalNAc:G_{M3} N-acetylgalactosaminyltransferase and UDPGal:G_{M2} galactosyltransferase and was not observed with lipid-specific glycosyltransferases using CMP sialic acid as sugar donor. As will be discussed further below, these authors also showed that the activities of both UDP sugar glycolipid glycosyltransferases are blocked by tunicamycin, which also blocks the incorporation of galactose from UDPGal to endogenous glycoproteins. These results indicate that both glycolipid glycosyltransferases are oriented similarly to the glycoprotein galactosyltransferase, that is, toward the lumen of the Golgi vesicles.

A luminal orientation for tyrosyl protein sulfotransferase is indicated by the inaccessibility of the enzyme to high molecular weight substrates added to intact Golgi vesicles (Lee and Huttner, 1985).

C. Implications of Functional Orientations

A major consequence of the luminal orientation of a number of Golgi glycosyltransferases which transfer sugars directly from nucleotide sugars to protein and lipid acceptors is that the donor nucleotide sugars must be accessible to the enzymes *in vivo*. In addition, the nucleotide products of the glycosyltransferases must not accumulate within the lumen of the Golgi, or product inhibition will occur. It was first proposed by Kuhn and White (1977), using isolated mammary gland Golgi, and later confirmed in our studies with rat liver Golgi (Brandan and Fleischer, 1982) that UDP formed by the luminally oriented galactosyltransferase is rapidly converted to the less inhibitory UMP by a nucleoside diphosphatase also luminally oriented. With liver Golgi vesicles we found that UMP appears to move rapidly out of the lumen and may be broken down further by 5'-nucleotidase on the cytoplasmic side of the vesicles. A similar fate is likely for CMP, the product of luminally oriented sialyltransferases. It is also possible that acid phosphatase activity, which is often seen cytochemically localized with areas of sialyltransferase activity, may function to cleave CMP intraluminally (Farquhar, 1986).

Similarly, for sulfation to occur intraluminally, as has been demonstrated so far only in the case of tyrosyl protein sulfation *in vitro,* 3'-phosphoadenosine 5'-phosphosulfate (PAPS) must be accessible to the luminally oriented transferase and the product, 3'-phosphoadenosine 5'-phosphate (PAP), must be removed. Analysis of rat liver Golgi vesicles prepared with a minimum of osmotic shock and a one-step flotation from a homogenate with no further washing reveals a significant level of AMP present in the Golgi but no detectable level of PAP (Fleischer, 1981a). It is likely that the product which accumulates as a result of sulfation reactions in the lumen is AMP.

Nucleotide sugars, with the exception of CMPNeuAc, are synthesized primarily in the cytoplasm (Coates *et al.*, 1980). CMPNeuAc synthesis occurs exclusively in the nucleus of hepatocytes (Kean, 1970; Coates *et al.*, 1980) while the synthesis of PAPS is probably cytoplasmic (Robbins and Lipmann, 1957). Thus it became clear that mechanisms must exist that allow the passage of nucleotide sugars and PAPS across the Golgi membrane into the lumen as well as the exit of nucleotide monophosphate products from the lumen into the cytoplasm.

IV. TRANSPORT PROPERTIES OF GOLGI MEMBRANES

A. Techniques for Measuring Transport

Three rather different techniques have been used in recent studies on the permeability of Golgi membranes to various molecules. The measurement of osmotic fragility has been utilized by White *et al.* (1980) to study the permeability of mammary gland Golgi to monosaccharides and electrolytes. The stability of the vesicles loaded *in vitro* with endogenously generated [^{14}C]lactose and suspended in lactose was tested after dilution into isosmolar solutions of the test solute by measuring the retention of the luminal [^{14}C]lactose after recovery of the vesicles by centrifugation or by filtration (White *et al.*, 1981a). This method has the advantage that only vesicles containing galactosyltransferase activity are measured so that heterogeneity of the vesicle preparation is irrelevant. It also allows measurement and comparison of solutes of high permeability. The disadvantage of this approach is that it may not discriminate between solutes of widely different affinities since only differences between rates of transport will similarly affect lysis rates.

A second approach is that utilized by Hirschberg and associates which measures the distribution of a radioactive solute between Golgi vesicles and the medium after recovery of the vesicles by centrifugation (Sommers and Hirschberg, 1982). Occluded medium in the pellet is corrected for by including radioactively labeled inulin in the assay which presumably does not penetrate or bind to the vesicles. In order to calculate the internal concentration of solute, the internal volume of the pellet must be determined. This is done in a parallel experiment using a radioactive pentrating solute, in this case labeled 2-deoxyglucose, which is assumed to equilibrate with the total internal volume of the vesicles and not be bound in any way. Incorporation of substrates into macromolecules is assessed by determining the fraction of solute not extractable from the pellet by cold trichloroacetic or perchloric acid. The method has been applied successfully to a number of subcellular organelles. Major disadvantages include the large amount of purified vesicles needed per assay tube (1–3 mg protein) since the internal volume of Golgi vesicles is quite small. In addition, binding of solutes to vesicles cannot be evaluated, and, unless uptake is quite slow, initial rates of uptake cannot be determined since the reaction is not efficiently stopped. To stop the reaction, the mix is put on ice, and incubation in the cold is continued for an hour during centrifugation.

A third method which we have applied successfully in our laboratory for studying transport in Golgi vesicles overcomes some of the limitations of the centrifugation method (Fleischer, 1983b). In this assay, concentrated vesicles are incubated with the substance to be tested in a small volume and the reaction stopped by diluting the samples 20- to 25-fold with cold isosmotic medium. The vesicles and their contents are recovered by rapid filtration using Millipore filters. Binding can be estimated with such a system either by permeabilizing the vesicles with filipin pretreatment or by including a low concentration of detergent in the stop solution. The level of detergent which is effective is determined by measuring the level necessary to release the bulk of the nucleotide or nucleotide sugar without solubilizing proteins from the vesicles. The assay uses less protein than the centrifugation method, can be used to determine initial kinetics for uptakes which are linear for at least 15 sec, and allows some estimate of binding. The main disadvantages are that high specific radioactivities of solutes must be used because the internal volume of the vesicles is small and thus the blank values of filters alone are high and errors large. If the affinity of the transport system is low and high concentrations of solutes must be used, initial kinetics again are difficult to achieve for simple facilitated diffusion because the small internal volume means a significant concentration of solute appears rapidly inside and impedes the rate of influx.

B. Permeability of Golgi Vesicles

1. Small Ions

Using the osmotic lysis approach, White *et al.* (1981b) showed that Golgi vesicles containing galactosyltransferase activity are quite permeable to chlorides of Li^+, Na^+, or K^+ and only somewhat less so to choline chloride or sodium citrate. Few specific transport mechanisms for inorganic ions have been described for Golgi. A proton pump driven by ATP hydrolysis has been described in isolated liver Golgi fractions (Glickmann *et al.*, 1983). Even greater specific activity, however, is found in ER. Other fractions, such as endosomes, also contain proton pumps. Since both ER and endosomes are contaminants of Golgi preparations, the significance of the finding is not clear. Immunocytochemical data on human fibroblasts support the view that the Golgi compartment is increasingly acidic from the cis to the trans side with secretory vesicles the most acidic Golgi compartment, although still much less acidic than lysosomes (Anderson and Pathak, 1985). The accumulation of the basic cytochemical probe is inhibited by treatment with monensin, an agent known to block specifically terminal glycosylation of secreted proteins (Tartakoff, 1983) and of glycosphingolipids (Saito et al., 1985) as well as sulfation of proteoglycans (Nishimoto et al., 1982). The ionophore also blocks transport of these products from the Golgi to the plasma membrane and causes swelling and vesiculation of the Golgi. Thus it appears that cationic gradients do exist across the Golgi membrane in vivo and that their perturbation by monensin causes physical swelling and disruption of a wide range of luminal Golgi functions. The exact nature and size of these gradients and how they are maintained are still not clear.

2. Monosaccharides and Nucleosides

Studies by White et al. (1980) have demonstrated that Golgi membranes from mammary gland are quite permeable to a wide range of monosaccharides, and expecially to either D- or L-glucose. The wide range of permeating solutes and the cutoff on the basis of molecular size above 300 led these authors to propose that Golgi membranes contain hydrophilic pores with a radius of about 0.54 nm (White et al., 1984).

Nucleotides are uncharged molecules with a molecular weight between that of mono- and disaccharides. We have found that uridine (244 Da) penetrates Golgi vesicles rapidly with a linear dependence on the outside concentration up to 5 mM (Fig. 2A). When efflux of uridine from Golgi vesicles is measured (Fig. 2B), a biphasic curve is obtained. About 50% of the uridine is released in about 1 min, and a second phase with a $t_{1/2}$ of 15 min is found for the remainder. These results may indicate heterogeneity in the total population of vesicles, i.e., about one-half very permeable to uridine and the other half much less permeable.

3. Disaccharides

In mammary gland, a major product of galactosyltransferase (in the presence of α-lactalbumin and glucose) is lactose, which has a molecular weight of 342 and is retained within Golgi vesicles from mammary gland (White et al., 1980). Similarly, exchange of [^{14}C]sucrose from inside passively loaded liver Golgi vesicles with outside sucrose (i.e., without net flux) is very slow compared to that measured using liver RER vesicles (Fig. 3).

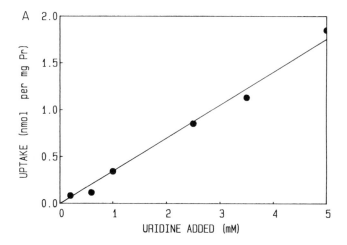

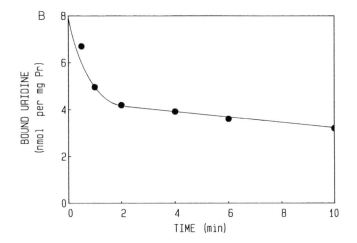

Fig. 2. (A) Uptake of uridine by Golgi vesicles. Golgi vesicles (3 mg protein/ml) were incubated with varying concentrations of [^{14}C]uridine at 37°C in 0.25 M sucrose containing 10 mM imidazole–HCl, pH 6.75, for 30 sec. The vesicles were then diluted with 20 volumes cold isosmotic buffered sucrose and filtered using 0.45-μm HA type Millipore filters and an Amicon filtration manifold. Values of filter blanks consisting of incubation mixtures without Golgi were subtracted from each experimental point. (B) Efflux of uridine from preloaded Golgi vesicles. Golgi vesicles (3 mg protein/ml) were incubated overnight in the cold in 0.25 M sucrose containing 10 mM imidazole–HCl, pH 6.75, and 4 mM [^{14}C]uridine. After 15 min at room temperature, the mixture was diluted 25-fold into isosmotic buffered sucrose, and aliquots were filtered as described for Fig. 2A. The filters are washed briefly with fresh buffered sucrose and counted to determine the amount of [^{14}C]uridine remaining in the vesicles.

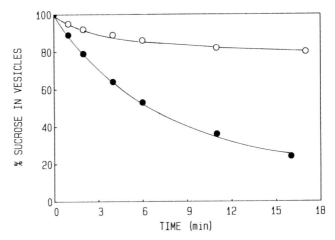

Fig. 3. [^{14}C]Sucrose exchange of rat liver Golgi or RER vesicles with the medium in the absence of net flux. Golgi (○) or RER (●) vesicles were equilibrated with 0.25 M [^{14}C]sucrose containing 10 mM imidazole–HCl, pH 7.4, overnight at 6°C. After warming to room temperature, the vesicles were diluted 250-fold into isosmotic unlabeled buffered sucrose, and the amount of [^{14}C]sucrose remaining in the vesicles was determined as a function of time as described in the legend to Fig. 2B.

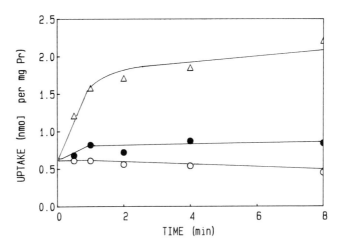

Fig. 4. Uptake of UMP by rat liver Golgi vesicles as a function of time. Golgi fractions (3 mg protein/ml) were incubated at 37°C for 3 min in 0.25 M sucrose containing 10 mM imidazole–HCl, pH 6.75, in a final volume of 100 μl. At various times after addition of [^{14}C]UMP, uptake was stopped by rapid addition of 2 ml ice-cold isosmotic buffered sucrose and aliquots filtered as described for Fig. 2A. Blank values, determined using Golgi vesicles pretreated with filipin (0.2 mg per mg protein) at 37°C for 15 min, were subtracted from each experimental value shown. ○, 0.15 mM UMP; ●, 1.0 mM UMP; △, 3.0 mM UMP.

4. Nucleotides and Nucleotide Sugars

Since the products that accumulate in the Golgi lumen during glycosylation and sulfation are probably nucleoside monophosphates, we have studied the permeability of rat liver Golgi vesicles to UMP and AMP directly, using the Millipore filtration assay. The uptake of UMP appears to be a two-stage process (Fig. 4). Below 1 mM, uptake is too rapid at room temperature for kinetic measurements with the Millipore filtration method. This uptake is abolished by disrupting the vesicles with filipin or by N_2 decompression, indicating that it is not due to simple binding to external sites on the vesicles. It is saturable with a K_b of about 20 μM at 37°C. The binding shows considerable specificity, with pyrimidine monophosphates competing more efficiently than purine monophosphates or UDPGal (Table I). The uptake is not detectable in RER or plasma membrane vesicles (Fleischer, 1983b). Above 1 mM UMP, a second slower accumulation of UMP can be demonstrated (Fig. 4). This uptake probably represents accumulation of free UMP within the vesicles.

Efflux of UMP shows similar behavior (Fig. 5). At the higher concentrations, efflux of both UMP and AMP from Golgi vesicles is much more rapid than from RER (Table II), even though the molecular weights of these compounds are equal to or greater than sucrose and the sucrose permeability of Golgi is much less than that of RER. Both the rate of uptake and efflux of UMP (measured above 1 mM) are saturable but appear cooperative with a Hill coefficient of 4 (Fig. 6). The efflux rate is not affected by the presence of an equal concentration of UDPGal outside the vesicles at either 1 or 4 mM UMP inside the vesicles. We can conclude from these observations that Golgi membranes are permeable to nucleoside monophosphates probably via carrier mechanisms as discussed further in Section IV,C.

TABLE I

Uptake of Mononucleotides by Golgi Vesicles[a]

Nucleotide	K_b (μM)	B_{max} (nmol/mg protein)	Type of inhibition versus UMP	K_i (μM)
UMP	17	0.90	—	—
CMP	21	0.75	Competitive	82
dTMP	13	0.60	Competitive	20
AMP	47	0.18	Competitive	550
GMP	150	0.42	Noncompetitive	—

[a] Uptake was measured for 30 sec at 37°C using Millipore filtration. In all cases, blanks were run using Golgi vesicles pretreated with filipin and values subtracted to correct for simple binding of nucleotides.

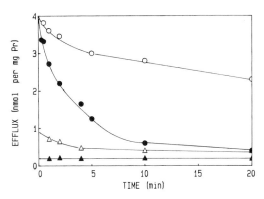

Fig. 5. Efflux of UMP from rat liver Golgi and RER vesicles as a function of time at 22°C. Vesicles were equilibrated at 22°C with various concentrations of [^{14}C]UMP for 3–4 hr, and the efflux rate was determined as described for Fig. 2A. ○, RER in 4 mM UMP; ●, Golgi in 4 mM UMP; △, Golgi in 1 mM UMP; ▲, Golgi in 0.1 mM UMP.

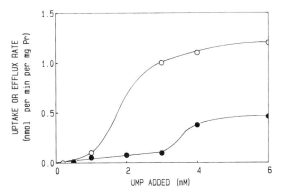

Fig. 6. Effect of UMP concentration on initial rate of UMP uptake (●) or efflux (○) from Golgi vesicles. Measurements made at 22°C as described for Figs. 4 and 5.

Uptake of a variety of nucleotide sugars has been demonstrated using either the Millipore filtration assay (Fleischer, 1983b) or the measurement of solute penetration in vesicles (Sommers and Hirschberg, 1982; Abeijon and Hirschberg, 1987). In addition, PAPS uptake has been demonstrated (Schwartz *et al.*, 1984) and been shown to be inhibited by atractylosides which also inhibit the ATP/ADP exchange transporter of mitochondria, although much more efficiently. Furthermore, PAPS is not transported into mitochondria (Capasso and Hirschberg, 1984a), indicating considerable difference between the two transport systems. Important recognition features for PAPS transport in Golgi are the purine base and the position

TABLE II

Comparison of Rates of Efflux of Nucleoside
Monophosphates from Golgi and RER Vesicles[a]

Nucleotide	Initial rate of efflux (nmol/min/mg protein)	
	RER	Golgi
UMP	0.3 ± 0.2	1.3 ± 0.3
AMP	0.2 ± 0.3	1.4 ± 0.2

[a] Efflux was measured at 22°C essentially as
described by Meissner and Allen (1981). The initial internal concentration of nucleotide in all
cases was 4 mM. Values are averages of duplicate
assays done on each of two preparations.

of the phosphate group on the ribose (Capasso and Hirschberg, 1984b).
Similarly, for CMPNeuAc transport, these studies show that UMP and
CMP are competitive inhibitors while AMP does not inhibit.

Studies in our laboratory on the uptake of UDPGal show that total
uptake is saturable, with more than one apparent K_m observed. The lowest apparent K_m (about 25 μM) is similar to that found using filipin-treated
Golgi, that is, for endogenous incorporation, in completely leaky vesicles.
This indicates that transport of UDPGal in intact vesicles is not rate
limiting for incorporation of galactose into endogenous acceptors. The
higher apparent K_m observed using higher UDPGal concentration reflects
accumulation of free UDPGal within the vesicles.

As shown by Yusuf et al. (1983) tunicamycin inhibits both ganglioside
biosynthesis and endogenous incorporation of UDPGal into proteins in
Golgi vesicles. We have confirmed these authors' suggestion that this
effect is due to inhibition of UDPGal transport in these vesicles. Total
accumulation of UDPGal within the vesicles as well as endogenous incorporation of [^{14}C]galactose is inhibited (Fig. 7A). The inhibition of total
uptake is mixed, with both apparent K_m and V_{max} affected (Fig. 7B). From
these data an apparent K_i of 0.65 mM can be estimated for tunicamycin
inhibition of UDPGal uptake under these assay conditions.

A summary of the apparent kinetic constants we have observed for the
accumulation of a number of sugar nucleotides is shown in Table III.
There is considerable specificity among the uridine nucleotide sugar
transport systems. UDPGal is a poor competitive inhibitor of UDPGlc
uptake ($K_i = 200 \mu M$) while UDPGlcNAc is a good competitive inhibitor
($K_i = 33 \mu M$). UDPGlc is a very poor competitive inhibitor of UDPGal

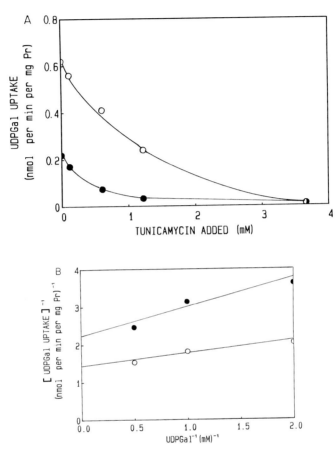

Fig. 7. (A) Effect of tunicamycin on uptake of UDPGal by Golgi vesicles. Golgi vesicles (3 mg protein/ml 0.25 M sucrose containing 100 mM imidazole–HCl, pH 6.9) are preincubated for 3 min at 37°C with various amounts of tunicamycin as indicated. UDP[^{14}C]Gal (2 mM in 0.25 M sucrose) is then added, and uptake is terminated after 3 min at 37°C by 20-fold dilution in cold isosmotic buffered sucrose followed by immediate Millipore filtration. To measure endogenous incorporation of [^{14}C]galactose, 0.015% Triton X-100 is included in the stop solution. ○, Total uptake of UDP[^{14}C]Gal; ●, endogenous incorporation of [^{14}C]galactose. (B) Effect of tunicamycin on the kinetic parameters of total UDPGal uptake in Golgi vesicles. Uptake was measured as described in (A). Tunicamycin, when present, was added together with UDP [^{14}C]Gal. ○, No tunicamycin added. An apparent K_m of 220 μM and a V_{max} of 0.7 nmol/min/mg protein for UDPGal uptake was observed. ●, Tunicamycin (0.6 mM) added. An apparent K_m of 330 μM and a V_{max} of 0.45 nmol/min/mg protein for UDPGal uptake was observed.

TABLE III

Uptake of Nucleotide Sugars by Golgi Vesicles[a]

Nucleotide sugar	K_m (μM)	V_{max} (nmol/min/mg protein)
UDPGal	420	0.80
UDPGlc	37	0.30
UDPGlcNAc	400	1.20
CMPNeuAc	200	0.50

[a] Initial rates of uptake were measured at 37°C using Millipore filtration. In all cases endogenous incorporation as well as binding has been corrected for by subtracting values obtained by releasing intravesicular nucleotide with Triton X-100.

uptake (K_i = 1.2 mM). From these studies we conclude there are at least two transport systems for uridine nucleotide sugars, one for UDPGal and a second for UDPGlc/UDPGlcNAc.

Strong evidence of the existence of specific protein carrier systems for sugar nucleotide translocation in Golgi membranes is the isolation of mutant CHO cells defective in such transport. Golgi vesicles from Lec 2 mutants have been shown to transport CMPNeuAc at 2% of the rate of wild-type cells (Deutscher et al., 1984). Partially purified Golgi preparations from Lec 8 and clone 13 cells, known to be deficient in the amount of galactose incorporated into glycoproteins and glycolipids (Briles et al., 1977; Stanley, 1983) were recently shown to have less than 5% of the uptake of UDPGal as wild-type Golgi (Deutscher and Hirschberg, 1986). In all cases the mutant cells contained normal levels of glycosyltransferases. The Lec 8 and clone 13 Golgi also had impaired uptake of CMP-sialic acid, but this may reflect lack of endogenous acceptors containing terminal galactose. Uptake as measured in these studies mainly reflects endogenous incorporation of the sugars, which is the rate-limiting step at the low concentrations of nucleotide sugars used.

C. Possible Mechanisms for Nucleotide Sugar Uptake

The suggestion that Golgi cisternae active in galactosylation may be acidic compartments raises the possibility that proton gradients could play a role in UDPGal uptake. We find, however, that addition of MgATP to Golgi vesicles, which should activate proton pumping, has no effect on the rate of UDPGal uptake. Uptake of UDPGal is also not affected by imposing a pH gradient across the vesicle membrane (Table IV) although uptake is decreased as the pH of the Golgi is increased during preincubation. As shown in Table IV, attempts to create a $\Delta\psi$ using a K^+ gradient in the presence of valinomycin also has no effect on the uptake.

The rate of UDPGal uptake can be stimulated if an acceptor for galactosyltransferase is introduced into the lumen of the Golgi vesicles. Preincubation of Golgi with N-acetylglucosamine results in a 4-fold enhancement of the rate of [^{14}C]galactose accumulation inside the vesicles (Fig. 8). Thus, UDPGal uptake appears to be a facilitated diffusion driven by the concentration gradient of the nucleotide sugar.

TABLE IV

Effect of ΔpH or $\Delta\psi$ on Uptake of UDPGal by Golgi Vesicles[a]

pH		UDPGal uptake (nmol/mg protein)	
Golgi	Medium	Nigericin present	Nigericin absent
6.0	7.0	0.82	0.81
7.0	7.0	0.65	0.73
8.0	7.0	0.52	0.55

0.1M KCl and valinomycin present (+) or absent (−)		
In Golgi	In Medium	UDPGal uptake
−	−	0.73
+	−	0.61
+	+	0.54
−	+	0.52

[a] Uptake was measured at 37°C for 3 min in the presence of 20 mM KCl and 2 mM UDPGal. To create a ΔpH, the Golgi vesicles were preincubated with 0.1 M imidazole–HCl at the indicated pH for 1 hr at 22°C, and uptake was initiated by addition of Golgi vesicles to UDPGal in 0.1 M imidazole buffer at the indicated pH. To create a $\Delta\psi$, Golgi was preincubated ±0.1 M KCl and valinomycin (10 μg/mg protein) in 0.1 M imidazole–HCl, pH 6.9, and was uptake measured as described above. Similar results are obtained if potassium gluconate is used instead of KCl to create a $\Delta\psi$.

clones of Chinese hamster ovary cells deficient in membrane sialic acid and galactose. *J. Biol. Chem.* **252,** 1107–1116.

Capasso, J. M., and Hirschberg, C. B. (1984a). Mechanisms of glycosylation and sulfation in the Golgi apparatus: Evidence for nucleotide sugar/nucleoside monophosphate and nucleotide sulfate/nucleoside monophosphate antiports in the Golgi apparatus membrane. *Proc. Natl. Acad. Sci. U.S.A.* **81,** 7051–7055.

Capasso, J. M., and Hirschberg, C. B. (1984b). Effect of nucleotides on translocation of sugar nucleotides and adenosine 3′-phosphate 5′-phosphosulfate into Golgi apparatus vesicles. *Biochim. Biophys. Acta* **177,** 133–139.

Coates, S. W., Gurney, T., Jr., Sommers, L. W., Yeh, M., and Hirschberg, C. B. (1980). Subcellular localization of sugar nucleotide synthetases. *J. Biol. Chem.* **255,** 9225–9229.

De Luca, S., Richmond, M. E., and Silbert, J. E. (1973). Biosynthesis of chondroitin sulfate—sulfation of the polysaccharide chains. *Biochemistry* **12,** 3911–3915.

Deutscher, S. L., and Hirschberg, C. B. (1986). Mechanism of galactosylation in the Golgi apparatus. A Chinese hamster ovary cell mutant deficient in translocation of UDP-galactose across Golgi vesicle membranes. *J. Biol. Chem.* **261,** 96–100.

Deutscher, S. L., Nuwayhid, N., Stanley, P., Briles, E. I. B., and Hirschberg, C. B. (1984). Translocation across Golgi vesicle membranes: A CHO glycosylation mutant deficient in CMP-sialic acid transport. *Cell* **39,** 295–299.

Dunphy, W. G., and Rothman, J. E. (1983). Compartmentation of asparagine-linked oligosaccharide processing in the Golgi apparatus. *J. Cell Biol.* **97,** 270–275.

Dunphy, W. G., and Rothman, J. E. (1985). Compartmental organization of the Golgi stack. *Cell* **42,** 13–21.

Farquhar, M. G. (1986). Progress in unraveling pathways of Golgi traffic. *Annu. Rev. Cell Biol.* **1,** 447–488.

Farquhar, M. G., and Palade, G. E. (1981). The Golgi apparatus (complex)—(1954–1981)—from artifact to center stage. *J. Cell Biol.* **91,** (No. 1, Pt. 2), 77s–103s.

Farquhar, M. G., Bergeron, J. J. M., and Palade, G. E. (1974). Cytochemistry of Golgi fractions prepared from rat liver. *J. Cell Biol.* **60,** 8–25.

Fleischer, B. (1974). Isolation and characterization of Golgi apparatus and membranes from rat liver. *Methods Enzymol.* **31,** 180–191.

Fleischer, B. (1977). Localization of some glycolipid glycosylating enzymes in the Golgi apparatus of rat kidney. *J. Supramol. Struct.* **7,** 79–89.

Fleischer, B. (1981a). The nucleotide content of rat liver Golgi vesicles. *Arch. Biochem. Biophys.* **212,** 602–610.

Fleischer, B. (1981b). Orientation of glycoprotein galactosyltransferase and sialyltransferase enzymes in vesicles derived from rat liver Golgi apparatus. *J. Cell Biol.* **89,** 246–255.

Fleischer, B. (1983a). Isolation of Golgi apparatus from rat liver using D_2O–sucrose gradients. *Methods Enzymol.* **98,** 60–67.

Fleischer, B. (1983b). Mechanism of glycosylation in the Golgi apparatus. *J. Histochem. Cytochem.* **31,** 1033–1040.

Fleischer, S., and Fleischer, B. (1977). Membrane diversity in the rat hepatocyte. *In* "Membrane Alterations as Basis of Liver Injury" (H. Popper, L. Bianchi, and W. Reutter, eds.), pp. 31–48. MTP Press, Lancaster, England.

Fleischer, B., and Zambrano, F. (1974). Golgi apparatus of rat kidney—preparation and role in sulfatide formation. *J. Biol. Chem.* **249,** 5995–6003.

Glickman, J., Croen, K., Kelly, S., and Al-Awqati, Q. (1983). Golgi membranes contain an electrogenic H^+ pump in parallel to a chloride conductance. *J. Cell Biol.* **97,** 1303–1308.

Goldfischer, S. (1982). The internal reticular apparatus of Camillo Golgi: A complex heterogeneous organelle, enriched in acid, neutral and alkaline phosphatases, and involved in glycosylation, secretion, membrane flow, lysosome formation, and intracellular digestion. *J. Histochem. Cytochem.* **30,** 717–733.

Goldfischer, S., Essner, E., and Novikoff, A. B. (1964). The localization of phosphatase activities at the level of ultrastructure. *J. Histochem. Cytochem.* **12,** 72–95.

Gordesky, S. E., Marinetti, G. V., and Love, R. (1975). The reaction of chemical probes with the erytyrocyte membrane. *J. Membr. Biol.* **20,** 111–132.

Grinna, L. S., and Robbins, P. W. (1979). Glycoprotein biosynthesis, rat liver microsomal glucosidases which process oligosaccharides. *J. Biol. Chem.* **254,** 8814–8818.

Hoshina, H., Glen, H., and Boime, I. (1982). Rat proopiomelanocortin contains sulfate. *Science* **217,** 63–64.

Kaplan, M. R., and Simoni, R. D. (1985a). Intracellular transport of phosphatidylcholine to the plasma membrane. *J. Cell Biol.* **101,** 441–445.

Kaplan, M. R., and Simoni, R. D. (1985b). Transport of cholesterol from the endoplasmic reticulum to the plasma membrane. *J. Cell Biol.* **101,** 446–453.

Kean, E. L. (1970). Nuclear cytidine 5'-monophosphosialic acid synthetase. *J. Biol. Chem.* **245,** 2301–2308.

Keenan, T. W., Morré, D. J., and Basu, S. (1974). Ganglioside biosynthesis. Concentration of glycosphingolipid glycosyltransferases in Golgi apparatus from rat liver. *J. Biol. Chem.* **249,** 310–315.

Kornfeld, R., and Kornfeld, S. (1985). Assembly of asparagine-linked oligosaccharides. *Annu. Rev. Biochem.* **54,** 631–664.

Kuhn, N. J., and White, A. (1977). The role of nucleoside diphosphatase in a uridine nucleotide cycle associated with lactose synthesis in rat mammary gland Golgi apparatus. *Biochem. J.* **168,** 423–433.

Lange, Y., and Ramos, B. V. (1983). Analysis of the distribution of cholesterol in the intact cell. *J. Biol. Chem.* **258,** 15130–15134.

Lee, R. W. H., and Huttner, W. B. (1985). (Glu^{62},Ala^{30},Tyr^8)$_n$ serves as high-affinity substrate for tyrosylprotein sulfotransferase: A Golgi enzyme. *Proc. Natl. Acad. Sci. U.S.A.* **82,** 6143–6147.

Lennarz, W. (1983). Overview: Role of intracellular membrane systems in glycosylation of proteins. *Methods Enzymol.* **98,** 91–97.

Lipsky, N. G., and Pagano, R. E. (1985). Intracellular translocation of fluorescent sphingolipids in cultured fibroblasts: Endogenously synthesized sphingomyelin and glucocerebroside analogues pass through the Golgi apparatus en route to the plasma membrane. *J. Cell Biol.* **100,** 27–34.

Lucocq, J. M., Brada, D., and Roth, J. (1986). Immunolocalization of the oligosaccharide trimming enzyme glucosidase II. *J. Cell Biol.* **102,** 2137–2146.

Lyon, E. S., and Jacoby, W. B. (1980). The identity of alcohol sulfotransferase and hydroxysteroid sulfotransferase. *Arch. Biochem. Biophys.* **202,** 474–481.

Meissner, G., and Allen, R. (1981). Evidence for two types of rat liver microsomes with differing permeability to glucose and other small molecules. *J. Biol. Chem.* **256,** 6413–6422.

Neutra, M., and Leblond, C. P. (1966). Radioautographic comparison of the uptake of galactose-^{3}H and glucose-^{3}H in the Golgi region of various cells secreting glycoproteins or mucopolysaccharides. *J. Cell Biol.* **30,** 137–150.

Nishimoto, S. K., Kajiwara, T., Ledger, P. W., and Tanzer, M. L. (1982). Effects of the ionophore monensin on type II collagen and proteoglycan synthesis and secretion by cultured chondrocytes. *J. Biol. Chem.* **257,** 11712–11716.

Op den Kamp, J. A. F. (1979). Lipid asymmetry in membranes. *Annu. Rev. Biochem.* **48,** 47–71.

Orci, L., Montesano, R., Meda, P., Malaisse-Lagae, F., Brown, D., Perrelet, A., and Vassalli, P. (1981). Heterogeneous distribution of filipin–cholesterol complexes across the cisternae of the Golgi apparatus. *Proc. Natl. Acad. Sci. U.S.A.* **78**, 293–297.

Palade, G. E. (1975). Intracellular aspects of the process of protein synthesis. *Science* **189**, 347–358.

Robbins, P. W., and Lipmann, F. (1957). Isolation and identification of active sulfate. *J. Biol. Chem.* **229**, 837–851.

Roth, J., and Berger, E. G. (1982). Immunocytochemical localization of galactosyltransferase in HeLa cells: Codistribution with thiamin pyrophosphatase in trans-Golgi cisternae. *J. Cell Biol.* **93**, 223–229.

Roth, J., Taatjes, D. J., Lucocq, J. M., Weinstein, J., and Paulson, J. C. (1985). Demonstration of an extensive *trans*-tubular network continuous with the Golgi apparatus stack that may function in glycosylation. *Cell* **43**, 287–295.

Saito, M., Saito, M., and Rosenberg, A. (1985). Influence of monovalent cation transport on anabolism of glycosphingolipids in cultured human fibroblasts. *Biochemistry* **24**, 3054–3059.

Schachter, H., and Roseman, S. (1980). Mammalian glycosyltransferases: Their role in the synthesis and function of complex carbohydrates and glycolipids. *In* "The Biochemistry of Glycoproteins and Proteoglycans" (W. J. Lennarz, ed.), pp. 85–180. Plenum, New York.

Schachter, H., Narasimhan, S., Gleeson, P., and Vella, G. (1983). Glycosyltransferases involved in elongation of N-glycosidically linked oligosaccharides of the complex or *N*-acetyllactosamine type. *Methods Enzymol.* **98**, 98–134.

Schwartz, J. K., Capasso, J. M., and Hirschberg, C. B. (1984). Translocation of adenosine 3'-phosphate 5'-phosphosulfate into rat liver Golgi vesicles. *J. Biol. Chem.* **259**, 3554–3559.

Sly, W. S., and Fischer, H. D. (1982). The phosphomannosyl recognition system for intracellular and intercellular transport of lysosomal enzymes. *J. Cell. Biochem.* **18**, 67–85.

Sommers, L. W., and Hirschberg, C. B. (1982). Transport of sugar nucleotides into rat liver Golgi, a new Golgi marker activity. *J. Biol. Chem.* **257**, 10811–10817.

Stanley, P. (1983). Selection of lectin-resistant mutants of animal cells. *Methods Enzymol.* **96**, 157–184.

Tabas, I., and Kornfeld, S. (1979). Purification and characterization of a rat liver Golgi α-mannosidase capable of processing asparagine-linked oligosaccharides. *J. Biol. Chem.* **254**, 11655–11663.

Tabas, I., and Kornfeld, S. (1982). *N*-Asparagine-linked oligosaccharides: Processing. *Methods Enzymol.* **83**, 416–429.

Tartakoff, A. M. (1983). The confined function model of the Golgi complex: Center for ordered processing of biosynthetic products of the rough endoplasmic reticulum. *Int. Rev. Cytol.* **85**, 221–252.

Tulsiani, D. R. P., Hubbard, S. C., Robbins, P. W., and Touster, O. (1982). α-D-Mannosidases of rat liver Golgi membranes. Mannosidase II is the GlcNAcMan$_5$-cleaving enzyme in glycoprotein biosynthesis and mannosidase IA and IB are the enzymes converting Man$_9$ precursors to Man$_5$ intermediates. *J. Biol. Chem.* **257**, 3660–3668.

van Deenen, L. L. M. (1981). Topology and dynamics of phospholipids in membranes. *FEBS Lett.* **123**, 3–15.

Voelker, D. R., and Kennedy, E. P. (1982). Cellular and enzymic synthesis of sphingomyelin. *Biochemistry* **21**, 2753–2759.

White, M. D., Kuhn, N. J., and Ward, S. (1980). Permeability of lactating-rat mammary gland Golgi membranes to monosaccharides. *Biochem. J.* **190**, 621–624.

White, M. D., Kuhn, N. J., and Ward, S. (1981a). Mannitol and glucose movement across the Golgi membrane of lactating-rat mammary gland. *Biochem. J.* **194,** 173–177.

White, M. D., Ward, S., and Kuhn, N. J. (1981b). Composition stability and electrolyte permeability of Golgi membranes from lactating-rat mammary gland. *Biochem. J.* **200,** 663–669.

White, M. D., Ward, S., and Kuhn, N. J. (1984). Pore properties of the Golgi membrane from lactating-rat mammary glands. *Biochem. J.* **217,** 297–301.

Widnell, C. C. (1972). Cytochemical localization of 5'-nucleotidase in subcellular fractions isolated from rat liver. *J. Cell Biol.* **52,** 542–548.

Yusuf, H. K. M., Pohlentz, G., and Sandhoff, K. (1983). Tunicamycin inhibits ganglioside biosynthesis in rat liver Golgi apparatus by blocking sugar nucleotide transport across the membrane vesicles. *Proc. Natl. Acad. Sci. U.S.A.* **80,** 7075–7079.

Yusuf, H. K. M., Pohlentz, G., and Sandhoff, K. (1984). Ganglioside biosynthesis in Golgi apparatus: New perspectives on its mechanism. *J. Neurosci. Res.* **12,** 161–178.

Zambrano, F., Fleischer, S., and Fleischer, B. (1975). Lipid composition of the Golgi apparatus of rat kidney and liver in comparison with other subcellular organelles. *Biochim. Biophys. Acta* **380,** 357–369.

Zil versmit, D. B. (1983). Lipid transfer proteins: Overview and application. *Methods Enzymol.* **98,** 565–573.

Zilversmit, D. B., and Hughes, M. E. (1977). Extensive exchange of rat liver microsomal phospholipids. *Biochim. Biophys. Acta* **469,** 99–110.

8

Protein Sorting and Biogenesis of the Lysosome-like Vacuole in Yeast

JOEL H. ROTHMAN AND TOM H. STEVENS

317

I. INTRODUCTION

The problem of eukaryotic protein localization, the process by which cytoplasmically synthesized proteins are directed to the appropriate subcellular compartments, has been intensively scrutinized in recent years. Progress in this area of research has accelerated not only as a result of the biochemical, immunological, and microscopic tools available to cell biologists but also as a result of classical and molecular genetic approaches that have been used more recently to assist in unveiling the mechanisms of protein localization. An emerging eukaryotic system in which protein localization studies have been undertaken is the unicellular organism yeast (Schekman, 1985). Due to its readily manipulated genetics and uniform cell type, this simple eukaryote appears to be ideal for addressing certain fundamental questions in cell biology. Classical genetic methods have allowed yeast cell biologists to identify the genes that encode products required for protein processing (Julius *et al.*, 1983, 1984b; Hemmings *et al.*, 1981; Yaffe and Schatz, 1984), glycosylation (Huffaker and Robbins, 1982), secretion (Novick *et al.*, 1980; Ferro-Novick *et al.*, 1984a), and sorting (Rothman and Stevens, 1986; Bankaitis *et al.*, 1986). Moreover, the recent techniques of reverse genetics, whereby the *in vivo* role of a biochemically identified gene product can be determined, have allowed researchers to study the roles of actin (Novick and Botstein, 1985), tubulin (Schatz *et al.*, 1986a,b), clathrin (Payne and Schekman, 1985), and other molecules in a number of cell biological processes. The localization determinants that direct proteins to the yeast nucleus (Hall *et al.*, 1984; Silver *et al.*, 1986), mitochondrion (van Loon *et al.*, 1986), endoplasmic reticulum (ER) (Kaiser and Botstein, 1986; Blachly-Dyson and Stevens, 1987), and lysosome-like vacuole (Valls *et al.*, 1987; Johnson *et al.*, 1987) have been identified using classical and molecular genetic techniques. Genetic studies, in conjunction with more traditional cell biological techniques of immunofluorescence, cell fractionation, and immunoelectron microscopy should allow a much deeper understanding of protein localization in yeast.

Advances in the study of yeast protein localization (Schekman, 1985), function and synthesis of vacuolar proteases (Jones, 1984; Achstetter and Wolf, 1985; Suarez Rendueles and Wolf, 1987b), and endocytosis in yeast (Riezman *et al.*, 1986) have been considered in recent reviews. In this chapter, we focus on the mechanism of assembly of the yeast lysosome-like vacuole. Our discussion includes a review of the pathway of delivery of vacuolar proteins and a brief discussion of the posttranslational modifications carried out during biosynthesis of these proteins. In addition, we describe the current level of understanding of vacuolar protein sorting and present a discussion of endocytosis in yeast and its relationship to vacuolar biogenesis. Whenever possible, we relate these processes in yeast to those in other eukaryotic systems, in particular mammalian cells.

II. BIOSYNTHESIS OF VACUOLAR PROTEINS

The yeast vacuole is considered to be equivalent to the lysosome of mammalian cells based primarily on its content of a large number of hydrolytic enzymes (Wiemken et al., 1979) and its acidic pH (Navon et al., 1979; Makarow and Nevalainen, 1987). Most vacuolar hydrolases have been identified by their enzymatic activities; however, several vacuolar proteins have been purified and the biosynthesis of these proteins examined. The soluble hydrolases, proteinases A (PrA) and B (PrB) and carboxypeptidase Y (CPY), are the best characterized proteins of the vacuole (Jones, 1984; Achstetter and Wolf, 1985; Suarez Rendueles and Wolf, 1987b). The biosynthesis of these three proteins appears to follow similar patterns; all are synthesized as larger precursors (Hasilik and Tanner, 1976, 1978a; Mechler et al., 1982a), and the precursors of all three proteins contain asparagine-linked (N-linked) carbohydrate (Hasilik and Tanner, 1978b; Mechler et al., 1982a). Numerous other soluble vacuolar proteins have been identified (Wiemken et al., 1979; Emter and Wolf, 1984). A number of vacuolar membrane activities, including several transport systems (Sato et al., 1984a,b; Ohsumi and Anraku, 1981, 1983), a proton-translocating ATPase (H^+-ATPase) (Uchida et al., 1985), a dipeptidyl aminopeptidase (DPAP B) (Bordallo et al., 1984), and an α-mannosidase (Opheim, 1978; Wiemken et al., 1979) have been described, but only the H^+-ATPase (Uchida et al., 1985) and the DPAP B (Garcia Alvarez et al., 1985; Roberts et al., 1988) proteins have been purified and characterized to any extent. The biosynthesis of the latter membrane protein has been examined: the protein is also modified with N-linked carbohydrate residues but does not appear to be synthesized as a zymogen (Roberts et al., 1988).

In our discussion of the biosynthesis of vacuolar proteins we shall contrast the modifications that these proteins undergo with those of secreted proteins such as invertase. A more complete consideration of post-translational modifications in the yeast secretory pathway can be found elsewhere in this volume (see Runge, Chapter 4, this volume).

A. Primary Structure of Vacuolar Polypeptides

The primary structures of a number of vacuolar proteins have been determined recently, in general by sequencing the genes encoding these proteins. In the case of PrA and CPY, the primary structures deduced from the sequences of the cloned structural genes (Ammerer *et al.,* 1986; Woolford *et al.,* 1986; Valls *et al.,* 1987) have been partially confirmed by protein sequencing (Dreyer *et al.,* 1986; Martin *et al.,* 1982; Svendsen *et al.,* 1982; Breddam and Svendsen, 1984). The sequence of the cloned genes encoding the vacuolar alkaline phosphatase (AlP) (Kaneko *et al.,* 1985; Y. Kaneko, personal communication), PrB (Jones *et al.,* 1986; Moehle *et al.,* 1987; C. Moehle and E. Jones, personal communication), and DPAP B (C. Roberts and T. Stevens, unpublished results) have been determined, but extensive amino acid sequencing of these proteins has not been performed. All of the protein sequences indicate the presence of from two to six potential sites of N-linked glycosylation. In the case of the membrane protein DPAP B, the translated DNA sequence predicts a single hydrophobic stretch that could serve to anchor this protein to the vacuolar membrane. The structural gene (*AMS1*) encoding another vacuolar membrane protein, α-mannosidase, has been cloned recently (Kuranda and Robbins, 1987), but its sequence has not yet been reported. None of the five vacuolar proteins for which sequences are known appears to share obvious amino acid sequence similarities; thus if vacuolar proteins carry localization determinants that are recognized by a common protein sorting apparatus, such determinants are not readily apparent at the primary sequence level. At least one vacuolar protein (PrA) is homologous to a mammalian lysosomal protein (cathepsin D) (Ammerer *et al.,* 1986; Woolford *et al.,* 1986). The significance of this observation will be addressed in a later section.

B. Translocation into the Endoplasmic Reticulum

It is probable that most or all vacuolar proteins reach the yeast vacuole via early stages (ER and Golgi apparatus) of the secretory pathway (Stevens *et al.,* 1982). An early step in biosynthesis of vacuolar proteins, then, is translocation of the proteins into the ER lumen either during or immediately following their synthesis. Consistent with this hypothesis is the finding that CPY resides in the lumen of the ER during the earliest stage of its synthesis (Ferro-Novick *et al.,* 1984b; Feldman *et al.,* 1987). Thus, like lysosomal proteins of mammalian cells (Erickson *et al.,* 1981; Rosenfeld *et al.,* 1982), it is likely that vacuolar proteins are synthesized with signal sequences, which are recognized by the ER protein translocating machinery (Walter *et al.,* 1984; Walter and Lingappa, 1986).

The precursors of the vacuolar proteins CPY (Valls *et al.,* 1987), PrA (Ammerer *et al.,* 1986; Woolford *et al.,* 1986), and PrB (C. Moehle and E. Jones, personal communication) contain at their amino termini hydrophobic sequences characteristic of signal sequences (von Heijne, 1985; Watson, 1984). In the case of CPY, it has been directly demonstrated that the amino-terminal sequence carried by the newly synthesized form of this protein is necessary for its efficient translocation into the ER (Blachly-Dyson and Stevens, 1987). In addition, the amino-terminal regions of CPY and PrA are sufficient to substitute for the signal sequence of the secretory protein invertase (Johnson *et al.,* 1987; D. Klionsky and S. Emr, personal communication).

Although mutations in the portion of the CPY structural gene encoding its amino-terminal signal sequence greatly diminish the efficiency of translocation of the protein, this sequence is not absolutely required for its entry into the lumen of the ER (Blachly-Dyson and Stevens, 1987). Yeast cells expressing CPY which completely lacks the amino-terminal signal sequence translocate over 10% of the protein. This surprising result appears to conflict with the hypothesis that signal sequences are absolutely required for translocation of proteins into the ER (Blobel and Dobberstein, 1975) and indicates that other portions of the newly translated CPY polypeptide may help to target this protein to the ER lumen. Deletion mutations that eliminate the CPY signal sequence and remove successively larger portions of its propeptide have addressed this hypothesis (Blachly-Dyson, 1987). This latter analysis demonstrated that an element between amino acids 53 and 63 of preproCPY is required for translocation of signal-peptideless CPY; amino-terminal deletions extending from the beginning of the signal sequence through this region reduce glycosylation (and hence, probably translocation) of CPY to less than 1%. This additional element on CPY that allows it to be translocated might be an internal signal sequence; however, this region contains a number of charged residues, which is uncharacteristic of a classical signal sequence. Alternatively, this portion of the protein might be required to keep the newly translated molecule in an unfolded state that is translocation-competent, in analogy to the apparent requirement for the lack of folding of maltose binding protein during its translocation in *Escherichia coli* (Randall and Hardy, 1986).

It has been observed that the form of CPY lacking the signal sequence undergoes posttranslational glycosylation and, therefore, probably posttranslational insertion into the ER lumen. This result is also at odds with the hypothesis of Blobel and Dobberstein (1975) but is analogous to the observations that the human glucose transporter (Mueckler and Lodish, 1986) and the yeast α-factor precursor (Hansen *et al.,* 1986; Waters and Blobel, 1986; Rothblatt and Meyer, 1986b) can be posttranslationally inserted into microsomes *in vitro* and that yeast killer preprotoxin is posttranslationally translocated *in vivo* (Lolle and Bussey, 1986).

CPY is synthesized with a 111 amino acid propeptide at its amino terminus which is not found on the mature vacuolar form of the protein (Valls *et al.*, 1987). It seemed feasible that proteolytic cleavage of the propeptide as a result of zymogen activation (Section II,D) would be sufficient to liberate CPY from its amino-terminal ER targeting information and that signal sequence cleavage might not occur. It has been reported, in fact, that the translocated, unglycosylated precursor form of CPY has the same apparent molecular weight as the *in vitro* translated molecule (Müller and Müller, 1981; Distel *et al.*, 1983), suggesting that the signal sequence is not cleaved from the polypeptide following its translocation. However, an amino-terminal analysis of the precursor performed recently indicates that the signal sequence is indeed removed from the propeptide of CPY (Blachly-Dyson and Stevens, 1987). It is as yet unknown whether the signal sequences of other soluble or membrane proteins of the vacuole are removed by signal peptidase.

In addition to the dependence of CPY on a signal sequence for efficient translocation, additional evidence suggests that the protein is inserted into the lumen of the ER via the same machinery as are secretory proteins. A mutation (*sec61*) that blocks translocation of the secretory proteins invertase and α-factor precursor also prevents translocation of CPY (Deshaies and Schekman, 1987). Moreover, an *in vitro* translation/translocation system derived from yeast translocates both secretory proteins and CPY when programmed with the appropriate RNA (Rothblatt and Meyer, 1987). It is apparent, then, that vacuolar proteins and other proteins that traverse the secretory pathway enter the ER by a common mechanism.

C. Glycosyl and Phosphoryl Modifications during Transport

During or immediately following translocation into the ER, vacuolar proteins such as CPY become glycosylated by transfer of oligosaccharides from dolichol pyrophosphate to the appropriate asparagine residues of the proteins. These oligosaccharide chains are subsequently modified in the ER by removal of the glucose residues to produce the "core" carbohydrate moiety. These early events, common to all eukaryotes, are reviewed in more detail elsewhere (Kornfeld and Kornfeld, 1985; Runge, Chapter 4, this volume). An average of four to five phosphates are added to the core mannose residues of each CPY molecule in the ER (Hashimoto *et al.*, 1981; Stevens *et al.*, 1982). Phosphorylation of CPY in the ER contrasts with phosphorylation of glycoproteins in animal cells, since the latter event appears to occur exclusively in the Golgi apparatus (Kornfeld and Kornfeld, 1985). Certain mannose residues of the yeast secretory protein invertase are phosphorylated (Lehle *et al.*, 1979; Trimble *et al.*,

1983) in the Golgi apparatus (Stevens *et al.*, 1982); phosphate groups are not detectable on the ER form of this protein. Thus, the earliest detectable landmark that distinguishes the vacuolar protein CPY and the secretory protein invertase is the differential intracellular site of phosphorylation of the two proteins. The significance of this difference is not clear. Unlike the case of lysosomal proteins of mammalian fibroblasts, however, addition of phosphate is apparently not required for sorting and delivery of CPY to the vacuole (Schwaiger *et al.*, 1982; Stevens *et al.*, 1982; J. Rothman and T. Stevens, unpublished observations). The difference between the sites of CPY and invertase phosphorylation might reflect either the segregation of these two molecules into distinct subcompartments of the ER or the specificity of the ER mannosylphosphotransferase for certain proteins.

Following the transport of CPY from the ER to the Golgi apparatus, the N-linked carbohydrate moieties are further modified to the form found on the mature protein (Hashimoto *et al.*, 1981; Stevens *et al.*, 1982). The passage of CPY from the ER to the Golgi apparatus can be monitored as an increase in apparent molecular weight of the protein from 67K to 69K as a result of the Golgi-mediated modifications of N-linked carbohydrate. The ER and Golgi precursor forms of CPY are referred to as p1 CPY and p2 CPY, respectively. The modifications of the oligosaccharides of CPY within the Golgi apparatus are distinct from those of the invertase oligosaccharides. Golgi-directed carbohydrate modifications of CPY result in an increase in the molecular weight of each chain by approximately 500, and the resulting glycosylated CPY molecules exhibit a fairly homogeneous distribution in molecular weight. In contrast, the oligosaccharide chains on invertase receive more extensive carbohydrate additions in the Golgi apparatus (an increase of approximately 2,000–6,000 per chain). Thus, the resulting invertase species exhibit a broad molecular weight distribution owing to this heterogeneous hyperglycosylation (Lehle *et al.*, 1979; Esmon *et al.*, 1981). These differences in oligosaccharide structure are clearly evident in the electrophoretic mobilities of the Golgi forms of the two proteins (Runge, Chapter 4, this volume).

It is likely that the distinction in glycosyl modification applies to all vacuolar and secretory proteins, since other vacuolar proteins that have been examined (PrA, PrB, aminopeptidase I, AIP, and DPAP B) do not appear to be hyperglycosylated (Jones, 1984; Achstetter and Wolf, 1985; Ohnishi *et al.*, 1979; Roberts, *et al.*, 1988), while other secretory proteins (acid phosphatase, α-factor precursor) contain the hyperglycosylated "invertase-like" modifications (Julius *et al.*, 1984a; Boer and Steyn-Parve, 1966; Haguenauer-Tsapis and Hinnen, 1984). Although both CPY and invertase are extensively glycosylated, neither of the proteins receives O-linked glycosylation (Hasilik and Tanner, 1978b; Lehle *et al.*, 1979), and no such modification has yet been detected on any vacuolar protein. A more complete discussion of CPY and invertase glycosyl modifications that occur during their transit can be found in the review by Runge (Chapter 4, this volume).

Treatment of yeast cells with tunicamycin, a drug that prevents addition of N-linked carbohydrate (Kuo and Lampen, 1974), does not perturb the efficient sorting of CPY or PrA to the vacuole (Schwaiger *et al.*, 1982; Stevens *et al.*, 1982; J. Rothman and T. Stevens, unpublished results), suggesting that, unlike lysosomal protein sorting in mammalian fibroblasts, vacuolar protein sorting does not rely on a carbohydrate determinant. Moreover, vacuolar AlP synthesized under these conditions is enzymatically active (Ohnishi *et al.*, 1979), implying that carbohydrate is not required for the activity of this protein. Thus, no role for the carbohydrate carried by yeast vacuolar proteins has yet been demonstrated.

Despite the observation that carbohydrate is probably not required for sorting of CPY from invertase during transit of these proteins through the secretory pathway, it is of interest to determine how secretory and vacuolar proteins are differentially modified in the Golgi apparatus. There are several possible explanations for this phenomenon. One hypothesis is that secretory and vacuolar proteins are segregated into different subcompartments of the Golgi apparatus and that the Golgi mannosyltransferases are differentially sequestered into the appropriate subcompartments. According to this model, proteins destined for delivery to the vacuole would not encounter the enzymes that carry out secretory protein carbohydrate modifications. This is not likely to be the case in yeast, however, since mislocalization of either CPY or PrA to the cell surface caused by a number of aberrant conditions (Section III) is not accompanied by a change in the glycosylation of these molecules from the normal vacuolar protein pattern (Stevens *et al.*, 1986; Rothman *et al.*, 1986; Rothman and Stevens, 1986; Bankaitis *et al.*, 1986; Johnson *et al.*, 1987; Valls *et al.*, 1987).

Alternatively, CPY and invertase might carry signals that are recognized by the modifying enzymes and that dictate the appropriate glycosyl modifications. Evidence for cis-acting elements that direct oligosaccharide modifications derives from studies of the glycosylation of CPY–invertase fusion proteins (Johnson *et al.*, 1987). Hybrid proteins consisting of the amino-terminal 95 or 433 amino acids of CPY fused to the amino terminus of the cytoplasmic form of invertase are faithfully localized to the vacuole, indicating that the vacuolar localization determinant on CPY is epistatic to that of invertase. When these two chimeric proteins are examined, they are found to exhibit different glycosylation patterns. The larger fusion protein resides in the vacuole in a discrete glycosylated form, suggesting that the 12–15 asparagine residues on the hybrid protein have received oligosaccharides of similar size and form to the 4 normally found on CPY. In contrast, the smaller fusion protein appears in the vacuole as a hyperglycosylated species characteristic of the secreted form of invertase. In mutants that fail to deliver this smaller CPY–invertase fusion proteins to the vacuole, and instead secrete the molecules (Sections III,C and III,D), the hybrid proteins are glycosylated in the same way as when they were delivered to the vacuole in wild-type cells. These findings suggest that the glycosylation state of these chimeric proteins is independent of their ultimate subcellular location.

It is possible that the two classes of fusion proteins are delivered to the vacuole via different routes and hence are exposed to different modifying enzymes; however, it seems more probable that the hybrids reach the vacuole by a common pathway, particularly since the kinetics of vacuolar delivery of these proteins is similar. The simplest interpretation of these results is that CPY carries a determinant, between amino acid 95 and amino acid 433, which suppresses elaboration of its oligosaccharide chains to the hyperglycosylated form. Consistent with this hypothesis is the existence of a mutation in the CPY structural gene which leads to hyperglycosylation of the CPY encoded by the mutant gene (E. Blachly-Dyson, C. Tachibana, and T. Stevens, unpublished results). A substantial fraction of this mutant form of CPY is not secreted and may reside in the vacuole. Unexpectedly, this mutation results in a change in the amino acid sequence near the carboxy terminus of CPY and beyond the region defined as important for suppression of hyperglycosylation in the fusion proteins. The relationship between the element defined by the mutation and by the studies with chimeric proteins is not currently understood.

The action of an element within CPY that regulates the global pattern of glycosylation of oligosaccharides well removed from this site (i.e., the chains linked to the invertase portion of the CPY–invertase hybrids) could be explained by at least two models. It is possible that this element interacts with the carbohydrate modifying enzymes directly and globally regulates their action over the entire protein. However, it is unclear how such an element could act to control the pattern of glycosylation over long ranges. Alternatively, this element may result in partitioning of the CPY polypeptide to a subcompartment of the Golgi apparatus. In such a case, the region of the Golgi to which the protein partitions would determine the type of carbohydrate modification it receives. This partitioning could not, however, determine the ultimate intracellular localization of the polypeptide, since both the CPY-like and invertase-like glycosylated forms of the CPY–invertase chimeras can appear either in the vacuole or at the cell surface. Further work with the fusion proteins and cis-acting CPY glycosylation mutants will help to clarify the role played by polypeptide chains in directing oligosaccharide modifications.

D. Proteolytic Activation of Vacuolar Zymogens

All soluble hydrolases of the yeast vacuole whose biosynthesis has been carefully examined appear to be synthesized as larger molecular weight precursors (Hasilik and Tanner, 1978a; Mechler *et al.*, 1982a). The two proteases, PrA and CPY, are known to be synthesized with amino-terminal extensions of 76 and 111 amino acids, respectively (Ammerer *et al.*, 1986; Woolford *et al.*, 1986; Valls *et al.*, 1987). In addition, PrB is synthesized as a polypeptide which is larger than the mature species (Mechler *et al.*, 1982a), perhaps indicative of an amino-terminal propeptide. The inactive CPY precursor (proCPY) is converted to the mature, active species (mCPY) immediately prior to or on arrival at the vacuolar lumen (Stevens *et al.*, 1982). Conversion of proCPY to mCPY is dependent on a wild-type copy of the *PEP4* gene (Hemmings *et al.*, 1981; Zubenko *et al.*, 1983). *PEP4* is also required for the activity of a number of other soluble vacuolar enzymes (Table I), including vacuolar species of ribonuclease, AlP, trehalase, and aminopeptidase (Jones *et al.*, 1982; Trumbly and Bradley, 1983; Harris and Cotter, 1987). Significantly, none of the vacuolar membrane activities that have been examined, i.e., α-

TABLE I

Perturbation of Vacuolar Protein Sorting and Biogenesis[a]

Condition	Proteins known to be affected	Phenotype
Multicopy expression of structural gene	PrA, CPY, AlP, (DPAP B?)	Secretion of the protein overproduced
prc1-linked sorting mutants	CPY only	Secretion of CPY only
vpl mutants	PrA, CPY, (PrB?)	Secretion of vacuolar proteins
vpt mutants	CPY, (PrB?)	Secretion of vacuolar proteins
pep mutants (some)	CPY, (others?)	Secretion of vacuolar proteins
end1 mutant	CPY, (others?)	Secretion of vacuolar proteins
Early *sec* mutants	CPY, PrA, DPAP B	Conditionally blocked in transport to the vacuole
pep4 mutant	CPY, PrA, PrB, trehalase, amino-peptidase I, ribo-nuclease, AlP	Failure to undergo zymogen activation

[a] Complete allelism tests between *vpl*, *vpt*, *pep*, and *end* mutants have not been performed. It is possible that many or all of these classes of mutants overlap substantially. Only the proteins that have been conclusively shown to be mislocalized in the mutants are noted—others may be affected but have not been examined.

mannosidase (Jones *et al.*, 1982), DPAP B (our unpublished observation), and the vacuolar H$^+$-ATPase (C. Yamashiro and T. Stevens, unpublished observation), is affected by a *pep4* mutation. In addition to these findings, the molecular weight of DPAP B does not change significantly following delivery to the vacuole, suggesting that it is not made as a larger precursor (Roberts *et al.*, 1988). Consequently, it seems feasible that none of the known vacuolar membrane proteins is synthesized with an inactivation propeptide.

The *PEP4* gene has been shown to encode the vacuolar enzyme PrA (Ammerer *et al.*, 1986; Woolford *et al.*, 1986). This was initially determined when the structural gene for PrA (Rothman *et al.*, 1986) and a gene capable of complementing a *pep4* mutation were cloned and found to have identical sequences (Ammerer *et al.*, 1986). Thus, PrA plays an essential role in the activation of vacuolar zymogens. The finding that *PEP4* encodes PrA was unexpected based on earlier studies of mutants deficient in PrA activity (Betz, 1979; Mechler and Wolf, 1981; Jones *et al.*, 1981). These *pra1* mutants were identified as cells that failed to exhibit *in vitro* PrA activity but which maintained approximately wild-type levels of other vacuolar hydrolases. Some *pra1* alleles eliminated detectable steady-state levels of PrA antigen as well as the activity of the protease (Mechler and Wolf, 1981). Since a number of *pra1* alleles fail to complement mutations in the *PEP4* gene with respect to the deficiency in PrA activity (Woolford *et al.*, 1986), it is not obvious how these *pra1* mutations eliminate the *in vitro* activity of PrA without abolishing its zymogen activating function. The most likely explanation for this apparent paradox is that the mutant forms of PrA are sufficiently active *in vivo* to catalyze conversion of vacuolar zymogens to their active forms, yet are unstable in extracts prepared from *pra1* cells because of degradation or denaturation. In this regard it is reassuring that all of the *pra1* mutations sequenced are missense mutations resulting in the substitution of a single amino acid in the mature portion of the protein, whereas all *pep4* alleles are nonsense mutations and consequently eliminate a large portion of the PrA polypeptide (Van Arsdell *et al.*, 1986). Of relevance to the contrasting phenotypes of the *pra1* and *pep4* alleles is the observation that at least one of the *pra1* alleles, when in combination with a mutation in the PrB structural gene (*PRB1*; Wolf and Ehmann, 1978; Zubenko, *et al.*, 1980; Mechler, *et al.*, 1982b) prevents conversion of proCPY to mCPY, and thus at least some aspects of vacuolar zymogen activation are blocked by this allele (Mechler, *et al.*, 1987). Further analysis of *pra1* missense mutants should provide insights into the structure, enzymatic specificity, and stability of PrA.

The biochemical characterization (Meussdoerffer *et al.*, 1980) and amino acid sequence of PrA indicate that it is a member of the aspartyl family of proteases (Tang, 1979). This class of proteins includes both secreted (e.g., pepsin) and lysosomal (e.g., cathepsin D) enzymes. Although PrA is homologous to proteins that are delivered to the surface of mammalian cells as well as to those targeted to the mammalian lysosome, lysosomal cathepsin D (Faust *et al.,* 1985) exhibits the greatest degree of similarity to PrA (Ammerer *et al.,* 1986). It is reasonable to suggest that, like PrA in yeast, lysosomal cathepsin D may play a role in activating lysosomal zymogens in animal cells.

A particularly conspicuous region of homology between human cathepsin D and yeast PrA is seen at the zymogen activation site: the two amino acids preceding and the one immediately following the site of proteolytic cleavage are identical, suggesting that the two proteins may be activated via similar mechanisms. Moreover, the similarities in the sequence of these proteins raises the possibility that cathepsin D and PrA may share common features in their localization determinants. However, recognition of cathepsin D by the lysosomal localization machinery in fibroblasts requires phosphorylation of N-linked carbohydrate moieties (Creek and Sly, 1984; Kornfeld, 1986; von Figura and Hasilik, 1986), whereas PrA is efficiently delivered to the vacuole even in the absence of glycosylation (J. Rothman and T. Stevens, unpublished observations). In order to investigate whether human cathepsin D is recognized by the yeast vacuolar localization machinery and is subsequently directed to the vacuole, the gene encoding this lysosomal enzyme has been expressed in yeast (our unpublished result). Cathepsin D was found not to be secreted when synthesized in yeast; it remains to be determined whether the protein is efficiently sorted to the vacuole. Expression of the human protein in a *pep4* mutant results in complementation of the defect in proCPY proteolytic processing. This result implies that at least a portion of cathepsin D can be delivered to the vacuole and activated, and suggests that lysosomal (vacuolar) zymogen activation is a related process in yeast and animal cells. Further studies of the human lysosomal protein in yeast will more clearly delineate similarities and differences between the mechanisms of protein sorting in these organisms and might provide a model system in yeast in which to study the localization determinants on lysosomal proteins of animal cells.

The finding that PrA, itself synthesized as a zymogen, is required for activation of nearly all soluble vacuolar enzymes examined raises a potential dilemma. How does proPrA itself become proteolytically activated? One reasonable hypothesis (Ammerer *et al.,* 1986; Woolford *et al.,* 1986) is that the precursor can activate itself under the appropriate environmental conditions such as the low pH state of the vacuole (Navon *et al.,* 1979;

Makarow and Nevalainen, 1987). Indeed, activation of the mammalian aspartyl protease pepsinogen is initiated by an intramolecular processing event triggered by low pH (Bustin and Conway-Jacobs, 1971; James and Sielecki, 1986). The structural similarities between proPrA and pepsinogen (40% amino acid identity between the mature proteins) leads one to speculate that PrA might also undergo pH-dependent autoactivation. Preliminary studies suggest that proPrA is converted to a molecule that comigrates with mature PrA on incubation in a low pH buffer (J. Rothman and T. Stevens, unpublished observations), consistent with a pH-induced autocatalytic activation of proPrA. However, it cannot be ruled out that an as yet unidentified enzyme carries out the proteolytic maturation of proPrA (Fig. 1, X). Additional *in vitro* studies will help to clarify the mechanism of proPrA activation.

Based on biochemical and genetic evidence (Hasilik and Tanner, 1978a; Zubenko, *et al.*, 1982; Mechler, *et al.*, 1987), PrB appears to be capable of catalyzing proteolytic activation of proCPY both *in vitro* and *in vivo*. As mentioned earlier, PrA is required to convert proPrB to mPrB, and therefore one model describing the activation pathway of proCPY in wild-type cells is to propose that PrA activates proPrB, and the active PrB then cleaves proCPY to mCPY (Fig. 1). However, two observations indicate that PrA can also act in a redundant pathway to activate proCPY independently of the PrB-dependent reaction. First, the presence of a processed

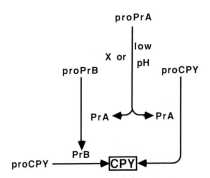

Fig. 1. Model for the vacuolar proCPY activation pathway. The proteolytic maturation of proCPY begins with activation of proPrA. The model proposes that proPrA either autoactivates owing to the low pH environment of the vacuole or is activated by an as yet unidentified protease (X). PrA activates proPrB, and PrB is capable of proCPY activation. In addition, PrA can directly activate proCPY by a redundant pathway. For the sake of simplicity, the ability of PrB to weakly activate proPrB, as evidenced by the phenomenon of *PRB1*-dependent phenotypic lag (Zubenko *et al.*, 1982), has not been included in the model. Active CPY is indicated in the box; the molecular weight of this active species differs depending on which of the two redundant pathways produced it.

and active form of CPY in cells carrying a nonsense mutation at the *PRB1* locus (Hemmings *et al.,* 1981) strongly suggests that a PrB-independent processing pathway can occur. In addition, recent work (Mechler *et al.,* 1987) has established that purified PrA can activate isolated proCPY *in vitro* in the presence of a vacuolar extract, high salt, or polyphosphate (a normal constituent of the yeast vacuole) but in the absence of PrB. Cleavage of proCPY by purified PrA yields a CPY species that, although active and indistinguishable in size from that produced *in vivo* in a *prb1* strain, is somewhat larger than CPY isolated from wild-type yeast cells. This implies that PrB always carries out the final cleavage of proCPY in wild-type cells, although it does not necessarily perform the initial cleavage that activates proCPY. Whether PrA can act directly on proCPY *in vivo* or instead acts in conjunction with additional vacuolar constituents (e.g., via a second cascade of proteolytic reactions) remains to be determined; however, in an effort to isolate a large number of mutants that do not activate vacuolar proCPY in a *prb1* genetic background, only mutations allelic to *PEP4* were identified (our unpublished results). This result suggests that the PrB-independent pathway of proCPY activation involves only PrA, or that there exist multiple pathways which are all redundant with the PrB-dependent processing reaction. Additional *in vivo* and *in vitro* studies should clarify whether the scheme depicted in Fig. 1 is complete.

III. SORTING OF VACUOLAR PROTEINS

A. Evidence for Golgi-Mediated Sorting of Vacuolar Proteins

It is well established that both lysosomal and secreted proteins of animal cells pass through the organelles (ER and Golgi) of the early secretory pathway en route to their ultimate destinations (von Figura and Hasilik, 1986; Kornfeld, 1986). Vacuolar delivery of CPY has been examined in yeast *sec* mutants to determine whether vacuolar protein biosynthesis requires any of the same machinery as is required for protein secretion (Stevens *et al.,* 1982). *sec* mutations conditionally block movement of the secretory proteins invertase, acid phosphatase, and α-factor to the cell surface at a restrictive temperature (Novick *et al.,* 1980, 1981; Ferro-Novick *et al.,* 1984a; Julius *et al.,* 1984a; Deshaies and Schekman, 1987). These mutations block secretion at distinct points in the pathway; certain of the "early" *sec* mutants fail to transport proteins beyond the ER, whereas others are blocked in movement of proteins beyond the Golgi. The "late" *sec* mutants allow secretory proteins to be packaged into secretory vesicles but do not permit fusion of these vesicles with the plasma membrane. It was found that early *sec* mutations not only block transport of secretory proteins through the ER or Golgi but also prevent delivery of newly synthesized CPY to the vacuole, resulting in the intracellular accumulation of the proenzyme forms of the protein in nonvacuolar compartments (Stevens *et al.*, 1982). In contrast, none of the late *sec* mutations prevents transport of CPY to the vacuole. The results of these experiments were taken as evidence that vacuolar proteins follow the same pathway as secretory proteins through the ER and Golgi apparatus. Within the Golgi apparatus, there appears to be a branch point in the pathway followed by vacuolar and secretory proteins, resulting in the transport of the former proteins to the vacuole and the latter to the periplasmic space surrounding the plasma membrane. The ability of yeast cells to distinguish proteins such as CPY and invertase and direct them to different subcellular compartments is defined as protein sorting.

For a number of animal cell types it is known that sorting of lysosomal proteins requires that mannose groups on the oligosaccharides of these proteins are modified to mannose 6-phosphate (M-6-P) (Creek and Sly, 1984; von Figura and Hasilik, 1986; Kornfeld, 1986). A human genetic disease (I cell disease), results in the failure to carry out phosphorylation of lysosomal protein mannose residues. Fibroblasts and other cell types from I cell individuals do not properly sort soluble lysosomal proteins as a result of this deficiency in the mannosylphosphotransferase activity, and thus secrete the unphosphorylated lysosomal glycoproteins. These findings led to the identification and purification of M-6-P recognition receptors (Sahagian *et al.*, 1981; Hoflack and Kornfeld, 1985), which appear to be an essential component of the lysosomal protein-sorting machinery in the Golgi apparatus (Brown and Farquhar, 1984; Geuze *et al.*, 1984, 1985). Detailed treatments of the M-6-P recognition system can be found elsewhere (Creek and Sly, 1984; Kornfeld, 1986; von Figura and Hasilik, 1986).

In contrast to the lysosomal protein-sorting system in many animal cell types, it appears that yeast cells do not rely on mannose phosphorylation nor even on carbohydrate for recognition of vacuolar proteins by the vacuolar protein-sorting apparatus (Schwaiger *et al.*, 1982; Stevens *et al.*, 1982; J. Rothman and T. Stevens, unpublished observations). This does not necessarily imply that the mechanisms of sorting of yeast vacuolar and mammlian lysosomal proteins are fundamentally different. Indeed, in some human cell types, as in yeast, lysosomal proteins are properly delivered even in the absence of the M-6-P modification (Owada and Neufeld, 1982; Waheed *et al.*, 1982). Moreover, the residual levels of certain lysosomal proteins in I cell fibroblast lysosomes can be a substantial fraction of the total amount synthesized (Hasilik and Neufeld, 1980; Lemansky *et al.*, 1985). These results suggest the existence of an alternative lysosomal protein-sorting mechanism in animal cells, perhaps not based on carbohydrate determinants. Finally, although the basis for recognition of lysosomal proteins in fibroblasts and vacuolar proteins in yeast appears to be different, the processes by which both types of proteins are sorted and packaged for delivery following recognition by the appropriate receptor might be quite similar. An example of the similarity between yeast vacuolar and animal cell lysosomal protein sorting is the observation that under conditions in which these proteins fail to be sorted they are aberrantly secreted (see below).

B. Overproduction-Induced Missorting of Vacuolar Proteins: A General Phenomenon in Yeast?

The aberrant phenomenon of secretion of a yeast vacuolar protein was initially observed as a consequence of overproduction of CPY (Stevens *et al.*, 1986) resulting from increased dosage of its structural gene (*PRC1*) (Wolf and Fink, 1975). In cells carrying a single copy of *PRC1*, only a low level of CPY is found at the cell surface. When the *PRC1* gene dosage is increased as the result of its presence on a multicopy plasmid, the total level of CPY synthesized is approximately 8-fold higher than in cells with a single copy of the gene, and over half of the total CPY is secreted. Thus, higher than normal levels of CPY lead to missorting of the protein into the late stages of the secretory pathway. This behavior of vacuolar proteins on overproduction may be a general phenomenon: overproduction of PrA (Rothman *et al.*, 1986) and vacuolar AlP (Y. Kaneko, personal communication) also lead to aberrant secretion of these proteins (Table I). It is possible that aberrant secretion of a mouse lysosomal protein in cells overexpressing the protein as a result of malignant transformation (Gals *et al.*, 1985) is a related phenomenon. It is not known, however, whether mislocalization of this lysosomal protein is a direct result of its overproduction or of some other aspects of the transformed phenotype.

When CPY or PrA is misdirected to the cell surface on overproduction, the protein does not become hyperglycosylated to a form that resembles the secreted form of invertase (Stevens *et al.*, 1986; Rothman *et al.*, 1986). This finding is consistent with the hypothesis presented above that the glycosylation characteristics of these proteins are determined by cis-acting elements carried by the proteins rather than by the ultimate destination of these proteins. It is likely that the mislocalized portion of these overproduced vacuolar proteins follows a pathway that is the same as that followed by secretory proteins during transit to the cell surface, since the late-acting *sec1* mutation prevents secretion of the missorted proteins. Moreover, immunoelectron microscopic analysis suggests that CPY resides in the secretory vesicles that accumulate in *sec1* mutants overproducing the protein (Fig. 2B), but not in the vesicles that accumulate in *sec1* mutants carrying a single copy of *PRC1* (Fig. 2A). Since all secretory vesicles appearing in this mutant contain the secretory enzyme acid phosphatase (Novick and Schekman, 1979), it is reasonable to suppose that misdirected CPY is delivered to the cell surface within the same secretory vesicles as are normal secretory proteins. Double-label immunoelectron microscopy, using both CPY and invertase antibodies, will allow this hypothesis to be tested.

One plausible explanation for the phenomenon of gene dosage-dependent secretion of vacuolar proteins is that overproduction of these proteins leads to saturation of some component which is required for vacuolar protein sorting. This component might be a sorting receptor that recognizes localization determinants on vacuolar proteins or, alternatively, an enzyme that carries out a modification recognized by the vacuolar protein sorting apparatus. However, overproduction of CPY does not lead to the mislocalization of PrA to the cell surface, nor does overproduction of PrA lead to CPY secretion (Stevens *et al.*, 1986; Rothman *et al.*, 1986). This finding is difficult to rationalize given the simple view that all vacuolar proteins are recognized by a single receptor, a model that is supported by the observation that trans-acting mutants that missort vacuolar proteins pleiotropically secrete PrA and CPY (Rothman and Stevens, 1986; see below). It is clear that further studies are required before an understanding of overproduction-induced mislocalization of yeast vacuolar proteins is attained.

C. Identification of Vacuolar Protein Localization Determinants

Gene dosage-dependent secretion of vacuolar proteins strongly suggested that a failure to sort vacuolar proteins would result in their delivery to the cell surface. Thus it seemed possible to identify mutants defective in sorting of CPY by identifying yeast cells which secrete CPY yet which express the protein at normal levels. Such an approach has allowed the isolation of both cis-acting (Valls *et al.*, 1987) and trans-acting mutations (Rothman and Stevens, 1986; Bankaitis *et al.*, 1986) that cause cells to be defective in vacuolar protein sorting (Table I).

Cis-acting mutations that lead to secretion of CPY were obtained by *in vitro* mutagenesis of the *PRC1* gene carried on a single-copy plasmid followed by screening for yeast transformants that exhibited extracellular

A

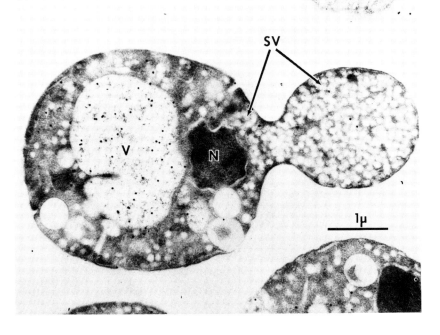

B

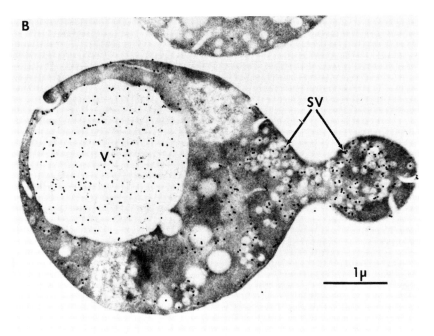

CPY (Valls *et al.*, 1987). An immunoblotting screen that simplifies identification of yeast colonies secreting CPY antigen among a large number of wild-type colonies (Rothman *et al.*, 1986) was used to isolate mutant transformants mislocalizing the protein to the cell surface. In-frame deletions in the propeptide-encoding region of *PRC1* which removed from 3 to 12 amino acids around amino acid 29 of proCPY were isolated, and all such mutations resulted in secretion of a substantial fraction (40–95%) of the total newly synthesized CPY (Valls *et al.*, 1987). Only mutations mapping near the extreme amino terminus of the proCPY polypeptide (following signal sequence cleavage) were identified. It is possible that other regions in the proCPY polypeptide are also required for normal localization of the protein; however, random chemical mutagenesis of the cloned *PRC1* gene produced a point mutation that also mapped within the same region delimited by the in-frame deletions. Further mutagenesis studies will be required to assess whether the amino-terminal region of the CPY propeptide is the only portion of the polypeptide required for its targeting to the vacuole.

Nearly all of the *PRC1* deletion mutants described above mislocalize CPY in proportion to the length of the deletion mutation (i.e., the larger the deletion, the more CPY secreted; Valls *et al.*, 1987). In contrast, the largest deletion, encompassing 12 amino acids of the propeptide, caused a significantly lower level of CPY missorting (42%) than did most of the smaller deletions. The anomalous behavior of this mutant was also seen in the kinetics of CPY secretion: newly synthesized CPY was found to be delivered to the cell surface at a much slower rate than in the other mutants. This increased time of residence of CPY in the secretory pathway might arise from a failure of the signal sequence to be removed from the mutant form of the protein, since this 12 amino acid deletion removes the signal sequence cleavage site (Blachly-Dyson and Stevens, 1987). Consistent with this argument is the observation that mutations in the invertase and acid phosphatase structural genes which prevent signal sequence cleavage result in a greatly diminished rate of transit of these

---

Fig. 2. Missorted CPY is contained in secretory vesicles. In *sec1* cells, secretory vesicles are seen to accumulate at the restrictive temperature. Thin sections were labeled with CPY antiserum followed by protein A–15 nm colloidal gold. Only background labeling is observed in a strain deleted for the CPY structural gene (not shown). (A) Yeast strain containing a single copy of the CPY structural gene. (B) Yeast strain containing a multicopy plasmid carrying the CPY structural gene. With normal levels of CPY expression (A), labeling is seen predominantly over the vacuole, with little or no labeling over the accumulated secretory vesicles. In contrast, overproduction of CPY (B) results in labeling over both the vacuole and secretory vesicles, indicating that the mislocalized fraction of CPY reaches the cell surface via secretory vesicles. N, Nucleus; SV, secretory vesicles; V, vacuole.

proteins through the secretory pathway (Schauer *et al.*, 1985; Haguenauer-Tsapis and Hinnen, 1984). The slower rate of movement of this mutant form of CPY through the secretory pathway might allow greater time for the protein to be recognized by the sorting apparatus, and consequently result in more efficient vacuolar delivery of the polypeptide compared with the other mutants.

One of the major caveats in interpreting the effects of altered primary protein structure on proper localization of a polypeptide is that such mutations might lead to gross alterations in the overall tertiary structure of the molecule rather than specifically perturbing only a localization determinant. For example, a mutant form of the *Dictyostelium* lysosomal protein α-mannosidase, which fails to be localized to the lysosome, apparently exists in an altered conformation (Woychik *et al.*, 1986). This appears not to be true in the case of the *PRC1*-linked CPY mislocalization mutants since the secreted form of the protein, isolated from the medium of mutant cells, can be converted to an enzymatically active species *in vitro*, and thus cannot be in a grossly unfolded or altered state (Valls *et al.*, 1987). The simplest interpretation of these cis-acting mutations is that they perturb a vacuolar sorting determinant on the CPY propeptide, resulting in the failure of the sorting apparatus to recognize the protein.

The *PRC1*-linked mutants demonstrate that residues near the amino terminus of the CPY propeptide (amino acids 21–31) are necessary for sorting of the protein to the vacuole; however, they do not allow one to determine whether this region is sufficient for vacuolar protein sorting. This complementary question has been considered by examining the behavior of genetically engineered chimeric proteins containing the cytoplasmic form of invertase linked to the carboxy terminus of CPY (Bankaitis *et al.*, 1986; Johnson *et al.*, 1987). These CPY–invertase fusion proteins express invertase activity, and the subcellular location of this activity is taken as an indicator of the location of the hybrid proteins. In fusions that include the first 50 or more residues of the amino terminus of CPY linked to the amino terminus of invertase, all plasmid-directed invertase activity is found in an intracellular compartment that cofractionates with vacuolar markers. In contrast, chimeric proteins containing only the first 20 residues of the CPY amino terminus (i.e., only the CPY signal sequence) are found to reside in the periplasm, and thus are not directed to the vacuole. These findings indicate that the amino-terminal 50 amino acids of CPY are sufficient to divert a secretory protein to the vacuole, in good agreement with the observations that amino acids 21 through 31 are essential for sorting CPY to the vacuole (Valls *et al.*, 1987). From these studies it is clear that the amino-terminal portion of the CPY propeptide is both necessary and sufficient as a vacuolar localization determinant. It is not yet known whether this amino-terminal element must reside at the amino terminus of a protein to direct that protein to the vacuole.

The amino-terminal propeptide of another soluble vacuolar protein also carries information sufficient for vacuolar localization. An invertase fusion protein carrying at its amino terminus the 76 amino acid propeptide of PrA is localized to the vacuole (D. Klionsky and S. Emr, personal communication). Thus, it is conceivable that the amino-terminal propeptides of vacuolar zymogens generally contain the information for vacuolar localization. Despite the common localization function of the CPY and PrA propeptides, no obvious homologies are observed between the region defined as essential for CPY sorting and any portion of the PrA polypeptide (J. Rothman and T. Stevens, unpublished observations). Mutational analysis of the *PEP4* gene will define more precisely the region(s) required for vacuolar localization and will allow a more complete determination of features common to localization determinants carried by vacuolar proteins.

It is reasonable to suggest that the element on vacuolar proteins which is recognized as an initial step in the sorting of these proteins is proteinaceous, since carbohydrate is not required for proper localization of these proteins. In mammalian cells, it also appears that the initial recognition step in lysosomal protein sorting is the binding of a protein determinant carried by these proteins. Specifically, the polypeptide component of lysosomal proteins appears to contain denaturable elements recognizable by the phosphotransferase that modifies mannose groups to the sorting-proficient M-6-P form (Reitman and Kornfeld, 1981; Lang *et al.*, 1984). As in yeast, the determinants required for the initial step in mammalian lysosomal protein sorting are present on the precursor form of the proteins (von Figura and Hasilik, 1986). Thus, it is possible that the tertiary structures of yeast vacuolar and mammalian lysosomal proteins carry qualitatively similar localization determinants, even though recognition of these determinants results in different biochemical consequences.

From studies with the *PRC1* gene, it is evident that the CPY propeptide (and perhaps the propeptides of other vacuolar proteins) carries out at least three functions: this propeptide contains the information for targeting of the nascent vacuolar proteins to the ER and for subsequent sorting to the vacuole, and it keeps the enzyme inactive during its transit to the vacuole. A summary of the various elements contained within CPY and its propeptide is presented in Fig. 3. The existence of two elements required for localization of a vacuolar protein (i.e., signal sequence and sorting determinant) would be predicted from the model of Blobel (1980) in which multiple topogenic sequences are proposed to be required for translocation and sorting of compartmentalized proteins in eukaryotes.

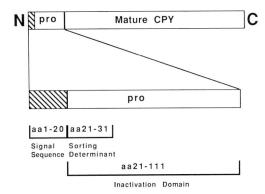

Fig. 3. Multiple functions of CPY propeptide. The DNA sequence of *PRC1* indicates that CPY is synthesized with a 111 amino acid prepropeptide. The first 20 amino acids are necessary and sufficient for efficient ER targeting and translocation. Amino acids 21–111 are required for inactivation of CPY, whereas the first 11 amino acids of the propeptide (21–31) are necessary for vacuolar targeting.

These ER and vacuolar targeting determinants are analogous to the multiple signals present on mitochondrial (van Loon *et al.*, 1986) and chloroplast proteins (van den Broeck *et al.*, 1985; Smeekens *et al.*, 1986) that are necessary for import and intraorganellar sorting of these proteins.

D. Genes Required for Sorting of Soluble Vacuolar Proteins

1. Mutants Pleiotropically Defective in Protein Sorting

A number of genes that are required for vacuolar protein sorting and that are therefore likely to encode components of the yeast protein-sorting apparatus have been identified. One set of such genes was identified by selecting for mutants that secrete CPY (Rothman and Stevens, 1986). The selection used to isolate these mutants exploited the observation that when CPY is secreted as a result of multicopy expression of *PRC1* a portion of the secreted protein is converted in the periplasm to an enzymatically active species that comigrates with mCPY (Stevens *et al.*, 1986). The extracellular conversion of proCPY to an active species occurs even in *pep4* cells, suggesting that the periplasmic peptidase carrying out this cleavage is unrelated to the vacuolar processing activity. Yeast cells lacking a leucine biosynthetic enzyme but expressing CPY activity can grow on the dipeptide carbobenzoxy-L-phenylalanyl-L-leucine as the only leucine supplement since CPY can cleave the peptide bond in this molecule (Kuhn *et al.*, 1974) thereby liberating leucine. *pep4* mutants that are defective for leucine biosynthesis will not grow on this dipeptide since they lack CPY activity (Jones *et al.*, 1982). When proCPY is mislocalized to the cell surface, however, it is converted to an active species even in *pep4* cells. Thus a *pep4* strain secreting CPY can utilize the dipeptide to relieve an auxotrophy for leucine. Demanding growth of *pep4 leu⁻* cells on the dipeptide as sole leucine source, then, allows a selection for cells that mislocalize CPY to the cell surface.

This selection has allowed the identification of a large number of mutants that fail to properly sort CPY to the vacuole, and these mutants have been classified into more than eight *VPL* complementation groups (Rothman and Stevens, 1986) required for vacuolar protein localization (Table I). Some of the mutants exhibit a dominant CPY secretion phenotype and consequently have not yet been placed into any of the *VPL* complementation groups. Although *vpl* mutants were selected for secretion of CPY, those mutants examined also secreted PrA. Moreover, they were deficient in internal PrB activity, perhaps indicating that these mutants secrete PrB as well as CPY and PrA. It seems reasonable to suggest that the pleiotropy of these mutations points to the existence of a single vacuolar protein-sorting apparatus that acts on multiple vacuolar proteins, and which can be perturbed by a defect in any one of a number of components. For the two *vpl* mutants examined in detail, it was found that this pleiotropy did not extend to defects in secretory protein localization since the rate of secretion and maturation of invertase was not substantially altered in the mutants. This was taken as evidence that these mutants do not perturb the general integrity of the secretory pathway, but instead act specifically on the vacuolar localization pathway.

The secretion of vacuolar hydrolases in *vpl* mutants appears not to result from a failure to assemble vacuoles; in the mutants examined (*vpl 1–vpl 8*), electron micrographs indicate the presence of vacuoles of approximately normal appearance. However, a number of the mutants do exhibit aberrant morphologies. For example, *vpl1* mutants accumulate large vesicular organelles, similar to those seen in certain of the *sec* mutants (Novick *et al.*, 1980, 1981), in addition to normal-looking vacuoles. These organelles might be related to intermediates in vacuolar assembly—a possibility that might be clarified by immunoelectron microscopic analysis.

A second approach to selecting for trans-acting CPY missorting mutants has been reported (Bankaitis *et al.*, 1986). This selection relies on the observation that cells which do not express invertase activity at the cell surface cannot utilize sucrose as a fermentable carbon source whereas cells expressing this enzyme extracellularly can ferment sucrose. By demanding growth of such cells on sucrose as a fermentable carbon source, Bankaitis *et al.* (1986) have identified mutants that secrete a CPY–invertase hybrid protein. This selection has resulted in the identification of over eight *VPT* genes required for vacuolar protein targeting (Table I). These *vpt* mutants secrete high levels of the fusion protein and are also found to mislocalize the wild-type CPY polypeptide to the cell surface. Because of the similar phenotypes of the *vpl* and *vpt* mutants it is likely that they represent the same general class of mutations. Consistent with this suggestion is the observation that at least one of the first eight *VPT* complementation groups is the same as one of the first eight *VPL* complementation groups (J. Rothman and V. Bankaitis, unpublished results). However, the number of complementation groups resulting in the secretion of vacuolar proteins has greatly expanded, since more than 33 *VPT* and 18 *VPL* complementation groups have now been identified (J. Robinson and S. Emr, personal communication; J. Rothman and T. Stevens, unpublished results). The degree of overlap of these two groups of genes has recently been determined to be extensive.

The *pep* mutants, originally identified by screening for yeast mutants deficient in CPY activity (see Section II), have been categorized into 17 complementation groups (Jones, 1977, 1984). Many *pep* mutants exhibited pleiotropic phenotypes with respect to deficiencies in several vacuolar hydrolases. As described in Section II,D, *PEP4* has been shown to encode the vacuolar protein PrA; however, the nature and function of the other *PEP* gene products is as yet unknown. It is possible that these other *PEP* genes are required for some stage in assembly of the vacuole. One of the phenotypes observed in the *vpl* mutants appears to be similar to that of the *pep* mutants, i.e., a deficiency in vacuolar hydrolase activities (Rothman and Stevens, 1986). Thus, it seemed feasible that the *VPL* and *PEP* complementation groups might overlap. An examination of the localization of CPY in several of the *pep* mutants has indicated that, indeed, all of the *pep* mutants that were analyzed show a *vpl* phenotype (i.e., secrete CPY; J. Rothman and T. Stevens, unpublished results). Preliminary data indicate that four of the *vpl* mutations are allelic to four of the *pep* mutations. Thus, it is likely that the relatively large number of genes which have been shown to be required for normal vacuolar function (*VPL*, *VPT*, and *PEP* genes) actually represents a somewhat smaller grouping, because of the overlaps among these groups of genes.

As summarized in Table I, there are three general conditions in which vacuolar proteins are missorted and delivered to the cell surface (i.e., overproduction of vacuolar proteins and cis- and trans-acting mutations), and in all three cases the missorting of CPY displays common features. First, CPY is always secreted in a form indistinguishable from the normal intracellular p2 precursor both in polypeptide and carbohydrate components (Stevens *et al.*, 1986; Rothman and Stevens, 1986; Valls *et al.*, 1987). Furthermore, the *sec1* mutation acts epistatically to secretion of CPY in cells overproducing the protein, in *vpl 1–8* mutants and in all of the *PRC1*-linked mislocalization mutants. Although these findings suggest that CPY may traverse the same pathway in all three aberrant situations, other interpretations are possible as will be discussed later. It appears from these studies that the vacuolar assembly pathway is a specialized branch of the secretory pathway, i.e., that in the absence of sorting signals or of a functional sorting apparatus, vacuolar proteins transit the Golgi and are mislocalized into the late secretory pathway by default. By this hypothesis, vacuolar proteins carry all the signals necessary for transport through the entire secretory pathway and, in addition, contain an element for sorting to the vacuole. This view is substantiated by the finding that vacuolar sorting signals are epistatic to signals on invertase that allow it to reach the cell surface. Moreover, the observation that lysosomal proteins are secreted in the absence of the M-6-P sorting tag suggests that this hypothesis may hold for all eukaryotic systems.

2. Analysis of Vacuolar Protein-Sorting Functions

The large number of genes required for sorting of vacuolar proteins suggests that many components are involved in this process. It seems likely that one of the components of the vacuolar protein-sorting system is a receptor which binds to newly synthesized vacuolar proteins during their transit through the Golgi apparatus and effects their sorting to the vacuole, in analogy to the M-6-P receptors of animal cells. If this receptor directly recognizes the vacuolar localization determinants at the amino terminus of proCPY (i.e., following signal sequence cleavage), it might be possible to identify the gene encoding such a component by isolating compensatory revertant mutants that properly sort the mutant forms of CPY. An alternative approach to identifying a putative CPY recognition molecule would be to biochemically isolate a protein from yeast membranes which binds to proCPY. The *PRC1*-linked mutants would provide useful controls for identification of such a component since it should bind to wild-type proCPY but not to the mutant forms of the proenzyme.

Studies with lysosomotropic agents (von Figura and Hasilik, 1986) and acidification mutants (Robbins *et al.*, 1983, 1984; Mellman *et al.*, 1986) in animal cells have strongly implicated a role for lysosomal or endosomal acidification in the biogenesis of lysosomes. It has been hypothesized that, in the absence of lysosomal acidification, receptor–lysosomal protein complexes formed in the Golgi apparatus fail to dissociate, resulting in saturation of the sorting receptors and a consequent secretion of lysosomal proteins. If this argument holds for yeast, it is reasonable that some of the *vpl, vpt,* or *pep* mutants might be defective in vacuolar acidification, as a result of lesions in the vacuolar H^+-ATPase (Uchida *et al.,* 1985) or in an accessory element required for vacuolar acidification. Two approaches have been taken in an attempt to identify components involved in vacuolar acidification, thereby allowing direct testing of this model. The first approach has been to isolate the genes encoding subunits of the vacuolar H^+-ATPase in an effort to examine the phenotypes of mutations in these genes generated *in vitro.* Antibody to the vacuolar H^+-ATPase purified from *Neurospora* (Bowman *et al.,* 1986) has been shown to cross-react with the yeast enzyme (C. Yamashiro and T. Stevens, unpublished observation). This antibody has been used to screen a yeast expression library (λ gt11) for clones that encode the protein. These H^+-ATPase structural gene clones will be useful for generating *in vitro* deletions and conditional mutations as has been done for many other yeast genes. Examination of the phenotypes of these H^+-ATPase mutants will allow an assessment of the involvement of the H^+-ATPase in vacuolar protein sorting.

A second approach to identifying mutants defective in vacuolar acidification involves the use of neutral red, a lysosomotropic agent that accumulates in the vacuoles of yeast cells (R. Preston, and E. Jones, personal communication). Vacuolar accumulation of this compound is probably dependent on the existence of a pH gradient across the vacuolar membrane. A selection has been developed based on the lethality of this agent to cells exposed to light. Survivors of such a treatment fail to accumulate neutral red and define at least four *NRA* (neutral red accumulation) linkage groups. At least one of these mutants appears to be defective in CPY activity, suggesting that vacuolar functions may be perturbed. Further analysis of these mutants will allow a determination of the acidification state of their vacuoles and an assessment of the role of acidification in vacuolar biogenesis.

A more complete understanding of the vacuolar protein-sorting reaction requires knowledge of the gene products that carry out this process. In yeast it is possible to clone a gene in which a mutation exists and to identify and characterize the product of that gene. The identification of the gene product allows one to undertake a biochemical analysis of its function and to attempt to understand this function in the context of the phenotypes caused by such mutations. With this motivation in mind, several of the *VPL* genes have been cloned (C. Raymond and T. Stevens, unpublished results). The *VPL1* gene product has been expressed in *E. coli* and has been used to generate antibody (C. Raymond, J. Rothman, and T. Stevens, unpublished results). This antibody has allowed a determination of the abundance and size of the *VPL1* protein and may provide clues to its intracellular site of residence and biochemical function. The cloned gene has also been used to determine the phenotype of a null mutation in *VPL1*. Since many of the *vpl1* alleles exhibited a temperature-sensitive lethal defect, it was expected that a deletion in the gene might be lethal. Unexpectedly, a *vpl1* null mutation leads to a greatly diminished growth rate but does not result in a lethal phenotype (J. Rothman and T. Stevens, unpublished observation). The *vpl1* deletion mutant also exhibits a thermosensitive lethal phenotype, suggesting that the *VPL1* gene is required for thermotolerance. Sequence analysis of the *VPL1* gene indicates that it encodes an approximately 700 amino acid protein which by standard criteria (Kyte and Doolittle, 1982) appears to be a soluble protein (J. Rothman, P. O'Hara, and T. Stevens, unpublished result). Based on this observation, the VPL1 protein may be analogous to yeast cytosolic factors required for transport of glycoproteins through the Golgi *in vitro* (Dunphy *et al.*, 1986).

E. Genetic Analysis of Vacuolar Membrane Protein Sorting

In our discussion thus far we have concentrated on the biogenesis of soluble proteins of the yeast vacuole. In mammalian cells, there is mounting evidence that soluble and membrane-bound lysosomal proteins are sorted via distinct mechanisms (Tager, 1985; von Figura and Hasilik, 1986; Lippincott-Schwartz and Fambrough, 1986; Barriocanal *et al.*, 1986). A number of lysosomal membrane proteins have been identified (Reggio *et al.*, 1984; Chen *et al.*, 1985; Lewis *et al.*, 1985; Lippincott-Schwartz and Fambrough, 1986; Barriocanal *et al.*, 1986), and none of those examined possesses the M-6-P moieties required for delivery of soluble lysosomal proteins (von Figura and Hasilik, 1986). Moreover, normal levels of a lysosomal membrane protein have been measured in mutant cells that fail to carry out the M-6-P modification. These results suggest that lysosomal membrane proteins utilize an alternative recognition system for sorting. The sketchy evidence from yeast seems to indicate that soluble and membrane vacuolar proteins are also sorted via distinct mechanisms. This argument is based on the finding that some *vpl* and *vpt* mutants which secrete high levels of multiple soluble vacuolar hydrolases do not exhibit extracellular activities of two vacuolar membrane proteins (DPAP B and α-mannosidase; Rothman and Stevens, 1986; J. Rothman and T. Stevens, unpublished observations) and seem to contain normal levels of α-mannosidase in their vacuoles (Bankaitis *et al.*, 1986). Therefore, it is of interest to pursue an analysis of vacuolar membrane protein delivery, in order to understand the differences and similarities between the pathways of sorting for these two classes of proteins.

The vacuolar membrane protein DPAP B appears to be well suited for studies of sorting and biosynthesis of a vacuolar membrane protein, since its activity is easily measured (Suarez Rendueles *et al.,* 1981), it is made in sufficient abundance to be studied using immunological procedures (J. Rothman and T. Stevens, unpublished observations), and the structural gene for this protein has been cloned (Julius *et al.,* 1983; Suarez Rendueles and Wolf, 1987a; our unpublished results) and sequenced (Roberts et al., 1988). DPAP B was originally identified as a heat-labile activity in yeast capable of cleaving an amino-terminal dipeptide from a chromogenic substrate (Suarez Rendueles *et al.,* 1981) and was subsequently shown to be localized to the vacuolar membrane (Bordallo *et al.,* 1984). Another activity capable of cleaving the same substrate has been described (Suarez Rendueles *et al.,* 1981; Julius *et al.,* 1983). The latter protein, DPAP A, is a heat-stable activity, is not localized to the vacuole, is encoded by a gene distinct from the DPAP B structural gene, and has been shown to be required for proteolytic maturation of the precursor of the α-cell-specific mating pheromone, α-factor. Thus, these two DPAP species are readily distinguished. DPAP B is probably an integral protein of the vacuolar membrane since it fails to be extracted from membranes by treatment with high-pH sodium carbonate (G. Pohlig and T. Stevens, unpublished result), a behavior generally associated with integral membrane proteins (Fujiki *et al.,* 1982). The translated sequence of the DPAP B structural gene predicts that the disposition of DPAP B in the vacuolar membrane is characteristic of a Group B (or II) membrane protein (Garoff, 1985; Wickner and Lodish, 1985), containing a short cytoplasmic domain at the amino terminus, the membrane-spanning segment, and a large lumenal domain which includes the carboxyl terminus.

DPAP B is found in the vacuolar membrane as a glycosylated species which can be converted by treatment with endoglycosidase F to a molecular weight (96 K) consistent with that predicted from the translated sequence. In accordance with this finding, the translated DNA sequence predicts six sites of N-linked glycosylation (i.e., Asn-x-Thr or -Ser). It is unknown whether the protein contains O-linked glycosylation. DPAP B appears to be glycosylated in two stages, in a manner similar to CPY and PrA. In the ER-blocked *sec18* mutant at a restrictive temperature, newly synthesized DPAP B accumulates as the glycosylated species seen at early times in a pulse–chase experiment, suggesting that this protein, like other vacuolar proteins, utilizes the early stages of the secretory pathway for its delivery to the vacuole. The form found in the vacuolar membrane is somewhat larger than this ER-accumulated species, consistent with subsequent modification of its oligosaccharides in the Golgi apparatus. The oligosaccharide chains found on the vacuolar species are apparently not hyperglycosylated like secretory proteins, but instead appear to be similar to those found on soluble vacuolar proteins.

The structural gene for DPAP B was fortuitously cloned by its ability to complement at high gene dosage the deficiency in mature α-factor production in a mutant lacking DPAP A (Julius et al., 1983). Since it is unlikely that the α-factor precursor passes through the vacuole, a reasonable interpretation for this observation is that DPAP B, like CPY, PrA, and vacuolar AlP, is mislocalized into the late secretory pathway on overproduction of the protein. This would allow the vacuolar DPAP to come in contact with the α-factor precursor during transit of the precursor to the cell surface. Once exposed to DPAP B, the α-factor precursor would then be matured by the reaction normally performed by DPAP A.

The observation that DPAP B, when overproduced, can suppress the α-factor production defect of a yeast cell lacking DPAP A suggests a means of screening for mutants defective in sorting of DPAP B to the vacuole. If it is indeed mislocalization of DPAP B that allows maturation of α-factor in the absence of DPAP A, then isolating yeast cells of α mating type that secrete active α-factor in the absence of DPAP A activity might identify mutants that missort DPAP B. Mutagenesis of the cloned DPAP B structural gene would allow one to assess which portions of the protein (e.g., transmembrane domain or cytoplasmic tail) are required for its localization to the vacuole. Such an approach has been successful in defining the important regions for the intracellular sorting of several cell surface integral membrane proteins (Adams and Rose, 1985; Doyle et al., 1986). Of relevance to these studies is the observation that DPAP B and DPAP A (apparently another Group B-like membrane protein) are substantially homologous in their lumenal domains but share no homology in their putative hydrophobic and cytoplasmic domains (J. Thorner, personal communication; our unpublished observations). This finding may indicate that the information that targets these two proteins to their respective destinations resides within the membrane spanning and cytoplasmic domains, a hypothesis that will be testable by exchanging these segments of the two proteins. By this approach, it has been possible to examine the regions important for sorting of membrane glycoproteins in polarized cells (Roth et al., 1987; Puddington et al., 1987).

Mutations unlinked to the DPAP B structural gene would be likely to define genes whose products are involved in vacuolar membrane protein sorting, similar to the genes (VPL, VPT, and PEP) that are required for soluble vacuolar protein sorting. Significantly, it has been shown that the α-factor processing defect of a strain lacking DPAP A is not suppressed by several of the vpl mutations, consistent with vacuolar membrane protein-sorting functions being distinct from the VPL gene products (J. Rothman, unpublished results). Moreover, a number of trans-acting mutations that suppress a deficiency in DPAP A (probably as a result of DPAP B mislocalization) have been isolated, and most of these do not result in CPY secretion (G. Pohlig, J. Rothman, and T. Stevens, unpublished results).

It will be of interest to extend studies of vacuolar membrane protein assembly to the H^+-ATPase, for example, to determine whether the various subunits of the complex (Uchida et al., 1985) are assembled prior to or on their arrival at the vacuole. It is possible that one or more of the H^+-ATPase subunits are synthesized on free ribosomes and are attached to the cytoplasmic face of the vacuolar membrane without passing through the secretory pathway. If this is true, it will be of great interest to determine how the cell coordinates the delivery and assembly of proteins that transit the secretory pathway with those synthesized and delivered directly through the cytoplasm.

IV. ENDOCYTOSIS IN YEAST AND ITS RELATIONSHIP TO VACUOLAR PROTEIN DELIVERY

A. Fluid-Phase Endocytosis and Isolation of Endocytosis Mutants

Recently it was demonstrated that materials can be delivered to the yeast vacuole not only from the early secretory pathway but also from the cell surface, in the process known as endocytosis. Fluid-phase endocytosis was first demonstrated in yeast through the use of the soluble fluorescent dyes fluorescein isothiocyanate (FITC)–dextran (Makarow, 1985a) and lucifer yellow (LY) (Riezman, 1985) as well as by internalization of virus particles into yeast spheroplasts (Makarow, 1985b). It was demonstrated that fluid-phase endocytosis in yeast is an energy-requiring process that is dependent on time and temperature. Uptake of these markers was taken to be a nonreceptor-mediated (i.e., fluid-phase) process, since the rate of uptake was not saturable with increasing marker concentrations. In the cases of the two fluorescent markers, it has been demonstrated that endocytic uptake results in delivery of these materials to the vacuole. Endocytosis of LY was shown to be blocked in a number of the sec mutants. Specifically, all of the vesicle-accumulating (late-acting) sec mutants were found to be deficient in LY uptake into the vacuole at the nonpermissive temperature, as were two of the ER-blocked mutants and a Golgi-blocked mutant (Riezman, 1985). It is perhaps not surprising that all of the genes required for fusion of secretory vesicles with the plasma membrane are also required for processes associated with internalization of portions of the plasma membrane during endocytosis. However, the finding that some of the early-acting (Golgi and ER) genes are also required for endocytosis suggests that their respective gene products may be required at multiple stages in the yeast secretory pathway.

Uptake of LY into yeast vacuoles has been used to identify thermosensitive mutants defective in endocytosis (Chvatchko *et al.*, 1986). Random temperature-sensitive lethal mutants were screened for a failure to take up the dye into the vacuole without also blocking secretion of invertase. Mutations in two complementation groups, *end1* and *end2*, were identified. The *end1* mutant exhibited a total block to LY uptake at the level of resolution of the LY assay, whereas the *end2* mutant was only partially defective in uptake. Surprisingly, these mutants were defective in endocytosis at the permissive as well as the restrictive temperature, indicating that efficient fluid-phase endocytosis is not essential for cell viability. The finding that mutations can be isolated which result in defective endocytosis but not in a failure to secrete seems to indicate that secretion is not absolutely dependent on a fully functional endocytic pathway.

The intracellular morphology of the two *end* mutants might be revealing with regard to the stage in endocytosis at which these mutants are blocked. The *end1* mutant accumulates invaginations of the plasma membrane at a higher frequency than is seen in wild-type cells, and these mutants also lack vacuoles of normal size. In contrast, the *end2* mutant contains vacuoles of normal appearance and exhibits a frequent accumulation of an organelle seen less often in wild-type cells. The latter structure was proposed to represent an intermediate in the endocytic pathway, such as an endosome.

Further evidence for the existence of an endosome-like compartment has been obtained by following the uptake of FITC–dextran in energy-depleted cells (Makarow and Nevalainen, 1987). Under such conditions the fluorescent label is taken up by yeast cells, but the fluorescence labeling is seen as a diffuse pattern throughout the cell—no vacuolar labeling is seen under these conditions. When a source of energy is resupplied to the cells this diffuse labeling pattern disappears, and the fluorescent dye is found to be localized specifically to the vacuole. This result suggests that the diffuse labeling pattern represents accumulation of the dye in an endocytic intermediate, perhaps the yeast equivalent of a mammalian endosome.

An analysis of the intensity and excitation spectrum of FITC fluorescence has allowed an assessment of the pH of the environment of the FITC–dextran during its uptake into the vacuole (Makarow and Nevalainen, 1987). Unlike mammalian cells, in which the endosomal pH is somewhat higher than the lysosomal pH (Mellman *et al.*, 1986), it was found that the pH of the putative yeast endosomal compartments is lower than the pH of the vacuole. The relationship between this low pH compartment, the nonvacuolar compartment in which endocytosed viruses partially accumulate (Makarow, 1985a), and the organelle accumulating in the *end2* mutants (Chvatchko *et al.*, 1986) has not yet been examined.

B. Receptor-Mediated Endocytosis

Once endocytosis had been demonstrated in yeast, the question arose as to its function. It is reasonable to suggest that endocytosis is used by yeast to gather extracellular nutrients (e.g., macromolecules) that cannot be transported into the cell via other pathways. Degradation of these materials could then proceed in the vacuole, liberating the nutrients in a metabolizable form. However, a requirement for endocytosis has not been demonstrated for any particular nutrient.

Another likely function of yeast endocytosis is its role in the yeast mating response, which has been described recently as a result of studies of receptor-mediated endocytosis (Chvatchko et al., 1986; Jenness and Spatrick, 1986). Haploid yeast cells exist in one of two mating types, called **a** or α, and each mating type secretes a small polypeptide hormone (**a**-factor or α-factor) that induces a mating response in cells of the opposite type (Sprague et al., 1983). α-Factor, secreted by α cells, has been shown to bind to specific receptors on the surface of **a** cells as the initial event in induction of the mating response (Jenness et al., 1983). It has been demonstrated that, following binding of the pheromone to **a** cells, bound α-factor cannot be dissociated from the cells by washing, consistent with its internalization. The apparent uptake of α-factor requires energy and an intact α-factor receptor, and is time and temperature dependent. This putative internalization of the pheromone parallels the loss of receptor binding sites from the cell surface and results in degradation of α-factor. These results have been taken as evidence for receptor-mediated endocytosis of the peptide pheromone and its receptor into yeast cells. Although it has not been directly demonstrated that α-factor is internalized by yeast cells, the finding that degradation of the presumably internalized factor does not occur in pep4 mutants (Riezman et al., 1986) implicates the role of vacuolar proteases in its breakdown. This observation is consistent with the hypothesis that α-factor is internalized and delivered to the vacuole following binding to its receptor, in analogy to receptor-mediated endocytosis of polypeptide hormones into lysosomes of animal cells (James and Bradshaw, 1984). It has been noted that, following binding and apparent uptake of α-factor and its receptor by **a** cells, pheromone binding sites do not reappear at the cell surface in the absence of protein synthesis (Jenness and Spatrick, 1986). Therefore, it is reasonable to suggest that loss of external receptors following binding to their ligand is a result of their degradation in the vacuole.

Receptor-mediated endocytosis of α-factor has been shown to be perturbed in the end mutants. The end1 mutant exhibits approximately a 10-fold reduction in the rate of α-factor uptake, consistent with a block in endocytosis. In this regard, it is interesting that the morphology of this mutant suggests it is defective in events occurring at the cell surface (Section IV,B). In contrast, *end2* cells do take up pheromone, but fail to degrade it. The latter phenotype is interpretable as a failure to deliver an endocytic intermediate containing the receptor and ligand to the compartment in which they are degraded (i.e., the vacuole), consistent with the observed accumulation of an intracellular organelle in this mutant. The action of *end1* and *end2* suggests that the former mutation exerts its effects earlier in the endocytic pathway. It will be of interest to test this model by examining epistatic interactions between these two mutations.

C. Role of Endocytosis and Vacuolar Functions in the Yeast Mating Response

A potential role for receptor-mediated endocytosis in yeast mating has been suggested by observations of the mating response of *end* mutants (Chvatchko *et al.*, 1986). All the parameters of mating response examined (cell cycle arrest, agglutination, and shmoo formation) were found to be deficient in both mutants. This finding suggests that endocytosis may be essential for many of the physiological manifestations associated with reception of mating pheromone. The finding that *end2* mutants may not deliver α-factor to the vacuole, but appear to take up the pheromone, suggests that some aspects of the mating response are dependent on vacuolar delivery of α-factor and/or its receptor. However, the normal proteolytic environment of the vacuole is not essential for eliciting the usual repertoire of mating responses, since *pep4* mutants, which do not degrade internalized α-factor, nonetheless respond normally to the pheromone. In contrast, the ability of cells to recover from the effects of high concentrations of α-factor is likely to be dependent on an intact vacuolar degradation system, since *pep4* mutants do not recover, or recover very slowly, from α-factor-induced cell cycle arrest when compared with wild-type cells (H. Riezman, personal communication).

The finding that the cell cycle of *end1* and *end2* mutants is not arrested by concentrations of α-factor that do cause arrest of wild-type cells suggested that it might be possible to identify additional *end* mutants by selecting for cells capable of growing in the presence of α-factor. Indeed, such a selection has resulted in the identification of a number of mutants that are deficient in endocytosis (H. Riezman, personal communication). Most of the α-factor-resistant mutants found by subsequent screening to be endocytosis deficient also exhibit a Sec⁻ phenotype. One of the complementation groups that show only an End⁻ phenotype is allelic to *cyr1*

ponent of the late secretory pathway (Fig. 4B). Alternatively, CPY might
be delivered to a post-Golgi compartment that is an intermediate in vacuolar biogenesis. If such a compartment is generated by the combined action
of endocytosis and Golgi-directed transport, then CPY might become
missorted to the cell surface via an endocytic recycling pathway (Fig.
4A). In both scenarios, the late *sec* mutations would be expected to act
epistatically to *vpl, vpt,* or *pep* mutations, since such *sec* mutations block
both pathways. It is conceivable that some mutations cause mislocalization of vacuolar proteins via one route, whereas others result in vacuolar
protein secretion via the other proposed route. Using immunological techniques, it should be possible to resolve these issues.

V. SUMMARY AND CONCLUSIONS

The problems of intracellular protein sorting, endocytosis, and vacuolar biogenesis in yeast have only begun to be addressed. It is clear that a
great deal of work is required before an understanding of these processes
at the molecular level is achieved. However, it is also apparent that genetic approaches to these problems have greatly assisted in identifying the
genes required for these cellular activities and should result in the identification of their respective gene products. The study of yeast as a cell
biological system has not only made it possible to identify components of
the protein localization pathways but has also allowed a direct assessment
of the role of proteins whose involvement in these processes has only
been implicated.

For example, clathrin, originally identified as a constituent of coated
pits and coated vesicles (Goldstein *et al.,* 1979), has been proposed to be
involved in mediating endocytosis (Goldstein *et al.,* 1979) and intracellular protein transport (Rothman and Fine, 1980). Once clathrin had been
identified in yeast (Mueller and Branton, 1984), it became possible to test
the *in vivo* role of this protein, by cloning the clathrin heavy and light
chain structural genes and examining the phenotypes associated with deletion of these genes. Because clathrin is ubiquitous, fairly conserved in a
number of diverse eukaryotic organisms, and was thought to play a critical role in protein localization (Goldstein *et al.,* 1979; Rothman *et al.,*
1980), it came as a surprise to cell biologists that this protein is apparently
dispensable—i.e., yeast cells that carry a deletion of the single clathrin
heavy chain gene (*CHC1*) are viable (Payne and Schekman, 1985). It is
even more surprising that endocytosis is not deficient in cells lacking
clathrin (G. Payne, personal communication), since endocytosis in mam-

malian cells is generally accompanied by the assembly of clathrin-coated pits and vesicles.

Clathrin has also been implicated in lysosomal biogenesis: coated regions of the Golgi that appear to contain newly synthesized lysosomal proteins also contain clathrin (Schulze-Lohoff *et al.*, 1985; Griffith and Simons, 1986). Yeast cells deleted for *CHC1* appear, however, to produce mCPY at a nearly normal rate and do not exhibit an obvious *vpl* phenotype (G. Payne, personal communication). Perhaps other proteins that are involved in generating coated membranes, such as those identified in an *in vitro* Golgi transport system (Balch *et al.*, 1984; Dunphy *et al.*, 1986; Orci *et al.*, 1986), are able to carry out protein localization functions in *chc1* mutants. Alternatively, coated membranes may not be essential intermediates in protein transport. The work with yeast clathrin clearly points to our profound ignorance of the molecular mechanisms of eukaryotic protein localization and underscores the utility of identifying the genes that are likely to be involved in protein localization based on the localization-defective phenotypes of lesions in these genes. Moreover, experiments similar to those done with yeast clathrin should help to clarify the roles that heretofore identified components such as cytoskeletal elements may play in protein localization and organellar biogenesis.

The identification of the genes required for secretion, organelle assembly, and endocytosis will allow isolation of the gene products that carry out these functions. However, an examination of the molecular details of the way in which these gene products operate will require that the reactions underlying these complex processes be reconstructed *in vitro*. In yeast, it has been shown that certain of these reactions can be examined in cell-free systems. For example, translocation of proteins across the ER membrane (Rothblatt and Meyer, 1986a,b; Hansen *et al.*, 1986; Waters and Blobel, 1986) and transport from ER to Golgi (Haselbeck and Schekman, 1986) have been reconstituted *in vitro* using yeast extracts. Furthermore, an *in vitro* reaction that mimics transport through stacks of the Golgi apparatus has been established using a combination of yeast and mammalian components (Dunphy *et al.*, 1986). Reactions of this sort will allow the determination of the precise biochemical roles of gene products originally identified via genetic methods.

ACKNOWLEDGMENTS

The authors would like to thank Ruskin Gould for excellent technical assistance with the immunoelectron microscopy experiments, Molly Rothman, Luis Valls, Beth Blachly-Dyson, Carl Yamashiro, Chris Roberts, and Chris Raymond for their comments on the manuscript, those individuals who communicated their results to us prior to publication, and Elizabeth Cooksey for her assistance in preparing the manuscript. The authors acknowledge the support of grants from the National Institute of General Medical Sciences (GM 32448) and the Chicago Community Trust/Searle Scholars Program. J.H.R. was supported by a National Institutes of Health predoctoral traineeship.

REFERENCES

Achstetter, T., and Wolf, D. H. (1985). Proteinases, proteolysis and biological control in the yeast *Saccharomyces cerevisiae. Yeast* **1,** 139–157.

Adams, G. A., and Rose, J. K. (1985). Structural requirements of a membrane spanning domain for protein anchoring and cell surface transport. *Cell* **41,** 1007–1015.

Ammerer, G., Hunter, C. P., Rothman, J. H., Saari, G. C., Valls, L. A., and Stevens, T. H. (1986). *PEP4* gene of *Saccharomyces cerevisiae* encodes proteinase A, a vacuolar enzyme required for processing of vacuolar precursors. *Mol. Cell. Biol.* **6,** 2490–2499.

Balch, W. E., Dunphy, W. G., Braell, W. A., and Rothman, J. E. (1984). Reconstitution of the transport of protein between successive compartments of the Golgi measured by the coupled incorporation of *N*-acetylglucosamine. *Cell* **39,** 405–416.

Bankaitis, V. A., Johnson, L. M., and Emr, S. D. (1986). Isolation of yeast mutants defective in protein targeting to the vacuole. *Proc. Natl. Acad. Sci. U.S.A.* **83,** 9075–9079.

Barriocanal, J. G., Bonifacino, J. S., Yuan, L., and Sandoval, I. V. (1986). Biosynthesis, glycosylation, movement through the Golgi system, and transport to lysosomes by an N-linked carbohydrate-independent mechanism of three lysosomal integral membrane proteins. *J. Biol. Chem.* **261,** 16755–16763.

Betz, H. (1979). Loss of sporulation ability in a yeast mutant with low proteinase A levels. *FEBS Lett.* **100,** 171–174.

Blachly-Dyson, E. G. (1987). Mutational analysis of the secretory signal sequence of yeast carboxypeptidase Y. Ph.D. Dissertation, University of Oregon.

Blachly-Dyson, E., and Stevens, T. H. (1987). Yeast carboxypeptidase Y is translocated and glycosylated without its amino-terminal signal sequence. *J. Cell Biol.* **104,** 1183–1191.

Blobel, G. (1980). Intracellular protein topogenesis. *Proc. Natl. Acad. Sci. U.S.A.* **77,** 1496–1500.

Blobel, G., and Dobberstein, B. (1975). Transfer of proteins across membranes. I. Presence of proteolytically processed and unprocessed immunoglobulin light chains on membrane-bound ribosomes of murine myeloma. *J. Cell Biol.* **67,** 835–851.

Boer, P., and Steyn-Parve, E. (1966). Isolation and purification of an acid phosphatase from baker's yeast. *Biochim. Biophys. Acta* **128,** 400–402.

Bordallo, C., Schwencke, J., and Suarez Rendueles, M. S. (1984). Localization of the thermosensitive X-prolyl dipeptidyl aminopeptidase in the vacuolar membrane of *Saccharomyces cerevisiae. FEBS Lett.* **173,** 199–203.

Bowman, E. J., Mandala, S., Taiz, L., and Bowman, B. J. (1986). Structural studies of the vacuolar membrane ATPase from *Neurospora crassa* and comparison with the tonoplast membrane ATPase from *Zea mays. Proc. Natl. Acad. Sci. U.S.A.* **83,** 48–52.

Breddam, K., and Svendsen, I. (1984). Identification of methionyl and cysteinyl residues in the substrate binding site of carboxypeptidase Y. *Carlsberg Res. Commun.* **49,** 639–645.

Brown, W. J., and Farquhar, M. G. (1984). The mannose 6-phosphate receptor for lysosomal enzymes in concentrated in cis Golgi cisternae. *Cell* **36,** 295–307.

Bustin, M., and Conway-Jacobs, A. (1971). Intramolecular activation of porcine pepsino-gen. *J. Biol. Chem.* **246**, 615–620.

Chen, J. W., Murphy, T., Willingham, M., Pastan, I., and August, J. T. (1985). Identifica-tion of two lysosomal membrane glycoproteins. *J. Cell Biol.* **101**, 85–95.

Chvatchko, Y., Howald, I., and Riezman, H. (1986). Two yeast mutants defective in endo-cytosis are defective in pheromone response. *Cell* **46**, 355–364.

Creek, K. E., and Sly, W. S. (1984). The role of the phosphomannosyl receptor in the transport of acid hydrolases to lysosomes. *In* "Lysosomes in Biology and Pathology" (J. T. Dingle, R. T. Dean, and W. Sly, eds.), Vol. 7, pp. 63–82. Elsevier, Amsterdam.

Deshaies, R. J., and Schekman, R. (1987). A yeast mutant defective at an early stage in import of secretory protein precursors into the endoplasmic reticulum. *J. Cell Biol.* **105**, 633–645.

Distel, B., Al, E. J. M., Tabak, H. F., and Jones, E. W. (1983). Synthesis and maturation of the yeast vacuolar enzymes carboxypeptidase Y and aminopeptidase I. *Biochim. Biophys. Acta* **741**, 128–135.

Doyle, C., Sambrook, J., and Gething, M.-J. (1986). Analysis of progressive deletions of the transmembrane and cytoplasmic domains of influenza hemagglutinin. *J. Cell Biol.* **103**, 1193–1204.

Dreyer, T., Halkjaer, B., Svendsen, I., and Ottesen, M. (1986). Primary structure of the aspartic proteinase A from *Saccharomyces cerevisiae*. *Carlsberg Res. Commun.* **51**, 27–41.

Dunphy, W. G., Pfeffer, S. R., Clary, D. O., Wattenberg, B. W., Glick, B. S., and Rothman, J. E. (1986). Yeast and mammals utilize similar cytosolic components to drive protein transport through the Golgi complex. *Proc. Natl. Acad. Sci. U.S.A.* **83**, 1622–1626.

Emter, O., and Wolf, D. H. (1984). Vacuoles are not the sole compartments of proteolytic enzymes in yeast. *FEBS Lett.* **166**, 321–325.

Erickson, A. H., Conner, G. E., and Blobel, G. (1981). Biosynthesis of a lysosomal enzyme: Partial structure of two transient and functionally distinct NH_2-terminal sequences in cathepsin D. *J. Biol. Chem.* **256**, 11224–11231.

Esmon, B., Novick, P., and Schekman, R. (1981). Compartmentalized assembly of oligosac-charides on exported glycoproteins in yeast. *Cell* **25**, 451–460.

Faust, P. L., Kornfeld, S., and Chirgwin, J. M. (1985). Cloning and sequence analysis of cDNA for human cathepsin D. *Proc. Natl. Acad. Sci. U.S.A.* **82**, 4910–4914.

Feldman, R. I., Bernstein, M., and Schekman, R. (1987). Product of *SEC53* is required for folding and glycosylation of secretory proteins in the lumen of the yeast endoplasmic reticulum. *J. Biol. Chem.* **262**, 9332–9339.

Ferro-Novick, S., Novick, P., Field, C., and Schekman, R. (1984a). Yeast secretory mu-tants that block the formation of active cell surface enzymes. *J. Cell Biol.* **98**, 44–53.

Ferro-Novick, S., Hansen, W., Schauer, I., and Schekman, R. (1984b). Genes required for completion of import of proteins into the endoplasmic reticulum in yeast. *J. Cell Biol.* **98**, 44–53.

Fujiki, Y., Hubbard, A. L., Fowler, S., and Lazarow, P. B. (1982). Isolation of intracellular membranes by means of sodium carbonate treatment: Application to endoplasmic reticulum. *J. Cell Biol.* **93**, 97–102.

Gals, S., Willingham, M. C., and Gottesman, M. M. (1985). Processing and lysosomal localization of a glycoprotein whose secretion is transformation stimulated. *J. Cell Biol.* **100**, 535–544.

Garcia Alvarez, N., Bordallo, C., Gascon, S., and Suarez Rendueles, P. (1985). Purification and characterization of a thermosensitive X-prolyl dipeptidyl aminopeptidase (dipepti-dyl aminopeptidase yscV) from *Saccharomyces cerevisiae*. *Biochem. Biophys. Acta* **832**, 119–125.

Garoff, H. (1985). Using recombinant DNA techniques to study protein targeting in the eucaryotic cell. *Ann. Rev. Cell Biol.* **1**, 403–445.

Geuze, H. J., Slot, J. W., Strous, G. J. A. M., Peppard, J., von Figura, K., Hasilik, A., and Schwartz, A. L. (1984). Intracellular receptor sorting during endocytosis: Comparative immunoelectron microscopy of multiple receptors in rat liver. *Cell* **37**, 195–204.

Geuze, H. J., Slot, J. W., Strous, G. J. A. M., Hasilik, A., and von Figura, K. (1985). Possible pathways for lysosomal enzyme delivery. *J. Cell Biol.* **101**, 2253–2262.

Goldstein, J. L., Anderson, R. G. W., and Brown, M. S. (1979). Coated pits, coated vesicles, and receptor-mediated endocytosis. *Nature (London)* **279**, 679–685.

Griffith, G., and Simons, K. (1986). The *trans* Golgi network: Sorting at the exit site of the Golgi complex. *Science* **234**, 438–443.

Haguenauer-Tsapis, R., and Hinnen, A. (1984). A deletion that includes the signal peptidase cleavage site impairs processing, glycosylation, and secretion of cell surface yeast acid phosphatase. *Mol. Cell. Biol.* **4**, 2668–2675.

Hall, M. N., Hereford, L., and Herskowitz, I. (1984). Targeting of *E. coli* β-galactosidase to the nucleus in yeast. *Cell* **36**, 1057–1065.

Hansen, W., Garcia, P. D., and Walter, P. (1986). *In vitro* protein translocation across the yeast endoplasmic reticulum: ATP-dependent posttranslational translocation of the prepro-α-factor. *Cell* **45**, 397–406.

Harris, S. D., and Cotter, D. A. (1987). Vacuolar (lysosomal) trehalase of *Saccharomyces cerevisiae*. *Curr. Micro.* **15**, 247–249.

Haselbeck, A., and Schekman, R. (1986). Interorganelle transfer and glycosylation of yeast invertase *in vitro*. *Proc. Natl. Acad. Sci. U.S.A.* **83**, 2017–2021.

Hashimoto, C., Cohen, R. E., Zhang, W., and Ballou, C. E. (1981). Carbohydrate chains on yeast carboxypeptidase Y are phosphorylated. *Proc. Natl. Acad. Sci. U.S.A.* **78**, 2244–2248.

Hasilik, A., and Neufeld, E. F. (1980). Biosynthesis of lysosomal enzymes in fibroblasts. Synthesis as precursors of higher molecular weight. *J. Biol. Chem.* **255**, 4937–4945.

Hasilik, A., and Tanner, W. (1976). Biosynthesis of carboxypeptidase Y in yeast. Evidence for a precursor form of the glycoprotein. *Biochem. Biophys. Res. Commun.* **72**, 1430–1436.

Hasilik, A., and Tanner, W. (1978a). Biosynthesis of the vacuolar yeast glycoprotein carboxypeptidase Y. Conversion of precursor into the enzyme. *Eur. J. Biochem.* **85**, 599–608.

Hasilik, A., and Tanner, W. (1978b). Carbohydrate moiety of carboxypeptidase Y and perturbation of its synthesis. *Eur. J. Biochem.* **91**, 567–575.

Hemmings, B. A., Zubenko, G. S., Hasilik, A., and Jones, E. W. (1981). Mutant defective in processing of an enzyme located in the lysosome-like vacuole of *Saccharomyces cerevisiae*. *Proc. Natl. Acad. Sci. U.S.A.* **78**, 435–439.

Hoflack, B., and Kornfeld, S. (1985). Lysosomal enzyme binding to mouse P388D1 macrophage membranes lacking the 215-kDa mannose 6-phosphate receptor: Evidence for the existence of a second mannose 6-phosphate receptor. *Proc. Natl. Acad. Sci. U.S.A.* **82**, 4428–4432.

Huffaker, T., and Robbins, P. (1982). Temperature-sensitive yeast mutants deficient in asparagine-linked glycosylation. *J. Biol. Chem.* **257**, 3203–3210.

James, M. N. G., and Sielecki, A. R. (1986). Molecular structure of an aspartic proteinase zymogen, porcine pepsinogen, at 1.8 Å resolution. *Nature (London)* **319**, 33–38.

James, R., and Bradshaw, R. A. (1984). Polypeptide growth factors. *Annu. Rev. Biochem.* **53**, 259–292.

Jenness, D. D., and Spatrick, P. (1986). Down regulation of the α-factor pheromone receptor in *S. cerevisiae*. *Cell* **46**, 345–353.

Jenness, D. D., Burkholder, A. C., and Harwell, L. H. (1983). Binding of α-factor pheromone to yeast **a** cells: Genetic evidence for an α-factor receptor. *Cell* **35**, 521–529.

Johnson, L. M., Bankaitis, V. A., and Emr, S. D. (1987). Distinct sequence determinants direct intracellular sorting and modification of a yeast vacuolar proteinase. *Cell* **48**, 875–885.

Jones, E. W. (1977). Proteinase mutants of *Saccharomyces cerevisiae*. *Genetics* **85**, 23–33.

Jones, E. W. (1984). The synthesis and function of proteases in *Saccharomyces:* Genetic approaches. *Annu. Rev. Genet.* **18**, 233–270.

Jones, E. W., Zubenko, G. S., Parker, R. R., Hemmings, B. A., and Hasilik, A. (1981). Pleiotropic mutations of *S. cerevisiae* which cause deficiency for proteinases and other vacuolar enzymes. *In* "Molecular Genetics in Yeast, Alfred Benzon Symposium" (D. von Wettstein, J. Friis, M. Kielland-Brandt, and A. Stenderup, eds.), pp. 182–198. Munksgaard, Copenhagen.

Jones, E. W., Zubenko, G. S., and Parker, R. R. (1982). *PEP4* gene function is required for expression of several vacuolar hydrolases in *Saccharomyces cerevisiae*. *Genetics* **102**, 665–677.

Jones, E. W., Moehle, C., Kolodny, M., Aynardi, M., Park, F., Daniels, L., and Garlow, S. (1986). Genetics of vacuolar proteases. *In* "Yeast Cell Biology" (J. Hicks, ed.), pp. 505–518. Alan R. Liss, New York.

Julius, D., Blair, L., Brake, A., Sprague, G., and Thorner, J. (1983). Yeast α-factor is processed from a larger precursor polypeptide: The essential role of a membrane-bound dipeptidyl aminopeptidase. *Cell* **32**, 839–852.

Julius, D., Schekman, R., and Thorner, J. (1984a). Glycosylation and processing of prepro-α-factor through the yeast secretory pathway. *Cell* **36**, 309–318.

Julius, D., Brake, A., Blair, L., Kunisawa, R., and Thorner, J. (1984b). Isolation of the putative structural gene for the lysine–arginine cleaving endopeptidase required for processing of yeast prepro-α-factor. *Cell* **37**, 1075–1089.

Kaiser, C. A., and Botstein, D. (1986). Secretion-defective mutations in the signal sequence for *Saccharomyces cerevisiae* invertase. *Mol. Cell. Biol.* **6**, 2382–2391.

Kaneko, Y., Tamai, Y., Toh-e, A., and Oshima, Y. (1985). Transcriptional and post-transcriptional control of *PHO8* expression by *PHO* regulatory genes in *Saccharomyces cerevisiae*. *Mol. Cell. Biol.* **5**, 248–252.

Kornfeld, R., and Kornfeld, S. (1985). Assembly of asparagine-linked oligosaccharides. *Annu. Rev. Biochem.* **54**, 631–664.

Kornfeld, S. (1986). Trafficking of lysosomal enzymes in normal and diseased states. *J. Clin. Invest.* **77**, 1–6.

Kuhn, R. W., Walsh, K. W., and Neurath, H. (1974). Isolation and partial characterization of an acid carboxypeptidase from yeast. *Biochemistry* **13**, 3871–3877.

Kuo, S., and Lampen, J. (1974). Tunicamycin, an inhibitor of yeast glycoprotein synthesis. *Biochem. Biophys. Res. Commun.* **58**, 287–295.

Kuranda, M. J., and Robbins, P. W. (1987). Cloning and heterologous expression of glycosidase genes from *Saccharomyces cerevisiae*. *Proc. Natl. Acad. Sci. U.S.A.* **84**, 2585–2589.

Kyte, J., and Doolittle, R. F. (1982). A simple method for displaying the hydropathic character of a protein. *J. Mol. Biol.* **157**, 105–132.

Lang, L., Reitman, M., Tang, J., Roberts, R. M., and Kornfeld, S. (1984). Lysosomal enzyme phosphorylation. Recognition of a protein-dependent determinant allows specific phosphorylation of oligosaccharides present on lysosomal protein. *J. Biol. Chem.* **259**, 14663–14671.

Lehle, L., Cohen, R. E., and Ballou, C. (1979). Carbohydrate structure of yeast invertase. *J. Biol. Chem.* **254,** 12209–12218.

Lemansky, P., Gieselmann, V., Hasilik, A., and von Figura, K. (1985). Synthesis and transport of lysosomal acid phosphatase in normal and I-cell fibroblasts. *J. Biol. Chem.* **260,** 9023–9030.

Lewis, V., Green, S. A., Marsh, M., Vihko, P., Helenius, A., and Mellman, I. (1985). Glycoproteins of the lysosomal membrane. *J. Cell Biol.* **100,** 1839–1847.

Lippincott-Schwartz, J., and Fambrough, D. M. (1986). Lysosomal membrane dynamics: Structure and interorganellar movement of a major lysosomal membrane glycoprotein. *J. Cell Biol.* **102,** 1593–1605.

Lolle, S. J., and Bussey, H. (1986). *In vivo* evidence for posttranslational translocation and signal sequence cleavage of the killer preprotoxin of *Saccharomyces cerevisiae. Mol. Cell. Biol.* **6,** 4274–4280.

Makarow, M. (1985a). Endocytosis in *Saccharomyces cerevisiae:* Internalization of enveloped viruses into spheroplasts. *EMBO J.* **4,** 1855–1860.

Makarow, M. (1985b). Endocytosis in *Saccharomyces cerevisiae:* Internalization of α-amylase and fluorescent dextran into cells. *EMBO J.* **4,** 1861–1866.

Makarow, M., and Nevalainen, L. T. (1987). Transport of a fluorescent macromolecule via endosomes to the vacuole in *Saccharomyces cerevisiae. J. Cell Biol.* **104,** 67–75.

Martin, B., Svendsen, I., Viswanatha, T., and Johansen, J. T. (1982). Amino acid sequence of carboxypeptidase Y. I. Peptides from cleavage with cyanogen bromide. *Carlsberg Res. Commun.* **47,** 1–13.

Matsumoto, K., Uno, I., and Ishikawa, T. (1984). Identification of the structural gene and nonsense alleles for adenylate cyclase in *Saccharomyces cerevisiae. J. Bacteriol.* **157,** 277–282.

Mechler, B., and Wolf, D. H. (1981). Analysis of proteinase A function in yeast. *Eur. J. Biochem.* **121,** 47–52.

Mechler, B., Müller, M., Müller, H., Meussdoerffer, F., and Wolf, D. H. (1982a). *In vivo* biosynthesis of the vacuolar proteinases A and B in the yeast *Saccharomyces cerevisiae. J. Biol. Chem.* **257,** 11203–11206.

Mechler, B., Müller, M., Müller, H., and Wolf, D. H. (1982b). *In vivo* biosynthesis of vacuolar proteinases in proteinase mutants of *Saccharomyces cerevisiae. Biochem. Biophys. Res. Commun.* **107,** 770–778.

Mechler, B., Müller, H., and Wolf, D. H. (1987). Maturation of vacuolar (lysosomal) enzymes in yeast: proteinase yscA and proteinase yscB are catalysts of the processing and activation of carboxypeptidase yscY. *EMBO J.* **6,** 2157–2163.

Mellman, I., Fuchs, R., and Helenius, A. (1986). Acidification of the endocytic and exocytic pathways. *Annu. Rev. Biochem.* **55,** 663–700.

Meussdoerffer, F., Tortora, P., and Holzer, H. (1980). Purification and properties of proteinase A from yeast. *J. Biol. Chem.* **255,** 12087–12093.

Moehle, C. M., Aynardi, M. W., Kolodny, M. R., Park, F. J., and Jones, E. W. (1987). Protease B of *Saccharomyces cerevisiae:* Isolation and regulation of the *PRB1* structural gene. *Genetics* **115,** 255–263.

Mueckler, M., and Lodish, H. F. (1986). The human glucose transporter can insert posttranslationally into microsomes. *Cell* **44,** 629–637.

Mueller, S., and Branton, D. (1984). Identification of coated vesicles in *Saccharomyces cerevisiae. J. Cell Biol.* **98,** 341–346.

Müller, M., and Müller, H. (1981). Synthesis and processing of *in vitro* and *in vivo* precursors of the vacuolar yeast enzyme carboxypeptidase Y. *J. Biol. Chem.* **256,** 11962–11965.

Navon, G., Shulman, R. G., Yamane, T., Eccleshall, T. R., Lam, K.-B., Baronofsky, J. J., and Marmur, J. (1979). Phosphorus-31 nuclear magnetic resonance studies of wild-type and glycolytic pathway mutants of *Saccharomyces cerevisiae*. *Biochemistry* **18**, 4487–4499.

Novick, P., and Botstein, D. (1985). Phenotypic analysis of temperature-sensitive yeast actin mutants. *Cell* **40**, 405–416.

Novick, P., and Schekman, R. (1979). Secretion and cell-surface growth are blocked in a temperature-sensitive mutant of *Saccharomyces cerevisiae*. *Proc. Natl. Acad. Sci. U.S.A.* **76**, 1858–1862.

Novick, P., Field, C., and Schekman, R. (1980). Identification of 23 complementation groups required for posttranslational events in the yeast secretory pathway. *Cell* **21**, 205–215.

Novick, P., Ferro, S., and Schekman, R. (1981). Order of events in the yeast secretory pathway. *Cell* **25**, 461–469.

Ohnishi, H. R., Tkacz, J. S., and Lampen, J. O. (1979). Glycoprotein nature of yeast alkaline phosphatase: Formation of active enzyme in the presence of tunicamycin. *J. Biol. Chem.* **254**, 11943–11952.

Ohsumi, Y., and Anraku, Y. (1981). Active transport of basic amino acids driven by a proton motive force in vacuolar membrane vesicles of *Saccharomyces cerevisiae*. *J. Biol. Chem.* **256**, 2079–2082.

Ohsumi, Y., and Anraku, Y. (1983). Calcium transport driven by a proton motive force in vacuolar membrane vesicles of *Saccharomyces cerevisiae*. *J. Biol. Chem.* **258**, 5614–5617.

Opheim, D. J. (1978). α-D-Mannosidase of *Saccharomyces cerevisiae* characterization and modulation of activity. *Biochim. Biophys. Acta* **524**, 121–130.

Orci, L., Glick, B. S., and Rothman, J. E. (1986). A new type of coated vesicular carrier that appears not to contain clathrin: Its possible role in protein transport within the Golgi stack. *Cell* **46**, 171–184.

Owada, M., and Neufeld, E. F. (1982). Is there a mechanism for introducing acid hydrolases into liver lysosomes that is independent of mannose 6-phosphate recognition? *Biochem. Biophys. Res. Commun.* **105**, 814–820.

Payne, G. S., and Schekman, R. (1985). A test of clathrin function in protein secretion and cell growth. *Science* **230**, 1009–1014.

Puddington, L., Woodgett, C., and Rose, J. K. (1987). Replacement of the cytoplasmic domain alters sorting of a viral glycoprotein in polarized cells. *Proc. Natl. Acad. Sci. U.S.A.* **84**, 2756–2760.

Randall, L. L., and Hardy, S. J. S. (1986). Correlation of competence for export with lack of tertiary structure of the mature species: A study *in vivo* of maltose-binding protein in *E. coli*. *Cell* **46**, 921–928.

Reggio, H., Bainton, D., Harms, E., Coudrier, E., and Louvard, D. (1984). Antibodies against lysosomal membranes reveal a 100,000-mol-wt protein that cross-reacts with purified H^+,K^+-ATPase from gastric mucosa. *J. Cell Biol.* **99**, 1511–1526.

Reitman, M. L., and Kornfeld, S. (1981). Lysosomal enzyme targeting. *N*-Acetyl-glucosaminylphosphotransferase selectively phosphorylates native lysosomal enzymes. *J. Biol. Chem.* **256**, 11977–11980.

Riezman, H. (1985). Endocytosis in yeast: Several of the yeast secretory mutants are defective in endocytosis. *Cell* **40**, 1001–1009.

Riezman, H., Chvatchko, Y., and Dulic, V. (1986). Endocytosis in yeast. *Trends Biochem. Sci.* **11**, 325–328.

Robbins, A. R., Peng, S. S., and Marshall, J. L. (1983). Mutant Chinese hamster ovary cells pleiotropically defective in receptor-mediated endocytosis. *J. Cell Biol.* **96**, 1064–1071.

Robbins, A. R., Oliver, C., Bateman, J. L., Krag, S. S., Galloway, C. J., and Mellman, I. (1984). A single mutation in Chinese hamster ovary cells impairs both Golgi and endosomal functions. *J. Cell Biol.* **99,** 1296–1308.

Roberts, C., Pohlig, G., Rothman, J., and Stevens, T. (1988). Dipeptidyl aminopeptidase B, an integral membrane protein of the yeast vacuole, utilizes early stages of the secretory pathway for transport. In preparation.

Rosenfeld, M. G., Kreibich, G., Popov, D., Kato, K., and Sabatini, D. D. (1982). Biosynthesis of lysosomal hydrolases: Their synthesis in bound polysomes and the role of co- and posttranslational processing in determining their subcellular distribution. *J. Cell Biol.* **93,** 135–143.

Roth, M. G., Gunderson, D., Patil, N., and Rodriguez-Boulan, E. (1987). The large external domain is sufficient for the correct sorting of secreted or chimeric influenza virus hemagglutinins in polarized monkey kidney cells. *J. Cell Biol.* **104,** 769–782.

Rothblatt, J. A., and Meyer, D. I. (1986a). Secretion in yeast: Reconstitution of the translocation and glycosylation of α-factor and invertase in a homologous cell-free system. *Cell* **44,** 619–628.

Rothblatt, J. A., and Meyer, D. I. (1986b). Secretion in yeast: Translocation and glycosylation of prepro-α-factor *in vitro* can occur via an ATP-dependent posttranslational mechanism. *EMBO J.* **5,** 1031–1036.

Rothblatt, J. A., and Meyer, D. I. (1987). Analysis of secretory protein translocation across yeast microsomes. *J. Cell. Biochem. Suppl.* **11A,** 263.

Rothman, J. E., and Fine, R. E. (1980). Coated vesicles transport newly synthesized membrane glycoproteins from endoplasmic reticulum to plasma membrane in two successive stages. *Proc. Natl. Acad. Sci. U.S.A.* **77,** 780–784.

Rothman, J. E., Bursztyn-Pettigrew, H., and Fine, R. E. (1980). Transport of the membrane glycoprotein of vesicular stomatitis virus to the cell surface in two stages by clathrin-coated vesicles. *J. Cell Biol.* **86,** 162–171.

Rothman, J. H., and Stevens, T. H. (1986). Protein sorting in yeast: Mutants defective in vacuole biogenesis mislocalize vacuolar proteins into the late secretory pathway. *Cell* **47,** 1041–1051.

Rothman, J. H., Hunter, C. P., Valls, L. A., and Stevens, T. H. (1986). Overproduction-induced mislocalization of a yeast vacuolar protein allows isolation of its structural gene. *Proc. Natl. Acad. Sci. U.S.A.* **83,** 3248–3252.

Sahagian, G. G., Distler, J., and Jourdian, G. W. (1981). Characterization of a membrane-associated receptor from bovine liver that binds phosphomannosyl residues of bovine testicular β-galactosidase. *Proc. Natl. Acad. Sci. U.S.A.* **78,** 4289–4293.

Sato, T., Ohsumi, Y., and Anraku, Y. (1984a). Substrate specificities of active transport systems for amino acids in vacuolar-membrane vesicles of *Saccharomyces cerevisiae*. *J. Biol. Chem.* **259,** 11505–11508.

Sato, T., Ohsumi, Y., and Anraku, Y. (1984b). An arginine/histidine exchange system in vacuolar-membrane vesicles of *Saccharomyces cerevisiae*. *J. Biol. Chem.* **259,** 11509–11511.

Schatz, P. J., Pillus, L., Grisafi, P., Solomon, F., and Botstein, D. (1986a). Two functional α-tubulin genes of the yeast *Saccharomyces cerevisiae* encode divergent proteins. *Mol. Cell. Biol.* **6,** 3711–3721.

Schatz, P. J., Solomon, F., and Botstein, D. (1986b). Genetically essential and nonessential α-tubulin genes specify functionally interchangeable proteins. *Mol. Cell. Biol.* **6,** 3722–3733.

Schauer, I., Emr, S., Gross, C., and Schekman, R. (1985). Invertase signal and mature sequence substitutions that delay intercompartmental transport of active enzyme. *J. Cell Biol.* **100**, 1664–1675.

Schekman, R. (1985). Protein localization and membrane traffic in yeast. *Annu. Rev. Cell Biol.* **1**, 115–143.

Schulze-Lohoff, E., Hasilik, A., and von Figura, K. (1985). Cathepsin D precursors in clathrin-coated organelles from human fibroblasts. *J. Cell Biol.* **101**, 824–829.

Schwaiger, H., Hasilik, A., von Figura, K., Wiemken, A., and Tanner, W. (1982). Carbohydrate-free carboxypeptidase Y is transferred into the lysosome-like yeast vacuole. *Biochem. Biophys. Res. Commun.* **104**, 950–956.

Silver, P. A., Keegan, L. P., and Ptashne, M. (1984). Amino terminus of the yeast *GAL4* gene product is sufficient for nuclear localization. *Proc. Natl. Acad. Sci. U.S.A.* **81**, 5951–5955.

Smeekens, S., Bauerle, C., Hageman, J., Keegstra, K., and Weisbeek, P. (1986). The role of the transit peptide in the routing of precursors toward different chloroplast compartments. *Cell* **46**, 365–375.

Sprague, G. F., Jr., Blair, L. C., and Thorner, J. (1983). Cell interactions and regulation of cell type in the yeast *Saccharomyces cerevisiae*. *Annu. Rev. Microbiol.* **37**, 623–660.

Stevens, T., Esmon, B., and Schekman, R. (1982). Early stages in the yeast secretory pathway are required for transport of carboxypeptidase Y to the vacuole. *Cell* **30**, 439–448.

Stevens, T. H., Rothman, J. H., Payne, G. S., and Schekman, R. (1986). Gene dosage-dependent secretion of yeast vacuolar carboxypeptidase Y. *J. Cell Biol.* **102**, 1551–1557.

Suarez Rendueles, M. P., Schwencke, J., Garcia Alvarez, N., and Gascon, S. (1981). A new X-prolyl-dipeptidyl aminopeptidase from yeast associated with a particulate fraction. *FEBS Lett.* **131**, 296–300.

Suarez Rendueles, P., and Wolf, D. H. (1987a). Identification of the structural gene for dipeptidyl aminopeptidase yscV (*DAP2*) of *Saccharomyces cerevisiae*. *J. Bacteriol.*, **169**, 4041–4048.

Suarez Rendueles, P., and Wolf, D. H. (1987b). Proteinase function in yeast: Biochemical and genetic approaches to a central mechanism of post-translational control in the eukaryote cell. *FEMS Microbiol. Rev.* (in press).

Svendsen, I., Martin, B. M., Viswanatha, T., and Johansen, J. T. (1982). Amino acid sequence of carboxypeptidase Y. II. Peptides from enzymatic cleavages. *Carlsberg Res. Commun.* **47**, 15–27.

Tager, J. M. (1985). Biosynthesis and deficiency of lysosomal enzymes. *Trends Biochem. Sci.* **10**, 324–326.

Tang, J. (1979). Evolution in the structure and function of aspartyl proteases. *Mol. Cell. Biochem.* **26**, 93–109.

Trimble, R. B., Maley, F., and Chu, F. K. (1983). Glycoprotein biosynthesis in yeast: Protein conformation affects processing of high mannose oligosaccharides on carboxypeptidase Y and invertase. *J. Biol. Chem.* **258**, 2562–2567.

Trumbly, R., and Bradley, G. (1983). Isolation and characterization of aminopeptidase mutants of *Saccharomyces cerevisiae*. *J. Bacteriol.* **156**, 36–48.

Uchida, E., Ohsumi, Y., and Anraku, Y. (1985). Purification and properties of H^+-translocating, Mg^{2+}-adenosine triphosphatase from vacuolar membranes of *Saccharomyces cerevisiae*. *J. Biol. Chem.* **260**, 1090–1095.

Valls, L. A., Hunter, C. P., Rothman, J. H., and Stevens, T. H. (1987). Protein sorting in yeast: The localization determinant of yeast vacuolar carboxypeptidase Y resides in the propeptide. *Cell* **48**, 887–897.

Van Arsdell, J. N., Woolford, C. A., Jones, E. W., and Innis, M. A. (1986). Characterization of genetic mutations in the *PEP4* locus. *Yeast* **2**, s399.

van den Broeck, G., Timko, M. P., Kausch, A. P., Cashmore, A. R., Van Montagu, M., and Herrera-Estrella, L. (1985). Targeting of foreign proteins to chloroplasts by fusion to the transit peptide from the small subunit of ribulose-1,5-bisphosphate carboxylase. *Nature (London)* **313**, 358–363.

van Loon, A. P. G. M., Brandli, A. W., and Schatz, G. (1986). The presequences of two imported mitochondrial proteins contain information for intracellular and intramitochondrial sorting. *Cell* **44**, 801–812.

von Figura, K., and Hasilik, A. (1986). Lysosomal enzymes and their receptors. *Annu. Rev. Biochem.* **55**, 167–193.

von Heijne, G. (1985). Signal sequences: The limits of variation. *J. Mol. Biol.* **184**, 99–105.

Waheed, A., Pohlmann, R., Hasilik, A., von Figura, K., van Elsen, A., and Leroy, J. G. (1982). Deficiency of UDP–*N*-acetylglucosamine:lysosomal enzyme *N*-acetylglucosamine-1-phosphotransferase in organs of I-cell patients. *Biochem. Biophys. Res. Commun.* **105**, 1052–1058.

Walter, P., and Lingappa, V. R. (1986). Mechanism of protein translocation across the endoplasmic reticulum membrane. *Annu. Rev. Cell Biol.* **2**, 499–516.

Walter, P., Gilmore, R., and Blobel, G. (1984). Protein translocation across the endoplasmic reticulum. *Cell* **38**, 5–8.

Waters, M. G., and Blobel, G. (1986). Secretory protein translocation in a yeast cell-free system can occur posttranslationally and requires ATP hydrolysis. *J. Cell Biol.* **102**, 1543–1550.

Watson, M. E. E. (1984). Compilation of published signal sequences. *Nucleic Acids Res.* **12**, 5145–5164.

Wickner, W. T., and Lodish, H. F. (1985). Multiple mechanisms of protein insertion into and across membranes. *Science* **230**, 400–407.

Wiemken, A., Schellenberg, M., and Urech, K. (1979). Vacuoles: The sole compartments of digestive enzymes in yeast (*Saccharomyces cerevisiae*)? *Arch. Microbiol.* **123**, 23–35.

Wolf, D. H., and Ehmann, C. (1978). Isolation of yeast mutants lacking proteinase B activity. *FEBS Lett.* **92**, 121–124.

Wolf, D. H., and Fink, G. R. (1975). Proteinase C (carboxypeptidase Y) mutant of yeast. *J. Bacteriol.* **123**, 1150–1156.

Woolford, C. A., Daniels, L. B., Park, F. J., Jones, E. W., van Arsdell, J. N., and Innis, M. A. (1986). The *PEP4* gene encodes an aspartyl protease implicated in the posttranslational regulation of *Saccharomyces cerevisiae* vacuolar hydrolases. *Mol. Cell. Biol.* **6**, 2500–2510.

Woychik, N. A., Cardelli, J. A., and Dimond, R. L. (1986). A conformationally altered precursor to the lysosomal enzyme α-mannosidase accumulates in the endoplasmic reticulum in a mutant strain of *Dictyostelium discoideum*. *J. Biol. Chem.* **261**, 9595–9602.

Yaffe, M. P., and Schatz, G. (1984). Two nuclear mutations which block mitochondrial protein import in yeast. *Proc. Natl. Acad. Sci. U.S.A.* **81**, 4819–4823.

Zubenko, G. S., Mitchell, A., and Jones, E. W. (1980). Mapping of the proteinase B structural gene, *PRB1*, in *Saccharomyces cerevisiae* and identification of nonsense alleles within the locus. *Genetics* **96**, 137–146.

Zubenko, G. S., Park, F. J., and Jones, E. W. (1982). Genetic properties of mutations at the *PEP4* locus in *Saccharomyces cerevisiae*. *Genetics* **102**, 679–690.

Zubenko, G. S., Park, F. J., and Jones, E. W. (1983). Mutations in *PEP4* locus of *Saccharomyces cerevisiae* block final step in maturation of two vacuolar hydrolases. *Proc. Natl. Acad. Sci. U.S.A.* **80**, 510–514.

9

Transport and Targeting of Lysosomal Enzymes in *Dictyostelium discoideum*

JAMES A. CARDELLI AND RANDALL L. DIMOND

363

PROTEIN TRANSFER AND ORGANELLE BIOGENESIS

I. INTRODUCTION

A great deal of information has accumulated over the past two decades concerning the biosynthesis, posttranslational processing, and intracellular transport of proteins destined to be secreted or targeted to various cellular compartments. General features of the secretory pathway have been known for some time (Palade, 1975; Farquhar and Palade, 1981), and details are now available concerning the localization and function of many of the glycoprotein-modifying enzymes residing in the endoplasmic reticulum (ER) and Golgi complex (Farquhar, 1985; Kornfeld and Kornfeld, 1985). Research in this area is now focused on defining the molecular mechanisms by which cells sort newly synthesized proteins and direct them to their proper intracellular location (Garoff, 1985).

An inherent assumption in many of the more recent studies is that proteins contain a recognition signal embodied in their structure and that by interacting with specific receptors they are transported to the proper compartment. Recombinant DNA techniques have proved particularly useful in defining the signals involved in the targeting of proteins to nuclei, mitochondria, and chloroplasts (Garoff, 1985; Wickner and Lodish, 1985). Classic biochemical approaches as well as recombinant DNA techniques have also helped define the signals and mechanisms involved in the translocation of proteins across the lipid bilayer into the lumen of the ER (Garoff, 1985). However, except for lysosomal enzymes, very little is presently known about how proteins which enter the secretory pathway are sorted and directed to the proper compartments.

Lysosomal enzymes in human fibroblasts are now known to be phosphorylated on asparagine-linked mannose-rich oligosaccharide side chains by an enzyme in the Golgi complex, followed by the binding of the mannose 6-phosphate recognition marker to receptors which direct the acid hydrolases to lysosomes (Sly and Fischer, 1982; Creek and Sly, 1984). In certain human cells, however, lysosomal enzymes lacking the phosphomannosyl recognition marker can still be localized to lysosomes (Owada and Neufeld, 1982; Waheed et al., 1982). This suggests that even for lysosomal enzymes a complete description of all the molecular mechanisms which transport and correctly localize these molecules is lacking.

Attention has recently been directed at investigations of protein transport and targeting in systems in which a wider range of techniques and approaches are available than are possible in animal cells. Because of the relative ease of application of genetics, molecular biology, and biochemistry techniques, yeast and the cellular slime mold Dictyostelium discoideum have become recognized as excellent systems in which to investigate basic cellular and developmental processes. In this chapter, we review what is currently known about the synthesis, modification, processing, localization, and secretion of lysosomal enzymes during growth and development of *Dictyostelium*. Where applicable, we will compare and contrast information about lysosomal enzyme biosynthesis and transport in *Dictyostelium* with other eukaryotic systems. A recent review summarizes many of the approaches used to study lysosomal enzyme synthesis and targeting in *Dictyostelium* (Cardelli *et al.*, 1987).

II. LIFE HISTORY OF A CELLULAR SLIME MOLD LYSOSOMAL ENZYME

Dictyostelium is a soil organism which feeds exclusively by phagocytosis and pinocytosis, and, as expected for an organism which feeds in this manner, it has a well-developed lysosomal system containing acid hydrolases 10-to 1000-fold higher in concentration than in animal cells. Many of the lysosomal enzymes in this organism have been purified and biochemically analyzed (Table I). As described below, group- and enzyme-specific polyclonal and monoclonal antibodies have also been generated and used to study the biosynthesis and modification of many of these enzymes.

A. Synthesis of Precursor Polypeptides: *In Vivo* Studies

The biosynthesis and maturation of lysosomal enzymes have been studied by incubating cells in culture with radioactive amino acids for brief periods of time (pulse) followed by a period (chase) in which the cells are incubated in the presence of a vast excess of the unlabeled amino acids. The radiolabeled proteins are immunoprecipitated with enzyme-specific antibodies and subjected to gel electrophoresis followed by fluorography. All the lysosomal enzymes examined by this approach in animal cells have been found to be synthesized as larger precursors which are processed to mature forms (Hasilik and von Figura, 1984; Skudlarek *et al.*, 1984). As described below, a similar obervation has also been made for three *Dictyostelium* lysosomal enzymes examined by radiolabel pulse–chase analysis.

Available evidence suggests that the precursors for α-mannosidase (M_r 140,000), β-glucosidase (M_r 105,000), and perhaps acid phosphatase (M_r 60,000) undergo proteolytic processing to generate the lower molecular weight mature forms of the enzyme (see Table I and Section II, D). The change in molecular weight is most dramatic in the case of the α-mannosidase precursor which is subject to at least one proteolytic cleavage event in order to generate the mature polypeptides of M_r 60,000 and M_r 58,000

TABLE I

Properties of *Dictyostelium* Lysosomal Enzymes

Enzyme[a]	Genetic locus	Molecular mass (kDa)					Proenzyme →Mature forms		Refs.
		Holoenzyme	Primary translation product	Pro-enzyme form	Inter-mediate form	Mature form	Half-life[b] (min)	Complete conversion (min)	
α-Mannosidase	manA	220–280	120	140	80	58 and 60	30–35	60–90	c
β-Glucosidase	gluA	220	94	105	103	100	10	35–40	d
Acid phosphatase	acpA	115–135	—	60	—	58	5	10–15	e
N-Acetylglucosaminidase	nagA	168	—	—	—	68 and 51?	—	—	f
α-Glucosidase	—	150	—	—	—	84	—	—	g
Proteinase B/I	—	38	—	—	—	30–34.5	—	—	h

[a] β-Galactosidase has been purified and characterized enzymatically (Dimond et al., 1976). For simplicity we are assuming that proteinases B and I are coded for by the same gene.

[b] Proenzyme half-lives were determined as described following a 10- to 15-min pulse with [35S]methionine.

[c] Free and Loomis (1974), Free et al. (1976), Livi et al. (1985a), Cardelli et al. (1986b), Mierendorf et al. (1983, 1985), Pannell et al. (1982).

[d] Dimond and Loomis (1976), Loomis, (1980b), Cardelli et al. (1986a), Golumbeski (1985), Cardelli (1987).

[e] Loomis and Kuspa (1984), Bennett and Dimond (1986).

[f] Dimond et al. (1973), Loomis (1978), Judelson (1985).

[g] Every and Ashworth (1973), Borts and Dimond (1981a, b).

[h] North (1985), North and Whyte (1984), Gustafson and Thon (1979).

(Mierendorf *et al.*, 1983; Pannell *et al.*, 1982) that are related but not identical in sequence (Mierendorf *et al.*, 1983). Intermediate molecular weight forms have also been observed *in vivo* during the maturation of α-mannosidase (Pannell *et al.*, 1982; Mierendorf *et al.*, 1985), β-glucosidase (Cardelli *et al.*, 1986a; Golumbeski, 1985), and acid phosphatase precursors (Bush and Cardelli, unpublished observations), a phenomenon originally observed during maturation of lysosomal enzymes in higher organisms (Hasilik and von Figura, 1984; Skudlarek *et al.*, 1984).

Precursors to lysosomal enzymes in *Dictyostelium* are processed to mature forms much faster than their mammalian counterparts (Mierendorf, *et al.*, 1985; Pannell *et al.*, 1982; Cardelli *et al.*, 1986a; Bennett and Dimond, 1986). For instance, the half-life for lysosomal enzyme precursors in mammalian cells ranges from 1 to 5 h (Skudlarek *et al.*, 1984), while in *Dictyostelium* precursor half-life ranges from 5 min for acid phosphatase (Bennett and Dimond, 1986; J. Bush and J. Cardelli, unpublished observations) to 30 min for α-mannosidase (Mierendorf *et al.*, 1985; Pannell, 1982). Moreover, the length of time for complete processing of all three *Dictyostelium* lysosomal enzymes to mature forms ranges from 15 to 90 min. In contrast, for many lysosomal enzymes in animal cells, it takes days to complete the conversion of intermediate forms to mature forms (Skudlarek *et al.*, 1984).

B. Synthesis of Precursor Polypeptides: *In Vitro* Studies

Erickson and Blobel (1979) were the first to show that lysosomal enzymes synthesized *in vitro* in the presence of dog pancreas microsomes were cotranslationally translocated into the lumen of the ER. Similar results have also been reported for β-hexosaminidase (Proia and Neufeld, 1982), β-glucuronidase (Rosenfeld *et al.*, 1982), and yeast carboxypeptidase Y (Mueller and Mueller, 1981). Further work by Erickson *et al.* (1981, 1983) revealed that the early events in the synthesis of the lysosomal enzyme cathepsin D were identical to those established for secretory proteins. In particular, the *in vitro* translation of cathepsin D mRNA generated preprocathespin D with the amino-terminal presequence being cleaved cotranslationally when microsomes were present. The translocation, core glycosylation, and segregation of these enzymes in the lumen of the ER were dependent on the presence of the signal recognition particle (Erickson *et al.*, 1983) originally described by Walter *et al.* (1984) and known to be required for the translocation of secretory proteins.

The primary translation products for the α-mannosidase and β-glucosidase mRNAs have been identified (Fig. 1) by *in vitro* translation and

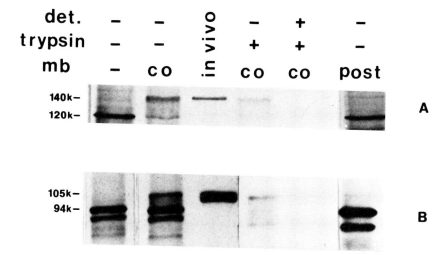

Fig. 1. Cotranslational translocation and modification of α-mannosidase and β-glucosidase by dog pancreas microsomal membranes. RNA prepared from growing cells was translated *in vitro* in the absence or presence (co) of 2 A_{260} units of microsomal membranes. Following translation, the indicated samples (30 μl) received 7.5 μg trypsin, 7.5 μg trypsin plus 1.5 μl 10% Triton X-100 or 2 A_{260} units of membranes (post). Following a 1-h incubation on ice, samples containing trypsin received 75 μg of α_1-antitrypsin. α-Mannosidase (A) and β-glucosidase (B) polypeptides were immunoprecipitated as previously described and subjected to sodium dodecyl sulfate–polyacrylamide gel electrophoresis (SDS–PAGE) followed by fluorography. One lane contains immunoprecipitates prepared from cells pulse-labeled for 30 min with [^{35}S]methionine.

immunoprecipitation (Golumbeski, 1985; Cardelli *et al.*, 1986b, 1987). Both mRNAs are predominantly associated with membrane-bound polysomes and code for polypeptides of M_r 120,000 (α-mannosidase) and M_r 94,000 (β-glucosidase). Two lines of evidence have indicated that N-glycosylation accounts for the difference in molecular weight between the *in vitro* and *in vivo* forms of the α-mannosidase and β-glucosidase precursors. First, the *in vitro* translation of the β-glucosidase and α-mannosidase mRNAs in the presence of dog pancreas microsomes resulted in the production of proteins identical in molecular weight to the cellular precursors (see Fig. 1) (Golumbeski, 1985; Cardelli *et al.*, 1986b, 1987). Under conditions where microsomes remain intact, the modified precursors are resistant to trypsin treatment, suggesting that the polypeptides have been translocated into the lumen of the ER. Second, polypeptides similar in molecular weight to the primary translation products were produced when the immunopurified *in vivo* and *in vitro* modified α-mannosidase and β-glucosidase precursors were treated with endo-β-N-acetylglucosaminidase H, an enzyme which cleaves N-linked mannose-rich side chains. In summary, these results reveal that β-glucosidase and α-mannosidase are synthesized as precursor polypeptides on membrane-bound polysomes, cotranslationally translocated into the lumen of the ER, and modified by the addition of asparagine-linked mannose-rich oligosaccharide side chains.

C. Posttranslational Modifications on N-Linked Oligosaccharide Side Chains

In animal cells, Golgi-localized enzymes promote the synthesis of mannose 6-phosphate on asparagine-linked oligosaccharide side chains which are recognized by a receptor thought to target the enzymes to lysosomes (Creek and Sly, 1984; von Figura and Hasilik, 1986). Recently, a second mannose-6-phosphate receptor has been described (Hoflack and Kornfeld, 1985) which depends on divalent cations for activity. The role this receptor plays in lysosomal enzyme targeting remains to be elucidated. The model for lysosomal enzyme localization in animal cells may not apply to *Dictyostelium* since no phosphomannosyl receptor has been identified, although *Dictyostelium* lysosomal enzymes contain mannose-6-phosphate residues (see Section III, C). Nevertheless, posttranslational modifications may be very important in the function and localization of *Dictyostelium* lysosomal enzymes. Radioactive pulse–chase experiments coupled with subcellular fractionation have, therefore, been done to determine the timing, nature, and location of postranslational modifications in *Dictyostelium*.

1. Sulfation

The precursor to α-mannosidase is sulfated immediately on arrival in the Golgi complex (Mierendorf *et al.*, 1985). Freeze (1985) has shown that the 6 position of mannose is sulfated and that mature secreted forms of lysosomal enzymes contain, on average, three sulfate groups and two phosphate groups per N-linked oligosaccharide side chain. The carbohydrate-linked sulfate and phosphate groups are thought to account for the extremely anionic character of *Dictyostelium* lysosomal enzymes (Knecht *et al.*, 1984; Freeze *et al.*, 1983a). The mannose 6-sulfate groups on N-linked carbohydrate side chains are also highly immunogenic and have been shown to be equivalent to the epitope termed common antigen-1 (CA-1) (Knecht and Dimond, 1981; Knecht *et al.*, 1984; Freeze *et al.*, 1984). CA-1 was originally described as a determinant shared by all lysosomal enzymes in *Dictyostelium* and recognized by the vast majority of rabbit polyclonal and mouse monoclonal antibodies produced in response to injected purified lysosomal enzymes (Knecht and Dimond, 1981; Knecht *et al.*, 1984; Dimond and Loomis, 1976; Judelson, 1985; Golumbeski and Dimond, 1986).

A *Dictyostelium* cell-free sulfation system has been described by Hohmann *et al.* (1985) which consists of membrane fractions, as the source of sulfotransferase, and acceptor proteins plus added 2-phosphoadenosine-5'-phospho [^{35}S] sulfate as the substrate. Subcellular fractionation experiments suggested that the sulfotransferase activity was localized to Golgi membranes, consistent with earlier findings based on *in vivo* ^{35}SO$_4$ labeling (Mierendorf *et al.*, 1985). Sulfated proteins are not unique to *Dictyostelium* and have been identified in developing sea urchin embryos (Heifetz and Lennarz, 1979), as part of a group of basement membrane proteins (Lemkin and Farquhar, 1981), and as members of the group of polypeptide hormones (Green *et al.*, 1984). Also, Waheed and van Etten (1985) have demonstrated that lysosomal arylsulfatase A is sulfated in the human cell line, WI-38.

2. Phosphorylation

Dictyostelium lysosomal enzymes are also phosphorylated after they arrive in the Golgi complex. Unlike the mammalian lysosomal enzymes (Goldberg *et al.*, 1984), however, the *Dictyostelium* mannose 6-phosphate is capped with an acid-stable methyl ester (Freeze and Wolgast, 1986; Gabel *et al.*, 1984). Freeze and Wolgast (1986) have reported that the methyl groups are derived *in vivo* from methylmethionine, although the route of biosynthesis of this unusual phosphodiester and the role it may play in enzyme localization remain to be determined. Gustafson and Milner (1980) have reported that *Dictyostelium* lysosomal enzymes may also be phosphorylated on serine residues.

Phosphorylated *Dictyostelium* lysosomal enzymes are efficiently recognized by fibroblast cell surface phosphomannosyl receptors and rapidly endocytosed (Freeze *et al.*, 1980, 1983b). In fact, in many cases *Dictyostelium* lysosomal enzymes were taken up more efficiently by animal cells than were fibroblast enzymes. Furthermore, *Dictyostelium* lysosomal enzymes have been used by a number of investigators as ligands during affinity chromatography to purify the phosphomannosyl receptor (Fischer *et al.*, 1982; Geuze *et al.*, 1984).

3. Other Modifications

Glycoproteins in vegetative amebas are not as extensively modified as mammalian glycoproteins during passage through the Golgi complex. The typical Golgi-processed *Dictyostelium* glycoprotein (including lysosomal enzymes) contains N-linked carbohydrate side chains constructed of eight to nine mannose residues and two *N*-acetylglucosamine residues (Henderson, 1984). Many of the N-linked oligosaccharides contain fucose in addition to sulfate and phosphate although fucose has not been detected on lysosomal enzymes. The Golgi complex in *Dictyostelium* remains poorly defined both morphologically and biochemically. For instance, characteristic Golgi membrane stacks are seldom seen in growing cells observed with the electron microscope, although these stacks have been detected in developing cells (see Section V, B). Furthermore, sulfotransferase is currently the only reliable Golgi membrane enzyme marker detected in growing cells (Mierendorf *et al.*, 1985; Hohmann *et al.*, 1985) while galactosyl transferase, sialyltransferase, and neutral α-mannosidase activities have not been detected (Wood and Kaplan, 1985; Henderson, 1984). Recent reports indicate the presence in growing cells of fucosyltransferase activity (Wood and Kaplan, 1985), but its utility as a Golgi membrane marker remains to be determined.

D. Maturation of Precursors

Lysosomal enzyme precursors in all types of cells are subject to a number of proteolytic events during their conversion to mature forms (Skudlarek *et al.*, 1984; Hasilik and von Figura, 1984). Direct protein sequence analysis has revealed that cathepsin D contains an amino-terminal localized presequence that is removed during translocation of the nascent polypeptide into the lumen of the ER (Erickson *et al.*, 1981). Rosenfeld *et al.* (1982) and Proia and Neufeld (1982) have also shown by more indirect methods that β-glucuronidase of rat hepatocytes and β-hexosaminidase of human fibroblasts contain cleavable presequences. In *Dictyostelium*, no direct evidence is available concerning the existence of presequences on lysosomal enzyme precursors. However, Pears *et al.* (1985) have determined by DNA sequence analysis of the cysteine proteinase-1 and -2 genes that a potential amino-terminal presequence, rich in hydrophobic amino acids, may exist for these acid proteinases.

Lysosomal enzyme precursors are subjected to additional proteolytic processing events following the removal of the presequence in the ER. This claim was originally inferred from pulse–chase labeling experiments which invariably indicated the disappearance of labeled precursor forms and the concomitant accumulation over time of smaller molecular weight intermediate and mature forms (Hasilik and von Figura, 1984; Skudlarek *et al.*, 1984). Below we review available data describing the cellular compartments where these proteolytic processing events occur.

1. Intracellular Sites of Proteolysis

Only precursor forms of lysosomal α-mannosidase and β-glucosidase are observed in the ER and Golgi membrane fractions following fractionation on sucrose density gradients of *Dictyostelium* cells pulse-labeled with [^{35}S]methionine (Mierendorf *et al.*, 1985; Cardelli *et al.*, 1986a). This suggests that the proteolytic events converting the precursor forms to intermediate and mature forms occur after these polypeptides exit the Golgi. Based on results from Percoll gradient fractionations, however, Wood and Kaplan (1985) have concluded that processing of the α-mannosidase precursor may initiate in the Golgi. Regardless of the exact intracellular location where processing begins, only mature forms are found compartmentalized in vesicles which cofractionate with lysosomes, indicating that complete proteolytic cleavage of the precursor occurs very rapidly either during transport to or immediately on arrival at lysosomes (Mierendorf *et al.*, 1985; Cardelli *et al.*, 1986a; Wood and Kaplan, 1985). Brown and Swank (1983) have also reported that conversion of β-glucuronidase and β-galactosidase precursors to mature forms is closely coupled to lysosomal localization. In contrast, β-hexosaminidase (Frisch and Neufeld, 1981) and cathepsin D (Gieselmann *et al.*, 1983) in fibroblasts as well as mouse spleen cathepsin D and rat preputial β-glucuronidase (Rosenfeld *et al.*, 1982) show relatively slow proteolytic maturation after arrival in lysosomes.

2. Proteinases Involved in Maturation

Although information is available concerning the kinetics of proteolytic maturation of lysosomal enzymes and the intracellular compartments in which these events occur (Skudlarek *et al.*, 1984; Hasilik and von Figura, 1984), very little is known about proteinases that are involved. However, cysteine proteinases in *Dictyostelium* and higher organisms probably play a major role in the conversion of intermediate forms of lysosomal enzymes to mature forms. Hentze *et al.* (1984) have determined that processing of cathepsin D to mature forms in human fibroblasts is prevented by treatment of cells with inhibitors of cysteine proteinases. Similar results have also been found for α-glucosidase, α-N-acetylglucosaminidase, and arylsulfatase (Steckel *et al.*, unpublished observations). Furthermore, cysteine proteinase inhibitors prevent the degradation of mutant arylsulfatase A in late onset forms of metachromatic leukodystrophy (von Figura *et al.*, 1983) and β-galactosidase in combined β-galactosidase/neuraminidase deficiency (Van Diggelen *et al.*, 1982). Our laboratory has also recently determined that treatment of Dictyostelium cells with the cysteine proteinase inhibitors E-64 and Z-phe-ala-CHN$_2$ results in the intracellular accumulation of intermediate forms of α-mannosidase and β-glucosidase (Richardson *et al.*, Submitted for publication).

Even less is known about the biochemistry and cellular location of the proteases in animal cells which process precursor forms to intermediate forms (Hasilik and von Figura, 1984; Skudlarek *et al.*, 1984). Gieselmann *et al.* (1985) have reported that the M_r 53,000 cathepsin D precursor in fibroblasts is cleaved to the intermediate M_r 47,000 form prior to transfer to lysosomes by an enzyme(s) insensitive to inhibitors of cysteine proteinases.

Leupeptin and antipain prevent the proteolytic processing *in vivo* of the precursors to *Dictyostelium* α-mannosidase and β-glucosidase (Richardson *et al.*, submitted for publication), implicating the involvement of cysteine or serine proteinases. Multiple forms of acid proteinases have in fact been identified in *Dictyostelium* and include aspartic (cathepsin D-like) as well as cysteine (cathepsin B-like) proteinases (North, 1985). Also, two cysteine proteinases have been purified from growing cells, proteinase B (North and Whyte, 1984) and proteinase I (Gustafson and Thon, 1979). The very similar properties of these two enzymes suggest that they may be the same protein although their identity and role in lysosomal enzyme maturation remain to be determined.

3. Proteolysis In Vitro

Many of the cellular proteolytic processing events occurring during lysosomal enzyme maturation can be duplicated *in vitro*. For example, Frisch and Neufeld (1981) reported that precursors of β-hexosaminidase were partially processed *in vitro* to intermediate and mature forms on treatment with trypsin, chymotrypsin, or a lysosomal extract. Gieselmann *et al.* (1985) have also demonstrated that mature cathepsin D (M_r 31,000) can be generated from the intermediate form (M_r 47,000) by incubating postnuclear supernatants of fibroblasts at 37°C. In both cases, the reaction was prevented by sulfhydryl blocking reagents, suggesting the involvement of a cysteine proteinase. Woychick *et al.* (1986) have observed that treatment of radiolabeled purified *Dictyostelium* α-mannosidase precursor with trypsin *in vitro* generates polypeptide chains identical in molecular weight to the cellular mature forms. More recent results from our laboratory indicate that detergent-treated cell extracts containing radiolabeled α-mannosidase and β-glucosidase precursors are capable of completely processing these polypeptides to mature forms of the enzymes (N. A. Woychick, unpublished data). Optimal conversion of the precursors to mature forms occurs under acidic pH conditions and can be prevented by the addition of leupeptin, antipain, and heavy metals (e.g.,

HgCl$_2$), which suggests that cysteine proteinases play a role in this process.

4. Inhibition of Maturation

As described above, *Dictyostelium* cells treated with leupeptin and antipain, inhibitors of cysteine and serine proteinases, are unable to process the α-mannosidase and β-glucosidase precursors. Interestingly, the precursor forms of these enzymes do not reach lysosomes or accumulate in cells but are instead rapidly secreted (Richardson *et al.*, submitted for publication) suggesting that in *Dictyostelium* proteolytic processing and localization may be a coupled process. This hypothesis has also been proposed for proinsulin in pancreatic β cells (Orci *et al.*, 1984).

Treatment of cells with proteinase inhibitors represents only one of a number of conditions that inhibit lysosomal enzyme maturation in *Dictyostelium* and animal cells. For instance, newly synthesized α-mannosidase precursors are preferentially secreted from cells treated with chloroquine (J. A. Cardelli, unpublished results), a result originally observed in human fibroblasts treated with NH$_4$ Cl (Hasilik and Neufeld, 1980). Chloroquine and NH$_4$Cl are weak bases which accumulate in lysosomes and other acidic compartments and raise the intravacuolar pH (Ohkuma and Poole, 1978). Thus, under near neutral pH conditions, the acid proteinases may be less effective in the processing of lysosomal enzyme precursors. Alternatively, weak bases may prevent the proper localization of lysosomal enzymes by blocking the reutilization of receptors necessary for targeting (Gonzales-Noriega *et al.*, 1980).

Chloroquine may also act in a pH-independent manner to prevent normal proteolytic processing of precursor polypeptides and to stimulate secretion of the uncleaved forms. Our laboratory has recently determined that certain concentrations of chloroquine can dramatically stimulate oversecretion of newly synthesized precursor polypeptides while increasing vacuolar pH only slightly (0.2 pH units). In contrast, increases in vacuolar pH (induced by NH$_4$Cl) as great as 0.7–0.8 pH units have no effect on the proteolytic processing and proper targeting of Dictyostelium lysosomal enzymes (Cardelli, Richardson, and Mieavs, submitted for publication). Interestingly, chloroquine has been reported to be a cysteine proteinase inhibitor (North, 1985) and therefore potentially may act by the same mechanism as leupeptin and antipain (see Section IID4) to prevent processing and localization of lysosomal enzymes.

Potassium cyanide and potassium cyanate have also been reported to prevent the processing of precursors of lysosomal enzymes in *Dictyostelium* (Mierendorf *et al.*, 1985) and fibroblasts (Hasilik *et al.*, 1983), respectively. The cyanate-induced block in processing is probably indirect because the unprocessed precursors accumulate predominantly in the Golgi complex. These results suggest, however, that the transport of lysosomal enzymes through or out of the Golgi complex may be an energy-dependent process, a finding recently reported for the prohormone form of insulin (Orci *et al.*, 1985).

Conflicting reports have appeared concerning the effects of monensin on the processing and localization of precursors to lysosomal enzymes. Monensin, a carboxylic proton ionophore, is known to interfere with the transport of secretory proteins and membrane proteins at the level of the Golgi complex (Tartakoff, 1983). Wood and Kaplan (1985) have reported that monensin did not inhibit the processing of *Dictyostelium* α-mannosidase precursor polypeptides but did delay the transport of the enzyme from the Golgi complex to lysosomes. A similar result has been reported by Gabel and Kornfeld (1984) for α-glucuronidase in a mouse myeloma cell line. In contrast, Pohlmann *et al.* (1984) have found that monensin inhibits the processing and the secretion of the precursor of cathepsin D in human fibroblasts.

Finally, a number of mutant *Dictyostelium* and animal cell lines have been described that no longer process lysosomal enzyme precursors. For instance, I cell fibroblasts isolated from patients with mucolipidosis II are severely deficient in the enzyme *N*-acetylglucosaminylphosphotransferase and are unable to phosphorylate newly synthesized acid hydrolases (reviewed in Goldberg *et al.*, 1984). The absence of the mannose 6-phosphate recognition signal prevents receptor-mediated localization of newly synthesized acid hydrolases to lysosomes and instead the enzymes are secreted into the medium in precursor form. As described in Section IV, a number of *Dictyostelium* mutants have been isolated that are defective in the processing and localization of acid hydrolase precursors. Some of the strains carry mutations in the α-mannosidase structural gene, and precursors to α-mannosidase accumulate in the ER while processing of β-glucosidase precursors continues at a normal rate (Woychick *et al.*, 1986). The remainder of the nonprocessing mutants contain mutations that lie outside known lysosomal enzyme structural genes and these mutants oversecrete precursor forms of α-mannosidase (Fig. 2) and β-glucosidase.

III. DUAL PATHWAYS FOR THE SECRETION OF LYSOSOMAL ENZYMES

The secretion of lysosomal enzymes is now known to be a property of many cells (Jessup *et al.*, 1985). Ashworth and Quance (1972) first described the secretion of lysosomal enzymes in *Dictyostelium discoideum* during axenic growth and development. A number of subsequent reports

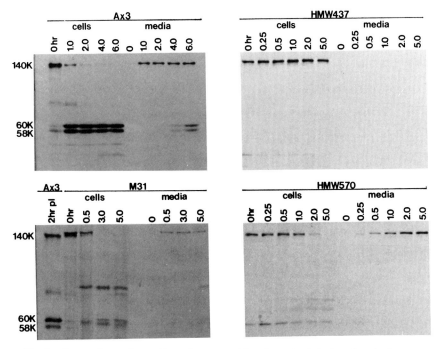

Fig. 2. Processing and secretion of α-mannosidase in mutant strains. Axenically grow-ing cells were pulse-labeled with [³⁵S]methionine and chased for the indicated times in growth medium lacking the isotope. Cells were separated from the medium by centrifuga-tion, and the labeled α-mannosidase was immunoprecipitated from both fractions as de-scribed in Mierendorf *et al.* (1985). Immunoprecipitates were subject to SDS—PAGE fol-lowed by fluorography. Strains were pulse-labeled for the following times: Ax3, 20 min; M31, 30 min; HMW 437, 60 min; HMW570, 60 min.

have characterized the secretion of lysosomal enzymes during growth and following suspension of cells in a nonnutrient buffer (Dimond *et al.*, 1981; Burns *et al.*, 1981; Rossomando *et al.*, 1978; Crean and Rossomando, 1979). As will be described in this section, lysosomal enzymes in *Dictyos-telium* are exocytosed by at least two different pathways, one of which appears to be regulated by a variety of genetic, developmental, and nutri-tional factors.

A. Constitutive Secretory Pathways

In *Dictyostelium*, a small percentage (5–10%) of newly synthesized α-mannosidase and β-glucosidase precursor polypeptides never reach lyso-somes and instead are rapidly secreted from cells still in precursor form

(Mierendorf *et al.*, 1985; Cardelli *et al.*, 1986a). Human fibroblasts (Hasilik and Neufeld, 1980), smooth muscle cells (Hasilik *et al.*, 1981), kidney cells (Erickson *et al.*, 1981), hepatocytes (Rosenfeld *et al.*, 1982), and endothelial cells (Hasilik *et al.*, 1981) also secrete a variable portion of newly synthesized precursor forms of acid hydrolases. A simple interpretation of these data is that the system which operates in each type of cell to direct newly synthesized acid hydrolases to lysosomes is not 100% efficient and thus a variable amount of lysosomal enzyme precursors escape the cell via the constitutive cell surface secretory pathway (Kelly, 1985). As described in Section II, D, 4, the percentage of newly synthesized acid hydrolases which exit the cell via this pathway is greatly increased under certain experimental conditions.

B. Regulated Secretion of Lysosomal Enzymes

Mature forms of lysosomally localized acid hydrolases are also secreted from *Dictyostelium* (Mierendorf *et al.*, 1985; Cardelli *et al.*, 1986a; Pannell *et al.*, 1982; Dimond *et al.*, 1981; Burns *et al.*, 1981) and a variety of other cell types (reviewed in Jessup *et al.*, 1985) including mast cells, kidney cells, macrophages, neutrophils, and the protozoan, *Tetrahymena*. During growth in axenic medium, mature forms of lysosomal enzymes in *Dictyostelium* begin to appear extracellularly within 3–4 hr after synthesis and accumulate to a level accounting for up to 50% of the total acid hydrolase activity found in the culture. These mature forms of the enzymes apparently pass through a presecretory pool (compartment) prior to reaching the efficiently secreted lysosomal compartment (Wood *et al.*, 1983).

Secretion of mature acid hydrolases in *Dictyostelium* is also a regulated response. After suspension of growing cells in nonnutrient buffer, acid hydrolase secretion is greatly stimulated and further enzyme synthesis stops (Dimond *et al.*, 1981). The secretory process is energy dependent, is distinct from phagosomal egestion, and is inhibited by the presence of external amino acids. One group of enzymes consisting of N-acetylglucosaminidase, β-glucosidase-1, α-mannosidase-1, β-galactosidase-1, and α-glucosidase-1 is very efficiently secreted such that 50–80% of the enzymes become extracellular in a few hours. In contrast, acid phosphatase is secreted to a much lesser extent with different secretion kinetics. In addition, a variety of substances (e.g., cycloheximide) inhibit the secretion of acid phosphatase to a different extent than the glycosidases, suggesting that these two groups of enzymes may be localized to different lysosomal vesicles. Finally, at least one lysosomal enzyme, β-galactosidase-2, is secreted very inefficiently and therefore may represent a third class of lysosomal vesicle in *Dictyostelium*. This potential heterogeneity in enzyme content in the population of lysosomal vesicles in *Dictyostelium* has also been observed in kidney cells and macrophages (Skudlarek et al., 1984). Alternatively, all lysosomal enzymes may reside in the same vesicles, and other factors (e.g., differential affinity for membranes) may account for differences in secretion (Hohman and Bowers, 1984).

The different secretory pathways for precursor and mature forms of lysosomal enzymes in *Dictyostelium* closely resemble the two intracellular secretory routes followed by precursor and mature forms of adrenocroticotropic hormone (Kelly, 1985). Thus, in *Dictyostelium* and some vertebrate cells, at least two exocytosis pathways, regulated and constitutive, can coexist in the same cell.

C. Role of Receptors in Intracellular Transport of Lysosomal Enzymes

Lysosomal enzyme localization in fibroblasts (Creek and Sly, 1984) and receptor-mediated endocytosis in a variety of cells (Wileman et al., 1985) are two well-characterized processes in which membrane-associated receptors have been shown to play a critical role. In both processes, receptors mediate the transport of proteins carrying the proper recognition markers to specific cellular compartments. Indirect evidence also supports the existence of additional receptors which may mediate the transport of proteins from the ER to the Golgi complex (Fitting and Kabat, 1982; Lodish et al., 1983), the sorting of proteins destined for storage in secretory granules of endocrine cells (Kelly, 1985), or the localization of enzymes in lysosomes in cells lacking phosphomannosyl receptors. (Cardelli et al., 1986c; Stevens et al., 1986). Below we review the evidence supporting the existence of these hypothetical receptors.

1. Lysosomal Enzyme Precursors Transported at Different Rates

Recent reports indicate that secretory and membrane proteins in animal cells are transported from the ER to the Golgi complex at inherently different rates (Fitting and Kabat, 1982; Lodish et al., 1983; Fries et al., 1984). A model accounting for these data postulates that ER receptors bind recognition markers on these newly synthesized proteins and that the proteins which most rapidly exit the ER are those which are bound most avidly by the receptor. Similarly, in *Dictyostelium*, the β-glucosidase precursor is transported to the Golgi complex and proteolytically processed much faster than the α-mannosidase precursor, a result consistent with transport being receptor mediated (Cardelli et al., 1986a). A comparable result has recently been reported for two other lysosomal enzymes in macrophages (Brown *et al.,* 1985).

2. Importance of Precursor Conformation and Amino Acid Sequence

A model proposing receptor-mediated transport of newly synthesized proteins from the ER to the Golgi also predicts that alterations in the conformation of these proteins or changes in the amino acid sequence may decrease their affinity for the receptor and thus lead to their retention in the ER. Results consistent with this prediction have been reported for a number of different proteins. For instance, the *Dictyostelium* strain HMW 437 which carries a mutation in the α-mannosidase structural gene (*manA$_5$* allele) synthesizes an apparently normal M_r 140,000 N-glycosylated α-mannosidase precursor which never exits the ER. Sedimentation velocity experiments and *in vitro* protease analysis indicated that the α-mannosidase precursor was altered in conformation, which may prevent receptor interaction and subsequent transport to the Golgi complex (Woychick *et al.,* 1986). Schauer *et al.* (1985) have reported that a mutation in the yeast invertase gene resulting in the substitution of the amino acid threonine for isoleucine at position +64 slows the rate of transport of the enzymatically active protein from the ER. These authors conclude that this mutation may interfere with the interaction of invertase with a receptor which mediates transport to the Golgi complex.

Additional examples of transport-defective mutant forms of secretory and membrane proteins include vesicular stomatitis virus G protein (Lodish and Kong, 1983), α_1-antitrypsin (Hercz *et al.,* 1978), major histocompatibility complex class I antigen (Miyazaki *et al.,* 1986), influenza virus hemagglutinin (Doyle *et al.,* 1985), and immunoglobulin A light chain (Wu *et al.,* 1983). Finally, Ronne *et al.* (1983) have shown that movement of the retinol-binding protein from the ER to the Golgi complex requires retinol, presumably to change the conformation of the protein in order to trigger transport.

Although the experiments discussed above are consistent with the existence of an ER-localized receptor mediating transport of proteins to the Golgi complex, no such receptor has been directly demonstrated. Moreover, the nonuniform transport rates described above might not be due to receptor interactions at all but rather to differences in physical properties of individual proteins that determine the rate at which they are included in transport vesicles. For example, some of the mutant proteins may fold in such a way that transport is prevented for physical reasons as opposed to an inability to be recognized by a receptor. Such mutant proteins may become insoluble (Gibson *et al.*, 1979), may be bound by ER localized proteins (Hendershot *et al.*, 1987), or be unable to oligomerize, (Gething *et al.*, 1986; Copeland *et al.*, 1986), a mechanism potentially important in the transport process (Garoff, 1985). Clearly, more work needs to be done in this area in order to settle this issue.

3. Sorting Mechanisms

Dictyostelium lysosomal enzymes are phosphorylated on their carbohydrate side chains and are excellent ligands for the cation-independent mammalian phosphomannosyl receptors; however, similar receptors have not been detected in the slime mold, suggesting the existence of an alternative localization mechanism for lysosomal enzymes. Mechanisms for lysosomal enzyme targeting that are not dependent on phosphomannosyl receptors have already been proposed for yeast (see Section IV, E), plants (Vitale and Chrispeels, 1984), and certain types of animal cells (Waheed *et al.*, 1982; Owada and Neufeld, 1982). For instance, parenchymal cells in cotyledons of the common bean (*Phaseolus vulgaris*) contain intracellular protein bodies (lysosomal compartment) rich in a number of acid hydrolases which are glycosylated but apparently not phosphorylated. Instead, the oligosaccharides contain terminal N-acetylglucosamine which is slowly removed after the acid hydrolases reach the lysosomal compartment (Vitale and Chrispeels, 1984).

The *Dictyostelium* α-mannosidase and β-glucosidase precursors are tightly bound to ER and Golgi membranes during transport to lysosomes while proteolytically processed mature forms of both enzymes are soluble in the lumen of lysosomes (Mierendorf *et al.*, 1985; Cardelli *et al.*, 1986a). The physical nature underlying the precursor–membrane association requires further study, but it is evident that the association is strong based on the inability to release the precursor following treatment of membranes with high salt and chaotropic agents (Cardelli *et al.*, 1987). Thus, the precursor polypeptides may be membrane associated because they are tightly bound to a receptor. Alternatively, they may be anchored in the membrane due to an uncleaved signal sequence or interaction with an integral membrane protein or glycolipid. Finally, the binding of the precursor polypeptides to a stationary ER component might also explain the nonuniform transport rates of lysosomal enzymes. For, instance, α-mannosidase might reside in the ER longer than β-glucosidase because it is more firmly attached to a protein in the lumen of this organelle. The membrane-bound protein egasyn, found in the ER in kidney cells, does in fact selectively retain a significant proportion of newly synthesized β-glucuronidase (Paigen, 1979).

IV. GENETIC APPROACHES TO DISSECT INTRACELLULAR TRANSPORT PATHWAYS

The distinct advantage of using systems like yeast and *Dictyostelium* to answer basic biological questions derives from their amenability to both genetic and biochemical approaches. *Dictyostelium* is normally haploid, and therefore the isolation of mutations affecting cell and developmental processes is relatively straightforward. In this section, we review what is currently known about *Dictyostelium* mutants altered in the biosynthesis, modification, processing, localization, and secretion of lysosomal enzymes.

A. Mutants Selected on the Basis of Secretion Defects

Our laboratory has isolated and completed the initial characterization of 74 independently isolated mutants defective in lysosomal enzyme secretion (Dimond *et al.*, 1983; D. Ebert unpublished results). Over three-fourths of the mutants fall into one of four defined secretory classes based on their behavior under standard secretion conditions. As described in Section III,B, assays done under standard secretion conditions measure only the release of preformed lysosomally localized enzyme. The first class (containing 20 mutants) undersecretes three glycosidases as well as acid phosphatase. This establishes that there are genes required for the secretion of all the studied lysosomal enzymes. The second class (consisting of 16 mutants) undersecretes the glycosidases while showing no abnormalities in the secretion of acid phosphatase. Nine strains make up the third class of mutants which oversecrete acid phosphatase. The existence of these last two classes of mutants indicates that genes can independently regulate the secretion of the glycosidases versus acid phosphatase, consistent with the proposed heterogeneity in the lysosomal system (Section III,B). Finally, the fourth class of mutants are normal in secretion of enzymes during standard secretions conditions, but all of the mutants are altered in secretion during growth. Secretion during growth is highly regulated (Dimond *et al.*, 1981) and involves the release of newly synthesized enzymes as well as enzymes already localized to lysosomes. This fourth class of mutants may therefore define genes involved in the regulation of secretion or in the proper localization of enzymes. The remainder of the mutants are defective in the secretion of a variety of multiple or single enzymes.

In summary, altered lysosomal enzyme secretion may be the result of mutations which (1) prevent the proper localization of lysosomal enzymes, (2) direct acid hydrolases to incorrect compartments in a heterogeneous lysosomal system, (3) affect the production, movement, and fusion of vesicles with each other or the plasma membrane, or (4) alter the regulation in any of the above processes. Research to distinguish between these possibilities is currently underway in our laboratory.

B. Modification Mutants

Dictyostelium mutants have also been isolated that are altered in the posttranslational modification of lysosomal enzymes and other glycoproteins. In fact, many of the secretory mutants themselves are altered in posttranslational modification of lysosomal enzymes as revealed by isoelectric focusing gel electrophoresis, suggesting a functional relationship between secretion and modification.

The recessive mutation, *modA* (strain M31), results in a reduction in the electrophoretic mobility on native gels of multiple lysosomal enzymes as well as a partial loss in the cellular activities of these enzymes (Free *et al.*, 1978; Free and Schimke, 1978). Strains carrying the *modA* mutation are deficient in α-1,3-glucosidase (Freeze *et al.*, 1983c), the enzyme responsible for removing the inner two glucose residues from N-linked carbohydrate side chains. As a result, the N-linked mannose-rich side chains in M31 retain their two innermost glucose residues. Moreover, oligosaccharides in these mutants are undersulfated and underphosphorylated (Freeze *et al.*, 1983c; Freeze and Miller, 1980), perhaps as a result of the inability of the transferases to efficiently recognize glucosylated oligosaccharide side chains. The *modB* locus was identified in mutant cells which failed to express a shared carbohydrate epitope recognized by a monoclonal antibody (Murray *et al.*, 1984). The molecular nature of the defect and its effect on lysosomal enzymes remain to be determined although a number of plasma membrane glycoproteins are localized normally in this mutant (West and Loomis, 1985).

Five mutants have recently been isolated that fail to express CA-1, an epitope shared by all lysosomal enzymes (Knecht *et al.*, 1984) and most likely consisting of mannose 6-sulfate residues in N-linked carbohydrate side chains (H. Freeze, personal communication). One of the mutants, HL-244, is almost completely deficient in sulfation while the other four mutants are reduced 2- to 5-fold (Knecht *et al.*, 1984). Strains HL-241 and HL-243 synthesize lysosomal enzymes that are N-glycosylated with truncated oligosaccharide side chains at least four hexoses smaller than those of the wild type (H. Freeze, personal communication). As discussed in Section IV,E and IV,F, these modification mutants are currently being

studied to determine if their lysosomal enzymes are processed and localized normally.

C. Structural and Regulatory Gene Mutants

Dictyostelium mutants have been isolated that carry defects in the
structural genes coding for the lysosomal enzymes α-mannosidase (Free
and Loomis, 1974; Free *et al.*, 1976; Livi *et al.*, 1985a), β-glucosidase
(Dimond and Loomis, 1976; Loomis, 1980a), N-acetylglucosaminidase
(Dimond *et al.*, 1973; Loomis, 1978), and acid phosphatase (Loomis and
Kuspa, 1984). Many of these are null mutations, and the cells synthesize
no detectable gene product. Others strains, such as HMW 437, HMW
433, and HMW 464 (carrying the *manA*$_5$, *manA*$_7$, and *manA*$_6$ alleles,
respectively) synthesize the α-mannosidase precursor at one-tenth to one-
third the rate observed in wild-type cells but accumulate less than 1% of
the wild-type level of functional enzyme (Livi *et al.*, 1985a). As described
in Section III,C,1, the conformation of the α-mannosidase precursors in
each of these mutants is altered, and newly synthesized polypeptides
accumulate inside the lumen of the ER.

Over the years, *Dictyostelium* has proved to be an excellent system in
which to investigate the molecular mechanisms regulating gene expression during development (Loomis, 1982). In fact, many of the genes coding for lysosomal enzymes are developmentally controlled; thus these
enzymes were originally studied to increase our understanding of the
genetic mechanisms regulating their synthesis (see Section V,A). Direct
screening has identified five aggregation-defective mutants that, unlike
wild-type strains, accumulate negligible amounts of α-mannosidase activity during axenic growth or development (Livi *et al.*, 1985a). Genetic
complementation analysis has revealed that all of the mutations in these
strains map outside the α-mannosidase structural gene, and thus they
have been classified as α-mannosidase regulatory mutants (Livi *et al.*,
1985a; C. Singleton, personal communication). These *mad* mutations (α-
mannosidase-1 deficient) have variable effects on the rate of both synthesis and processing of the α-mannosidase precursor. For instance, the
mutation *mad*-404 results in a complete lack of functional α-mannosidase
mRNA during development (Livi *et al.*, 1984, 1985b) and the misregula-
tion of greater than half of the genes expressed early in development
(Cardelli *et al.*, 1984, 1985). The mutation *mad*-426 causes a complete lack
of α-mannosidase precursor processing while strains carrying *mad*-414
and *mad*-417 mutations process abnormally modified precursors at re-

duced rates. Thus, a small number of developmentally essential genes also regulate the synthesis, modification, and processing of α-mannosidase.

D. Classes of Mutants Deficient in Cellular α-Mannosidase Activity

The mutants described above represent a valuable source to aid in the determination of the molecular mechanisms involved in lysosomal enzyme biosynthesis, localization, and secretion. These mutants are currently being analyzed at the molecular level to identify those that are altered in the localization and processing of lysosomal enzymes as these mutants may help define the recognition system used in targeting. Mutants deficient in cellular α-mannosidase activity are grouped in Table II into classes with similar phenotypes.

Mutants in class 5 (Fig. 2, Table II), which oversecrete lysosomal enzymes in precursor form, are particularly intriguing because they may be defective in the generation of (1) a recognition marker on enzymes destined for lysosomes, (2) a functional receptor to mediate transport, (3) a functional proteinase involved in maturation, or (4) acidic pH conditions in lysosomal compartments.

E. Role of Carbohydrate Side Chains in Enzyme Localization

No α-mannosidase or β-glucosidase precursor polypeptides were detected following immunoprecipitation of pulse-labeled proteins from cells treated with tunicamycin, an inhibitor of N-linked glycosylation (J. A. Cardelli, unpublished results). This suggests that the unglycosylated precursors are very unstable or that their synthesis is inhibited. Examples of tunicamycin affecting both synthesis and stability of glycoproteins have been previously reported (reviewed in Olden et al., 1985). Our results using this drug are thus inconclusive in regard to the role of the carbohydrate side chain in the targeting of Dictyostelium lysosomal enzymes.

Failure to remove the innermost glucose residues from N-linked mannose-rich side chains results in a delay in the transport of secretory proteins (Lodish and Kong, 1984; Gross et al., 1983) and lysosomal enzymes (Lemansky et al., 1984) from the ER to the Golgi complex. In Dictyostelium, the rate of transport and processing of the α-mannosidase precursor is only slightly impaired in both strain M31 (Free et al., 1978) which lacks

TABLE II

Mutants Deficient in Cellular α-Mannosidase Activity

Class of mutant	Representative		Steady-state location of precursor (P) or mature (M) forms[a]			Comments	Ref.
	Allele	Strain	ER	Lysosome	Extracellular		
1	$manA_1$	M_1				No functional mRNA	b
2	$manA_5$	HMW 437	P++			Conformationally altered precursor	c
2 (variant)	$manA_6$	HMW 464	P++		P+/-	Conformationally altered precursor slowly secreted	d
3	$manA_8$	HMW 452		M		Mature forms unstable	e
4	mad-417	HMW 417		M		Precursor abnormal molecular weight, mature forms inactive	f
5	Not determined	HMW 570	P+/-		P+++	Precursor inefficiently processed	g

[a] Plus and minus superscripts represent the relative abundance of the indicated form of the enzyme in that particular cellular compartment.

[b] Free and Loomis (1974), Free et al. (1976), Mierendorf et al. (1983), Cardelli et al. (1986b).
[c] Livi et al. (1985a), Woychick et al. (1986).
[d] N. A. Woychick et al. (unpublished data), Livi et al. (1985a).
[e] Livi et al. (1985a), C. Singleton (personal communication).
[f] Livi et al. (1985a).
[g] D. Ebert (unpublished results), Cardelli et al. (1986c).

ER-localized α-1,3-glucosidase activity (Fig. 2) and in cells treated with the α-glucosidase inhibitor, deoxynorjirimycin, (D. Ebert, Bush, Dimond, and Cardelli, submitted for publication). Thus, in *Dictyostelium,* the retention of glucose residues on carbohydrate side chains of lysosomal enzymes does not prevent exit from the ER or greatly influence intracellular movement of the precursors.

The localization and processing of α-mannosidase and β-glucosidase are normal in strains HL-241 and HL-243 which synthesize lysosomal enzymes containing the truncated carbohydrate structure Glc-NAc$_2$Man$_6$ (Cardelli and Freeze, unpublished results). This suggests that the outermost mannose residues do not play an essential role in the localization process.

It is conceivable that carbohydrate does not play a direct role in the localization of *Dictyostelium* lysosomal enzymes and that instead the sorting signal is part of the structure of the precursor polypeptide itself. This has recently been shown to be the case in the targeting of carboxypeptidase Y (CPY) to the lysosomelike vacuole in yeast, a compartment which contains hydrolytic enzymes that are synthesized as larger molecular weight glycosylated proenzyme forms (Schwaiger et al., 1982). Recombinant DNA approaches have in fact revealed that the sorting signal for CPY appears to lie within the region of 10–40 amino acids from the amino terminus of the protein (Johnson *et al.,* 1987; Valls *et al.,* 1987). A recent report has also indicated that the overproduction of CPY in yeast hosting a multiple copy plasmid containing the *PRC1* gene (encoding CPY) leads to the secretion of 50% of the precursor form of this enzyme (Stevens *et al.,* 1986). These findings suggest that targeting of CPY in yeast is a saturable process consistent with the existence of a CPY-sorting receptor.

F. Role of Sulfate and Phosphate in Enzyme Localization

The mutant HL-244, which is almost completely deficient in sulfation of N-linked oligosaccharides (Knecht *et al.,* 1984), proteolytically processes the α-mannosidase precursor at an apparently normal rate, suggesting that carbohydrate-linked sulfate groups do not play an essential role in this process (C. Singleton, unpublished results). Mutants unable to phosphorylate N-linked carbohydrate side chains have not been isolated; thus the role of phosphate in lysosomal enzyme localization in *Dictyostelium* remains unknown. However, although the oligosaccharide side chains on lysosomal enzymes in strain M31 (*modA* mutation) contain reduced amounts of phosphate, the enzymes are still localized and processed with near normal kinetics, (D. Ebert, *et al.*, in press). Therefore, the wild-type number of carbohydrate-linked phosphate groups are unnecessary for proper targeting.

V. DEVELOPMENTAL REGULATION OF THE SYNTHESIS, MODIFICATION, AND LOCALIZATION OF LYSOSOMAL ENZYMES

On starvation, *Dictyostelium* undergoes a developmental process which results in the genesis of two major differentiated cell types, spore cells and stalk cells (Loomis, 1975). This process is characterized by defined morphological changes that are accompanied by stage-specific (temporal) changes in the expression of many developmentally regulated genes (Alton and Lodish, 1977a,b; Cardelli *et al.*, 1984, 1985; Livi *et al.*, 1984, 1985b; Loomis, 1982) including the genes coding for lysosomal enzymes (Loomis, 1975). In this section, we review what is currently known about the developmental control of the synthesis, modification, and localization of *Dictyostelium* lysosomal enzymes.

A. Regulation of Lysosomal Enzyme Synthesis

α-Mannosidase is one of the earliest developmentally controlled gene products in *Dictyostelium* (Loomis, 1970). Cellular activity of the enzyme is very low in amebas grown on bacteria, begins to increase immediately on starvation, and accumulates linearly over the next 20–22 hr of development (Loomis, 1970; Livi *et al.*, 1985b). The accumulation of α-mannosidase activity is the result of an increase in the relative rate of synthesis of the α-mannosidase precursor form which can be accounted for by a parallel increase in the concentration of functional mRNA (Livi *et al.*, 1984; 1985b). As discussed in Section IV, a number of aggregation-essential genes have been identified which regulate the developmentally induced accumulation of α-mannosidase either by affecting its rate of synthesis or by controlling posttranslational modification and/or processing (Livi *et al.*, 1985a).

β-glucosidase activity increases prior to aggregation, decreases during the slug stage, and increases again during culmination (Coston and Loomis, 1969; Golumbeski, 1985). These changes in enzyme activity are paralleled by changes in the concentration of functional β-glucosidase mRNA and in the relative rate of synthesis of the β-glucosidase precursor (Golumbeski, 1985). Thus, like α-mannosidase, developmental synthesis of β-glucosidase is controlled pretranslationally. The activities of other

VII. SUMMARY AND FUTURE PERSPECTIVES

The process involving the biosynthesis, modification, processing, localization, and secretion of lysosomal enzymes in *Dictyostelium* is similar in many aspects to that occurring in higher organisms. Lysosomal enzymes are synthesized as precursor forms on membrane-bound polysomes, translocated into the lumen of the ER, and N-glycosylated with mannose-rich oligosaccharide side chains. The membrane-associated precursor forms are transported with nonuniform kinetics to the Golgi where they are phosphorylated and sulfated on their carbohydrate side chains. A small percentage of the precursors escapes further processing and is rapidly secreted from cells, while the vast majority is routed to lysosomes where proteolytic processing generates soluble mature forms of the enzymes. The enzymes localized to lysosomes are also secreted from cells by a regulated, energy-dependent mechanism.

The distinct advantage in studying *Dictyostelium* is that the various processes pertaining to the lysosomal system including phagocytosis, pinocytosis, autodigestion, enzyme localization, and secretion can be studied using both biochemical and genetic approaches. Thus, the biochemical analysis of mutants defective in the modification, synthesis, localization, secretion, and processing of lysosomal enzymes will greatly increase our knowledge concerning a number of different aspects of the lysosomal system.

Future studies will rely on recombinant DNA approaches that will include the examination of cloned *Dictyostelium* lysosomal enzyme genes to determine the molecular nature of the targeting signal on lysosomal enzymes as well as the isolation of genes involved in the biogenesis and function of the lysosomal system. An area of research with the potential for industrial application will be the development of *Dictyostelium* as a eukaryotic expression system for foreign genes. *Dictyostelium* does not have a cell wall to trap secreted proteins, and the modification of proteins in the slime mold are much more like animal cells than is yeast. Moreover, the availability of mutants that secrete greater than 90% of proteins that enter the secretory pathway coupled with the ability to culture large numbers of cells makes *Dictyostelium* an excellent choice as a biological system for the production of large amounts of proteins with applications to medicine and industry.

ACKNOWLEDGMENTS

Research by the authors was supported by grants from the National Institutes of Health (AM 36447 and GM 29156). The authors would like to thank David Ebert, George Golumbeski, Jan Richardson, and John Bush for critical reading of the manuscript; Steve Free, Charles Singleton, and Hud Freeze for communicating their unpublished work; and Cynthia Harrison for her exceptional word processing skills. A special tribute is extended to all the past members of the laboratory who contributed so greatly to various aspects of the research presented in this chapter.

REFERENCES

Alton, T. A., and Lodish, H. F. (1977a). Developmental changes in mRNAs and protein synthesis in *Dictyostelium discoideum*. *Dev. Biol.* **60**, 180–206.

Alton, T. A., and Lodish, H. F. (1977b). Translational control of protein synthesis during early stages of differentiation of the slime mold, *Dictyostelium discoideum*. *Cell* **12**, 301–310.

Ashworth, J. M., and Quance, J. (1972). Enzyme synthesis in myxamoebae of the cellular slime mold, *Dictyostelium discoideum*. *Biochem. J.* **126**, 601–608.

Bennett, V. D., and Dimond, R. L. (1986). Biosynthesis of two developmentally distinct acid phosphatase isozymes in *Dictyostelium discoideum*. *J. Biol. Chem.* **261**, 5355–5362.

Borts, R. H., and Dimond, R. L. (1981a). The α-glucosidases of *Dictyostelium discoideum*. I. Identification and characterization. *Dev. Biol.* **87**, 176–184.

Borts, R. H., and Dimond, R. L. (1981b). The α-glucosidases of *Dictyostelium discoideum*. II. Developmental regulation and cellular localization. *Dev. Biol.* **87**, 185–192.

Brown, J. A., and Swank, R. T. (1983). Subcellular redistribution of newly synthesized macrophage lysosomal enzymes. *J. Biol. Chem.* **258**, 15323–15328.

Brown, J. A., Novak, E. K., and Swank, R. T. (1985). Effects of ammonia on processing and secretion of precursor and mature lysosomal enzyme from macrophages of normal and pale ear mice: Evidence for two distinct pathways. *J. Cell Biol.* **100**, 1894–1904.

Burns, R. A., Livi, G. P., and Dimond, R. L. (1981). Regulation and secretion of early developmentally controlled enzymes during axenic growth in *Dictyostelium discoideum*. *Dev. Biol.* **84**, 407–416.

Cardelli, J. A., Livi, G. P., Mierendorf, R. C., Knecht, D. A., and Dimond, R. L. (1984). Gene regulation during early development of the cellular slime mold *Dictyostelium discoideum*. *In* "Molecular Biology of Development" (UCLA Symposia on Molecular and Cellular Biology, Vol. 19, E. H. Davidson and R. A. Firtel, ed., pp. 427–435. Alan R. Liss, New York.

Cardelli, J. A., Knecht, D. A., Wunderlich, R., and Dimond, R. L. (1985). Major changes in gene expression occur during at least four stages of development of *Dictyostelium discoideum*. *Dev. Biol.* **110**, 147–156.

Cardelli, J. A., Golumbeski, G. S., and Dimond, R. L. (1986a). Lysosomal enzymes in *Dictyostelium discoideum* are transported at distinctly different rates. *J. Cell Biol.* **102**, 1264–1270.

Cardelli, J., Mierendorf, R. C., and Dimond, R. L. (1986b). Initial events involved in the synthesis of the lysosomal enzyme α-mannosidase in *Dictyostelium discoideum*. *Arch. Biochem. Biophys.* **244**, 338–345.

Cardelli, J. A., Golumbeski, G. S., Woychick, N. A., Ebert, D. L., Mierendorf, R. C., and Dimond, R. L. (1987). Defining the intracellular localization pathways followed by lysosomal enzymes in *Dictyostelium discoideum*. *In* "Methods in Cell Biology," Vol. 28, pp. 139–155 (J. Spudich, ed.), Academic Press, Orlando.

Cardelli, J. A., Richardson, J., and Mieavs, D. (1987). Weak bases differentially affect the proteolytic processing and proper targeting of lysosomal enzyme precursors in (*Dictyostelium discoideum*. *J. Cell Biol.* (Submitted for publication).

Catterall, J. F., and Leary, S. L. (1983). Detection of early changes in androgen-induced mouse renal β-glucuronidase messenger ribonucleic acid using cloned complementary deoxyribonucleic acid. *Biochemistry,* **22**, 6049–6053.

Copeland, C., Doms, R. W., Bolzau, E. M., Webster, R. G., and Helenius, A. (1986). Assembly of influenza hemagglutinin trimers and its role in intracellular transport. *J. Cell Biol.* **103**, 1179–1191.

Coston, M. B., and Loomis, W. F. (1969). Isozymes of β-glucosidase in *Dictyostelium discoideum*. *J. Bacteriol.* **100**, 1208–1217.

Crean, E. V., and Rossomando, E. F. (1979). Effects of sugars on glycosidase secretion in *Dictyostelium discoideum*. *J. Gen. Microbiol* **110**, 315–322.

Creek, K. E., and Sly, W. S. (1984). The role of the phosphomannosyl receptor in the transport of acid hydrolases to lysosomes. *In* "Lysosomes in Biology and Pathology" (J. T. Dingle, R. T. Dean, and W. Sly, eds.), pp. 63–82. Elsevier, Amsterdam.

Dimond, R. L., and Loomis, W. F. (1976). Structure and function of β-glucosidases in *Dictyostelium discoideum*. *J. Biol. Chem.* **251**, 2680–2687.

Dimond, R. L., Brenner, M., and Loomis, W. F. (1973). Mutations affecting N-acetylglucosaminidase in *Dictyostelium discoideum*. *Proc. Natl. Acad. Sci. U.S.A.* **70**, 3356–3360.

Dimond, R. L., Mayer, M., and Loomis, W. F. (1976). Characterization and developmental regulation of β-galactosidase isozymes in *Dictyostelium discoideum*. *Dev. Biol.* **52**, 74–82.

Dimond, R. L., Burns, R. A., and Jordan, K. B. (1981). Secretion of lysosomal enzymes in the cellular slime mold, *Dictyostelium discoideum*. *J. Biol. Chem.* **256**, 6565–6572.

Dimond, R. L., Knecht, D. A., Jordan, K. B., Burns, R. A., and Livi, G. P. (1983). Secretory mutants in cellular slime mold, *Dictyostelium discoideum*. *Methods Enzymol.* **96**, 815–828.

Doyle, C., Roth, M. G., Sambrook, J., and Gething, M. J. (1985). Mutations in the cytoplasmic domain of the influenza virus hemagglutinin affect different stages of intracellular transport. *J. Cell Biol.* **100**, 704–714.

Ebert, D. L., Bush, J., Dimond, R. L., and Cardelli, J. A. (1987). α1,3-linked glucose removal not required for maturation and localization of two lysosomal enzymes in *Dictyostelium discoideum*. *J. Biol. Chem.* (submitted).

Emr, S. D., Bankaitis, V. A., Garrett, J. M., and Douglass, M. G. (1985). Gene fusions to study intracellular protein traffic and organelle assembly in yeast. *In* "Protein Transport and Secretion, Current Communications in Molecular Biology" (M. J. Gething, ed.), pp. 184–189. Cold Spring Harbor Laboratory, Cold Spring Harbor, New York.

Erickson, A., and Blobel, G. (1979). Early events in the biosynthesis of lysosomal enzyme cathepsin D. *J. Biol. Chem.* **254**, 11771–11774.

Erickson, A. H., Conner, G. E., and Blobel, G. (1981). Biosynthesis of a lysosomal enzyme. Partial structure of two transient and functionally distinct NH2-terminal sequences in cathepsin D. *J. Biol. Chem.* **256**, 11224–11231.

Erickson, A. H., Walter, P., and Blobel, G. (1983). Translocation of a lysosomal enzyme across the microsomal membrane requires signal recognition particle. *Biochem. Biophys. Res. Commun.* **115**, 275–280.

Every, D., and Ashworth, J. M. (1973). The purification and properties of the extracellular glycosidases of the cellular slime mold, *Dictyostelium discoideum*. *Biochem. J.* **133**, 37–47.

Farquhar, M. G. (1985). Progress in unraveling pathways of Golgi traffic. *Annu. Rev. Cell Biol.* **1**, 447–488.

Farquhar, M. G., and Palade, G. E. (1981). The Golgi apparatus (Complex)—(1954–1981)—from artifact to center stage. *J. Cell Biol.* **91**, 77S–103S.

Fischer, H. D., Creek, D. E., and Sly, W. S. (1982). Binding of phosphorylated oligosaccharides to immobilized phosphomannosyl receptors. *J. Biol. Chem.* **257**, 9938–9943.

Fitting, T., and Kabat, D. (1982). Evidence for a glycoprotein "signal" involved in transport between subcellular organelles. *J. Biol. Chem.* **257**, 14011–14017.

Fong, D., Calhoun, D. H., Hsieh, W.-T., Lee, B., and Wells, R. D. (1986). Isolation of a cDNA clone for the human lysosomal proteinase cathepsin B. *Proc. Natl. Acad. Sci. U.S.A.* **83**, 2909–2913.

Free, S. J., and Loomis, W. F. (1974). Isolation of mutations in *Dictyostelium discoideum.* affecting α-mannosidase. *Biochimie* **56**, 1525–1528.

Free, S. J., and Schimke, R. T. (1978). Effects of a posttranslational modification mutation on different developmentally regulated glycosidases in *Dictyostelium. J. Biol. Chem.* **253**, 4107–4111.

Free, S. J., Schimke, R. T., and Loomis, W. F. (1976). The structural gene for α-mannosidase in *Dictyostelium discoideum. Genetics* **84**, 159–174.

Free, S., Schimke, R. T., Freeze, H., and Loomis, W. F. (1978). Characterization and genetic mapping of *modA*. A mutation in the posttranslational modification of the glycosidases of *Dictyostelium discoideum. J. Biol. Chem.* **253**, 4102–4106.

Freeze, H. H. (1985). Mannose 6-sulfate is present in the N-linked oligosaccharides of lysosomal enzymes of *Dictyostelium. Arch. Biochem. Biophys.* **243**, 690–693.

Freeze, H., and Miller, A. L. (1980). *modA:* A posttranslational mutation affecting phosphorylated and sulfated glycoproteins in *Dictyostelium discoideum. Mol. Cell Biochem.* **35**, 17–27.

Freeze, H. H., and Wolgast, D. (1986). Biosynthesis of methylphosphomannosyl residues in the oligosaccharides of *Dictyostelium discoideum* glycoproteins. *J. Biol. Chem.* **261**, 135–141.

Freeze, H., Miller, A. L., and Kaplan, A. (1980). Acid hydrolases from *Dictyostelium discoideum* contain phosphomannosyl recognition markers. *J. Biol. Chem.* **255**, 11081–11084.

Freeze, H. H., Yek, R., Miller, A. L., and Kornfeld, S. (1983a). Structural analysis of the asparagine-linked oligosaccharides from three lysosomal enzymes of *Dictyostelium discoideum. J. Biol. Chem.* **258**, 14874–14879.

Freeze, H. H., Yeh, R. Y., and Miller, A. L. (1983b). Uptake of α-D-mannosidase and β-D-glucosidase from *Dictyostelium discoideum* via phosphohexosyl receptor on normal human fibroblasts. *J. Biol. Chem.* **258**, 8928–8933.

Freeze, H. H., Yeh, R., Miller, A. L., and Kornfeld, S. (1983c). The *modA* mutant of *Dictyostelium discoideum* is missing the α-1,3-glucosidase involved in asparagine-linked oligosaccharide processing. *J. Biol. Chem.* **258**, 14880–14884.

Freeze, H. H., Mierendorf, R. C., Wunderlich, R., and Dimond, R. L. (1984). Sulfated oligosaccharides block antibodies to many *Dictyostelium discoideum* acid hydrolases. *J. Biol. Chem.* **259**, 10641–10643.

Fries, E., Gustafsson, L., and Peterson, P. A. (1984). Four secretory proteins synthesized by hepatocytes are transported from endoplasmic reticulum to Golgi complex at different rates. *EMBO J.* **3**, 147–152.

Frisch, A., and Neufeld, E. F. (1981). Limited proteolysis of the β-hexosoaminidase precursor in a cell-free system. *J. Biol. Chem.* **256**, 8242–8246.

Fukushima, H., deWet, J. R., and O'Brien, J. S. (1985). Molecular cloning of a cDNA for human α-fucosidase. *Proc. Natl. Acad. Sci. U.S.A.* **82**, 1262–1265.

Gabel, C. A., and Kornfeld, S. (1984). Targeting of β-glucuronidase to lysosomes in mannose 6-phosphate receptor-deficient MOPC 315 cells. *J. Cell Biol.* **99**, 296–306.

Gabel, C. A., Costello, C. E., Reinhold, V. N., Kurz, L., and Kornfeld, S. (1984). Identification of methylphosphomannosyl residues as components of the high mannose oligosaccharides of *Dictyostelium discoideum* glycoproteins. *J. Biol. Chem.* **259**, 13762–13769.

Garoff, H. (1985). Using recombinant DNA techniques to study protein targeting in the eukaryotic cell. *Annu. Rev. Cell Biol.* **1**, 403–406.

Gething, M.-J., McCammon, K., and Sambrook, J. (1986). Expression of wild-type and mutant forms of influenza hemagglutinin: the role of folding in intracellular transport. *Cell* **46**, 939–950.

Geuze, H. J., Slot, J. W., Strous, G. J. A. M., Hasilik, A., and von Figura, K. (1984). Ultrastructural localization of the mannose 6-phosphate receptor in rat liver. *J. Cell Biol.* **98**, 2047–2054.

Gibson, R., Schlesinger, S., and Kornfeld, S. (1979). The nonglycosylated glycoprotein of vesicular stomatitis virus is temperature-sensitive and undergoes intracellular aggregation at elevated temperatures. *J. Biol. Chem.* **254**, 3600–3607.

Gieselmann, V., Pohlmann, R., Hasilik, A., and von Figura, K. (1983). Biosynthesis and transport of cathepsin D in cultured human fibroblasts. *J. Cell Biol.* **97**, 1–5.

Gieselmann, V., Hasilik, A., and von Figura, K. (1985). Processing of human cathepsin D in lysosomes *in vitro*. *J. Biol. Chem.* **260**, 3215–3220.

Goldberg, D., Gabel, C., and Kornfeld, S. (1984). Processing of lysosomal enzyme oligosaccharide units. *In* "Lysosomes in Biology and Pathology" (J. T. Dingle, R. T. Dean, and W. Sly, eds.), pp. 45–62. Elsevier, Amsterdam.

Golumbeski, G. S. (1985). Cellular events and developmental controls involved in the biosynthesis of lysosomal β-glucosidase in *Dictyostelium discoideum*. Ph.D. Thesis, University of Wisconsin, Madison.

Golumbeski, G. S., and Dimond, R. L. (1986). The use of tolerization in the production of monoclonal antibodies against minor antigenic determinants. *Anal. Biochem.* **154**, 373–381.

Gonzales-Noriega, A., Grubb, J. H., Talkad, V., and Sly, W. S. (1980). Chloroquine inhibits lysosomal enzyme pinocytosis and enhances lysosomal enzyme secretion by impairing receptor recycling. *J. Cell Biol.* **85**, 839–852.

Green, E. D., Gruenebaum, J., Bielinska, M., Baenziger, J. V., and Boime, I. (1984). Sulfation of leutropin oligosaccharides with a cell-free system. *Proc. Natl. Acad. Sci. U.S.A.* **81**, 5320–5324.

Gross, V., Andrus, T., Tran-Thi, T.-A., Schwarz, R. T., Decker, K., and Heinrich, P. C. (1983). l-Deoxnorjirimycin impairs oligosaccharide processing of α-$_1$-proteinase inhibitors and inhibits its secretion in primary cultures of rat hepatocytes. *J. Biol. Chem.* **258**, 12203–12209.

Gustafson, G. L., and Milner, L. A. (1980). Occurrence of N-acetylglucosamine 1-phosphate in proteinase I from *Dictyostelium discoideum*. *J. Biol. Chem.* **255**, 7208–7210.

Gustafson, G. L., and Thon, L. A. (1979). Purification and characterization of a proteinase from *Dictyostelium discoideum*. *J. Biol. Chem.* **254**, 12471–12478.

Hasilik, A., and Neufeld, E. F. (1980). Biosynthesis of lysosomal enzymes in fibroblasts. *J. Biol. Chem.* **255**, 4937–4945.

Hasilik, A., and von Figura, K. (1984). Processing of lysosomal enzymes in fibroblasts. *In* "Lysosomes in Biology and Pathology" (J. T. Dingle, R. T. Dean, and W. Sly, eds.), pp. 3–16. Elsevier, Amsterdam.

Hasilik, A., Voss, B., and von Figura, K. (1981). Transport and processing of lysosomal enzymes by smooth muscle cells and endothelial cells. *Exp. Cell Res.* **133**, 23–30.

Hasilik, A., Pohlmann, R., and von Figura, K. (1983). Inhibition by cyanate of the process-ing of lysosomal enzymes. *Biochem. J.* **210,** 795–802.

Heifetz, A., and Lennarz, W. J. (1979). Biosynthesis of N-glycosidically linked glycopro-teins during gastrulation of sea urchin embryos. *J. Biol. Chem.* **254,** 6119–6127.

Hendershoot, L., Bole, D., Köhler, G., and Kearney, J. F. (1987). Assembly and secretion of heavy chains do not associate post-translationally with immunoglobulin heavy chain-binding protein. *J. Cell Biol.* **104,** 761–767.

Henderson, E. J. (1984). The role of glycoproteins in the life cycle of the cellular slime mold *Dictyostelium discoideum. In* "The Biology of Glycoproteins" pp. 371–443. (R. Ivatt, ed.), Plenum, New York.

Hentze, M., Hasilik, A., and von Figura, K. (1984). Enhanced degradation of cathepsin D synthesized in the presence of the threonine analog β-hydroxynorvaline. *Arch. Bio-chem. Biophys.* **230,** 375–382.

Hercz, A., Katona, E., Cutz, E., Wilson, J., and Barton, M. (1978). α_1-Antitrypsin: The presence of excess mannose in the Z variant isolated from liver. *Science* **201,** 1229–1231.

Hieber, V. C. (1982). Cloning of a cDNA complementary to rat preputial gland β-glucuroni-dase mRNA. *Biochem. Biophys. Res. Commun.* **104,** 1271–2178.

Hoflack, B., and Kornfeld, S. (1985). Lysosomal enzyme binding to mouse P388D, macro-phage membranes lacking the 215-kDa mannose-6-phosphate receptor: Evidence for the exisence of a second mannose 6-phosphate receptor. *Proc. Natl. Acad. Sci. U.S.A.* **82,** 4428–4432.

Hohmann, H., Gerish, G., Lee, R., and Huttner, W. B. (1985). Cell-free sulfation of the contact site A glycoprotein of *Dictyostelium discoideum* and of a partially glycosylated precursor. *J. Biol. Chem.* **260,** 13869–13878.

Hohman, T. C., and Bowers, B. (1984). Hydrolase secretion is a consequence of membrane recycling. *J. Cell Biol.* **98,** 246–252.

Ikeda, T. (1981). Subcellular distributions of UDPgalactose:polysaccharide transferase and UDPglucose pyrophosphorylase involved in biosynthesis of prespore-specific acid mu-copolysaccharide in *Dictyostelium discoideum. Biochim. Biophys. Acta* **675,** 69–76.

Jessup, W., Leoni, P., and Dean, R. T. (1985). Constitutive and triggered lysosomal enzyme secretion. *In* "Developments in Cell Biology I. Secretory Processes"(Dean and Stahl, eds.), pp. 38–57. Butterworths, London.

Johnson, L. M., Bankaitis, V. A., and Emr, S. D. (1987). Distinct sequence determinants direct intracellular sorting and modification of a yeast vacuolar protease. *Cell* **48,** 875–885.

Judelson, H. J. (1985). Synthesis and posttranslational-modification of lysosomal proteins in *Dictyostelium discoideum.* Ph.D. Thesis, University of Wisconsin, Madison.

Kelly, R. B. (1985). Pathways of protein secretion in eukaryotes. *Science* **230,** 25–31.

Knecht, D. A., Cohen, S. M., Loomis, W. F., and Lodish, H. F. (1986). Developmental regulation of *Dictyostelium discoideum* actin gene fusions carried on low-copy and high-copy transformation vectors. *Molec. Cell. Biol.* **6,** 3973–3983.

Knecht, D. A., and Dimond, R. L. (1981). Lysosomal enzymes possess a common antigenic determinant in the cellular slime mold, *Dictyostelium discoideum. J. Biol. Chem.* **256,** 3564–3575.

Knecht, D. A., Dimond, R. L., Wheeler, S., and Loomis, W. F. (1984). Antigenic determi-nants shared by lysosomal proteins of *Dictyostelium discoideum. J. Biol. Chem.* **259,** 10633–10640.

Knecht, D. A., Green, E. D., Loomis, W. F., and Dimond, R. L. (1985). Developmental changes in the modification of lysosomal enzymes in *Dictyostelium discoideum. Dev. Biol.* **107,** 490–502.

Konings, A., Hupkes, P., Versteeg, R., Grosweld, G., Reuser, A., and Galjaard, H. (1984). Cloning a cDNA for the lysosomal α-glucosidase. *Biochem. Biophys. Res. Commun.* **119**, 252–258.

Kornfeld, R., and Kornfeld, S. (1985). Assembly of asparagine-linked oligosaccharides. *Annu. Rev. Biochem.* **54**, 631–634.

Lemansky, P., Gieselmann, V., Hasilik, A., and von Figura, K. (1984). Cathepsin D and β-hexosaminidase synthesized in the presence of 1-deoxynorjirimycin accumulate in the endoplasmic reticulum. *J. Biol. Chem.* **259**, 10129–10135.

Lemkin, M., and Farquhar, M. G. (1981). Sulfated and nonsulfated glycosaminoglycans and glycopeptides are synthesized by kidney *in vivo* and incorporated into glomerular basement membranes. *Proc. Natl. Acad. Sci. U.S.A.* **78**, 1726–1730.

Livi, G. P., Cardelli, J. A., Mierendorf, R. C., and Dimond, R. L. (1984). Developmental control of α-mannosidase-1 synthesis in *Dictyostelium discoideum*. In "Molecular Biology of Development" (UCLA Symposia on Molecular and Cellular Biology, Vol 19, E. H. Davidson and R. A. Firtel ed.), pp. 447–457. Alan R. Liss, New York.

Livi, G. P., Cardelli, J. A., and Dimond, R. L. (1985a). α-Mannosidase-1 mutants of *Dictyostelium discoideum* and early aggregation-essential genes regulate enzyme precursor synthesis, modification, and processing. *Differentiation* **29**, 207–215.

Livi, G. P., Cardelli, J. A., Mierendorf, R. C., and Dimond, R. L. (1985b). Regulation of lysosomal α-mannosidase-1 synthesis during development in *Dictyostelium discoideum*. *Dev. Biol.* **110**, 514–520.

Livi, G. P., Woychick, N. A., and Dimond, R. L. (1985c). Lysosomal enzyme inactivation associated with defects in posttranslational modification during development in *Dictyostelium discoideum*. *Differentiation* **30**, 83–91.

Lodish, H. F., and Kong, N. (1983). Reversible block in intracellular transport and budding of mutant vesicular stomatitis virus glycoproteins. *Virology* **125**, 335–348.

Lodish, H. F., and Kong, N. (1984). Glucose removal from N-linked oligosaccharides is required for efficient maturation of certain secretory glycoproteins from the rough endoplasmic reticulum to the Golgi complex. *J. Cell Biol.* **98**, 1720–1729.

Lodish, H. F., Kong, N., Snider, M., and Strous, G. J. A. M. (1983). Hepatoma secretory proteins migrate from rough endoplasmic reticulum to Golgi at characteristic rates. *Nature (London)* **304**, 80–83.

Loomis, W. F. (1970). Developmental regulation of α-mannosidase in *Dictyostelium discoideum*. *J. Bacteriol.* **103**, 375–381.

Loomis, W. F. (1975). "*Dictyostelium discoideum*: A Developmental System." Academic Press, New York.

Loomis, W. F. (1978). Genetic analysis of the gene for *N*-acetylglucosaminidase in *Dictyostelium discoideum*. *Genetics* **88**, 277–284.

Loomis, W. F. (1980a). Genetic analysis of development in *Dictyostelium*. In "The Molecular Genetics of Development" (T. Leighton and W. F. Loomis, eds.). Academic Press, New York.

Loomis, W. F. (1980b). A β-glucosidase gene of *Dictyostelium discoideum*. *Dev. Genet.* **1**, 241–246.

Loomis, W. F. (1982). "The Development of *Dictyostelium discoideum*." Academic Press, New York.

Loomis, W. F., and Kuspa, A. (1984). Biochemical and genetic analysis of pre-stalk specific acid phosphatase in *Dictyostelium*. *Dev. Biol.* **102**, 498–503.

Mierendorf, R. C., Cardelli, J. A., Livi, G. P., and Dimond, R. L. (1983). Synthesis of related forms of the lysosomal enzyme α-mannosidase in *Dictyostelium discoideum*. *J. Biol. Chem.* **258**, 5878–5884.

Mierendorf, R. C., Cardelli, J. A., and Dimond, R. L. (1985). Pathways involved in targeting and secretion of a lysosomal enzyme in *Dictyostelium discoideum*. *J. Cell Biol.* **100**, 1777–1787.

Miyazaki, J., Appella, E., and Ozato, K. (1986). Intracellular transport blockade caused by disruption of the disulfide bridge in the third external domain of major histocompatibility complex I antigen. *Proc. Natl. Acad. Sci. U.S.A.* **83**, 757–761.

Moore, B. R., Gossels, J. D., and Free, S. J. (1985). Changes in the posttranslational modification of lysosomal enzymes during development of *Dictyostelium Differentiation* **29**, 7–13.

Mueller, M., and Mueller, H. (1981). Synthesis and processing of *in vitro* and *in vivo* precursor of the vacuolar yeast enzyme carboxypeptidase Y. *J. Biol. Chem.* **256**, 11962–11965.

Murray, B. A., Wheeler, S., Jongens, T., and Loomis, W. F. (1984). Mutations affecting a surface glycoprotein, gp80, of *Dictyostelium discoideum*. *Mol. Cell. Biol.* **4**, 514–519.

Nellin, W., Silan, C., and Firtel, R. A. (1984). DNA-mediated transformation in *Dictyostelium discoideum*: regulated expression of an actin gene fusion. *Mol. Cell. Biol.* **4**, 2890–2898.

North, M. J. (1985). Cysteine proteinases of cellular slime molds. *Biochem. Soc. Trans.* **13**, 288–290.

North, M. J., and Whyte, A. (1984). Purification and characterization of two acid proteinases from *Dictyostelium discoideum*. *J. Gen. Microbiol.* **130**, 123–134.

O'Dowd, B. F., Quan, F., Willard, H. F., Lamhonwah, A., Korneluk, R. G., Lowden, J. A., Gravel, R. A., and Mahuran, D. J. (1985). Isolation of cDNA clones coding for the β subunit of human β-hexosaminidase. *Proc. Natl. Acad. Sci. U.S.A.* **82**, 1184–1188.

Ohkuma, S., and Poole, B. (1978). Fluorescence probe measurement of intralysosomal pH in living cells and the perturbation of pH by various agents. *Proc. Natl. Acad. Sci. U.S.A.* **75**, 3327–3331.

Olden, K., Bernard, B. A., Humphries, M. J., Yeo, K-T., White, S. L., Newton, S. A., Bauer, H. C., and Parent, J. B. (1985). Function of glycoprotein glycans. *Trends Biochem. Sci.* **10**, 76–82.

Orci, L., Halban, P., Amherdt, M., Ravazzola, M., Vassali, J.-D., and Perrelet, A. (1984). Nonconverted, amino acid analog modified proinsulin stays in a Golgi derived clathrin coated membrane compartment. *J. Cell Biol.* **99**, 2187–2192.

Orci, L., Ravazzola, M., Amherdt, M., Madsen, O., Vassalli, J.-D., and Perrelet, A. (1985). Direct identification of prohormone conversion site in insulin secreting cells. *Cell* **42**, 671–681.

Owada, M., and Neufeld, E. F. (1982). Is there a mechanism for introducing acid hydrolases into liver lysosomes that is independent of mannose t-phosphate recognition? Evidence from I-cell disease. *Biochem. Biophys. Res. Commun.* **105**, 814–820.

Paigen, K. (1979). Acid hydrolases as models of genetic control. *Annu. Rev. Genet.* **13**, 417–466.

Palade, G. (1975). Intracellular aspects of the process of protein secretion. *Science* **189**, 347–358.

Pannell, R., Wood, L., and Kaplan, A. (1982). Processing and secretion of α-mannosidase forms by *Dictyostelium discoideum*. *J. Biol. Chem.* **257**, 9861–9865.

Pears, C. J., Mahbugani, H. M., and Williams, J. G. (1985). Characterization of two highly diverged but developmentally coregulated cysteine proteinase genes in *Dictyostelium discoideum*. *Nucleic Acids Res.* **13**, 8853–8866.

Pohlmann, R., Kruger, S., Hasilik, A., and von Figura, K. (1984). Effect of monensin on intracellular transport and receptor mediated endocytosis of lysosomal enzymes. *Biochem. J.* **217**, 649–658.

Proia, R. L., and Neufeld, E. F. (1982). Synthesis of β-hexosaminidase in cell-free translation and in intact fibroblasts: An insoluble precursor chain in a rare form of Tay–Sachs disease. *Proc. Natl. Acad. Sci. U.S.A.* **79**, 6360–6364.

Richardson, J. M., Woychik, N. A., Dimond, R. L., and Cardelli, J. A. (1987). Primary but not secondary proteolytic processing events may be essential for targeting of lysosomal enzymes in *Dictyostelium discoideum*. *J. Cell Biol.* (submitted).

Ronne, H., Ocklind, C., Wiman, K., Rask, L., Obrink, B., and Peterson, P. A. (1983). Ligand-dependent regulation of intracellular protein transport: Effect of vitamin A on the secretion of the retinol-binding protein. *J. Cell Biol.* **96**, 907–910.

Rosenfeld, M. G., Kreibich, G., Popov, D., Kato, K., and Sabatini, D. D. (1982). Biosynthesis of lysosomal hydrolases: Their synthesis in bound polysomes and the role of co- and posttranslational processing in determining their subcellular distribution. *J. Cell Biol.* **93**, 135–143.

Rossomando, E. F., Maldonado, B., Crean, E. V., and Kollar, E. J. (1978). Protease secretion during onset of development in *Dictyostelium discoideum*. *J. Cell Sci.* **30**, 305–318.

Schauer, I., Emr, S., Gross, C., and Schekman, R. (1985). Invertase signal and mature sequence substitutions that delay intercompartmental transport of active enzyme. *J. Cell. Biol.* **100**, 1664–1675.

Schwaiger, H., Hasilik, A., von Figura, K., Wiemken, A., and Tanner, W. (1982). Carbohydrate free carboxypeptidase is transferred into the lysosome-like yeast vacuole. *Biochem. Biophys. Res. Commun.* **104**, 950–956.

Segundo, B. S., Chan, S. J., and Steiner, D. F. (1985). Identification of cDNA clones encoding a precursor of rat liver cathepsin B. *Proc. Natl. Acad. Sci. U.S.A.* **82**, 2320–2324.

Skudlarek, M. D., Novak, E. K., and Swank, R. T. (1984). Processing of lysosomal enzymes in macrophages and kidney. *In* "Lysosomes in Biology and Pathology" (J. T. Dingle, R. T. Dean, W. Sly, ed.), pp. 17–43. Elsevier, Amsterdam.

Sly, W. S., and Fischer, H. D. (1982). The phosphomannosyl recognition system for intracellular and intercellular transport of lysosomal enzymes. *J. Cell Biochem.* **18**, 67–85.

Stevens, T. H., Rothman, J. H., Payne, G. S., and Schekman, R. (1986). Gene dosage-dependent secretion of yeast vacuolar carboxypeptidase Y. *J. Cell Biol.* **102**, 1551–1557.

Tartakoff, A. M. (1983). Pertubation of vesicular traffic with the carboxylic ionophore monensin. *Cell* **32**, 1026–1028.

Tsuji, S., Choudary, P. V., Martin, B. M., Winfield, S., Barranger, J. A., and Ginns, E. I. (1986). Nucleotide sequence of cDNA containing the complete coding sequence for human lysosomal glucocerebrosidase. *J. Biol. Chem.* **261**, 50–53.

Valls, L. A., Hunter, C. P., Rothman, J. H., and Stevens, T. H. (1987). Protein sorting in yeast: the localization determinant of yeast vacuolar carboxypeptidase Y resides in the propeptide. *Cell* **48**, 887–897.

Van Diggelen, D. P., Hoogeveen, A. T., Smith, P. J., Reuser, A. J. J., and Galjaard, H. (1982). Enhanced proteolytic degradation of normal β-galactosidase in the lysosomal storage disease with combined β-galactosidase and neuraminidase deficiency. *Biochim. Biophys. Acta* **703**, 69–76.

Vitale, A., and Chrispeels, M. J. (1984). Transient *N*-acetylglucosamine in the biosynthesis of phytohemagglutin: Attachment in the Golgi apparatus and removal in protein bodies. *J. Cell Biol.* **99**, 133–140.

von Figura, K., and Hasilik, A. (1986). Lysosomal enzymes and their receptors. *Annu. Rev. Biochem.* **55,** 167–194.

von Figura, K., Steckel, F., and Hasilik, A. (1983). Juvenile and adult metachromatic leukodystrophy: Partial restoration of arylsulfatase A (cerebroside sulfatase) activity by inhibition of thiol proteinases. *Proc. Natl. Acad. Sci. U.S.A.* **80,** 6066–6070.

Waheed, A., and Van Etten, R. L. (1985). Phosphorylation and sulfation of arylsulfatase-A accompanies biosynthesis of the enzyme in normal and carcinoma cell-lines. *Biochim. Biophys. Acta* **847,** 53–61.

Waheed, A., Pohlmann, R., Hasilik, A., von Figura, K., van Elsen, A., and Leroy, J. G. (1982). Deficiency of UDP–N-acetylglucosamine:lysosomal enzyme N-acetylglucosamine-1-phosphotransferase in organs of I-cell patients. *Biochem. Biophys. Res. Commun.* **105,** 1052–1058.

Walter, P., Gilmore, R., and Blobel, G. (1984). Protein translocation across the endoplasmic reticulum. *Cell* **38,** 5–8.

West, C. M., and Loomis, W. F. (1985). Absence of a carbohydrate modification does not affect the level or subcellular localization of three membrane glycoproteins in *mod* B mutants of *Dictyostelium discoideum*. *J. Biol. Chem.* **260,** 13803–13809.

Wickner, W. T., and Lodish, H. F. (1985). Multiple mechanisms of protein insertion into and across membranes. *Science* **230,** 400–407.

Wileman, T., Harding, C., and Stahl, P. (1985). Receptor-mediated endocytosis. *Biochem. J.* **232,** 1–14.

Wood, L., and Kaplan, A. (1985). Transit of α-mannosidase during its maturation in *Dictyostelium discoideum*. *J. Cell Biol.* **101,** 2063–2069.

Wood, L., Pannell, R. N., and Kaplan, A. (1983). Linked pools of processed α-mannosidase in *Dictyostelium discoideum*. *J. Biol. Chem.* **258,** 9426–9430.

Woychick, N. A., Cardelli, J. A., and Dimond, R. L. (1986). A conformationally altered precursor to the lysosomal enzyme α-mannosidase accumulates in the endoplasmic reticulum in a mutant strain of *Dictyostelium discoideum*. *J. Biol. Chem.* **261,** 9595–9602.

Wu, G. E., Hozumi, N., and Murialdo, H. (1983). Secretion of a λ_2 immunoglobulin chain is prevented by a single amino acid substitution in its variable region. *Cell* **33,** 77–83.

10

Organelles of Endocytosis and Exocytosis

JOHN A. HANOVER AND ROBERT B. DICKSON

I. INTRODUCTION

The study of endocytosis (engulfment of materials) and exocytosis (secretion) has historical roots in a number of subfields of cell biology. Probably the secretory pathway was the first to be appreciated; initial studies visualized electron-dense granules in specialized secretory cells (reviewed in, Palade, 1975; Kelly, 1985). Some of the earliest experiments on endocytosis in eukaryotic cells distinguished phagocytosis (engulfment of large particles) (Metchnikoff, 1893) from pinocytosis (engulfment of fluid) (Lewis, 1931). The first clear examples of receptor-mediated endocytosis

401

were viruses (Dales, 1973), followed several years later by lipoproteins (Goldstein *et al.,* 1979). Both of these structures are large and easily visualized by electron microscopy.

Work by Palade (reviewed in Farquhar and Palade, 1981) in the late 1950s on cells with regulated secretion processes led to the realization that endocytosis and secretion might be coupled. During secretion, internal vesicle membranes were inserted into the plasma membrane. To maintain a constant size, it was proposed that cells must recapture or endocytose the membrane to their interior. By the 1970s, electron-dense ferritin membrane tracers had been developed, initially by Bessis (Bessis, 1963) and subsequently by Fawcett (Fawcett, 1964) and Farquhar (reviewed in Farquhar and Palade, 1981). Such tracers could be applied from the cell exterior and, following membrane adsorption, could be observed traveling into the cell. These studies emphasized that sometimes membrane tracers could reemerge from the cell following endocytosis (Ottosen *et al.,* 1980). Together, these studies emphasized the dynamic aspects of membrane movement in and out of cells.

In 1964 Roth and Porter identified the coated pit on the surface of oocytes as a site of uptake of yolk proteins. Then in the 1970s Pearse isolated "coated vesicles" and their principle proteinaceous component clathrin (Pearse, 1976), and the stage was set for the full definition of receptor-mediated endocytosis and secretion pathways. Much of this recent work has focused on the intracellular movement of biologically relevant, receptor-bound macromolecules. The early pioneering studies of Ashwell and Morell (1974) on the asialogycoprotein receptor pointed out the biological significance of the endocytic pathway. The role of the receptor in these processes has now been facilitated because several receptors have been purified, molecularly cloned, and sequenced. This kind of information, coupled with availability of genetically altered receptors, particularly for low density lipoprotein (LDL), has led to a greater appreciation of the role of receptor domains and biochemical modifications (Brown and Goldstein, 1986).

Independent investigations, using pharmacologic agents or ionic manipulations to inhibit intracellular transport as well as the selection of cell and viral mutants *in vitro,* have begun to define other important components regulating membrane protein movement. This chapter summarizes information from these diverse approaches to define the organelles (or classes of cellular membranes) which mediate endocytosis and secretion. Where available, direct biochemical characterization of these membrane classes will be presented. Two themes will recur: (1) consideration of endocytosis and secretion as partially coupled phenomena and (2) definition of the biochemical signals expressed by receptors and other proteins which allow their routing, sorting, and transfer through these dynamic membrane systems.

II. ENDOCYTOSIS

A. Ligands and Their Receptors

1. Ligands Taken up by Receptor-Mediated Endocytosis

Numerous proteins are selectively removed from the aqueous environment of cells for degradation and/or utilization. In the case of LDL and transferrin, the nutritional requirements for cholesterol and iron, respectively, provide a straightforward rationale for their cellular entrapment (Brown and Goldstein, 1986; Hanover and Dickson, 1985a). Scavenger receptors, such as those mediating α_2-macroglobulin–protease uptake, have also been described (Dickson *et al.,* 1981a; Hanover *et al.,* 1982, 1983a,b, 1985b). In the case of various hormones and growth factors, the biological consequences of endocytosis are much less certain. Although the fate of the hormone and sometimes the receptor is degradation in lysosomes, it is not certain whether this breakdown terminates or initiates receptor signaling. In addition, many toxins (Olsnes and Sandvig, 1985) and viruses (Marsh *et al.,* 1983) appear to have opportunistically adopted endocytosis into an acidic intracellular environment to trigger penetration into the cytoplasm. Finally, extracellular fluid is also taken into cells via receptor-mediated endocytosis (Ryser *et al.,* 1982; Gonatas *et al.,* 1984). This process might not be as efficient, however, as other mechanisms of fluid entry (osmotic pumps, pinocytosis, etc.)

2. Biochemical and Genetic Analysis of Receptor Molecules

The most simple view of receptors must allow for at least four important functional domains encoded within their structure. First, receptors must bind with high avidity and specificity the ligands for which they have evolved. Second, receptors must concentrate within the plane of the plasma membrane in sites of endocytosis (in most cases the clathrin-coated pits). Third, receptors (or the receptor–ligand complex) may contain additional routing signals to ensure their subsequent fate. Fourth, in cases where signal transduction occurs, some means for transmission of that signal from the extracellular binding site to an intracellular (cytoplasmic) site is required.

Currently at least half a dozen receptors have been sequenced. Some are monomeric, others homo- or heterodimeric. Most contain cysteine-rich amino-terminal regions which extend outside the cell and contain the ligand binding site(s) (Brown and Goldstein, 1986). In many cases, receptors are glycosylated through Asn or Ser/Thr residues in the primary amino acid structure (Brown and Goldstein, 1986; Hanover and Lennarz, 1981). In one case described so far a receptor is also covalently linked to ubiquitin (Siegelman *et al.*, 1986). In some instances, glycosylation is required for high affinity ligand binding (Sleiker and Lane, 1985). All receptors appear to have a short (20–24 amino acid) hydrophobic segment, the transmembrane domain. Sometimes fatty acid residues are also covalently linked (via esterification) to the primary sequence which may also contribute to proper transmembrane positioning (Omary and Trowbridge, 1981). Finally, receptors usually contain significant intracellular domains; for growth factor receptors, this domain may contain a protein kinase (Carpenter and Cohen, 1979; Cohen *et al.*, 1982). Mechanisms of transmembrane signaling by receptors are largely unknown; this chapter concentrates on the localization of receptors and other membrane proteins.

Probably the most complete genetic analysis of any receptor system is available for LDL. This is the result of the extensive catalog of naturally occurring diseases of cholesterol metabolism (reviewed in Brown and Goldstein, 1986; Krieger *et al.*, 1985). Four major classes of aberrant LDL receptor function have been described, and essentially all receptor mutations are in ligand binding and receptor localization. First, the largest group of mutant receptor alleles codes for proteins that are undetectable by antibody methods and are never transferred from the endoplasmic reticulum to the Golgi complex. One of these mutations which has been analyzed contains a large deletion.

The second most common class of mutations is characterized by essentially normal precursor proteins (100–135 kDa) and initial glycosylation events. However, further processing of carbohydrate residues (trimming of N-linked sugars to endo-β-N-acetylglucosaminidase H-resistant forms and elongation of O-linked sugar chains) has not occurred. Most of these mutants also accumulate in the endoplasmic reticulum. The molecular defects giving rise to this class of mutants in humans are unknown at present. A related mutant phenotype in the rabbit model system has been sequenced (the Watanabe heritable hyperlipidemic, or WHHL, rabbit) and contains a small in-frame deletion in the cysteine-rich extracellular domain (Yamamoto *et al.*, 1986). This mutation apparently leads to a "degradation shunt" pathway prior to localization of the receptor in the Golgi. Krieger and co-workers have characterized a mutant altering Ser/Thr O-glycosylation in Chinese hamster ovary (CHO) cells which can yield a similar phenotype (Kingsley *et al.*, 1986; Krieger, 1983; Kingsley and Krieger, 1984).

In the third class of receptor mutations, receptor molecules are normally processed and reach the cell surface, but they are unable to bind LDL. These are all synthesized from different sized precursors which are 40 kDa smaller than the mature form (normally 160 kDa). Presumably, these receptors are also defective in some portion of the extracellular domain (reviewed in Brown and Goldstein, 1986).

Finally, a class of mutants exists in which receptor is present on the cell surface, binds LDL normally, but does not concentrate in coated pits. The altered receptor is unable to deliver cholesterol to the cell (Lehrman *et al.*, 1985; Davis *et al.*, 1986). The three mutants in this class which have been sequended all have defective cytoplasmic receptor "tails." In one case, a tryptophan codon has been converted to a stop codon truncating the receptor just beneath the plasma membrane. In another case, a frameshift insertion leads to a similar phenotype. Perhaps the most interesting of all is the JD mutant, which has a missense mutation leading to a tyrosine to cysteine substitution 18 amino acids below the presumed plasma membrane-spanning domain on the cytoplasmic tail. This suggests that the detailed structure of the cytoplasmic tail near the membrane is intimately involved in initial entry to the endocytic pathway. This region of the receptor has a relatively high density of positively changed residues (Lys and Arg) which may be involved in association with plasma membrane proteins and anionic phospholipids. A likely candidate for the initial membrane structure which would normally interact with the receptor tail is the clathrin-coated pit. The biochemical nature of this proposed interaction is unknown at present.

3. Two Prototypic Receptors: Transferrin Receptor and EGF Receptor

As a model system to compare the intracellular routes taken by receptors following ligand binding, we have chosen transferrin and epidermal growth factor (EGF). These receptors are found together on a wide variety of cells, and the ligands are easily labeled for simultaneous studies in microscopy and cell fractionation. The EGF receptor is a monomer containing a glycosylated, cysteine-rich external binding domain, a transmembrane domain, and a long cytoplasmic tail containing an EGF-induced tyrosine kinase activity. There is also a phosphorylated threonine residue on the cytoplasmic face the plasma membrane (Hunter, 1984; Mayes and Waterfield, 1984), which will be addressed in a later section (Section II,F.) of this chapter. The transferrin receptor is a dimeric, glycosylated molecule, but with a shorter cytoplasmic tail. The tail contains fatty acid modifications and a serine residue just under the plasma membrane which can be phosphorylated (Newman *et al.*, 1982; reviewed in Hanover and Dickson, 1985b).

B. Pathways of Receptor-Mediated Endocytosis

1. Endocytosis Leading to Lysosomal Delivery

The most widely studied endocytic pathway is the one terminating in lysosomes (Fig. 1, pathway 2). A variety of polypeptide hormones and other plasma proteins, including insulin, EGF, prolactin, human chori-

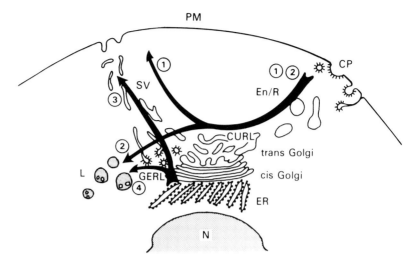

Fig. 1. Endocytic–recycling pathways. Polypeptide receptors enter cells via endocytosis from the plasma membrane or biosynthesis in the endoplasmic reticulum. Four major pathways are as follows: (1) Endocytosis–recycling pathway (for many receptors, and receptor–transferrin, receptor–immunoglobulin complexes); (2) endocytosis–lysosome pathway (some receptors, most ligands); (3) biosynthesis–secretion pathway (presumably all receptors); and (4) biosynthesis–lysosome pathway (phosphomannosyl receptor–biosynthesized acid hydrolase complexes). CP, Coated pits; PM, plasma membrane; SV, secretory vesicles (different from concentrative secretory granules in stimulus–secretion systems); En/R, endosome–receptosome; ER, endoplasmic reticulum; L, lysosome; N, nucleus; GERL, Golgi–endoplasmic reticulum–lysosome region; CURL, compartment for uncoupling of receptor and ligand.

onic gonadotropin (HCG), LDL, α_2-macroglobulin, asialoglycoproteins, and lysosomal hydrolases, follow this route (Pastan and Willingham, 1981a). By using density gradient centrifugation of cell homogenates on colloidal silica, this route is correlated with a smooth transition of ligands (for instance, EGF) from light endocytic vesicles to a sense, lysosomal enzyme-containing compartment (Dickson *et al.*, 1983a). Electron and fluorescence microscopies have demonstrated that this delivery proceeds through Golgi-associated membranes. Lipid is also transferred from the cell surface to the Golgi (Sleight and Pagano, 1984). Uncoated and clathrin-coated vesicles and tubules have been observed (Willingham and Pastan, 1980, 1982; Geuze *et al.*, 1983) in this sequence. Golgi-associated, endocytic marker-containing membranes have been termed endosomes II and III (Helenius *et al.*, 1983), juxtanuclear endosomes (Hopkins, 1983), transreticular Golgi (TRG) (Willingham *et al.*, 1984), and CURL (compartment for uncoupling of receptor and ligand; see below). While all of these terms may refer to analogous or partially overlapping membrane compartments, the identification of functionally distinct Golgi-associated subregions will probably proceed for many years. At some as yet poorly defined transition, endocytic vesicle-associated ligands and receptors may enter a proposed Golgi subregion termed GERL (defined by Novikoff and Novikoff, 1977, as Golgi–endoplasmic reticulum–lysosome region, an acid phosphatase-rich membrane network with a lysosomal terminus). As mentioned above, clathrin-coated pits in the Golgi have been observed to concentrate ligands destined for lysosomes (Farquhar and Palade, 1981; Willingham and Pastan, 1982; Hanover *et al.*, 1984, 1985b).

Some receptors undergo down-regulation or removal from the plasma membrane for very extensive periods of time (hours) following endocytosis. Biochemical studies of the insulin receptor in A431 carcinoma cells (Krupp *et al.*, 1982) have suggested that, although down-regulation occurs after ligand binding, it is spared lysosomal degradation. The insulin receptor appears to be sequestered inside the cell and slowly degraded by a nonlysosomal proteolytic system. Endocytosis of the EGF–receptor complex also leads to down-regulation of the receptor as well as to a rapid decrease in receptor half-life. Here, the mechanism of this down-regulation is better understood than for the insulin receptor. Biochemical and morphologic studies in KB carcinoma cells have recently shown that the EGF receptor is degraded in lysosomes (Stoscheck and Carpenter, 1984; Beguinot *et al.*, 1984). Evidence for lysosomal degradation of endocytosed prolactin and acetylcholine receptors, and gonadotropin receptors, has been provided (Ascoli, 1984; reviewed in Dickson, 1985). No biochemical recognition signal for lysosomal routing of receptors themselves has yet been established.

A phosphomannose receptor-mediated, lysosome-terminating, biosynthetic route for lysosomal proteins has also been described (Fig. 1, pathway 4); however, others may exist (Hoflack and Kornfeld, 1985a). Mannose 6-phosphate, a modification present on lysosomal hydrolases, allows binding to a specific receptor involved in targeting to lysosomes (Neufeld and Ashwell, 1980; see Robbins, Chapter 11, this volume). Phosphomannosyl receptor has been localized in Golgi-associated membranes, using electron microscopy (Willingham *et al.*, 1983; Brown and Farquhar, 1984). Receptor-bound acid hydrolases traversing the Golgi in a biosynthetic pathway bind to this receptor (Gabel *et al.*, 1982). Using density gradient fractionation of fibroblasts on colloidal silica, Sahagian and Neufeld (1983) have observed that, while the endocytosed ligand accumulated in lysosomes, the receptor was sorted into a prelysosomal compartment and recycled. It is not yet known if this biosynthetic pathway converges with receptor-bound, endocytically scavenged acid hydrolases, hormones, and receptors to enter a common membrane pathway (GERL?) terminating in lysosomes.

2. Recycling of Ligands and Their Receptors

Receptors can be inserted into the plasma membrane as a consequence both of biosynthesis and of reutilization (recycling) following endocytosis (Fig. 1, pathway 1). Examples of proteins that undergo recycling are LDL, α_2-macroglobulin, asialoglycoproteins, immunoglobulins, lysosomal hydrolases, and transferrin (reviewed in Brown *et al.*, 1983; Steinman *et al.*, 1983). Some of these appear to have large intracellular pools of receptors which may contribute to recycling. Transferrin and immunoglobulin receptors appear to be unique in that an intact ligand–receptor complex is reinserted into the plasma membrane following endocytosis (Iacopetta and Morgan, 1984). A double-label study employing both fluorescence microscopy and colloidal silica gradient fractionation of cell homogenates has demonstrated that transferrin and EGF enter cells in the same vesicles but diverge in a prelysosomal compartment (Dickson *et al.*, 1983b; Willingham *et al.*, 1984). In contrast to endocytic vesicles and lysosomes, the recycling compartment for transferrin has only a mildly acidic pH (6.5) (Yamashiro *et al.*, 1984).

The recycling pathway for transferrin and its receptor may be similar or identical to a general recycling pathway (Hanover and Dickson, 1985a). In hepatocytes, internalized asialoglycoprotein has been localized by electron microscopy in endocytic vesicles and lysosomes, whereas its receptor appeared uniquely concentrated in tubular extensions of these endocytic vesicles and in an extended tubular network in the Golgi system. The receptor-rich, tubular network (CURL) was proposed as an intermediate in receptor recycling (Geuze, *et al.*, 1983). A double-label study was subsequently carried out, comparing distributions of receptors with different intracellular fates. The phosphomannosyl receptor and asialoglycoprotein receptors (which recycle, but whose ligands are largely lysosome directed) were compared to the immunoglobulin receptor. Its ligand, polymeric immunoglobulin A (IgA), is directed along with the receptor by "transcytosis" to bile cannicular domains of the plasma membrane (see Section II,B,4). All three receptors were localized to a system of peripheral Golgi-associated tubules (CURL), but a strikingly different localization was observed for the immunoglobulin receptor and the other two receptors (Geuze *et al.*, 1984).

Time-course electron microscopic studies in cultured cells have been able to directly visualize similar CURL vesicles and tubules involved in transferrin–receptor recycling. A number of groups have shown that the bulk of the transferrin receptor and endocytosed transferrin is located in peripheral or trans-Golgi elements in epidermoid carcinoma, CHO, and other cell types. Farquhar and co-workers, in contrast, have reported an additional localization of transferrin receptor in cis-Golgi elements in myeloma cells (Woods *et al.*, 1986; Hopkins, 1983; Willingham and Pastan, 1985). The additional cis localization in myeloma cells was interpreted to result from the more extensive plasma membrane to Golgi trafficking in these specialized, secretory cells (Woods *et al.*, 1986).

Interestingly, resialation of both transferrin and its receptor can occur during recycling (Regoeczi *et al.*, 1982; Snider and Rogers, 1985). In addition, Snider and Rogers (1986) have also shown that recycling transferrin receptor can be transported to a mannosidase I-containing Golgi subcompartment but with a very slow half-time (>6 hr) compared with the half-time of internalization itself (10–20 min). Clearly, recycling receptors can gain access to various Golgi compartments though they may preferentially or initially accumulate at the trans face.

3. Endocytosis in Vivo Not Resulting in Degradation of Ligand or Receptor (Diacytosis)

One of the initial demonstrations that endocytosis of ligands *in vivo* can lead to a nondegradative pathway was for the asialoglycoprotein system. Regoeczi and co-workers (1982) have shown that human asialotransferrin type 3 is endocytosed by the liver and returned to the circulation partially resialated. The process was termed diacytosis (Fig. 1, pathway 1). A plasma membrane fraction, the ''diacytosome,'' was isolated which was enriched in the asialoglycoprotein receptor and distinct from Golgi plasma membrane and lysosome membrane fractions (Regoeczi *et al.*, 1982; Debanne *et al.*, 1982). Presumably the initial uptake was mediated by the asialoglycoprotein receptor, but, following partial replacement of sialic acid residues in a Golgi-like membrane fraction, recycling of the protein (by the transferrin receptor?) to the plasma membrane occurred.

4. Transcellular Endocytosis (Transcytosis) in Polarized Cells

In polarized cells, the possibility exists for exocytic and endocytic pathways to originate and be directed to specific plasma membrane domains. In the case of IgA (Kuhn and Kraehenbuhl, 1982; Geuze *et al.*, 1984; Hoppe *et al.*, 1985), endocytosis occurs on the basolateral surface of epithelial cells such as liver parenchymal cells. The receptor (originally called secretory component) mediates this translocation from blood to the apical surface as well as ''secretion'' into bile channels. In liver, galactose-terminal receptor-mediated endocytosis occurs in the same endocytic vesicles as for IgA, but then diverges, transferring galactose-terminal proteins to lysosomes (Fig. 1, pathway 2) (Courtoy *et al.*, 1985; Limet *et al.*, 1985). Abrahamson and Rodenwald (1981) have shown that IgG is transported across intestinal epithelial cells, being sorted away from soluble peroxidase used as a fluid-phase marker. Peroxidase and, presumably, the bulk of the endocytosed fluid are transferred to lysosomes. The process of transcytosis has been recently recognized to be quite common in polarized endothelial and epithelial cells (Mostov and Simister, 1985).

5. Summary

Clearly, a variety of fates exists for receptors and ligands (Fig. 1). The initial uncoated endocytic compartments appear to be key elements in ligand–receptor sorting events. It is not yet known whether recycled receptors converge with other newly biosynthesized, secreted proteins in a common secretory pathway. The whole system, however, is reminiscent of early work by Farquhar and others on the commonality of the secretory pathways for newly synthesized immunoglobulins and endocytosed membrane tracers in cultured myeloma cells.

C. Organelles of Endocytosis: Analysis by Inhibitors of Ligand Internalization

From the previous discussion, it should be clear that endocytosis and secretion are dynamic processes. One approach to defining different membrane "organelles" or functional semiautonomous domains in membrane trafficking has been the inhibitor approach. By modulating temperature and ionic environment or applying pharmacologic agents, attempts have been made to dissect these complex processes. One hope was to define "restriction points" or sites of important membrane transfers between functional compartments.

1. Temperature and K⁺ Depletion: Lesions Primarily at the Cell Surface

One of the simplest reversible manipulations of cells in culture which arrests endocytosis and secretion is the lowering of temperature. At 4°C, essentially all membrane flow is arrested (reviewed in Steinman *et al.,* 1983). Receptor binding occurs, and some (but not all) receptor–ligand complexes localize to coated pits. Most of the coated pits in such experiments appear open and in clear communication with the extracellular fluid environment (Willingham *et al.,* 1981). Similarly, endocytosis is arrested (Steinman *et al.,* 1983) if cells are depleted of their ATP content through inhibitors of the glycolytic and oxidative phosphorylation pathways. This suggests an ATP requirement in an early step in endocytosis, presumably in formation of endocytic vesicles.

Interestingly, depletion of the cellular potassium content over a period of several hours (Larkin *et al.*, 1983) also leads to an arrest in endocytosis. In this case, coated pits are depleted from the plasma membrane. The process can be accelerated with addition of nigericin, a potassium proton ionophore (Larkin *et al.*, 1985). K^+ depletion protects cells from toxicity of diphtheria toxin (Sandvig *et al.*, 1985; Moya *et al.*, 1985). The biochemical mechanism of these effects is unknown, but it has been hypothesized that K^+ is required inside the cell for clathrin attachment, polymerization, or some other coated pit assembly process. These data provide support for a functional role of clathrin in organizing the position and structure of the coated pit.

However, other experiments utilizing microinjected polyclonal anticlathrin light chain antibodies in endocytosing fibroblasts may cast doubt on a dynamic role for the clathrin coating. Wehland and co-workers observed that such microinjected antibodies heavily decorated the cytoplasmic surface of coated pits but had no effect on their function in endocytosis. Microinjected coated vesicles in antibody-injected cells formed cross-linked aggregates. No such aggregates were observed in cells microinjected with antibody alone. If coated vesicles existed, they were either too low in concentration or were functionally isolated from each other, preventing their cross-linking (Wehland *et al.*, 1981). More recently, high concentration of monoclonal anticlathrin heavy chain antibodies were shown to decrease fluid phase endocytosis by 50% (Doxsey *et al.*, 1987). Phenothiazines (such as chlorpromazine and trifluoperazine) prevent endocytosis by a calmodulin-independent mechanism. These drugs alter both cell surface binding (DiPaola *et al.*, 1984) and receptor aggregation (Peng, 1984; Dickson *et al.*, 1982a).

If the temperature of cells is limited to 18°C, endocytosis occurs but is aberrant. Ligand–receptor dissociation events do not occur, limiting ligand uptake in receptor recycling systems (Wolkoff *et al.*, 1984). A second step is also blocked at 18°C: transfer of endocytic vesicle contents to lysosomes (Marsh *et al.*, 1983).

2. Ionophores, Ions, and Cytoskeletal Disrupting Compounds: Inhibition of Intracellular Movement and Recycling

The monovalent ionophores are lipid-soluble molecules which collapse Na^+, K^+, and/or H^+ gradients across cellular membranes. Na^+/H^+- or H^+-specific ionophores block receptor-mediated endocytosis at several points. At micromolar concentrations, they have effects similar to experiments carried out at 18°C (and similar to treatment with primary amines such as NH_4Cl or chloroquine): ligand–receptor disassociation events are prevented, and Golgi membranes are vacuolated and swollen (Tartakoff,

receptors to the coated pit while vesicle formation takes place. The exact early stages in this endocytic process are unknown; coated pits on long intracytoplasmic stalks have been proposed to give rise to uncoated vesicles, the pit returning to the cell surface (Willingham et al., 1981; Willingham and Pastan, 1983). Alternatively, isolated short-lived cytoplasmic coated vesicles may exist (Anderson et al., 1977). Rothman and co-workers have purified a 70 k Da coated vesicle-uncoating ATPase which they propose has a role in the hypothetical vesicle-uncoating process. Interestingly, this enzyme is a previously described heat shock protein (reviewed in Rothman and Schmid, 1986).

   c. **Methods of Isolation.** Initial purification schemes for coated vesicles utilize bulk gradient and differential centrifugation technique (Pearse, 1976). Subsequently, advances have been made in column chromatography, agar electrophoresis, and immunoprecipitation to allow greater purity with less effort and time (Merisko et al., 1982; Rubenstein et al., 1981). Figures 2E–H and 3 show coated vesicles purified by the last of these techniques, immunoprecipitation. Anticlathrin antibodies are tightly adsorbed (via protein A) to formaldehyde-fixed *Staphylococcus aureus* bacteria. The adsorbed coated vesicles are washed free of contamining debris and then studied as is or after elution from the fixed bacteria (Hanover et al., 1984). Purifications such as these have allowed detailed biochemical characterization of coated vesicles.

   d. **Biochemical Characterization.** In addition to the predominant, 180 kDa component known as clathrin, isolated coated vesicles contain two 30–36 kDa species which are closely associated with clathrin, together forming a unit structure known as a triskelion (reviewed in Keen, 1985). There are also multiple proteins of 100–120 kDa (Pearse and Robinson, 1984; Robinson and Pearse, 1986) and 48–55 kDa. Triskelions may be polymerized into intact polyhedral, 3-dimensional closed baskets *in vitro*. Of particular interest is a 50 kDa polypeptide which is a substrate of a coated vesicle kinase; it is related structurally and immunologically to microtubule-associated polypeptides. The existence of such a component suggests that coated membranes are capable of interactions with microtubules or soluble tubulin. Coated vesicles also contain a phosphatidylinositol kinase whose function at present is unknown (Campbell et al., 1985). Coated vesicles contain calmodulin and numerous receptors and are high in cholesterol content (Steer et al., 1984; Hanover, et al., 1985a,b). Interestingly, the Golgi-associated coated pit subclass appears to have a much lower cholesterol content than the subclass involved in endocytosis (Helmy et al., 1986). A variety of studies have also shown that isolated

Fig. 3.   Electron micrograph of bovine brain coated vesicles (CV) adsorbed to *Staphylococcus aureus* anticlathrin (SAC) (Hanover et al., 1984).

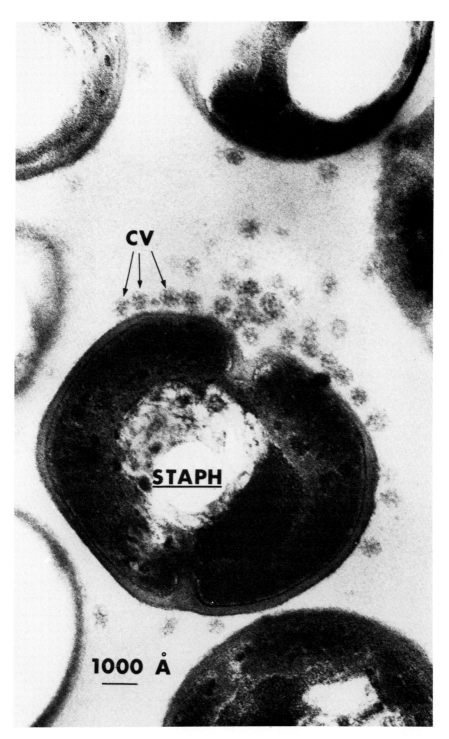

CV

STAPH

1000 Å

coated vesicles are capable of ATP-dependent acidification of their lumen (Van Dyke *et al.*, 1984; Xie *et al.*, 1984; Forgac and Cantley, 1984). Racker and co-workers (Stone *et al.*, 1984) have described an inhibitor for this proton pump (duramycin).

The isolated clathrin components from coated vesicles have long been known to be capable of reassembly into polyhedral cages under appropriate conditions *in vitro* (reviewed in Keen, 1985). In addition, membrane binding sites for clathrin have been described in coated vesicle membranes (Hanspal *et al.*, 1984). Finally, since both soluble and assembled pools of clathrin exist in the cell (Goud *et al.*, 1985) and since an uncoating ATPase has been isolated (Rothman and Schmid, 1986; Chappell *et al.*, 1986), a dynamic, clathrin assembly–disassembly process has been proposed as part of an endocytosis cycle. Schmid and co-workers (1984) have shown that the uncoating ATPase binds to the clathrin light chains and may thus target to coated vesicles for disassembly. It is not yet known if clathrin can undergo multiple cycles of assembly—disassembly *in vivo*. It is also not yet proven that clathrin coat disassembly is relevant to the endocytosis process (Wehland *et al.*, 1981).

   e. Kinetics of Transit of Ligands through Clathrin-Coated Membranes. The development of the immunoadsorption technique (Fig. 3) for rapid purification of coated vesicles has made possible rapid, kinetic studies of ligand-receptor entry and exit from this compartment. Figure 4 shows an example of this. In the absence of EGF little of the EGF receptor is clathrin associated, but following ligand addition, within as little as 2 min, the EGF–receptor complexes are concentrated in clathrin-coated pits (Hanover *et al.*, 1985a,b). These kinds of studies confirm previous morphological notions of the rapidity of early events in endocytosis and allow multiple label studies to compare endocytic pathways.

   Figure 5 shows a comparison of transferrin and EGF based on the kinetics of clathrin association. As shown in Fig. 4, EGF rapidly enters a clathrin compartment but transiently moves to another compartment before reentering another clathrin-coated membrane (the Golgi-associated class) prior to lysosomal degradation and release by 30 min. In contrast, transferrin associates only with the early clathrin compartment and is rapidly released from the cell in an undegraded form. Clearly, multiple signals must be involved (Hanover *et al.*, 1984) to allow receptor targeting to these two pathways. Section II,F addresses the possible nature of these signals.

2. Endosomes (Receptosomes): Sites of Ligand–Receptor Dissociation

   a. Morphologic Characteristics. Probably the first description of endocytic vesicles for receptor-mediated endocytosis was provided by Dales

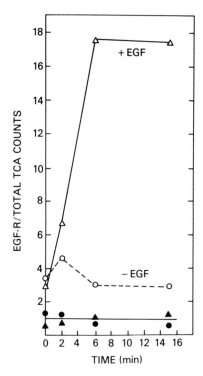

Fig. 4. EGF dependence of the appearance of EGF-R in clathrin-coated membranes. ▲, Accumulation of EGF-R in clathrin-coated membranes in the presence of EGF; ●, amount of receptor in clathrin-coated membranes in the absence of EGF; △ and ○, amount of receptor in the SAR pellets in the presence and absence, respectively, of EGF.

(1973) in studies with viropexis (entry of viruses into the cell). More complete descriptions have been recently presented in studies of liver and fibroblast model systems. Hubbard and co-workers, and Geuze and co-workers have described in the liver a network of anastamosing tubules and vesicles transporting asialoglycoproteins and their receptors from coated pits to the Golgi region (Helenius *et al.*, 1983; Geuze *et al.*, 1983). This compartment has been termed the endosome. Pastan, Willigham, and co-workers have described *in vivo* studies with cultured cells, fluorescently labeled ligands, and time-lapse photomicroscopy. These studies demonstrated a class of saltating, ligand-containing vesicles (termed receptosomes). The vesicles associated with microtubule tracks, sometimes fused with each other, and moved bidirectionally between the plasma membrane and the perinuclear Golgi region. Subsequent studies visualized heteromorphic endocytic vesicles and tubules, sometimes seen as partially coated (Fig. 6) (Willingham and Pastan, 1980; Herman and Al-

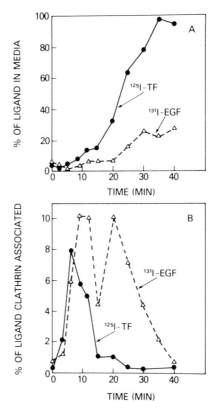

Fig. 5. (A) Kinetics of release of ^{125}I-labeled TF and ^{131}I-labeled EGF from KB cells. The labeled compounds were bound to cell monolayers at 4°C. After various periods of warming at 37°C, the amount of ligand released into the media was assessed. (B) Kinetics of association of ^{125}I-labeled TF and ^{131}I-labeled EGF with clathrin-coated membranes.

bertini, 1984; Valberg and Albertini, 1985). Double-label studies with various ligand pairs or antireceptor antibodies have suggested that no selectivity exists in coated pits and vesicles: a common pathway seems to exist for at least most polypeptides and their receptors (Willingham and Pastan, 1984; Pastan and Willingham, 1985; Dickson *et al.*, 1981b).

In endocytic vesicles, various ligands and their receptors begin to respond to the acid environment. Many viruses (Semliki Forest virus, VSV) begin to fuse with endocytic vesicle membranes, invading the cytosol. Some receptor–ligand complexes dissociate (asialoglycoprotein) while others (IgG, polymeric IgA, and transferrin) remain intact. Transferrin appears to rapidly release its iron in this compartment. A key property of endocytic vesicles is their acid pH, first demonstrated by Tycko and

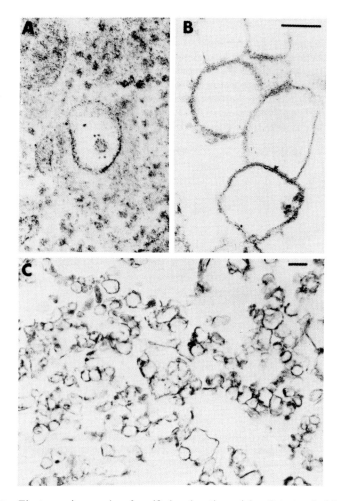

Fig. 6. Electron micrographs of purified endocytic vesicles. Intact cells labeled with ferritin–EGF (A) show a typical structure. (B and C) Appearance of membranes in the most highly purified fraction after isolation. (A and B) ×107,900 (bar = 0.1 μm), (C) ×24,000 (bar = 0.2 μm).

Maxfield (1982) using fluorescent, endocytosed probes. The volume of endocytic vesicles appears to be much larger than that of the coated pits. Decreasing pH and increasing volume may act in concert to promote these dissociation events (Maxfield, 1985; Mellman *et al.*, 1986; Murphy *et al.*, 1984; Seth *et al.*, 1984).

b. Methods of Isolation. A highly enriched preparation of endocytic vesicles has been characterized after a four-step purification procedure involving equilibrium density gradient centrifugation on colloidal silica and sucrose and column chromatography on Sephacryl S-1000 (Fig. 7). The morphologic appearance of this membrane fraction is shown in Fig. 6. While this preparation was homogeneous by rate velocity sedimentation and free flow electrophoresis, it was contaminated by microvilli (Dickson *et al.*, 1983b, 1985). More recent studies have improved on this purification (Evans 1985; Evans and Flint, 1985; Quintart *et al.*, 1984).

c. Biochemical Characterization. Figure 8 shows the polypeptide composition of purified endocytic vesicles. All of the major protein bands observed by Coomassie blue staining were also observed by [^{35}S]methionine labeling and autoradiography, indicating that the major proteins were cellular and not serum in origin. Figure 8 shows the major bands of M_r of 120,000, 92,000, 70,000–75,000, 68,000, 55,000, 43,000 (actin), and 36,000. Some of these proteins appear to be accessible to *in situ* iodination after endocytosis of lactoperoxidase–ligand conjugates (Watts, 1984). Actin was found to a variable extent in different preparations: its presence may be due to its very high concentration in contaminating microvilli (see Fig. 6). The polypeptide composition of the endocytic vesicle fraction was different from that of a plasma membrane-containing or lysosome-containing fraction. In addition, clathrin could not be detected in the endosome fraction (Dickson *et al.*, 1983).

These studies directly confirmed previous suggestions that the endocytic vesicle compartment is biochemically distinct from coated pits and lysosomes. Endocytic vesicles are a light membrane fraction and appear to be enriched in cholesterol and transferrin receptors, slightly enriched in galactosyltransferase (usually considered to be a marker for trans Golgi), and depleted of marker enzymes for plasma membranes and lysosomes. Studies by others have confirmed these conclusions in other cell types (Hornick *et al.*, 1985; Rodman *et al.*, 1986; Quintart *et al.*, 1984; Wall and Hubbard, 1985). Further studies by Galloway and others have shown that isolated endocytic vesicles can acidify *in vitro* by an ATP-dependent process (Geisow and Evans, 1984; Van Dyke *et al.*, 1985; Galloway *et al.*, 1983; Saermark *et al.*, 1985). Recently, fusion between endocytic vesicles has been obtained in a cell-free system (Braell, 1987).

3. Perinuclear Endosomes; Transreticular Golgi Elements

a. Morphological Characteristics. Using membrane-adsorbed, cationized ferritin "tracers" and horseradish peroxidase (HRP) as a fluid-phase marker, Farquhar was among the first to demonstrate, in suspensions of cultured pituitary cells, elements of the complete endocytic

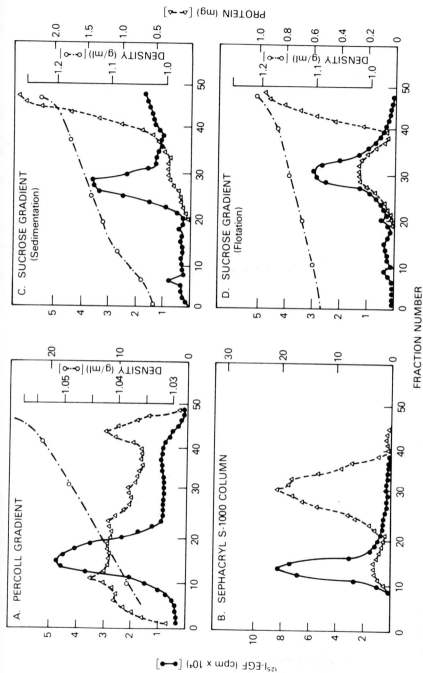

Fig. 7. Summary of purification steps in endosome fractionation. Endocytic vesicles of KB cells were marked with ^{125}I-labeled EGF and the cells homogenized. Membranes were fractionated and ^{125}I-labeled EGF and protein (Bradford method) determined. In each purification step the peak of ^{125}I-labeled EGF was pooled. In (B), V of the column was at fraction 12, whereas V_i was at fraction 31.

423

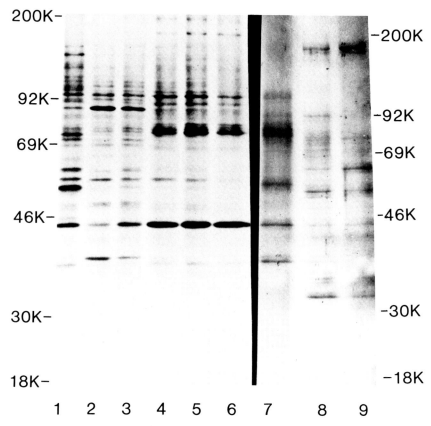

Fig. 8. Sodium dodecyl sulfate–gel electrophoresis of membrane fractions at various stages of purification and comparison with plasma membrane- and lysosome-containing fractions. Membranes were fractionated, and SDS–gel electrophoresis was carried out on a 5–17% gradient gel. Lane 1, homogenate; lane 2, low-speed supernatant; lane 3, Percoll pool; lane 4, Sephacryl S-1000 pool; lane 5, sucrose gradient 1 pool; lane 6, sucrose gradient 2 (flotation) pool (lanes 1–6, [^{35}S]methionine labeling and autoradiography); lane 7, sucrose gradient 2 pool; lane 8, plasma membrane fraction; lane 9, lysosome fraction (lanes 7–9, Coomassie blue staining).

pathway. Ferritin and HRP were visualized by electron microscopy in common endocytic vesicles, Golgi cisternae, condensing vacuoles, the trans-Golgi face, and in lysosomes. HRP selectively concentrated in lysosomes, while ferritin was selectively sequestered in Golgi-associated membranes. Subsequent studies in immunoglobulin-secreting myeloma cells demonstrated that the membrane-adsorbed cationized ferritin endocytic pathway and the secretion pathway for biosynthesized immunoglobulins converged at the trans Golgi; both proteins were secreted from cells by small exocytic vesicles. HRP and ferritin were taken up by common endocytic vesicles, but sorting occurred in the trans Golgi. Fluid was transferred to lysosomes, and bulk membrane was apparently recycled to the plasma membrane (Ottosen, 1981; Farquhar and Palade, 1981).

Many morphological and cell fractionation studies have focused on identification of polypeptide ligands and their receptors in Golgi-associated membrane systems of many cell types. A large number of cell fractionation and morphologic studies have localized α_2-macroglobulin, EGF, insulin, prolactin, nerve growth factor (NGF), and melanocyte-stimulating hormone (MSH) to the Golgi area of the cell (but not in Golgi stacks) and in Golgi-containing membrane fractions. In addition, receptors for insulin, asialoglycoproteins, and lysosomal hydrolases have also been detected in Golgi fraction membranes (Willingham and Pastan, 1982; Posner et al., 1985; Evans, 1985). A variety of ligands and receptors (e.g., NGF, insulin, and EGF) have also been localized in the cell nucleus or the nuclear envelope (Goldfine, 1981). While this may be a very minor destination for most ligands, it may be more important for others, such as NGF. It is possible that a slow but significant membrane exchange occurs between Golgi and nuclear membrane compartments.

Figure 9 shows the complete visualized pathway for transferrin–peroxidase. In Fig. 9E endosomes appear to be continuous with a host of tubules in the Golgi region, ultimately transferring contents to exocytic vesicles lying in association with microtubules. Finally, in Fig. 9L, dumbbell-shaped secretion vesicles are seen just under the plasma membrane (Hopkins, 1983; Willingham et al., 1984).

b. Role in Sorting and Lysosomal Delivery. Similar sorts of perinuclear, Golgi-associated sorting events have recently been observed by Geuze and co-workers in study of the asialoglycoprotein receptor (Geuze et al., 1983). While endocytosed asialoglycoprotein was localized in endocytic vesicles and lysosomes, only its receptor appeared to be highly concentrated in tubular extensions of some of these endocytic vesicles and in an extended tubular system. The receptor-rich, tubular network (termed CURL, for compartment of uncoupling of receptor and ligand) was proposed as an intermediate in receptor recycling. Since the endocytic vesicles are also known to eventually transfer some of their contents to lysosomes (Goldstein et al., 1979; Pastan and Willingham, 1981), it was proposed that the endocytic vesicles with receptor-rich tubular extensions represent evidence of a sorting compartment. It is possible that sorting of transferrin from EGF occurs through a similar system in KB cells. CURL appears very similar to a system of vesicles and tubules containing transferrin–HRP prior to plasma membrane delivery (Hopkins, 1983; Willingham et al., 1984). The exocytic vesicles mediating

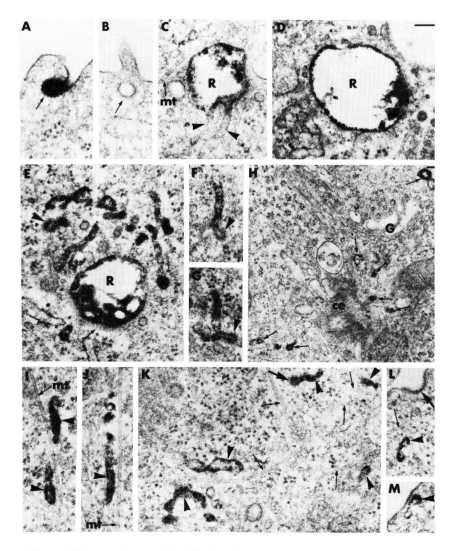

Fig. 9. Electron microscopic localization of TF–HRP during endocytosis and exocytosis in KB cells. Cells incubated in TF–HRP along with a large excess of unlabeled transferrin showed no significant labeling (B). At 1 min, TF–HRP could be found clustered in plasma membrane coated pits (A, arrow). At 5 min after warming, most of the TF–HRP was found in receptosomes (R) with their characteristic connections to adjacent tubular elements (C–E, arrowheads), some of which also showed small clathrin-coated regions characteristic of the coated pits of the Golgi (E, arrowhead). Between 5 and 12 min, images such as those shown in F–M were seen. In F and G the Golgi coated pits (arrowheads) show TF–HRP in the adjacent tubular elements of the transreticular Golgi but very little concentration in the pits themselves relative to the tubular elements, unlike that previously shown for EGF–

receptor recycling remain to be characterized biochemically. They do, however, appear to have a weakly acid pH by fluorescence quenching studies with fluorescein–transferrin (Yamashiro *et al.*, 1984).

4. Lysosomes

a. Morphological Characteristics. Lysosomes are electron-dense, perinuclear structures which receive both newly synthesized enzymes and endocytosed ligands and receptors. A full description of their properties is beyond the scope of this chapter. Many excellent reviews are available (deDuve, 1983; Kornfeld, 1986; Dingle, 1977; Neufeld and Ashwell, 1980). Interestingly, Abertini and co-workers (Herman and Albertini, 1983) have shown that both lysosomes and endocytic vesicles appear to move by saltatory motion toward the Golgi region. This suggests that a mechanism exists to coordinate accumulation of both kinds of endocytic organelle to this sorting–transfer region. Presumably, such a coordinated effort would require interaction with the cytoskeleton.

b. Enzyme Activities. Owing to the ease of purification of lysosomes by a variety of differential, equilibrium centrifugation and free-flow electrophoresis methods, their biochemical composition is well established (Dingle, 1977). Lysosomes are acidified by an ATP-dependent proton pump, which genetic analysis suggests is a different pump than that involved in endocytic vesicle acidification (Section II,E,2,a). A very large number of hydrolytic enzymes, predominantly of acid pH optimum and containing the receptor recognition mannose 6-phosphate linkage, have been identified. The reader is referred to one of several exhaustive reviews on the subject (Neufeld and Ashwell, 1980).

c. Kinetics of Transfer to Lysosomes. The exact sequence of events either in lysosome genesis or in transfer of biosynthesized components to this compartment is not completely clear. Proposed lysosomal precursor structures have been isolated by some (Rome *et al.*, 1979; Helmy *et al.*, 1986) and also suggested by Farquhar and others to be the Golgi-associated coated pits (Farquhar and Palade, 1981; Willingham and Pastan,

---

HRP. (H) Noticeable concentration of labeled tubular elements (arrows) occurred in the vicinity of centrioles (ce) in an interphase cell. (I and J) Unusual vesicular profiles that contain TF–HRP (arrowheads) were associated with microtubules (mt). (K) Various shapes of these elements (arrowheads) associated with microtubules (small arrows), which also appear near Golgi coated pits (large arrow), were observed. (L and M) What appear to be later stages in the process of exocytic delivery back to the cell surface were seen, in which small tubular profiles (arrowhead), often dumbbell shaped, were found near the plasma membrane, usually close to a microtubule (small arrow) (note the plasma membrane coated pit, large arrow, L). No significant label was detected in lysosomes during this initial stage of entry and exocytosis. (A–G and I–M) ×72,000, (H) ×42,000, bar = 0.1 μm.

1982; Hanover *et al.*, 1984). Figure 10 shows a cell fractionation density gradient analysis of the transfer of EGF to lysosomes. In contrast to transferrin, substantial amounts of EGF are associated with lysosomal enzyme-containing membranes within 15 min. Degradation was extensive by 40 min. Transfer of EGF and other ligands from endocytic vesicles proceeds as a smooth transition from light to dense lysosomal membranes. This suggests that while lysosomes can be isolated, and characterized as such, in reality a graded progression of maturation steps or membrane transfers may exist between endosomes and final mature lysosomes (Dickson *et al.*, 1983a).

5. Recycling Pathway: Morphological Considerations

Figure 9 shows the elements of the recycling pathway for a diferric transferrin–peroxidase conjugate (TF–HRP). From the perinuclear Golgi-associated membrane stage on, TF–HRP had a very different distribution

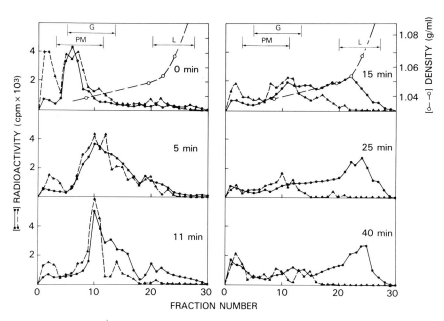

Fig. 10.   Kinetics of transfer of ^{125}I-labeled transferrin and ^{131}I-labeled EGF between intracellular compartments as determined by equilibrium density centrifugation. In a double-label study, cells were incubated with EGF or transferrin at 4°C. The radioactivity associated with each ligand is plotted. Density marker beads were used to determine gradient density. The solid line indicates EGF while the dashed line indicates transferrin. Abbreviations for the gradient regions are as follows: PM, plasma membrane (composite of the results obtained for Na^+,K^+-ATPase activity, 5'-nucleotidase activity, surface lactoperoxidase-catalyzed iodination, and 4°C ^{125}I-labeled EGF binding); G, Golgi (galactosyltransferase activity); L, lysosomes (β-hexosaminidase and β-galactosidase activities).

from that previously shown for EGF–HRP. EGF–HRP accumulated in lysosomes; in contrast, TF–HRP was found in tubular elements closely associated with microtubules (Fig. 9I–K). These structures were around 600 Å in width and up to 5,000 Å in length. The relationship of the structures to microtubules was particularly evident in interphase cells, in which the transferrin-containing structures were found near microtubules around the centriolar region (Fig. 9H). At later times (8–12 min), TF–HRP was found progressively closer to the plasma membrane. Eventually, some TF–HRP was found in small dumbbell-shaped profiles near the plasma membrane (Fig. 9L and M). Beyond 20 min after entry, very little TF–HRP could be detected in the cell. Only very small amounts of TF–HRP were ever detected in lysosomes (Willingham et al., 1984). The sorting of TF–HRP and EGF–HRP into different membrane systems is clearly evident but remains mysterious at the biochemical level.

F. Phosphorylation as a Receptor Sorting Signal

Various receptors are known to be phosphorylated on serine, threonine, and tyrosine residues. Some mitogenic hormones (EGF; insulin-like growth factor I, IGF-I; insulin; platelet-derived growth factor, PDGF) induce extremely rapid phosphorylations of tyrosine residues in the primary sequence of their own receptors (reviewed in Heldin and Westermark, 1984, and Sibley et al., 1987). A cascade of subsequent phosphorylations occurs, involving serine/threonine residues of receptors and resulting in attenuation of their tyrosine kinase activity. Recent evidence has strongly suggested that receptor phosphorylation may play a crucial role in inducing receptor entry into the cell and modulating receptor function (Hanover and Dickson, 1985b; Beguinot et al., 1985; Lin et al., 1986). The EGF receptor has now been cloned and sequenced; site-directed mutagenesis has been used to identify a critical phosphorylation-induced sorting signal. The EGF receptor, after treatment with EGF, enters first the plasma membrane clathrin compartment (termed coated pits) and is rapidly transferred to a nonclathrin-coated compartment (endosome, receptosome). The receptor and the ligand then enter a second, Golgi-associated and partially clathrin-coated membrane compartment just prior to lysosomal delivery and degradation (Hanover et al., 1984, 1985b).

4β-Phorbol 12-acetate 13-myristate (PMA), an activator of protein kinase C, induces EGF receptor movement into the first clathrin compartment prior to its recycling through a noncoated (endosome) compartment (see Figs. 5 and 11). It appears that PMA treatment is sufficient to induce association with clathrin-coated pits and cellular entry, but not subsequent lysosomal delivery. It is possible that lysosomal delivery could

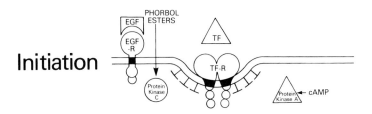

Initiation

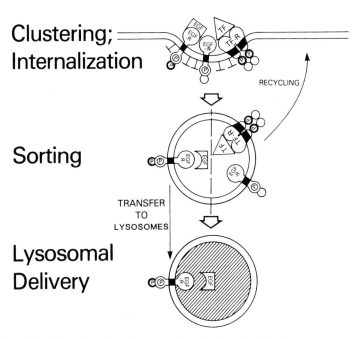

Clustering;
Internalization

Sorting

Lysosomal
Delivery

Fig. 11. Possible modulation of receptor function by phosphorylation. For illustrative purposes the epidermal growth factor (EGF) and transferrin (TF) receptors are shown (EGF-R and TF-R, respectively). Initiation: The signal for receptor phosphorylation may be ligand binding, elevation of cAMP levels (protein kinase A), or activation of protein kinase C by Ca^{2+}, diacylglycerol, or phorbol esters. Clustering and internalization: In response to phorbol esters, the EGF-R clusters in coated pits and is internalized (heterologous down-regulation). Phorbol esters can also trigger TF-R entry. Sorting: EGF-R internalized in response to phorbol esters and TF-R appear to be recycled rapidly back to the cell surface. The acid pH of the endocytic vesicle promotes EGF dissociation from its receptor and release of iron from TF. Lysosomal delivery: EGF-R proceeds to lysosomes when internalized in response to EGF (homologous down-regulation).

depend on other phosphorylation events, such as receptor (or other protein) phosphorylations on tyrosine. In contrast, the transferrin receptor (TF-R) is constitutively phosphorylated on serine/threonine on submembrane residues and appears to be "preclustered" in coated pits. TF-R entry into cells does not appear to depend on ligand occupancy (in nonhematopoietic cell types) (Watts, 1985). In addition, endocytosis of the transferrin receptor requires the cytoplasmic domain but not cytoplasmic serine phosphorylation (Rothenberger et al., 1987). The transferrin receptor is not a tyrosine kinase and is not delivered to lysosomes (or the Golgi-associated clathrin-coated compartment) during its life cycle (Fig. 11) (Beguinot et al., 1985). As previously mentioned (Section II,B), the LDL receptor is not a kinase and is not phosphorylated, yet it possesses a submembranous clathrin localization signal which is disrupted in the JD mutant (Davis et al., 1986).

In site-directed mutagenesis studies, Rosenfeld and co-workers (Lin et al., 1986) mutated threonine 659 of the EGF receptor (a submembranous residue mentioned in Section II,C) to alanine (a nonphosphorylatable amino acid). Phorbol esters were unable to induce internalization of the mutant receptor or attenuate its tyrosine kinase activity. In contrast, EGF still induced the mutant EGF receptor to undergo other phosphorylations and full endocytosis terminating in lysosomal degradation. This indicates that, while a single submembranous phosphorylation might be sufficient to trigger clathrin-coated pit association of EGF and transferrin receptors and their transient internalization prior to delivery to recycling membranes, other signals must exist. These other signals can act independently (in response to EGF) to promote several different membrane transfers terminating in lysosomes. Recent evidence also suggests multiple internalization signals for PDGF receptors (Olashaw et al., 1986) and β-adrenergic receptor (Nambi et al., 1985; Sibley et al., 1985; Sibley and Lefkowitz, 1985). Interestingly, dephosphorylation of the IGF-II and transferrin receptors in response to insulin treatment of cells leads to their pooling on the cell surface (Oka et al., 1984, 1985). In the LDL receptor system, acid dependent ligand and recycling has been shown to be mediated by the EGF-precursor homology region (Davis et al., 1987). Future studies utilizing both biochemical and genetic approaches will undoubtedly further identify these trafficking signals.

III. EXOCYTOSIS

The secretory pathway of eukaryotic cells provides the dual functions of protein export and membrane biogenesis (Palade, 1975). Although originally described in specialized cells (epithelial cells), the secretory pathway is now known to operate in all cell types, including yeast. In this section, the organelles participating in the secretory pathway will be examined. Emphasis will be placed on the genetically and biochemically defined distinctions between the various exocytic organelles. Recognition of these distinctions has allowed major advances in our understanding of the nature of membrane biogenesis and the biosynthesis and processing of membrane and secretory proteins.

The exocytic process was first visualized in cells which formed secretory vesicles (Farquhar and Palade, 1981; Jamieson and Palade, 1971; Palade, 1975). The process was suggested to proceed in an orderly and sequential fashion beginning with elements of the rough endoplasmic reticulum (RER), proceeding to the smooth endoplasmic reticulum (lacking ribosomes) then to the Golgi apparatus, and finally to secretory vesicles which fuse with the cell surface. The recognition that morphologically distinct compartments were involved led rapidly to attempts to fractionate and characterize these subcompartments. The methodological details of the many subcellular fractionation schemes are beyond the scope of this chapter and have been exhaustively reviewed previously (see Fleischer and Packer, 1974). The focus here is on recent model systems that have allowed examination of the organelles of the secretory pathway at a molecular level.

A. Assembly and Exocytosis of a Model Membrane Glycoprotein: VSV G Protein

Perhaps the single most important model system for the analysis of membrane and secretory biogenesis has been that of the glycoproteins of enveloped RNA viruses (Lenard and Compans, 1974; Lodish *et al.*, 1981). The most widely studied, the G protein of vesicular stomatitis virus (VSV), is the sole glycoprotein encoded by the virus. Since it is made in large quantities in virally infected cells, the biosynthesis and processing of the protein are conveniently followed. Mutants defective in the various stages of intracellular transport of the G protein have been selected and characterized (Flamand, 1970). In addition, it has been possible to obtain cDNA clones derived from the RNA genome of VSV (Gallione and Rose, 1983). As a result, a great deal is now known about the structure and exocytosis of the membrane glycoprotein of VSV. A description of the maturation of the G protein serves as a useful introduction to the process of export of a membrane component by the exocytic pathway.

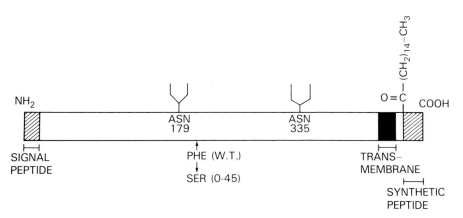

Fig. 12.   Primary structure of vesicular stomatitis virus G protein.

1. Domain Structure of VSV G Protein

As suggested in Fig. 12, the primary sequence of the G protein is now known in detail from analysis of the cDNA (Gallione and Rose, 1983, 1985). The molecule is composed of 511 amino acids. Two glycosylation sites (residues 179 and 335) have been identified; the oligosaccharide chains are of the complex type. In the various strains of VSV so far examined, there is considerable variation in the amino acid sequence. For example, the Orsay strain differs in 11 amino acids from the San Juan strain (Gallione and Rose, 1983, 1985). G protein can be divided into four primary structural domains: the signal sequence, the main body of the protein, the transmembrane domain, and the cytoplasmic tail. As Fig. 12 indicates, a signal peptide of 16 amino acids is present which is cleaved from the molecule cotranslationally after translocation into the lumen of the RER. A transmembrane domain is present containing hydrophobic amino acid residues which separates the body of the protein from the small carboxy-terminal cytoplasmic domain. A cysteine residue very close to the transmembrane domain carries palmitic acid in some strains of VSV.

A very useful mutant of VSV G (O-45) has been isolated which exhibits a temperature-sensitive phenotype with regard to viral maturation and G protein exocytosis (Flamond, 1970; Bergman and Singer, 1983). A recent examination of the primary sequence of this variant G protein suggests that a single amino acid replacement results in the temperature-sensitive phenotype. As indicated in Fig. 12, the mutant was shown to have a Ser instead of Phe at residue 204 near the first glycosylation site (Gallione and Rose, 1985).

2. Morphological Characterization of G Protein Exocytosis

Because the G protein of VSV strain O-45 is exported to the cell surface only at the permissive temperature (32°C), it is possible to follow the transport of the G protein in a very synchronous fashion. This has allowed the movement of G protein to the various organelles at the secretory pathway to be followed morphologically. Figure 13 shows the time course of exocytosis of the G protein of strain O-45 in BHK cells shifted to 32°C for the indicated time as determined by indirect immunofluorescence. Initially, the immunoreactive G protein is found only in the RER and nuclear envelope. After approximately 5 min, the protein appears in a perinuclear organelle which has been shown to correspond to the Golgi apparatus. Nearly all of the G protein can be found in the Golgi by 15 min at 32°C. The protein appears at the cell surface by 20 min. This time

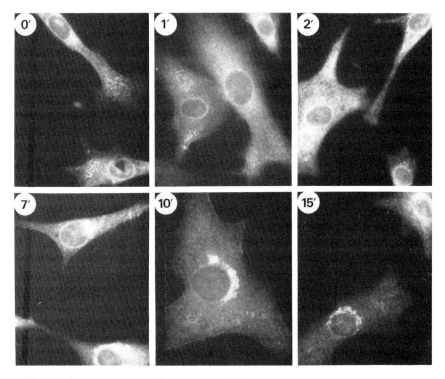

Fig. 13. Temperature-sensitive transport of VSV strain O-45 G protein as assessed by immunofluorescence. The time (in minutes) at the permissive temperature (32°C) is indicated on each panel.

course has also been followed in living cells by microinjection of fluorescently labeled antibody against a peptide corresponding to the carboxy terminus of VSV G protein (Fig. 13) (Arnheiter et al., 1984).

Temperature-dependent transport of VSV has also been examined by electron microscopy (Bergmann and Singer, 1983). The G protein accumulates in O-45-infected cells in elements of the RER and the nuclear envelope. After switching to the permissive temperature for 3 min, the G protein appears in saccules at one face of the Golgi apparatus. Interestingly, there was a suggestion that the G protein was concentrated at blebbed sites on the nuclear envelope at these early time points. An additional 3 min was required for appearance of the G protein in all stacks of the Golgi. At later times, the protein was localized in vesicles with diameters of 200–300 nm. These recent data support and extend previous morphological studies describing G protein maturation (Wehland et al., 1982; Bergman et al., 1981; Levine et al., 1981).

3. Posttranslational Modification of VSV G Protein

As is apparent from Fig. 12, numerous posttranslational modifications must occur during VSV G maturation. The earliest involves interaction of the amino-terminal signal peptide with the RER membrane and translocation of the body of the protein into the RER lumen, followed by cleavage of the signal peptide (Chatis and Morrison, 1979; Irving et al., 1979; Lingappa et al., 1978; Rose, 1977). After translocation into the lumen, the two glycosylation sites at Asn residues 179 and 335 are glycosylated by direct transfer to Asn from oligosaccharidylpyrophosphoryldolichol (Rothman and Lodish, 1977). The glycosylation enzyme is present within the lumen of the endoplasmic reticulum (Rothman and Lodish, 1977; Hanover et al., 1980; Glabe et al., 1980; Hanover and Lennarz, 1979, 1980, 1981, 1982). The glucose residues and one mannose are removed in the RER (Kornfeld and Kornfeld, 1985). Once in the Golgi apparatus, the carbohydrate is further processed to a $Man_5GlcNAc_2$—Asn structure. In the medial stack of the Golgi apparatus, GlcNAc transferases I and II and α-mannosidase 2 act to produce a GlcNAc-terminal structure which is further elongated in the trans stack with β-1-4 Gal and N-acetylneuraminic acid to produce the final glycosylated species inserted on the cell surface. The G protein may also be modified with palmitic acid at a cysteine residue in the cytoplasmic domain of the G protein (see Fig. 12).

4. Structural Features of the G Protein Required for Intracellular Transport

Recently, the technique of site-directed mutagenesis combined with mediated gene transfer has been applied to define the requirements for

VSV G transport. This has led to a wealth of information which is only briefly summarized here.

a. Domains of the Polypeptide Chain.

i. The Transmembrane Domain. As shown in Fig. 12, the G protein is composed of several domains. The membrane-spanning domain contains 20 uncharged and mostly hydrophobic amino acids. When this domain was shortened to 18, 16, or 14 residues by site-directed mutagenesis, the resultant G protein still maintained its proper transmembrane configuration and was transported to the cell surface. Shortening this domain to 12 or 8 residues, however, blocked intracellular transport at what appeared to be elements of the Golgi apparatus. A G protein lacking the transmembrane domain entirely was shown to accumulate in the RER and to be slowly secreted (Adams and Rose, 1985a). More subtle changes in the transmembrane domain also resulted in a G protein with altered properties. Replacement of an isoleucine in the center of the transmembrane region with an arginine resulted in a G protein that still spanned intracellular membranes but was not transported properly to the cell surface (Adams and Rose, 1985b).

ii. The Cytoplasmic Domain. A series of mutants have also been engineered which alter parts of the carboxy-terminal cytoplasmic domain (Rose and Bergman, 1983). Absence of this domain or addition of amino acids to the domain resulted in a G protein unable to exit the RER or Golgi. Partial deletions of this segment greatly retard the rate of intracellular transport.

iii. The Amino-Terminal "Body" of the G Protein. Several laboratories have constructed hybrid proteins consisting of coding regions of other membrane and secretory proteins and the carboxy-terminal anchor region of VSV G protein. One such hybrid was made between growth hormone (GH) and the carboxy terminus of the G protein (Rizzolo *et al.*, 1985). The GH–G hybrid could be detected only in transfected cells treated with chloroquine. A similar fusion protein was engineered by Guan and Rose (1984). This fusion protein was anchored in microsomal membranes in the expected orientation and was found to be transported to the Golgi apparatus. However, it was not transported to the cell surface (Guan and Rose, 1984). In a similar set of experiments, a chimeric influenza virus hemagglutinin containing either the amino terminus or the carboxy terminus of G protein was found to be defective in transport to the cell surface (McQueen *et al.*, 1984). Thus it appears that the body of the G protein may also be a determinant of whether the molecule moves normally through the secretory pathway. This point is perhaps most clearly made by the previously discussed O-45 mutant of VSV G protein which is blocked in movement to the Golgi at the permissive temperature. This defect appears to be attributable to a single amino acid change in a hydrophobic domain in the body of the protein (Gallione and Rose, 1985). Recently, oligomerization has been suggested to be an absolute requirement for proper transport of the G protein (Kreis and Lodish, 1986).

b. The Requirement for Posttranslational Modifications. The G protein is modified with two complex oligosaccharide chains and, in some strains, with covalently attached palmitate. Recently, the role of these moieties in the assembly and transport of the G protein has been examined. Inhibitors which interfere with the various steps in the assembly and processing of the carbohydrate chains of N-linked oligosaccharides (see Elbein, 1984) have been used to treat VSV-infected cells. Deoxynorjirimycin and castanospermine (inhibitors of glucosidases) affect virus growth in the San Juan strain but not the Orsay strain. The inhibition of VSV synthesis was thought to be due to inability of the G protein to adopt the correct conformation. This may be due to the fact that the excision of glucose and mannose can occur cotranslationally on nascent G protein (Atkinson and Lee 1984). Interestingly, the inhibition of virus production observed with these glucosidase inhibitors was temperature sensitive and did not result from inability of the G protein to be transported to the cell surface (Schlesinger *et al.*, 1984).

In another study, deoxynorjirimycin and deoxymannojirimycin (a mannosidase inhibitor) were shown to alter the carbohydrate chain of the G protein without altering its expression on the cell surface (Burke *et al.*, 1984). Using site-directed mutagenesis, Guan *et al.*, 1985 produced G proteins have two, one, or no glycosylation sites and examined the ability of these proteins to be transported to cell surface. Unglycosylated G protein was not efficiently transported to the cell surface from the Golgi, while the addition of one or two oligosaccharide chains allowed movement to the cell surface. In addition, a hybrid protein consisting of the amino terminus of growth hormone and the membrane-spanning and carboxy-terminal domains of G protein was engineered to contain two of the consensus signals for core glycosylation, Asn-X-Ser/Thr. This fusion protein was glycosylated and transported to the cell surface, whereas the unglycosylated hybrid was blocked at the Golgi apparatus (Guan *et al.*, 1985). Taken together, these data suggest that glycosylation can serve as a signal for transport to the cell surface. Recently a mutant VSV (Kotwal *et al.*, 1986) was found to synthesize an underglycosylated G protein. At 32°C the G protein was found to contain one oligosaccharide and was transported properly. At 39°C, the virus was unglycosylated and was not transported.

Site-directed mutagenesis has been also used to remove the cysteine to which palmitate is covalently attached in some forms of G protein. This resulted in a G protein which lacked palmitate but was normally transported to the cell surface (Rose *et al.*, 1984). The role of the covalently attached palmitate of G protein has not been identified.

5. Reconstitution of VSV G Intracellular Transport

To study the details of the intracellular transport of VSV G, recent efforts have been directed toward reconstituting the various transfer steps involved. Much of this work has focused on the movement of G protein between the compartments of the Golgi apparatus. Cell fusion studies (Rothman *et al.*, 1984a,b), using donor cells defective in oligosaccharide assembly and wild-type acceptor cells, demonstrated that inter-Golgi transfers could be relatively efficient; this suggested a vesicular transfer model. These studies have been extended by examining transport between Golgi compartments in a cell-free system (Fries and Rothman, 1980; Balch *et al.*, 1984a,b; Wattenberg and Rothman 1986). In this reconstituted system, movement is marked by the addition of GlcNAc on the G protein. Transport activity in this system has been shown to be dependent on ATP and a high-speed supernatant fraction from cells. Further analysis of this cytosolic fraction suggests that many soluble factors appear to be required. One of these factors is a 25 kDa protein which seems to be involved in vesicle processing leading to fusion. Reconstitution of transport of VSV G protein from the RER to the Golgi has been recently accomplished (Balch *et al.*, 1987). Such *in vitro* approaches should result eventually in a detailed understanding of the mechanism of membrane fusion as well as intracellular transport and targeting.

B. Pathways of Exocytosis: Constitutive and Regulated Secretion

Secretory cells can be separated into two classes: constitutive and secretagogue-dependent (regulated) secretory cells (see Kelly, 1985). This distinction was suggested by Tartakoff and Vassalli (1977) and points out that while some cells form dense secretory vesicles others do not. An additional distinction is that secretion of newly synthesized proteins from the constitutive pathway occurs shortly after synthesis whereas in specialized cells capable of regulated secretion these proteins are stored and released many hours after synthesis. For constitutive secretion to occur there is no requirement for an external stimulus. It is imagined that in these cells the rate of protein synthesis regulates the rate of secretion.

Regulated secretory cells accumulate secretory vesicles which fuse with the plasma membrane only when the level of a cytoplasmic component (usually calcium) is altered. Thus, although the basic pathway of protein export appears to be similar in organisms as diverse as yeast and humans this important distinction should be made.

C. Inhibitors of the Exocytic Pathway

1. Ionophores

The ionophore monensin disrupts traffic across the stack of the Golgi apparatus and terminates processing of the oligosaccharides on many proteins (Tartakoff and Vassalli 1977; Tartakoff 1983). Monesin was shown to cause the accumulation of Semliki Forest virus envelope glyco-proteins in the medial Golgi of BHK cells. This stack of the Golgi became dilated (Griffiths *et al.*, 1983; Quinn *et al.*, 1983). Monensin may also block galactosylation, sialylation, and sulfation of glycoproteins (Alonso-Caplen and Compans, 1983). It appears that the drug may therefore be useful for trapping proteins in the Golgi apparatus. The precise mechanism for this trapping is unknown.

2. Inhibitors which Alter Intracellular ATP Levels

Use of uncouplers of oxidative phosphorylation has established that movement from the RER to the Golgi and movement from the Golgi apparatus to the cell surface requires ATP production (Tartakoff and Vassalli, 1977). Of course, these are relatively nonspecific inhibitors of protein movement through the secretory pathway.

3. Inhibitors of Oligosaccharide Assembly and Processing

Inhibitors of glycoprotein assembly and processing have been widely used to examine whether the carbohydrate moiety of glycoproteins is required for export. These studies have been widely reviewed (Schwartz and Datema, 1982). One of the most widely used inhibitors is tunicamy-cin, a nucleoside analog which inhibits the transfer of *N*-acetylgluco-samine 1-phosphate from the sugar nucleotide to dolichyl phosphate (Tkacz and Lampen, 1975; Takatsuki *et al.*, 1975). It has become clear that certain proteins are correctly assembled and transported in the presence of tunicamycin (Siuta-Mangano *et al.*, 1982; Budarf and Herbert, 1982). Other proteins appear to be blocked in their ability to reach the cell surface after treatment with tunicamycin (Sidman, 1981; Reed *et al.*, 1981; Firestone, 1983).

Inhibitors of oligosaccharide processing have also been examined for their ability to alter intracellular transport. The mannosidase II inhibitor swainsonine was recently shown to accelerate intracellular transport in human hepatoma cells (Yeo *et al.*, 1985). Cells infected with VSV when treated with the glucosidase inhibitor deoxynorjirimycin become temperature sensitive for virus production (Schlesinger *et al.*, 1984). On the other hand, the mannosidase inhibitors deoxymannojirimycin and swainsonine were shown to have no effect on the biosynthesis and infectivity of Rous sarcoma virus (Bosch *et al.*, 1985). No generalizations are possible regarding the effects of these inhibitors on intracellular transport.

D. Organelles of Exocytosis: Genetic Analysis

1. Yeast Secretory Mutants

A significant advance in our knowledge of the organelles of the exocytic pathway has come from analysis of yeast mutants pleiotropically defective in protein export (Scheckman, 1982). Although these mutants are more fully discussed in other chapters in this volume, several points should be made here. The mutants were originally selected by screening a collection of temperature-sensitive growth mutants for colonies that failed to export active invertase and acid phosphatase at the nonpermissive temperature but were normal at the permissive temperature. These cells became dense at the nonpermissive temperature, and this finding suggested that additional mutants could be isolated using a high-density enrichment protocol. Adopting this strategy, Schekman and colleagues have defined 23 complementation groups required for the posttranslational events in the yeast secretory pathway. Electron microscopic study of these mutants revealed that they accumulated membrane-enclosed vesicles. When these structures were analyzed, they were found to correspond to endoplasmic reticulum (ER), Golgi-like structures, and 80–100 nm vesicles. The sequence of events was then ordered using haploid double mutant strains. The model which emerged suggested that secretory proteins enter the ER, where initial glycosylation takes place. At least nine gene products and energy were found to be required to transfer material to a Golgi-like structure where additional glycosylation occurs. Two additional gene products and energy were then required to package nearly fully glycosylated proteins into vesicles. Movement to the plasma membrane and fusion was shown to be dependent on energy and a minimum of ten gene products. Although the nature of the numerous gene products involved in the exocytic process is still under investigation, the yeast mutants have provided a means of dissecting this complex problem into its component parts.

2. Somatic Cell Mutants with Pleiotropic Exocytosis Defects

In the past, efforts to obtain mammalian cell mutants with pleiotropic defects in exocytosis have been thwarted by lack of appropriate selection and screening procedures. Recently, a new procedure was developed, using diphtheria toxin as a selection agent and nitrocellulose filters as a screening method for defective protein secretion (Nakano *et al.*, 1985). These procedures have allowed the isolation of a temperature-sensitive mutant clone of CHO cells which shows pleiotropic defects in protein export. This mutant was found to be temperature sensitive for growth. When the mutant cell line was infected with VSV, the G glycoprotein remained in an endoglycosidase H-resistant form, suggesting that the temperature-sensitive lesion may be in the RER or cis-Golgi apparatus. Isolation of many such mutants defective in the various steps in secretion will provide a powerful tool for determining the precise role of each of the organelles in the secretory pathway of mammalian cells.

3. Analysis of Viral Membrane Protein Mutants

A source of numerous proteins that follow the secretory pathway is found in the surface glycoproteins of enveloped viruses (Rindler *et al.*, 1985). These proteins have been particularly useful in examining this process since mutants with a temperature-sensitive phenotype can be produced.

For VSV the temperature-sensitive mutations have been classified in five complementation groups. The group V mutations involve the single transmembrane glycoprotein of the virus (G protein). One such mutant, O-45, is blocked in its transport from the RER (discussed in Section III,A). Temperature-sensitive transport mutants have also been characterized for other viruses. For two of the Sindbis virus mutants (*ts10* and *ts23*), oligosaccharide assembly and fatty acylation are interrupted at the nonpermissive temperature (Schmidt and Schlesinger, 1979; Kääriäinen *et al.*, 1980). Transport mutants have been identified for other viruses including the Friend and Rauscher erythroleukemia viruses (Fitting *et al.*, 1981; Kabat *et al.*, 1980) and Rous sarcoma virus (Wills *et al.*, 1984; also see Hunter, Chapter 3, this volume).

E. Organelles of Exocytosis: Biochemical Characterization

The previous sections have suggested that compartmentalization of enzyme activities might account for the sequential nature of the posttranslational modifications which occur during exocytosis. In this section the various intracellular compartments are discussed in terms of their morphology, enzyme activities, and the nature of the molecules which reside there.

1. The Rough Endoplasmic Reticulum and Nuclear Envelope

When thin sections of cells are examined by electron microscopy, the RER and nuclear envelope appear to be studded with numerous membrane-bound ribosomes. It is this morphological feature which first led investigators to suggest that these organelles were the initial sites of synthesis of membrane and secretory proteins. The elements of the RER appear throughout the cytoplasm and appear to be in continuity with the outer membrane of the nuclear envelope.

Identification of the proteins which are normal components of the RER has only recently become the subject of investigation (see Green and Mazzarella, Chapter 6, this volume). The ribophorins are two transmembrane glycoproteins which are thought to be involved in the binding of ribosomes to the RER membrane. These proteins have been studied in detail, and their biosynthetic pathways have been examined (Rosenfeld *et al.*, 1984). Direct analysis of the oligosaccharides of ribophorin I showed that they do not contain the terminal sugars characteristic of complex oligosaccharides and therefore may not require passage to the Golgi apparatus.

Another component of the RER which is involved in the early stages of synthesis of secretory proteins is the signal recognition particle (SRP) receptor. This protein is an integral membrane protein which, in conjunction with the signal recognition particle itself, ensures the correct targeting of nascent secretory proteins to this membrane system. The complete primary structure of the SRP receptor has been deduced from the cDNA. The amino-terminal region of this protein is anchored in the membrane of the RER (Lauffer *et al.*, 1985). In addition to being the location of the proteins involved in protein translocation across the membrane, the RER is the site of attachment of oligosaccharides from oligosaccharide–lipid precursors to the sequence Asn-X-Ser/Thr in the polypeptide chain (see Lennarz, 1980; Hanover and Lennarz 1980, 1981; Kornfeld and Kornfeld, 1985). The oligosaccharide transferase has been identified using an active site-directed photoaffinity probe (Welply *et al.*, 1986). The molecular weight of the protein was found to be approximately 60,000.

Numerous other enzymes are present in the RER, including many of the enzymes of detoxication (Jakoby, 1982) and those involved in the assembly of the oligosaccharide chain of N-linked glycoproteins (Hanover and Lennarz, 1981; Kornfeld and Kornfeld, 1985) as well as removal of glucose and at least one mannose residue from the oligosaccharide. The nuclear envelope has been shown to share many of the enzyme activities present in the RER (Franke *et al.*, 1981). A recent study demonstrated the involvement of the nuclear envelope in the synthesis, processing, and transport of membrane glycoproteins (Puddington *et al.,* 1985). It has become clear, however, that the nuclear envelope must be considered more than simply an extension of the ER. The polypeptide composition of the nuclear envelope is distinct; the nuclear lamina are well-characterized proteins which appear to specifically associate with the inner membrane of the nuclear envelope. In addition, components of the nuclear porecomplex have been identified (Gerace *et al.,* 1982; Park *et al.,* 1987; Snow *et al.,* 1987; Hanover *et al.,* 1987, Davis and Blobel, 1986). The nuclear envelope has been suggested to be a critical component in the maturation of the *env* glycoprotein of the Rous sarcoma virus (Wills *et al.,* 1984). Another finding of great interest is that the elements of the smooth ER which contain the enzyme hydroxymethylglutaryl-CoA reductase have been shown to emerge from the nuclear envelope (Pathak *et al.,* 1986).

The elements which shuttle proteins between the RER–nuclear envelope compartment and the Golgi apparatus have recently come under investigation. Several reports had suggested that such transfer might occur via clathrin–coated vesicles (Rothman and Fine, 1980). However, recently it has been possible to construct yeast strains which have the gene for the clathrin heavy chain deleted and thus cannot assemble coated vesicles. In such a strain secretion appeared normal (Payne and Schekman, 1985). These data, coupled with previous morphological data regarding the exocytosis of viral glycoproteins (Wehland *et al.,* 1982; Saraste and Hedman, 1983), suggest that coated vesicles are not involved in this transfer step. Several recent studies have shed some light on the possible nature of these transport vesicles. In studying the intracellular transport of a hybrid protein made up of the carboxy-terminal domain of influenza virus hemagglutinin and growth hormone, Rizzolo *et al.* (1985) found the hybrid accumulated in perinuclear region in modified ER cisternae. These cisternae were found very close to the Golgi apparatus and were suggested to represent overdeveloped transitional ER element natural way stations between the ER and Golgi apparatus. In another study, using the O-45 temperature-sensitive strain of VSV (see Section III,A), blebbed regions of the nuclear envelope were suggested to serve as possible transition elements (Bergman and Singer, 1983). Characterization of the vesicles mediating RER–Golgi transport may soon be possible since methods are becoming available to arrest or slow protein export at this stage. In addition, the biochemical requirements for ER–Golgi transport may be defined by *in vitro* assays such as those used to analyze the transfer of yeast invertase from the RER to the Golgi (Haselbeck and Scheckman, 1986).

2. The Golgi Apparatus

The Golgi is placed strategically at the crossroads of the pathways of exocytosis and endocytosis. The membrane system itself consists of 4–10 cisternae with associated vesicles, some of which are coated. Membrane thickness increases from the proximal to the distal cisternae and these cisternae sometimes display a clear polarity. The Golgi stack is functionally as well as morphologically polarized. Proteins enter at the entry face (cis stack), proceed to the medial, and leave at the trans stack. Numerous reviews have appeared which describe the compartmental organization of the Golgi stack (Dunphy and Rothman, 1985; Kornfeld and Kornfeld, 1985). These data are derived from studies involving subfractionation of Golgi membranes, *in situ* enzyme localization, and the distribution of oligosaccharide products within the stacks.

Although a complete treatment of this compartmentalization is not possible here, several points should be made concerning the distribution of the enzymes involved in oligosaccharide assembly. The cis stack is suggested to serve as the site of glucose removal from oligosaccharide precursors, initial mannose removal, and phosphorylation of lysosomal enzymes. The medial stack is presumed to be the site of GlcNAc addition and removal of mannose residues. The trans stack is the site of addition of galactose and sialic acids (Kornfeld and Kornfeld, 1985; Roth, 1983). In addition, O-glycosidically linked oligosaccharides having a core *N*-acetylgalactosamine appear to be entirely assembled in the Golgi complex (Hanover and Lennarz, 1980, 1981, 1982; Elhammer and Kornfeld, 1984). Although the precise compartmentalization of O-linked oligosaccharide assembly is not known in detail, it is likely that the elongation of GalNAc by galactose and sialic acid (and peripheral GalNAc) occur in different cisternae.

The observed compartmentalization of the Golgi suggests that proteins must traverse membrane barriers to travel from compartment to compartment. Recent evidence suggests that this intracisternal movement is mediated by a series of transport vesicles (see Rothman and Fine, 1980; and Dunphy and Rothman, 1985). These vesicles may have a coat which is different from clathrin. Much of this work stems from studies on the VSV G protein and is summarized in Section III,A.

3. Secretory Vesicles

Once they have left the Golgi complex, membrane and secretory proteins proceed to the cell surface. In many cell types in which regulated secretion occurs, secretory vesicles are present in the cytoplasm that appear to have an electron-opaque core in electron micrographs. Secretory proteins appear to be selectively targeted to these vesicles (Moore and Kelly, 1985). Such vesicles have been partially purified from the rat parotid gland (Cameron and Castle, 1984) and found to have a unique polypeptide composition. The internal pH of the vesicles was found to be 6.8 (Arvan *et al.*, 1984). Attempts have recently been made to define antigens specific for secretory vesicles. A transmembrane glycoprotein of 100 kDa has been shown to be present in all endocrine and neural cell lines studied (Buckley and Kelly, 1985). Sulfated proteoglycans and sulfated glycoproteins have been demonstrated to be a common feature of zymogen granules, chromaffin granules, and pituitary granules (Burgess and Kelly, 1984). Interestingly, perturbation of the level of sulfated proteoglycan chains did not alter either sorting or secretion. The mechanism of secretory vesicle fusion with the plasma membrane is not known although a metalloendoprotease has been suggested to be required for exocytosis in mast cells and adrenal chromaffin cells (Mundy and Strittmatter, 1985).

One system which may provide an approach toward understanding the function of secretory granules is trichocyst discharge in *Paramecium*. Constitutive discharge mutants have been described (Cohen and Beisson, 1980; LeFort-Tran *et al.*, 1981). Approximately 1100–1300 secretory trichocysts are normally docked to the cell membrane and can be discharged within seconds, using a secretogogue. It has been demonstrated that synchronous exocytosis in *Paramecium* cells involves a reversible dephosphorylation of a 65 kDa phosphoprotein. This dephosphorylation occurs only in exocytosis-competent strains and not in mutant strains defective in vesicle fusion, docking, or transport to the surface. The suggestion has been made that this dephosphorylation event may trigger the exocytic event (Zieseniss and Plattner, 1985). The combination of biochemical and genetic analysis may provide a detailed understanding of secretory vesicle structure and function.

F. Relationship between Exocytosis and Endocytosis

After surveying the organelles involved in endocytosis and exocytosis, it is clear that some relationship may exist between the two pathways. Mutants in CHO cells which are defective in endocytosis have been shown to have defects in Golgi function (Robbins *et al.*, 1984). In addition, yeast secretion mutants defective in exocytosis were shown to be defective in uptake of fluid-phase markers (Riezman, 1985). These mutants should provide a means for determining where the pathways of exocytosis and endocytosis intersect.

One question which has recently been addressed is the question of whether the pathway of receptor recycling is part of the exocytic pathway. Using the O-45 mutant of VSV, the exit of newly synthesized membrane proteins from the trans cisternae of the golgi to the cell surface was examined. Double-label experiments were performed in which G protein localization was combined with staining for the fluid-phase marker HRP taken up by endocytosis (Griffiths *et al.*, 1985). The results suggest that the G protein did not colocalize with internalized HRP. Similar results were obtained when G protein exit was compared to transferrin recycling; no colocalization was observed (Hedman *et al.*, 1986). However, in the perfused rat liver, exocytic vesicles have been shown to contain both newly synthesized cholinesterase and internalized transferrin (Fishman, 1984).

IV. SUMMARY AND FUTURE PROSPECTS

The focus of this chapter is the interrelationship between the organelles of the exocytic and endocytic pathways. The combination of site-directed mutagenesis and the ability to introduce foreign genes into cultured cells should allow a detailed analysis of the sorting signals which operate to determine the ultimate intracellular location of proteins. In addition, the analysis of mutants which accumulate secretory and endocytic organelles should aid in the biochemical characterization of the membrane-limited compartments of the eukaryotic cell.

ACKNOWLEDGMENTS

We would like to thank all of the many scientists who forwarded reprints and preprints of their published work for inclusion in this chapter. The assistance of Ms. Dee Goodrich in the preparation of the manuscript is also gratefully acknowledged.

REFERENCES

Abrahamson, D. R., and Rodenwald, R. (1981). Evidence for the sorting of endocytic vesicle contents during the receptor-mediated transport of IgG across the newborn rat intestine. *J. Cell Biol.* **91,** 270–280.
Adams, G. A., and Rose, J. K. (1985a). Structural requirements of a membrane-spanning domain for protein anchoring and cell surface transport. *Cell* **41,** 1007–1015.
Adams, G. A., and Rose, J. K. (1985b). Incorporation of a charged amino acid into the membrane-spanning domain blocks cell surface transport but not membrane anchoring of a viral glycoprotein. *Mol. Cell. Biol.* **5,** 1442–1448.
Alonso-Caplen, F. V., and Compans, R. W. (1983). Modulation of glycosylation and transport of viral membrane glycoproteins by a sodium ionophore. *J. Cell Biol.* **97,** 659–668.
Anderson, R. G., Brown, M. S., and Goldstein, J. L. (1977). Role of the coated endocytic vesicle in the uptake of receptor-bound low density lipoprotein in human fibroblasts. *Cell* **10,** 351–364.

Arnheiter, H., Dubois-Dalcq, M., and Lazzarini, R. A. (1984). Direct visualization of protein transport and processing in the living cell by microinjection of specific antibodies. *Cell* **39**, 99–109.

Arvan, P., Rudnick, G., and Castle, J. D. (1984). Osmotic properties and internal pH of isolated rat parotid secretory granules. *J. Biol. Chem.* **259**, 13567–13572.

Ascoli, M. (1984). Lysosomal accumulation of the hormone–receptor complex during receptor-mediated endocytosis of human choriogonadotropin. *J. Cell Biol.* **99**, 1242–1250.

Ashwell, G., and Morell, A. G. (1974). The role of surface carbohydrate in the hepatic recognition and transport of circulating glycoproteins. *Adv. Enzymol.* **41**, 99–128.

Atkinson, P. H., and Lee, J. T. (1984). Cotranslational excision of α-glucose and α-mannose in nascent vesicular stomatitis virus G protein. *J. Cell Biol.* **98**, 2245–2249.

Baenziger, J. A., and Fiete, D. (1982). Recycling of the hepatocyte asialoglycoprotein receptor does not require delivery of ligand to lysosomes. *J. Biol. Chem.* **257**, 6007–6009.

Balch, W. E., Dunphy, W. G., Baell, W. D., and Rothman, J. E. (1984a). Reconstitution of the transport of protein between successive compartments of the Golgi measured by the coupled incorporation of *N*-acetylglucosamine. *Cell* **39**, 511–525.

Balch, W. E., Glick, B. S., and Rothman, J. E. (1984b). Sequential intermediates in the pathway of intercompartmental transport in a cell-free system. *Cell* **39**, 525–536.

Balch, W. E., Wagner, K. R., and Keller, D. S. (1987). Reconstitution of transport of vesicular stomatitis G protein from the endoplasmic reticulum to the Golgi apparatus using a cell-free system. *J. Cell Biol.* **104**, 749–760.

Basu, S. K., Goldstein, J. L., Anderson, R. G. W., and Brown, M. S. (1981). Monensin interrupts the recycling of low density lipoprotein receptors in human fibroblasts. *Cell* **24**, 493–502.

Beguinot, L., Lyall, R. M., Willingham, M. C., and Pastan, I. H. (1984). Down-regulating of the EGF receptor in KB cells is due to receptor internalization and subsequent degradation in lysosomes. *Proc. Natl. Acad. Sci. U.S.A.* **81**, 2384–2388.

Beguinot, L., Hanover, J. A., Willingham, M. C., and Pastan, I. (1985). Phorbol esters induce transient internalization of unoccupied epidermal growth factor receptors without receptor degradation. *Proc. Natl. Acad. Sci. U.S.A.* **82**, 2774–2778.

Bergmann, J. E., and Singer, S. J. (1983). Immunoelectron microscopic studies of the intracellular transport of the membrane glycoprotein (G) of vesicular stomatitis virus in infected Chinese hamster ovary cells. *J. Cell Biol.* **97**, 1777–1787.

Bergmann, J. E., Tokuyasu, K. T., and Singer, S. J. (1981). Passage of an integral membrane protein, the VSV glycoprotein, through the Golgi apparatus en route to the plasma membrane. *Proc. Natl. Acad. Sci. U.S.A.* **78**, 1746–1750.

Bessis, M. (1963). Cytologic aspects of hemoglobin production. In "The Harvey Lecture Series," Vol. 58, pp. 125–156. Academic Press, New York.

Bosch, J. V., Tlusty, A., McDowell, W., Legler, G., and Schwarz, R. T. (1985). The mannosidase inhibitors 1-deoxymannojirimycin and swainsonine have no effect on the biosynthesis and infectivity of Rous sarcoma virus. *Virology* **143**, 342–346.

Braell, W. A. (1987). Fusion between endocytic vesicles in a cell-free system. *Proc. Natl. Acad. Sci. U.S.A.* **84**, 1137–1141.

Brown, M. S., and Goldstein, J. L. (1986). A receptor-mediated pathway for cholesterol homostasis. *Science* **232**, 34–47.

Brown, M. S., Anderson, R. G. W., and Goldstein, J. L. (1983). Recycling receptors: The round trip itinerary of migrant membrane proteins. *Cell* **32**, 663–667.

Brown, W. J., and Farquhar, M. G. (1984). The mannose 6-phosphate receptor for lysosomal enzyme is concentrated in cis Golgi cisternae. *Cell* **36**, 295–307.

Buckley, K., and Kelly, R. B. (1985). Identification of a transmembrane glycoprotein specific for secretory vesicles of neural and endocrine cells. *J. Cell Biol.* **100**, 1284–1294.

Budarf, M. L., and Herbert, E. (1982). Effect of tunicamycin on the synthesis processing and secretion of proopipmelanocortin peptides in mouse pituitary cells. *J. Biol. Chem.* **257**, 10128–10135.

Burgess, T. L., and Kelly, R. B. (1984). Sorting and secretion of adrenocorticotropin in a pituitary tumor cell line after perturbation of the level of a secretory granule-specific proteoglycan. *J. Cell Biol.* **99**, 2223–2230.

Burke, B., Matlin, K., Bause, E., Legler, G., Peyrieras, N., and Pleogh, H. (1984). Inhibition of N-linked oligosaccharide trimming does not interfere with surface expression of certain integral membrane proteins. *EMBO J.* **3**, 551–556.

Cameron, R. S., and Castle, J. D. (1984). Isolation and composition analysis of secretory granules and their membrane subtraction from parotid glad. *J. Membr. Biol.* **79**, 127–144.

Campbell, C. R., Fishman, J. F., and Fine, R. E. (1985). Coated vesicles contain a phosphatidylinositol kinase. *J. Biol. Chem.* **260**, 10948–10951.

Carpenter, G., and Cohen, S. (1979). Epidermal growth factor. *Annu. Rev. Biochem.* **48**, 193–216.

Chappell, T. G., Welch, W. J., Schlossman, D. M., Palter, K. B., Schlesinger, M. J., and Rothman, J. E. (1986). Uncoating ATPase is a member of the 70 kilodalton family of stress proteins. *Cell* **45**, 3–13.

Chatis, P. A., and Morrison, T. G. (1979). Vascular stomatitis virus glycoprotein is anchored to intracellular membranes near its carboxy end and is proteolytically cleaved at its amino terminus. *J. Virol.* **29**, 957–963.

Chvatchko, Y., Howard, I., and Riezman, H. (1986). Two yeast mutants defective in endocytosis are defective in pheromone response. *Cell* **46**, 355–364.

Cohen, J., and Beisson, J. (1980). Genetic analysis of the relationships between the cell surface and the nuclei in *Paramecium tetrahymena*. *Genetics* **95**, 797–818.

Cohen, S., Fava, R. A., and Sawyer, S. T. (1982). Purification and characterization of epidermal growth factor receptor/protein kinase from normal mouse lines. *Proc. Natl. Acad. Sci. U.S.A.* **79**, 6237–6241.

Courtoy, P. J., Quintart, J., Limet, J. N., deRoe, C., and Baudhuin, P. (1985). Polymeric IgA and galactose-specific pathways in hepatocytes: Evidence for intracellular ligand sorting. In "Endocytosis" (I. H. Pastan, and M. C. Willingham, eds.), pp. 163–194. Plenum, New York.

Dales, S. (1973). Early events in cell animal virus interactions. *Bacteriol. Rev.* **37**, 103–135.

Davis, C. C., Lehrman, M. A., Russell, D. W., Anderson, R. G. W., Brown, M. S., and Goldstein, J. L. (1986). The J. D. mutation in familial hypercholesterolemia: Amino acid substitution in cytoplasmic domain impedes internalization of LDL receptor. *Cell* **43**, 15–24.

Davis, C. C., Goldstein, J. L., Saudhol, I. C., Anderson, R. G., Russell, D. W., and Brown, M. S. (1987). Acid-dependent ligand dissociation and recycling of LDL receptor mediated by growth factor nomology region. *Nature* **326**, 760–765.

Davis, I., and Blobel, G. (1986). Identification and characterization of a nuclear pore complex protein. *Cell* **45**, 699–709.

deDuve, C. (1983). Lysosomes revisited. *Eur. J. Biochem.* **137**, 391–397.

Debanne, M. T., Evans, W. H., Flint, N., and Regoeczi, E. (1982). Receptor-rich intracellular membrane vesicles transporting asialotransferrin and insulin in liver. *Nature (London)* **298**, 398–400.

Dickson, R. B. (1985). Endocytosis of polypeptides and their receptors. *Trends Pharmacol. Sci.* **6**, 164–167.

Dickson, R. B., Willingham, M. C., and Pastan, I. (1981a). Binding and internalization of ^{125}I-α_2-macroglobulin by cultured fibroblasts. *J. Biol. Chem.* **256**, 3454–3459.

Dickson, R. B., Willingham, M. C., and Pastan, I. (1981b). α_2-Macro globulin absorbed to colloidal gold: A new probe in the study of receptor-mediated endocytosis. *J. Cell Biol.* **89**, 29–34.

Dickson, R. B., Schlegel, R., Willingham, M. C., and Pastan, I. H. (1982a). Reversible and irreversible inhibitors of clustering of α_2M in clathrin-coated pits on the surface of fibroblasts. *Exp. Cell. Res.* **140**, 215–225.

Dickson, R. B., Schlegel, R., Willingham, M. C., and Pastan, I. H. (1982b). Binding and internalization of α_2-macroglobulin by cultured fibroblasts: Effects of monovalent ionophores. *Exp. Cell. Res.* **142**, 124–140.

Dickson, R. B., Schlegel, R., Willingham, M. C., and Pastan, I. H. (1982c). Involvement of Na$^+$ and HCO$_3^-$ in receptor-mediated endocytosis of α_2-macroglobulin, epidermal growth factor, and vesicular stomatitis virus. *J. Cell. Physiol.* **113**, 353–358.

Dickson, R. B., Hanover, J. A., Willingham, M. C., and Pastan, I. (1983a). Prelysosomal divergence of EGF and transferrin during receptor-mediated endocytosis. *Biochemistry* **22**, 5667–5674.

Dickson, R. B., Beguinot, L., Hanover, J. A., Richert, N. D., Willingham, M C., and Pastan, I. (1983b). Isolation and characterization of a highly enriched preparation of receptosomes (endosomes) from a human cell line. *Proc. Natl. Acad. Sci. U.S.A.* **80**, 5335–5339.

Dickson, R. B., Hanover, J. A., Willingham, M. C., and Pastan, I. (1985). Isolation of receptosomes (endosomes) from human KB cells. *Methods Enzymol.* **109**, 257–271.

Dingle, J. T., ed. (1977). "Lysosomes." North Holand Publ. New York.

DiPaola, M., Keith, C. H., Feldman, D., Tycko, B., and Maxfield, F. R. (1984). Loss of α_2-macroglobulin and epidermal growth factor surface binding induced by phenothiazines and naphthalene sulfonamides. *J. Cell Physiol.* **118**, 193–202.

Doxsey, S. J., Brodsky, F. M., Blank, G. S., and Helenius, A. (1987). Inhibition of endocytosis by anticlathrin antibodies. *Cell* **50**, 453–463.

Dunphy, W. G., and Rothman, J. E. (1985). Compartmental organization of the Golgi stack. *Cell* **42**, 13–21.

Elbein, A. D. (1984). Inhibitors of the biosynthesis and processing of N-linked oligosaccharides. *CRC Crit. Rev. Biochem.* **16**, 21–49.

Elhammer, A., and Kornfeld, S. (1984). Two enzymes involved in the synthesis of O-linked oligosaccharides are localized on membranes of different densities in mouse lymphoma BW5147 cells. *J. Cell Biol.* **99**, 327–331.

Evans, W. H. (1985). Preparation of low-density "endosome" and "endosome"-depleted Golgi fractions from rat liver. *Methods Enzymol.* **109**, 246–257.

Evans, W. H., and Flint, N. (1985). Subfractionation of hepatic endosomes in Nycodenz gradients and by free-flow electrophoresis. Separation of ligand-transporting and receptor-enriched membranes. *Biochem. J.* **232**, 25–32.

Farquhar, M. G. (1985). Progress in unraveling pathways of Golgi traffic. *Annu. Rev. Cell Biol.* **1**, 447–488.

Farquhar, M. G., and Palade, G. B. (1981). The Golgi apparatus (complex)—(1954–1981)—from artifact to center stage. *J. Cell Biol.* **91**, 775–1035.

Fawcett, D. W. (1964). Local specialization of the plasmalemma in micropinocytosis vesicles of erythroblasts. *Anat. Rec.* **148**, 370 (Abstr.).

Fitting, T., Ruta, M., and Kabat, D. (1981). Mutant cells that abnormally process plasma membrane glycoproteins encoded by murine leukemia virus. *Cell* **24**, 847–858.

Flamand, A. (1970). Etude génétique du virus do la specie vesiculaire: Classement de mutants thermosensible specie engroupe de compleméntetion. *J. Gen. Virol.* **8**, 187–195.

Firestone, G. L. (1983). The role of G protein glycosylation in the compartmentalization and processing of mouse mammary tumor virus glycoproteins in mouse mammary tumor virus-infected rat hepatoma cells. *J. Biol. Chem.* **258**, 6155–6161.

Fishman, J. B., and Fine, R. E. (1987). A trans Golgi derived exocytic coated vesicle can contain both newly synthesized cholinesterase and internalized transferrin. *Cell* **48**, 157–164.

Fleischer, S., and Packer, L. (1974). Biomembranes. *Methods Enzymol.* **31**, 1–889.

Forgac, M., and Cantley, L. (1984). Characterization of the ATP-dependent proton pump of clathrin-coated vesicles. *J. Biol. Chem.* **259**, 8101–8105.

Franke, W. W., Schear, U., Krohne, G., and Jarasch, E. D. (1981). The nuclear envelope and the architecture of the nuclear periphery. *J. Cell Biol.* **91**, 395–505.

Fries, E., and Rothman, J. E. (1980). Transient activity of Golgi-like membranes as donors of vesicular stomatitis viral glycoprotein *in vitro*. *J. Cell Biol.* **90**, 697–704.

Gabel, C. A., Goldberg, D. E., and Kornfeld, S. (1982). Lysosomal enzyme oligosaccharide phosphorylation in mouse lymphoma cells: Specificity and kinetics of binding to the mannose 6-phospate receptor *in vivo*. *J. Cell Biol.* **95**, 536–542.

Gallione, C. J., and Rose, J. K. (1983). Nucleotide sequence of a cDNA clone encoding the glycoprotein from the New Jersey serotype of vesicular stomatitis virus. *J. Virol.* **46**, 162–169.

Gallione, C. J., and Rose, J. K. (1985). A single amino acid substitution in a hydrophobic demain causes temperature-sensitive cell-surface transport of a mutant viral glycoprotein. *J. Virol.* **54**, 374–382.

Galloway, C. D., Dean, G. F., Marsh, M., Rudnick, G., and Mellman, I. (1983). Acidification of macrophage and fibroblast endocytic vesicles *in vitro*. *Proc. Natl. Acad, Sci. U.S.A.* **80**, 3334–3338.

Geisow, M. J., and Evans, W. H. (1984). pH in the endosome. Measurements during pinocytosis and receptor-mediated endocytosis. *Exp. Cell. Res.* **150**, 36–46.

Gerace, L., Ottaviano, Y., and Kondorkoch, C. (1982). Identification of a major protein of the nuclear pore complex. *J. Cell Biol.* **95**, 826–837.

Geuze, H. J., Slot, J. W., Straus, G. J., Lodish, H. F., and Schwartz, A. L. (1983). Intracellular site of asialoglycoprotein receptor ligand uncoupling: Double-label immunoelectron microscopy during receptor-mediated endocytosis. *Cell* **32**, 277–287.

Geuze, H. J., Slot, J. W., Straus, G. J. A. M., von Figura, K., Hasilik, A., and Schwartz, A. L. (1984). Intracellular receptor sorting during endocytosis: Comparative immunoelectron microscopy of multiple receptors in rat liver. *Cell* **37**, 195–204.

Glabe, C. G., Hanover, J. A., and Lennarz, W. J. (1980). Glycosylation of ovalbumin nascent chains. The spatial relationship between translation and glycosylation. *J. Biol. Chem.* **255**, 9236–9242.

Goldfine, I. D. (1981). Interaction of insulin, polypeptide hormones and growth factors with intracellular membranes. *Biochim. Biophys. Acta* **650**, 53–67.

Goldstein, J. L., Anderson, R. G., and Brown, M. S. (1979). Coated pits, coated vesicles, and receptor-mediated endocytosis. *Nature (London)* **279**, 679–685.

Gonatas, N. K., Stieber, A., Hickey, W. F., Herbert, S. H., and Gonatas, J. O. (1984). Endosomes and Golgi vesicles in absorptive and fluid phase endocytosis. *J. Cell Biol.* **99**, 1379–1390.

Goud, B., Huet, C., and Louvard, D. (1985). Assembled and unassembled pools of clathrin: A quantitative study using an enzyme immunoassay. *J. Cell Biol.* **100**, 521–527.

Greenspan. P., and St. Clair, W. (1984). Retroendocytosis of low density lipoprotein. *J. Biol. Chem.* **259**, 1703–1713.

Griffiths, G., Quinn, P., and Warren, G. (1983). Dissection of the Golgi complex I. Monensin inhibits the transport of viral membrane proteins from medial to trans Golgi cisternae in BHK cells infected with Semliki Forest virus. *J. Cell Biol.* **96**, 835–850.

Griffiths, G., Pfeiffer, S., Simons, K., and Matlin, K. (1985). Exit of newly synthesized membrane proteins from the trans cisterna of the Golgi complex to the plasma membrane. *J Cell Biol.* **101**, 949–964.

Guan, J. L., and Rose, J. K. (1984). Conversion of a secretory protein into a transmembrane protein results in its transport to the Golgi complex but not to the cell surface. *Cell* **37**, 779–787.

Guan, J. L., Machamer, C. E., and Rose, J. K. (1985). Glycosylation allows cell-surface transport of an anchored secretory protein. *Cell* **42**, 489–496.

Hanover, J. A., and Dickson, R. B. (1985a). Transferrin: Receptor-mediated endocytosis and iron delivery. *In* "Receptor-Mediated Endocytosis" (M. C. Willingham and I. Pastan, eds.) Plenum, New York, pp. 131–156.

Hanover, J. A., and Dickson, R. B. (1985b). The possible link between receptor phorphorylation and internalization. *Trends Pharmacol. Sci.* **6**, 457–459.

Hanover, J. A., and Lennarz, W. J. (1979). The topological orientation of N,N'-diacetylchitobiosylpyrophosphoryldolichol in artificial and natural membranes. *J. Biol. Chem.* **254**, 9237–9446.

Hanover, J. A., and Lennarz, W. J. (1980). N-Linked glycoprotein assembly: Evidence that oligosaccharide attachment occurs within the lumen of the endoplasmic reticulum. *J. Biol. Chem.* **255**, 3600–3604.

Hanover, J. A., and Lennarz, W. J. (1981). Transmembrane assembly of membrane and secretory glycoproteins. *Arch. Biochem. Biophys.* **24**, 1–19.

Hanover, J. A., and Lennarz, W. J. (1982). Transmembrane assembly of N-linked glycoproteins: Studies on the topology of saccharide–lipid synthesis. *J. Biol. Chem.* **257**, 10172–10177.

Hanover, J. A., Lennarz, W. J., and Young, J. D. (1980). Synthesis of N- and O-linked glycopeptides in oviduct membrane preparations. *J. Biol. Chem.* **255**, 6713–6716.

Hanover, J. A., Cheng, S.-Y., Willingham, M. C., and Pastan, I. H. (1982). α_2-Macroglobulin binding to cultured fibroblasts: Solubilization and partial characterization of binding sites. *J. Biol. Chem.* **258**, 370–377.

Hanover, J. A., Willingham, M. C., and Pastan, I. H. (1983a). Receptor-mediated endocytosis of α_2-macroglobulin. *Ann. N.Y. Acad. Sci.* **421**, 310–423.

Hanover, J. A., Rudick, J., Willingham, M. C., and Pastan, I. H. (1983b). α-Macroglobulin binding to cultured fibroblasts: Identification by affinity chromatography of high affinity binding sites. *Arch. Biochem. Biophys.* **2257**, 570–579.

Hanover, J. A., Willingham, M. C., and Pastan, I. (1984). Kinetics of transit of transferrin and epidermal growth factor through clathrin-coated membranes. *Cell* **39**, 283–293.

Hanover, J. A., DeSousa, P., Murphy, T., Willingham, M. C. August, T., and Patan, I. (1985a). Monoclonal antibodies which inhibit α_2-macroglobulin binding and uptake in cultured fibroblasts. *J. Biol. Chem.* **261**, 16732–16737.

Hanover, J. A., Beguinot, L., Willingham, M. C., and Pastan, I. H. (1985b). Transit of receptors for epidermal growth factor and transferrin through clathrin coated pits: Analysis of the kinetics of receptor entry. *J. Biol. Chem.* **260**, 15938–15945.

Hanover, J. A., Cohen, C., Willingham, M. C., and Park, M. K. (1987). O-linked N-acetylglucosamine is attached to proteins of the nuclear pore: evidence for cytoplasmic and nucleoplasmic glycoproteins. *J. Biol. Chem.* **262**, 9887–9895.

Hanspal, M., Luna, E., and Branton, D. (1984). The association of clathrin fragments with coated vesicle membranes. *J. Biol. Chem.* **259**, 11075–11082.

Haselbeck, A., and Scheckman, R. (1986). Intraorganelle transfer and glycosylation of yeast invertase *in vitro*. *Proc. Natl. Acad. Sci. U.S.A.* **83**, 2017–2020.

Hedman, K., Pastan, I., and Willingham, M. C. (1986). VSV G protein export and transferrin recycling occur in different vesicle fraction. *J. Histochem. Cytochem.* **35**, 233–243.

Helenius, A., Mellman, I., Wall, D., and Hubbard, A. (1983). Endosomes. *Trends Biochem. Sci.* **8**, 245–250.

Helenius, A., Kielian, M., Wellsteed, J., Mellman, I., and Rudnick, G. (1985). Effects of monovalent cations on Semliki Forest virus entry into BHK-21 cells. *J. Biol. Chem.* **260**, 5691–5697.

Heldin, C. H., and Westermark, B. (1984). Growth factors: Mechanism of action and relation to oncogenes. *Cell* **37**, 9–20.

Helmy, S., Porter-Jordan, K., Dawidowica, E. A., Pitch, P., Schwartz, A. L., and Fine, R. T. (1986). Separation of endocytic from exocytic coated vesicles using a novel cholinesterase mediated density shift technique. *Cell* **44**, 497–506.

Herman, B., and Albertini, D. F. (1982). The intracellular movement of endocytic vesicles in cultured granulosa cells. *Cell Motility* **2**, 583–597.

Herman, B., and Albertini, D. F. (1983). Ligand-induced rapid redistribution of lysosomes is temporally distinct from endosome translocation. *Nature (London)* **304**, 738–740.

Herman, B., and Albertini, D. F. (1984). A time-lapse video image intensification analysis of cytoplasmic organelle movements during endosome translocation. *J. Cell Biol.* **98**, 565–576.

Hoflack, B., and Kornfeld, S. (1985a). Lysosomal enzyme binding to mouse P388D1 macrophage membranes lacking the 215-kDa mannose 6-phosphate receptor: Evidence for the existence of a second mannose 6-phosphate receptor. *Proc. Natl. Acad. Sci. U.S.A.* **82**, 4428–4432.

Hoflack, B., and Kornfeld, S. (1985b). Purification and Characterization of a Cation-dependent Mannose 6-Phosphate Receptor from Murine P388D1 Macrophages and Bovine Liver. *J. Biol. Chem.* **260**, 12008–12014.

Hopkins, C. R. (1983). Intracellular routing of transferrin and transferrin receptor in epidermoid carcinoma A 451 cells. *Cell* **35**, 321–330.

Hopkins, C. R. (1985). Coated pits and their role in membrane receptor internalization. *In* "Molecular Mechanisms of Transmembrane Signalling" pp. 337–357. (Cohen and Housley, eds.), Elsevier, Amsterdam.

Hoppe, C. A., Connolly, T. P., and Hubbard, A. L. (1985). Transcellular transport of polymeric IgA in the rat hepatocyte: Biochemical and morphological characterization of the transport pathway. *J. Cell Biol.* **101**, 2113–2123.

Hornick, C. A., Hamilton, R. L., Spaziani, E., Enders, G. H., and Havel, R. J. (1985). Isolation and characterization of multivesicular bodies from rat hepatocytes: an organelle distinct from secretory vesicles of the Golgi apparatus. *J. Cell Biol.* **100**, 1558–1569.

Hunter, T. (1984). The epidermal growth factor receptor gene and its product. *Nature (London)* **311**, 414–416.

Iacopetta, B., and Morgan, E. (1984). Heme inhibits transferrin endocytosis in immature erythroid cells. *Biochim. Biophys. Acta* **805**, 211–216.

Irving, R. A., Toneguzzo, F., Rhee, S. H., Hofmann, T., and Ghosh, H. P. (1979). Synthesis and assembly of membrane glycorproteins: Presence of leader peptide in nonglycosylated precursor of membrane glycoprotein of vesicular stomatitis virus. *Proc. Natl. Acad. Sci. U.S.A.* **76**, 570–574.

Jakoby, W. J., ed. (1982). "Enzymatic Basis of Detoxication," p. 415. Academic Press, New York.

Jamieson, J., and Palade, G. (1971). Synthesis intracellular transport and discharge of secretory proteins in stimulated exocrine cells. *J. Cell Biol.* **50,** 135–158.

Kääriäinen, L., Hashimoto, K., Saraste, J., Virtanen, I., and Penttinen, K. (1980). Monensin and FCCP inhibit the intracellular transport of alphavirus membrane glycoproteins. *J. Cell Biol.* **87,** 783–791.

Kabat, D., Ruta, M., Murray, M., and Polonoff, E. (1980). Immunoselection of mutant deficient in all surface glycoprotein encoded by murine erythroleukemia viruses. *Proc. Natl. Acad. Sci. U.S.A.* **77,** 57–61.

Keen, J. H. (1985). The structure of clathrin-coated membranes: Assembly and disassembly. *In* "Endocytosis" (I. H. Pastan and M. C. Willingham, eds.), pp. 85–130. Plenum, New York.

Kelly, R. B. (1985). Pathways of protein secretion in eukaryotes. *Science* **230,** 25–32.

Kingsley, D. M., and Krieger, M. (1984). Receptor-mediated endocytosis of low density lipoprotein: Somatic cell mutants define multiple genes required for expression of surface-receptor activity. *Proc. Natl. Acad. Sci. U.S.A.* **81,** 5454–5458.

Kingsley, D. M., Kozarsky, K. F., Hobbie, L., and Krieger, M. (1986). Reversible defects in O-linked glycosylation and LDL receptor expression in a UDP-Gal/UDP-GalNAc 4-epimerase deficient mutant. *Cell* **44,** 749–759.

Klausner, R. D., van Renswonde, J., Kempf, C., Rao, K., Bateman, J. L., and Robbins, A. R. (1984). Failure to release iron from transferrin in a Chinese hamster ovary cell mutant pleiotropically defective in endocytosis. *J. Cell Biol.* **98,** 1098–1101.

Kornfeld, S. (1986). Trafficking of lysosomal enzymes in normal and disease states. *J. Clin. Invest.* **77,** 1–6.

Kornfeld, R., and Kornfeld, S. (1985). Assembly of asparagine-linked oligosaccharides. *Annu. Res. Biochem.* **54,** 631–664.

Kotwal, G., Buller, R. M. L., Wvnnerj, W. H., Pringle, C. R., and Ghosh, H. P. (1986). Role of glycosylation in transport of vesicular stomatitis envelope glycoprotein. A new class of mutant defective in glycosylation and transport of G protein. *J. Biol. Chem.* **261,** 8936–8953.

Kreis, T. T., and Lodish, H. F. (1986). Oligomerization is essential for transport of vesicular stomatitis glycoprotein to the cell surface. *Cell.* **46,** 929–937.

Krieger, M. (1983). Complementation of mutations in the LDL pathway of receptor-mediated endocytosis by cocultivation of LDL receptor-defective hamster cell mutants. *Cell* **33,** 413–422.

Krieger, M., Kingsley, D., Sege, R., Hobbie, L., and Kozarsky, K. (1985). Genetic analysis of receptor-mediated endocytosis. *Trends Biochem. Sci.* **10,** 442–452.

Krupp, M. N., Connolly, D. T., and Lane, M. D. (1982). Synthesis, turnover and down-regulation of epidermal growth factor receptors in human A431 epidermoid carcinoma cells and skin fibroblasts. *J. Biol. Chem.* **257,** 11489–11496.

Kuhn, L. C., and Kraehenbuhl, J. P. (1982). The sacrificial receptor-translocation of polymeric IgA across epithelia. *Trends Biochem. Sci.* **7,** 299–302.

Larkin, J. M., Brown, M. S., Goldstein, J. L., and Anderson, R. G. W. (1983). Depletion of intracellular potassium arrests coated pit formation and receptor mediated endocytosis in fibroblasts. *Cell* **33,** 273–285.

Larkin, J. M., Donzell, W. C., and Anderson, R. C. (1985). Modulation of intracellular potassium and ATP: Effects on coated pit function in fibroblasts and hepatocytes. *J. Cell. Physiol.* **124,** 372–378.

Lauffer, L., Garcia, P. D., Harkins, R. N., Coussens, L., Ullrich, A., and Walter, P. (1985). Topology of signal recognition particle receptor in endoplasmic reticulum membrane. *Nature (London)* **318,** 334–338.

LeFort-Tran, M., Aufderhude, K., Pouphile, M., Rossignol, M., and Beisson, J. (1981). Control of exocytic processes. Cytological and physiological studies of trichocyst mutants in *Paramecium tetrahymena. J. Cell Biol.* **88,** 301–311.

Lehrman, M. A., Goldstein, J. L., Brown, M. S., Russell, D. W., and Schweiden, W. J. (1985). Internalization-defective LDL receptors produced by genes with deletion and frameshift mutations that truncate the cytoplasmic domain. *Cell* **41,** 735–743.

Lenard, J., and Compans, R. W. (1974). The membrane structure of lipid-containing viruses. *Biochim. Biophys. Acta* **344,** 51–94.

Lennarz, W. J. (1980). "The Biochemistry of Glycoproteins and Proteoglycans." Plenum, New York.

Levine, G., Kotwal, G., Hamilton, M. C., Rachubinski, R., Bilan, P., Shore, G. C., Ghosh, H., and Bergerson, J. J. M. (1981). Transport of VSV "G" through the Golgi apparatus of cultured fibroblasts as visualized by radioautography. *J. Cell Biol.* **91,** 413a.

Lewis, W. H. (1931). Pinocytosis. *Bull. Johns Hopkins Hosp.* **49,** 17–36.

Limet, J. N., Quintart, J., Schneider, Y. J., and Courtoy, P. J. (1985). Receptor-mediated endocytosis of polymeric IgA and galactosylated serum albumin in rat liver. Evidence for intracellular ligand sorting and identification of distinct endosomal compartments. *Eur. J. Biochem.* **146,** 539–548.

Lin, C. R., Chen, W. S., Lazar, C. S., Carpenter, C. D., Gill, G. N., Evans, R. M., and Rosenfeld, M. G. (1986). Protein kinase C phosphorylation at Thr 654 of the unoccupied EGF receptor and EGF binding regulate functional receptor loss by independent mechanisms. *Cell* **44,** 839–848.

Lingappa, V. R., Katz, F. N., Lodish, H. F., and Blobel, G. (1978). A signal sequence for the insertion of a transmembrane glycoprotein. Similarities to the signals of secretory proteins in primary sequence and function. *J. Biol. Chem.* **253,** 8667–8670.

Lodish, H. F., Braell, W. A., Schwartz, A. L., Straus, G. J. S. M., and Zilberstein, A. (1981). Synthesis and assembly of membrane and organelle proteins. *Int. Rev. Cytol. Suppl.* **12,** 247–307.

McQueen, N. L., Nayak, D. P., Jones, L. V., and Compans, R. W. (1984). Chimeric influenza virus hemagglutinin containing either the NH$_2$ terminus or the COOH terminus of G protein of vesicular stomatitis virus is defective in transport to the cell surface. *Proc. Natl. Acad. Sci. U.S.A.* **81,** 395–399.

Makarow, M. (1985). Endocytosis in *Saccharomyces cerevisiae:* Internalization of α-amylase and fluorescent dextran into cells. *EMBO J.* **4,** 1861–1866.

Marnell, M. H., Mathis, L. S., Stookey, M., Shia, S. P., Stone, O. K., and Draper, R. K. (1984). *J. Cell Biol.* **99,** 1907–1916.

Marsh, M., Bolzau, E., and Helenius, A. (1983). Penetration of Semliki Forest virus from acidic prelysosomal vacuoles. *Cell* **32,** 931–940.

Mayes, E. L., and Waterfield, M. D. (1984). Biosynthesis of the epidermal growth factor receptor in A431 cells. *EMBO J.* **3,** 531–537.

Maxfield, F. R. (1985). Acidification of "endocytic vesicles and lysosomes." *In* "Endocytosis" (I. H. Pastan and M. C. Willingham, eds.), pp. 235–257. Plenum, New York.

Mellman, I., Fuchs, R., and Helenius, A. (1986). Acidification of the endocytic and exocytic pathways *Annu. Rev. Biochem.* **55,** 663–700.

Merion, M., Schlesinger, L. P., Brooks, R. M., Moehring, J. M., Moehring, T. J., and Sly, W. S. (1983). Defective acidification of endosomes in Chinese hamster ovary cell mutants "cross resistant" to toxins and viruses. *Proc. Natl. Acad. Sci. U.S.A.* **80**, 5315–5319.

Merisko, E. M., Farquhar, M. G., and Palade, G. E. (1982). Coated vesicle isolation by immunoadsorption on *Staphylococcus aureus* cells. *J. Cell Biol.* **92**, 846–857.

Metchnikoff, E. (1893). "Lectures on Comparative Pathology of Inflammation." Paul, Kegan, Trench, Trabner, and Co., London.

Moore, H. H., and Kelly, R. B. (1985). Secretory protein targeting in a pituitary cell line: Differential transport of foreign secretory proteins to distinct secretory pathways. *J. Cell Biol.* **101**, 1773–1781.

Mostov, K. E., and Simister, N. E. (1985). Transcytosis. *Cell* **43**, 389–390.

Moya, M., Dautry-Varsat, A., Goud, B., Louvard, D., and Boquet, P. (1985). Inhibition of coated pit formation in Hep2 cells blocks the cytotoxicity of diphtheria toxin but not that of ricin toxin. *J. Cell Biol.* **101**, 548–559.

Mueller, S. C., and Branton, D. (1984). Identification of coated vesicles in *Saccharomyces cerevisiae*. *J. Cell Biol.* **98**, 341–346.

Mundy, D. I., and Strittmatter, W. J. (1985). Requirement for metalloendoprotease in exocytosis: Evidence in mast cells and adrenal chromaffin cells. *Cell* **40**, 645–656.

Murphy, R. F., Powers, S., and Cantor, C. R. (1984). Endosome pH measured in single cells by dual fluorescence flow cytometry: Rapid acidification of insulin to pH 6. *J. Cell Biol.* **98**, 1757–1762.

Nakano, A., Nishijima, M., Maeda, M., and Akamatsu, Y. (1985). A temperature-sensitive Chinese hamster ovary cell mutant pleiotropically defective in protein export. *Biochim. Biophys. Acta* **845**, 324–332.

Nambi, P., Peters, J. R., Sibley, D. R., and Lefkowitz, R. J. (1985). Desensitization of the turkey erythrocyte β-adrenergic receptor in a cell-free system: Evidence that multiple protein kinases can phosphorylate and desensitize the receptor. *J. Biol. Chem.* **260**, 2165–2171.

Neufeld, E. F., and Ashwell, G. (1980). *In* "The Biochemistry of Glycoproteins and Proteoglycans" (W. J. Lennarz, ed), pp. 241–266. Plenum, New York.

Newman, R., Schneider, C., Sutherland, R. Vodinelich, L., and Greaves, M. (1982). The transferrin receptor. *Trends Biochem. Sci.* **7**, 397–400.

Novikoff, A. B., and Novikoff, P. M. (1977). Cytochemical contributions to differentiating GERL from the Golgi apparatus. *Histochem. J.* **9**, 525–551.

Oka, Y., Mottola, C., Oppenheimer, C. L., and Czech, M. P. (1984). Insulin activates the appearance of insulin-like growth factor II receptors on the adipocyte cell surface. *Proc. Natl. Acad. Sci. U.S.A.* **81**, 4028–4032.

Oka, Y., Rozek, L. M., and Czech, M. P. (1985). Direct demonstration of rapid insulin-like growth factor II receptor internalization and recycling in rat adipocytes. *J. Biol. Chem.* **260**, 9435–9442.

Olashaw, N. E., O'Keefe, E. J., and Pledger, W. J. (1986). Platelet-derived growth factor modulates epidemal growth factor receptors by a mechanism distinct from that of phorbol esters. *Proc. Natl. Acad. Sci. U.S.A.* **83**, 3834–3838.

Olsnes, S., and Sandvig, K. (1985). Toxins. *In* "Receptor-Mediated Endocytosis" (M. C. Willingham and I. H. Pastan, eds.), Plenum, New York, pp. 195–234.

Omary, M. B., and Trowbridge, I. S. (1981). Covalent attachment of fatty acid to the transferrin receptor in cultured human cells. *J. Biol. Chem.* **256**, 4715–4718.

Ottosen, P. D., Courtoy, P. J., and Farquhar, M. G. (1980). Pathways followed by membrane recovered from the surface of plasma cells and myeloma cells. *J. Exp. Med.* **152,** 1–19.

Palade, G. B. (1975). Intracellular aspects of the process of protein secretion. *Science* **189,** 347–351.

Park, M. K., D'Onofrio, M., Willingham, M. C., and Hanover, J. A. (1987). A monoclonal antibody against a family of nuclear pore proteins: O-linked GlcNAc is part of the immunodeterminant. *Proc. Natl. Acad. Sci. U.S.A.,* **84,** 6462–6466.

Pastan, I., and Willingham, M. C. (1981). Receptor-mediated endocytosis of hormones in cultured cells. *Annu. Rev. Physiol.* **43,** 239–250.

Pastan, I., and Willingham, M. C. (1985). The pathway of endocytosis. *In* "Endocytosis" (I. H. Pastan and M. C. Willingham, eds.), pp. 1–44. Plenum, New York.

Paterson, S., and Morgan, E. H. (1980). Effect of changes in the ionic environment of reticulocytes on the uptake of transferrin-bound iron. *J. Cell. Physiol.* **105,** 489–502.

Pathak, R. K., Luskey, K. L., and Anderson, R. A. W. (1986). Biogenesis of the crystalloid endoplasmic reticulum in UT 1 cells: Evidence that newly formed endoplasmic reticulum emerges from the nuclear envelope. *J. Cell Biol.* **102,** 2158–2168.

Payne, G. S., and Schekman, R. (1985). A test of clathrin function in protein secretion and cell growth. *Science* **230,** 1009–1014.

Pearse, B. M. (1976). Clathrin: A unique protein associated with intracellular transfer of membrane by coated vesicles. *Proc. Natl. Acad. Sci U.S.A.* **73,** 1255–1259.

Pearse, B. M. F., and Robinson, M. S. (1984). Purification and properties of 100 Kd proteins from coated vesicles and their reconstitution with clathrin. *EMBO J.* **3,** 1951–1957.

Peng, H. B. (1984). Participation of calcium and calmodulin in the formation of acetylcholine receptor clusters. *J. Cell Biol.* **98,** 550–557.

Posner, B., Kahn, M. N., and Bergeson, J. M. (1985). Peptide hormone receptors in intracellular structures from rat liver. *Methods Enzymol.* **109,** 219–231.

Puddington, L., Lively, M. O., and Lyles, D. S. (1985). Role of the nuclear envelope in synthesis processing and transport of membrane glycoproteins. *J. Biol. Chem.* **260,** 5641–5647.

Quinn, P., Griffiths, G., and Warren, G. (1983). Dissection of the Golgi complex II. Density separation of specific Golgi functions in virally infected cells treated with monensin. *J. Cell Biol.* **96,** 851–856.

Quintart, J., Courtoy, P. J., and Baudhuin, P. (1984). Receptor-mediated endocytosis in rat liver: Purification and enzymic characterization of low density organelles involved in uptake of galactose-exposing proteins. *J. Cell Biol.* **98,** 877–884.

Reed, B. C., Bonnet, G. U., and Lane, M. D. (1981). Roles of glycosylation and protein synthesis in insulin receptor metabolism by 3T3-LI mouse adipocytes. *Proc. Natl. Acad. Sci. U.S.A.* **78,** 2908–2912.

Regoeczi, E., Chindemi, P. A., Debanne, M. T., and Hatton, M. W. C. (1982). Dual nature of the hepatic lectin pathway for human asialotransferrin type 3 in the rat. *J. Biol. Chem.* **257,** 5431–5436.

Riezman, H. (1985). Endocytosis in yeast: Several of the yeast secretory mutants are defective in endocytosis. *Cell* **40,** 1001–1009.

Riezman, H., Chvatchko, Y., and Howald, I. (1986). Endocytosis in yeast: Relationship to other cellular pathways. *In* "Yeast Cell Biology" (UCLA Symposia on Molecular and Cellular Biology. Vol. 33, J. Hichks, ed.). Alan R. Liss, New York.

Rindler, M. J., Ivanov, I. E., Plesken, H., Rodriguez-Boulan, E., and Sabatini, D. D. (1984). Viral glycoproteins destined for apical or basolateral plasma membrane domains traverse the same Golgi apparatus during their intracellular transport in doubly infected Madin–Darby canine kidney cells. *J. Cell Biol.* **98,** 1304–1319.

Rindler, M. J., Ivanov, I. E., Plesken, H., and Sabatini, D. D. (1985). Polarized delivery of viral glycoproteins to the apical and basolateral plasma membranes of Madin–Darby canine kidney cells infected with temperature-sensitive viruses. *J. Cell Biol.* **100,** 136–151.

Rizzolo, L. J., Finidori, J., Gonzalez, A., Arpin, M., Ivanov, I. E., Adesnik, M., and Sabatini, D. C. (1985). Biosynthesis and intracellular sorting of growth hormone–viral envelope glycoprotein hybrids. *J. Cell Biol.* **101,** 1351–1362.

Robbins, A. R., Peng, S. S., and Marshall, J. L. (1983). Mutant Chinese hamster ovary cells pleiotypically defective in receptor-mediated endocytosis. *J. Cell Biol.* **96,** 1064–1071.

Robbins, A. R., Oliver, C., Bateman, J. L., Krag, S. S., Galloway, C. J., and Mellman, I. (1984). A single mutation in Chinese hamster ovary cells impairs both Golgi and endosomal functions. *J. Cell Biol.* **99,** 1296–1308.

Robinson, M. S., and Pearse, B. M. F. (1986). Immunofluorescent localization of 100K coated vesicle proteins. *J. Cell Biol.* **102,** 48–54.

Rodman, J. S., Seidman, L., and Farquhar, M. G. (1986). The membrane composition of coated pits microvilli endosomes is distinctive in the rat kidney proximal tubule cell. *J. Cell Biol.* **102,** 77–87.

Rome, L. H., Garvin, A. J., Allietta, M. M., and Neufeld, E. F. (1979). Two species of lysosomal organelles in cultured human fibroblasts. *Cell* **17,** 143–153.

Rose, J. K. (1977). Nucleotide sequence of ribosome recognition site in mRNAs of vesicular stomatitis virus. *Proc. Natl. Acad. Sci. U.S.A.* **74,** 3672–3676.

Rose, J. K., and Bergman, J. E. (1983). Altered cytoplasmic domains affect ultracellular transport of the vesicular stomatitis virus glycoprotein. *Cell* **34,** 513–524.

Rose, J. K., Adams, G. A., and Gallione, C. J. (1984). The presence of cysteine in the cytoplasmic domain of the vesicular stomatitis virus glycoprotein is required for palmitate addition. *Proc. Natl. Acad. Sci. U.S.A.* **81,** 2050–2054.

Rosenfeld, M. G., Marcantonio, E. E., Hakimi, J., Ort, V. M., Atkinson, P. H., Sabatini, D., and Kreibich, G. (1984). Biosynthesis and processing of ribophorins in the endoplasmic reticulum. *J. Cell Biol.* **99,** 1076–1082.

Roth, J. (1983). Application of lectin–gold complexes for electron microscopic localization of glycoconjugates on thin sections. *J. Histochem. Cytochem.* **31,** 987–999.

Roth, T. F., and Porter, K. R. (1964). Yolk protein uptake in the oocyte of the mosquito *Aedes aegypti. J. Cell Biol.* **20,** 313–332.

Rothenberger, S. Iacopelta, B. J., and Kühn, L. C. (1987). Endocytosis of the transferrin receptor requires the cytoplasmic domain but not its phosphorylation site. *Cell* **49,** 423–431.

Rothman, J. E., and Fine, R. E. (1980). Coated vesicles transport newly synthesized membrane glycoproteins from endoplasmic reticulum to plasma membrane in two successive stages. *Proc. Natl. Acad. Sci. U.S.A.* **77,** 780–784.

Rothman, J. E., and Lodish, H. F. (1977). Synchronized transmembrane insertion and glycosylation of a nascent membrane protein. *Nature (London)* **269,** 775–780.

Rothman, J. E., and Schmid, S. L. (1986). Enzymatic recycling of clathrin from coated vesicles. *Cell* **46,** 5–9.

Rothman, J. E., Bursztn-Pettegrew, H., and Fine, R. E. (1980). Transport of the membrane glycoprotein of vesicular stomatitis virus to the cell surface in two stages by clathrin coated vesicles. *J. Cell Biol.* **86,** 162–171.

Rothman, J. E., Miller, R. L., and Urbani, L. J. (1984a). Intercompartmental transport in the Golgi complex is a constituitive process: Facile transfer of membrane protein between two Golgi populations. *J. Cell Biol.* **99**, 260–271.

Rothman, J. E., Urbanii, L. J., and Brdands, R. (1984b). Transport of protein between cytoplasmic membranes of fused cells: Correspondence to processes reconstituted in a cell-free system. *J. Cell Biol.* **99**, 248–259.

Rubenstein, J. L. R., Fine, R. E., Lusky, B. D., and Rothman, J. E. (1981). Purification of coated vessicles by agarose gel electrophoresis. *J. Cell Biol.* **89**, 357–361.

Ryser, H. J., Drummond, I., and Shen, W. C. (1982). The cellular uptake of horseradish peroxidase and its poly(lysine) conquests by cultured fibroblasts is qualitatively similar despite a 900-fold difference in rate. *J. Cell. Physiol.* **113**, 167–178.

Saermark, T., Flint, N., and Evans, W. H. (1985). Hepatic endosome fractions contain an ATP-driven proton pump. *Biochem. J.* **225**, 51–58.

Sahagian, G. G., and Neufeld, E. F. (1983). Biosynthesis and turnover of the mannose 6-phosphate receptor in cultured Chinese hamster ovary cells. *J. Biol. Chem.* **258**, 7121–7128.

Sandvig, K., Sundan, A., and Olsnes, S. (1985). Effect of potassium depletion of cells on their sensitivity to diphtheria toxin and *Pseudomonas* toxin. *J. Cell Physiol.* **124**, 54–60.

Saraste, J., and Hedman, K. (1983). Intracellular vesicles involved in the transport of Semliki Forest virus membrane proteins to the cell surface. *EMBO J.* **2**, 2001–2006.

Scheckman, R. (1982). The secretory pathway in yeast. *Trends Biochem. Sci.* **7**, 243–246.

Schlegel, R., Willingham, M. C., and Pastan, I. (1981). Monensin blocks endocytosis of vesicular stomatitis virus. *Biochem. Biophys. Res. Commun.* **102**, 992–998.

Schlesinger, S., Malfer, C., and Schlesinger, M. J. (1984). The formation of vesicular stomatitis virus (San Juan strain) becomes temperature-sensitive when glucose residues are retained on the oligosaccharides of the glycoprotein. *J. Biol. Chem.* **259**, 7597–7601.

Schmid, S. L., Braell, W. A., Schlossman, D. M., and Rothman, J. E. (1984). A role for clathrin light chains in the recognition of clathrin cages by "uncoating ATPase." *Nature (London)* **311**, 228–231.

Schmidt, M., and Schlesinger, M. (1979). Fatty acid binding to vesicular stomatitis virus glycoprotein: A new type of posttranslational modification of the viral glycoprotein. *Cell* **17**, 813–819.

Schwartz, R. T., and Datema, R. (1982). The lipid pathway of protein glycosylation and its inhibitors: The biological significance of protein-bound carbohydrates. *Adv. Carbohydr. Chem. Biochem.* **40**, 287–379.

Seth, P., Fitzgerald, D. J., Willingham, M. C., and Pastan, I. (1984). Role of a low-pH environment in adenovirus enhancement of the toxicity of a *Pseudomonas* exotoxin–epidermal growth factor conjugate. *J. Virol.* **51**, 650–655.

Sibley, D. R., and Lefkowitz, R. J. (1985). Molecular mechanisms of receptor desensitization using the β-adrenergic receptor-coupled adenylate cyclase system as a model. *Nature (London)* **317**, 124–129.

Sibley, D. R., Strasser, R. H., Caron, M. G., and Lefkowitz, R. J. (1985). Homologous desensitization of adenylate cyclase is associated with phosphorylation of the β-adrenergic receptor. *J. Biol. Chem.* **260**, 3883–3886.

Sibley, D. R., Benovic, J. L., Caron, M. G., and Lefkowitz, R. J. (1987). Regulation of transmembrane signalling by receptor phosphorylation. *Cell* **48**, 913–922.

Sidman, C. (1981). Differing requirement for glycosylation in the secretion of related glycoproteins in neither by the producing cell nor by the relative number of oligosaccharide units. *J. Biol. Chem.* **256**, 9374–9376.

Siegelman, M., Bond, M. W., Gallatin, M., St. John, T., Smith, H. T., Fried, V. A., and Weissman, I. L. (1986). Cell surface molecule associated with lymphocyte activation is ubiquitinated branched-chain glycoprotein. *Science* **231**, 823–829.

Siuta-Mangano, P., Janero, D. R., and Lane, M. D. (1982). Association and assembly of triglyceride and phospholipid with glycosylated and unglycosylated apoproteins of very low density lipoproteins in the intact liver cell. *J. Biol. Chem.* **257**, 11463–11467.

Sleight, R. G., and Pagano, R. E. (1984). Transport of a fluorescent phosphatidylcholine analog from the plasma membrane to the Golgi apparatus. *J. Cell Biol.* **99**, 742–751.

Slieker, L. J., and Lane, M. D. (1985). Glycosylation dependent acquisition of ligand binding capacity. Posttranslational processing of the epidermal growth factor receptor. *J. Biol. Chem.* **260**, 687–690.

Snider, M. D., and Rogers, O. C. (1985). Intracellular movement of cell surface receptors after endocytosis: Resialation of asialotransferrin receptor in human erythroleukemia cells. *J. Cell Biol.* **100**, 826–834.

Snider, M. D., and Rogers, O. C. (1986). Membrane traffic in animal cells: Cellular glycoproteins return to the site of Golgi mannosidase I. *J. Cell Biol.* **103**, 265–275.

Snow, C. M., Senior, A., and Gerace, L. (1987) Monoclonal antibodies identify a group of nuclear pore complex glycoproteins. *J. Cell Biol.* **104**, 1143–1156.

Steer, C. J., Bisher, M., Blumenthal, R., and Steven, A. C. (1984). Detection of membrane cholesterol by filipin in isolated rat liver coated vesicles is dependent upon removal of the clathrin coat. *J. Cell Biol.* **99**, 315–319.

Steinman, R. M., Mellman, I. S., Mullter, W. A., and Cohn, Z. A. (1983). Endocytosis and the recycling of plasma membrane. *J. Cell Biol.* **96**, 1–27.

Stone, D. K., Xie, X. S., and Racker, E. (1984). Inhibition of clathrin-coated vesicle acidification by duramycin. *J. Biol. Chem.* **259**, 2701–2703.

Stoscheck, E. M., and Carpenter, G. (1984). Down-regulation of epidermal growth factor receptors: Direct demonstration of receptor degradation in human fibroblasts. *J. Cell Biol.* **98**, 1048–1053.

Takatsuki, A., Kohno, K., and Tamura, G. (1975). Inhibition of biosynthesis of polysopenol sugars in chick embryo microsome by tunicamycin. *J. Biol. Chem.* **39**, 2089–2091.

Tartakoff, S. M. (1983). Perturbation of vesicular traffic with the carboxylic ionophore monensin. *Cell* **32**, 1026–1028.

Tartakoff, S. M., and Vassalli (1977). Plasma cell immunoglobulin secretion. Arrest is accompanied by alteration of the Golgi complex. *J. Exp. Med.* **146**, 1332–1345.

Tkacz, J. S., and Lampen, J. B. (1975). Tunicamycin, inhibitor of polyisoprenyl *N*-acetylglucosaminyl pyrophosphate formation in calf liver microsome. *Biochem. Biophys. Res. Commun.* **65**, 248–257.

Tycko, B., and Maxfield, F. R. (1982). Rapid acidification of endocytic vesicles containing α_2-macroglobulin. *Cell* **28**, 643–651.

Valberg, P. A., and Albertini, D. F. (1985). Cytoplasmic motions, rheology, and structure probed by a novel magnetic particle method. *J. Cell Biol.* **101**, 130–140.

Van Dyke, R. W., Steer, C. J., and Scharschmidt, B. F. (1984). Clathrin-coated vesicles from rat liver: Enzymatic profile and characterization of ATP-dependent proton transport. *Proc. Natl. Acad. Sci. U.S.A.* **81**, 3108–3112.

Van Dyke, R. W., Hornick, C. A., Belcher, J., Scharschmidt, B. F., and Havel, R. J. (1985). Identification and characterization of ATP-dependent proton transport by rat liver multivesicular bodies. *J. Biol. Chem.* **260**, 11021–11026.

Wall, D. A., and Hubbard, A. L. (1985). Receptor mediated endocytosis of asialoglycoproteins by rat liver hepatocytes: Biochemical characterization of the endosomal compartments. *J. Cell Biol.* **101**, 2104–2112.

Wattenberg, B. W., and Rothman, J. E. (1986). Multiple cytosolic components promoter intra-Golgi protein transport. *J. Biol. Chem.* **261**, 2208–2213.

Watts, C. (1984). *In situ* ^{125}I-labelling of endosome proteins with lactoperoxidase conjugates. *EMBO J.* **3**, 1965–1970.

Watts, C. (1985). Rapid endocytosis of the transferrin receptor in the absence of bound transferrin. *J. Cell Biol.* **100**, 633–637.

Wehland, J., Willingham, M. C., Dickson, R. B., and Pastan, I. (1981). Microinjection of anticlathrin antibodies into fibroblasts does not interfere with the receptor-mediated endocytosis of α_2-macroglobulin. *Cell* **25**, 105–120.

Wehland, J., Willingham, M. C., Galt, M. G., and Pastan, I. (1982). The morphologic pathway of exocytosis of the VSV G protein in cultured fibroblasts. *Cell* **28**, 831–841.

Welply, J. K., Kaplan, H. A., Shenbagamurthi, P., Naider, F., and Lennarz, W. J. (1986). Studies on properties of membrane-associated oligosaccharyltransferase using an active site-directed photoaffinity probe. *Arch. Biochem. Biophys.* **246**, 808–819.

Willingham, M. C., and Pastan, I. (1980). The receptosome: An intermediate organelle or receptor-mediated endocytosis in cultured fibroblasts. *Cell* **21**, 67–77.

Willingham, M. C., and Pastan, I. H. (1982). The transit of epidermal growth factor through coated pits of the Golgi system. *J. Cell Biol.* **94**, 207–212.

Willingham, M. C., and Pastan, I. (1983). Formation of receptosomes from plasma membrane coated pits during endocytosis: Analysis by serial sections with improved membrane labeling and preservation techniques. *Proc. Natl. Acad. Sci. U.S.A.* **80**, 5617–5621.

Willingham, M. C., and Pastan, I. (1984). Endocytosis and exocytosis: current concepts of vesicle traffic in animal cells. *Int. Rev. Cytol.* **92**, 51–92.

Willingham, M. C., and Pastan, I. (1985). Ultrastructural immunocytochemical localization of the transferrin receptor using a monoclonal antibody in human KB cells. *J. Histochem. Cytochem.* **33**, 59–64.

Willingham, M. W., Rutherford, A. V., Gallo, M. G., Wehland, J., Dickson, R. B., Schlegel, R., and Pastan, I. (1981). Receptor-mediated endocytosis in cultured fibroblasts: Cryptic coated pits and the formation of receptosomes. *J. Histochem. Cytochem.* **29**, 1003–1013.

Willingham, M. C., Pastan, I. H., and Sahagian, G. G. (1983). Ultrastructural immunocytochemical localization of the phosphomannosyl receptor in Chinese hamster ovary (CHO) cells. *J. Histochem. Cytochem.* **31**, 1–11.

Willingham, M. C., Hanover, J. A., Dickson, R. B., and Pastan, I. (1984). Morphologic characterization of the pathway of transferrin endocytosis and recycling in human KB cells. *Proc. Natl. Acad. Sci. U.S.A.* **81**, 175–179.

Wills, J. W., Srinivas, R. V., and Hunter, E. (1984). Mutations of the Rous sarcoma virus env gene that affect the transport and subcellular location of the glycoprotein products. *J. Cell Biol.* **99**, 2011–2023.

Wilcox, D. K., Kitson, P. P., and Widnell, C. C. (1982). Inhibition of pinocytosis in rat embryo fibroblasts treated with monensin. *J. Cell Biol.* **92**, 859–864.

Wolkoff, A. W., Klausner, R. D., Ashwell, G., and Harford, J. (1984). Intracellular segregation of asialoglycoproteins and their receptor: a prelysosomal event subsequent to dissociation of the ligand-receptor complex. *J. Cell Biol.* **98**, 375–381.

Woods, J. W., Doriaux, M., and Farquhar, M. G. (1986). Transfer in receptors recycle to cis and middle as well as trans Golgi cisternae in Ig-secreting myeloma cells. *J. Cell Biol.* **103**, 277–286.

Xie, X. S., Stone, D. K., and Racker, E. (1984) Activation and partial purification of the

ATPase of -coated vesicles and reconstitution of the proton pump. *J. Biol. Chem.* **259,** 11676–11678.

Yamamato, T., Bishop, R. W., Brown, M. S., Goldstein, J. L., and Russell, D. W. (1986). Deletion in cysteine-rich region of LDL receptor impedes transport to cell surface in WHHL rabbit. *Science* **232,** 1230–1239.

Yamashiro, D. J., Tycko, B., Fluss, S. R., and Maxfield, F. R. (1984). Segregation of transferrin to a mildly acidic (pH 6.5) para-Golgi compartment in the recycling pathway. *Cell* **37,** 789–800.

Yeo, K. T., Yeo, T. K., Parent, J. B., Olden, K. (1985). Swainsonine treatment accelerates intracellular transport and secretion of glycoproteins in human hepatoma cells. *J. Biol. Chem.* **260,** 2565–2569.

Zieseniss, E., and Plattner, H. (1985). Synchronous exocytosis in Paramecium cells involves very rapid (less than or equal to 1 s) reversible dephosphorylation of a 65-kD phosphoprotein in exocytosis-competent strains. *J. Cell Biol.* **101,** 2028–2035.

11

Endocytosis and Compartmentalization of Lysosomal Enzymes in Normal and Mutant Mammalian Cells: Mannose 6-Phosphate-Dependent Pathways

APRIL R. ROBBINS

PROTEIN TRANSFER AND ORGANELLE BIOGENESIS

7.0 (Glaser *et al.*, 1975). Urinary α-L-iduronidase was also separable into low and high uptake forms (Shapiro *et al.*, 1976), but in this case the two differed significantly in molecular weight (67,000 and 87,000, respectively).

Consistent with both the anionic and carbohydrate nature of the recognition marker was the subsequent finding that sugar phosphates, specifically mannose 6-phosphate (Man 6-P) and fructose 1-phosphate, were far more potent inhibitors of β-glucuronidase uptake than were nonphosphorylated sugars (Kaplan *et al.*, 1977a). Conversion to the low uptake form was observed after treatment of the enzyme with alkaline phosphatase (Kaplan *et al.*, 1977a). In accord with the hypothesis of a common recognition marker, uptake of β-galactosidase and β-hexosaminidase also was inhibited by Man 6-P or treatment with phosphatase (Kaplan *et al.*, 1977b), as was uptake of α-L-iduronidase (Sando and Neufeld, 1977), α-N-acetylglucosaminidase, arylsulfatase A, and α-mannosidase (Ullrich *et al.*, 1978). K_{uptake} values for all of these enzymes ranged from 3 to 6 nM, whereas the K_i values measured for Man 6-P were from 20 to 60 μM (Kaplan *et al.*, 1977a,b; Sando and Neufeld, 1977; Ullrich *et al.*, 1978).

Direct demonstration of Man 6-P on lysosomal enzymes was accomplished both chemically (Distler *et al.*, 1979a) and enzymatically (Natowicz *et al.*, 1979). For the former a bovine testicular glycoprotein fraction, enriched in lysosomal hydrolases and very inhibitory for β-galactosidase uptake (Distler *et al.*, 1979b), was employed. Sugars and sugar phosphates were released by acid hydrolysis, radiolabeled by reduction with NaB^3H_4, then identified by electrophoresis against authentic standards, both before and after treatment with phosphatase. The position of phosphate substitution was established by periodate oxidation and identification of the resultant fragments (Distler *et al.*, 1979a). Enzymatic analyses were of sugars and sugar phosphates released by acid hydrolysis from purified human spleen β-glucuronidase. Man 6-P was quantitated via generation of NADPH (Man 6-P $\rightarrow$ Fru 6-P $\rightarrow$ Glc 6-P + $NADP^+$ $\rightarrow$ 6-P-gluconate + NADPH), amplified by $NADP^+$–NADPH cycling. The activity of β-glucuronidase fractions in uptake assays increased linearly with the Man 6-P content (Natowicz *et al.*, 1979).

Treatment of α-N-acetylglucosaminidase with endo-β-N-acetylglucosaminidase H (endo H) markedly reduced uptake of that enzyme; of the released oligosaccharides $\sim$12% were acidic on electrophoresis, and treatment with phosphatase converted these to neutral species (von Figura and Klein, 1979). Similarly, treatment of β-glucuronidase with endo H decreased both uptake and Man 6-P content of the enzyme by 90%, without affecting enzymatic activity; Man 6-P was recovered from the released oligosaccharides (Natowicz *et al.*, 1979). Thus, based on the specificity of endo H (Tarentino *et al.*, 1974; Yamashita *et al.*, 1978), the Man 6-P recognition marker is on asparagine-linked oligosaccharides of the high-mannose and/or hybrid type (discussed further below).

Completing the circle of evidence regarding the common recognition marker, it was shown that I-cells do not phosphorylate their lysosomal enzymes (Hasilik and Neufeld, 1980b; Bach *et al.*, 1979). Cathepsin D, β-hexosaminidase α and β chains, and α-glucosidase immunoprecipitated from normal fibroblasts after metabolic labeling with $^{32}P_i$ all contained radiolabel; electrophoretic and enzymatic analyses as described above demonstrated that the label was in Man 6-P. No ^{32}P was incorporated into these enzymes by I cells (Hasilik and Neufeld, 1980b). Similar results were obtained with β-hexosaminidase purified by affinity chromatography from I-cell versus normal fibroblast secretions (Bach *et al.*, 1979).

B. Biosynthesis of the Recognition Marker

1. Phosphotransferase

Formation of Man6-P on lysosomal enzymes involves an α-*N*-acetylglucosamine-1-phospho-6-mannose diester intermediate. The majority of anionic oligosaccharides released by endo H from β-glucuronidase of murine lymphoma cells (Tabas and Kornfeld, 1980), or from β-hexosaminidase and cathepsin D of human fibroblasts (Hasilik *et al.*, 1980) were not susceptible to phosphatase unless previously subjected to mild acid hydrolysis, indicating that the phosphate was blocked. After metabolic labeling with glucose or glucosamine, analysis of the sugar released from the enzymes by mild acid showed that it was *N*-acetylglucosamine (GlcNAc); enzymatic digestion showed that it was the α anomer.

The phosphodiester is generated by direct transfer of GlcNAc-P from UDPGlcNAc to mannose(s) on protein-linked oligosaccharides, without involvement of a lipid-linked intermediate. [β-^{32}P]UDP[^{3}H]GlcNAc was incubated with Chinese hamster ovary (CHO) cell extracts (as a source of both transferase and acceptors), and transfer of [^{3}H]GlcNAc-^{32}P was measured (Reitman and Kornfeld, 1981a). No inhibition of transfer to protein was observed under conditions that abolished transfer to lipid (i.e., in the presence of tunicamycin), nor was any increase obtained under conditions that markedly stimulated formation of labeled GlcNAc-P-P-lipid. Characterization of the proteins labeled *in vitro* showed they contained GlcNAc-1-P-6-Man on high-mannose oligosaccharides.

Formation of the phosphodiester was also demonstrated on exogenously added thyroglobulin glycopeptides (Reitman *et al.*, 1981a). Subsequent studies, using partially purified transferase (DEAE fractions of proteins extracted from rat liver microsomes by 0.3% Lubrol), showed that nonlysosomal glycoproteins such as ribonuclease B, high-mannose oligo-saccharides, and even α-methylmannoside could serve as acceptors of GlcNAc-P; however, the efficiency of transfer (V_{max}/K_m) to lysosomal enzymes was at least 100-fold higher (Reitman and Kornfeld, 1981b). Heat denaturation of α-N-acetylglucosaminidase greatly diminished its efficiency as an acceptor; thus, some feature beyond the primary structure appears to determine the efficacy of lysosomal enzymes as acceptors. Characterization of GlcNAc-P-transferase* activity in these studies (Reitman and Kornfeld, 1981a,b) showed that the enzyme required Mg^{2+} or Mn^{2+} (preferring a combination of both), had a pH optimum of 7–8 (with <1% activity at pH ≤5), and exhibited an apparent K_m for UDPGlcNAc of 21 μM.

I-cell fibroblasts lack GlcNAc-P-transferase activity (Hasilik et al., 1981; Reitman et al., 1981). Using as acceptor either deblocked and de-phosphorylated purified human β-hexosaminidase (Hasilik et al., 1981), or endogenous glycoproteins and thyroglobulin glycopeptides (Reitman et al., 1981), no transfer of ^{32}P from [^{32}P]UDPGlcNAc was detected. Mixing experiments with membranes (Hasilik et al., 1981) or extracts (Reitman et al., 1981) from normal fibroblasts gave no indication of an inhibitor in the I-cell preparations. These results plus the lack of activity with exogenous acceptors indicate that I-cells have a defect in the transferase enzyme itself. Later studies showed absence of transferase activity in organs of I-cell patients, including liver (Owada and Neufeld, 1982; Waheed et al., 1982), of note because activities of lysosomal enzymes in liver appear normal in this disorder (Tondeur et al., 1971).

Depending on the sensitivity of the assay system, residual GlcNAc-P-transferase activity could be detected in fibroblasts from patients with mucolipidosis III [pseudo-Hurler polydystrophy, thought to be a clini-cally milder form of I-cell disease (Neufeld and McKusick, 1983)]. With endogenous glycoproteins as acceptors, transferase activity ranged from I-cell levels (i.e., undetectable) to about 25% of normal; with thyroglobulin glycopeptides, activities were 0–10% of normal (Reitman et al., 1981). Metabolic incorporation of ^{32}P into β-hexosaminidase by some of these same cell lines showed a similar range (Robey and Neufeld, 1982). Using α-methylmannoside as exogenous acceptor, activity as low as 0.4% of normal could be detected; in this assay I-cell and mucolipidosis III fibro-blasts were distinguishable (Varki et al., 1981). Four of five I-cell lines showed no activity (the fifth was ≤1% of normal), whereas all cell lines from the latter disorder exhibited transferase activity. In addition, two siblings with mucolipidosis III were discovered, extracts of whose fibro-

* UDP-N-Acetylglucosamine : glycoprotein lysosomal enzyme N-acetylglucosamine-1-phosphotransferase.

blasts exhibited normal levels of transfer to α-methylmannoside but less than 5% of normal activity with endogenous acceptors, and no detectable phosphorylation of exogenously added purified β-hexosaminidase (Varki *et al.*, 1981). Analysis of the kinetics of *in vitro* GlcNAc-P transfer indicated two distinct groups among mucolipidosis III patients, one exhibiting less than one-fourth the normal V_{max} for phosphorylation of α-methylmannoside and lysosomal enzymes, with near normal K_m values for both these acceptors, and a second group, including but not limited to the siblings described above, showing no effect on V_{max}, with a greater than 8-fold increase in K_m for lysosomal enzymes (Lang *et al.*, 1985).

Based on clinical presentation and these biochemical studies, it appears that both I-cell disease and mucolipidosis III reflect defects in the GlcNAc-P-transferase. Among individuals with the latter disorder, those showing normal activity with α-methylmannoside but depressed activity with a *bona fide* lysosomal enzyme seem to have a defect specifically affecting the ability of the transferase to recognize lysosomal enzymes. Consistent with separation of recognition and catalytic sites (subunits?) is the observation that partially deglycosylated lysosomal enzymes inhibited phosphorylation of intact lysosomal enzymes (with K_i values similar to the K_m values of phosphorylation of intact enzymes) but did not inhibit phosphorylation of α-methylmannoside (Lang *et al.*, 1984).

Results of genetic complementation between mucolipidosis III cells also suggest that different subunits of the transferase mediate recognition and GlcNAc-P transfer. First two complementation groups (Gravel *et al.*, 1981), then three (designated MLIII$_A$, MLIII$_B$, and MLIII$_C$) were obtained; the group MLIII$_C$ consisted of five cell lines all of which showed normal levels of GlcNAc-P transfer to α-methylmannoside (Honey *et al.*, 1982). Does I-cell disease reflect more debilitating mutations at each of these loci? In one study two complementation groups were defined with I-cells (Shows *et al.*, 1982), although at least two I-cell lines failed to complement with the prototypical members of either group. Crosses with mucolipidosis III cells suggested that most I-cell lines belonged to the group MLIII$_A$ (Mueller *et al.*, 1983); exceptions observed were dismissed as resulting from intragenic complementation. Interpretation of negative results in the above studies is problematic: complementation was monitored by the increase in cellular lysosomal enzyme activity; signal to noise ratios were quite low. Use of a more sensitive assay (perhaps metabolic labeling with ^{32}P, followed by immunoprecipitation of a lysosomal enzyme, or measurement of *in vitro* GlcNAc-^{32}P transfer to either endogenous acceptors or added lysosomal enzyme) as well as some method for enrichment of heterokaryons over like–like fusions might eliminate some of the present confusion.

2. *GlcNAc Phosphodiesterase*

In order to phosphorylate lysosomal enzymes the cell must transfer GlcNAc-P; in order to form the Man 6-P recognition marker GlcNAc must subsequently be removed. Two preliminary studies using rat liver showed cosedimentation of GlcNAc phosphodiesterase* activity with that of galactosyltransferase, suggesting localization in the Golgi (Varki and Kornfeld, 1980a; Waheed *et al.,* 1981a); GlcNAc-P-transferase was also enriched in this fraction (Waheed *et al.,* 1981a). GlcNAc phospho-diesterase, assayed with oligosaccharides containing blocked phosphates, could be distinguished from lysosomal α-N-acetylglucosaminidase by cell fractionation, susceptibility to inhibitors, and through genetics: fibro-blasts from patients with the Sanfilippo B lysosomal storage disorder lacked α-N-acetylglucosaminidase but contained normal levels of GlcNAc phosphodiesterase activity (Varki and Kornfeld, 1980a).

GlcNAc phosphodiesterase activity was characterized in partially puri-fied (1000- to 2000-fold) preparations from human placenta (Waheed *et al.,* 1981b) and rat liver (Varki and Kornfeld, 1981). Activity from either source was unaffected by divalent cations or the lack thereof, was inhib-ited in the presence of both UDPGlcNAc, which appeared to act as a substrate of the enzyme, and phosphate; N-acetylglucosamine inhibited enzyme from liver but not that from placenta. Both liver and placental activities exhibited broad pH optima, although enzyme from the latter source retained significantly more activity below pH 5.0. The liver en-zyme removed blocking GlcNAc residues from all of the positions (see Section II,C) in which they appeared on the oligosaccharide, albeit there was a preference for those on the branch linked α-1,3 to the β mannose (see structure **I,** below). This was accentuated in deblocking of oligosac-charides with two diesters: a distinct lag was seen in removal of the second GlcNAc; analysis of the partially deblocked intermediate showed that hydrolysis had occurred almost exclusively on the α-1,3 branch. Neither the size of the underlying oligosaccharide (Man_{6-9}) nor the attach-ment of that oligosaccharide to a peptide affected GlcNAc phosphodies-terase activity.

Enzymatic removal of the blocking GlcNAc groups from lysosomal enzymes actually proceeds by a glycosidase rather than a phosphodies-terase mechanism (Varki *et al.,* 1983). Using $H_2{}^{18}O$ enrichment, enzyme purified from rat liver was shown to cleave the C—O bond, rather than the O—P bond, in hydrolysis of GlcNAc from both UDPGlcNAc and GlcNAc-P-(α-methyl)mannoside.

---

* α-N-Acetyl-D-glucosamine-1-phosphodiester N-acetylglucosaminidase; formerly called α-N-acetylglucosaminyl phosphodiesterase, see below.

A genetic defect in the phosphodiesterase might result in a phenotype similar to that observed in I-cell disease or mucolipidosis III, i.e., multiple enzyme deficiencies within the cell accompanied by elevated secretion of those enzymes. No such defect has been observed to date; phosphodiesterase activity in I-cell and mucolipidosis III fibroblast lines was normal or somewhat elevated (Reitman *et al.*, 1981). The possibility of a phosphodiesterase defect was suggested but not tested as an explanation for the elevation of lysosomal enzymes in plasma in members of one family (Alexander *et al.*, 1984).

3. Kinetics

Phosphorylation and removal of the GlcNAc residues were examined in endo H-sensitive oligosaccharides of β-glucuronidase, immunoprecipitated from the mouse macrophage line P388D$_1$* after pulses with [2-^{3}H]mannose and varying periods of chase (Goldberg and Kornfeld, 1981). Phosphorylation began at or about 15 min of labeling. Although 20–30 min is often required for complete trimming of glucose from the Glc$_3$ Man$_9$GlcNAc$_2$ oligosaccharide that is transferred *en bloc* to protein (for example, see Kornfeld *et al.*, 1978), even the earliest phosphorylated oligosaccharide was without glucose. Oligosaccharides containing one blocked phosphate (i.e., phosphodiester) were observed first; radioactivity in this species continued to increase through 40 min of chase. From 20 to 40 min of chase, increases in the radioactivity representing two blocked phosphates and one and two unblocked phosphates were seen. From 80 to 160 min, label in oligosaccharides with blocked phosphates decreased, while it increased in those with unblocked phosphates.

Phosphorylated oligosaccharides on β-glucuronidase were more processed (i.e., contained fewer mannoses) than were the neutral oligosaccharides (examined up to 80 min of chase); oligosaccharides with two diesters were more processed than those with one, and oligosaccharides with phosphomonoesters contained fewer mannoses than those with diesters (Goldberg and Kornfeld, 1981). A small degree of selectivity was observed with respect to phosphorylation of oligosaccharides on different glycosylation sites of β-glucuronidase. One of the three sites [examined as tryptic glycopeptides separable on high-performance liquid chromatography (HPLC)] contained about half the phosphorylated oligosaccharide observed at each of the other two; this same site contained a lower fraction of phosphorylated oligosaccharide in monoester form, and it was the only site that never contained complex oligosaccharides, suggesting that

---

* As described in Section III,B, this cell line is without the 215 kDa, cation-independent mannose 6-phosphate receptor.

its oligosaccharide may be generally less accessible to modifying enzymes.

C. Effects of Oligosaccharide Structure

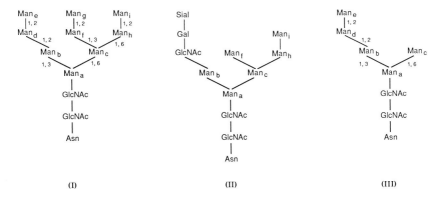

(I)   (II)   (III)

The structures deduced for the major phosphorylated asparagine-linked oligosaccharides are described with reference to structure I above. In the first study murine lymphoma BW5147 cells were labeled for 3 hr with [2-^{3}H]mannose, then total endo H-releasable phosphorylated oligosaccharides were examined (Varki and Kornfeld, 1980b). In the second, β-glucuronidase was purified from human spleen, oligosaccharides were generated with endo H, and then oligosaccharides were labeled by reduction with NaB^{3}H$_4$ (Natowicz *et al.,* 1982). The major anionic oligosaccharide observed in both studies (>50% of the radioactivity) contained a single phosphate in diester linkage. In BW5147 phosphorylation occurred at any of five positions on the prototypical Man$_9$ structure, with only the three core mannoses (a, b, and c) and Man$_g$ not phosphorylated. Spleen β-glucuronidase also contained no phosphate on the core; because of the position of the radiolabel it was not possible to distinguish between Man$_g$ and Man$_i$. Whereas from BW5147 approximately 50% of the oligosaccharides containing a single phosphodiester were phosphorylated on Man$_h$ or Man$_f$, oligosaccharides from spleen β-glucuronidase with one phosphodiester were not phosphorylated at either of these positions; instead a significant portion of these oligosaccharides were phosphorylated at Man$_d$. The majority of the oligosaccharides from both sources that contained two phosphates had one on the α-1,3 branch and the second on either of the other two branches.

Sizing of the phosphorylated oligosaccharides indicated that they ranged from Man$_6$ to Man$_9$; the lack of phosphorylated Man$_5$ in the structures from BW5147 is puzzling: phosphorylation at Man_f and/or Man_h would not be expected to interfere with trimming of any of the four α-1,2-linked mannose residues. Also, recalling the lag observed *in vitro* with respect to removal of the second GlcNAc by GlcNAc phosphodiesterase (Varki and Kornfeld, 1981; discussed in Section II,B,2), oligosaccharides containing both mono- and diesters might have been expected, but they were not seen.

In a subsequent study additional phosphorylated structures were identified among total endo H-releasable anionic oligosaccharides from P388D₁ macrophages after a 14-hr pulse with [2-³H]mannose (Varki and Kornfeld, 1983). A very minor species containing both monoester and diester was observed. A novel oligosaccharide had a hybrid structure (II), in which the α-1,3 branch contained the sugars characteristic of complex oligosaccharides [GlcNAc, galactose (Gal), and one or two sialic acids (Sial)], with Man_h or Man_i phosphorylated. This hybrid oligosaccharide, with phosphate present as the diester, was a very minor species within the cell but was a major species (with phosphate as monoester) in material obtained from the medium.

Although a large amount of the phosphorylation described above involves the mannoses emanating from the α-1,6 side of the Man_3 core (i.e., $Man_{f,g,h,i}$), these mannoses appear unnecessary for either phosphorylation or uptake of lysosomal enzymes. In synthesis of the lipid-linked $Man_9GlcNAc_2$ intermediate, these four mannoses are transferred to the growing oligosaccharide from mannosylphosphoryldolichol (Man-P-dol) (Rearick *et al.*, 1981). Defects in Man-P-dol synthase result in formation of a truncated oligosaccharide of structure III. Lysosomal enzymes from two mutants defective in the synthesis of Man-P-dol [B4-2-1, a CHO cell mutant (Stoll *et al.*, 1982), and Thy-1⁻, class E, a mutant of the lymphoma line BW5147 (Gabel and Kornfeld, 1982)] were taken up by recipient human fibroblasts via a Man 6-P-inhibitable pathway at a rate similar to that observed with enzymes from parent cells. Acid hydrolysis of anionic oligosaccharides from Thy-1⁻ demonstrated the presence of Man 6-P; structural analysis showed that phosphorylated oligosaccharides contained four mannoses with phosphate on Man_d. Immunoprecipitation of β-glucuronidase secreted by B4-2-1 after metabolic labeling showed that it was phosphorylated and that its oligosaccharides were resistant to hydrolysis by endo H (Robbins *et al.*, 1987), characteristic of structures with terminal Man_c (Kobata, 1978).

In contrast to the apparent irrelevance of mannoses on the α-1,6 side of the oligosaccharide, alterations on the α-1,3 linked branch affect phosphorylation and uptake of the lysosomal enzymes. The CHO mutant B211 fails to glucosylate lipid-linked oligosaccharide (Krag, 1979); thus, the oligosaccharide transferred to nascent protein has an exposed Man on the α-1,3 branch. B211 was shown to be pleiotropically deficient in intracellular lysosomal enzymes; α-L-iduronidase and β-hexosaminidase from B211 were not phosphorylated and instead contained complex oligosaccharides (Krag and Robbins, 1982). Lysosomal enzymes secreted by I-cell fibroblasts also contain elevated levels of complex oligosaccharides (Miller *et al.*, 1981), but, unlike I-cells, B211 exhibited normal levels of GlcNAc-P-transferase activity, with exogenous human β-hexosaminidase used as acceptor (Krag and Robbins, 1982). It has been suggested that the oligosaccharides that are processed to complex structures are those located at relatively exposed sites on the glycoprotein (Hsieh *et al.*, 1983); oligosaccharides that are acceptors for GlcNAc-P may be similarly located, based on the efficiency of their hydrolysis from native β-glucuronidase by endo H (Natowicz *et al.*, 1979; also see Section II,B,3). In I-cells or B211, without phosphorylation to block further processing, complex oligosaccharides would be formed. But how in B211 mutant does the absence of glucose residues, destined to be removed from the oligosaccharide minutes after transfer to protein, cause loss of phosphorylation? One possibility is that without glucoses to impede it, the α-mannosidase in the endoplasmic reticulum (Bischoff and Kornfeld, 1983) removes too many α-1,2-linked mannoses for phosphorylation to occur. This is pure speculation.

Lending credence to the adage that "you can't live with 'em and you can't live without 'em", failure to remove glucoses from the protein-linked oligosaccharide also inhibits phosphorylation. The lymphoma cell mutant PhaR2.7 is defective in glucosidase II and thus is impaired in removal of the inner two (α-1,3-linked) glucoses (Reitman *et al.*, 1982). After metabolic labeling with [2-^{3}H]mannose, phosphorylated oligosaccharides from parental cells (isolated by cleavage with endo H) contained four times as much radioactivity as did phosphorylated oligosaccharides from this mutant (Gabel and Kornfeld, 1982). No glucose-containing oligosaccharide from PhaR2.7 had more than a single phosphate; phosphorylation occurred only on Man_f, Man_h, or Man_i, never on the glucosylated branch.

Similarly, on incubation with deoxynojirimycin, an inhibitor of glucosidase I (Saunier *et al.*, 1982), B4-2-1, the CHO mutant transferring the Man_5 structure (see above), synthesized and secreted β-glucuronidase that was almost devoid of phosphate, as judged by ^{32}P incorporation (Robbins *et al.*, 1987); thus, the presence of glucose inhibits phosphorylation on the α-1,3 branch. Levels of ^{32}P in enzyme secreted by deoxynojirimycin-treated wild-type CHO cells were decreased approximately 50%; Man 6-P-dependent uptake of these secreted enzymes was decreased 6-fold (Robbins *et al.*, 1987), suggesting that the phosphates present were either blocked or located on the oligosaccharides in positions unattractive to the Man 6-P receptor (see below). In mutant PhaR2.7 radioactivity in oligosaccharides containing phosphomonoesters was reduced even more than in those with diesters (Gabel and Kornfeld, 1982); phosphodiesterase activity may be inhibited by the presence of glucose either on the oligosaccharide substrate or on its own oligosaccharides. A complication to be borne in mind when considering secondary phenomena in glycosylation mutants (or on treatment with inhibitors of glycosylation) is that the enzymes mediating those events may themselves be glycoproteins whose activities are affected by changes in oligosaccharide structure.

The effectiveness of the various phosphorylated oligosaccharides as ligands was compared measuring either uptake by recipient fibroblasts (Creek and Sly, 1982; Natowicz et al., 1983) or retardation on columns of immobilized Man 6-P receptor* (Varki and Kornfeld, 1983; Fischer et al., 1982). In all cases oligosaccharides containing two phosphomonoesters were superior with respect to binding and/or uptake; the K_{uptake} for these oligosaccharides was about 40 nM (Natowicz et al., 1983), and elution from the affinity column required Man 6-P (Varki and Kornfeld, 1983; Fischer et al., 1982). Oligosaccharides with a single phosphomonoester also acted as ligands (K_{uptake} 300 nM) and were retarded by the column. Oligosaccharides with a single phosphodiester were not taken up (Creek and Sly, 1982) and coeluted with neutral oligosaccharides. The sole disparity observed between in vivo and in vitro assays came with oligosaccharides containing two phosphodiesters; although some uptake was observed (Creek and Sly, 1982), no retardation was seen on passage over the affinity column. This discrepancy appears to result from removal of the blocking groups by an enzyme present either in the medium or on the surface of recipient human fibroblasts (Natowicz et al., 1983).

Both the position of the phosphomonoester and the size of the underlying oligosaccharide were of significance with respect to binding to the receptor affinity column (Varki and Kornfeld, 1983). The results may be summarized as follows: among oligosaccharides containing a single phosphomonoester, phosphorylation on the α-1,3 branch (Man$_d$) effected higher affinity than phosphorylation on the α-1,6 side of the oligosaccharide (Man$_f$, Man$_h$, or Man$_i$); the affinity of oligosaccharides phosphorylated on the α-1,6 side was increased when they contained fewer mannoses, as indicated both by comparison of oligosaccharide size with

---

* As described in Section III, there are two distinct Man 6-P receptors. However, the method used for purification of receptor for these affinity columns (i.e., by adsorption to lysosomal enzymes from Dictyostelium discoideum), and the absence of divalent cations from the binding studies dictate that the binding measured was to the 215 kDa cation-independent receptor.

relative affinity and by the increase in affinity obtained after treatment with α-mannosidase. Even among oligosaccharides containing two phosphomonoesters, a lower affinity fraction was observed whose interaction with the column was increased by enzymatic removal of mannoses.

Comparing the relative receptor-binding abilities of the various oligosaccharides with the predominant structures observed either in BW5147 after 3 hr of labeling (Varki and Kornfeld, 1980b) or in steady-state β-glucuronidase (Natowicz et al., 1982), it seems that what can be bound cannot be found and vice versa. Do the major phosphorylated oligosaccharides obtained reflect cellular synthetic activities, or do they accumulate because they are poor ligands for the receptor, and thus are not carried forward to the lysosomes where phosphatase(s) could render them neutral? In this regard BW5147 lymphoma cells were an unlucky choice for structural analyses; phosphorylated oligosaccharides decreased in these cells with a $t_{1/2}$ of 1.4 hr (Gabel et al., 1982), whereas in human fibroblasts phosphate on oligosaccharides of β-hexosaminidase α and β chains, cathepsin D, α-glucosidase (Hasilik and Neufeld, 1980b), and α-L-iduronidase (Myerowitz and Neufeld, 1981) appeared stable for at least 22 hr. If cell types differ with respect to the oligosaccharides found on their lysosomal enzymes, does this reflect differences in synthesis, or in the subsequent action of lysosomal hydrolases on those oligosaccharides, or in the selectivity of the receptor(s) that divert the enzymes to lysosomes?

III. THE RECEPTORS

A. The 215 kDa Cation-independent Receptor

Uptake of lysosomal enzymes from the growth medium implied the presence of receptor on the plasma membrane. Incubation at 0°C of human fibroblasts deficient in α-L-iduronidase with that enzyme resulted in approximately 14,000 enzyme molecules (measured by activity) bound per cell (Rome et al., 1979a); binding was saturable (K_D 1.1 nM) and was inhibited in the presence of Man 6-P (K_i 100 μM). Release of bound enzyme was greatly accelerated by addition of Man 6-P. In contrast to other carbohydrate-binding proteins, cell surface binding of α-L-iduronidase was independent of Ca^{2+} and Mg^{2+}. A study of β-glucuronidase binding to fibroblasts (Gonzalez-Noriega et al., 1980) confirmed these results; in addition, rapid release of bound enzyme was demonstrated when extracellular pH was decreased below 6.0. Total fibroblast membranes, following removal of nuclei, bound about six times more β-glucuronidase than did intact cells; characteristics of binding to intracellular and surface sites were very similar (Fischer et al., 1980a). Membranes prepared from liver and a variety of other tissues also showed a very similar binding activity (Fischer *et al.,* 1980b), although the fraction of receptors occupied with endogenous ligands was much higher in membranes from liver (80%) than from fibroblasts (10%).

Man 6-P receptor was first purified from bovine liver (Sahagian *et al.,* 1981). Membrane proteins, solubilized with 1% Triton X-100, were incubated with β-galactosidase, then the receptor–enzyme complex was immunoprecipitated with antiserum directed against the enzyme. Bound enzyme was released by addition of Man 6-P, receptor was iodinated, adsorbed to a β-galactosidase affinity column, and then eluted with Man 6-P. On sodium dodecyl sulfate (SDS)–gel electrophoresis the purified receptor had an M_r of 215 K. Similar proteins were obtained from human fibroblasts and CHO cells [previously shown to have Man 6-P-inhibitable uptake of lysosomal enzymes (Robbins, 1979; Robbins *et al.,* 1981; Robbins and Myerowitz, 1981)]. Binding to purified receptor appeared identical to that measured with intact human fibroblasts or the membranes thereof: binding of high uptake β-galactosidase had a K_D of 20 nM, was inhibited by Man 6-P (K_i 60 μM), was dependent on the presence of phosphorylated oligosaccharide on the β-galactosidase, showed a pH optimum between 5.7 and 6.5 with a precipitous decrease in binding activity below pH 5.7, and did not require divalent cations. Subsequent purifications of the receptor (e.g., Sahagian *et al.,* 1982; Steiner and Rome, 1982) utilized immobilized yeast phosphomannan as the affinity matrix.

The natural history of the receptor was examined in CHO cells by metabolic labeling, immunoprecipitation, and SDS–gel electrophoresis (Sahagian and Neufeld, 1983). Receptor was first detected (50 min after initiation of labeling) in a form sensitive to endo H (treatment reduced the M_r by ~10K); sensitivity was lost over 2.5 hr. Phosphorylation of receptor, predominantly on serine residues, occurred late in maturation, i.e., on the endo H-resistant form. In monolayer cultures turnover of phosphate on the receptor was significantly faster than turnover of the protein itself ($t_{1/2}$ 2 and 16 hr, respectively), whereas in suspended cultures turnover was more coordinate (4 and 6 hr, respectively). Receptor present on the cell surface for at least a portion of its lifetime, identified by iodination of intact cells at 0°C, had a life span similar to that of total metabolically labeled receptor ($t_{1/2}$ 7 hr in suspended cultures).

Structure–function relationships were also examined (Sahagian and Neufeld, 1983). Both endo H-sensitive and -resistant forms of the receptor bound to phosphomannan, indicating that neither complex oligosaccharides nor phosphorylated serines were required; removal of oligosaccharides from the endo H-sensitive form did not inhibit binding. Although these posttranslational modifications appeared unnecessary for ligand binding, acquisition of binding activity occurred only 70 min after receptor synthesis. A potential problem with respect to this last observation is that, if newly synthesized receptors were occupied by endogenous lysosomal hydrolases, then those receptors would not bind phosphomannan.

In a subsequent study (Goldberg *et al.*, 1983) of receptor biosynthesis in the murine lymphoma BW5147* using an independent preparation of antiserum, two forms of receptor of slightly different electrophoretic mobilities were detected at early times (20 min pulse, 10 min chase). Differences between these forms could not be accounted for by oligosaccharide processing. In contrast to receptor from CHO cells, BW5147 receptor retained some endo H-sensitive oligosaccharides even 24 hr after synthesis.

Under some circumstances proteins of M_r 40K–90K have been found associated with, or reacting to antisera prepared against, the 215 kDa Man 6-P receptor. Solubilization from bovine liver in Zwittergent yielded a 57kDa polypeptide, as well as the 215 kDa species, on elution from a phosphomannan column with Man 6-P (Mitchell *et al.*, 1983). Storage of the purified 215 kDa form in Zwittergent resulted eventually in its complete conversion to the 57 kDa form. In a second study antiserum prepared against electrophoretically purified 215 kDa receptor decorated almost exclusively a 43–49 kDa polypeptide in Western blots of homogenates of Syrian hamster liver. This polypeptide, present also in hamster serum, appeared to be soluble rather than membrane bound and was associated in S—S bonded complexes (M_r ~190K) (Maler *et al.*, 1985). No S—S bonded complexes had been seen with the CHO cell Man 6-P receptor (Sahagian and Neufeld, 1983); however, several polypeptides of M_r 50K–85K cross-reacting with anti-215 kDa antiserum were found to be secreted from CHO cells under conditions that effected loss of Man 6-P-binding activity at the cell surface (Sahagian, 1984).

B. Binding-deficient Cells

CHO cell mutants defective in binding and uptake of lysosomal enzymes were isolated by selecting for cells resistant to a conjugate composed of 6-phosphopentamannose covalently linked to ricin (Robbins *et al.*, 1981). Two of the conjugate-resistant mutants showed about 10% of parental uptake of bovine β-galactosidase (Robbins *et al.*, 1981) and markedly reduced uptake of β-hexosaminidase secreted by human fibroblasts (Robbins and Myerowitz, 1981). In mutant A10-2-4 β-galactosidase appeared to be less tightly bound at the cell surface, i.e., dissociation with-

---

* A peculiarity of BW5147 cells is that they lack endocytosis of lysosomal enzymes as well as Man 6-P-dependent binding at the cell surface (Gabel *et al.*, 1983).

out added Man 6-P was more rapid than normal; in mutant B4-2-1 *cell surface binding exhibited sigmoidal rather than Michaelis–Menten kinetics, and saturation required about 10 times more ligand than did saturation of binding to parental cells (Robbins et al., 1981). No differences were observed in the 215 kDa receptors immunoprecipitated from mutants and parent after electrophoresis on SDS–polyacrylamide gels (G. G. Sahagian, unpublished). Man 6-P-dependent binding to total membranes from these mutants was not examined. Both CHO cell mutants secreted lysosomal enzymes in elevated amounts; enzymes were secreted as high uptake, precursor forms. But, with the exception of α-mannosidase, the major fraction of the activity of each enzyme examined was intra- rather than extracellular (Robbins and Myerowitz 1981).

Membranes of a number of cell types were tested for Man 6-P receptor activity by measurements of binding (in the absence of divalent cations) of purified, iodinated *D. discoideum* β-hexosaminidase (Gabel et al., 1983). No binding was detected with membranes from two murine macrophage lines, $P388D_1$ and J774.2, or with L cells.[†] Lack of receptor activity cannot be ascribed to the murine origin of these three lines, since the murine lymphoma BW5147 contained about two-thirds of the binding activity measured in CHO and human fibroblast membranes; nor is it a general phonomenon of mononuclear phagocytes, since a human monocyte exhibited significant binding activity. Also, primary rabbit alveolar macrophages showed both binding and cation-independent uptake of *D. discoideum* α-mannosidase (Shepherd et al., 1983). Absence of Man 6-P receptor activity in $P388D_1$ and J774.2 correlated with secretion of the majority of newly synthesized β-hexosaminidase and β-glucuronidase (based on enzyme activity); however, these macrophages retained about 10 times more enzyme than did the receptor-positive murine lymphoma. No 215 kDa receptor was found in $P388D_1$ on metabolic labeling with mannose or methionine followed by immunoprecipitation (Goldberg et al., 1983).

The murine myeloma MOPC 315 contained very low (2% of CHO) levels of binding activity, yet lysosomal hydrolases were not secreted from MOPC 315 in elevated amounts (Gabel et al., 1983). Biosynthetic studies showed synthesis of the 215 kDa receptor followed by rapid degradation (Goldberg et al., 1983). Among lymphoid cells it appears that T cell lines (e.g., BW5147) have 215 kDa receptor activity, whereas B cell lines do not (Gabel et al., 1984a).

* B4-2-1 lacks Man-P-dol synthase activity, as discussed in Section II,C.

† Absence of Man 6-P-dependent binding is peculiar to this L cell line (Gabel et al., 1984a); other L cells have been shown to have Man 6-P binding and uptake (Gabel and Foster, 1986) as well as the 215 kDa receptor (S. M. Laurie and A. R. Robbins, unpublished results).

C. The 46 kDa Cation-Dependent Receptor

Because P388D$_1$ macrophages contained high levels of intracellular acid hydrolases that sedimented on Percoll gradients at the density of lysosomes (Gabel *et al.*, 1983), it seemed that those cells must have an alternate receptor for enzyme compartmentalization. Using as ligand β-galactosidase purified from its own secretions, membranes from P388D$_1$ were shown to have significant binding activity in the presence of 10 mM Ca^{2+}, Mn^{2+} and Mg^{2+} (Hoflack and Kornfeld, 1985a); Mn^{2+} caused maximal stimulation. Binding in the absence of divalent cations was decreased to approximately 20%. Surprisingly, these binding sites appeared to recognize Man 6-P—among the sugars and sugar phosphates tested only Man 6-P was a potent inhibitor of enzyme binding (K_i 10 μM); also, treatment of β-galactosidase with phosphatase significantly decreased its binding activity. As was observed with the 215 kDa receptor, bound ligand could be released by lowering the pH ($<$5.5). Membranes from P388D$_1$ bound little or not *D. discoideum* β-hexosaminidase at concentrations of that enzyme manyfold higher than those required for maximal binding to membranes of human fibroblasts. The cation-dependent Man 6-P receptor appears unable to recognize the methylphosphomannosyl diesters present on the *D. discoideum* enzyme (Gabel *et al.*, 1984b).

The cation-dependent receptor was purified not only from P388D$_1$ but also from bovine liver (Hoflack and Kornfeld, 1985b). Membrane proteins, solubilized in 1% Triton X-100, 0.1% sodium deoxycholate, were applied to a phosphomannan affinity column in the presence of 10 mM Mn^{2+}; proteins eluted with Man 6-P were then passed over an affinity column of *D. discoideum* lysosomal enzymes in the absence of divalent cations; the cation-dependent receptor was recovered from the flowthrough, and the 215 kDa receptor was subsequently eluted with Man 6-P. The cation-dependent receptor had an M_r of 46K on SDS–gel electrophoresis.

Binding properties of purified, iodinated 46 kDa receptor were examined using various affinity matrices (Hoflack and Kornfeld, 1985b). In the presence of 10 mM Mn^{2+} receptor bound phosphomannose and could be eluted at pH 4.6, or with Man 6-P, but not with Glu 6-P or Man. About 50% of the receptor applied to a *D. discoideum* lysosomal enzyme column bound, in contrast to results obtained with membranes from P388D$_1$ (Hoflack and Kornfeld, 1985a). In the absence of divalent cations, receptor bound only weakly to phosphomannose and did not bind to *D. discoideum* enzymes.

Treatment of the 46 kDa receptor with endo H reduced the M_r to 42 K, treatment with endo F, which cleaves complex as well as high-mannose oligosaccharides (Elder and Alexander, 1982), to 28K. Binding to phosphomannan was reduced but not abolished on treatment with the latter enzyme. On gel filtration the endo F-treated receptor chromatographed with a molecular weight of 76K, suggesting that it may exist as a trimer. No cross-reactivity between the 215 kDa and 46 kDa receptors was observed on radioimmunoassays employing antisera prepared against each of the receptors (Hoflack and Kornfeld, 1985b).

IV. ENDOCYTOSIS

A. The Road Taken

The endocytic pathway has been examined by both electron microscopy and cell fractionation. Because cells also possess an intracellular pathway for delivery of lysosomal enzymes, confusion may arise between these pathways when other than an extracellularly added tracer is used. With this precautionary note, I have included in this section results that were interpreted by their authors as representing the endocytic pathway.

CHO cells incubated at 4°C with β-galactosidase followed by anti-β-galactosidase–peroxidase showed that label accumulated in coated pits on the cell surface (Willingham *et al.*, 1981). Double labeling with β-galactosidase and α_2-macroglobulin–colloidal gold showed these ligands together in coated pits. Surface labeling of pre-fixed CHO cells with antibody directed against the 215 kDa Man 6-P receptor primarily decorated coated pits; some label was observed on other areas of the plasma membrane (Willingham *et al.*, 1983). On Clone 9 hepatocytes the majority of surface 215 kDa receptor also was in coated pits, as determined by immunoperoxidase localization (Brown *et al.*, 1984). In cryosections of liver parenchymal cells, gold immunostaining showed the 215 kDa receptor randomly distributed on the sinusoidal plasma membrane and in coated pits invaginating from this membrane (Geuze *et al.*, 1984a); in double-labeling studies, the asialoglycoprotein and Man 6-P receptors were colocalized in both these areas (Geuze *et al.*, 1984b).

Warming CHO cells to 37°C for 8 min, after incubation at 4°C with β-galactosidase then anti-β-galactosidase–peroxidase, resulted in movement of the tracer to intracellular vesicles (Willingham *et al.*, 1981), some of which had tubular projections and resembled the structures currently known as CURL [compartment of uncoupling of receptors and ligands (Geuze *et al.*, 1983)]. Immunostaining of liver cryosections showed the 215 kDa receptor in CURL tubules (Geuze *et al.*, 1984a), again accompanied by the asialoglycoprotein receptor (Geuze *et al.*, 1984b). In cryosections of Hep G_2 (human hepatoma) cells, the 215 kDa receptor and a lysosomal enzyme were found together in CURL tubules, whereas only the enzyme was observed in large CURL vesicles (Geuze *et al.*, 1985).

In these last three studies there could be confusion between the endocytic and intracellular pathways: There is a striking resemblance between CURL and GERL (the trans-Golgi reticulum, or TGR), where large quantities of asialoglycoprotein and 215 kDa receptors as well as of lysosomal enzymes are located (Geuze *et al.*, 1984a,b, 1985); the authors contend that CURL and GERL are distinguishable by relative proximity to the cell surface and by the paucity of albumin in CURL. The former could be influenced by plane and depth of section; with regard to the latter, lysosomal enzymes have to be segregated from the secretory pathway (and albumin) at some point. Also, do hepatocytes in liver and Hep G_2 cells have Man 6-P-dependent endocytosis? Isolated hepatocytes exhibited Man 6-P-inhibitable binding but little or no endocytosis of β-galactosidase (Sahagian, 1981).

After 15 min at 37°C most of the endocytosed β-galactosidase was in lysosomes in CHO cells (Willingham *et al.*, 1981). Warming CHO cells after incubation with anti-215 kDa receptor antibody, followed by anti-antibody–peroxidase, resulted in the appearance of antibody in the lysosomes (Willingham *et al.*, 1983); whether antibody was still associated with receptor is unknown. No 215 kDa receptor was seen in lysosomes on staining permeabilized cells with antibody (Willingham *et al.*, 1983); with gold immunostaining procedures, lysosomes in cryosections of liver (Geuze *et al.*, 1984a) or Hep G_2 cells (Geuze *et al.*, 1985) showed lysosomal enzymes but little or no receptor. Immunoperoxidase staining of rat liver (Brown and Farquhar, 1984a), Clone 9 hepatocytes (Brown *et al.*, 1984), and human fibroblasts (Brown and Farquhar, 1984b) did show the 215 kDa receptor in structures resembling lysosomes; however, the authors pointed out that many organelles which stained for a lysosomal antigen were without receptor and that the lysosome-like structures containing receptor might be endosomes.

With Percoll gradients (8.2%) designed to separate lysosomes from light membranes (e.g., plasma membrane, endosomes, Golgi), fractionation of CHO cells incubated with ^{125}I-labeled β-galactosidase at 0°C then warmed for varying periods showed label exclusively in light membranes after 2 min at 34°C, approximately equal distribution in light membranes and lysosomal fractions after 18 min, and predominantly lysosomal localization after 30 min (Sahagian and Neufeld, 1983). Fractionation of parallel cultures in which the cells were iodinated at 0°C, then incubated with saturating levels of β-galactosidase at 34°C, showed immunoprecipitable, iodinated 215 kDa receptor remaining with the light membrane fraction through 30 min. Fractionation of CHO cells after metabolic labeling also showed little or no immunoprecipitable receptor in lysosomes (Sahagian and Neufeld, 1983).

Gradients of higher density separated two lysosomal fractions (Rome *et al.*, 1979b) in homogenates of human fibroblasts, the lighter of which cosedimented with marker enzyme activities for Golgi and plasma membranes. Endocytosed lysosomal enzyme was found in both lysosomal fractions. Using sequential gradients whereby dense lysosomes were first separated from light lysosomes and light membranes, then the latter activities were separated on a second gradient, it was shown in pulse–chase experiments that lysosomal enzymes endocytosed by CHO cells moved from light membranes through light lysosomes into the dense lysosomal fraction (Roff *et al.*, 1986). Assuming that fluid-phase and receptor-mediated endocytosis follow the same pathway, gradient fractions prepared after incubation of CHO cells with horseradish peroxidase were examined by electron microscopy: in the light membrane fraction peroxidase activity was in small vesicles, short tubules, and multivesicular-like bodies; in the light lysosome fraction it was in multivesicular-like bodies and lysosomes; and in the dense lysosome fraction it was in residual bodies (C. F. Roff and C. Oliver, unpublished observations).

B. Disruption of Endocytosis

1. Weak Bases

Incubation of human fibroblasts with chloroquine (2–5 μM) caused both increased secretion of endogenously synthesized lysosomal enzymes and decreased endocytosis of exogenously added enzymes (Wiesmann *et al.*, 1975). Inhibition of enzyme endocytosis was subsequently demonstrated for other amines (e.g., NH_4Cl, procaine, cyclohexylamine); these inhibitors altered V_{max} not K_{uptake} (Sando *et al.*, 1979). The effect of the amines on endocytosis appeared somehow related to their ability to rapidly increase lysosomal pH, as demonstrated in macrophages (Ohkuma and Poole, 1978).

Comparison of the effects of weak bases on I-cells versus β-glucuronidase-deficient fibroblasts led to the conclusion that internalized enzyme–Man 6-P receptor complexes normally dissociate in an acidic organelle, whence the receptor recycles back to the cell surface. By raising intraorganellar pH, the weak bases prevent dissociation, thus blocking recycling and resulting in depletion of surface receptors (Gonzalez-Noriega *et al.*, 1980). The supporting data may be summarized as follows: (1) Incubation for 3 hr with cycloheximide did not inhibit enzyme uptake.

Comparison of the V_{max} of uptake with the number of cell surface binding sites indicated that receptors must be replaced every 5 min; however, the total number of Man 6-P binding sites (surface plus intracellular) was not sufficient to allow uptake to continue for 3 hr without protein synthesis unless receptors were reutilized, i.e., recycled. (2) Preincubation with the weak base for increasing periods of time caused increased inhibition of uptake (provided that chloroquine was also included during the uptake assay) only if high uptake enzymes were present; i.e., I-cells behaved like the β-glucuronidase-deficient cells only when high uptake enzyme was provided to the I-cells. (3) Surface binding activity, measured at 4°C, decreased in cells preincubated with chloroquine or NH_4Cl at 37°C, to 60 and 35% of control values, respectively. Inclusion of Man 6-P during incubation with the weak bases prevented some of the loss of binding activity. (4) As described in Section III, release of cell surface-bound β-glucuronidase was demonstrated below pH 6.

Incubation of CHO cells with NH_4Cl also caused loss of cell surface-associated Man 6-P binding activity (<1% of controls after 12 hr); however, loss of receptor molecules from the cell surface did not correspond to loss of activity. Iodination at 0°C followed by immunoprecipitation of the 215 kDa Man 6-P receptor and electrophoresis showed that cells preincubated with NH_4Cl for 12 hr had 30–60% as much radioactive receptor as did control cells (Robbins et al., 1983). Similar results were obtained in measurements of ^{125}I-labeled anti-215 kDa receptor antibody binding to Hep G_2 cells pretreated with primaquine (Geuze et al., 1985). Absence of surface binding in the NH_4Cl-treated CHO cells did not reflect normal occupancy of the receptors with endogenous lysosomal enzymes: prior to addition of labeled ligand, cells were incubated with Man 6-P to release bound enzymes (Robbins et al., 1983); also, preincubation at pH 4.5 at 4°C did not increase surface binding activity of NH_4Cl-treated CHO cells (A. R. Robbins, unpublished observations).

Presumably, incubation of the cells with NH_4Cl effects some alteration in the receptor not detected on SDS gels. The Man 6-P binding site could be modified, or hidden from ligand. Alternatively, the aggregation state of the receptor may be changed. A number of observations indicate that interactions between ligands and Man 6-P receptors at the cell surface are polyvalent: binding and uptake of a multivalent phosphomannan fragment were about 100 times more efficient than that of monophosphopenta-mannose, as compared on a per mole of PO_4 basis (Fischer et al., 1980c). "Substrate inhibition" of binding and uptake was observed with β-galactosidase (Robbins et al., 1981) and the multivalent phosphomannan fragment (Fischer et al., 1980c), respectively, at ligand concentrations greater than 20 times those required for half-maximal activity. Release of receptor-bound lysosomal enzymes, of itself extremely slow, was markedly accelerated on addition of Man 6-P (Rome *et al.*, 1979a; Fischer *et al.*, 1980a). Finally, human fibroblasts internalized a (Man 6-P)$_{20}$ neoglycoprotein about 50 times more efficiently than would have been predicted by the ability of the neoglycoprotein to inhibit lysosomal enzyme uptake (Karson *et al.*, 1980), suggesting that the distribution of receptors on the human fibroblast surface allowed multivalent interaction with this ligand.

Regardless of the mechanism by which the weak bases diminish Man 6-P binding at the cell surface, they most likely inhibit endocytosis by raising the pH of an acidic organelle, thus preventing dissociation of ligand from receptor. Initially the lysosome was thought to be this organelle; however, as described above, little or no 215 kDa Man 6-P receptor was found in lysosomes. This contradiction was reconciled with the discovery of acidic, prelysosomal endocytic vesicles, through use of fluoresceinated macromolecules that served simultaneously as endocytic tracers and pH probes (Ohkuma and Poole, 1978). Fluoresceinated α_2-macroglobulin was observed to enter an acidic compartment in BALB/c 3T3 cells within 20 min of its internalization (Tycko and Maxfield, 1982), by which time in these cells less than about 10% of the ligand had entered lysosomes. Addition of chloroquine, NH$_4$Cl, or other weak bases rapidly (within 2 min) increased the pH of this compartment above 6.0 (Maxfield, 1982). Fractionation of fibroblasts and macrophages after incubation with fluoresceinated dextran, a marker of fluid-phase endocytosis, showed tracer in both lysosomes and a light membrane fraction, separated on Percoll gradients, each of which was acidified on addition of ATP *in vitro* (Galloway *et al.*, 1983).

2. *Mutants*

A variety of cell types (many of which cannot properly be referred to as mutants) do not show Man 6-P-dependent endocytosis of lysosomal enzymes. In no case is this completely understood, and in most cases we know precious little.

P388D$_1$ macrophages lack the 215 kDa receptor (Goldberg *et al.*, 1983); membranes of P388D$_1$ contain a 46 kDa, cation-dependent Man 6-P receptor that binds lysosomal enzymes secreted by these cells (Hoflack and Kornfeld, 1985a,b; see Sections III,B and III,C). No binding or uptake of enzymes was observed with intact P388D$_1$ (Gabel *et al.*, 1983), suggesting that the 215 kDa receptor may have sole responsibility for Man 6-P-dependent endocytosis. Consistent with this we have recently isolated mutant LTk$^-$ cells that are without detectable 215 kDa receptor and exhibit less than 10% of parental levels of Man 6-P-dependent endocytosis (S. M. Laurie and A. R. Robbins, unpublished).

BW5147 lymphoma cells lack uptake and surface binding activity (Gabel *et al.*, 1983), although the 215 kDa receptor is present in total membranes from those cells. Whether or not receptor is present on the plasma membrane has not been reported.

Decreases observed in surface binding activity in both B4-2-1 and A10-2-4 [the Man 6-P–ricin-resistant CHO cell mutants (see Section III,B)] are not entirely consistent with those in endocytosis (Robbins *et al.*, 1981). With B4-2-1, greater than parental binding activity was obtained at elevated concentrations (0.3 μM) of β-galactosidase, yet at this concentration uptake was still only 20% of normal. In A10-2-4, reduced binding correlated with accelerated dissociation of bound enzyme, the consequences of which are unclear with respect to uptake.

Hepatocytes, either freshly isolated from rat liver or kept in primary culture for 18 hr, showed high levels of Man 6-P-dependent cell surface binding of β-galactosidase but little or no Man 6-P-inhibitable uptake of that enzyme; parallel populations showed both binding and uptake of asialoglycoprotein (Sahagian, 1981). If the 215 kDa and asialoglycoprotein receptors of hepatocytes are colocalized in coated pits and CURL, as was seen with cryosections of rat liver (Gueze *et al.*, 1984b), then how to explain productive endocytosis of one ligand but not the other? One difference between these systems is the pH requirement for ligand dissociation from receptor: at Ca^{2+} concentrations below about 1 mM asialoglycoproteins are released at pH values greater than those required for release of lysosomal enzymes (Harford *et al.*, 1984). Depending on intravesicular pH and $[Ca^{2+}]$, efficient endocytosis of asialoglycoproteins but retroendocytosis (futile in–out shuttling) of lysosomal enzymes might occur. This is mere speculation.

CHO cell mutants pleiotropically defective in endocytosis have been isolated by selection (1) for resistance to one toxin with screening or selection for resistance to an unrelated toxin(s) (Moehring and Moehring, 1983; Didsbury *et al.*, 1983; Marnell *et al.*, 1984), (2) for toxin resistance then screening for loss of Man 6-P-dependent endocytosis (Roff *et al.*, 1986; Robbins *et al.*, 1983; Robbins and Roff, 1987), and (3) for resistance to a poly(L-lysine)–methotrexate conjugate (Mandel *et al.*, 1985). Mutants with temperature-sensitive (*ts*) defects and with nonconditional defects have been reported. All of the mutants tested to date fall into one of two genetic complementation groups (see Table I), that we have called End1 and End2 (Roff *et al.*, 1986; Robbins and Roff, 1987).

These mutants are aberrant in a variety of parameters, exhibiting increased resistance to modeccin (Roff *et al.*, 1986; Marnell *et al.*, 1984; Mandel *et al.*, 1985; Robbins *et al.*, 1984), diphtheria (Roff *et al.*, 1986; Robbins *et al.*, 1983, 1984; Moehring and Moehring, 1983; Didsbury *et al.*,

TABLE I

Genetic Complementation

Mutant	Hybridization results[a]			Group
	× DTG1-5-4-122[b]	× M311[b]	× P1R2.P50[c]	
DTG1-5-4[d]		+		End1
DTF1-5-1[e]	+[f]	−[f]		End2
RPE.44[g]	−	+	−	End1
RPE.28[g]			−	End1
RPE.51[g]			−	End1
RPE.1A[h]	+	−		End2
P1R2.P50[g]				End1
MPL^{R}3-4[i]	+	−		End2
ts B3853[j]	−	+		End1
ts M311[j]	+			End2
ts I223[j]	+	−		End2
ts G.7.1[k]				?

[a] All mutants were recessive in hybrids with parental cells. In crosses with the ts mutants the normal phenotype was always observed at the permissive temperature, regardless of the other member of the cross.

[b] Ouabain[res], thioguanine[res]. Hybrids were selected in hypoxanthine–aminopterin–thymidine (HAT) medium plus ouabain and were screened for Man 6-P uptake and modeccin resistance (Roff et al., 1986; Robbins et al., 1984).

[c] HPRT−, proline+. Hybrids were selected in HAT medium without proline and were screened for Pseudomonas toxin resistance (Moehring and Moehring, 1983; Didsbury et al., 1983).

[d] Robbins et al. (1984).

[e] Robbins et al. (1983).

[f] Hybrids were screened for Man 6-P uptake and diphtheria toxin resistance (Roff et al., 1986; Robbins et al., 1984).

[g] Moehring and Moehring (1983).

[h] J. M. Moehring and T. J. Moehring (unpublished results).

[i] Mandel et al. (1985).

[j] Roff et al. (1986).

[k] Marnell et al. (1984).

1983; Marnell et al., 1984; Mandel et al., 1985), and Pseudomonas toxin (Roff et al., 1986; Moehring and Moehring, 1983; Marnell et al., 1984; Robbins et al., 1984) and to Sindbis virus (Robbins et al., 1983, 1984; Moehring and Moehring, 1983), hypersensitivity to ricin (Roff et al., 1986; Robbins et al., 1983, 1984; Moehring and Moehring, 1983; Mandel et al., 1985), decreased accumulation of iron from diferric transferrin (Roff et al., 1986; Klausner et al., 1984), decreased α_2-macroglobulin uptake (Roff et al., 1986), decreased Man 6-P-dependent uptake (Roff et al., 1986;

Robbins *et al.*, 1983, 1984; Sly *et al.*, 1984), increased secretion of endogenously synthesized lysosomal enzymes (Roff *et al.*, 1986; Robbins *et al.*, 1983, 1984; Sly *et al.*, 1984), and decreased terminal glycosylation activities (Roff *et al.*, 1986; Robbins *et al.*, 1984); these last two features are discussed in Section V,B,2. Most, but not all, of these aspects of phenotype have been tested in each of the mutants listed in Table I; the few exceptions observed [e.g., DTF1-5-1 was not resistant to modeccin or *Pseudomonas* toxins (Robbins *et al.*, 1983)], appear to correspond to relative mildness of the lesion.

These mutants bear a striking resemblance to cells treated with agents that inhibit organelle acidification. In those cases tested, endocytic vesicles from the mutants were found to be defective in ATP-dependent acidification *in vitro* (Roff *et al.*, 1986; Marnell *et al.*, 1984; Robbins *et al.*, 1984; Merion *et al.*, 1983). The affected compartment was characterized by its sedimentation on Percoll gradients (including sequential gradients, where it cosediments with light membranes), by its labeling with endocytic tracer during a 5-min pulse, and by its labeling with fluoresceinated transferrin, a ligand for receptor-mediated endocytosis that does not go to lysosomes (van Renswoude *et al.*, 1982). Depending on the mutant, ATP-dependent acidification of lysosomes was either normal or significantly less inhibited than endosomal acidification (Robbins *et al.*, 1984; Merion *et al.*, 1983).

Restoration of endocytic activities in revertants isolated from DTG1-5-4 was accompanied by normal acidification of endosomes *in vitro* (Robbins *et al.*, 1984). Also, with mutants exhibiting temperature-sensitive endocytic defects, *in vitro* acidification was greater in endosomes isolated from cells grown at the permissive than at the nonpermissive temperature (Roff *et al.*, 1986; Marnell *et al.*, 1984). Thus, endocytic defects appear to correlate qualitatively with acidification defects; however, endosomes from End2 mutants tested to date have always exhibited some residual acidification activity, irrespective of the severity of these mutants' endocytic defects.

In the nonconditional endocytosis mutants, inhibition of Man 6-P uptake was accompanied by marked decreases in surface Man 6-P binding activity but maintenance of the 215 kDa Man 6-P receptor at the cell surface, measured by immunoprecipitation after iodination at 0°C (Robbins *et al.*, 1983; A. R. Robbins, unpublished). As was observed with NH$_4$Cl-treated CHO cells, lack of surface receptor activity could not be explained by normal receptor occupancy—attempts to release bound ligand by preincubation at 4°C with Man 6-P or at pH 4.5 did not increase surface binding activity.

With the temperature-sensitive mutants it was possible to see different stages in the loss of Man 6-P endocytosis by varying the length of incubation at the nonpermissive temperature. The earliest effect detected was a block in movement of endocytosed lysosomal enzymes to lysosomes (Roff et al., 1986). Shifting the mutants to the nonpermissive temperature 15 min before addition of radiolabeled ligand resulted in endocytosed lysosomal enzymes sedimenting almost exclusively with a light membrane compartment, whence very little could be chased; in the parent cells most of the radioactivity moved through to lysosomes. This block was not general; at the nonpermissive temperature whatever ^{125}I-labeled α_2-macroglobulin could be endocytosed was delivered to the mutants' lysosomes.

After 1.5 hr at the nonpermissive temperature uptake of lysosomal enzymes was reduced to less than about 10%; cell-surface Man 6-P binding at this time was between 70 and 90% of normal. Longer incubation at the nonpermissive temperature did effect loss of surface binding activity in the temperature-sensitive mutants, once again without loss of surface receptor (215 kDa) molecules (Roff et al., 1986).

Although endocytosis of lysosomal enzymes was a bit more sensitive, the various endocytic activities were lost quite coordinately after shifting to the nonpermissive temperature; onset of inhibition required at most 2 hr, and maximal inhibition was attained within 4 hr, depending on the parameter under study (Roff et al., 1986). This was in marked contrast to the resumption of endocytic activities on return of the mutants from the nonpermissive to the permissive temperature. Although most functions were restored within a few hours (e.g., normal sensitivity to diphtheria toxin in 2 hr; 50% normal accumulation of iron from transferrin in 2 hr, 100% in 8 hr), recovery of both Man 6-P binding activity at the cell surface and of Man 6-P uptake required 30 hr (Roff and Robbins, 1987). The only other activity observed to require this prolonged recovery period was endocytosis of α_2-macroglobulin. It is unlikely that the rate of recovery of endocytosis of either lysosomal enzymes or α_2-macroglobulin is limited by recovery of endosomal acidification—the pH requirement for dissociation of ligands from the 215 kDa Man 6-P receptor is no lower than that for diphtheria toxin penetration (Donovan et al., 1985; Draper and Simon, 1980; Sandvig and Olsnes, 1980) or release of both iron atoms from transferrin (Dautry-Varsat et al., 1983), and α_2-macroglobulin dissociates from its receptor at or near neutral pH (Maxfield, 1982).

The slow recovery of cell surface lysosomal enzyme binding activity may explain some apparently contradictory findings with respect to protection of that binding activity by addition of Man 6-P. As had been observed on treatment of human fibroblasts for 3 hr with NH_4Cl (see Section IV,B,1), inclusion of Man 6-P decreased the loss observed in surface binding activity after shifting the temperature-sensitive End1 mutant to the nonpermissive temperature (Roff and Robbins, 1987). Incubation of the nonconditional End1 and End2 mutants for 12 hr with Man 6-P, however, effected no increase in binding activity (A. R. Robbins, unpublished). Apparently, once cell surface binding activity has been depleted, restoration (as observed with the temperature-sensitive mutants) is very slow.

V. THE INTRACELLULAR PATHWAY

A. Biosynthesis and Maturation of Lysosomal Enzymes

Like other secretory proteins, lysosomal enzymes are synthesized on membrane-bound polysomes and cotranslationally inserted into endoplasmic reticulum (ER) membranes; only mRNAs from bound polysomes synthesized β-glucuronidase and cathepsin D (Rosenfeld et al., 1982). Translation in the presence of microsomal membranes resulted in protection of newly synthesized enzymes from added proteases (Rosenfeld et al., 1982; Erickson and Blobel, 1979) and glycosylation of the enzymes, as demonstrated by susceptibility to endo H (Erickson et al., 1981). Lysosomal enzymes contain signal sequences; cathepsin D (Rosenfeld et al., 1982; Erickson et al., 1981), β-glucuronidase (Rosenfeld et al., 1982), and β-hexosaminidase α and β chains (Proia and Neufeld, 1982) obtained from in vitro translation without membranes were about 2 kDa larger than the corresponding newly synthesized and deglycosylated enzymes made in vivo. In cathepsin D an amino-terminal stretch of 20 amino acids (containing 7 leucines) was present following translation in the absence but not the presence of membranes (Erickson et al., 1981). Comparison of the cDNA sequence of β-hexosaminidase α chain (Myerowitz et al., 1985) with the amino acid sequence of the protein newly synthesized in vivo also indicated a cleaved amino-terminal signal (Little and Neufeld, 1986).

After metabolic labeling, lysosomal enzymes were found in precursor forms, 2–20 kDa larger than the corresponding mature enzymes (Myerowitz and Neufeld, 1981; Rosenfeld et al., 1982; Erickson and Blobel, 1979; Skudlarek and Swank, 1979, 1981; Hasilik and Neufeld 1980a). These precursors are glycosylated and phosphorylated, as shown by metabolic labeling with sugars and phosphate (Hasilik and Neufeld 1980a,b; Myerowitz and Neufeld, 1981). All of the precursors examined to date appear to contain more than one oligosaccharide. With the exception of macrophages (Skudlarek and Swank, 1981), cells secrete enzymes in precursor form (Robbins and Myerowitz, 1981; Rosenfeld et al., 1982;

Erickson *et al.*, 1981; Hasilik and Neufeld, 1980a). The fraction of enzyme that is secreted varies with both enzyme and cell type: human fibroblasts secrete about 20% of β-hexosaminidase and barely detectable levels of cathepsin D (Hasilik and Neufeld, 1980a), whereas Clone 9 hepatocytes secrete about 50% of cathepsin D and β-glucuronidase (Rosenfeld *et al.*, 1982). The excess secretion of enzymes by I-cells (Hasilik and Neufeld, 1980a) or cells treated with weak bases (Rosenfeld *et al.*, 1982; Hasilik and Neufeld, 1980a), including macrophages (Imort *et al.*, 1983), is of the precursor forms. Some precursors exhibit enzymatic activity (e.g., β-hexosaminidase), others do not (e.g., cathepsin D).

Generation of mature lysosomal enzyme can occur in a single step, e.g., β-galactosidase of peritoneal macrophages (Skudlarek and Swank, 1979, 1981), or in several discrete steps, e.g., α-L-iduronidase and cathepsin D of human fibroblasts (Myerowitz and Neufeld, 1981; Hasilik and Neufeld, 1980a). In pulse–chase experiments the first processing intermediates were detected 1–3 hr after biosynthesis. Lag times vary with both enzyme [cathepsin D versus β-hexosaminidase in human fibroblasts (Hasilik and Neufeld, 1980a)] and cell type [β-hexosaminidase in CHO cells (Robbins and Myerowitz, 1981) versus human fibroblasts (Hasilik and Neufeld, 1980a)].

The proteolytically processed form(s) appear asynchronously. Following a 10-min pulse with [^{35}S]methionine, in human fibroblasts the intermediate form of cathepsin D could first be detected after a 60-min chase; after a 120-min chase, precursor and this intermediate were found in approximately equal amounts (Gieselmann *et al.*, 1983). Fractionation of human fibroblasts on Percoll gradients after a 20-min pulse with [^{35}S]methionine and varying chase times showed precursor and intermediate forms of cathepsin D in light membranes–light lysosomes after a 30-min chase; after 60 min precursor, intermediate, and mature forms all were in dense lysosomes (Gieselmann *et al.*, 1983). Some of the asynchrony seen in processing may result from different rates of transport to different compartments as well as different rates of proteolysis within those compartments. Similarly, some of the differences observed between lysosomal enzymes with respect to initiation of processing may depend on whether a particular enzyme can be cleaved in a prelysosomal compartment.

B. The Road Taken

The Man 6-P recognition marker and its receptors provide a means for segregating lysosomal enzymes from soluble, secreted glycoproteins. Several approaches have been taken toward defining where segregation

occurs: localization of the GlcNAc-P-transferase and GlcNAc phosphodiesterase by cell fractionation to fix the earliest point at which enzyme
would be competent to bind to receptor; examination of when binding
occurs in pulse–chase studies; oligosaccharide analysis to determine
which glycosylation enzymes the lysosomal enzyme has encountered;
and electron microscopy and cell fractionation to ascertain the cellular
compartments containing enzyme and/or receptor.

As described in Section II,B,2, on fractionation of rat liver the GlcNAc
phosphodiesterase sedimented with the Golgi (Varki and Kornfeld, 1980a;
Waheed *et al.*, 1981a). Newly synthesized secretory proteins are thought
to move through the Golgi in the direction cis $\rightarrow$ medial $\rightarrow$ trans $\rightarrow$
GERL–TGR. Modification of asparagine-linked oligosaccharides pro-
ceeds by a stepwise series of reactions (Fig. 1). Some of the glycosyltransferases responsible have been localized within the Golgi by immunocytochemical procedures: *N*-acetylglucosaminyltransferase I was found in the
medial cisternae of rat liver and of a rabbit kidney cell line (Dunphy *et al.*,
1985), galactosyltransferase in the trans cisternae of HeLa cells (Roth and
Berger, 1982), and sialyltransferase in the trans cisternae and in GERL–
TGR of rat liver hepatocytes (Roth *et al.*, 1985). When CHO cell homogenates were fractionated on sucrose density gradients, the majority of
mannosidase I activity sedimented at slightly greater density than that of
galactosyltransferase (Dunphy *et al.*, 1981). The gradient of cholesterol
observed across the Golgi apparatus (Orci *et al.*, 1981), increasing pro-

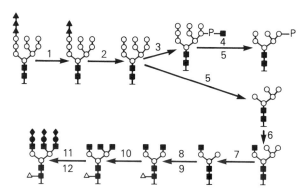

Fig. 1.   Steps in the processing of asparagine-linked oligosaccharides. 1, Glucosidase I;
2, glucosidase II; 3, *N*-acetylglucosaminylphosphotransferase; 4, phosphodiester glycosidase; 5, mannosidase I; 6, *N*-acetylglucosaminyltransferase I; 7, mannosidase II; 8, *N*-
acetylglucosaminyltransferase II; 9, core fucosyltransferase; 10, *N*-acetylglucosaminy-
ltransferase IV; 11, galactosyltransferase; 12, sialyltransferase. Symbols: ▲, glucose; ■,
GlcNAc; ○, mannose; ●, galactose; △, fucose; ◆, sialic acid; P, phosphate. From Goldberg
and Kornfeld (1983).

gressively from cis to trans, plus the early position of α-mannosidase I in oligosaccharide processing, suggested that the density gradients had separated cis and/or medial saccules from trans; subsequent experiments showed cosedimentation of mannosidases I and II with GlcNAc transferases I and II (Dunphy and Rothman, 1983).

Fractionation of rat liver homogenates by sucrose density gradient centrifugation showed GlcNAc-P-transferase, GlcNAc phosphodiesterase, and mannosidase I activities together at a density higher than that of galactosyltransferase activity (Pohlmann et al., 1982), suggesting that synthesis of the Man 6-P recognition marker could take place in the cis or medial saccules. This apparent separation may have resulted, however, from destruction of galactosyltransferase substrate in the denser fractions by a pyrophosphatase (Dunphy and Rothman, 1983).

Fractionation of membranes from BW5147 lymphoma cells (Goldberg and Kornfeld, 1983) showed the following, from dense to light: (1) GlcNAc-P-transferase sedimenting with glucosidase I and II activities [the last two enzymes are thought to be in ER (Lucocq et al., 1986)], (2) GlcNAc phosphodiesterase, (3) α-mannosidase II, GlcNAc transferase I and/or II, and fucosyltransferase, and (4) galactosyltransferase. These results suggested that in lymphoma cells GlcNAc phosphodiesterase is somewhat to the cis side of most early and middle oligosaccharide processing activities. It also was shown, however, that a large fraction of protein containing newly synthesized *sialylated*, phosphorylated hybrid oligosaccharide sedimented at the density of GlcNAc phosphodiesterase (Goldberg and Kornfeld, 1983).

GlcNAc-P-transferase and GlcNAc phosphodiesterase activities from rat liver were separated from each other by refractionating isolated Golgi on Percoll gradients, but in this system neither enzyme was separated from galactosyl- or sialyltransferases, both of which exhibited bimodal distributions (Deutscher et al., 1983). Thus, placement of the earliest point at which lysosomal enzymes would be competent to leave the secretory pathway has not been unequivocally established.

How far in the pathway can enzymes travel without being committed to secretion? About 20% of the mature (therefore lysosomal) form of cathepsin D in human fibroblasts contained an endo H-resistant oligosaccharide (Hasilik and von Figura, 1981), indicating that these molecules had at least passed mannosidase II. In another study 50% of the intracellular β-hexosaminidase activity from mucolipidosis I fibroblasts (defective in lysosomal sialidase activity) was shown to bind the galactose-specific lectin ricin following treatment with neuraminidase (Vladutiu, 1983). Although the form of the β-hexosaminidase was not determined, the lifetime of this enzyme is such that most of the activity would have represented mature enzyme. Thus, at least 50% of newly synthesized β-hexosaminidase appears to have passed through the trans Golgi en route to lysosomes.

The degree of oligosaccharide heterogeneity on lysosomal enzymes is intriguing; assuming that only one of the oligosaccharides on the enzyme precursor has to be of the phosphorylated high-mannose structure to mediate receptor binding, one might expect that the other(s) would always be modified to a uniform extent (i.e., galactosylated in the above case of β-hexosaminidase). If lysosomal enzymes bound receptors in various areas of the Golgi–GERL–TGR and if binding to receptor blocked further oligosaccharide processing on the lysosomal enzyme, either by removing the bound enzyme from Golgi–GERL–TGR or by making the oligosaccharides on bound enzyme inaccessible, this could explain the heterogeneity observed.

To time binding of newly synthesized lysosomal enzymes to the Man 6-P (215 kDa) receptor, BW5147 cells were pulsed for 20 min with [2-³H] mannose, then label was chased for varying intervals; membranes were permeabilized with 0.5% saponin (without divalent cations) to release soluble intravesicular proteins, then 10 mM Man 6-P was added to displace receptor-bound lysosomal enzymes (Gabel et al., 1982). Oligosaccharides cleaved from the Man 6-P-displaced proteins by endo H were analyzed. Oligosaccharides containing two phosphomonoesters were the major phosphorylated form, in contrast to their low abundance among total phosphorylated oligosaccharides (Varki and Kornfeld, 1980b). Results obtained with the Man 6-P receptor affinity column (see Section II,C) also suggested that this form of oligosaccharide is responsible for binding enzyme to the receptor. In the Man 6-P-displaced proteins, radioactivity in the various phosphorylated oligosaccharides increased and decreased coordinately with time, that is, radioactivity did not appear to chase from oligosaccharides with phosphodiesters into those with phosphomonoesters. Again this suggests that oligosaccharides on receptor-bound enzymes are not accessible to further processing, in this case by the GlcNAc phosphodiesterase.

Based on the kinetics of labeling of phosphorylated oligosaccharides derived from the Man 6-P-displaced proteins (Gabel et al., 1982), it appears that a major fraction of lysosomal enzymes binds to receptor 30–50 min after glycosylation. At the end of the 20-min pulse little or no radioactivity was found in phosphorylated oligosaccharides; after a 30-min chase radioactivity in those oligosaccharides was greater than one-half the maximum, attained after a 60-min chase. But, based on the radioactivity in neutral oligosaccharides from the Man 6-P-displaced proteins (~100 times greater than that in phosphorylated oligosaccharides), binding to the receptor was half-maximal at the end of the 20-min pulse. It is difficult to decide whether these neutral oligosaccharides represent contaminating proteins (and thus should be ignored) or "fellow traveler" oligosaccharides on lysosomal enzymes bound to receptors through other phosphorylated oligosaccharides: the relatively high level of radioactivity in the neutral oligosaccharides suggests that they are contaminants; on the other hand, the comparatively low levels of radioactive neutral oligosaccharides displaced by Man 6-P from receptor-deficient macrophages suggests that they are relevant. If the Man 6-P-eluted proteins were freed of contaminants by passage over a Man 6-P receptor affinity column prior to treatment with endo H, it might be possible to fix the timing of enzyme binding to the receptor a bit more firmly.

At no time was more than 50% of the radioactive oligosaccharide containing two phosphomonoesters found bound to receptor (Gabel *et al.*, 1982). This led the authors to suggest that most of the receptors in BW5147 cells were occupied. As noted in Section II, the degree of receptor occupancy varies greatly with cell type [~80% in rat liver homogenates (Fischer *et al.*, 1980b), ~10% in human fibroblasts (Fischer *et al.*, 1980a)], based on comparisons of binding activities without and with displacement of endogenous ligands with Man 6-P. These data and those from BW5147 may not be reporting on total occupied receptor or bound enzyme, but on multivalently bound material; enzymes bound univalently may have been released on the preparatory washes. It is unclear whether multivalent binding is necessary for segregation of newly synthesized enzymes; depending on the volume of the intracellular compartment in which receptor encounters ligand, univalent interaction may be sufficient.

Table II summarizes the results compiled from electron microscopic studies in which lysosomal enzymes and/or the 215 kDa Man 6-P receptor were localized using specific antisera. In liver and Hep G_2 * cells distribution was also compared to that of albumin, a secreted protein. With the exception of the endocytic compartment CURL (discussed in Section IV,A above) no prelysosomal compartment was without albumin; on the other hand, no secretory compartment was defined by the presence of only albumin. Thus, the point of divergence of secretory and lysosomal pathways could not be defined; the gold immunostaining procedure may lack the sensitivity necessary to detect low abundance antigens in other than confined, defined areas such as Golgi and GERL–TGR.

Distribution of the 215 kDa receptor among the Golgi cisternae and in GERL–TGR is a matter of controversy. Indirect immunoperoxidase

* Hep G_2 cells secrete $\geq 50\%$ of their newly synthesized β-hexosaminidase (S. M. Laurie, unpublished observations). If this is also true for other lysosomal enzymes then discrimination between lysosomal and secretory routes may be difficult in these cells.

Table II

Immunocytochemical Localization[a]

Tissue/cell	ER	Golgi		GERL–TGR
		Cis	Trans	
Rat liver[b]	E			E
Rat liver[c]		RE	RE	
		RA	RA	RA
Rat liver[d]		R	R	
Rat liver[e]		R		
Hepatocytes[f]		R[g]		
clone 9				
Hepatoma[h]		A	A	RA
Hep G$_2$		±E	±E	RE
		A±E	A±E	AE
Hep G$_2$[i]		E	E	
Human fibroblast[j]	E			
Human fibroblast[k]		R[g]		
CHO cells[l]	R			R

[a] E, Lysosomal enzyme; R, 215 kDa receptor; A, albumin. Paired symbols indicate colocalization studies.

[b] Novikoff et al. (1983).

[c] Geuze et al. (1984a).

[d] Geuze et al. (1984b).

[e] Brown and Farquhar (1984a).

[f] Brown et al. (1984).

[g] For these instances used to indicate presence in only two of cisternae visualized.

[h] Geuze et al. (1985).

[i] Slot and Geuze (1983).

[j] van Dongen et al. (1985); only the localization of soluble lysosomal enzymes in this study is included.

[k] Brown and Farquhar (1984b).

[l] Willingham et al. (1983).

staining of rat liver (Brown and Farquhar, 1984a), Clone 9 hepatocytes (Brown et al., 1984), and human fibroblasts (Brown and Farquhar, 1984b) showed receptor in only two of the cisternae observed, and also in small coated vesicles adjacent to and among the labeled cisternae. In rat liver the labeled cisternae were identified as cis, owing to their location on the side of the Golgi stack opposite the secretory granules. Receptor was not seen in GERL–TGR. On the other hand, gold immunostaining of rat liver cryosections (Geuze et al., 1984a,b) showed 215 kDa receptor, in low amounts, distributed throughout the Golgi cisternae, with most of the receptor in GERL–TGR. In cryosections of Hep G$_2$ cells (Geuze *et al.*, 1985) receptor was almost exclusively in the latter compartment. Receptor was also found in coated regions budding from GERL–TGR.

On neither side of the controversy does the localization obtained appear to reflect either newly synthesized 215 kDa receptor or newly synthesized lysosomal enzymes. Following treatment with cycloheximide, the Clone 9 hepatocytes still showed immunoperoxidase only in cis cisternae and coated vesicles (increased greatly in number on inhibition of protein synthesis); similar effects were observed after treatment with tunicamycin (Brown *et al.*, 1984), which would block biosynthesis of the Man 6-P recognition marker. Gold particles were seen across the cisternae and in GERL–TGR in cryosections of rat liver prepared after perfusion with cycloheximide (Geuze *et al.*, 1984c).

To this nonmorphologist, it seems worthy of note that while the two groups find receptor concentrated on opposite sides of the Golgi stack, both find it present in coated structures associated with their respective side of the stack. In resolving this controversy, it might be useful to determine whether these coated structures are comprised of clathrin. Gold immunostaining of endocrine pancreas and pituitary cells showed clathrin predominantly associated with the trans side of the Golgi (Orci *et al.*, 1985), whereas coated buds without detectable clathrin were found associated with multiple cisternae of the CHO cell Golgi, both *in vivo* and *in vitro* (Orci *et al.*, 1986). Two types of coated structures were also observed in BHK-21 cells infected with vesicular stomatitis virus (Griffiths *et al.*, 1985); clathrin-coated buds were restricted to GERL–TGR, coated structures without clathrin were found both in GERL–TGR and at the rims of all Golgi cisternae.

The clathrin-coated vesicle is an attractive site for dissociation of enzyme from the Man 6-P receptor in that it contains an ATP-dependent proton pump (Forgac *et al.*, 1983; Stone *et al.*, 1983). Clathrin-coated vesicles isolated from bovine brain and rat liver (Campbell *et al.*, 1983; Sahagian and Steer, 1985) contained the 215 kDa Man 6-P receptor. [It should be noted that coated vesicles are also an early, transient structure in the endocytic pathway (e.g., Anderson *et al.*, 1977); analysis of preparations of coated vesicles from rat liver suggested that the ratio of endocytic to exocytic vesicles was 2 : 1 (Helmy *et al.*, 1986)]. Man 6-P binding activity was observed with coated vesicle preparations [in one case (Campbell *et al.*, 1983) using α-mannosidase from *D. discoideum*, thus binding was probably to the 215 kDa, rather than the 46 kDa, receptor]. Binding was stimulated 5- to 20-fold by treatment with low concentrations of detergent, suggesting that most of the binding sites were cryptic. Consistent with this, pretreatment of intact vesicles with protease did not alter binding activity in subsequently permeabilized vesicles (Campbell *et al.*, 1983 Sahagian and Steer, 1985), and reduced the observed M_r of the receptor by only 12 kD (Sahagian and Steer, 1985). Exposure of intact vesicles to carboxypeptidase Y caused a similar reduction, indicating that the carboxy terminus of the receptor is exposed on the exterior of the vesicles, in agreement with the orientation observed with total membrane vesicles prepared from human fibroblasts (von Figura *et al.*, 1985).

Lysosomal enzymes also were found in coated vesicles (Sahagian and Steer, 1985; Campbell and Rome, 1983; Tümmers *et al.*, 1983; Schulze-Lohoff *et al.*, 1985). The degree to which enzymes were receptor-bound varied. Permeabilization with detergent plus Man 6-P released 90% of the β-hexosaminidase activity from bovine brain coated vesicles in one study (Sahagian and Steer, 1985) but less than 5% from either bovine brain or rat liver coated vesicles in another (Campbell and Rome, 1983). With human fibroblast coated vesicles, about 70% of radiolabeled precursor cathepsin D was released by permeabilization with detergent alone; inclusion of Man 6-P increased release by only about 10% (Schulze-Lohoff *et al.*, 1985). Different detergents at different concentrations were employed in the three studies, possibly accounting for the varying results. Also, during purification the coated vesicles may not maintain an acidic interior; thus, receptor-bound enzyme may reflect rebinding as the proton gradient is dissipated.

Differences were also observed with respect to the presence of mature versus precursor forms of the enzymes in the coated vesicles. Both forms of β-hexosaminidase α chain and cathepsin D were obtained from coated vesicles of rat liver (Campbell and Rome, 1983) and human placenta (Tümmers *et al.*, 1983), respectively; the intermediate form of cathepsin D was not detected. About 90% of the radioactive cathepsin D in coated vesicles from human fibroblasts, metabolically labeled for 3 hr with [^{35}S]methionine, was in precursor form (Schulze-Lohoff *et al.*, 1985). The authors of the last study concluded that only the precursor form of the enzyme was a constituent of the coated vesicle, the mature forms observed were deemed contaminants, based on the fact that all of the mature form could be extracted in the absence of Man 6-P, and that the amount of mature form in the vesicles decreased rather than increased after a 10-hr chase of the radiolabel. The first argument rests on the assumption that enzyme in coated vesicles *should be* receptor-bound, albeit one of the roles of the vesicle would seem to be dissociation of enzyme from receptor; with regard to the second, a residence time of enzyme in vesicles less than 10 hr would be expected, based on the time of appearance of cathepsin D in lysosomes (Gieselmann *et al.*, 1983). Actually, the decrease in mature form associated with the vesicles after the chase seems to argue against its being a contaminant—radiolabel in total mature form increased significantly after the 10-hr chase; radioactivity in a contaminant would be expected to increase proportionately.

C. Disruption of the Pathway

1. Inhibitors

As would be predicted from the behavior of I-cells, agents interfering with synthesis of the Man 6-P recognition marker on lysosomal enzymes prevent the segregation of those enzymes from the secretory pathway. Thus, on treatment with tunicamycin, an inhibitor of N-linked glycosylation, all newly synthesized cathepsin D was secreted from Clone 9 hepatocytes (Rosenfeld *et al.*, 1982) in unglycosylated precursor form. Absence of glycosylation can also result in increased turnover of lysosomal enzymes as evidenced by β-glucuronidase of Clone 9 hepatocytes (Rosenfeld *et al.*, 1982) and cathepsin D of porcine kidney cells (Erickson *et al.*, 1981) after treatment of those cells with tunicamycin; in these cases also, the unglycosylated enzymes recovered were in the medium.

Treatment with weak bases also effects oversecretion of newly synthesized lysosomal enzymes (Rosenfeld *et al.*, 1982; Hasilik and Neufeld, 1980a); enzymes already proteolytically processed are secreted in negligible amounts. Weak bases do not interfere with biosynthesis of the Man 6-P recognition marker; secreted enzymes were shown to be phosphorylated (Hasilik and Neufeld, 1980b) and high uptake (Robbins and Myerowitz, 1981). Nor do they alter synthesis of the 215 kDa Man 6-P receptor, although they do accelerate its turnover, 2-fold in CHO cells (Sahagian, 1984; Robbins *et al.*, 1983) and up to 1.3-fold in human fibroblasts (Creek and Sly, 1983). By analogy to their effects on endocytosis, weak bases are assumed to block compartmentalization of enzymes in lysosomes through prevention of acidification of some compartment(s) in which enzymes dissociate from receptors.

After treatment for 30 min with NH_4Cl or primaquine, gold immunostaining of the 215 kDa Man 6-P receptor in Hep G_2 cells increased in coated regions of GERL–TGR, due primarily to an accumulation of coated buds on this structure (Geuze *et al.*, 1985). Colocalization experiments showed lysosomal enzymes present in these regions. Also, enzymes, Man 6-P receptor, and albumin all were observed together in smooth vesicles. This last observation is of interest because treatment with weak bases markedly inhibited secretion of newly synthesized albumin from Hep G_2 cells (Strous *et al.*, 1985); secretion of lysosomal enzymes was not examined in this study.

Chloroquine caused Clone 9 hepatocytes to secrete essentially all newly synthesized cathepsin D (Rosenfeld *et al.*, 1982). Immunoperoxidase staining of these cells after treatment with chloroquine for 3 hr showed depletion of 215 kDa Man 6-P receptor from the Golgi and its associated coated vesicles and accumulation of receptor in very large, vacuolated vesicles (Brown *et al.*, 1984). A subsequent study identified these structures as endocytic, prelysosomal multivesicular bodies, based on colocalization of receptor with endocytic tracers, lack of colocalization with a lysosomal antigen, and morphology (Brown *et al.*, 1986).

Do any of these sites where receptor accumulates in the presence of weak base correspond to compartments that are acidified under normal conditions? Acidic compartments have been identified at the electron microscopic level using antibodies directed against weak bases. In human fibroblasts these antibodies stained, in addition to endosomes, multivesicular bodies and lysosomes (Anderson *et al.*, 1984), the trans-Golgi cisternae, and GERL–TGR (Anderson and Pathak, 1985). In Hep G_2 cells gold immunostaining for primaquine showed the weak base in what appeared to be either swollen Golgi cisternae or GERL–TGR (Schwartz *et al.*, 1985).

An acidic pH makes the GERL–TGR an attractive candidate for the site of segregation of enzyme from receptor in the intracellular pathway; consistent with this is the occasional observation in Hep G_2 cells that receptor is limited to tubular elements of GERL–TGR whereas lysosomal enzymes are in both tubular and vesicular areas of this organelle (Geuze *et al.*, 1985). To serve as the compartment for release of lysosomal enzymes from receptor, however, the intraorganellar pH should be below 6. While the actual pH cannot be determined by immunostaining procedures, a cogent argument has been put forth against a pH below 6 in Golgi–GERL (Anderson and Pathak, 1985): membrane proteins from some viruses are powerful fusogens below pH 6, yet these viral membrane proteins pass through the Golgi–GERL en route to the plasma membrane without causing membrane fusion. If lysosomal enzyme–receptor complexes have access to some region of the Golgi–GERL from which other newly synthesized proteins are excluded, then the possibility of a pH below 6 in that region is still viable. The clathrin-coated buds of the GERL–TGR [shown to contain 215 kDa receptor (Geuze *et al.*, 1985)] are a possibility; in BHK-21 cells infected with vesicular stomatitis virus then shifted to 20°C, both clathrin-coated buds and viral membrane protein accumulated in the GERL–TGR, but clathrin-coated buds were devoid of the viral protein (Griffiths *et al.*, 1985).

Treatment of human fibroblasts with 10 m*M* potassium cyanate, a reversible sulfhydryl reagent, caused inhibition of transport of β-hexosaminidase and cathepsin D (Hasilik *et al.*, 1983). In untreated cells after a 16-hr pulse the majority of the radiolabeled enzymes were found in mature forms in a dense fraction on Percoll gradients, whereas in cells pretreated with cyanate most of the labeled enzymes were in precursor forms sedimenting with light membranes. Some stimulation of secretion of cathepsin D but not of β-hexosaminidase was observed. On removal of cyanate the accumulated precursor of β-hexosaminidase was secreted. Phosphorylation of enzyme was not inhibited.

Incubation of human fibroblasts with affinity-purified antibodies, directed against the 215 kDa Man 6-P receptor and capable of blocking uptake of lysosomal enzymes via that receptor, effected a signficant increase in the secretion of newly synthesized (precursor) lysosomal enzymes (von Figura *et al.*, 1984): 70% of β-hexosaminidase and 40% of cathepsin D synthesized during a 24-hr pulse were secreted in the presence of antibody, whereas 21 and 12% were secreted by control cells. Similar results were obtained with Fab fragments of the antibody (Gartung *et al.*, 1985). This is the strongest evidence to date that the 215 kDa receptor functions in the intracellular pathway, at least in human fibroblasts. Whereas after incubation for 3 hr with antibody less than 20% of control levels of receptor could be recovered on detergent solubilization (owing to insolubility of the cross-linked receptor–antibody complex), Fab fragments did not cause receptor disappearance.

2. Mutants*

I-cells are the prototypical mutant with regard to disruption of the intracellular pathway for lysosomal enzymes. The defect in these cells has been described in Section II,B,1. One point to note is the differential effect of lack of phosphorylation on the various lysosomal enzymes: essentially all of the newly synthesized β-hexosaminidase but only about 50% of cathepsin D and α-glucosidase were secreted; proteolytic processing of retained cathepsin D appeared complete, that of α-glucosidase stopped at an intermediate form (Hasilik and Neufeld, 1980a); and no phosphorylation of any of these enzymes was observed (Hasilik and Neufeld, 1980b). Thus, even in human fibroblasts significant lysosomal compartmentalization of some soluble enzymes occurs without biosynthesis of the recognition marker.

P388D$_1$ macrophages, which lack the 215 kDa Man 6-P receptor, secreted the majority of their newly synthesized β-hexosaminidase and β-

---

* As in Section IV,B,2, some of the cells included in this section cannot properly be referred to as mutants.

glucuronidase, albeit the amount of those enzymes retained by P388D$_1$ was more than that found in many "normal" cell types. Macrophages can secrete lysosomal enzymes in mature form (Skudlarek and Swank, 1981), that is, from a compartment beyond the segregation of lysosomal enzymes from the secretory pathway. It would be of interest to know if secreted enzymes from P388D$_1$ had passed the point of segregation. Lysosomal enzymes secreted by this cell line were shown to be high uptake (Hoflack and Kornfeld, 1985a), but that does not preclude their being mature forms; mature β-hexosaminidase isolated from human fibroblasts was high uptake (Frisch and Neufeld, 1981).

Mutant LTk$^-$ cells lacking detectable 215 kDa Man 6-P receptor contained only 25% of parental levels of α-L-iduronidase and α-mannosidase activities. β-Hexosaminidase, β-galactosidase, and β-glucuronidase activities in these cells were at least 75% of normal (S. M. Laurie and A. R. Robbins, unpublished).

The above mutant cells present a common problem, namely, determination of the system(s) responsible for Man 6-P-independent and/or 215 kDa Man 6-P receptor-independent compartmentalization of lysosomal enzymes. For P388D$_1$ macrophages a candidate protein, the 46 kDa cation-dependent Man 6-P receptor, has been identified, but even in this case a role for this binding protein in enzyme compartmentalization has yet to be established. What is needed are mutants of mutants (e.g., derivatives of the 215 kDa receptorless LTk$^-$ mutants which oversecrete β-hexosaminidase, or 46 kDa receptorless variants of P388D$_1$) in order to approach the nature of these second systems.

Clone 9 hepatocytes secreted about 50% of their newly synthesized β-glucuronidase and cathepsin D (Rosenfeld et al., 1982); Hep G$_2$ cells secreted at least 50% of newly synthesized β-hexosaminidase (S. M. Laurie, unpublished). Both these lines have 215 kDa receptor, as shown in the immunocytochemical studies described in Section V,B. Perhaps in hepatic cells more membrane area in GERL–TGR is devoted to formation of secretory vesicles, and enzyme or receptor–enzyme complexes are simply swept passively into those vesicles. If, like freshly isolated hepatocytes (see Section IV,B,2), the hepatocyte lines lack endocytosis via the 215 kDa receptor, then secreted enzyme would not be recaptured.

The CHO cell mutants A10-2-4 and B4-2-1 selected with the monophosphopentamannose–ricin conjugate (see Section III,B) exhibited 2- to 6-fold increased secretion of lysosomal enzymes (α-mannosidase, β-glucuronidase, α-L-fucosidase, α-L-iduronidase, and β-hexosaminidase); the last two were tested and shown to be high uptake (Robbins and Myerowitz, 1981). A large fraction of these enzymes, however, remained cell-associated. These cells may use an alternate receptor for segregation of newly synthesized hydrolases, or the low affinity Man 6-P binding exhibited by these mutants at the cell surface may be sufficient within the intracellular pathway for segregation of enzymes.

The End1 and End2 mutants, pleiotropically defective in receptor-mediated endocytosis, also secreted newly synthesized lysosomal enzymes in elevated amounts (Roff *et al.*, 1986; Robbins *et al.*, 1983, 1984; Sly *et al.* 1984). With the nonconditional mutants DTF1-5-1 (Robbins *et al.*, 1983) and DTG1-5-4 (Robbins *et al.*, 1984) the fraction of newly synthesized enzymes secreted ranged from 60 to 80% and 90 to 95%, respectively.* In the mutants with temperature-sensitive endocytic defects, enzymes were secreted at parental levels at the permissive temperature, and oversecreted only at the nonpermissive temperature (Roff *et al.*, 1986). Nearly complete secretion of enzyme could be effected by shifting to the nonpermissive temperature at or before the initiation of a 15-min pulse with radiolabeled amino acid; in the case of MEP [Major Excreted Protein, a thiol-dependent acid protease (Gal and Gottesman, 1986), the secretion and/or compartmentalization of which in CHO cells is very rapid], 1.25 hr at the nonpermissive temperature was sufficient to obtain complete secretion. Enzymes secreted by End1 and End2 mutants were in precursor, high uptake form (Roff *et al.*, 1986; Sly *et al.*, 1984).

Activity of the cation-independent Man 6-P receptor was decreased in total (minus nuclear) membranes of DTF1-5-1 and DGT1-5-4, albeit to a lesser degree than cell surface binding activity (Robbins *et al.*, 1983; A. R. Robbins, unpublished data). Results with the *ts* End1 and End2 mutants, however, suggest that decreased binding activity is not responsible for oversecretion of newly synthesized lysosomal enzymes: after incubation at the nonpermissive temperature for 4 hr binding activity remained 50–100% of normal, varying with the membrane preparation (Roff *et al.*, 1986). In the nonconditional End mutants turnover of the 215 kDa Man 6-P receptor was about twice as fast as turnover in parental cells, similar to results obtained on treating CHO cells with NH_4Cl.

Because endosomes from End1 and End2 mutants are defective in ATP-dependent acidification *in vitro*, it is tempting to speculate that compartmentalization of newly synthesized acid hydrolases proceeds through the endosomes; i.e., that dissociation of endogenous as well as endocytosed enzyme from receptor occurs in this compartment. But (it is my unrealized dream to isolate a But$^-$ mutant), all of the End1 and End2

* Always with the exception of β-glucuronidase, whose secretion never exceeded 50%. CHO cells exhibit a phenomenon resembling the egasyn binding of β-glucuronidase in rat and mouse liver (reviewed in Paigen, 1979); i.e., about 50% of the newly synthesized enzyme associates with light membranes and chases from this fraction very slowly, if at all. This fraction of β-glucuronidase is always found inside the CHO cell mutants.

mutants tested to data also exhibit a defect localized to Golgi–GERL; sialylation of some (predominantly secreted) proteins is decreased in these mutants (Roff *et al.*, 1986). It is unlikely that sialylation is the primary defect, because onset of diminution of sialylation required incubation of the *ts* End1 mutant for 2–3 hr at the nonpermissive temperature, whereas decreased endocytic activity and oversecretion of lysosomal enzymes required only 1 hr.

Recalling that sialyl transferase is localized in those regions of the Golgi–GERL shown to be acidic, decreased sialylation may be reporting on the increasing pH of that compartment. This was suggested also by observations made with the End1 mutant DTG1-5-4 after infection with Sindbis virus: a dramatic accumulation of cisternae decorated with viral nucleocapsid was seen adjacent to the Golgi apparatus (Robbins *et al.*, 1984), very similar to what has been reported after treatment of normal cells with the ionophore monensin (Johnson and Schlesinger, 1980). If acidification of endosomes and Golgi–GERL has a common dependence, whether it is on the proton pump itself, on an ion channel required for maintenance of a pH gradient, or on some modification that regulates pumping activity, then coated vesicles or some other acidic compartment may also share this dependence, and the defect in the End1 and End2 mutants could affect lysosomal enzyme segregation at any of these locations.

As mentioned in Section IV,B,2, recovery of Man 6-P-dependent binding and uptake in the *ts* End1 and End2 mutants on return to the permissive temperature was extremely slow (24–30 hr). In contrast, oversecretion of lysosomal enzymes had diminished by 50% in cells returned to the permissive temperature at the time of the radioactive pulse; secretion was restored to normal levels after 4 hr. Sialylation of newly synthesized proteins had returned to normal within 4–6 hr (Roff and Robbins, 1987). These results are discussed in the next section.

VI. JUNCTION OF THE PATHWAYS

A. Mixing of Components

Correction of mucopolysaccharide storage in mutant fibroblasts (see Section I) demonstrated that endocytosed lysosomal enzymes must be colocalized with endogenous enzymes in lysosomes. To date there are no studies directly determining whether endocytosed enzymes meet enzymes from the intracellular pathway in prelysosomal compartments. I am confident that these studies are underway: required is addition of a purified high uptake enzyme, at levels saturating for the surface Man 6-P receptors, to mutant human fibroblasts CRM⁻ for that enzyme, preparation of cryosections, then gold immunostaining, using different sizes of gold particles to decorate antibodies against the added enzyme versus antibodies against an endogenous lysosomal enzyme. All the reagents, cells, and techniques presently exist.

Although the 215 kDa receptor does not accompany enzyme to lysosomes, it appears that endocytic and intracellular receptors also mix. In the absence of protein synthesis, uptake of β-glucuronidase into chloroquine-treated I-cells continued at a linear albeit reduced rate for 4 hr, during which period the cells internalized an amount of enzyme about 4 times greater than the number of surface binding sites (Gonzalez-Noriega et al., 1980). Assuming that chloroquine quickly and completely blocked recycling of surface receptors by preventing release of endocytosed enzyme, then receptors must have moved to the cell surface from within.

Movement of intracellular receptor to the cell surface also occurs in cells synthesizing high uptake enzymes. On incubation of CHO cells (Sahagian, 1984) and human fibroblasts (von Figura et al., 1984) at 35°C with immunoglobulin G (IgG) against the 215 kDa receptor, 75–80% of the total receptor in those cells was antibody bound after 1 hr and 90–95% after 2 hr. Binding to Fab fragments followed a similar time course (Gartung et al., 1985). The simplest interpretation of these results is that most of the receptors make an appearance at the cell surface within an hour. The amount of antibody employed and the rapidity of the effect obviate antibody entering by fluid-phase pinocytosis then binding to receptors inside the cell. It would be reassuring to know that the antibody remained bound to receptor at pH 5, thus excluding the possibility of antibody entering the cells bound to surface receptor, dissociating in the endosome, then rebinding to intracellular receptors in this or some subsequent compartment.

Colocalization of receptors from the endocytic and intracellular pathways will not be as straightforward as colocalization of lysosomal enzymes. Needed is some means of labeling surface receptor that is specific for the receptor, does not interfere with, or alter the route of, subsequent endocytosis via that receptor, remains stable at acid pH, and prevents the receptor from acquiring any additional label when sections of the cells are subsequently stained to locate intracellular receptors. This presents a difficult, albeit not impossible, task.

Demonstration of rapid receptor mixing led to an elegant model in which receptor–enzyme complexes from both endocytic and intracellular pathways meet in a single compartment, acidification of which results in dissociation, then segregation of receptors from enzyme (Sahagian, 1984).

However, given the plethora of acidic compartments in which enzyme and/o. receptor have been found (endosomes, multivesicular bodies, coated vesicles, GERL–TGR), it seems equally likely that mixing occurs after dissociation and segregation, that is, receptors meet in one compartment and enzymes in another. What evidence exists vis à vis any of the compartments of either pathway serving as a meeting place?

Colocalization of endocytosed ligands with endogenous lysosomal enzymes has been examined cytochemically. In hepatocytes following perfusion, endocytosed lactosaminated ferritin (a ligand for the asialoglycoprotein receptor) was initially located in small peripheral endocytic vesicles and tubules that lacked arylsulfatase activity (Wall et al., 1980). Similarly, after 3 min of uptake into CHO cells, horseradish peroxidase (internalized by fluid-phase pinocytosis) did not colocalize with acid phosphatase activity (Storrie et al., 1984). The significance of these studies with respect to lysosomal enzyme mixing is the apparent absence of lysosomal enzyme in early endocytic vesicles. One caveat with respect to interpretation of these results is that they depend on activity of the endogenous lysosomal enzyme; colocalization of endocytosed protein with an inactive early form of enzyme would have gone undetected. Precursor forms of acid phosphatase from human fibroblasts were reported to lack enzymatic activity (Lemansky et al., 1985).

After 10 min of uptake greater than 90% of the endocytosed horseradish peroxidase in the CHO cells was in acid phosphatase-positive structures (Storrie et al., 1984), some of which appeared to be multivesicular bodies (MVBs). Endocytosed asialoglycoprotein in rat hepatocytes also appeared to move from peripheral endosomes to more centrally located MVBs between 2 and 5 min of internalization (Wall and Hubbard, 1985). Although the results of the CHO study indicate that MVBs contain at least one lysosomal enzyme, degradation of the asialoglycoprotein was not detected until 25 min. With rat liver homogenates early endocytic vesicles could be separated from MBVs on sucrose density gradients; characterization of the isolated fractions showed that asialoglycoprotein receptor cosedimented with early vesicles but was very much diminished in the fractions containing MVBs (Mueller and Hubbard, 1986).

Assuming that MVBs participate in endocytosis of lysosomal enzymes, do they also function in the intracellular pathway? In a study of Clone 9 hepatocytes treated with NH_4Cl or chloroquine, depletion of Man 6-P receptors from Golgi–GERL and simultaneous accumulation of receptors in MVBs was noted (Brown et al., 1986). The key issue is whether receptors moved directly to the MVBs from Golgi–GERL, or went first to the plasma membrane (thus entering the endocytic pathway) and only then to the MVBs. The kinetics of receptor relocation observed in this study are not inconsistent with the latter route: after addition of weak base there was a 1-hr lag, then receptor relocation increased linearly to 100% over a 2-hr period. Comparing these results to those of the antibody labeling experiments described above, all of the intracellular receptor could have moved to the surface then into the MVBs during this time. Uptake of an endocytic tracer, cationized ferritin, into the MVBs was not blocked in the presence of weak bases.

The Golgi–GERL is another possible site for mixing of receptors from the intracellular and endocytic pathways. In K562 cells recycling of an endocytic receptor through the Golgi–GERL was indicated by the resialylation of transferrin receptor following removal of sialic acid from surface receptors with neuraminidase (Snider and Rogers, 1985). In a second study recycling of this receptor through even earlier regions of the Golgi (i.e., the compartment(s) containing α-mannosidase I) was demonstrated (Snider and Rogers, 1986). However, a large disparity between the time required for the receptor to recycle to the surface (10–20 min) and the time required for resialylation ($t_{1/2}$ 2–3 hr) or mannose trimming ($t_{1/2}$ 6 hr) was noted. Reglycosylation may be a relatively inefficient process, or recycling of surface receptor through the Golgi–GERL may be a minor, secondary route.

B. Disruption of Mixing

Following incubation for 4 hr at the nonpermissive temperature, cation-independent Man 6-P binding activity in intracellular membranes of the *ts* End1 mutant* was 50–100% of normal (Roff *et al.*, 1986). On return to the permissive temperature (Roff and Robbins, 1987), recovery of the intracellular pathway for lysosomal enzymes was rapid (100% in 4 hr), whereas recovery of cell-surface binding and endocytosis of lysosomal enzymes was slow (lag of 4–8 hr, 100% recovery in 24–30 hr). The 8-hr lag is consistent with the time required for complete restoration of endosomal acidification, as indicated by other endocytic activities. Inclusion of Man 6-P during incubation at both the nonpermissive then permissive temperatures increased the basal level of enzyme uptake after the lag (from ≤10 to 30% of control values) but did not affect the rate of recovery. Although several scenarios are consistent with our findings, they all require that mixing of intracellular and cell surface Man 6-P receptors be curtailed during recovery.

If the intracellular pool of active Man 6-P receptors is rescued on recovery, then those receptors must repopulate *only* the intracellular pathway.

* Specifying the *ts* End1 mutant in this context is not meant to imply that the *ts* End2 mutants behaved differently; where examined they behaved similarly. Discussion is restricted to the End1 mutant simply because our studies of this mutant have been more thorough.

But this pool may be a red herring: unlike cells recovering after treatment with weak base (Gonzalez-Noriega *et al.*, 1980; Brown *et al.*, 1986), recovery of the *ts* End1 mutant requires new protein synthesis (Roff and Robbins, 1987), presumably of the *ts* component. On removal of weak base, acidification of existing compartments resumes (Maxfield, 1982); lysosomal enzyme–receptor complexes trapped in those compartments can then dissociate, and receptor and enzyme each proceed to its appropriate next compartment (Brown *et al.*, 1986). In contrast, the newly synthesized *ts* component may be incorporated only into newly forming compartments, that is, there may be no rescue of trapped Man 6-P receptors. Thus, at the start of recovery the mutants may for all practical purposes be receptorless cells, needing new synthesis of receptor as well as of the *ts* component.

Whether the source of receptors for the intracellular pathway is newly synthesized or rescued receptors, the level of receptor in this pathway is sufficient for normal function after 4 hr. If restoration of the intracellular pathway requires a full complement of 215 kDa Man 6-P receptor, then these receptors must replenish the cell surface very slowly. What if the nature or number of receptors required by the intracellular and endocytic pathways are different, that is, enzyme segregation can proceed via an alternate receptor or at much lower levels of 215 kDa receptor? This could explain the disparity between rates of recovery of the two pathways. Under these circumstances, however, Man 6-P receptor from the cell surface must be restricted from mixing with the intracellular pathway, otherwise endocytosis during recovery would result in diminution of surface receptor, which is not observed. Preincubation of the recovering *ts* End1 mutant with high uptake β-galactosidase did not diminish subsequent endocytosis of lysosomal enzymes; also, uptake of enzymes increased linearly with time in the recovering mutant.

Thus, it appears that whatever mechanism is invoked, old receptors, new receptors, or alternate receptors, receptor mixing must be curtailed on resumption of activity in the *ts* End1 mutant. Elucidation of the mechanism by which the pathways are kept separate under these peculiar circumstances may contribute to our understanding of how they mix under normal conditions.

ACKNOWLEDGMENTS

I am grateful to my friends and colleagues Clara W. Hall, Sharon S. Krag, Susan M. Laurie, Rachel Myerowitz, Richard L. Proia, and Calvin F. Roff for their careful, thoughtful consideration of the manuscript. I especially thank Sharon Krag, without whose advice, encouragment, and good example it would not have been accomplished.

REFERENCES

Alexander, D., Dudin, G., Talj, F., Bitar, F., Deeb, M., Khudr, A., Abboud, M., and Der Kaloustian, V. M. (1984). Five related Lebanese individuals with high plasma lysosomal hydrolases: A new defect in mannose 6-phosphate receptor recognition? *Am. J. Hum. Genet.* **36**, 1001–1014.

Anderson, R. G. W., and Pathak, R. K. (1985). Vesicles and cisternae in the trans Golgi apparatus of human fibroblasts are acidic compartments. *Cell* **40**, 635–643.

Anderson, R. G. W., Brown, M. S., and Goldstein, J. L. (1977). Role of the coated endocytic vesicle in the uptake of receptor-bound low density lipoprotein in human fibroblasts. *Cell* **10**, 351–364.

Anderson, R. G. W., Falck, J. R., Goldstein, J. L., and Brown, M. S. (1984). Visualization of acidic organelles in intact cells by electron microscopy. *Proc. Natl. Acad. Sci. U.S.A.* **81**, 4838–4842.

Bach, G., Friedman, R., Weissmann, B., and Neufeld, E. F. (1972). The defect in the Hurler and Scheie syndromes: Deficiency of α-L-iduronidase. *Proc. Natl. Acad. Sci. U.S.A.* **69**, 2048–2051.

Bach, G., Bargal, R., and Cantz, M. (1979). I-Cell disease: Deficiency of extracellular hydrolase phosphorylation. *Biochem. Biophys. Res. Commun.* **91**, 976–981.

Barton, R. W., and Neufeld, E. F. (1971). The Hurler corrective factor. Purification and some properties. *J. Biol. Chem.* **246**, 7773–7779.

Barton, R. W., and Neufeld, E. F. (1972). A distinct biochemical deficit in the Maroteaux–Lamy syndrome (mucopolysaccharidosis VI). *J. Pediat.* **80**, 114–116.

Bischoff, J., and Kornfeld, R. (1983). Evidence for an α-mannosidase in endoplasmic reticulum of rat liver. *J. Biol. Chem.* **258**, 7907–7910.

Brot, F. E., Glaser, J. H., Roozen, K. J., and Sly, W. S. (1974). *In vitro* correction of deficient human fibroblasts by β-glucuronidase from different human sources. *Biochem. Biophys. Res. Commun.* **57**, 1–8.

Brown, W. J., and Farquhar, M. G. (1984a). The mannose 6-phosphate receptor for lysosomal enzymes is concentrated in cis Golgi cisternae. *Cell* **36**, 295–307.

Brown, W. J., and Farquhar, M. G. (1984b). Accumulation of coated vesicles bearing mannose 6-phosphate receptors for lysosomal enzymes in the Golgi region of I-cell fibroblasts. *Proc. Natl. Acad. Sci. U.S.A.* **81**, 5135–5139.

Brown, W. J., Constantinescu, E., and Farquhar, M. G. (1984). Redistribution of mannose 6-phosphate receptors induced by tunicamycin and chloroquine. *J. Cell Biol.* **99**, 320–326.

Brown, W. J., Goodhouse, J., and Farquhar, M. G. (1986). Mannose 6-phosphate receptors for lysosomal enzymes cycle between the Golgi complex and endosomes. *J. Cell Biol.* **103**, 1235–1247.

Campbell, C. H., and Rome, L. H. (1983). Coated vesicles from rat liver and calf brain contain lysosomal enzymes bound to mannose 6-phosphate receptors. *J. Biol. Chem.* **258**, 13347–13352.

Campbell, C. H., Fine, R. E., Squicciarini, J., and Rome, L. H. (1983). Coated vesicles from rat liver and calf brain contain cryptic mannose 6-phosphate receptors. *J. Biol. Chem.* **258**, 2628–2633.

Cantz, M., Crambach, A., and Neufeld, E. F. (1970). Characterization of the factor deficient in the Hunter syndrome by polyacrylamide gel electrophoresis. *Biochem. Biophys. Res. Commun.* **39**, 936–942.

Creek, K. E., and Sly, W. S. (1982). Adsorptive pinocytosis of phosphorylated oligosaccharides by human fibroblasts. *J. Biol. Chem.* **257**, 9931–9937.

Creek, K. E., and Sly, W. S. (1983). Biosynthesis and turnover of the phosphomannosyl ₁eceptor in human fibroblasts. *Biochem. J.* **214**, 353–360.

Danes, B. S., and Bearn, A. G. (1965). Hurler's syndrome: Demonstration of an inherited disorder of connective tissue in cell culture. *Science* **149**, 987–989.

Dautry-Varsat, A., Ciechanover, A., and Lodish, H. F. (1983). pH and the recycling of transferrin during receptor-mediated endocytosis. *Proc. Natl. Acad. Sci. U.S.A.* **80**, 2258–2262.

Deutscher, S. L., Creek, K. E., Merion, M., and Hirschberg, C. B. (1983). Subfractionation of rat liver Golgi apparatus: Separation of enzyme activities involved in the biosynthesis of the phosphomannosyl recognition marker in lysosomal enzymes. *Proc. Natl. Acad. Sci. U.S.A.* **80**, 3938–3942.

Didsbury, J. R., Moehring, J. M., and Moehring, T. J. (1983). Binding and uptake of diphtheria toxin by toxin-resistant Chinese hamster ovary and mouse cells. *Mol. Cell. Biol.* **3**, 1283–1294.

Distler, J., Hieber, V., Sahagian, G., Schmickel, R., and Jourdian, G. W. (1979a). Identification of mannose 6-phosphate in glycoproteins that inhibit the assimilation of β-galactosidase by fibroblasts. *Proc. Natl. Acad. Sci. U.S.A.* **76**, 4235–4239.

Distler, J., Hieber, V., Schmickel, R., and Jourdian, G. W. (1979b). The role of glycosidically bound mannose in the cellular assimilation of β-D-galactosidase. *In* "Carbohydrate-Protein Interaction," (I. Goldstein, ed.), pp. 163–180. ACS Symposium Series, Am. Chem. Soc., Washington, D.C.

Donovan, J. J., Simon, M. I., and Montal, M. (1985). Requirements for the translocation of diphtheria toxin fragment across lipid membranes. *J. Biol. Chem.* **260**, 8817–8823.

Draper, R. K., and Simon, M. I. (1980). The entry of diphtheria toxin into the mammalian cell cytoplasm: Evidence for lysosomal involvement. *J. Cell Biol.* **87**, 849–854.

Dunphy, W. G., and Rothman, J. E. (1983). Compartmentation of asparagine-linked oligosaccharide processing in the Golgi apparatus. *J. Cell Biol.* **97**, 270–275.

Dunphy, W. G., Fries, E., Urbani, L. J., and Rothman, J. E. (1981). Early and late functions associated with the Golgi apparatus reside in distinct compartments. *Proc. Natl. Acad. Sci. U.S.A.* **78**, 7453–7457.

Dunphy, W. G., Brands, R., and Rothman, J. E. (1985). Attachment of terminal *N*-acetylglucosamine to asparagine-linked oligosaccharides occurs in central cisternae of the Golgi stack. *Cell* **40**, 463–472.

Elder, J. H., and Alexander, S. (1982). *endo*-β-*N*-Acetylglucosaminidase F: Endoglycosidase from *Flavobacterium meningosepticum* that cleaves both high-mannose and complex glycoproteins. *Proc. Natl. Acad. Sci. U.S.A.* **79**, 4540–4544.

Erickson, A. H., and Blobel, G. (1979). Early events in the biosynthesis of the lysosomal enzyme cathepsin D. *J. Biol. Chem.* **254**, 11771–11774.

Erickson, A. H., Conner, G. E., and Blobel, G. (1981). Biosynthesis of a lysosomal enzyme. Partial structure of two transient and functionally distinct NH_2-terminal sequences in cathepsin D. *J. Biol. Chem.* **256**, 11224–11231.

Fischer, H. D., Gonzalez-Noriega, A., and Sly, W. S. (1980a). β-Glucuronidase binding to human fibroblast membrane receptors. *J. Biol. Chem.* **255**, 5069–5074.

Fischer, H. D., Gonzalez-Noriega, A., Sly, W. S., and Morre, D. J. (1980b). Phosphomannosyl-enzyme receptors in rat liver. Subcellular distribution and role in intracellular transport of lysosomal enzymes. *J. Biol. Chem.* **255**, 9608–9615.

Fischer, H. D., Natowicz, M., Sly, W. S., and Bretthauer, R. K. (1980c). Fibroblast receptor for lysosomal enzymes mediates pinocytosis of multivalent phosphomannan fragment. *J. Cell Biol.* **84**, 77–86.

Fischer, H. D., Creek, K. E., and Sly, W. S. (1982). Binding of phosphorylated oligosaccharides to immobilized phosphomannosyl receptors. *J. Biol. Chem.* **257**, 9938–9943.

Forgac, M., Cantley, L., Wiedenmann, B., Alstiel, L., and Branton, D. (1983). Clathrin-coated vesicles contain an ATP-dependent proton pump. *Proc. Natl. Acad. Sci. U.S.A.* **80**, 1300–1303.

Fratantoni, J. C., Hall, C. W., and Neufeld, E. F. (1968a). The defect in Hurler's and Hunter's syndromes: Faulty degradation of mucopolysaccharide. *Proc. Natl. Acad. Sci. U.S.A.* **60**, 699–706.

Fratantoni, J. C., Hall, C. W., and Neufeld, E. F. (1968b). Hurler and Hunter syndromes: Mutual correction of the defect in cultured fibroblasts. *Science* **162**, 570–572.

Fratantoni, J. C., Hall, C. W., and Neufeld, E. F. (1969). The defect in Hurler and Hunter syndromes, II. Deficiency of specific factors involved in mucopolysaccharide degradation. *Proc. Natl. Acad. Sci. U.S.A.* **64**, 360–366.

Frisch, A., and Neufeld, E. F. (1981). Limited proteolysis of the β-hexosaminidase precursor in a cell-free system. *J. Biol. Chem.* **256**, 8242–8246.

Gabel, C. A., and Foster, S. A. (1986). Lysosomal enzyme trafficking in mannose 6-phosphate receptor-positive mouse L-cells: Demonstration of a steady-state accumulation of phosphorylated acid hydrolases. *J. Cell Biol.* **102**, 943–950.

Gabel, C. A., and Kornfeld, S. (1982). Lysosomal enzyme phosphorylation in mouse lymphoma cell lines with altered asparagine-linked oligosaccharides. *J. Biol. Chem.* **257**, 10605–10612.

Gabel, C. A., Goldberg, D. E., and Kornfeld, S. (1982). Lysosomal enzyme oligosaccharide phosphorylation in mouse lymphoma cells: Specificity and kinetics of binding to the mannose 6-phosphate receptor *in vivo*. *J. Cell Biol.* **95**, 536–542.

Gabel, C. A., Goldberg, D. E., and Kornfeld, S. (1983). Identification and characterization of cells deficient in the mannose 6-phosphate receptor: Evidence for an alternate pathway for lysosomal enzyme targeting. *Proc. Natl. Acad. Sci. U.S.A.* **80**, 775–779.

Gabel, C. A., Goldberg, D. E., and Kornfeld, S. (1984a). Evidence for a mannose 6-phosphate-independent pathway for lysosomal enzyme targeting. *In* "Molecular Basis of Lysosomal Storage Disorders" (J. A. Barranger and R. O. Brady, eds.), pp. 175–193. Academic Press, New York.

Gabel, C. A., Costello, C. E., Reinhold, V. N., Kurz, L., and Kornfeld, S. (1984b). Identification of methylphosphomannosyl residues as components of the high mannose oligosaccharides of *Dictyostelium discoideum* glycoproteins. *J. Biol. Chem.* **259**, 13762–13769.

Gal, S., and Gottesman, M. M. (1986). The major excreted protein of transformed fibroblasts (MEP) is an activatable acid-protease. *J. Biol. Chem.* **260**, 1760–1765.

Galloway, C. J., Dean, G. E., Marsh, M., Rudnick, G., and Mellman, I. (1983). Acidification of macrophage and fibroblast endocytic vesicles *in vitro*. *Proc. Natl. Acad. Sci. U.S.A.* **80**, 3334–3338.

Gartung, C., Braulke, T., Hasilik, A., and von Figura, K. (1985). Internalization of blocking antibodies against mannose 6-phosphate specific receptors. *EMBO J.* **4**, 1725–1730.

Geuze, H. J., Slot, J. W., Strous, G. J. A. M., Lodish, H. F., and Schwartz, A. L. (1983). Intracellular site of asialoglycoprotein receptor–ligand uncoupling: Double-label immunoelectron microscopy during receptor-mediated endocytosis. *Cell* **32**, 277–287.

Geuze, H. J., Slot, J. W., Strous, G. J. A. M., Hasilik, A., and von Figura, K. (1984a). Ultrastructural localization of the mannose 6-phosphate receptor in rat liver. *J. Cell Biol.* **98**, 2047–2054.

Geuze, H. J., Slot, J. W., Strous, G. J. A. M., Peppard, J., von Figura, K., Hasilik, A., and Schwartz, A. L. (1984b). Intracellular receptor sorting during endocytosis: Comparative immunoelectron microscopy of multiple receptors in rat liver. *Cell* **37**, 195–204.

Geuze, H. J., Slot, J. W., Strous, G. J., Luzio, J. P., and Schwartz, A. L. (1984c). A cycloheximide-resistant pool of receptors for asialoglycoproteins and mannose 6-phosphate residues in the Golgi complex of hepatocytes. *EMBO J.* **3**, 2677–2685.

Geuze, H. J., Slot, J. W., Strous, G. J. A. M., Hasilik, A., and von Figura, K. (1985). Possible pathways for lysosomal enzyme delivery. *J. Cell Biol.* **101**, 2253–2262.

Gieselmann, V., Pohlmann, R., Hasilik, A., and von Figura, K. (1983). Biosynthesis and transport of cathepsin D in cultured human fibroblasts. *J. Cell Biol.* **97**, 1–5.

Glaser, J. H., Roozen, K. J., Brot, F. E., and Sly, W. S. (1975). Multiple isoelectric and recognition forms of human β-glucuronidase activity. *Arch. Biochem. Biophys.* **166**, 536–542.

Goldberg, D. E., and Kornfeld, S. (1981). The phosphorylation of β-glucuronidase oligosaccharides in mouse P388D$_1$ cells. *J. Biol. Chem.* **256**, 13060–13067.

Goldberg, D. E., and Kornfeld, S. (1983). Evidence for extensive subcellular organization of asparagine-linked oligosaccharide processing and lysosomal enzyme phosphorylation. *J. Biol. Chem.* **258**, 3159–3165.

Goldberg, D. E., Gabel, C. A., and Kornfeld, S. (1983). Studies of the biosynthesis of the mannose 6-phosphate receptor in receptor-positive and -deficient cell lines. *J. Cell Biol.* **97**, 1700–1706.

Gonzalez-Noriega, A., Grubb, J. H., Talkad, V., and Sly, W. S. (1980). Chloroquine inhibits lysosomal enzyme pinocytosis and enhances lysosomal enzyme secretion by impairing receptor recycling. *J. Cell Biol.* **85**, 839–852.

Gravel, R. A., Gravel, Y. L., Miller, A. L., and Lowden, J. A. (1981). Genetic complementation analysis of I-cell disease and pseudo-Hurler polydystrophy. *In* "Lysosomes and Lysosomal Storage Diseases" (J. W. Callahan and J. A. Lowden, eds.), pp. 289–298. Raven, New York.

Griffiths, G., Pfeiffer, S., Simons, K., and Matlin, K. (1985). Exit of newly synthesized membrane proteins from the trans cisterna of the Golgi complex to the plasma membrane. *J. Cell Biol.* **101**, 949–964.

Harford, J., Klausner, R. D., Wolkoff, A. W., Bridges, K. R., and Ashwell, G. (1984). Asialoglycoprotein catabolism by hepatocytes: Insights from perturbation of the endocytic pathway. *In* "Molecular Basis of Lysosomal Storage Disorders" (J. A. Barranger and R. O. Brady, eds.), pp. 149–162. Academic Press, New York.

Hasilik, A., and Neufeld, E. F. (1980a). Biosynthesis of lysosomal enzymes in fibroblasts. Synthesis as precursors of higher molecular weight. *J. Biol. Chem.* **255**, 4937–4945.

Hasilik, A., and Neufeld, E. F. (1980b). Biosynthesis of lysosomal enzymes in fibroblasts. Phosphorylation of mannose residues. *J. Biol. Chem.* **255**, 4946–4950.

Hasilik, A., and von Figura, K. (1981). Oligosaccharides in lysosomal enzymes. Distribution of high-mannose and complex oligosaccharides in cathepsin D and β-hexosaminidase. *Eur. J. Biochem.* **121**, 125–129.

Hasilik, A., Klein, U., Waheed, A., Strecker, G., and von Figura, K. (1980). Phosphorylated oligosaccharides in lysosomal enzymes: Identification of α-N-acetylglucosamine(1)phospho(6)mannose diester groups. *Proc. Natl. Acad. Sci. U.S.A.* **77**, 7074–7078.

Hasilik, A., Waheed, A., and von Figura, K. (1981). Enzymatic phosphorylation of lysosomal enzymes in the presence of UDP-N-acetylglucosamine. Absence of the activity in I-cell fibroblasts. *Biochem. Biophys. Res. Commun.* **98**, 761–767.

Hasilik, A., Pohlmann, R., and von Figura, K. (1983). Inhibition by cyanate of the processing of lysosomal enzymes. *Biochem. J.* **210**, 795–802.

Helmy, S., Porter-Jordan, K., Dawidowicz, E. A., Pilch, P., Schwartz, A. L., and Fine, R. E. (1986). Separation of endocytic from exocytic coated vesicles using a novel cholinesterase mediated density shift technique. *Cell* **44**, 497–506.

Hickman, S., and Neufeld, E. F. (1972). A hypothesis for I-cell disease: Defective hydrolases that do not enter lysosomes. *Biochem. Biophys. Res. Commun.* **49**, 992–999.

Hickman, S., Shapiro, L. J., and Neufeld, E. F. (1974). A recognition marker required for uptake of a lysosomal enzyme by cultured fibroblasts. *Biochem. Biophys. Res. Commun.* **57**, 55–61.

Hoflack, B. and Kornfeld, S. (1985a). Lysosomal enzyme binding to mouse P388D$_1$ macrophage membranes lacking the 215-kDa mannose 6-phosphate receptor: Evidence for the existence of a second mannose 6-phosphate receptor. *Proc. Natl. Acad. Sci. U.S.A.* **82**, 4428–4432.

Hoflack, B. and Kornfeld, S. (1985b). Purification and characterization of a cation-dependent mannose 6-phosphate receptor from murine P388D$_1$ macrophages and bovine liver. *J. Biol. Chem.* **260**, 12008–12014.

Honey, N. K., Mueller, O. T., Little, L. E., Miller, A. L., and Shows, T. B. (1982). Mucolipidosis III is genetically heterogeneous. *Proc. Natl. Acad. Sci. U.S.A.* **79**, 7420–7424.

Hsieh, P., Rosner, M. R., and Robbins, P. W. (1983). Selective cleavage by endo-β-N-acetylglucosaminidase H at individual glycosylation sites of Sindbis virion envelope glycoproteins. *J. Biol. Chem.* **258**, 2555–2561.

Imort, M., Zühlsdorf, M., Feige, U., Hasilik, A., and von Figura, K. (1983). Biosynthesis and transport of lysosomal enzymes in human monocytes and macrophages. *Biochem. J.* **214**, 671–678.

Johnson, D. C., and Schlesinger, M. J. (1980). Vesicular stomatitis virus and Sindbis virus glycoprotein transport to the cell surface is inhibited by ionophores. *Virology* **103**, 407–424.

Kaplan, A., Achord, D. T., and Sly, W. S. (1977a). Phosphohexosyl components of a lysosomal enzyme are recognized by pinocytosis receptors on human fibroblasts. *Proc. Natl. Acad. Sci. U.S.A.* **74**, 2026–2030.

Kaplan, A., Fischer, D., Achord, D., and Sly, W. (1977b). Phosphohexosyl recognition is a general characteristic of pinocytosis of lysosomal glycosidases by human fibroblasts. *J. Clin. Invest.* **60**, 1088–1093.

Karson, E. M., Neufeld, E. F., and Sando, G. N. (1980). p-Isothiocyanatophenyl 6-phospho-α-D-mannopyranoside coupled to albumin. A model compound recognized by the fibroblast enzyme uptake system. 2. Biological properties. *Biochemistry* **19**, 3856–3860.

Klausner, R. D., van Renswoude, J., Kempf, C., Rao, K., Bateman, J. L., and Robbins, A. R. (1984). Failure to release iron from transferrin in a Chinese hamster ovary cell mutant pleiotropically defective in endocytosis. *J. Cell Biol.* **98**, 1098–1101.

Kobata, A. (1978). Endo-β-N-Acetylglucosaminidase C$_I$ and C$_{II}$ from *Clostridium perfringens*. *Methods Enzymol.* **50**, 567–574.

Kornfeld, S., Li, E., and Tabas, I. (1978). The synthesis of complex-type oligosaccharides. II. Characterization of the processing intermediates in the synthesis of the complex oligosaccharide units of the vesicular stomatitis virus G protein. *J. Biol. Chem.* **253**, 7771–7778.

Krag, S. S. (1979). A concanavalin A-resistant Chinese hamster ovary cell line is deficient in the synthesis of [^{3}H]glucosyl oligosaccharide–lipid. *J. Biol. Chem.* **254**, 9167–9177.

Krag, S. S., and Robbins, A. R. (1982). A Chinese hamster ovary cell mutant deficient in glucosylation of lipid-linked oligosaccharide synthesizes lysosomal enzymes of altered structure and function. *J. Biol. Chem.* **257**, 8424–8431.

Kresse, H., and Neufeld, E. F. (1972). The Sanfilippo A corrective factor. Purification and mode of action. *J. Biol. Chem.* **247**, 2164–2170.

Kresse, H., Wiesmann, U., Cantz, M., Hall, C. W., and Neufeld, E. F. (1971). Biochemical heterogeneity of the Sanfilippo syndrome: Preliminary characterization of two deficient factors. *Biochem. Biophys. Res. Commun.* **42**, 892–898.

Lagunoff, D., Nicol, D. M., and Pritzl, P. (1973). Uptake of β-glucuronidase by deficient human fibroblasts. *Lab. Invest.* **29**, 449–453.

Lang, L., Reitman, M., Tang, J., Roberts, R. M., and Kornfeld, S. (1984). Lysosomal enzyme phosphorylation. Recognition of a protein-dependent determinant allows specific phosphorylation of oligosaccharides present on lysosomal enzymes. *J. Biol. Chem.* **259**, 14663–14671.

Lang, L., Takahashi, T., Tang, J., and Kornfeld, S. (1985). Lysosomal enzyme phosphorylation in human fibroblasts. Kinetic parameters offer a biochemical rationale for two distinct defects in the uridine diphospho-*N*-acetylglucosamine:lysosomal enzyme precursor *N*-acetylglucosamine-1-phosphotransferase. *J. Clin. Invest.* **76**, 2191–2195.

Lemansky, P., Gieselmann, V., Hasilik, A., and von Figura, K. (1985). Synthesis and transport of lysosomal acid phosphatase in normal and I-cell fibroblasts. *J. Biol. Chem.* **260**, 9023–9030.

Lightbody, J., Wiesmann, U., Hadorn, B., and Herschkowitz, N. (1971). I-cell disease: Multiple lysosomal-enzyme defect. *Lancet* **1**, 451.

Little, L. E., and Neufeld, E. F. (1986). The amino terminus of precursor and mature forms of the α chain of β-hexosaminidase. *Fed. Proc., Fed. Am. Soc. Exp. Biol.* **45**(6), 1711 (Abstr.).

Lucocq, J. M., Brada, D., and Roth, J. (1986). Immunolocalization of the oligosaccharide trimming enzyme glucosidase II. *J. Cell Biol.* **102**, 2137–2146.

Maler, T., Rosenblum, B. B., and Jourdian, G. W. (1985). Properties of the Syrian hamster phosphomannosyl receptor: An aggregate of low molecular weight proteins. *Proc. Natl. Acad. Sci. U.S.A.* **82**, 8379–8383.

Mandel, R., Ryser, H. J.-P., Hacobian, A., and Shen, W.-C. (1985). Patterns of sensitivity and cross resistance to diphtheria toxin, ricin, modeccin and methotrexate–poly(lysine) in mutants of Chinese hamster ovary cells. *J. Cell Biol.* **101**(5, Pt. 2), 289a (Abstr.).

Marnell, M. H., Mathis, L. S., Stookey, M., Shia, S.-P., Stone, D. K., and Draper, R. K. (1984). A Chinese hamster ovary cell mutant with a heat-sensitive, conditional-lethal defect in vacuolar function. *J. Cell Biol.* **99**, 1907–1916.

Maxfield, F. R. (1982). Weak bases and ionophores rapidly and reversibly raise the pH of endocytic vesicles in cultured mouse fibroblasts. *J. Cell Biol.* **95**, 676–681.

Merion, M., Schlesinger, P., Brooks, R. M., Moehring, J. M., Moehring, T. J., and Sly, W. S. (1983). Defective acidification of endosomes in Chinese hamster ovary cell mutants "cross-resistant" to toxins and viruses. *Proc. Natl. Acad. Sci. U.S.A.* **80**, 5315–5319.

Miller, A. L., Kress, B. C., Stein, R., Kinnon, C., Kern, H., Schneider, J. A., and Harms, E. (1981). Properties of *N*-acetyl-β-D-hexosaminidase from isolated normal and I-cell lysosomes. *J. Biol. Chem.* **256**, 9352–9362.

Mitchell, D. C., Maler, T., and Jourdian, G. W. (1983). Detergent dissociation of bovine liver phosphomannosyl binding protein. *J. Cell. Biochem.* **24**, 319–330.

Moehring, J. M., and Moehring, T. J. (1983). Strains of CHO-K1 cells resistant to *Pseudomonas* exotoxin A and cross-resistant to diphtheria toxin and viruses. *Infect. Immun.* **41**, 998–1009.

Mueller, O. T., Honey, N. K., Little, L. E., Miller, A. L., and Shows, T. B. (1983). Mucolipidosis II and III. The genetic relationships between two disorders of lysosomal enzyme biosynthesis. *J. Clin. Invest.* **72,** 1016–1023.

Mueller, S. C., and Hubbard, A. L. (1986). Receptor-mediated endocytosis of asialoglycoproteins by rat hepatocytes: Receptor-positive and receptor-negative endosomes. *J. Cell Biol.* **103,** 932–942.

Myerowitz, R., and Neufeld, E. F. (1981). Maturation of α-L-iduronidase in cultured human fibroblasts. *J. Biol. Chem.* **256,** 3044–3048.

Myerowitz, R., Piekarz, R., Neufeld, E. F., Shows, T. B., and Suzuki, K. (1985). Human β-hexosaminidase α chain: Coding sequence and homology with the β chain. *Proc. Natl. Acad. Sci. U.S.A.* **82,** 7830–7834.

Natowicz, M. R., Chi, M. M.-Y., Lowry, O. H., and Sly, W. S. (1979). Enzymatic identification of mannose 6-phosphate on the recognition marker for receptor-mediated pinocytosis of β-glucuronidase by human fibroblasts. *Proc. Natl. Acad. Sci. U.S.A.* **76,** 4322–4326.

Natowicz, M., Baenziger, J. U., and Sly, W. S. (1982). Structural studies of the phosphorylated high mannose-type oligosaccharides on human β-glucuronidase. *J. Biol. Chem.* **257,** 4412–4420.

Natowicz, M., Hallett, D. W., Frier, C., Chi, M., Schlesinger, P. H., and Baenziger, J. U. (1983). Recognition and receptor-mediated uptake of phosphorylated high mannose-type oligosaccharides by cultured human fibroblasts. *J. Cell Biol.* **96,** 915–919.

Neufeld, E. F., and Cantz, M. J. (1971). Corrective factors for inborn errors of mucopolysaccharide metabolism. *Ann. N.Y. Acad. Sci.* **179,** 580–587.

Neufeld, E. F., and McKusick, V. A. (1983). Disorders of lysosomal enzyme synthesis and localization: I-cell disease and pseudo-Hurler polydystrophy. *In* "The Metabolic Basis of Inherited Disease" (J. B. Stanbury, J. B. Wyngaarden, D. S. Fredrickson, J. L. Goldstein, and M. S. Brown, eds.), 5th Ed., pp. 778–787. McGraw-Hill New York.

Nicol, D. M., Lagunoff, D., and Pritzl, P. (1974). Differential uptake of human β-glucuronidase isoenzymes from spleen by deficient fibroblasts. *Biochem. Biophys. Res. Commun.* **59,** 941–946.

Novikoff, P. M., La Russo, N. F., Novikoff, A. B., Stockert, R. J., Yam, A., and Le Sage, G. D. (1983). Immunocytochemical localization of lysosomal β-galactosidase in rat liver. *J. Cell Biol.* **97,** 1559–1565.

Ohkuma, S., and Poole, B. (1978). Fluorescence probe measurement of the intralysosomal pH in living cells and the perturbation of pH by various agents. *Proc. Natl. Acad. Sci. U.S.A.* **75,** 3327–3331.

Orci, L., Montesano, R., Meda, P., Malaisse-Lagae, F., Brown, D., Perrelet, A., and Vassalli, P. (1981). Heterogeneous distribution of filipin–cholesterol complexes across the cisternae of the Golgi apparatus. *Proc. Natl. Acad. Sci. U.S.A.* **78,** 293–297.

Orci, L., Ravazzola, M., Amherdt, M., Louvard, D., and Perrelet, A. (1985). Clathrin-immunoreactive sites in the Golgi apparatus are concentrated at the trans pole in polypeptide hormone-secreting cells. *Proc. Natl. Acad. Sci. U.S.A.* **82,** 5385–5389.

Orci, L., Glick, B. S., and Rothman, J. E. (1986). A new type of coated vesicular carrier that appears not to contain clathrin: Its possible role in protein transport within the Golgi stack. *Cell* **46,** 171–184.

Owada, M., and Neufeld, E. F. (1982). Is there a mechanism for introducing acid hydrolases into liver lysosomes that is independent of mannose 6-phosphate recognition? Evidence from I-cell disease. *Biochem. Biophys. Res. Commun.* **105,** 814–820.

Paigen, K. (1979). Acid hydrolases as models of genetic control. *Annu. Rev. Genet.* **13,** 417–466.

Pohlmann, R., Waheed, A., Hasilik, A., and von Figura, K. (1982). Synthesis of phosphory-lated recognition marker in lysosomal enzymes is located in the cis part of Golgi apparatus. *J. Biol. Chem.* **257**, 5323–5325.

Proia, R. L., and Neufeld, E. F. (1982). Synthesis of β-hexosaminidase in cell-free transla-tion and in intact fibroblasts: An insoluble precursor α chain in a rare form of Tay-Sachs disease. *Proc. Natl. Acad. Sci. U.S.A.* **79**, 6360–6364.

Rearick, J. I., Fujimoto, K., and Kornfeld, S. (1981). Identification of the mannosyl donors involved in the synthesis of lipid-linked oligosaccharides. *J. Biol. Chem.* **256**, 3762–3769.

Reitman, M. L., and Kornfeld, S. (1981a). UDP–*N*-acetylglucosamine:glycoprotein *N*-ace-tylglucosamine:glycoprotein *N*-acetylglucosamine-1-phosphotransferase. Proposed enzyme for the phosphorylation of the high mannose oligosaccharide units of lysoso-mal enzymes. *J. Biol. Chem.* **256**, 4275–4281.

Reitman, M. L., and Kornfeld, S. (1981b). Lysosomal enzyme targeting. *N*-acetylgluco-saminylphosphotransferase selectively phosphorylates native lysosomal enzymes. *J. Biol. Chem.* **256**, 11977–11980.

Reitman, M. L., Varki, A., and Kornfeld, S. (1981). Fibroblasts from patients with I-cell disease and pseudo-Hurler polydystrophy are deficient in uridine 5′-diphosphate-*N*-acetylglucosamine:glycoprotein *N*-acetylglucosaminylphosphotransferase activity. *J. Clin. Invest.* **67**, 1574–1579.

Reitman, M. L., Trowbridge, I. S., and Kornfeld, S. (1982). A lectin-resistant mouse lym-phoma cell line is deficient in glucosidase II, a glycoprotein-processing enzyme. *J. Biol. Chem.* **257**, 10357–10363.

Robbins, A. R. (1979). Isolation of lysosomal α-mannosidase mutants of Chinese hamster ovary cells. *Proc. Natl. Acad. Sci. U.S.A.* **76**, 1911–1915.

Robbins, A. R., and Myerowitz, R. (1981). The mannose 6-phosphate receptor of Chinese hamster ovary cells. Compartmentalization of acid hydrolases in mutants with altered receptors. *J. Biol. Chem.* **256**, 10623–10627.

Robbins, A. R., and Roff, C. F. (1987). Isolation of mutant Chinese hamster ovary cells defective in endocytosis. *Methods Enzymol.* **138**, 458–470.

Robbins, A. R., Myerowitz, R., Youle, R. J., Murray, G. J., and Neville, D. M., Jr. (1981). The mannose 6-phosphate receptor of Chinese hamster ovary cells. Isolation of mu-tants with altered receptors. *J. Biol. Chem.* **256**, 10618–10622.

Robbins, A. R., Peng, S. S., and Marshall, J. L. (1983). Mutant Chinese hamster ovary cells pleiotropically defective in receptor-mediated endocytosis. *J. Cell Biol.* **96**, 1064–1071.

Robbins, A. R., Oliver, C., Bateman, J. L., Krag, S. S., Galloway, C. J., and Mellman, I. (1984). A single mutation in Chinese hamster ovary cells impairs both Golgi and endosomal functions. *J. Cell Biol.* **99**, 1296–1308.

Robbins, A. R., Herscovics, A., and Krag, S. S. (1987). Manuscript in preparation.

Robey, P. G., and Neufeld, E. F. (1982). Defective phosphorylation and processing of β-hexosaminidase by intact cultured fibroblasts from patients with mucolipidosis III. *Arch. Biochem. Biophys.* **213**, 251–257.

Roff, C. F., and Robbins, A. R. (1987). Manuscript in preparation.

Roff, C. F., Fuchs, R., Mellman, I., and Robbins, A. R. (1986). Chinese hamster ovary cell mutants with temperature-sensitive defects in endocytosis. I. Loss of function on shifting to the non-permissive temperature. *J. Cell Biol.* **103**, 2283–2297.

Rome, L. H., Weissmann, B., and Neufeld, E. F. (1979a). Direct demonstration of binding of a lysosomal enzyme, α-L-iduronidase, to receptors on cultured fibroblasts. *Proc. Natl. Acad. Sci. U.S.A.* **76**, 2331–2334.

Rome, L. H., Garvin, A. J., Allietta, M. M., and Neufeld, E. F. (1979b). Two species of lysosomal organelles in cultured human fibroblasts. *Cell* **17**, 143–153.

Rosenfeld, M. G., Kreibich, G., Popov, D., Kato, K., and Sabatini, D. D. (1982). Biosynthesis of lysosomal hydrolases: Their synthesis in bound polysomes and the role of co- and posttranslational processing in determining their subcellular distribution. *J. Cell Biol.* **93**, 135–143.

Roth, J., and Berger, E. G. (1982). Immunocytochemical localization of galactosyltrans- ferase in HeLa cells: Codistribution with thiamine pyrophosphatase in trans-Golgi cisternae. *J. Cell Biol.* **93**, 223–229.

Roth, J., Taatjes, D. J., Lucocq, J. M., Weinstein, J., and Paulson, J. C. (1985). Demonstra- tion of an extensive trans-tubular network continuous with the Golgi apparatus stack that may function in glycosylation. *Cell* **43**, 287–295.

Sahagian, G. G. (1981). Receptor-mediated uptake of lysosomal enzymes. Ph.D. disserta- tion. Univ. of Michigan, Ann Arbor.

Sahagian, G. G. (1984). The mannose 6-phosphate receptor: Function, biosynthesis and translocation. *Biol. Cell* **51**, 207–214.

Sahagian, G. G., and Neufeld, E. F. (1983). Biosynthesis and turnover of the mannose 6- phosphate receptor in cultured Chinese hamster ovary cells. *J. Biol. Chem.* **258**, 7121– 7128.

Sahagian, G. G., and Steer, C. J. (1985). Transmembrane orientation of the mannose 6-phosphate receptor in isolated clathrin-coated vesicles. *J. Biol. Chem.* **260**, 9838– 9842.

Sahagian, G. G., Distler, J., and Jourdian, G. W. (1981). Characterization of a membrane- associated receptor from bovine liver that binds phosphomannosyl residues of bovine testicular β-galactosidase. *Proc. Natl. Acad. Sci. U.S.A.* **78**, 4289–4293.

Sahagian, G. G., Distler, J. J., and Jourdian, G. W. (1982). Membrane receptor for phospho- mannosyl residues. *Methods Enzymol.* **83**, 392–396.

Sando, G. N., and Neufeld, E. F. (1977). Recognition and receptor-mediated uptake of a lysosomal enzyme, α-L-iduronidase, by cultured human fibroblasts. *Cell* **12**, 619–627.

Sando, G. N., Titus-Dillon, P., Hall, C. W., and Neufeld, E. F. (1979). Inhibition of recep- tor-mediated uptake of a lysosomal enzyme into fibroblasts by chloroquine, procaine and ammonia. *Exp. Cell Res.* **119**, 359–364.

Sandvig, K. and Olsnes, S. (1980). Diphtheria toxin entry into cells is facilitated by low pH. *J. Cell Biol.* **87**, 828–832.

Saunier, B., Kilker, R. D., Jr., Tkacz, J. S., Quaroni, A., and Herscovics, A. (1982). Inhibition of N-linked complex oligosaccharide formation by 1-deoxynojirimycin, an inhibitor of processing glucosidases. *J. Biol. Chem.* **257**, 14155–14161.

Schulze-Lohoff, E., Hasilik, A., and von Figura, K. (1985). Cathepsin D precursors in clathrin-coated organelles from human fibroblasts. *J. Cell Biol.* **101**, 824–829.

Schwartz, A. L., Strous, G. J. A. M., Slot, J. W., and Geuze, H. J. (1985). Immunoelectron microscopic localization of acidic intracellular compartments in hepatoma cells. *EMBO J.* **4**, 899–904.

Shapiro, L. J., Hall, C. W., Leder, I. G., and Neufeld, E. F. (1976). The relationship of α-L- iduronidase and Hurler corrective factor. *Arch. Biochem. Biophys.* **172**, 156–161.

Shepherd, V. L., Freeze, H. H., Miller, A. L., and Stahl, P. D. (1983). Identification of mannose 6-phosphate receptors in rabbit alveolar macrophages. *J. Biol. Chem.* **259**, 2257–2261.

Shows, T. B., Mueller, O. T., Honey, N. K., Wright, C. E., and Miller, A. L. (1982). Genetic heterogeneity of I-cell disease is demonstrated by complementation of lysoso- mal enzyme processing mutants. *Am. J. Med. Gen.* **12**, 343–353.

Skudlarek, M. D., and Swank, R. T. (1979). Biosynthesis of two lysosomal enzymes in macrophages. Evidence for a precursor of β-galactosidase. *J. Biol. Chem.* **254**, 9939– 9942.

Skudlarek, M. D., and Swank, R. T. (1981). Turnover of two lysosomal enzymes in macro- phages. *J. Biol. Chem.* **256**, 10137–10144.

Slot, J. W., and Geuze, H. J. (1983). Immunoelectron microscopic exploration of the Golgi complex. *J. Histochem. Cytochem.* **31**, 1049–1056.

Sly, W. S., Grubb, J. H., Moehring, J. M., and Moehring, T. J. (1984). The role of acidification in transport of acid hydrolases to lysosomes. In "Molecular Basis of Lysosomal Storage Disorders" (J. A. Barranger and R. O. Brady, eds.), pp. 163–174. Academic Press, New York.

Snider, M. D., and Rogers, O. C. (1985). Intracellular movement of cell surface receptors after endocytosis: Resialylation of asialo-transferrin receptor in human erythroleukemia cells. *J. Cell Biol.* **100**, 826–834.

Snider, M. D., and Rogers, O. C. (1986). Membrane traffic in animal cells: Cellular glycoproteins return to the site of Golgi mannosidase I. *J. Cell Biol.* **103**, 265–275.

Steiner, A., and Rome, L. H. (1982). Assay and purification of a solubilized membrane receptor that binds the lysosomal enzyme α-L-iduronidase. *Arch. Biochem. Biophys.* **214**, 681–687.

Stoll, J., Robbins, A. R., and Krag, S. S. (1982). Mutant of Chinese hamster ovary cells with altered mannose 6-phosphate receptor activity is unable to synthesize mannosylphosphoryldolichol. *Proc. Natl. Acad. Sci. U.S.A.* **79**, 2296–2300.

Stone, D. K., Sie, X-S, and Racker, E. (1983). An ATP-driven proton pump in clathrin-coated vesicles. *J. Biol. Chem.* **258**, 4059–4062.

Storrie, B., Pool, R. R., Jr., Sachdeva, M., Maurey, K. M., and Oliver, C. (1984). Evidence for both prelysosomal and lysosomal intermediates in endocytic pathways. *J. Cell Biol.* **98**, 108–115.

Strous, G. J., Du Maine, A., Zijderhand-Bleekemolen, J. E., Slot, J. W., and Schwartz, A. L. (1985). Effect of lysosomotropic amines on the secretory pathway and on the recycling of the asialoglycoprotein receptor in human hepatoma cells. *J. Cell Biol.* **101**, 531–539.

Tabas, I., and Kornfeld, S. (1980). Biosynthetic intermediates of β-glucuronidase contain high mannose oligosaccharides with blocked phosphate residues. *J. Biol. Chem.* **255**, 6633–6639.

Tarentino, A. L., Plummer, T. H., Jr., and Maley, F. (1974). The release of intact oligosaccharide from specific glycoproteins by endo-β-N-acetylglucosaminidase H. *J. Biol. Chem.* **249**, 818–824.

Tondeur, M., Vamos-Hurwitz, E., Mockel-Pohl, S., Dereume, J. P., Cremer, N., and Loeb, H. (1971). Clinical, biochemical, and ultrastructural studies in a case of chondrodystrophy presenting the I-cell phenotype in tissue culture. *J. Pediat.* **79**, 366–378.

Tümmers, S., Zühlsdorf, M., Robenek, H., Hasilik, A., and von Figura, K. (1983). Molecular forms of cathepsin D in coated vesicle preparations. *Hoppe Seyler's Z. Physiol. Chem.* **364**, 1287–1295.

Tycko, B., and Maxfield, F. R. (1982). Rapid acidification of endocytic vesicles containing α₂-macroglobulin. *Cell* **28**, 643–651.

Ullrich, K., Mersmann, G., Weber, E., and von Figura, K. (1978). Evidence for lysosomal enzyme recognition by human fibroblasts via a phosphorylated carbohydrate moiety. *Biochem. J.* **170**, 643–650.

van Dongen, J., Willemsen, R., Ginns, E. I., Sips, H. J., Tager, J. M., Barranger, J. A., and Reuser, A. J. J. (1985). The subcellular localization of soluble and membrane-bound lysosomal enzymes in I-cell fibroblasts: A comparative immunocytochemical study. *Eur. J. Cell Biol.* **39**, 179–189.

van Renswoude, J., Bridges, K. R., Harford, J. B., and Klausner, R. D. (1982). Receptor-mediated endocytosis of transferrin and the uptake of Fe in K562 cells: Identification of a nonlysosomal acidic compartment. *Proc. Natl. Acad. Sci. U.S.A.* **79**, 6186–6190.

Varki, A., and Kornfeld, S. (1980a). Identification of a rat liver α-N-acetylglucosaminyl phosphodiesterase capable of removing "blocking" α-N-acetylglucosamine residues from phosphorylated high mannose oligosaccharides of lysosomal enzymes. *J. Biol. Chem.* **255**, 8398–8401.

Varki, A., and Kornfeld, S. (1980b). Structural studies of phosphorylated high mannose-type oligosaccharides. *J. Biol. Chem.* **255**, 10847–10858.

Varki, A., and Kornfeld, S. (1981). Purification and characterization of rat liver α-N-acetylglucosaminyl phosphodiesterase. *J. Biol. Chem.* **256**, 9937–9943.

Varki, A., and Kornfeld, S. (1983). The spectrum of anionic oligosaccharides released by endo-β-N-acetylglucosaminidase H from glycoproteins. Structural studies and interactions with the phosphomannosyl receptor. *J. Biol. Chem.* **258**, 2808–2818.

Varki, A. P., Reitman, M. L., and Kornfeld, S. (1981). Identification of a variant of mucolipidosis III (pseudo-Hurler polydystrophy): A catalytically active N-acetylglucosaminyl phosphotransferase that fails to phosphorylate lysosomal enzymes. *Proc. Natl. Acad. Sci. U.S.A.* **78**, 7773–7777.

Varki, A., Sherman, W., and Kornfeld, S. (1983). Demonstration of the enzymatic mechanisms of α-N-acetyl-D-glucosamine-1-phosphodiester N-acetylglucosaminidase (formerly called α-N-acetylglucosaminylphosphodiesterase) and lysosomal α-N-acetylglucosaminidase. *Arch. Biochem. Biophys.* **222**, 145–149.

Vladutiu, G. D. (1983). Effect of the coexistence of galactosyl and phosphomannosyl residues on β-hexosaminidase on the processing and transport of the enzyme in mucolipidosis I fibroblasts. *Biochim. Biophys. Acta* **760**, 363–370.

von Figura, K., and Klein, U. (1979). Isolation and characterization of phosphorylated oligosaccharides from α-N-acetylglucosaminidase that are recognized by cell-surface receptors. *Eur. J. Biochem.* **94**, 347–354.

von Figura, K., and Kresse, H. (1972). The Sanfilippo B corrective factor: A N-acetyl-α-D-glucosaminidase. *Biochem. Biophys. Res. Commun.* **48**, 262–269.

von Figura, K., Gieselmann, V., and Hasilik, A. (1984). Antibody to mannose 6-phosphate specific receptor induces receptor deficiency in human fibroblasts. *EMBO J.* **3**, 1281–1286.

von Figura, K., Gieselmann, V., and Hasilik, A. (1985). Mannose 6-phosphate-specific receptor is a transmembrane protein with a C-terminal extension oriented towards the cytosol. *Biochem. J.* **225**, 543–547.

Waheed, A., Pohlmann, R., Hasilik, A., and von Figura, K. (1981a). Subcellular location of two enzymes involved in the synthesis of phosphorylated recognition markers in lysosomal enzymes. *J. Biol. Chem.* **256**, 4150–4152.

Waheed, A., Hasilik, A., and von Figura, K. (1981b). Processing of the phosphorylated recognition marker in lysosomal enzymes. Characterization and partial purification of a microsomal α-N-acetylglucosaminyl phosphodiesterase. *J. Biol. Chem.* **256**, 5717–5721.

Waheed, A., Pohlmann, R., Hasilik, A., von Figura, K., van Elsen, A., and Leroy, J. G. (1982). Deficiency of UDP-N-acetylglucosamine:lysosomal enzyme N-acetylglucosamine-1-phosphotransferase in organs of I-cell patients. *Biochem. Biophys. Res. Commun.* **105**, 1052–1058.

Wall, D. A., and Hubbard, A. L. (1985). Receptor-mediated endocytosis of asialoglycoproteins by rat liver hepatocytes: Biochemical characterization of the endosomal compartments. *J. Cell Biol.* **101**, 2104–2112.

Wall, D. A., Wilson, G., and Hubbard, A. L. (1980). The galactose-specific recognition system of mammalian liver: The route of ligand internalization in rat hepatocytes. *Cell* **21**, 79–93.

Wiesmann, U. N., Lightbody, J., Vassella, F., and Herschkowitz, N. N. (1971). Multiple lysosomal enzyme deficiency due to enzyme leakage. *N. Engl. J. Med.* **284,** 109–110.

Wiesmann, U. N., DiDonato, S., and Herschkowitz, N. N. (1975). Effect of chloroquine on cultured fibroblasts: Release of lysosomal hydrolases and inhibition of their uptake. *Biochem. Biophys. Res. Commun.* **66,** 1338–1343.

Willingham, M. C., Pastan, I. H., Sahagian, G. G., Jourdian, G. W., and Neufeld, E. F. (1981). Morphologic study of the internalization of a lysosomal enzyme by the mannose 6-phosphate receptor in cultured Chinese hamster ovary cells. *Proc. Natl. Acad. Sci. U.S.A.* **78,** 6967–6971.

Willingham, M. C., Pastan, I. H., and Sahagian, G. G. (1983). Ultrastructural immunocytochemical localization of the phosphomannosyl receptor in Chinese hamster ovary (CHO) cells. *J. Histochem. Cytochem.* **31,** 1–11.

Yamashita, K., Tachibana, Y., and Kobata, A. (1978). The structures of the galactose-containing sugar chains of ovalbumin. *J. Biol. Chem.* **253,** 3862–3869.

12

Biogenesis of Secretory Vesicles

HSIAO-PING HSU MOORE, LELIO ORCI, AND GEORGE F. OSTER

I. INTRODUCTION

Most cells secrete continuously. The process of excreting substances into the extracellular milieu is accomplished by packaging materials into membranous vesicles and transporting them to the cell surface. There they fuse with the plasma membrane and dump their contents to the outside world. Historically, only specialized secretory cells, such as endocrine, exocrine, or neuronal cells, were classified as "secretory," for only in these types of cells could secretory vesicles be unequivocally identified. We realize now that all cells are equipped to secrete, but they have different types of secretory vesicles. The "professional" secretory cells store their products in morphologically distinguishable vesicles, which accumulate in the cytoplasm. Other cells do not present obviously distinguishable secretory vesicles, despite the fact that biochemical assays reveal that they are secreting quite prodigiously. For example, Fig. 1 shows two kinds of secretory cells. The insulin cell contains numerous

521

Insulin cell

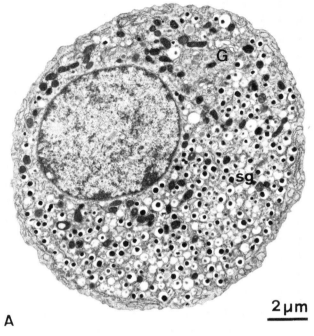

G

sg

2 μm

A

Plasma cell

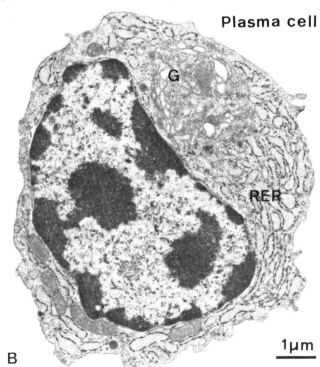

G

RER

1 μm

B

dense-core vesicles, while the plasma cell—which secretes quite vigorously—shows few discernible cytoplasmic vesicles. This is because the lifetime of a vesicle may be only a few minutes, so that not many vesicles accumulate within the cell. Thus we can distinguish between two types of secretion mechanism: (i) vesicles secreted as soon as they are created, the "constitutive vesicles," and (ii) "storage vesicles" which accumulate in the cytoplasm before release.

As we shall discuss, the two types of vesicles differ not only in their secretion properties but also in their mechanism of biogenesis. Thus it is important to bear in mind that the classic definition of "secretory granules" refers to only the storage-type secretory vesicles present in professional secretory cells. For discussion of this particular type of secretory vesicles, readers may consult several reviews (Palade, 1975; Winkler, 1977; Holtzman, 1977; Hand and Oliver, 1981; Orci, 1982; Steiner, 1984; Farquhar and Palade, 1981; Whittaker, 1986). In this chapter we shall devote our attention to the distinctions between the different pathways of protein secretion, and we shall discuss the biogenesis of both constitutive and storage vesicles.

An important advance in our understanding of the assembly of secretory granules came from the observation that a single cell can sometimes manufacture more than one type of vesicle. For example, recent experiments have shown that endocrine cells possess both constitutive and storage vesicles, and that each type of vesicle contains different sets of proteins (Gumbiner and Kelly, 1982; Moore *et al.,* 1983b). Moreover, epithelial cells manufacture at least two different types of constitutive vesicles: some are routed to the apical surface, and others find their way to the basolateral surface. These vesicles provide the vehicles for transporting different proteins to distinct domains on the cell surface (for a review, see Simons and Fuller, 1985).

The realization that vesicles in the same cell may have different fates and contain different proteins raises an important question: How do proteins find their way to the appropriate vesicle? This question is stimulating much research on how secretory vesicles are formed and how the cell

Fig. 1. Differences in pathways of protein secretion among higher eukaryotes: examples of storage secretory cells (A) and constitutive secretory cells (B). Both cells secrete large amounts of protein products, but only the storage cells contain morphologically identifiable secretory granules. (A) Classic endocrine secretory cell from the pancreas. The cell stores insulin and releases it when the plasma glucose level increases. The morphological hallmark of this type of secretory cell is the numerous dense-core secretory granules (sg) in the cytoplasm. (B) Plasma cell which constitutively synthesizes and exports immunoglobulins. In these cells, there is no accumulation of secretory vesicles, although the compartments of protein synthesis and membrane insertion (the rough endoplasmic reticulum, RER) and of protein processing and sorting (the Golgi complex, G) are conspicuous.

sorts and routes its proteins to the right vesicle. In this chapter we shall emphasize some recent progress in understanding the protein-sorting problem. However, we shall not attempt to cover sorting into all types of secretory vesicles; rather we shall focus mostly on a particular model system—the adrenocorticotropic hormone (ACTH)-secreting pituitary cell line (AtT-20)—under the assumption that the principles governing these cells can be applied to sorting and secretory activities in other neural and endocrine cells. For sorting in polarized epithelial cells, readers are referred to several outstanding recent reviews (Simons and Fuller, 1985; Rodriguez-Boulan *et al.*, 1985; Matlin, 1986). In Section II we discuss the molecular structure and function of secretory organelles. Section III discusses how the components of the vesicles are sorted and assembled. Finally, we shall present our view on the most productive avenues of inquiry for future research.

II. STRUCTURE AND FUNCTION OF SECRETORY ORGANELLES

Proteins are passengers on a transport pathway destined for the cell surface. The Golgi plays the role of the train station, with stationmaster proteins that help direct the passenger proteins to the correct departure sites. The transport machines—the secretory vesicles—must also contain specific molecules that enable it to propel itself to, and fuse with, the plasma membrane. Thus the secretory vesicles must contain at least two varieties of molecules: the passenger proteins themselves as well as transport and recognition machinery. Since the mid-1970s vesicles from a number of systems have been isolated and their composition characterized. While the studies have greatly clarified the nature of the passenger proteins, our knowledge of the transport and recognition proteins remains quite sketchy. This is because good assay systems for these proteins have not yet been developed. Therefore, in this section we shall try to formulate a conceptual framework within which one can incorporate future knowledge about the various molecular components found in secretory vesicles. To appreciate the roles of the transport and recognition machinery we need to first discuss the life cycle of a typical secretory vesicle.

A. Types of Secretory Vesicles

1. Life Cycle of Secretory Vesicles

In most cases, the amount of membrane inserted into the cell surface is far greater than that required for cell growth. Therefore, much of the membrane material comprising the secretory vesicles must be recycled continuously. Figure 2 shows a schematic view of the life cycle of a secretory vesicle. For descriptive purposes, we can break down this cycle into the following steps: (1) assembly in the Golgi region; (2) transport to the cell periphery; (3) anchoring to the release site at the cell surface; (4) fusion with the plasma membrane and release of contents; and (5) retrieval of membrane components by endocytosis and recycling of membrane components back to newly formed secretory vesicles.

2. Classification of Secretory Vesicles

While all vesicles share this general life cycle, there are many types of vesicles, and their components differ according to their specific roles in the cell. One can classify secretory vesicles in several ways. The most obvious classification is by their physical appearance, which reflects differing contents. Many nerve terminals contain two types of secretory vesicles with distinct morphologies. Those vesicles containing small neurotransmitter molecules are typically translucent, while those containing peptide hormones have electron-dense cores (Fig. 3a). Another difference between vesicle types is their destination. Exocrine cells discharge the vesicle contents vectorially from their apical surfaces to the inner or outer surfaces of the organism, while endorine cells secrete into the blood. Many epithelial cells can direct vesicles to either their apical or

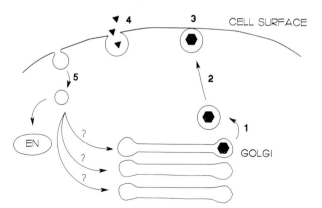

Fig. 2. Life cycle of a secretory vesicle. (1) Proteins destined for secretion accumulate in the trans Golgi where they are packaged into secretory vesicles. (2) The vesicles are transported to the cell periphery. Storage vesicles remain in the cortex (3) until an external signal triggers membrane fusion and exocytosis (4). Constitutive vesicles do not require any external signal to fuse with the plasma membrane. Membrane proteins destined for recycling are reinternalized by endocytosis (5). They are then routed back to the Golgi for incorporation into newly formed secretory vesicles. EN, Endosome.

Finally, it is worth noting that secretory vesicles containing large proteins typically acquire their cargo at the Golgi, while vesicles containing smaller molecules, such as nuerotransmitters, may ingest them directly from the cytoplasm. In some cases it appears that biosynthesis of neurotransmitters continues in the vesicles, since biosynthetic enzymes are found there, e.g., dopamine β-hydroxylase (dopamine β-monooxygenase) in chromaffin granules.

2. Proton Pumps

In addition to the passenger proteins and their processing enzymes, secretory vesicles possess the molecular equipment required to perform the other functions suggested by Fig. 2. Important components of storage vesicles are proton pumps, which generate an acidic environment inside the granules (Fig. 4). These serve at least two functions. In vesicles containing dopamine or acetylcholine, the proton pumps provide the proton-motive force for the uptake of neurotransmitters from the cytoplasm. Indeed, proton pumps and the transporter which performs this uptake reside in the vesicle membrane (Parsons *et al.*, 1986; Henry *et al.*, 1986; Johnson 1986; Kirshner *et al.*, 1986; Apps and Percy, 1986). In vesicles containing peptide hormones the proton pump appears to provide a pH environment conducive to enzymatic processing. Lowering the internal pH may also play a role in sorting of peptide hormones, as we shall discuss below.

3. Transport Machinery

Once the secretory vesicle has formed, it must be transported to the cell surface. This could be accomplished either by random diffusion or via active transport along cytoskeletal elements. In cells where vesicles must travel long distances there is good evidence that they are actively transported. In neurons, vesicles move along microtubules, driven by the mechanochemical molecule kinesin (Vale *et al.*, 1985). In pancreatic B cells disruption of microtubules also slows the intracellular transport of newly synthesized proinsulin and its conversion to insulin (Malaisse-Lagae *et al.*, 1979). However, microtubule-dependent transport has not been demonstrated in cells where vesicles do not travel far to reach the cell surface. In yeast cells, for example, disruption of the gene encoding for microtubules does not affect secretion rate nor the polarized membrane growth that accompanies budding. Instead, it appears that actin filaments play a role in transporting vesicles to the cell surface: disruption of actin structural genes slows the rate of secretion, leading to accumulation of vesicles in the cytoplasm (Novick and Botstein, 1985). In fibroblasts, following disruption of the microtubules, the overall rate of secretion is also unaf-

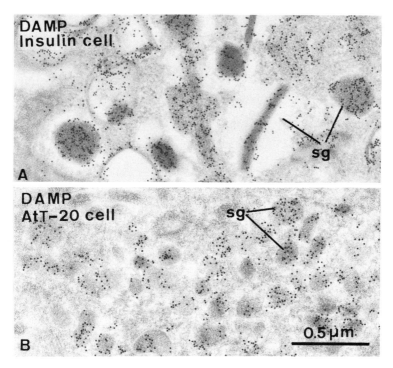

Fig. 4. Storage granules of endocrine cells are characteristically acidic. The acidic compartments inside a cell can be identified by uptake of a weak base, 3-(2,4-dinitroanilino)-3'-amino-*N*-methyldipropylamine (DAMP) (Anderson *et al.,* 1984). DAMP contains a dinitrophenol moeity which can be labeled with antibodies against this hapten and visualized by protein A–gold techniques. DAMP immunostaining of an insulin (A) and an AtT-20 cell (B) are shown. In both cell types, gold particles are distinctly concentrated on the core of secretory granules (sg). In the insulin cell the core can appear as an elongated crystalloid, or a round mass, while in the AtT-20 cell all cores are roundish.

fected (Rogalski *et al.,* 1984). This suggests that vesicles may not travel on microtubule tracks in these cells. Disrupting microtubules, however, does destroy the preferential insertion of secretory vesicles into the plasma membrane closest to the Golgi. This may be a simple consequence of dispersing the Golgi apparatus, which depends on the presence of microtubules to maintain its morphological integrity. In Madin–Darby canine kidney (MDCK) epithelial cells, disrupting microtubules does not affect the directed transport of specific viral proteins to the apical or basolateral surfaces (Salas *et al.,* 1986). In this case, either other filament types are involved, or vesicles simply diffuse until they are captured by a selective surface. In summary, there appear to be several mechanisms cells employ to transport their secretory vesicles from the Golgi to the cell surface. At least in those cases where microtubule transport is involved, however, there must be specific recognition molecules on the secretory vesicle that bind to the transport motor.

4. Targeting

To date, no specific molecules which target secretory vesicles to the plasma membrane have been identified. However, there is suggestive evidence in several cell types that such molecules do exist. In neurons, for example, there are specific "zones of release," near which vesicles attach to the plasma membrane. It is possible that this involves interactions involving components of the cortical cytoskeleton, such as spectrin. In this regard, it is important to note that synapsin-1, a major protein found in brain synaptic vesicles, is homologous to the band 4.1 protein in red blood cells, which links spectrin and actin (Huttner et al., 1983; Llinas et al., 1985; Goldenring et al., 1986; Baines and Bennett, 1985). In chromaffin cells, there is also evidence that the subplasmalemmal microfilament network is modified during exocytosis (Marie-France et al., 1986).

5. Fusion and Regulation of Release

In storage cells, exocytosis of the secretory vesicle is often triggered by an external signal, such as membrane depolarization, binding of a hormone, or a neurotransmitter. Is there a common feature shared by these stimuli? Most triggered exocytosis is accompanied by a rise in intracellular calcium. Raising of intracellular calcium concentrations can be brought about in several ways (Fig. 5). In neuromuscular junctions, the membrane depolarization accompanying action potentials triggers a rise in internal calcium by opening voltage-dependent calcium channels (Fig. 5a). In other situations, intracellular calcium is controlled by different second messengers. For example, in corticotrophs, binding of corticotropin-releasing factor triggers the influx of extracellular calcium by first increasing the level of intracellular cAMP (Luini et al., 1985; Reisine et al., 1985; Fig. 5b). In many cell types the rise of intracellular calcium results from the release of internal stores, rather than influx from the surroundings. In these cases, binding of ligands triggers the hydrolysis of phosphatidylinositol to inositol trisphosphate (IP_3) and diacylglycerol (DG). IP_3 is the proximal trigger releasing calcium from membranous stores (Streb et al., 1983; Fig. 5c). Recent evidence suggests that IP_4 may also play a role in regulating the influx of external calcium (Houslay, 1987).

How does a rise in intracellular calcium promote exocytosis? There are two physical barriers that prevent fusion of lipid bilayers: the electrostatic repulsion between the negatively charged surfaces and the hydration

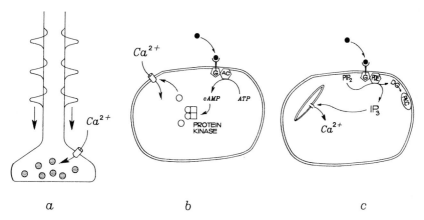

Fig. 5. Various pathways leading to a rise in intracellular calcium levels during stimulated exocytosis. (a) In nerve terminals, an incoming depolarization wave opens voltage-dependent calcium channels which allow influx of extracellular calcium ions. (b) In cortico-trophs, an external ligand binds to a receptor which activates adenylate cyclase (AC) via an intermediate G protein (G). AC catalyzes the production of cAMP from ATP, which then acts on a protein kinase which, in turn, acts on ligand-dependent calcium channels in the plasma membrane. (c) A third pathway for calcium release involves the inositol lipids (Berridge, 1986). A ligand activates a phospholipase (PDE) via a G protein that cleaves the inositol lipid, phosphatidylinositol 4,5-bisphosphate (PEP_2) into diacylglycerol (DG), which remains in the membrane, and inositol trisphosphate (IP_3), which is released into the cytoplasm. IP_3 induces release of sequestered calcium from intracellular membrane stores.

layer that prevents the layers from coming in contact (Lucy and Ahkong, 1986). *In vitro* experiments on fusion of lipid bilayers indicate that divalent cations are particularly effective in shielding the electrostatic forces, and an osmotic pressure difference across the membrane compartments is effective in removing the hydration barrier. However, as we shall discuss below, while calcium is implicated in regulating secretion in storage cells, the levels of calcium involved are not nearly as high as those required *in vitro* bilayer fusion. It is more likely, therefore, that calcium is a regulatory agent in controlling the activity of specific fusagens which overcome the electrostatic and hydration barriers via their hydrophobic structure. The best example of a protein-mediated membrane fusion is that produced by viral surface glycoproteins, such as the hemagglutinin of influenza virus (White *et al.*, 1983). The activity of these fusagens is triggered by a drop in pH, which exposes a hydrophobic fusion peptide. It is possible that the fusion of vesicles with the plasma membrane in storage cells is mediated by a protein which responds to calcium in a similar fashion.

The fact that calcium appears to trigger exocytosis has stimulated the search for regulatory intermediates. Several candidates are currently under investigation. Synexin is a protein that promotes aggregation of chromaffin vesicles in the presence of calcium (Scott *et al.*, 1985). Calmodulin is a common intermediate in calcium-regulated responses; therefore, it is natural to seek a role for it in regulating exocytosis. Indeed, both calmodulin and calmodulin binding sites have been detected on synaptic vesicles isolated from electric fish (Hooper and Kelly, 1984a,b). Also, injection of calmodulin antibodies into chromaffin cells diminishes amine release (Trifaro, 1986).

Depite its apparent ubiquity, exocytosis is not always accompanied by a rise in intracellular calcium. Rink *et al.* (1983) found that diacylglycerol (DG) and phorbol esters—both activators of protein kinase C (PKC)—when applied to platelet cells can increase the release rate of serotonin without an associated rise in intracellular calcium. Several examples of such heightened secretion in storage cells, unaccompanied by elevated cytosolic calcium, have since been observed. It is interesting to note that constitutive secretion also does not appear to require elevated calcium levels (Tartakoff and Vassalli, 1978). It will be important to determine whether these types of calcium-independent exocytosis employ a different mechanism from calcium-dependent release. It is possible that they use the same mechanism, albeit with modified calcium sensitivity. In this regard, it is interesting to note that PKC can reduce the activation threshold for calcium dependence on exocytosis (Knight and Scrutten, 1984).

It is apparent from this abbreviated discussion that, aside from the passenger proteins and their processing enzymes, much work lies ahead before we fully understand the molecular components of secretory vesicles.

III. FORMATION OF SECRETORY VESICLES

The composition of the secretory vesicles is, as we have seen, quite complex. In addition to the peptide hormones, these vesicles must contain molecules for hormone processing, targeting, transport of the vesicle, fusion with the plasma membrane, and recycling of the membrane components. How are these components brought together during their biosynthesis? Unfortunately, most of the components of the secretory vesicles have not been explicitly isolated, and their mechanisms of action are only vaguely understood. Thus our understanding of the assembly of secretory vesicles has derived mostly from the study of the only component that has been well characterized: the peptide hormone itself. For the purposes of discussion, we shall divide the formation of the secretory vesicles into three steps: (a) sorting, assembly, and packaging of its constituents in the Golgi region; (b) budding of the vesicles; and (c) transport of the vesicle to the cell surface and the subsequent recycling of the membrane components. This breakdown may not correspond to the actual sequence of events. For example, sorting and budding may not be strictly sequential events (see below, and Fig. 8).

A. Molecular Sorting of Proteins: General Principles

1. Why Is Sorting Necessary?

Most proteins are manufactured by cytoplasmic ribosomes. Although many proteins remain in the cytoplasm, many others are inserted into membrane organelles, such as the rough endoplasmic reticulum, (RER), nucleus, and mitochondria. Therefore, there must be subsequent sorting events to direct the different proteins to their appropriate target organelles.

2. Sorting of Proteins Is a Hierarchical Procedure

Sorting occurs in multiple steps, much like a zip code system, as shown schematically in Fig. 6 and 7. From the ribosomes the proteins are routed to four major destinations: the cytoplasm, nucleus, mitochondria, and endoplasmic reticulum (ER) (Fig. 6). Proteins that possess a specific "signal peptide" are inserted into the ER membrane and will thence be routed to the secretory pathway. Similarly, proteins equipped with mitochondrial- or nuclear-specific tags are routed to the various mitochondrial or nuclear compartments (for reviews, see Schatz and Butow, 1983; Hurt and Loon, 1986; Dingwall and Laskey, 1986). Proteins with no organelle-specific tag remain, by default, in the cytoplasm. (Here we are using the term "tag" to refer to either a particular primary sequence or to a post-translational modification.)

Within each of these pathways further sorting takes place (Fig. 7). For example, the ER is the entry port for at least five classes of proteins: those retained by the ER itself, and those destined for the Golgi, lysosomes, secretory vesicles, and plasma membrane. Similarly, mitochondria contain four subcompartments: the outer membrane, the intermembrane space, the inner membrane, and the matrix. Proteins must be further sorted into these subcompartments. These subsequent sorting processes work on the same principle as the primary sorting event at the ribosomes: there is a "default" destination that will be followed unless secondary molecular tags intervene. In the secretory pathway, there are two types of secondary molecular tags that affect the destination of a particular protein

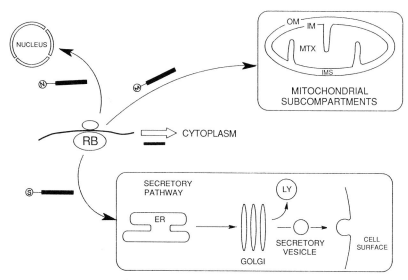

Fig. 6. Sorting of secretory proteins is a hierarchical process. Proteins assembled on the ribosomes (RB) are targeted to four major cellular compartments, the nucleus, mitochondria, the endoplasmic reticulum (ER), or the cytosol, according to their targeting sequences. We have indicated the targeting sequence by small circle with enclosed letters: N, nucleus; M, mitochondria; S, secretory pathway. Proteins with no signal sequence remain in the cytosol. It should be noted that the targeting signal for the secretory pathway and the mitochondria usually resides at the amino terminus of the polypeptide, whereas the targeting for the nucleus is usually not restricted to the amino terminus. After routing to their major compartments, many proteins are further routed to subcompartments. For example, mitochondrial proteins may be inserted into the outer membrane (OM), the intermembrane space (IMS), the inner membrane (IM), or the luminal matrix (MTX). Proteins in the secretory pathway may be retained in the ER, be shunted to the Golgi or lysosomes (LY), or pass to the cell surface.

(cf. Fig. 7A). A "retention tag" will cause a protein to be retained by an organelle (e.g., the ER or Golgi). A "branching tag" will cause the protein to be diverted from the secretory pathway to another compartment, such as the lysosomes. Those proteins that are neither retained nor diverted (e.g., plasma membrane and secretory proteins) will default to the cell surface via the "constitutive" secretory pathway, discussed below. A similar scheme operates to sort proteins to the various mitochondrial compartments (Fig. 7B). Proteins without retention tags for the outer membrane or the intermembrane space end up in the matrix (Hurt and Loon, 1986).

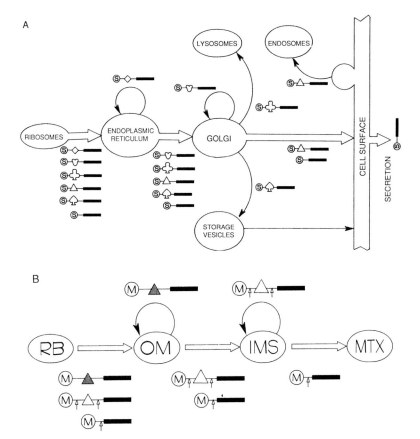

Fig. 7. The second stage of sorting is accomplished by retention and/or branching tags. (A) Sorting in the secretory pathway. All proteins with the signal sequence (S) are inserted into the lumen of the rough endoplasmic reticulum. A subset of these contains retention tags which are captured by the ER ($\triangle$), or the Golgi ($\heartsuit$). Another subset contains branching signals. Those containing the lysosomal tag ($\clubsuit$) are diverted to lysosomes, and, in storage cells, those containing localization signals for storage vesicles ($\spadesuit$) are diverted into secretory granules. In all cells, those proteins not retained or diverted into the above compartments are, by default, transported to the cell surface along the constitutive pathway. At the cell surface, soluble proteins are discharged into the extracellular environment. Membrane proteins containing an addition signal ($\triangle$) are collected into coated pits for recycling, whereas those without such signals remain on the cell surface. We should emphasize that, for illustrative purposes, we have indicated a signal sequence on all proteins along the entire secretory pathway; however, most of them are removed at the rough endoplasmic reticulum. Note that most of the retention and branching tags have not been explicitly identified; therefore, their locations on the polypeptide have been drawn arbitrarily. (B) Sorting in the mitochondrial compartment. All proteins with the mitochondrial targeting sequence are captured by binding sites on the mitochondrial membrane. Some are retained on the outer membrane (○), while others containing additional cleavage signals (denoted by small arrows) are retained in the intermembrane space. Those not retained by the intervening compartments reach the luminal matrix. In contrast to the case of sorting within the secretory pathway, all of the mitochondrial targeting signals are found at the amino terminus.

3. Sorting Is Mediated by Sorting Domains and Receptors

How is the zip code system decoded by the cell's machinery? The retention and diversion tags we have referred to are "sorting domains" (Blobel, 1980); the cell appears to use specific receptors to recognize and bind these domains. The two best characterized receptors are (a) the signal recognition particle and its receptor that route proteins to the secretory pathway (Walter and Blobel, 1981; Walter *et al.*, 1981; Gilmore *et al.*, 1982a,b; Meyer and Dobberstein, 1980) and (b) the mannose 6-phosphate receptor that targets lysosomal hydrolases to the lysosomes (for reviews, see Sly and Fischer, 1982; Kornfeld and Kornfeld, 1984; Farquhar, 1985).

B. Sorting of Secretory Proteins in Storage Cells

1. Are Secretory Proteins Sorted?

As we have discussed above, not all proteins require positive sorting. In the secretory pathway, it seems that the cell surface is the "default" destination for proteins not retained by or diverted to other organelles. That is, many proteins end up on the plasma membrane without any apparent signal for the cell surface—other than the signal peptide that routed them to the secretory pathway back at the ribosomes. For example, globin, a cytoplasmic protein, can be fused to the signal peptide of β-lactamase and threaded into the lumen of the ER (Lingappa *et al.*, 1984). On cleavage of the signal peptide, it is efficiently secreted along the constitutive pathway (D. Quinn and H.-P. H. Moore, unpublished). Moreover, Weiland *et al.* (1987) recently demonstrated that a derivative of the tripeptide, Asn- Tyr- Thr, can be introduced into a number of constitutive cells, where it is glycosylated and transported to the cell surface with kinetics faster than any known secretory protein. Finally, yeast mutants defective in vacuole biogenesis shunt their vacuolar proteins into the secretory pathway (Rothman and Stevens, 1986).

These experiments demonstrate that a protein equipped only with the signal sequence will eventually reach the cell surface by the "default" constitutive pathway. Indeed, in many cell types secretory and plasma membrane proteins do not appear to be sorted and thus reach the surface by the same pathway. For example, in hepatocyctes the plasma membrane glycoprotein G and the secretory proteins albumin and transferrin are found in the same secretory vesicles. Moreover, yeast mutants defective in secretion are also defective in export of plasma membrane proteins, supporting the view that they are transported via the same pathway (for review, see Schekman, 1985). Thus it appears that constitutive cells do not sort their secretory proteins. However, recent experiments indicate that storage cells do sort their secretory and plasma membrane proteins. This is necessary because these cells contain more than one secretory pathway.

2. Storage Pathway Coexisting with the Constitutive Pathway

As discussed earlier, professional secretory cells have a specialized secretory pathway. Does this imply that the constitutive pathway is discarded? Tartakoff and Vassalli (1978) noticed that secretory cells could be classified into two categories: those which could respond quickly to secretagogues and those which did not respond. He called the fast responding cells "regulated secretory cells," since their secretion could be regulated by an external signal. This nomenclature was widely adopted, and appeared to define two distinct cell types. Moreover, it became generally accepted that each of these cell types possessed only one or the other secretory pathway. It has recently become clear, however, that the two pathways could coexist in the same cell. This makes sense, for the evolution of a specialized function does not preclude the necessity of serving the cell's previous secretory tasks, such as supplying the cell surface with new membrane proteins.

3. Experimental Evidence Supporting the Coexistence of the Two Pathways

Gumbiner and Kelly (1982) examined the secretory pathways in the pituitary cell line AtT -20. They used antibodies against an endogenous viral membrane protein, gp70, to follow the secretory pathway of a plasma membrane protein and compared it with the pathway taken by the hormone ACTH. They concluded that there are at least two distinct routes to the cell surface in this cell. While ACTH is packaged into storage secretory vesicles and its secretion is accelerated by secretagogues, the viral protein is externalized by a different pathway. This protein is transported to the cell surface as soon as it is synthesized, i.e., without first being stored in secretory vesicles. Furthermore, its externalization is constitutive and not influenced by secretagogues. Does the constitutive pathway also provide a general route for secreting nonviral proteins made by AtT -20 cells? Moore et al. (1983b) examined the pathways taken by all endogenous secretory products after labeling the cells with [^{35}S] methionine. A number of radiolabeled molecules were found in the tissue culture medium, and they all fell into one of two classes: those externalized by the constitutive pathway and those by the regulated pathway. Thus, in addition to gp70 and ACTH, other products in the secretory apparatus are also sorted into these two pathways. The presence of these same two secretory pathways has also been found in other endocrine and neuronal cell lines. These include the growth hormone-producing pituitary tumor GH3 cells (H.-P. H. Moore, unpublished), the pheochromacytoma-derived PC-12 cells (Schweitzer and Kelly, 1985), and the pancreatic insulin-producing HIT cell line (Moore *et al.,* 1983a).

4. Are Proteins Selectively Sorted into the Two Pathways?

The coexistence of these two secretory routes is important, for it raises the question of whether proteins are selectively sorted into the two pathways. In principle, one could imagine accomplishing fast response by simply shunting, without discrimination, some of the constitutively secreted material into the storage pathway. Experimental evidence indicates, however, that it does not work this way: proteins are not randomly distributed between the two pathways. Experimentally, it is not easy to demonstrate this, for there are few endogenously manufactured proteins that can be quantitated accurately. However, it is possible to transfect foreign DNA coding for known markers into AtT-20 cells, and their fate can be followed using specific antibodies. When these cells are transfected with human growth hormone (HGH)—a protein that is normally stored in secretory vesicles of somatotrophs—the foreign hormone is packaged efficiently into the ACTH vesicles. In contrast, a soluble form of a viral membrane protein (the VSV G protein with its membrane anchor truncated, VSV TG) is exported via the constitutive pathway by the same cell (Moore and Kelly, 1985).

5. What Kinds of Proteins Are Sorted?

The unequal distribution of proteins between the two pathways implies that at least one class of proteins is actively sorted. Which one? To address this question Moore and Kelly (1986) fused a constitutively secreted protein (TG) with a storage protein (HGH) and observed how the hybrid molecule was sorted by the cell. If the constitutive protein contained a sorting domain, then some fraction of the fusion product should be diverted to the constitutive pathway. Conversely, if the storage fragment contained the sorting domain, the hybrid protein should be directed to storage vesicles. It turned out that the viral protein TG was diverted from its normal route to the storage vesicles. These results strongly support the view that proteins are actively sorted into storage vesicles; that is, it appears that the storage pathway is the recipient of sorted proteins. Figure 8 summarizes the current view: the constitutive pathway is a "bulk flow" route to the cell surface, whereas the storage pathway is a shunt

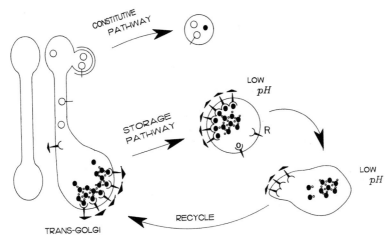

Fig. 8. Hypothetical scheme for sorting secretory proteins in storage cells. Storage proteins (●) are actively captured by receptors (R) which segregate into clathrin-coated buds (▲). Electron microscopy shows dense material collecting into the forming vesicle, suggesting that there may be linking proteins (○) that help aggregate the secretory protein. In the newly formed clathrin-coated vesicles the pH drops, releasing the ligand from its receptor. The receptors are presumably recycled back to the Golgi. Proteins not actively sorted into this pathway flow out along the bulk flow constitutive pathway. Constitutive buds are coated with some nonclathrin substance (—). Sorting of molecules into these two pathways may not be perfect: some storage proteins find their way into constitutive vesicles.

which receives selectively sorted proteins destined for storage vesicles to be used for controlled, rapid exocytosis.

6. Where Does Sorting Take Place?

Conceivably, segregation of proteins into the two pathways could take place in the RER, where different proteins are inserted into separate domains of the ER cisternae. Alternatively, they could be sorted at a later stage of transport, i.e., within a given Golgi cisterna or in some post-Golgi station. The intracellular sorting compartment has now been identified via immunoelectron microscopy (Orci *et al.*, 1987). When transfected AtT -20 cells producing rat proinsulin were infected with the influenza virus, the constitutive marker hemagglutinin was found to be mixed together with the regulated protein proinsulin in all Golgi cisternae. Segregation appears to take place at the time the two proteins exit the Golgi: hemagglutinin was packaged into clear vesicles (i.e., with no electron-dense contents), whereas proinsulin became concentrated in the dense core of budding granules (Fig. 9 and 10). Thus, the two pathways appear to diverge at the trans-most Golgi cisternae, which is acid phosphatase positive.

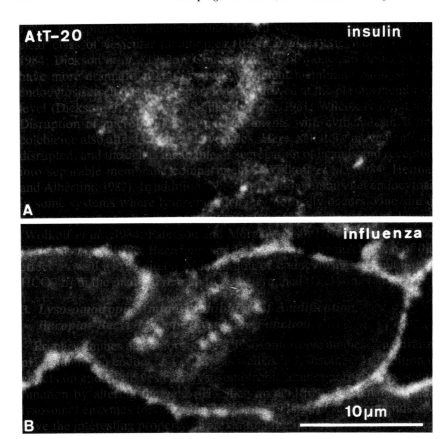

Fig. 9. Influenza virus hemagglutinin and proinsulin are segregated into different secretory vesicles in AtT-20 cells. Consecutive thick serial sections of an insulin-transfected, influenza virus-infected AtT-20 cell following immunofluorescent staining of insulin (A) and influenza virus hemagglutinin (B) are shown. Insulin immunofluorescence appears mostly as a spotty perinuclear (Golgi region) staining; influenza virus hemagglutinin antibody also elicits a perinuclear reaction but in addition an intense cell surface staining.

Fig. 10. Immunocytochemical localization of influenza virus hemagglutinin (A and B) and insulin (C) in insulin-transfected, influenza virus-infected AtT-20 cells. (A) In the Golgi area (G) situated in the cytocenter (attested by the presence of a centriole, C), gold particles revealing influenza virus hemagglutinin are found in the Golgi cisternae and in clear vesicles (v), but they are absent from core-containing secretory granules (sg). (B) At the cell periphery, the gold particles decorate the cell surface, which is thrown into numerous microvilli (several particles are indicated by circles), but are absent from the numerous secretory granules beneath. (C) In the thin serial section consecutive to that shown in (B) and stained with antiinsulin antibodies, gold particles appear virtually restricted to the dense-core secretory granules.

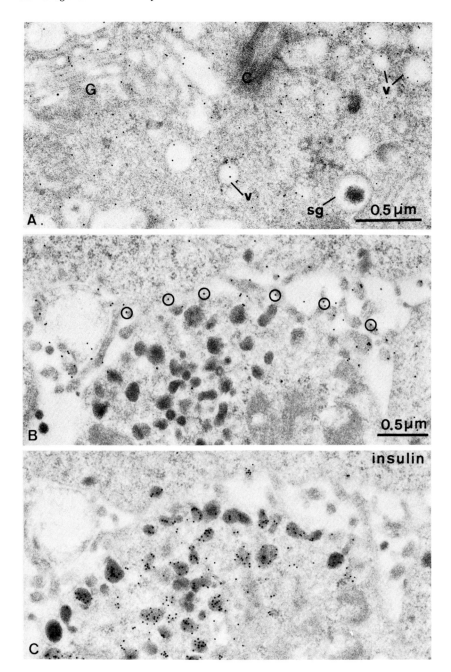

7. What Is the Molecular Mechanism of Sorting?

The mechanisms the cell uses to divert proteins into the storage pathway remain obscure. However, some clues may be gleaned from examining related sorting systems, such as that which localizes the lysosomal hydrolases. In fibroblasts, sorting of lysosomal enzymes to the lysosomes is accomplished by the mannose 6-phosphate receptor, which binds to a common recognition marker on lysosomal hydrolases and diverts them to lysosomes. The hybrid protein experiment discussed above showed that storage proteins are positively sorted into secretory vesicles. This suggests that storage proteins may have a specialized sorting domain which is recognized by a specialized receptor. If this hypothesis is correct, what is the nature of the sorting domain and its receptor?

8. Molecular Structure of the Sorting Domain

Most peptide hormones are synthesized as a larger precursor consisting of a pre-, or signal, sequence followed by the prohormone sequence. During transit through the cell, the preprohormones undergo two sets of proteolytic processing events to yield the final mature hormones. The first cleavage event takes place in the RER where the presequence is removed by the signal peptidase as the nascent polypeptides are translocated through the membrane of the RER. A later processing step then cleaves the prohormones to separate the prosequences from the mature products. It is appealing to hypothesize that targeting signals for granular localization may reside on one of these cleavable sequences.

A clear advantage of using such sequences for address information is that, once a protein is correctly localized in the target organelle, removal of these signals could render the process irreversible. Recent results obtained from *in vitro* mutagenesis of storage proteins, however, do not support such a sorting function for the pre- or propeptide of hormones. Two lines of evidence suggest that the prepeptide is not involved in sorting into storage granules. First, identification of the trans-Golgi compartment as the sorting site rules out the possible involvement of this sequence, since it is removed much earlier in the RER. Second, as mentioned earlier, the entire human growth hormone with its preprosequence is able to divert the VSV G fragment into storage granules. The presequence alone, however, is insufficient in rerouting the constitutive protein (L. Lehmicke and H.-P. H. Moore, unpublished results). Thus, the targeting signal does not reside on this hormone's presequence. Moreover, the propeptide, or C peptide, of insulin can be either completely deleted or replaced with a much shorter sequence from the insulin-like growth factor without affecting its granule localization (S. Powell and H.-

P. H. Moore, unpublished results). Taken together, these analyses suggest that the information responsible for granular localization is contained within the mature peptide hormone sequences, rather than in the pre- or prosequence.

9. Cellular Sorting Receptors

How does the cell decode the sorting information on mature hormone sequences? A likely mechanism is that the sequence is recognized by cellular carriers, or receptors, which concentrate and localize the hormones within selective regions of the Golgi. These receptors would then serve as a molecular filter: they would bind only to proteins destined for packaging into storage granules and leave those destined for the constitutive pathway untouched. There is preliminary evidence for such receptor proteins. Using affinity chromatography, K.-N. Chung, P. Walter, and H.-P. H. Moore (unpublished results) have isolated two proteins with molecular weights of 15K and 24K–29K from detergent-solubilized Golgi fractions of dog pancreas. These proteins selectively bind to storage proteins such as prolactin but do not show significant affinity for constitutive proteins. Interestingly, the pH dependence of ligand binding is similar to that of the mannose 6-phosphate receptor: binding occurs at neutral pH but is abolished at acidic pH. Since the internal pH of the storage granules and their precursor compartments are acidic (Orci et al., 1986a) (cf. Fig. 4), the dissociation of ligands from receptors at reduced pH may facilitate the unloading of transported proteins at the target compartments, much like the unloading of hydrolases in prelysosomal compartments (cf. Moore et al., 1983c).

10. Does Clathrin Play a Role in Intracellular Protein Sorting?

If sorting of storage proteins is indeed mediated by receptor proteins situated in the Golgi compartments, how then are the receptor proteins segregated to the storage granules? An intriguing observation came from studies of the intracellular distribution of clathrin. Orci et al. (1985b) and Tooze and Tooze (1986) found that, in peptide hormone-secreting cells, a significant fraction of the clathrin occurs in the Golgi areas. Electron microscopic examination of AtT-20 cells revealed that the forming secretory granules are conspicuously coated with clathrin (Fig. 11). These clathrin molecules—and perhaps other associated proteins—may serve to aggregate receptors carrying ligands for granular localization, much like the recruitment of receptors by coated pits during endocytosis. Since transport via constitutive vesicles does not appear to require positive

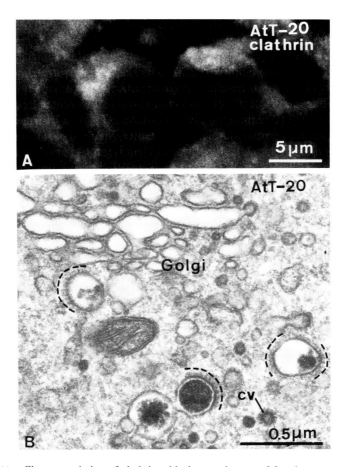

Fig. 11. Close association of clathrin with the membranes of forming secretory granules. (A) Immunofluorescent detection of clathrin in AtT-20 cells reveals a positive perinuclear (Golgi) region. (B) Thin section of the Golgi of an AtT-20 cell shows several core-containing secretory granules with a distinct clathrin coat on segments of their limiting membrane (dashed line). A typical clathrin-coated vesicle (cv) is also indicated.

sorting signals (see above), and hence need not be mediated by receptors, one would expect that the formation of constitutive vesicles would not involve clathrin, if the latter functions exclusively as a scavenger for receptors. In this regard, it is interesting to note that clathrin is not associated with budding microvesicles, which may mediate intro-Golgi bulk transport (Orci *et al.*, 1986), nor with budding trans-Golgi vesicles containing constitutive proteins (Griffiths *et al.*, 1985). Moreover, disruption of the clathrin gene in yeast does not have a significant effect on protein transport and secretion (Payne and Schekman, 1985).

11. Is There a Ubiquitous Sorting Mechanism in Storage Secretory Cells?

There are many types of storage cells. Endocrine cells typically secrete predominantly one hormone type: pancreatic β cells secrete insulin, somatotrophs in pituitary tissue secrete growth hormone, and corticotrophs secrete ACTH. Is it possible that each cell type has evolved a unique sorting mechanism, that is, a unique receptor to sort its hormone? In principle, there are two possibilities: (a) there is a common "zip code" that is recognized by a universal receptor and (b) each protein has a unique receptor. Transfection experiments have demonstrated that a number of endocrine and exocrine polypeptides can be packaged in the storage vesicles of heterologous cells, which seems to support the universal receptor model (Moore et al., 1983a; Burgess et al., 1985; Schweitzer and Kelly, 1985; Fig. 12). Examination of the molecular structure of these proteins, however, fails to reveal any common sequence. This may indicate that the recognition site resides in the 3-dimensional structure. This suspicion has several precedents, for example, in the lysosomal enzymes, where there appears to be a universal recognition system with no obvious consensus sequence. More work is required to identify the sorting domain and receptors before we can resolve this issue.

C. Budding from the Golgi: Mechanical Considerations

1. How Do Vesicles Bud from the Golgi?

Aside from the machinery to concentrate proteins in a particular region of the Golgi, there is another problem the cell must solve, namely, how to physically separate the condensed proteins from the Golgi itself into a secretory vesicle. While, much is known about the process of vesicle fusion (Wilschut and Hoekstra, 1984), vesicle budding—both from the Golgi and from the plasma membrane—is still quite mysterious. Although probably requiring ATP, receptor-mediated endocytosis continues in the presence of cytoskeletal inhibitors; moreover, despite its suspected role, endocytosis appears to proceed independently of clathrin in yeast cells (Payne and Schekman, 1985).

The fundamental mechanical requirement for budding a vesicle or a tube from the Golgi is the following: somehow a bending moment must be generated across the lipid bilayer. This is a mechanical problem that can be solved in a number of ways; in Cheng et al. (1987) we discuss the various physical mechanisms for generating this force. Here we shall summarize some of our investigations into the mechanics of budding that may shed light on this crucial process (cf. also Israelachvili et al., 1980; Tartakoff, 1987; Cevc and Marsh, 1987).

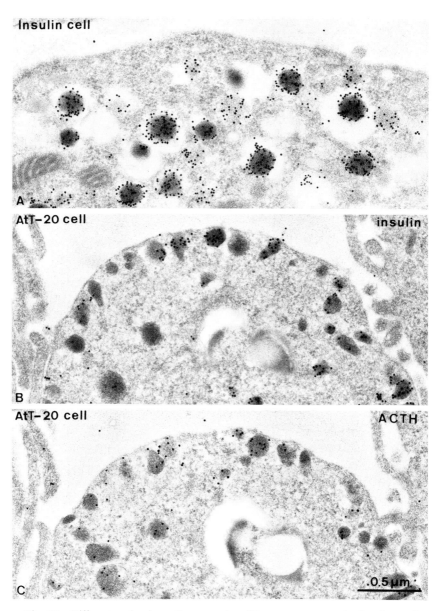

Fig. 12. Different endocrine cells appear to utilize a common mechanism for sorting regulated secretory proteins into their storage granules. This is shown by transfecting a proinsulin cDNA into a pituitary cell line, AtT-20. The latter normally synthesizes the pituitary hormone, ACTH, and packages it in dense-core secretory granules. The pituitary cell is perfectly capable of targeting the foreign pancreatic hormone into its dense-core secretory granules. (A) Peripheral part of the cytoplasm of an insulin cell following protein A–gold immunostaining with anti-insulin antibodies, showing the hormone's normal intracellular location in its native environment. The gold particles stain heavily the dense core of

2. *The Mechanical Equilibrium of the Lipid Bilayer as a Balance between Interfacial Tension and Surface Pressure*

Associated with each leaflet of a lipid bilayer is a "surface stress," which can be characterized in two equivalent ways: (1) the force per unit length (say, dyne/cm) tending to pull the surface apart* or (2) the energy per unit area (erg/cm^2 = dyne/cm) stored in the surface. The surface stress in a lipid bilayer is the result of an imbalance between opposing forces, the surface pressure and the interfacial tension, both of which act within each leaf of the bilayer. The net surface stress in each lipid layer (cf. Fig. 13a) can be written as

$$\text{Surface stress = surface pressure - interfacial tension} \qquad (1)$$

The surface pressure arises from several contributions, including steric repulsions between the lipid head groups and hydrocarbon tails, as well as from electrostatic and hydration forces between the head groups (Israelachvili *et al.*, 1980). These forces are counted as positive because they tend to expand the membrane surface area.

An alternative view of the surface pressure is to treat it as a "surface osmotic pressure" (Defay and Prigogine 1966; Davies and Rideal, 1961; Evans and Skalak, 1980). Since water can interpenetrate into the surface phase among the hydrophilic head groups, this region can be viewed as a nearly 2-dimensional solution. That is, the activity of the water is lower in the interfacial region; water molecules tend to diffuse into this region, thus increasing the local osmotic pressure.†

The interfacial (surface) tension arises from the hydrophobic interactions between the lipid tails and the aqueous solvent (Tanford, 1980). It is counted as is negative because it tends to shrink the surface area of each leaflet. In a planar membrane patch at mechanical equilibrium, these surface forces just balance, and the net surface stress vanishes.

---

* Imagine cutting a small slit in the surface and measuring the force tending to make the slip "gape."

† It is customary to include in this surface osmotic pressure the hydration forces and the electrostatic repulsions between the polar head groups (since the counterions induced by the fixed charges also must be counted as osmotically active particles; Hill, 1960).

---

secretory granules. (B and C) Consecutive thin serial sections of the periphery of an insulin-transfected AtT-20 cell, ummunostained with anti-insulin and anti-ACTH antibodies, respectively. Both antibodies stain the core of secretory granules. Thus insulin in this foreign environment is still targeted to the same location as in the pancreas. Most probably, it is targeted by the cell's machinery for sorting ACTH.

These conclusions do not depend on the mechanism for generating the membrane bending moment, but are simply geometric properties of any deforming elastic surface.* In order to say more about how proteins might exit the trans Golgi, one must inspect specific molecular models for generating the membrane bending moment.

Finally, surface tension considerations suggest that tubular membrane structures are not stable without some form of support (Israelachvili *et al.*, 1975)†. This might explain why tubular membrane organelles, such as the tubular ER (and perhaps the trans-Golgi network and tubular lysosomal structures), appear to be colinear with cytoskeletal elements, and disappear when the cell is treated by agents which disrupt the cytoskeleton (Terasaki *et al.*, 1986). In this connection it is interesting to note that heliozoan axopodia, which are tubular organelles, vesiculate when microtubules are disrupted (Schliwa, 1976).

4. How Might the Membrane Bending Moment Be Generated?

A possible clue to the mechanism of membrane budding is the observation that a molecular coat appears to form on the cytoplasmic face of constitutive Golgi vesicles (Orci *et al.*, 1986b). This suggests the following scenario for vesicle and tube formation from the Golgi (Cheng *et al.*, 1987). If a molecular species dissolves in the interfacial environment of one leaflet, the attractive forces may be reduced, and/or the surface osmotic pressure may increase. This will cause the leaflet to expand, and the membrane will buckle, as shown in Fig. 13b. As the curvature of the vesicle increases, the circumferential tensions created by the membrane curvature will pinch the neck of the protrusion until the lipid bilayers contact. This "surfactant theory" is only one of a number of theoretical possibilities (cf. Cheng *et al.*, 1987; Israelachvili *et al.*, 1976, 1980).

Any model for vesicle formation must address the issue of how the vesicle actually pinches off once the membrane surfaces contact. This is a complicated process involving the details of the transient structure of lipids in the bilayer (Wilschut and Hoekstra, 1984; Nagle, 1980; Gruner *et al.*, 1985). As we mentioned in Section II,B, there is evidence that osmotic forces play an important role in membrane fusion (Lucy and Ahkong, 1986; Fisher and Parker, 1984; Horn, 1984). A macromolecular fusagen could surmount both the hydration layer barrier and electrostatic

---

* Recall that a membrane bilayer is a "surface liquid," i.e., with very low resistance to shear forces. The above conclusions are not altered by this property, however.

† The calculations discussed above are not contradictory to these conclusions, for the tubes generated by membrane bending moments alone are not indefinitely extruded.

repulsion. Therefore, isolation of a vesicle-specific fusagen would be an important step in elucidating the mechanism of vesicle budding from the Golgi.

5. Intracellular Membrane Morphology May Be Controlled by Local Ionic Conditions

The above discussion of membrane forces suggests some further speculation about the role of physical factors in the functioning of the ER–Golgi system. Membranous organelles assume a variety of geometric shapes, including spherical vesicles, flattened sheets, fenestrated cisternae, and tubular reticulums. A clue to the function of the Golgi–ER system may be found in the differences in their respective shapes. Since the geometry of a membrane system is controlled by the balance of surface forces in the bilayers, these diverse geometries must reflect differences in local surface stress. These differences, in turn, can arise from differences in membrane composition and from local changes in ionic conditions:

1. It is known that different organelles are characterized by distinct molecular constituents, e.g., reticuloplasmins in the ER (Koch, 1987). Different compositions almost certainly imply different surface stresses, especially owing to differing molecular interactions between polar head groups (Shinada and Friberg, 1986) and different packing constraints (Israelachvili *et al.*, 1980; Mitchell and Ninham, 1981). Thus one can expect that organelle shapes will correlate with their compositions. (Indeed, it may well be that the separation mechanism of differing membrane components *defines* the organelle geometry.)
2. Solutions of mixed surfactants, when subjected to varying ionic environments, exhibit a sequence of phases that is very suggestive of biological structures (Longley and McIntosh, 1983; Tabony *et al.*, 1987). Thus one candidate for controlling the surface stress of intracellular membranes is the local ionic and/or pH conditions.

As one adds salt to a micelle solution, the electrostatic repulsions between micelles is progressively screened, allowing them to contact and fuse. Increasing ionic strength also screens repulsions between polar head groups. According to Eqs. (1) and (2), this will cause a decrease in the local curvature, favoring the emergence of a lamellar phase. Sufficiently high ionic strength can even locally reverse the curvature from the micelle phase. Since the interfacial tension acts to minimize the total surface area, the global geometry of the membranous system is a complex compromise between the components of the surface stress. One solution to this problem is the formation of an interpenetrating minimal surface that has many

of the features of the fenestrated cisternae and tubular networks characteristic of the Golgi stacks and the ER (Tabony *et al.*, 1987). This similarity suggests that organelle geometry may be a reflection of local ionic and/
or pH conditions that regulate membrane surface stress. It is worth noting
that since a fluid membrane must flow down a gradient in surface stress,
such gradients could play a role in transporting membrane-bound constituents within interconnected compartments.

D. Recycling of Membrane Components

Following exocytosis of the transported proteins, much of the transport
equipment, including many membrane proteins, is probably recycled for
subsequent reuse. This notion is supported by the observation that the
turnover rate of the membrane proteins in secretory vesicles is much
slower than that of the soluble contents. This retrieval process must be
selective, since the composition of secretory vesicle membranes is quite
different from that of the plasma membrane. Farquhar (1978) explored the
route of recycled membrane by following the fate of cationized ferritin.
When added to the extracellular fluid, cationized ferritin associates with
the membrane, is internalized, and subsequently shows up in the Golgi
and in newly formed secretory vesicles, rather than in its usual site, the
lysosomes. In insulin cells, it has been shown that the externally applied
tracer is found in the medial Golgi cisternae (Orci *et al.*, 1986c; see Fig.
14). These experiments clearly demonstrate that the membrane of secretory granules is recycled, but they do not address the mechanism of
recycling.

Before we can understand the mechanism of recycling, we must first
understand the route followed by recycled membrane molecules. An important—and unresolved—question in this regard is whether membrane
components first disperse over the plasma membrane before retrieval, or
whether they are retrieved as a group immediately following exocytosis.
Two experiments suggest that dispersal may precede recycling. First,
Heuser and Reese (1981) examined a time sequence of sections of neuromuscular junctions that had been rapidly frozen and freeze–fractured just
following stimulation. They found that intramembranous particles appeared to disperse away from the site of exocytosis before retrieval.
Second, Pfeffer and Kelly (1985) isolated coated vesicles from bovine
brain, under the assumption that these were the vehicles of recycling.
They found that two vesicle membrane components—defined by monoclonal antibodies—tended to segregate into different vesicles. Their
results suggest that membrane components are recycled independently
after dispersing over the plasma membrane.

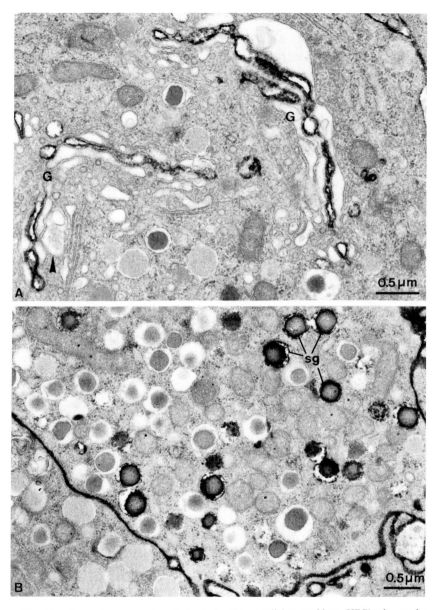

Fig. 14. Examples of an extracellular tracer (horseradish peroxidase, HRP) taken up by endocytosis in the insulin cell (A). The tracer can be found in the Golgi stack (G), where cisternae situated in an intermediate (medial) position are preferentially labeled (Orci *et al.,* 1986c), as well as in secretory granules (sg) (B). Arrowhead in (A) indicates the lack of HRP labeling of a trans-condensing cisterna. Pulse–chase labeling with [³H]leucine coupled to the cytochemical demonstration of HRP showed that HRP-labeled granules are not those which are newly synthesized during the pulse–chase experiment (Sawano *et al.,* 1986).

If membrane components are indeed dispersed before recycling, where do the dispersed components reassociate? This issue could be resolved by labeling two membrane markers and directly observing their distribution on the cell surface, and their subsequent route following endocytosis. Patzak and Winkler (1986) have made some progress in this direction by following a single vesicle antigen in chromaffin cells. If substances are indeed dispersed before retrieval, then the question arises of how the cell selectively recognizes these different components. This raises once again the specter of specific recognition molecules for each component. If true, this would open up an exciting avenue for further research.

IV. CONCLUSION AND PERSPECTIVES

It is clear that the molecular mechanisms underlying assembly of secretory vesicles are still largely *terra incognita*. The major point that emerges from studies of the secretory process is that there are many types of secretory vesicles, each with a characteristic ontogeny. Table I summarizes the major distinctions we have drawn between the constitutive and storage vesicles. It is likely that these, and perhaps additional, differences also exist between other vesicle types, such as those shuttling between apical and basolateral cell surfaces. We close with a list of issues that promise resolution in the foreseeable future.

1. What are the molecules that direct the sorting of storage proteins, and how are they routed to the correct locations?
2. How is the budding of vesicles from the Golgi accomplished? Are the molecular components involved in budding the same in constitutive and storage vesicles?
3. What is the function, if any, of clathrin in the assembly of storage vesicles?
4. How does intracompartmental pH influence the sorting and secretion process?
5. What is the molecular machinery for transporting and targeting vesicles between the different stations of the secretory pathways?
6. What are the mediators of membrane recognition and fusions?
7. After exocytosis, membrane proteins must be collected and rerouted to new secretory vesicles. Are the signals that guide recycling of these proteins the same as those previously used to target them to the secretory vesicles? That is, are addressing signals employed repetitively, or does each stage have its unique routing machinery?

TABLE I

Comparison between Characteristics of Constitutive and Storage Vesicles

Mechanism	Constitutive vesicles	Storage vesicles
Secretion	Independent of external signals	Exocytosis triggered by external signals
	Does not require extracellular calcium	Usually accompanied by rise in intracellular calcium
	Rapid transit from Golgi to cell surface ($t_{1/2} \approx 10$ min)	Long residence time within the cytoplasm ($t_{1/2} \approx 7$–10 hr)
Assembly	No sorting signal required for entry into vesicles	Specific sorting signals necessary for targeting to vesicles
	Transported proteins not concentrated before vesiculation	Proteins concentrated before packaging
	Nonclathrin coat associated with vesiculation	Clathrin associated with vesicle formation
	Assembly relatively insensitive to pH	Assembly impaired by elevation of luminal pH

ACKNOWLEDGMENTS

This work is supported by National Institutes of Health Grant GM 35239 (to H.-P. H. M.), National Science Foundation Presidential Young Investigator Award DCB-8451636 (to H.-P. H. M.), Swiss National Science Foundation Grant 3.404.86 (to L.O.), and NSF Grant MCS-8110557 (to G.F.O.). H.-P. H. M. is a recipient of a Juvenile Diabetes Career Development Award (284087) and an Alfred P. Sloan Research Fellowship (BR2497).

REFERENCES

Anderson, R. G. W., Falck, J. R., Goldstein, J. L., and Brown, M. J. (1984). Visualization of acidic organelles in intact cells by electron microscopy. *Proc. Natl. Acad. Sci. U.S.A.* **81,** 4838–4842.

Apps, D. K., and Percy, J. M. (1987). The H^+-translocating ATPase of chromaffin granules membranes. *Ann. N.Y. Acad. Sci.* **493,** 178–188.

Baines, A. J., and Bennett, V. (1985). Synapsin-I is a spectrin-binding protein immunologically related to erythrocyte protein 4.1. *Nature (London)* **315,** 410–413.

Berridge, M. (1986). Cell signalling through phospholipid metabolism. *J. Cell Sci.* **4,** 137–153.

Blobel, G. (1980). Intracellular protein topogenesis. *Proc. Natl. Acad. Sci. U.S.A.* **77,** 1496–1500.

Burgess, T. L., Craik, C. S., and Kelly, R. B. (1985). The exocrine protein trypsinogen is targeted into the secretory granules of an endocrine cell line: Studies by gene transfer. *J. Cell Biol.* **101,** 630–645.

Cevc, G., and Marsh, D. (1987). "Phospholipid Bilayers." Wiley, New York.

Cheng, L., Moore, H.-P. H., Oster, G., and Perelson, A. (1987). Vesicles, tubes and membrane organelles. In preparation.

Davies, J., and Rideal, E. (1961). "Interfacial Phenomena." Academic Press, New York.

Defay, R., and Prigogine, I. (1966). "Surface Tension and Absorption." Wiley, New York.

Dingwall, C., and Laskey, R. (1986). Protein import into the cell nucleus. *Annu. Rev. Cell Biol.* **2**, 367–390.

Evans, E., and Skalak, R. (1980). "Mechanics and Thermodynamics of Biomembranes." CRC Press, Boca Raton, Florida.

Farquhar, M. G. (1978). Recovery of surface membrane in anterior pituitary cells *J. Cell Biol.* **77**, R35–R42.

Farquhar, M. G. (1985). Progress in unraveling pathways of Golgi traffic *Annu. Rev. Cell Biol.* **1**, 447–488.

Farquhar, M. G., and Palade, G. E. (1981). The Golgi apparatus (complex)—(1954–1981)—from artifact to center stage. *J. Cell Biol.* **91 (3 Pt2)**, 77s–103s.

Fisher, L., and Parker, M. (1984). Osmotic control of bilayer fusion. *Biophys. J.* **46**, 253–258.

Gilmore, R., Blobel, G., and Walter, P. (1982a). Protein translocation across the endoplasmic reticulum. I. Detection in the microsomal membrane of a receptor *J. Cell Biol.* **95**, 463–469.

Gilmore, R., Walter, P., and Blobel, G. (1982b). Protein translocation across the endoplasmic reticulum. II. Isolation and characterization of the signal recognition particle receptor *J. Cell Biol.* **95**, 470–477.

Goldenring, J. R., Lasher, R. S., Vallano, M. L., Ueda, T., Naito, S., Sternberger, N. H., Sternberger, L. A., and DeLorenzo, R. J. (1986). Association of synapsin-I with neuronal cytoskeleton. *J. Biol. Chem.* **261**, 8495–8504.

Griffiths, G., and Simons, K. (1986). The trans Golgi network: Sorting at the exit site of the Golgi complex. *Science* **234**, 438–443.

Griffiths, G., Pfeiffer, S., Simons, K., and Matlin, K. (1985). Exit of newly synthesized membrane proteins from the trans-cisternae of the Golgi complex to the plasma membrane. *J. Cell Biol.* **101**, 949–964.

Gruner, S., Cullis, P., Hope, M., Tilcok, C. (1985). Lipid polymorphism: The molecular basis of nonbilayer phases. *Am. Rev. Biophys. Chem.* **14**, 211–238.

Gumbiner, B., and Kelly, R. B. (1982). Two distinct intracellular pathways transport secretory and membrane glycoproteins to the surface of pituitary tumor cells. *Cell* **28**, 51–59.

Hand, A. R., and Oliver, C. (1981). Transport and packaging in the Golgi region. *Methods Cell Biol.* **23**, 137–153.

Henry, J. P., Gasnier, B., Roisin, M. P., Isanbert, M. F., and Scherman, D. (1987). Molecular pharmacology of monoamine transporter. *Ann. N.Y. Acad. Sci.* **493**, 194–206.

Heuser, J. E., and Reese, T. S. (1981). Structural changes after transmitter release at the frog neuromuscular junction. *J. Cell Biol.* **88**, 564–580.

Hill, T. (1960). "Introduction to Statistical Thermodynamics." Addison-Wesley, Reading, Massachusetts.

Holcomb, C., Hansen, W., Etcheverry, T., and Schekman, R. (1987). Plasma membrane protein intermediates are present in the secretory vesicles of yeast. *J. Cell. Biochem.* (in press).

Holtzman, E. (1977). The origin and fate of secretory packages, especially synaptic vesicles. *Neuroscience* **2**, 327–355.

Hooper, J. E., and Kelly, R. B. (1984a). Calcium dependent calmodulin binding to cholinergic synaptic vesicles. *J. Biol. Chem.* **259**, 141–147.

Hooper, J. E., and Kelly, R. B. (1984b). Calmodulin is tightly associated with synaptic vesicles independent of calcium. *J. Biol. Chem.* **259**, 148–153.

Horn, R. (1984). Direct measurement of the force between two lipid bilayers and observation of their fusion. *Biochim. Biophys. Acta* **778**, 224–228.

Houslay, M. (1987). Egg activation unscrambles a potential role for IP₄. *Trends Biochem. Sci.* **12**(1), 1–2.

Hurt, E., and Loon, A. (1986). How proteins find mitochondria and intramitochondrial compartments. *Trends Biochem. Sci.* **11**, 204–207.

Huttner, W. B., Schiebler, W., Greengard, P., and DeCamilli, P. (1983). Synapsin-I, a nerve terminal specific phosphoprotein. III. Its association with synaptic vesicles studied in a highly purified synaptic vesicle preparation. *J. Cell Biol.* **96**, 1374–1388.

Israelachvili, J., and Mitchell, D. (1975). A model for the packing of lipid in bilayer membranes. *Biochim. Biophys. Acta* **389**, 13–19.

Israelachvili, J., Mitchell, J., and Ninham, B. (1976). Theory of self-assembly of hydrocarbon amphiphiles into micelles and bilayers. *J. Chem. Soc. Faraday Trans. II* **72**, 1525–1568.

Israelachvili, J., Marčelja, and Horn, R. (1980). Physical principles of membrane organization. *Quart. Rev. Biophys.* **13**, 121–200.

Johnson, R. G. (1987). Proton pumps and chemosmotic coupling as a generalized mechanism for neurotransmitter and hormone transport. *Ann. N.Y. Acad. Sci.* **493**, 162–177.

Julius, D., Schekman, R., and Thorner, J. (1984a). Glycosylation and processing of prepro-α-factor through the yeast secretory pathway. *Cell* **36**, 309–318.

Julius, D., Brake, A., Blair, L., Kunisawa, R., and Thorner, J. (1984b). Isolation of the putative structural gene for the lysine–arginine-cleaving endopeptidase required for processing of yeast prepro-α-factor. *Cell* **37**, 1075–1089.

Kelly, R. B. (1985). Pathways of protein secretion in eukaryotes. *Science* **230**, 25–32.

Kirshner, N., Corcoran, J. J., Caughey, B., and Korner, M. (1987). Chromaffin vesicles function in intact cells. *Ann. Rev. N.Y. Acad. Sci.* **493**, 207–219.

Knight, D. E., and Scrutten, M. C. (1984). Cyclic nucleotides control a system which regulates Ca^{2+} sensitivity of platelet secretion. *Nature (London)* **309**, 66–68.

Koch, G. (1987). Reticuloplasmins: A novel group of proteins in the endoplasmic reticulum. *J. Cell Sci.* **87**, 491–492.

Kornfeld, R., and Kornfeld, S. (1985). Assembly of asparagine-linked oligosaccharides. *Annu. Rev. Biochem.* **54;** 631–664.

Lingappa, V. R., Chaidez, J., Yost, C. S., and Hedgpeth, J. (1984). Determinants for protein localization: β-Lactamase signal sequence directs globin across microsomal membranes. *Proc. Natl. Acad. Sci. U.S.A.* **81**, 456–460.

Llinas, R., McGuinness, T. L., Leonard, C. S., Sugimori, M., and Greengard, P. (1985). Intraterminal injection of synapsin-I or calcium/calmodulin-dependent protein kinase II alters neurotransmitter release at the squid giant synapse. *Proc. Natl. Acad. Sci. U.S.A.* **82**, 3035–3039.

Loh, Y. P., Brownstein, M. J., and Gainer, H. (1984). Proteolysis in neuropeptide processing and other neural functions *Annu. Rev. Neurosci.* **7**, 189–222.

Longley, W., and McIntosh, T. (1983). A bicontinuous tetrahedral structure in a liquid–crystalline lipid. *Nature (London)* **303**, 612–614.

Lucy, J., and Ahkong, Q. (1986). An osmotic model for the fusion of biological membranes. *FEBS Letters* **199**, 1–11.

Luini, A., Lewis, D., Guild, S., Corda, D., and Axelrod, J. (1985). Hormone secretagogues increase cytosolic calcium by increasing cAMP in corticotropin-secreting cells. *Proc. Natl. Acad. Sci. U.S.A.* **82,** 8034–8038.

Malaisse-Lagae, F., Amherdt, M., Ravazzola, M., Sener, A., Hutton, J., Orci, L., and Malaisse, W. (1979). Role of microtubules in the synthesis, conversion and release of (pro)insulin. *J. Clin. Invest.* **63,** 1284–1296.

Marie-France, B., Trifaro, J. N., Langley, O. K., Thierse, D., and Aunis, D. (1986). Secretory cell actin-binding proteins: Identification of a gelsolin-like protein in chromaffin cells. *J. Cell Biol.* **102,** 636–646.

Matlin, K. S. (1986). The sorting of proteins to the plasma membrane in epithelial cells. *J. Cell Biol.* **103,** 2565–2568.

Meyer, D. I., and Dobberstein, B. (1980). Identification and characterization of a membrane component essential for the translocation of nascent proteins across the membrane of the endoplasmic reticulum *J. Cell Biol.* **87,** 503–508.

Mitchell, D., and Ninham, B. (1981). Micelles, vesicles and microemulsions. *J. Chem. Soc., Faraday Trans. 2,* **77,** 601–629.

Moore, H.-P. H. (1987). Factors controlling packaging of peptide hormones into secretory granules. *Ann. N.Y. Acad. Sci.* **493,** 50–61.

Moore, H.-P. H., and Kelly, R. B. (1985). Secretory protein targeting in a pituitary cell line: Differential transport of foreign secretory proteins to distinct secretory pathways. *J. Cell Biol.* **101,** 1773–1781.

Moore, H.-P. H., and Kelly, R. B. (1986). Rerouting of a secretory protein by fusion with human growth hormone sequences. *Nature (London)* **321,** 443–446.

Moore, H.-P. H., Walker, M., Lee, F., and Kelly, R. B. (1983a). Expressing a human proinsulin cDNA in a mouse ACTH secretory cell. Intracellular storage, proteolytic processing and secretion on stimulation. *Cell* **35,** 531–538.

Moore, H.-P. H., Gumbiner, B., and Kelly, R. B. (1983b). A subclass of proteins and sulfated macromolecules secreted by AtT-20 (mouse pituitary tumor) cells is sorted with adrenocorticotropin into dense secretory granules. *J. Cell Biol.* **97,** 810–817.

Moore, H.-P. H., Gumbiner, B., and Kelly, R. B. (1983c). Chloroquine diverts ACTH from the regulated secretory pathway to a constitutive one in AtT-20 cells. *Nature (London)* **302,** 434–436.

Mostov, K. E., and Simister, N. E. (1985). Transcytosis. *Cell* **43,** 389–390.

Nagle, J. (1980). Theory of the main lipid bilayer phase transition. *Annu. Rev. Phys. Chem.* **31,** 157–195.

Novick, P., and Botstein, D. (1985). Phenotypic analysis of temperature sensitive yeast actin mutants. *Cell* **40,** 405–416.

Orci, L. (1982). Macro- and micro-domains in the endocrine pancreas. *Diabetes* **31,** 538–565.

Orci, L., Ravazzola, M., Amherdt, M., Madsen, O., Vassalli, J.-D., and Perrelet, A. (1985a). Direct identification of prohormone conversion site in insulin-secreting cells. *Cell* **42,** 671–681.

Orci, L., Ravazzola, M., Amherdt, M., Louvard, D., and Perrelet, A. (1985b). Clathrin-immunoreactive sites in the Golgi apparatus are concentrated at the trans pole in polypeptide hormone-secreting cells. *Proc. Natl. Acad. Sci. U.S.A.* **82,** 5385–5389.

Orci, L., Ravazzola, M., Amherdt, M., Madsen, O., Perrelet, A., Vassalli, J. D., and Anderson, R. G. W. (1986a). Conversion of proinsulin to insulin occurs coordinately with acidification of maturing secretory vesicles. *J. Cell Biol.* **103,** 2273–2282.

Orci, L., Glick, B. S., and Rothman, J. E. (1986b). A new type of coated vesicular carrier that appears not to contain clathrin: Its possible role in protein transport within the Golgi stack. *Cell* **46**, 171–184.

Orci, L., Ravazzola, M., Amherdt, M., Brown, D., and Perrelet, A. (1986c). Transport of horseradish peroxidase from the cell surface to the Golgi in insulin-secreting cells: Preferential labelling of cisternae located in an intermediate position in the stack. *EMBO J.* **5**, 2097–2101.

Orci, L., Ravazzola, M., Amherdt, M., Perrelet, A., Powell, S., Quinn, D., and Moore, H.-P. H. (1987). The trans-most cisternae of the Golgi complex: A compartment for sorting of secretory and plasma membrane proteins. Submitted for publication.

Palade, G. (1975). Intracellular aspects of the process of protein secretion. *Science* **189**, 347–358.

Parsons, S. M., Bahr, B. A., Gracz, L. M., Kaufman, R., Kornreich, W., Nilsson, L., and Rogers, G. A. (1987). Acetylcholine transport: Fundamental properties and effects of phamacological agents. *Ann. N.Y. Acad. Sci.* **493**, 220–232.

Patzak, A., and Winkler, H. (1986). Exocytotic exposure and recycling of membrane antigens of chromaffin granules: Ultrastructural evaluation after immunolabeling *J. Cell Biol.* **102**, 510–515.

Payne, G. S., and Schekman, R. (1985). A test of clathrin function in protein secretion and cell growth. *Science* **230**, 1009–1014.

Pfeffer, S., and Kelly, R. (1985). The subpopulation of brain coated vesicles that carries synaptic vesicle proteins contains two unique polypeptides. *Cell* **40**, 949–957.

Reisine, T., Rougon, G., Barbet, J., and Affolter, H. U. (1985). Corticotropin-releasing factor-induced adrenocorticotropin hormone release and synthesis in blocked by incorporation of the inhibitor of cAMP-dependent protein kinase into anterior pituitary tumor cells by liposomes. *Proc. Natl. Acad. Sci. U.S.A.* **82**, 8261–8265.

Rink, T. J., Sanchez, A., and Hallam, T. J. (1983). Diacyglycerol and phorbol ester stimulate secretion without raising cytoplasmic free calcium in human platelets *Nature (London)* **305**, 317–319.

Rodriguez-Boulan, E., Misek, D., De Salas, D. V., Salas, P. J. I., and Bard, E. (1985). Protein sorting in the secretory pathway. *Curr. Topics Membranes Transport* **24**, 251–285.

Rogalski, A. A., Bergmann, J. E., and Singer, S. J. (1984). Effect of microtubule assembly status on the intracellular processing and surface expression of an integral protein of the plasma membrane. *J. Cell Biol.* **99**, 1101–1109.

Rothman, J. H., and Stevens, T. H. (1986). Protein sorting in yeast: Mutants defective in vacuole biogenesis mislocalize vacuolar proteins into the late secretory pathway. *Cell* **47**, 1041–1051.

Salas, P. J. I., Misek, D. E., Vega-Salas, D. E., Gundersen, D., Cereijido, M., and Rodriguez-Boulan, E. (1986). Microtubules and actin filaments are not critically involved in the biogenesis of epithelial cell surface polarity. *J. Cell Biol.* **102**, 1853.

Sawano, F., Ravazzola, M., Amherdt, M., Perrelet, A., and Orci, L. (1986). Horseradish peroxidase uptake and crinophagy in insulin-secreting cells. *Exp. Cell. Res.* **164**, 174–182.

Schatz, G., and Butow, R. A. (1983). How are proteins imported into mitochondria? *Cell* **32**, 316–318.

Schekman, R. (1985). Protein localization and membrane traffic in yeast. *Annu. Rev. Cell Biol.* **1**, 115–144.

Schliwa, M. (1976). The role of divalent cations in the regulation of microtubule assembly. *In vivo* studies on microtubules of the heliozoan axopodium using the ionophore A23187. *J. Cell Biol.* **70**, 527–540.

Schweitzer, E. S., and Kelly, R. B. (1985). Selective packaging of human growth hormone into synaptic vesicles in a rat neuronal (PC12) line. *J. Cell Biol.* **101**, 667–675.

Scott, J. H., Keiner, K. L., and Pollard, H. B. (1985). Purification of synexin by pH step elution from chromatofocusing methods in the absence of ampholytes. *Anal. Biochem.* **149**, 163–165.

Siegel, D. (1986). Inverted micellar intermediates and the transitions between lamellar, cubic, and inverted hexagonal lipid phases. *Biophys. J.* **49**, 1155–1183.

Shinada, K., and Friberg, S. (1986). "Emulsions and Solubilization." Wiley, New York.

Simons, K., and Fuller, S. D. (1985). Cell surface polarity in epithelia. *Annu. Rev. Cell Biol.* **1**, 243–288.

Sly, W. S., and Fischer, H. D. (1982). The phosphomannosyl recognition system for intracellular and intercellular transport of lysosomal enzymes. *J. Cell. Biochem.* **18**, 67–85.

Steiner, D. (1984). The biosynthesis of insulin: Genetic, evolutionary, and pathophysiologic aspects. *Harvey Lect. Ser.* **78**, 191–228.

Streb, H., Irvine, R. F., Berridge, M. J., and Schultz, I. (1983). Release of Ca^{2+} from a nonmitochondrial intracellular store in pancreatic acinar cells by inositol 1,4,5-triphosphate. *Nature (London)* **306**, 67–69.

Strous, G. J. A. M., Willemsen, P. K., Slot, J. W., Geuze, H. J., and Lodish, H. F. (1983). Vesicular stomatitis virus glycoprotein, albumin, and transferrin are transported to the cell surface via the same Golgi vesicles. *J. Cell Biol.* **97**, 1815–1822.

Tabony, J., de Geyer, A., and Braganza, L. (1987). Assembly of micelles into higher-order structures with increasing salt concentration. *Nature (London)* **327**, 321–324.

Tanford, C. (1980). "The Hydrophobic Effect." Wiley, New York.

Tartakoff, A. (1987). "Secretory and Endocytic Paths." Wiley, New York.

Tartakoff, A., and Vassalli, P. (1978). Comparative studies of intracellular transport of secretory proteins. *J. Cell Biol.* **79**, 694–707.

Terasaki, M., Chen, L., and Fujiwara, K. (1986). Microtubules and the endoplasmic reticulum are highly interdependent structures. *J. Cell Biol.* **103**, 1557–1586.

Tooze, J., and Tooze, S. A. (1986). Clathrin-coated vesicular transport of secretory proteins during the formation of ACTH-containing secretory granules in AtT-20 cells. *J. Cell Biol.* **103**, 839–850.

Trifaro, J.-M., and Fournier, S. (1987). Calmodulin and the secretory vesicle. *Ann. N.Y. Acad. Sci.* **493**, 417–434.

Vale, R. D., Schnapp, B. J., Mitcheson, T., Steuer, E., Reese, T. S., and Sheetz, M. P. (1985). Differential axoplasmic proteins generate movement in opposite directions along microtubules *in vitro*. *Cell* **43**, 623–632.

Walter, P., and Blobel, G. (1981). Translocation of proteins across the endoplasmic reticulum. II. Signal recognition protein (SRP) mediates the selective binding to microsomal membranes of *in vitro* assembled polysomes synthesizing secretory protein. *Cell Biol.* **91**, 551–556.

Walter, P., Ibrahimi, I., and Blobel, G. (1981). Translocation of proteins across the endoplasmic reticulum I. Signal recognition protein (SRP) binds to *in-vitro*-assembled polysomes synthesizing secretory protein. *J. Cell Biol.* **91**, 545–561.

Weiland, F., Gleason, M., Serafini, T., and Rothman, J. (1987). The rate of bulk flow from the endoplasmic reticulum to the cell surface. *Cell* **50**, 289–300.

White, J., Kielian, M., and Helenius, A. (1983). Membrane fusion proteins of enveloped animal viruses. *Quart. Rev. Biophys.* **16**, 151–195.

Whittaker, V. P. (1987). Cholinergic synaptic vesicles from the electromotor nerve terminals of *Torpedo:* Composition and life cycle. *Ann. N.Y. Acad. Sci.* **493,** 77–91.

Wilschut, J. and Hoekstra, D. (1984). Membrane fusion: From liposomes to biological membranes. *Trends Biochem. Sci.* **9,** 479–483.

Winkler, H. (1977). The biogenesis of adrenal chromaffin granules. *Neuroscience* **2,** 657–683.

Woods, J. W., Doriaux, M., and Farquhar, M. G. (1986). Transferrin receptors recycle to cis and middle as well as trans Golgi cisternae in Ig-secreting myeloma cells. *J. Cell Biol.* **103,** 277–286.

13

Expression of Extracellular Matrixlike Glycoproteins by Macrophages and Other Leukocytes

ARTHUR M. MERCURIO

I. INTRODUCTION

The tissue macrophage represents one biological end point of the diverse pathways of hematopoietic cell differentiation. Arguably, the biochemical complexity of this differentiated mammalian cell exceeds that of most other cell populations discussed in this volume. This complexity is reflected in the many physiological and pathological functions mediated by such macrophage populations as blood monocytes, alveolar macrophages, Kupffer cells, and peritoneal macrophages (Van Furth, 1980). Moreover, the complexity of these cells is compounded by their ability to

563

PROTEIN TRANSFER AND ORGANELLE BIOGENESIS

be "modulated" by stimuli such as inflammatory agents and lympho-kines. These modulation pathways alter the biochemical and, conse-quently, functional phenotype of the macrophage (Cohn, 1986). In other terms, populations of inflammatory and activated macrophages are char-acterized by their enhanced participation in specific functions such as wound repair, tissue remodeling, and host defense (Cohn, 1978; Karnovsky and Lazdins, 1978; North, 1978).

Work from many laboratories has identified specific quantitative and qualitative changes that occur on the macrophage surface in response to stimuli such as inflammatory agents and interferon γ. Importantly, several surface molecules have been implicated in a number of macrophage func-tions (Cohn, 1986). In addition, macrophages secrete many different types of molecules including proteases, growth factors, complement compo-nents, and lymphokines that participate in a multitude of macrophage-mediated functions (Nathan, 1987; Takemura and Werb, 1984).

One major class of glycoproteins that is not often dealt with are those glycoproteins associated with extracellular matrices. This group includes laminin, fibronectin, thrombospondin, the collagens, and the proteogly-cans (Hay, 1981). These are largely secreted glycoproteins that often associate with cell surfaces through specific receptors, and they are present in basement membranes that separate epithelial and endothelial cells from mesenchymal cells (Bernfield, 1984). In recent years, however, it has become evident that these glycoproteins are also synthesized by, and expressed on, populations of leukocytes. Although they probably do not assemble into well-defined extracellular matrices or basement mem-branes characteristic of epithelial and endothelial cells, they probably contribute significantly to many leukocyte functions. Unfortunately, the data on extracellular matrix glycoproteins in leukocytes are fragmentary, and major gaps exist in our understanding of their structure, biosynthesis, and localization in these cells.

This chapter focuses on this aspect of leukocyte biochemistry, and, more specifically, it is concerned largely with monocytes and macro-phages. This decision is based on our interest in the biosynthesis, pro-cessing, and function of macrophage glycoproteins, as well as on recent evidence implicating important roles for glycoproteins such as laminin and fibronectin in macrophage physiology (see below). One specialized function of macrophages in which we are interested is their capacity for the nonspecific lysis of oncogenically transformed cells (Hibbs et al., 1980). To date, good data are not available on the possible involvement of extracellular matrixlike glycoproteins in this specialized cellular interac-tion, although the possibliity has been proposed (Huard et al., 1985; Perri et al., 1982, 1985). However, other leukocytes such as large granular lymphocytes or natural killer (NK) cells also have the capacity to kill tumor cells (Herberman *et al.*, 1975), and recent reports suggest that laminin may participate in NK–tumor cell interactions (Hiserodt *et al.*, 1985; Laybourn *et al.*, 1986). For this reason, relevant data on the expression and function of extracellular matrixlike glycoproteins by NK cells are also discussed.

Several points relating to the study of extracellular matrix molecules in macrophages and related cells should be emphasized. First, it is important to ascertain the capacity of the macrophages themselves to synthesize and secrete specific matrix glycoproteins. These glycoproteins can also be deposited on the macrophage surface from exogenous sources including sera and the secretory products of other cell types, and the contribution of these glycoproteins to specific macrophage functions could be different from that of those which are synthesized and secreted by the macrophages themselves.

Second, much of our knowledge on the structure and function of matrix glycoproteins has been derived from studies on a few, specialized forms of basement membrane, or, in the case of fibronectin, from plasma fibronectin and from fibronectin derived from fibroblasts. Most of the antibodies currently used to identify and characterize matrix glycoproteins in leukocytes have been generated against components purified from these sources. Although many of these antibodies exhibit significant cross-reactivity with matrix glycoproteins synthesized by other cell populations, it is likely that differences in structure, as well as function, exist, for example, between matrix components synthesized by macrophages and those synthesized by epithelial cells. Also, adherent leukocytes probably do not assemble well-defined extracellular matrices characteristic of other cell types. It would seem appropriate, therefore, to use the term "extracellular matrixlike" when describing glycoproteins that have been identified using reagents generated against heterologous cells.

Finally, as mentioned above, the biochemical and functional phenotype of macrophages can be modulated by a number of physiological stimuli. The classic stimuli are inflammatory agents such as thioglycolate broth (TG), endotoxin, and lymphokines such as interferon γ (Cohn, 1978; Karnovsky and Lazdins, 1978). Interferon γ, for example, has the capacity to activate macrophages to express specialized functions including the nonspecific lysis of tumor cells (Schreiber *et al.*, 1983). Inflammatory macrophages differ in a number of parameters from both resident tissue macrophages and tumoricidally activated macrophages. [For a discussion of macrophage activation, see Adams and Hamilton (1984).] Probably, modulations in the expression of matrixlike glycoproteins occur among resident, inflammatory, and activated macrophages, as well as during the

pathway of monocyte/macrophage differentiation. Indeed, such alterations in the expression of matrixlike glycoproteins may correlate with
their involvement in specific macrophage functions. Thus, it is important
to define the particular macrophage population(s) being studied for their
expression of extracellular matrixlike glycoproteins.

This chapter emphasizes the importance of studying extracellular matrixlike glycoproteins within the context of macrophage cell biology. This
is probably the first review article that attempts to integrate the biochemistry of extracellular matrixlike glycoproteins with macrophage cell biology. The limited and fragmentary data that exist in this area preclude the
writing of a definitive and authoritative review. Rather, it is our intent that
this chapter should summarize existing data and highlight key research
areas, thus providing a framework and direction for future work.

One issue which should be stressed and one which is directly related to
the theme of this volume is that virtually nothing is known about the
biosynthesis, secretion, and surface association of these large, multisu-
bunit glycoproteins in macrophages or other leukocytes. In fact, as will
become apparent in the subsequent discussions of individual matrix glycoproteins, only limited data are available on these processes in cell types
that have been well studied with respect to these glycoproteins. The
assembly of an extracellular matrix in such cells certainly constitutes a
problem in organelle biogenesis, and the situation with macrophages and
other leukocytes can be considered a variation on this theme.

II. BIOSYNTHESIS OF EXTRACELLULAR MATRIXLIKE
GLYCOPROTEINS BY MACROPHAGES

A. Fibronectin

The term fibronectin refers to a family of glycoproteins that are found
on cell surfaces and associated with extracellular matrices (cellular fibronectin) and that are also present in plasma and other body fluids
(plasma fibronectin). [For reviews on fibronectin structure and function,
see Hynes and Yamada (1982) and Hynes (1985).] The fibronectin molecule is a dimer consisting of two related, but not identical subunits (M_r
220K $\pm$ 20K), that are held together at their carboxy termini by disulfide
bonds. Complex N-linked carbohydrates comprise 5–9% of the fibronectin molecule, and several studies have analyzed these oligosaccharides in
some detail [e.g., Zhu et al. (1984), Nichols et al. (1986)]. The primary
function of this carbohydrate may be to protect the molecule from proteolysis and abnormal rates of turnover (Hynes and Yamada, 1982). Fi-

bronectin contains binding sites for several molecules of physiological importance (e.g., heparin, collagen, and fibrin), and it also contains a binding site that is involved in its association with specific receptors on cell surfaces (see below). The recent generation of cDNA and genomic clones has provided significant insight into fibronectin structure–function relationships (Hynes, 1985). Indeed, the number of cellular functions in which fibronectin participates is large and includes cell attachment and spreading, maintenance of cell shape, cell migration, and, as will be discussed below, phagocytosis.

There have been several reports on the ability of macrophages to synthesize and secrete fibronectin, but some discrepancies exist in the data that have been accumulated. Alitalo *et al.* (1980) reported that human monocyte-derived macrophages synthesize and secrete fibronectin but that this expression is observed only on monocyte differentiation to macrophages. Fibronectin may be the major secreted protein of human lung alveolar macrophages in culture (Villiger *et al.*, 1981). Whether these *in vitro* levels of secretion reflect the situation *in vivo* is difficult to ascertain, particularly in view of the possibility that the *in vitro* culture of certain cells markedly induces fibronectin synthesis (L. B. Chen, personal communication). Mouse peritoneal macrophages also have the capacity to synthesize and secrete fibronectin (Johansson *et al.*, 1979). It is interesting to note that in all of these studies it was observed that little fibronectin could be detected on the macrophage surface and that an extensive fibronectin-containing extracellular matrix was not deposited by the macrophages. Such matrices are frequently observed in cultured fibroblasts and other cells (Hynes and Yamada, 1982).

In contrast to this group of studies is the report by Wicha and Huard (1983) that 25% of resident mouse macrophages and 35% of TG-elicited macrophages express cell surface fibronectin as determined by indirect immunofluorescent staining. This staining could have resulted, however, from exogenous fibronectin deposition on the cell surface, a possibility not excluded in the study. A more recent report suggests that TG-elicited mouse peritoneal macrophages synthesize significantly more fibronectin than resident macrophages, and that some of this fibronectin associates with the adherent surface of the macrophage (Cofano *et al.*, 1984). Taken together, these observations suggest that fibronectin expression may be a differentiated function of monocytes and macrophages, although the ability of this fibronectin to associate with the macrophage surface is less clear.

The possibility that fibronectin expression may be a differentiated function of macrophages is also supported by the observation that interferon γ stimulates fibronectin expression by TG-elicited macrophages (Cofano *et al.*, 1984). Experiments performed recently in our laboratory, however, indicate that TG-elicited mouse macrophages synthesize and secrete considerably less fibronectin than resident macrophages. More importantly, and in contrast to Cofano *et al.* (1984), we find that interferon γ markedly *decreases* fibronectin synthesis by TG-elicited macrophages. Clearly, more detailed studies on the regulation of fibronectin biosynthesis and secretion in macrophages are needed. In fact, it is not even known whether macrophages synthesize and secrete cellular or plasma forms of fibronectin (L. Van DeWater, personal communication). Moreover, it should be determined more carefully whether populations of monocytes, resident macrophages, elicited macrophages, and activated macrophages differ in their ability to express fibronectin, and, if so, whether these differences correlate with specific fibronectin-mediated functions.

Human macrophages also synthesize a 95K gelatin-binding protein that is distinct from fibronectin (Vartio *et al.*, 1982). Although the function of this protein is not known, its expression correlates with macrophage adherence to substrata, and it can be stimulated by agents that promote adherence such as phorbol esters.

The capacity of other types of leukocytes to synthesize fibronectin has not been critically evaluated. Rat natural killer (NK) cells, however, may synthesize and secrete a 450K protein that can be identified immunologically as fibronectin (A. Santoni, personal communication).

The identification and characterization of specific cell surface receptors for fibronectin has been a major goal of extracellular matrix biochemistry. Such receptors presumably serve as sites of attachment for fibronectin to cell surfaces, and also as mediators of transmembrane linkages between fibronectin and cytoskeletal elements that are known to occur (Hynes and Yamada, 1982). Monoclonal antibodies which inhibit the fibronectin-mediated adhesion of cells have been utilized to identify specific surface proteins that have the properties of a putative fibronectin receptor. Using this approach, several groups have reported recently that a complex of 140K glycoproteins exhibit the properties expected of a fibronectin receptor (Horowitz *et al.*, 1984; Hasegawa *et al.*, 1985; Knudsen *et al.*, 1985; Giancotti *et al.*, 1985; Pytela *et al.*, 1985; Brown and Juliano, 1986). This complex of glycoproteins has been designated by a number of different terms, including "integrin" (Tamkun *et al.*, 1986). Recently, the structure of Band III of integrin has been deduced from analyses of cDNA clones (Tamkun *et al.*, 1986).

Although a detailed discussion of fibronectin receptors is beyond the scope of this chapter, several points are worth mentioning. (1) The 140K complex can bind both fibronectin and laminin in chickens, but apparently it binds only fibronectin in mammalian cells (Brown and Juliano, 1986). (2) A specific peptide sequence in the cell-binding domain of fibronectin, Arg-Gly-Asp-Ser, mediates fibronectin attachment to cell surface receptors (Pierschbacher and Ruoslahti, 1984). (3) The 140K complex may be part of a larger family of adhesion molecules which includes the IIb/IIIa complex on platelets (Ginsberg et al., 1985) as well as the LFA-1/Mac-1/p150,95 glycoprotein family present on lymphoid and myeloid cells (Sanchez-Madrid et al., 1986; Cosgrove et al., 1986). This latter group of proteins shares a common 95K β subunit and have related but distinct α subunits. Recent data suggest that Band 3 of the 140K complex and protein IIIa may also exhibit homology with the β subunits of the LFA-1 family (Tamkun et al., 1986). All these β subunits exhibit extensive intrachain disulfide bonding as evidenced by changes in electrophoretic mobility on reduction.

Monocytes and macrophages do express cell surface receptors for fibronectin. Hosein and Bianco (1985) reported on a monoclonal antibody which blocked the interaction between monocytes and substrate-bound fibronectin. This antibody recognized a 110K polypeptide on the surfaces of human monocytes and macrophages, but this antigen was not present on fibroblasts, lymphocytes, or platelets. Human macrophages bind to surfaces coated with the Arg-Gly-Asp-Ser tetrapeptide (Wright and Meyer, 1985), and this interaction promotes complement-mediated phagocytosis (see below), suggesting that the macrophage fibronectin receptor interacts with the same fibronectin domains as in other cell types. We have shown that antibodies directed against a 140K glycoprotein on Chinese hamster ovary (CHO) cells which may identify a fibronectin receptor on these cells, as well as other mammalian cells (Brown and Juliano, 1986), can inhibit macrophage adherence to fibronectin-coated surfaces (Barlozzari et al., 1987b). This macrophage fibronectin receptor exhibits an Mr of 115K under reducing conditions, while under nonreducing conditions two bands of 110K and 130K are observed. Interestingly, the surface expression of this receptor can be up-regulated by interferon γ and LPS. These agents stimulate increased spreading of macrophages on a variety of substrata, and, in combination they can induce macrophage tumoricidal activity. The possible relationship between these processes and increased receptor expression is being investigated. It is interesting to note that neither interferon γ nor LPS alter the rate of receptor biosynthesis. This suggests that the observed increase in receptor surface expression is probably a posttranslational event.

The structural and functional relationship between this macrophage fibronectin receptor and the LFA-1/Mac-1/p150,95 glycoprotein family merits investigation. These molecules are all present on the macrophage surface, and, as discussed above, their β subunits share homology with Band III of integrin.

B. Thrombospondin

Thrombospondin is another member of the family of glycoproteins that mediate cell–cell and cell–matrix interactions. This 420K glycoprotein was first identified in platelets, where it associates with the cell surface on activation and becomes incorporated into fibrin clots (Lawler, 1986). Thrombospondin is also synthesized by a variety of cells in culture including monocytes and macrophages (Jaffe et al., 1985). In many of the cells, secreted thrombospondin is incorporated into the extracellular matrix. In human monocytes and mouse macrophages, however, thrombospondin in not present in an extracellular matrix, and very little can be detected intracellularly by indirect immunofluorescent staining (Jaffe et al., 1985). Thus, thrombospondin is largely a secreted protein in these cells, perhaps similar to the situation with fibronectin described above. In fact, thrombospondin binds to fibronectin as well as to Type V collagen, fibrinogen, and heparin (Lawler, 1986), and it also possesses a lectinlike activity that may mediate its involvement in cell–cell interactions (Jaffe et al., 1982). The synthesis and secretion of thrombospondin are markedly depressed in elicited mouse macrophages compared to resident macrophages (Jaffe et al., 1985). This observation is in agreement with our results on both fibronectin synthesis (see above) and laminin synthesis (see below) in mouse macrophage populations. Taken together, these findings suggest a coordinate regulation of extracellular matrix glycoprotein synthesis as a function of the state of macrophage activation and differentiation.

C. Laminin

The matrix glycoprotein laminin was originally isolated and characterized in cells that secrete copious amounts of basement membrane components (Timpl et al., 1979; Chung et al., 1979). Subsequent studies have demonstrated that laminin is a major component of all basement membranes and that it is an adhesive glycoprotein which mediates cell attachment, expecially to Type IV collagen [see Bernfield (1984) for recent reviews on basement membranes]. Most structural studies have utilized laminin purified from either the Englebreadth-Holm-Swarm (EHS) mouse tumor or from parietal endoderm cells. This laminin is comprised of an α subunit (M_r 440K) and two distinct β subunits (M_r 230K and 210K). The intact molecule exhibits extensive disulfide bonding, and its molecular weight is approximately 900K. All of the laminin subunits contain oligosaccharides, most of which are N-linked, and the intact molecule is approximately 15–18% carbohydrate. Specific structural domains of the laminin molecule have been implicated in several laminin-mediated functions including binding to Type IV collagen, cell attachment and spreading, and the promotion of neurite outgrowth (Liotta, 1986). It is interesting to note that the laminin synthesized by a number of different cell types in culture does not contain an α subunit. This defect is apparently not the result of degradation because α chain mRNA cannot be detected in these cells using specific cDNA probes (H. Kleinman, personal communication).

Wicha and Huard (1983) reported that mouse macrophages express a lamininlike molecule on their surfaces. More specifically, they observed that a substantially greater number of TG-elicited macrophages expressed cell surface laminin than resident macrophages, as determined by indirect immunofluorescent staining. The ability of the macrophages themselves, however, to synthesize and secrete laminin was not demonstrated in this study. Because of our interest in macrophage glycoprotein biosynthesis and because of the potential importance of laminin in macrophage physiology (see below), we initiated a study of laminin biosynthesis, secretion, and localization in populations of mouse peritoneal macrophages (Barlozzari et al., 1987a). Resident, TG-elicited, and interferon γ-activated macrophages radiolabeled with [^{35}S]methionine secrete into their culture media two polypeptides of M_r 220K on sodium dodecyl sulfate (SDS) reducing gels. These polypeptides are the major high molecular weight (i.e., $M_r > 220$K), newly synthesized proteins secreted by mouse macrophages in culture. Immunoprecipitation experiments using affinity purified rabbit antisera against mouse laminin indicate the presence in culture media of the laminin β subunits which comigrate with the high molecular weight polypeptides observed by direct, SDS gel analysis of culture media. Moreover, the 220K polypeptides present in the culture media can be removed specifically by treating the supernatants with laminin antisera and protein A–agarose. These data demonstrate that the β subunits of laminin are secretory products of mouse macrophages maintained in culture.

As with our results on fibronectin synthesis (see above), we observed that TG-elicited macrophages synthesize less laminin than resident macrophages, and that interferon-gamma decreases laminin synthesis in TG-elicited macrophages. Pulse–chase radiolabeling experiments demonstrate that laminin is synthesized and secreted rapidly in mouse macrophages. An interesting finding of these pulse–chase experiments is that the M_r of the secreted form of laminin is significantly larger than the corresponding cellular forms. Perhaps, a posttranslational modification such as terminal glycosylation occurs immediately prior to laminin secretion. This finding also raises the possibility that laminin is retained in an intracellular compartment that has the enzymatic capacity for such a posttranslational modification.

To our knowledge, there have been no published reports demonstrating the synthesis and expression of laminin by human monocytes or macrophages. Human macrophages in culture secrete a major protein that exhibits an M_r of 220K on SDS reducing gels (Alitalo et al., 1980). This protein had been identified as fibronectin, but it is possible, in view of the findings discussed above, that this band is comprised, at least in part, of laminin.

Germane to a discussion on laminin expression by macrophages are preliminary reports from two laboratories demonstrating the presence of a 67K laminin-binding protein on the surface of mouse macrophages (Wicha et al., 1983; Brown, 1986). This protein is similar in size to the laminin receptor on human tumor cells which has been purified and for which cDNA clones have been obtained (Wewer et al., 1986). From a functional standpoint, the observation that macrophages may express a laminin receptor is quite plausible. Such a receptor could serve as a site of attachment for laminin synthesized by the macrophages themselves, as well as laminin that may be deposited on the macrophage surface from other sources. This receptor could also participate in the interaction of macrophages with laminin-containing structures such as basement membranes, and it may function in cell–cell interactions that involve laminin. Surely, the further characterization of the structure and functions of this receptor is an important task. In this connection, recent work from our laboratory indicates that a 67K protein and a 77K protein on the macrophage surface have the ability to bind specifically to laminin-Sepharose (A. M. Mercurio and L. M. Shaw, in preparation). The surface expression of these proteins can be up-regulated by agents such as PMA and interferon γ suggesting that laminin binding proteins play key roles in the functions of activated macrophages (Mercurio and Barlozzari, 1987; A. M. Mercurio and L. M. Shaw, unpublished observation).

In addition, the possibility that macrophages express other receptors for laminin should be investigated. This point is raised because several different receptors for laminin may exist, including a specific type of sulfated glycolipid (Roberts et al., 1986). In this connection, we have demonstrated, in collaboration with Dr. Gary Schwarting, that sulfated glycolipids are present in macrophage membranes. Their potential role as laminin-binding molecules on the macrophage surface is being investigated.

Rodent NK cells also express laminin on their surfaces, an observation based largely on immunofluorescent analyses of purified NK populations (Hiserodt *et al.*, 1985; Laybourn *et al.*, 1986). Surprisingly, none of these initial studies reported on the ability of NK cells themselves to synthesize laminin. A recent report (Santoni *et al.*, 1987), however, does provide data indicating that rat NK cells do have this synthetic capacity.

D. Proteoglycans

Proteoglycans are a heterogeneous group of glycoproteins that contain a core protein and glycosaminoglycan side chains. Proteoglycan biochemistry was once viewed as an arcane endeavor by many cell biologists and immunologists. This perspective has changed radically in recent years as a result of the number of important cellular and immune functions in which specific proteoglycans have been implicated (Hascall, 1986). These functions range from exocytosis and secretion to basement membrane scaffolding. The invariant γ chain of Class II major histocompatibility complex (MHC) molecules has been identified as the core protein of the Ia-associated chondroitin sulfate proteoglycan, and it may facilitate the biosynthesis and targeting of Class II antigens to the plasma membrane (Sant *et al.*, 1985). Quite surprising is the report that the primary gene product responsible for the biological clock in *Drosophila* is a proteoglycan (Kolata, 1985). These exciting examples are mentioned, in part, to emphasize the need for more studies on the structural and functional characterization of proteoglycans in macrophages and related cells.

Studies on mast cells have provided good data demonstrating that sulfated proteoglycans are major components of secretory granules (Stevens, 1986). The anionic nature of these molecules enables them to complex with cationic proteins that are often stored in granules and that are often of functional importance. This complex formation may assist in the intracellular transport of such molecules and protect them from proteolytic degradation. Cloned populations of NK cells contain chondroitin sulfate proteoglycan in their secretory granules (MacDermott *et al.*, 1985). Interestingly, contact of NK cells with tumor target cells results in the fusion of secretory granules with the plasma membrane and the release of proteoglycan into the extracellular milieu (MacDermott *et al.*, 1985). This contact-dependent exocytosis of proteoglycan appears to be related directly to the ability of NK cells to kill tumor target cells, although the contribution of these molecules to this process is not well established. They may, for example, protect the NK cells from being damaged by the lytic molecules that are also stored in the secretory granules.

Proteoglycans have not been implicated in specific macrophage functions. In fact, not much is known about the structure and function of macrophage proteoglycans. Human peripheral blood monocytes and lymphocytes synthesize and secrete a chondroitin sulfate proteoglycan (Levitt and Ho, 1983). This chondroitin sulfate proteoglycan differs markedly from the well-studied cartilage chondroitin sulfate. The mononuclear proteoglycan is a relatively small molecule (M_r 150K) in comparison to the cartilage molecule (M_r 200K–300K), and it does not form the large aggregates observed in cartilage. Whether this chondroitin sulfate is contained within a macrophage secretory granule and whether it functions in a manner analogous to its role in NK cells remains to be determined. It should be pointed out, however, that the concept of a granule-type structure which fuses with the cell surface in response, for example, to tumor cell binding is not as well developed for the macrophage as it is for NK cells and cytotoxic T cells (Hibbs *et al.*, 1980). Takasu *et al.* (1982) reported that the majority of proteoglycans synthesized by mouse peritoneal macrophages are secreted in culture and that a highly sulfated proteodermatan sulfate is a major constituent of the secreted proteoglycans.

The characterization of *cell surface* proteoglycans on macrophages and NK cells may provide insight into many different functions associated with these cells. It is known from studies on other cell types that proteoglycans such as heparan sulfate occur on cell surfaces, as integral membrane structures (Hook *et al.*, 1984). These cell surface proteoglycans are thought to participate in cell adhesion and binding to extracellular matrix components, and they may interact with the cytoskeleton (Hook *et al.*, 1984).

III. FUNCTIONS OF EXTRACELLULAR MATRIXLIKE GLYCOPROTEINS IN MACROPHAGES AND RELATED CELLS

A. Phagocytosis

As stated quite elegantly in a recent review (Nathan, 1987), "Phagocytosis by macrophages is essential to metazoan life. Without it, organs would go unmolded, wounds unmended. Senescent cells would stagnate in the vessels, invading microbes flourish in the tissues." The involvement of extracellular matrixlike glycoproteins in phagocytosis is well documented, although the actual mechanism of their involvement is uncertain (Brown, 1986).

One common mechanism of phagocytosis involves the opsonization of foreign particles by antibody or complement components. Such opsonized particles interact with the macrophage surface through specific receptors (Brown, 1986; VanDeWater, 1985). Perhaps, the most potent opsonin for phagocytosis is immunoglobulin G (IgG). Foreign particles which are coated with IgG bind, via their Fc domain, to a class of Fc receptors on the macrophage surface and are rapidly internalized (Ravetch et al., 1986).

Somewhat more complex is the situation with complement-coated particles. Phagocytosis of complement-coated particles is mediated by two different receptors on the macrophage surface (Brown, 1986): CR1 (C4b/C3b receptor) and CR3 (C3bi receptor). Complement-coated particles will bind to, but not be ingested by, resident macrophages (Bianco et al., 1975). They will be ingested by resident macrophages treated with T cell derived lymphokines, by TG-elicited macrophages, or by macrophages exposed to either fibronectin, laminin, or serum amyloid P [reviewed in Brown (1986)]. Presumably, all of these treatments serve to "activate" the complement receptors enabling them to ingest the foreign particles.

The mechanism by which this activation occurs is not known, but the contribution of fibronectin and laminin to this process is intriguing. In fact, fibronectin (Wright et al., 1983; Pommier et al., 1983) and laminin (Bohnsack et al., 1985) can stimulate both complement-mediated and IgG-mediated phagocytosis. This stimulation requires a direct interaction between these matrix glycoproteins and the macrophage, although there is some controversy regarding the manner in which they are presented. Pommier et al., (1983) observed that both soluble and substrate-bound fibronectin will stimulate phagocytosis, but Wright et al., (1983) reported that soluble fibronectin is inefficient in this regard. Taken together, these observations suggest that matrix glycoproteins have a global effect on the macrophage surface resulting in an overall stimulation of phagocytosis. Perhaps, this effect occurs, at least in part, by a matrix glycoprotein-mediated modulation of the macrophage cytoskeleton. It is known that the extracellular matrix and the cytoskeleton are intimately related (Hynes, 1981). For example, extracellular matrix fibrils containing fibronectin align with intracellular microfilament bundles (Hynes, 1981). Moreover, the role of the cytoskeleton in phagocytosis is well documented (Silverstein et al., 1977). These data have also led to the suggestion that the interaction of phagocytic cells with matrix glycoproteins may act to signal their emigration from the bloodstream into areas of tissue inflammation and damage where their phagocytic function is required (Brown, 1986).

One point difficult to reconcile at present is, as discussed above, the constitutive synthesis and secretion of laminin and fibronectin by murine macrophages in vitro. Does this endogenous secretion affect the phago-

Hasegawa, T., Hasegawa, E., Chen, W. T., and Yamada, K. M. (1985). Characterization of a membrane-associated glycoprotein complex implicated in cell adhesion to fibronectin. *J. Cell. Biochem.* **28**, 307–318.

Hay, E. D. (1981). "Cell Biology of the Extracellular Matrix." Plenum, New York.

Herberman, R. B., Nunn, M., and Lavin, D. (1975). Natural cytotoxic reactivity of mouse lymphoid cells against syngeneic and allogeneic tumors. *Int. J. Cancer* **16**, 216.

Hibbs, J. B., Chapman, H. A., Jr., and Weinberg, J. B. (1980). Regulation of macrophage nonspecific tumoricidal capability. *In* "Mononuclear Phagocytes", (R. Van Furth, ed.), pp. 1681–1720. Nijhoff, Amsterdam.

Hiserodt, J. C., Laybourn, K. A., and Varani, J. (1985). Expression of a laminin-like substance on the surface of murine natural killer lymphocytes and its role in NK recognition of tumor target cells. *J. Immunol.* **135**, 1484–1487.

Hook, M., Kjellen, L., Johansson, S., and Robinson, J. (1984). Cell surface glycosaminoglycans. *Annu. Rev. Biochem.* **53**, 847–869.

Hosein, B., and Bianco, C. (1985). Monocyte receptors for fibronectin characterized by a monoclonal antibody that interferes with receptor activity. *J. Exp. Med.* **162**, 157–170.

Horowitz, A., Duggan, K., Greggs, R., Decker, C., and Buck, C. (1984). The cell substrate attachment (CSAT) antigen has properties of a receptor for laminin and fibronectin. *J. Cell Biol.* **101**, 2134–2144.

Huard, T. K., Baney, J. L., Wood, J. F. B., and Wicha, M. S. (1985). A potential role for the extracellular matrix glycoprotein laminin in macrophage–tumor cell interactions. *Int. J. Cancer* **36**, 511–517.

Hynes, R. O. (1981). Relationships between fibronectin and the cytoskeleton. *In* "Cytoskeletal elements and Plasma Membrane Organization" (Cell Surface Reviews, Vol. 71; G. Poste and G. L. Nicholson, eds.). Elsevier, New York, pp. 97–139.

Hynes, R. O. (1985). Molecular biology of fibronectin. *Annu. Rev. Cell Biol.* **1**, 67–90.

Hynes, R. O., and Yamada, K. M. (1982). Fibronectins: Multifunctional modular glycoproteins. *J. Cell Biol.* **95**, 369–377.

Jaffe, E. A., Leung, L. L. K., Nachman, R. L., Levin, R. I., and Mosher, D. F. (1982). Thrombospondin is the endogenous lectin of human platelets. *Nature (London)* **295**, 246.

Jaffe, E. A., Ruggiero, J. T., and Falcone, D. J. (1985). Monocytes and macrophages synthesize and secrete thrombospondin. *Blood* **65**, 79–84.

Johansson, S., Rubin, K., Hook, M., Ahlgren, T., and Seljilid, R. (1979). *In vitro* biosynthesis of cold insoluble globulin (fibronectin) by mouse peritoneal macrophages. *FEBS Lett.* **105**, 313–316.

Karnovsky, M. L., and Lazdins, J. K. (1978). Biochemical criteria for activated macrophages. *J. Immunol.* **121**, 809.

Knudsen, K. A., Horwitz, A. F., and Buck, C. A. (1985). A monoclonal antibody identifies a glycoprotein complex involved cell-substratum adhesion. *Exp. Cell Res.* **157**, 218–226.

Kolata, G. (1985). Genes and biological clocks. *Science* **230**, 1151.

Lawler, J. (1986). Review: The structural and functional properties of thrombospondin. *Boood* **67**, 1197–1209.

Laybourn, K. A., Hiserodt, J. C., Abruzzo, L. V., and Varani, J. (1986). *In vitro* and *in vivo* interaction between murine fibrosarcoma cells and natural killer cells. *Cancer Res.* **46**, 3407–3412.

Levitt, D., and Ho, P.-L. (1983). Induction of chondroitin sulfate proteoglycan synthesis and secretion in lymphocytes and monocytes. *J. Cell Biol.* **97**, 351.

Liotta, L. A. (1986). Tumor invasion and metastases—role of the extracellular matrix. *Cancer Res.* **46**, 1–7.

MacDermott, R. P., Schmidt, R. E., Caulfield, J. P., Hein, A., Bartley, G. T., Ritz, J., Schlossman, S. F., Austen, K. F., and Stevens, R. L. (1985). Proteoglycans in cell-mediated cytotoxicity: Identification, localization, and exocytosis of a chondroitin sulfate proteoglycan from human cloned natural killer cells during target cell lysis. *J. Exp. Med.* **162,** 1771–1787.

Mercurio, A. M. (1986). Disruption of oligosaccharide processing in murine tumor cells inhibits their susceptibility to lysis by activated mouse macrophages. *Proc. Natl. Acad. Sci. U.S.A.* **83,** 2609–2613.

Mercurio, A. M., and Barlozzari, T. (1987). Macrophage adherence to laminin-coated surfaces and expression of gp67 are augmented by interferon-gamma and PMA. *J. Cell Biol.* (in press). (abst.).

Mercurio, A. M., and Robbins, P. W. (1985). Activation of mouse peritoneal macrophages alters the structure and surface expression of protein-bound lactosaminoglycans. *J. Immunol.* **135,** 1305–1312.

Nathan, C. (1987). Secretory products of macrophages. *J. Clin. Invest.* **79,** 319–326.

Nichols. E. J., Fenderson, B. A., Carter, W. G., and Hakomori, S. (1986). Domain specific distribution of carbohydrates in human fibronectins and the transformation-dependent translocation of branched type 2 chain defined by monoclonal antibody. *J. Biol. Chem.* **261,** 11295–11301.

North, R. J. (1978). The concept of the activated macrophage. *J. Immunol.* **121,** 809.

Perri, R. T., Kay, N. E., McCarthy, Vessella, R. L., Jacob, H. S., and Furcht, L. T. (1982). Fibronectin enhances *in vitro* monocyte–macrophage mediated tumoricidal activity. *Blood* **60,** 430–435.

Perri, R. T., Vercellotti, G., McCarthy, J., Vessella, R. L., and Furcht, L. T. (1985). Laminin selectively enhances monocyte–macrophage mediated tumoricidal activity. *J. Lab. Clin. Med.* **105,** 30–35.

Pierschbacher, M. D., and Ruoslahti, E. (1984). The cell attachment activity of fibronectin can be duplicated by small synthetic fragments of the molecule. *Nature (London)* **309,** 30–33.

Pommier, C. G., Inada, S., Fries, L. F., Takahasi, T., Frank, M. M., and Brown, E. J. (1983). Plasma fibronectin enhances phagocytosis of opsonized particles by human peripheral blood monocytes. *J. Exp. Med.* **157,** 1844.

Pytela, R., Pierschbacher, M. D., and Ruoslahti, E. (1985). Identification and isolation of a 140Kd cell surface glycoprotein with properties expected of a fibronectin receptor. *Cell* **40,** 191–198.

Ravetch, J. V., Luster, A. D., Weinshank, R., Kochan, J., Pavlovec, A., Portnoy, D. A., Hulmes, J., Pan, Y. C. E., and Unkeless, J. C. (1986). Structural heterogeneity and functional domains of murine immunoglobulin G Fc receptors. *Science* **234,** 718.

Roberts, D. D., Rao, C. N., Liotta, L. A., Gralnick, H. R., and Ginsburg, V. (1986). Comparison of the specificities of laminin, thrombospondin, and von Willebrand factor for binding to sulfated glycolipids. *J. Biol. Chem.* **261,** 6872–6877.

Sanchez-Madrid, F., Nagy, J. A., Robbins, E., Simon, P., and Springer, T. A. (1986). A human leukocyte differentiation antigen family with distinct α subunits and a common β subunit: The lymphocyte function associated antigen (LFA-1), the C3bi complement receptor (OKM1/Mac-1), and the p150,195 molecule. *J. Exp. Med.* **158,** 1785–1803.

Sant, A., Cullen, S. E., and Schwartz, B. D. (1985). Biosynthetic relationships of the chondroitin sulfate proteoglycan with Ia and invariant chain glycoproteins. *J. Immunol.* **135,** 416–422.

Santoni, A., Scarpa, S., Testi, R., Morrone, S., Piccoli, M., Frati, L., and Modesti, A. (1987). Natural killer cells synthesize laminin. Submitted for publication.

Schreiber, R. D., Pace, J. L., Russell, S. W. Altman, A., and Katz, D. H. (1983). Macrophage activating factor produced by a T cell hybridoma: Physiochemical and biosynthetic resemblance to γ-interferon. *J. Immunol.* **131,** 826.

Silverstein, S. C., Steinman, R. M., and Cohn, Z. A. (1977). Endocytosis. *Annu. Rev. Biochem.* **46,** 699–722.

Stevens, R. L. (1986). Secretory granule proteolglycans of mast cells and natural killer cells. *In* "Functions of the Proteoglycans" (V. Hascall, ed.), pp. 272–285. Pitman, London.

Takasu, Y., Hasumi, F., and Mori, F. (1982). Biosynthesis of glycosaminoglycans in peritoneal macrophages from the guinea pig. *Biochim. Biophys. Acta* **716,** 316–323.

Takemura, R., and Werb, Z. (1984). Secretory products of macrophages and their physiological functions. *Am. J. Physiol.* **246** (*Cell Physiol.* **15**), C1–C9.

Tamkun, J. W., DeSimone, D. W., Fonda, D., Patel, R. S., Buck, C., Horwitz, A. F., and Hynes, R. O. (1986). Structure of integrin, a glycoprotein involved in the transmembrane linkage between fibronectin and actin. *Cell* **46,** 271–282.

Terranova, V. P., DiFlorio, R., Hujanen, E. S., Lyall, R. M., Liotta, L. A., Thorgeirsson, U., Siegal, G. P., and Schiffmann, E. (1985). Laminin promotes rabbit neutrophil motility and attachment. *J. Clin. Invest.* **77,** 1180–1186.

Timpl, R., Rohde, H., Robey, P. G., Rennard, S. I., Foidart, J. M., and Martin, G. R. (1979). Laminin—A glycoprotein from basement membranes. *J. Biol. Chem.* **254,** 9933–9937.

Trinkaus, J. P. (1984). "Cells into Organs. The Forces That Shape the Embryo," Prentice-Hall, Englewood Cliffs, New Jersey.

VanDeWater, L. (1985). Phagocytosis. *In* "Plasma Fibronectin. Structure and Function" (J. McDonagh, ed.), pp. 175–196. Dekker, New York.

VanDeWater, L., Destree, A. T., and Hynes, R. O. (1983). Fibronectin binds to some bacteria but does not promote their uptake by phagocytic cells. *Science* **220,** 201–204.

Van Furth, R. (1980). "Mononuclear Phagocytes. Functional Aspects. Parts I and II." Martinus Nijhoff, the Hague.

Vartio, T., Hovi, T., and Vaheri, A. (1982). Human macrophages synthesize and secrete a major 95,000-dalton gelatin-binding protein distinct from fibronection. *J. Biol. Chem.* **257,** 8862–8866.

Villiger, B., Kelley, D. G., Engleman, W., Kuhn, III, C. and McDonald, J. A. (1981). Human alveolar machophage fibronectin: Synthesis, secretion, and ultrastructural localization during gelatin-coated latex particle binding. *J. Cell Biol.* **90,** 711–720.

Wewer, U. M., Liotta, L. A., Jaye, M., Ricca, G. A., Drohan, W. N., Claysmith, A. P., Rao, C. N., Wirth, P., Coligan, J. E., Albrechtsen, R., Mudryj, M., and Sobel, M. E. (1986). Altered levels of laminin receptor mRNA in various human carcinoma cells that have different abilities to bind laminin. *Proc. Natl. Acad. Sci. U.S.A.* **83,** 7137–7141.

Wicha, M. S., and Huard, T. K. (1983). Macrophages express cell surface laminin. *Exp. Cell Res.* **143,** 475–479.

Wicha, M. S., Malinoff, H., and Huard, T. K. (1983). Laminin receptors on macrophages— a mechanism for metastatic tumor cell recognition. *Proc. Am. Asso. Cancer. Res.* **24,** 35.

Wright, S. D., and Meyer, B. C. (1985). Fibronectin receptor of human macrophages recognizes the sequence Arg-Gly-Asp-Ser. *J. Exp. Med.* **162,** 763–767.

Wright, S. D., Craigmyle, L. S., and Silverstein, S. C. (1983). Fibronectin and serum amyloid P component stimulate C3b- and C3bi-mediated phagocytosis in cultured human monocytes. *J. Exp. Med.* **158,** 1338.

Zhu, B. C. R., Fisher, R. F., Pande, H., Calacay, J., Shively, J., and Laine, R. A. (1984). *J. Biol. Chem.* **259,** 3962–3970.

IV

Transfer across Bacterial Membranes

14

Synthesis and Export of Lipoproteins in Bacteria

MIGUEL REGUE AND HENRY C. WU

I. INTRODUCTION

Among all the proteins synthesized in bacteria, there is a group whose final localization is extracytoplasmic. There are four extracytoplasmic

587

compartments in gram-negative bacteria: the inner membrane, the periplasmic space, the outer membrane, and the extracellular medium. There are only two such extracytoplasmic compartments in gram-positive bacteria, the cytoplasmic membrane and the extracellular medium. The extracytoplasmic or secreted proteins are synthesized in the cytoplasm and through a complex series of steps transverse the cytoplasmic membrane to reach their ultimate cellular locations (Oliver, 1985).

Much of our understanding of the secretion of proteins in bacteria has been obtained through the study of the secretion of periplasmic and outer membrane proteins of *Escherichia coli*. The murein lipoprotein is one of the major outer membrane proteins in *E. coli*, and much effort has been directed toward the study of this protein because of its unique structure, its abundance, and its usefulness as a model of protein secretion. The progress in this field during the last few years has been facilitated by the discovery of the antibiotic globomycin, which specifically blocks the processing of prolipoprotein (Hussain *et al.*, 1980), and by the realization that the murein lipoprotein is a prototype of a group of membrane lipoproteins widely distributed among both gram-negative and gram-positive bacteria (Nielsen *et al.*, 1981; Ichihara *et al.*, 1981; Huang *et al.*, 1983) and *Mycoplasma* (Dahl *et al.*, 1983). In this chapter we summarize current knowledge about the pathway for the biosynthesis and secretion of lipoproteins in bacteria.

II. LIPOPROTEIN STRUCTURE

The common feature of all lipoproteins is the unique amino-terminal residue of the mature protein. This structure, first shown to be present in murein lipoprotein by Hantke and Braun (1973), consists of a lipid-modified cysteine, with a glycerol moiety covalently attached to the sulfhydryl group of cysteine by a thioether linkage, and two fatty acid molecules ester-linked and one amide-linked to the glycerylcysteine. The murein lipoprotein is located in the outer membrane with about one-third of the lipoprotein covalently linked to the peptidoglycan through the carboxy-terminal lysine (Braun and Rehn, 1969), while two-thirds remain free (Inouye *et al.*, 1972). The facts that *E. coli* murein lipoprotein is nonessential (Hirota *et al.*, 1977) and that *E. coli* cells containing mutant lipoproteins show increased ethylenediaminetetraacetic acid (EDTA) sensitivity and outer membrane blebs (Yem and Wu, 1978) suggest a structural function for this lipoprotein, i.e., the anchorage of the outer membrane to the peptidoglycan via the murein-bound lipoprotein. This unique modification of the amino-terminal cysteine with glycerol and fatty acids has been demonstrated in other lipoproteins, such as *Bacillus licheniformis* penicillinase (Lai *et al.*, 1981; Nielsen *et al.*, 1981; Smith *et al.*, 1981) and new *E. coli* lipoproteins including lipoprotein-28 (Yu *et al.*, 1986).

III. BIOSYNTHESIS

The discovery of a precursor form of the murein lipoprotein, the prolipoprotein, with an extra 20 amino acid residues at the amino terminus while lacking glyceride and fatty acids (Inouye *et al.*, 1977b) indicates the existence of a number of posttranslational modification and processing reactions prior to the translocation of murein lipoprotein into the outer membrane.

A. Modification Reactions

Current understanding of the biosynthesis of the lipid moiety of the lipoprotein, as revealed by *in vivo* studies, can be summarized as follows: (1) the glycerol moiety of the mature lipoprotein is derived from the nonacylated glycerol of phosphatidylglycerol (Chattopadhyay and Wu, 1977; Chattopadhyay *et al.*, 1979); (2) both ester- and amide-linked fatty acids in lipoprotein are transferred from phospholipids to lipoproteins (Lai *et al.*, 1980); (3) the finding of a lipoprotein structural gene mutant (*lpp*C77) deficient in ester-linked fatty acids but with a normal amount of glycerylcysteine (Inouye *et al.*, 1977a; Rotering and Braun, 1977) suggests a sequential transfer of glycerol and fatty acid groups. This same modification pathway appears to be used by *Mycoplasma capricolum* (Dahl *et al.*, 1983; Dahl and Dahl, 1984).

B. Processing Reaction

The discovery that the cyclic antibiotic globomycin blocks the processing of prolipoprotein (Inukai *et al.*, 1978) and leads to the accumulation of glyceride-modified prolipoprotein, instead of unmodified prolipoprotein (Hussain *et al.*, 1980), was the first indication that modification reactions may precede the processing reaction. This notion was supported by the transient detection of unmodified prolipoprotein followed by modified prolipoprotein and mature lipoprotein in pulse–chase experiments (Tokunaga and Wu, 1984). *In vitro* studies have shown that unmodified prolipoprotein cannot be cleaved by either signal peptidase I (M13 procoat protein signal peptidase) or signal peptidase II (prolipoprotein signal peptidase) (Tokunaga *et al.*, 1982a; Dev and Ray, 1984). In addition, Tokunaga *et al.*, (1982b) showed that the processing of prolipoprotein by

signal peptidase II requires prior modification, whereas Hussain *et al.*
(1982) and Inukai *et al.* (1984) demonstrated that signal peptidase II is able
to fully process modified prolipoprotein to give signal peptide and apolipoprotein containing glyceride cysteine at its amino terminus but lacking
an amide-linked fatty acid. Taken together, these results indicate that the
glycerol and ester-linked fatty acid modifications precede and are in fact a
prerequisite for the processing of modified prolipoprotein by signal peptidase II. The overall pathway for the lipoprotein biogenesis is schematically shown in Fig. 1.

IV. DISTRIBUTION OF LIPOPROTEINS AMONG BACTERIA

The number of known lipoproteins is increasing rapidly, indicating that
the modifying and processing enzymes analogous to those found in *E. coli*
are widely distributed among bacteria.

A. *Escherichia coli* Non-Murein Lipoproteins

The addition of globomycin to growing *E. coli* cells inhibits the processing of modified murein prolipoprotein and leads to the accumulation of
lipid-modified murein prolipoprotein as well as the accumulation of precursors of other minor lipid-modified proteins (Ichihara *et al.*, 1981). The
accumulation of modified non-murein prolipoproteins is also observed in
a conditionally lethal signal peptidase II mutant after shifting to the nonpermissive condition (Yamagata *et al.*, 1982). Some of these so-called
new lipoproteins are located in the outer membrane while others are
found in the inner membrane (Ichihara *et al.*, 1981). These lipoproteins
are not immunologically related to murein lipoprotein (Ichihara *et al.*,
1981) nor is there extensive DNA sequence homology between genes
encoding these new lipoproteins and those encoding the *lpp* gene (Nakamura and Inouye, 1979; Nakamura *et al.*, 1979). One of the new lipoproteins, lipoprotein-28, has been characterized and its DNA sequence determined (Yu *et al.*, 1986). Lipoprotein-28, which has a molecular weight of
27,800, is located in the inner membrane; its function, however, is not yet
known (Yu *et al.*, 1986).

B. *Bacillus licheniformis* Penicillinase

Bacillus licheniformis penicillinase has been shown to exist in three
forms, one linked to the cytoplasmic membrane and two extracellular
forms, ExoF and ExoS (Yamamoto and Lampen, 1976, Simons *et al.*,
1978). The amino acid sequence of penicillinase, as deduced from the

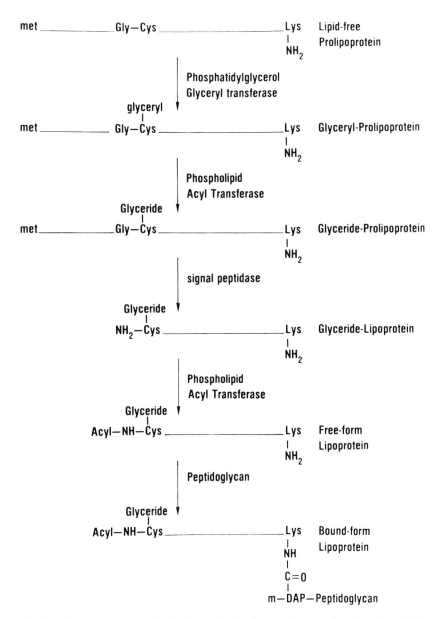

Fig. 1. Postulated pathway for the biosynthesis of murein lipoprotein in *E. coli.* *m*-DAP, *meso*-Diaminopimelic acid.

nucleotide sequence of its structural gene *penP*, has revealed the presence of a typical signal peptide containing a tetrapeptide sequence, -Leu-Ala-Gly-Cys-, identical to that found in the vicinity of the modification and processing site of murein prolipoprotein.

This finding first suggested the possibility of *B. licheniformis* penicillinase being a lipoprotein. Indeed, it was shown that *B. licheniformis* penicillinase synthesized both in *E. coli* and in *B. licheniformis* is modified with glycerol and fatty acids (Lai *et al.*, 1981; Nielsen *et al.*, 1981; Smith *et al.*, 1981). Hayashi and Wu (1983) demonstrated a globomycin-sensitive processing of the *B. licheniformis* prepenicillinase.

C. Other Lipoproteins

Lipoproteins are widely distributed among bacteria. They are found in Enterobacteriaceae, such as *Serratia marcescens* (Nakamura and Inouye, 1980), *Proteus mirabilis* (Ching and Inouye, 1986), *Morganella morganii* (Huang *et al.*,1983), *Erwinia amylovora* (Yamagata *et al.*, 1981), and *Klebsiella pneumoniae* (Chapon and Raibaud, 1985). They are also been found in gram-positive bacteria such as *B. licheniformis* and *Staphylococcus aureus* (McLaughlin *et al.*, 1981). Lipoproteins have also been found to be encoded by plasmids, such as the F factor (Ogata *et al.*, 1982) and several bacteriocinogens (Oka *et al.*, 1979; Hakkaart *et al.*, 1981; Watson *et al.*, 1984; Cole *et al.*, 1985).

Thus far all lipoproteins for which the DNA sequence is known share a common tetrapeptide -Leu-X-Y-Cys- around the modification and processing site; X and Y are usually small neutral amino acids (Fig. 2). The functions of these lipoproteins are widely different, ranging from the structural role of murein lipoproteins to the enzymatic activities of membrane-bound penicillinase in *B. licheniformis* and extracellular pullulanase in *K. pneumoniae*. Although lipoproteins have different structures and functions, the existence of the common tetrapeptide containing the modification and processing site suggests that they have evolved to preserve a common feature in their signal sequence. From an evolutionary point of view, it would be interesting to know if lipoproteins exist among the archaebacteria.

V. SIGNAL PEPTIDASES

Most, if not all, of the proteins secreted through the inner membrane are synthesized as precursor forms containing a signal peptide. The signal peptides in all secreted proteins have common general characteristics: (1)

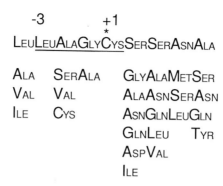

Fig. 2. Consensus sequence for the modification and processing of prolipoproteins in bacteria. The consensus sequence is derived from the signal sequences of lipoprotein precursors in bacteria, including murein lipoproteins (*E. coli, Serratia marcescens, Erwinia amylovora, Morganella morganii*, and *Proteus mirabilis*); new lipoprotein-28 in *E. coli*; penicillin-binding protein 3 (PBP-3) in *E. coli*; penicillinases (*B. licheniformis* and *Staphylococcus aureus*); TraT protein in *E. coli*; lysis proteins encoded by pCloDF13, pColE1, pColE2, pColE3, and pColA; and pullulanase from *Klebsiella pneumoniae*. Symbols: *, glyceride modification of cysteine; +1 and −3 refer to the amino acid residues from the cleavage site.

an amino terminus rich in basic residues; (2) a hydrophobic core region of 10–12 amino acids; and (3) a processing site at the junction between the signal peptide and the mature protein. The signal peptides are essential for the secretion process that leads to the translocation of proteins across the inner membrane. Signal peptidases are the proteolytic enzymes whose function is to cleave off the signal peptide from the mature protein. Cleavage or processing of the signal peptide is not necessary for the transmembrane export of secreted proteins (Lin *et al.*, 1978; Meek *et al.*, 1982). Probably the main role of the signal peptidase is to allow a rapid release of the mature protein from the inner membrane. Once the mature protein is free from the cytoplasmic membrane, it folds into its proper conformation and reaches its final localization, either the periplasmic space or the outer membrane. Additional steps are probably necessary for the secretion of proteins across the outer membrane.

A. Signal Peptidase I

Signal peptidase I (or M13 procoat protein leader peptidase) was the first signal peptidase to be discovered in *E. coli*. In addition to M13 procoat protein, signal peptidase I processes precursor forms of several periplasmic and outer membrane proteins (Ito, 1982; Wolfe *et al.*, 1982).

Signal peptidase I cannot process lipoprotein precursors (Tokunaga *et al.*, 1982b; Dev and Ray, 1984), except in special cases as discussed below.

Signal peptidase I contains 323 amino acids and has a molecular weight of 37,000 (Wolfe *et al.*, 1983). This enzyme is localized in about equal amounts in the inner and outer membranes (Zwizinski *et al.*, 1981). On the other hand, about 90% of signal peptidase I activity is found in the inner membrane when the enzyme level is amplified 75-fold (Wolfe *et al.*, 1983). The accessibility of signal peptidase I to trypsin digestion in spheroplasts (Wolfe *et al.*, 1983) suggests a transmembrane localization, with the bulk of the protein facing the periplasmic space. By cloning the *lep* gene under the control of an inducible *ara* promoter and by replacement insertion in the chromosome, it has been shown that the *lep* gene is essential to the viability of *E. coli* cells (Dalbey and Wickner, 1985). A statistical analysis of the cleavage sites of secreted proteins processed by a signal peptidase such as signal peptidase I indicates a preference for Ala-X sites (Von Heijne, 1983).

B. Signal Peptidase II

There is much evidence suggesting that signal peptidase II (or prolipoprotein signal peptidase) is specific for prolipoprotein processing (Tokunaga *et al.*, 1984a): (1) signal peptidase II cleaves modified prolipoprotein precursors but is not able to process unmodified precursors (Tokunaga *et al.*, 1982b; Dev and Ray, 1984); (2) signal peptidase II activity is specifically inhibited by globomycin (Inukai *et al.*, 1978), and the kinetics of inhibition suggests that globomycin binds to signal peptidase II in a noncompetitive manner (Dev *et al.*, 1985); (3) and *E. coli* mutant strain with temperature-sensitive signal peptidase II activity accumulates modified prolipoprotein under the nonpermissive condition (Yamagata *et al.*, 1982).

Signal peptidase II consists of 164 amino acid residues and has a molecular weight of 18,144 (Innis *et al.*, 1984; Yu *et al.*, 1984). There are four hydrophobic segments in this protein, each containing at least 10 hydrophobic amino acids with adjacent charged residues. These hydrophobic segments in β-sheet conformations are long enough to span the cytoplasmic membrane. A hypothetical model for the transmembrane structure of signal peptidase II (Innis *et al.*, 1984) is shown in Fig. 3. *In vivo* studies have shown that signal peptidase II is exclusively localized in the inner membrane (Tokunaga *et al.*, 1984b) and that a hybrid protein consisting of isoleucyl-tRNA synthetase–signal peptidase II is localized in the cell en-

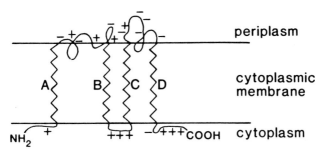

Fig. 3. Transmembrane structure of signal peptidase II according to Innis *et al.* (1984). The segments A, B, C, and D correspond to hydrophobic segments, and + and − symbols indicate the position of charged amino acids.

velope (Tokunaga *et al.*, 1985), probably owing to the high overall hydrophobicity of signal peptidase II. In addition to this hybrid protein encoded by the *ileS–lsp* gene, two other *lsp* mutant have been constructed *in vitro*, resulting in the removal and substitution of the three positively charged amino acid residues at the carboxy terminus of the enzyme (Table I). Both mutations decrease the signal peptidase II activity by 70–90% (Innis *et al.*, 1984), suggesting an important role for the positively charged carboxy terminus, for example, a putative interaction of this domain with the negatively charged inner face of the cytoplasmic membrane (Fig. 3).

C. Signal Pepidase III

A third signal peptidase has been recently discovered in *E. coli*. Signal peptidase III, which has an apparent molecular weight of 60,000, is able to process both the wild-type and mutant maltose-binding protein precursors more efficiently than signal peptidase I (Ray *et al.*, 1986).

In addition to the signal pepidases, two more proteases have been found to be involved in the hydrolysis of signal peptides. Signal peptide peptidase and protease IV hydrolyze the signal peptide from M13 procoat protein (Zwizinski and Wickner, 1980) and the prolipoprotein signal peptide (Pacaud, 1982; Ichihara *et al.*, 1984), respectively, after their release from the precursor proteins. Recently, Novak *et al.* (1986) have shown that most of the signal peptide degradative activity is due to two cytoplasmic proteases with apparent molecular weights of 68,000 and 81,000. These proteases probably play an important role in preventing the possible competition for the limited number of export sites between the cleaved signal peptides and the precursor proteins.

TABLE I

Carboxy-Terminal Sequences of Wild-Type and Mutant *lsp* Gene Products[a]

Plasmid								Sequence								
	154	155	156	157	158	159	160	161	162	163	164					
pMT521 (wt)	Glu	Gly	Phe	Leu	Pro	Ser	Arg	Ala	Lys	Lys	Gln—COOH					
pSYC902	Glu	Gly	Phe	Leu	Pro	Ser	Ser	Arg	Ala	Ala	Leu	Gly	Pro	Gly	His	Gly	Cys	Ala—COOH
pSYC988	Glu	Gly	Phe	Leu	Pro	Ser	Ser—COOH									

[a] Construction and characterization of the *lsp* mutants are described in Innis *et al.* (1984).

VI. GENOMIC ORGANIZATION OF THE *lsp* GENE

Using two different approaches, the *lsp* (signal peptidase II) gene has been cloned (Tokunaga *et al.*, 1983; Yamagata *et al.*, 1983a), mapped (Regue *et al.*, 1984; Yamagata *et al.*, 1983b), and its DNA nucleotide sequence determined (Innis *et al.* 1984; Yu *et al.*, 1984). The DNA sequence data confirmed the location of the *lsp* gene adjacent to the *ileS* (isoleucyl-tRNA synthetase) gene and revealed an unexpected finding: the stop codon of the *ileS* gene overlaps with the initiation codon of the *lsp* gene, suggesting that these two genes with no apparent functional relationship may share a common promoter. Deletion and *Tn5* insertion analysis of a plasmid containing both *ileS* and *lsp* genes supported the hypothesis of a polycistronic mRNA for the *ileS* and *lsp* genes and demonstrated the existence of a weak internal promoter for the *lsp* gene near the end of the *ileS* coding region (Tokunaga *et al.*, 1985). The subsequent DNA sequence analysis of the *ileS* upstream region led to the discovery of a new gene (Kamio *et al.*, 1985). This *x* gene is located 45 base pairs (bp) upstream of the *ileS* gene and codes for a cytoplasmic protein containing 312 amino acids and having a molecular weight of 34,716. The function of the X protein in unknown (Kamio *et al.*, 1985).

The three genes *x*, *ileS*, and *lsp* are transcribed in the same direction without obvious transcriptional stop signals in between (Kamio *et al.*, 1985), indicating that they constitute an operon. RNA polymerase protection experiments have demonstrated the existence of a promoter upstream of the *x* gene, while another predicted promoter located between *x* and *ileS* genes was not protected by RNA polymerase (Kamio *et al.*, 1985). Definitive proof of an *in vivo* transcriptional unit including both *ileS* and *lsp* genes was recently otained by Miller and Wu (1987). Northern blotting hybridization analysis of mRNA has shown that these two genes are in fact cotranscribed on common high molecular weight mRNAs of 5,000–6,000 nucleotides.

The genomic organization of the *lep* gene (encoding signal peptidase I) is somewhat similar to that of the *lsp* gene, since the *lep* gene is separated from its promoter by an upstream gene coding for a 76,000 molecular weight inner membrane protein (March and Inouye, 1985). The function of this upstream gene is not essential and is not related to protein secretion (Dibbs and Wolfe, 1986). The fact that both *lsp* and *lep* are part of operons with genes not related to the secretion process raises questions about how the cell coregulates the expression of these genes and about possible functional interactions of signal peptidases with other cellular functions.

VII. PROLIPOPROTEIN STRUCTURAL GENE MUTANTS
AFFECTING MODIFICATION AND/OR PROCESSING

Our current understanding of the roles that different domains in prolipoprotein play in the modification and processing reactions has emerged from the analysis of structural gene mutations obtained by both *in vivo* or *in vitro* methods in the *E. coli* murein lipoprotein *(lpp)* gene and in the *B. licheniformis* penicillinase *(penP)* gene. These studies have shown the expected crucial role played by the tetrapeptide consensus sequence (Fig. 2), as well as the importance of (1) a positively charged amino-terminal segment of the signal peptide, (2) the absence of charged residues in the hydrophobic region of the signal peptide, and (3) a unique local as well as overall conformation of the prolipoprotein allowing the potential cleavage sites to be recognized by the modification/processing enzymes.

A. Amino-Terminal Basic Domain in Signal Peptide

There are two positively charged amino acids in the amino-terminal region of murein prolipoprotein signal peptide. Sequential changes of the charge in this region from +2 to 0 by oligonucleotide site-directed mutagenesis lead to a decrease in the rate of prolipoprotein synthesis (Inouye *et al.*, 1982). The introduction of a net negative charge in this region results in an accumulation of soluble unmodified prolipoprotein that can be converted to mature lipoprotein, albeit at a very slow rate (Vlasuk *et al.*, 1983). These results indicate an important role of the positively charged amino-terminal domain in prolipoprotein signal peptide in the synthesis and translocation of prolipoprotein.

B. Central Hydrophobic Domain in Signal Peptide

The introduction of charged amino acids into the central hydrophobic region leads to an accumulation of unmodified prolipoprotein as shown in mutants *lpp*D9 (Pollitt *et al.*, 1985b) and *lpp*D14 (Wu *et al.*, 1977). Pseudorevertants partially restoring modification and processing were found to contain substitutions of Asp14 → Asn14 *(lpp*N14) and Thr16 → Ile16 *(lpp*D14I16), respectively (Giam *et al.*, 1984; Tokunaga and Wu, 1984). The importance of residues 14 to 18 is further stressed by the isolation of mutants *lpp*ΔG14 (Inouye *et al.*, 1984), *lpp*A15, and *lpp*A15A16 (Vlasuk *et al.*, 1984), all of which accumulate both modified and unmodified prolipoprotein. These results suggest important roles of the predicted coil region in residues 14 and 15 and the β-sheet structure in residues 16 to 18 in the modification and processing reactions (Giam *et al.*,

1984). Further deletion analysis has shown a high degree of flexibility in this domain, since mutant $lpp\Delta$V7ΔG9ΔL13ΔG14 is still functional; unexpectedly, a higher level of lipoprotein synthesis was observed in mutant $lpp\Delta$G9ΔL13ΔG14 than in wild-type lpp or mutants $lpp\Delta$V7ΔG9ΔG14 and $lpp\Delta$V7ΔG9ΔL13ΔG14 (Pollitt et al., 1984a).

C. Cleavage Region

The cleavage region is contained in the common tetrapeptide -Leu-Ala-Gly-Cys-. Since the cysteine residue is necessary for the modification process, it is not surprising that mutant lppG21 accumulates unmodified prolipoprotein (Inouye et al., 1983a). The residue Gly20 appears to be also important since its deletion results in the accumulation of unmodified prolipoprotein (Inouye et al., 1983b). Substitution of the Gly20 residue with amino acid residues of increasing side-chain size has shown that serine is the upper limit in the size of the side chain for the residue adjacent to the cleavage site; this size limit is analogous to that found in secreted proteins processed by signal peptidase I (Pollitt et al., 1986).

The predicted β turn from residues 19 to 24 (Giam et al., 1984) is another important structural feature. This is shown in mutant lppI23I24, which is defective in modification and processing because of a decrease in β-turn structure in this region. An additional substitution of Gly20 by Ala20, further slows the modification and processing kinetics of mutant lppA20I23I24 (Inouye et al., 1986).

The construction of a hybrid erythromycin–murein lipoprotein fusion, in which the common tetrapeptide sequence in internalized, leads to the formation of mature lipoprotein and two erythromycin–murein lipoprotein hybrid proteins containing 45 and 22 amino acid long extensions preceding the amino terminus of prolipoprotein (Hayashi et al., 1985). In both E. coli and B. subtilis, the precursor forms of the hybrid proteins are modified and processed to mature lipoprotein in vivo (Hayashi et al., 1985), indicating that the prolipoprotein signal sequence can be recognized even when it is not located at the amino terminus.

D. Other Structural Gene Mutations with No Modifiable Cysteine but Allowing Alternative Processing Sites

A mutant with a five amino acid deletion (Ala23-Leu24-Ala25-Gly26-Cys27) in the signal peptide of B. licheniformis prepenicillinase, which spans the modification and processing site, contains about 90% unmodified prepenicillinase and about 8% glyceride-modified but unprocessed prepenicillinase (Hayashi et al, 1984). The modification presumably occurs at Cys21, indicating that, in the absence of the normal modification site at Cys27, the modifying enzymes can recognize, albeit much less efficiently, the sequence -Leu18-Phe19-Ser20-Cys21-. Neither signal peptidase II nor signal peptidase I is able to process this Cys21-modified prepenicillinase.

The lack of processing of unmodified prolipoprotein in mutant *lpp*G21 is somewhat surprising as potential cleavage sites Ala-X for signal peptidase I are present in the mutant prolipoprotein. By contrast, a hybrid protein containing the *lpp*G21 mutant signal sequence plus nine amino acids joined to the mature TEM–β-lactamase is not modified but is processed in a globomycin-insensitive manner between Ala25 and Lys26 (Ghrayeb *et al.*, 1985). The same construction using the wild-type *lpp* is modified and processed at Cys21 to form lipo-β-lactamase; as expected, this processing is inhibited by globomycin (Ghrayeb and Inouye, 1984). These results indicate that the alternative processing site Ala25–Lys26, while present in both wild-type and mutant hybrid proteins, is only recognized when the normal modification site is absent. On the other hand, the same site, which is present in the *lpp*G21 mutant prolipoprotein as well, is not accessible to the alternative processing enzyme.

The substitution of Cys27 → Ser27 in mutant *penP*Ser27 results in the synthesis of unmodified but processed forms of penicillinases, with the processing taking place between Ala28 and Asn29 and between Ala34 and Ser35 (Hayashi *et al.*, 1984). Both processed forms of penicillinase are located in the periplasm. The additional substitution of Ala28 → Pro28 (mutant *penP*Ser27Pro28) leads to the synthesis of two unmodified but processed penicillinases, with the processing taking place between Ala34 and Ser35 as in *penP*Ser27, and at a new site between Ala25 and Ser26 (Hayashi *et al.*, 1986). The addition of Pro-Asp between Ala23 and Leu24 [mutant *penP*Ser27,23'(Pro-Asp)24'] results in the complete abolishment of any processing of the mutant prepenicillinase (Hayashi *et al.*, 1986).

These results indicate that, in the absence of the normal modification and processing site for lipoproteins, the precursor forms of mutant lipoproteins can be processed at alternative sites, as long as there are potential cleavage sites of Ala-X in the signal sequences and the overall conformation of the mutant prolipoproteins makes these sites accessible to alternative signal peptidases.

VIII. EARLY STEPS IN LIPOPROTEIN SECRETION

The process leading to the secretion of proteins is very complex and requires the orderly and interrelated functions of many cellular components. So far we have considered the role of signal peptide and signal peptidases in the secretion of lipoproteins. Using genetic techniques, based on the selection of extragenic suppressors of signal sequence mutations and defective export of hybrid proteins, it has been possible to identify several *E. coli* loci whose functions are necessary for the secretion of several exported proteins. This discovery led to the notion of a general secretory machinery for exported proteins (Oliver, 1985). Since lipoproteins as a group require several modification reactions and have their own signal peptidase, it is not clear, a priori, whether or not they utilize the same initial secretory pathway as the nonlipoproteins. Studies of the fate of murein lipoprotein in conditionally lethal mutants impaired in general secretion have shown that lipoprotein secretion is affected in *secA*, *secY* (Hayashi and Wu, 1985), *secD* and *secC* mutants (Watanabe *et al.*, 1988). Since the SecY protein appears to interact directly with signal peptides, and the *secA* and *secC* gene products may participate in the coupling of protein synthesis and secretion (Oliver, 1985), one can conclude that both prolipoproteins and other precursor proteins share common early steps in protein secretion.

IX. CONCLUDING REMARKS

The progress made in this field has led to the recognition of the widespread distribution of membrane lipoproteins in bacteria. The biosynthetic pathway for murein lipoprotein has been elucidated, and the lipoprotein-specific signal peptidase has been characterized. The secretion of lipoprotein appears to share the same early steps as those of nonlipoprotein exported proteins, the diverging point being the lipoprotein modification and processing reactions. The roles of different prolipoprotein domains in modification, processing, and export are being explored through the use of structural gene mutants of *lpp* and *penP* genes. Future work will be needed for the isolation and characterization of the modifying enzymes and for a comparative study of these enzymes among gram-positive and gram-negative bacteria.

ACKNOWLEDGMENTS

This work was supported by Public Health Service Grant GM-28811 from the National Institute of General Medical Sciences and American Heart Association Grant 84-606.

REFERENCES

Braun, V., and Rehn, K. (1969). Chemical characterization, spatial distribution and function of a lipoprotein (murein lipoprotein) of the *E. coli* cell wall. The specific effect of trypsin on the membrane structure. *Eur. J. Biochem.* **10**, 426–438.

Chapon, C., and Raibaud, O. (1985). Structure of two divergent promoters located in front of the gene encoding pullulanase in *Klebsiella pneumoniae* and positively regulated by *malT* product. *J. Bacteriol.* **164**, 639–645.

Chattopadhyay, P. K., and Wu, H. C. (1977). Biosynthesis of the covalently linked diglyceride in the murein lipoprotein of *Escherichia coli*. *Proc. Natl. Acad. Sci. U.S.A.* **74**, 5318–5322.

Chattopadhyay, P. K., Lai, J. S., and Wu, H. C. (1979). Incorporation of phosphatidylglycerol into murein lipoprotein in intact cells of Salmonella typhimurium by phospholipid vesicle fusion. *J. Bacteriol.* **137**, 309–312.

Ching, G., and Inouye, M. (1986). Expression of the *Proteus mirabilis* lipoprotein gene in *Escherichia coli*. Existence of tandem promoters. *J. Biol. Chem.* **261**, 4600–4606.

Cole, S. T., Saint-Joanis, B., and Pugsley, A. P. (1985). Molecular characterization of the colicin E-2 operon and identification of its products. *Mol. Gen. Genet.* **198**, 465–472.

Dahl, C. E., and Dahl, J. S. (1984). Phospholipids as acyl donors to membrane proteins of *Mycoplasma capricolum*. *J. Biol. Chem.* **259**, 10771–10776.

Dahl, C. E., Dahl, J. S., and Bloch, K. (1983). Proteolipid formation in *Mycoplasma capricolum*. *J. Biol. Chem.* **258**, 11814–11818.

Dalbey, R. E., and Wickner, W. (1985). Leader peptidase catalyzes the release of exported proteins from the outer surface of the *Escherichia coli* plasma membrane. *J. Biol. Chem.* **260**, 15925–15931.

Dev, I. K., and Ray, P. H. (1984). Rapid assay and purification of a unique signal peptidase that processes the prolipoprotein from *Escherichia coli*. *J. Biol. Chem.* **259**, 11114–11120.

Dev, I. K., Harvey, R. J., and Ray, P. H. (1985). Inhibition of prolipoprotein signal peptidase by globomycin. *J. Biol. Chem.* **260**, 5891–5894.

Dibbs, N. J., and Wolfe, P. B. (1986). *lep* operon proximal gene is not required for growth or secretion by *Escherichia coli*. *J. Bacteriol.* **166**, 83–87.

Ghrayeb, J., and Inouye, M. (1984). Nine amino acid residues at the NH$_2$-terminal of lipoprotein are sufficient for its modification, processing and localization in the outer membrane of *Escherichia coli*. *J. Biol. Chem.* **259**, 463–467.

Ghrayeb, J., Lunn, C. A., Inouye, S., and Inouye, M. (1985). An alternate pathway for the processing of prolipoprotein signal peptide in *Escherichia coli*. *J. Biol. Chem.* **260**, 10961–10965.

Giam, C. Z., Chai, T., Hayashi, S., and Wu, H. C. (1984). Prolipoprotein modification and processing in *Escherichia coli*. A unique secondary structure in prolipoprotein signal sequence for the recognition by glyceryl transferase. *Eur. J. Biochem.* **141**, 331–337.

Hakkaart, M., Veltkamp, E., and Nijkamp, H. J. J. (1981). Protein H encoded by plasmid Clo DF13 involved in lysis of the bacterial host. I. Localization of the gene and identification and subcellular localization of the gene H product. *Mol. Gen. Genet.* **183**, 318–325.

Hantke, K., and Braun, V. (1973). Covalent binding of lipid to protein. Diglyceride and amide-linked fatty acid at the N-terminal end of the murein lipoprotein of the *Escherichia coli* outer membrane. *Eur. J. Biochem.* **34**, 284–296.

Hayashi, S., and Wu, H. C. (1983). Biosynthesis of *Bacillus licheniformis* penicillinase in *Escherichia coli* and in *Bacillus subtilis*. *J. Bacteriol.* **156**, 773–777.

Hayashi, S., and Wu, H. C. (1985). Accumulation of prolipoprotein in *Escherichia coli* mutants defective in protein secretion. *J. Bacteriol.* **161,** 949–954.

Hayashi, S., Chang, S. Y., Chang, S., and Wu, H. C. (1984). Modification and processing of *Bacillus licheniformis* prepenicillinase in *Escherichia coli*. Fate of mutant penicillinase lacking lipoprotein modification site. *J. Biol. Chem.* **259,** 10448–10454.

Hayashi, S., Chang, S. Y., Chang, S., Giam, C. Z., and Wu, H. C. (1985). Modification and processing of internalized signal sequences of prolipoprotein in *Escherichia coli* and in *Bacillus subtilis*. *J. Biol. Chem.* **260,** 5753–5759.

Hayashi, S., Chang, S. Y., Chang, S., and Wu, H. C. (1986). Processing of *Bacillus licheniformis* penicillinases lacking a lipoprotein modification site in *Escherichia coli*. *J. Bacteriol.* **165,** 678–681.

Hirota, Y., Suzuki, H., Nishimura, Y., and Yasuda, S. (1977). On the process of cellular division in *Escherichia coli:* A mutant of *E. coli* lacking a murein-lipoprotein. *Proc. Natl. Acad. Sci. U.S.A.* **74,** 1417–1420.

Huang, Y. X., Ching, G., and Inouye, M. (1983). Comparison of the lipoprotein gene among Enterobacteriaceae. DNA sequence of *Morganella morganii* lipoprotein gene and its expresssion in *Escherichia coli*. *J. Biol. Chem.* **258,** 8139–8145.

Hussain, M., Ichihara, S., and Mizushima, S. (1980). Accumulation of glyceride-containing precursor of the outer membrane lipoprotein in the cytoplasmic membrane of *Escherichia coli* treated with globomycin. *J. Biol. Chem.* **255,** 3707–3712.

Hussain, M., Ichihara, S., and Mizushima, S. (1982). Mechanism of signal peptide cleavage in the biosynthesis of the major lipoprotein of the *Escherichia coli* outer membrane. *J. Biol. Chem.* **257,** 5177–5182.

Ichihara, S., Hussain, M., and Mizushima, S. (1981). Characterization of new membrane lipoproteins and their precursors of *Escherichia coli*. *J. Biol. Chem.* **256,** 3125–3129.

Ichihara, S., Beppu, N., and Mizushima, S. (1984). Protease IV, a cytoplasmic membrane protein of *Escherichia coli,* has signal peptide peptidase activity. *J. Biol. Chem.* **259,** 9853–9857.

Innis, M. A., Tokunaga, M., Williams, M. E., Loranger, J. M., Chang, S. Y., Chang, S., and Wu, H. C. (1984). Nucleotide sequence of the *Escherichia coli* prolipoprotein signal peptidase (*lsp*) gene. *Proc. Natl. Acad. Sci. U.S.A.* **81,** 3708–3712.

Inouye, M., Shaw, J., and Shen, C. (1972). The assembly of a structural lipoprotein in the evelope of *Escherichia coli*. *J. Biol. Chem.* **247,** 8154–8159.

Inouye, S., Lee, N., Inouye, M., Wu, H. C., Suzuki, H., Nishimura, Y., Iketani, H., and Hirota, Y. (1977a). Amino acid replacement in a mutant lipoprotein of the *Escherichia coli* outer membrane. *J. Bacteriol.* **132,** 308–313.

Inouye, S., Wang, S. S., Sekizawa, J., Halegoua, S., and Inouye, M. (1977b). Amino acid sequence for the peptide extension on the prolipoprotein of the *Escherichia coli* outer membrane. *Proc. Natl. Acad. Sci. U.S.A.* **74,** 1004–1008.

Inouye, S., Soberon, X., Franceschini, T., Nakamura, K., Itakura, K., and Inouye, M. (1982). Role of positive change at the amino-terminal region of the signal peptide for protein secretion across the membrane. *Proc. Natl. Acad. Sci. U.S.A.* **79,** 3438–3441.

Inouye, S., Franceschini, T., Sato, M., Itakura, K., and Inouye, M. (1983a). Prolipoprotein signal peptidase of *Escherichia coli* requires a cysteine residue at the cleavage site. *EMBO J.* **2,** 87–91.

Inouye, S., Hsu, C. P. S., Itakura, K., and Inouye, M. (1983b). Requirement for signal peptide cleavage of *Escherichia coli* prolipoprotein. *Science* **221,** 59–61.

Inouye, S., Vlasuk, G., Hsiung, H., and Inouye, M. (1984). Effects of mutations at glycine residues in the hydrophobic region of the *Escherichia coli* prolipoprotein signal peptide on the secretion across the membrane. *J. Biol. Chem.* **259,** 3729–3733.

Inouye, S., Duffaud, G., and Inouye, M. (1986). Structural requirement at the cleavage site for efficient processing of the lipoprotein secretory precursor of *Escherichia coli*. *J. Biol. Chem.* **261**, 10970–10975.

Inukai, M., Takeguchi, M., Shimizu, K., and Arai, M. (1978). Mechanism of action of globomycin. *J. Antibiot.* **31**, 1203–1205.

Inukai, M., Ghrayeb, J., Nakamura, K., and Inouye, M. (1984). Apolipoprotein, an intermediate in the processing of the major lipoprotein of the *Escherichia coli* outer membrane. *J. Biol. Chem.* **259**, 757–760.

Ito, K. (1982). Purification of the precursor form of maltose-binding protein, a periplasmic protein of *Escherichia coli*. *J. Biol. Chem.* **257**, 9895–9897.

Kamio, Y., Lin, C. K., Regue, M., and Wu, H. C. (1985). Characterization of the *ileS–lsp* operon in *Escherichia coli*. *J. Biol. Chem.* **260**, 5616–5620.

Lai, J. S., Philbrick, W., and Wu, H. C. (1980). Acyl moieties in phospholipids are the precursors of the fatty acids in murein lipoprotein of *Escherichia coli*. *J. Biol. Chem.* **255**, 5384–5387.

Lai, J. S., Sarvas, M., Brammar, W. J., Neugebauer, K., and Wu, H. C. (1981). *Bacillus licheniformis* penicillinase synthesized in *Escherichia coli* contains covalently linked fatty acid and glyceride. *Proc. Natl. Acad. Sci. U.S.A.* **78**, 3506–3510.

Lin, J. J., Kanazawa, H., Ozols, J., and Wu, H. C. (1978). An *Escherichia coli* mutant with an amino acid alteration within the signal sequence of outer membrane prolipoprotein. *Proc. Natl. Acad. Sci. U.S.A.* **75**, 4891–4895.

McLaughlin, J. R., Murray, C. L., and Rabinowitz, J. C. (1981). Unique features in the ribosome binding site sequence of the gram positive *Staphylococcus aureus* β-lactamase gene. *J. Biol. Chem.* **256**, 11283–11291.

March, P. E., and Inouye, M. (1985). Characterization of the *lep* operon of *Escherichia coli*. Identification of the promoter and the gene upstream of the signal peptidase I gene. *J. Biol. Chem.* **260**, 7206–7213.

Meek, R. L., Walsh, K. A., and Palmiter, R. D. (1982). The signal sequence of ovalbumin is located near the NH_2 terminus. *J. Biol. Chem.* **257**, 12245–12251.

Miller, K. W., and Wu, H. C. (1987). Cotranscription of the *Escherichia coli* isoleucyl-tRNA synthetase (*ileS*) and prolipoprotein signal peptidase (*lsp*) genes: Fine-structure mapping of the *lsp* internal promoter. *J. Biol. Chem.* **262**, 389–393.

Nakamura, K., and Inouye, M. (1979). DNA sequence of the gene for the outer membrane lipoprotein of *Escherichia coli:* An extremely A–T rich promoter. *Cell* **18**, 1109–1117.

Nakamura, K., and Inouye, M. (1980). DNA sequence of the *Serratia marcescens* lipoprotein gene. *Proc. Natl. Acad. Sci. U.S.A.* **77**, 1369–1373.

Nakamura, K., Katz-Wurtzel, E. T., Pirtle, R. R., and Inouye, M. (1979). Restriction enzyme cleavage sites surrounding the structural gene for the lipoprotein of the *Escherichia coli* outer membrane. *J. Bacteriol.* **138**, 715–720.

Nielsen, J. B., Caulfield, M. P., and Lampen, J. O. (1981). Lipoprotein nature of *Bacillus licheniformis* membrane penicillinase. *Proc. Natl. Acad. Sci. U.S.A.* **78**, 3511–3515.

Novak, P., Ray, P. H., and Dev, I. K. (1986). Localization and purification of two enzymes from *Escherichia coli* capable of hydrolyzing a signal peptide. *J. Biol. Chem.* **261**, 420–427.

Ogata, R. T., Winters, C., and Levine, R. P. (1982). Nucleotide sequence analysis of the complement resistance gene from plasmid R100. *J. Bacteriol.* **151**, 819-827.

Oka, A., Nomura, N., Morita, M., Sugisaki, H., Sugimoto, K., and Takanami, M. (1979). Nucleotide sequence of small Col E1 derivatives: Structure of the regions essential for autonomous replication and colicin E1 immunity. *Mol. Gen. Genet.* **127**, 151–159.

Oliver, D. (1985). Protein secretion in *Escherichia coli*. *Annu. Rev. Microbiol.* **39**, 615–648.

Pacaud, M. (1982). Purification and characterization of two novel proteolytic enzymes in membranes of *Escherichia coli*. *J. Biol. Chem.* **257,** 4333–4339.

Pollitt, S., Inouye, S., and Inouye, M. (1985a). A functional prolipoprotein signal peptide with deletion of four amino acid residues from the hydrophobic region. *J. Biol. Chem.* **260,** 7965–7969.

Pollitt, S., Inouye, S., and Inouye, M. (1985b). Functional analysis of signal peptide mutants. *In* "Microbiology—1985", (L. Leive, P. F. Bonventre, J. A. Morello, S. Schlesinger, S. D. Silver, and H. C. Wu, eds.), pp. 308–311. American Society for Microbiology, Washington, D.C.

Pollitt, S., Inouye, S., and Inouye, M. (1986). Effect of amino acid substitutions at the signal peptide cleavage site of the *Escherichia coli* outer membrane lipoprotein. *J. Biol. Chem.* **261,** 1835–1837.

Ray, P., Dev, I., MacGregor, C., and Bassford, P., Jr. (1986). Signal peptidases. *In* "Protein Secretion and Export in Bacteria" (H. C. Wu and P. C. Tai, eds.), pp. 75–102. Springer-Verlag, Berlin, Heidelberg, New York, Tokyo.

Regue, M., Remenick, J., Tokunaga, M., Mackie, G. A., and Wu, H. C. (1984). Mapping of the lipoprotein signal peptidase gene (*lsp*). *J. Bacteriol.* **158,** 632–635.

Rotering, H., and Braun, V. (1977). Lipid deficiency in a lipoprotein mutant of *Escherichia coli*. *FEBS Lett.* **83,** 41–44.

Simons, K., Sarvas, M., Garoff, H., and Helenius, A. (1978). Membrane-bound and secreted forms of penicillinase from *Bacillus licheniformis*. *J. Mol. Biol.* **126,** 673–690.

Smith, W. P., Tai, P. C., and Davis, B. C. (1981). *Bacillus licheniformis* penicillinase cleavage and attachment of lipid during cotranslational secretion. *Proc. Natl. Acad. Sci. U.S.A.* **78,** 3501–3505.

Tokunaga, H., and Wu, H. C. (1984). Studies on the modification and processing of prolipoprotein in *Escherichia coli*. Effects of structural alterations in prolipoprotein on its maturation in wild-type and *lpp* mutants. *J. Biol. Chem.* **259,** 6098–6104.

Tokunaga, M., Loranger, J. M., Wolfe, P. B., and Wu, H. C. (1982a). Prolipoprotein signal peptidase in *Escherichia coli* is distinct from the M13 procoat protein signal peptidase. *J. Biol. Chem.* **257,** 9922–9925.

Tokunaga, M., Tokunaga, H., and Wu, H. C. (1982b). Posttranslational modification and processing of *Escherichia coli* prolipoprotein *in vitro*. *Proc. Natl. Acad. Sci. U.S.A.* **79,** 2255–2259.

Tokunaga, M., Loranger, J. M., and Wu, H. C. (1983). Isolation and characterization of an *Escherichia coli* clone overproducing prolipoprotein signal peptidase. *J. Biol. Chem.* **258,** 12102–12105.

Tokunaga, M., Loranger, J. M., and Wu, H. C. (1984a). A distinct signal peptidase for prolipoprotein in *Escherichia coli*. *J. Cell. Biochem.* **24,** 113–120.

Tokunaga, M., Loranger, J. M., and Wu, H. C. (1984b). Prolipoprotein modification and processing enzymes in *Escherichia coli*. *J. Biol. Chem.* **259,** 3825–3830.

Tokunaga, M., Loranger, J. M., Chang, S. Y., Regue, M., Chang, S., and Wu, H. C. (1985). Identification of prolipoprotein signal peptidase and genomic organization of the *lsp* gene in *Escherichia coli*. *J. Biol. Chem.* **260,** 5610–5615.

Vlasuk, J. P., Inouye, S., Ito, H., Itakura, K., and Inouye, M. (1983). Effects of the complete removal of basic amino acid residues from the signal peptide on secretion of lipoprotein in *Escherichia coli*. *J. Biol. Chem.* **258,** 7141–7148.

Vlasuk, G. P., Inouye, S., and Inouye, M. (1984). Effects of replacing serine and threonine residues within the signal peptide on the secretion of the major outer membrane lipoprotein of *Escherichia coli*. *J. Biol. Chem.* **259,** 6195–6200.

Von Heijne, G. (1983). Patterns of amino acids near signal sequence cleavage sites. *Eur. J. Biochem.* **133,** 17–21.

Watanabe, T., Hayashi, S., and Wu, H. C. (1988). Synthesis and export of the outer membrane lipoprotein in *Escherichia coli* mutants defective in generalized protein export. *J. Bacteriol.* (in press).

Watson, R. J., Lau, P. C. K., Vernet, T., and Visentin, T. (1984). Characterization and nucleotide sequence of a colicin-release gene in the *hic* region of plasmid ColE3-CA38. *Gene* **29**, 175–184.

Wolfe, P. B., Silver, P., and Wickner, W. (1982). The isolation of homogenous leader peptidase from a strain of *Escherichia coli* which overproduces the enzyme. *J. Biol. Chem.* **257**, 7879–7902.

Wolfe, P. B., Wickner, W., and Goodman, J. M. (1983). Sequence of the leader peptidase gene of *Escherichia coli* and the orientation of the leader peptidase in the bacterial envelope. *J. Biol. Chem.* **258**, 12073–12080.

Wu, H. C., Hou, C., Lin, J. J., and Yem, D. W. (1977). Biochemical characterization of a mutant lipoprotein of *Escherichia coli*. *Proc. Natl. Acad. Sci. U.S.A.* **74**, 1388–1392.

Yamagata, H., Nakamura, K., and Inouye, M. (1981). Comparison of the lipoprotein gene among the Enterobacteriaceae. DNA sequence of *Erwinia amylovora* lipoprotein gene. *J. Biol. Chem.* **256**, 2194–2198.

Yamagata, H., Ippolito, C., Inukai, M., and Inouye, M. (1982). Temperature-sensitive processing of outer membrane lipoprotein in an *Escherichia coli* mutant. *J. Bacteriol.* **152**, 1163–1168.

Yamagata, H., Daishima, K., and Mizushima, S. (1983a). Cloning and expression of a gene coding for the prolipoprotein signal peptidase of *Escherichia coli*. *FEBS Lett.* **158**, 301–304.

Yamagata, H., Taguchi, N., Daishima, K., and Mizushima, S. (1983b). Genetic characterization of a gene for prolipoprotein signal peptidase in *Escherichia coli*. *Mol. Gen. Genet.* **192**, 10–14.

Yamamoto, S., and Lampen, J. O. (1976). Membrane penicillinase of *Bacillus licheniformis* 748/C: Sequence and possible repeated tetrapeptide structure of the prolipoprotein region. *Proc. Natl. Acad. Sci. U.S.A.* **73**, 1457–1461.

Yem, D. W., and Wu, H. C. (1978). Physiological characterization of an *Escherichia coli* mutant altered in the structure of murein lipoprotein. *J. Bacteriol.* **133**, 1419–1426.

Yu, F., Yamada, H., Daishima, K., and Mizushima, S. (1984). Nucleotide sequence of the *lspA* gene, the structural gene for lipoprotein signal peptidase of *Escherichia coli*. *FEBS Lett.* **173**, 264–268.

Yu, F., Inouye, S., and Inouye, M. (1986). Lipoprotein-28, a cytoplasmic membrane lipoprotein from *Escherichia coli*. *J. Biol. Chem.* **261**, 2284–2288.

Zwizinski, C., and Wickner, W. (1980). Purification and characterization of leader (signal) peptidase from *Escherichia coli*. *J. Biol. Chem.* **255**, 7973–7977.

Zwizinski, C., Date, T., and Wickner, W. (1981). Leader peptidase is found in both the inner and the outer membranes of *Escherichia coli*. *J. Biol. Chem.* **256**, 3593–3597.

15

Protein Secretion across the Outer Membrane of Gram-Negative Bacteria

ANTHONY P. PUGSLEY

I. INTRODUCTION

A great deal of research has been directed toward the study of two different aspects of the relocation of cytoplasmically synthesized proteins in bacteria. Fundamental studies, conducted mainly in the gram-negative bacterium *Escherichia coli,* have considerably increased our understanding of how exported proteins are inserted into or translocated across the

607

PROTEIN TRANSFER AND ORGANELLE BIOGENESIS

cytoplasmic (inner) membrane and directed to their final location (outer membrane or periplasmic space between the two membranes; Fig. 1). The reader is referred to reviews (Randall and Hardy, 1984; Benson *et al.*, 1985; Wickner and Lodish, 1985; Pugsley and Schwartz, 1985) for details of specific features of these aspects of bacterial protein export, which will not be covered in depth in this chapter.

The secretion of proteins into the medium (hereafter referred to simply as secretion) has been extensively studied in gram-positive bacteria, which do not have an outer membrane (Fig. 1). These studies are motivated, to a large extent, by potential industrial applications of the secreted enzymes, and, more recently, by the possibility of secreting hybrid proteins comprising part of the naturally secreted protein coupled to other proteins of commercial interest. Much of the research in this area is concerned with increasing the yields, quality, and stability of the secreted protein. Protein secretion by gram-positive bacteria is mechanistically similar to the process by which gram-negative bacteria export proteins to the periplasm (Mézes and Lampen, 1985; Pugsley and Schwartz, 1985; Section III,A,1).

Although it has long been recognized that gram-negative bacteria release proteins (hydrolytic enzymes and toxins) into the growth medium, the mechanisms involved have, until recently, received only scant attention. This was in part due to the belief that these proteins were released either by nonspecific blebbing of outer membrane vesicles from the cell surface or by lysis, in which case the process could not justifiably be referred to as secretion. The purpose of this chapter is to review the growing evidence for the existence of several novel mechanisms for translocating proteins across the gram-negative outer membrane (and in some cases the cytoplasmic membrane) and into the growth medium.

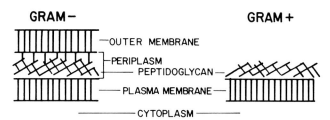

Fig. 1. Simplified representation of the envelope structure of gram-negative and gram-positive bacterial cells. Note that both types of bacteria have a cytoplasmic membrane surrounded by a peptidoglycan gel but that only gram-negative bacteria have an outer membrane. Secreted proteins are those which are fully exposed on the surface of the cell (surface appendages) or are free in the extracellular medium.

II. LYSIS OR SECRETION

Before going into details on how proteins are secreted by gram-negative bacteria, it seems pertinent to begin by reviewing the evidence that bona fide secretion does occur. The periplasm and outer membrane contain specific subsets of cellular proteins which are found in only trace amounts in the medium (generally <1% of that remaining cell associated), except when outer membrane permeability is compromised. The outer membrane is, therefore, a macropermeability barrier across which large polypeptides would not normally be expected to diffuse. This does not exclude the possibility that small polypeptides could diffuse through nonspecific pores in the outer membrane or could transiently partition into the outer membrane from the inner, periplasmic face and then adopt another conformation which would result in their release into the medium from the outer face. Possible examples of this method of protein secretion may include such small proteins as heat-stable enterotoxins I and II (So and McCarthy, 1980; Guzmán-Verduzco et al., 1983; Lee et al., 1984), microcins (Baquero and Moreno, 1984), and OmpF–β-endorphin and alkaline phosphatase–hirudin hybrid polypeptides (Nagahari et al., 1985; Dodt et al., 1986), all of which are released into the medium by E. coli.

In an earlier review, we presented a list of "secreted" proteins produced by different species of gram-negative bacteria (Pugsley and Schwartz, 1985). In most cases, the proteins were considered to be secreted because they were found in the culture medium and because their presence there was compatible with their function and mode of action. In an increasing number of cases, it has been unambiguously demonstrated that protein release is efficient (i.e., the bulk of the particular protein is released into the medium) and that release occurs without concomitant release of other proteins which are not normally extracellular. Figure 2 shows the sort of evidence presented in support of specific secretion of bacterial proteins. In most cases, only a limited number of proteins are found in the growth medium, which apparently rules out the possibility of wholesale lysis or nonspecific release of outer membrane vesicles (the exception, colicins, is discussed in detail in Section IV,B,1).

It will also be obvious from the examples shown in Fig. 2 that gram-negative bacteria can secrete proteins in amounts similar to those secreted by gram-positive bacteria under comparable conditions. Yields of secreted proteins from gram-positive bacteria have been substantially improved by the use of more suitable culture media, high cell densities, and genetically engineered strains. Similar strategies could presumably be used to increase yields of secreted proteins from gram-negative bacteria.

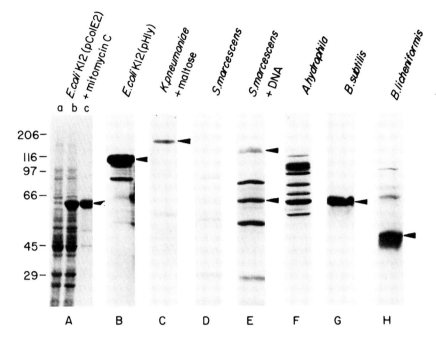

Fig. 2. Proteins secreted by gram-negative and gram-positive bacteria. Samples were as follows: (A) Total cell proteins of a strain of *E. coli* K12 carrying the colicin plasmid pColE2-P9 growing exponentially in rich medium under conditions where colicin production was not induced; b, same as a except that the culture was induced for 2 hr with 0.5 µg/ml mitomycin C (see Pugsley, 1984a); c, proteins in 0.1 ml of medium from culture used for sample b after 5 hr induction (arrow indicates colicin E2, the band in c being equivalent to 5 µg of protein). (B) Proteins in 1 ml of medium from an exponentially growing culture of *E. coli* K12 carrying the recombinant, multiple copy number α-hemolysin plasmid pSF4000 (a gift from R. Welch; Felmlee *et al.*, 1985a) in rich medium (arrow indicates α-hemolysin). (C) Proteins in 1 ml of medium from a 24-hr culture of *Klebsiella pneumoniae* ATCC 15050 (Michaelis *et al.*, 1985) grown in minimal medium with 0.4% maltose to induce expression of the maltose regulon, which includes the pullulanase structural gene (arrow indicates pullulanase). (D) Proteins in 1 ml of medium from an exponentially growing culture of *Serratia marcescens* (a gift from R. Barker) in minimal medium containing 1% glycerol as carbon source. (E) Same as D except that the medium was supplemented with 0.2% DNA (arrows indicate protein bands not present in D and therefore assumed to be nucleases whose production was induced by DNA; note also that culture E grew faster than D and therefore secreted more of the other extracellular proteins). (F) Proteins in 1 ml of medium of an exponentially growing culture of *Aeromonas hydrophila* (a gift from R. Barker) in rich medium (note that the pattern of extracellular proteins changed as incubation continued, probably due to proteolytic degradation). (G) Proteins present in 1 ml of medium of an exponentially growing culture of *Bacillus subtilis* (a gift from G. Rapoport) in rich medium (arrow indicates protein tentatively identified as a protease). (H) Proteins present in 1 ml of medium from an exponentially growing culture of *Bacillus licheniformis* (a gift from J. O. Lampen) in rich medium (arrow indicates penicillinase; the multiple banding pattern is typical of this protein). All samples other than

III. EARLY STAGES IN THE SECRETION PATHWAY

Gram-negative bacteria export numerous proteins to their two membranes and into the periplasmic space between them. The pathway(s) by which these proteins reach their destinations could also be used to initiate protein secretion, with one or more additional steps being necessary to translocate the polypeptide across the outer membrane and into the medium. Alternatively, the normal protein export route(s) could be bypassed by the evolution of independent pathways for protein secretion. In fact, there are numerous examples of both types of secretion pathway among the better characterized protein secretion systems in gram-negative bacteria.

A. Signal Peptide Route

Almost all of the secreted proteins of gram-positive bacteria, as well as the outer membrane, periplasmic, and some cytoplasmic membrane proteins of gram-negative bacteria, are made as precursors with amino-terminal signal sequences (signal peptides) which are normally cleaved off before the polypeptide reaches its final destination. All signal peptides share several basic characteristics including length (15–30 residues), a positively charged amino terminus, a hydrophobic and generally α-helical core, helix-breaking or turn-inducing residues (proline, glycine, or threonine + serine) at the end of the hydrophobic core, and one of two specific cleavage sites for signal peptidases (Pugsley and Schwartz, 1985). Signal peptides are thought to initiate the early stages of protein export across the cytoplasmic membrane, possibly via interactions with specific envelope proteins deemed to be involved in protein export (Pugsley and Schwartz, 1985; Benson et al., 1985). In the absence of any membrane anchor or "membrane targeting sequences," the protein should be released into the periplasm (gram-negative bacteria) or into the medium (gram-positive bacteria). It is still not known whether the translocation of the polypeptide through the membrane occurs through an energetically favorable interaction with membrane lipids, or whether it is catalyzed by specific protein translocators in the cytoplasmic membrane. Results from in vivo experiments and from in vitro reconstitution experiments require-ment for energy (ATP hydrolysis and transmembrane potential) in the

---

Aa and Ab were obtained by precipitating proteins in cell-free medium with 1% Triton X-100 plus 10% trichloroacetic acid. The samples were dissolved in 5% SDS, and proteins were separated by sodium dodecyl sulfate–polyacrylamide gel electrophoresis. The gels were stained with Coomassie blue.

early stages of protein export (Gellert *et al.*, 1986; Bakker and Randall, 1984).

1. Protein Secretion by Gram-Positive Bacteria

Figure 3 shows the two pathways by which proteins are secreted by gram-positive bacteria: the major pathway (left-hand side of Fig. 3), as used for the secretion of most proteins, and the alternate pathway (right), used for the secretion of lipoproteins. Many of these proteins seem to be subject to secondary proteolysis following cleavage of the signal peptide. This secondary processing step may be required to release a membrane-bound form of the protein into the medium, with the amino terminus acting as an anchor. There is strong evidence that this might be the case for staphylococcal nucleases A and B (Shortle, 1983) and for subtilisin (Power *et al.*, 1986). In most other cases, the secreted polypeptide is thought to undergo further proteolysis after its release, and any delay in releasing the protein into the medium probably reflects the posttranslational nature of the export-secretion process or, more likely limited diffusion of the polypeptide across the cell wall (Gould *et al.*, 1974).

The evidence for a membrane-bound intermediate is more convincing in the case of the penicillinases (β-lactamase) secreted by the lipoprotein

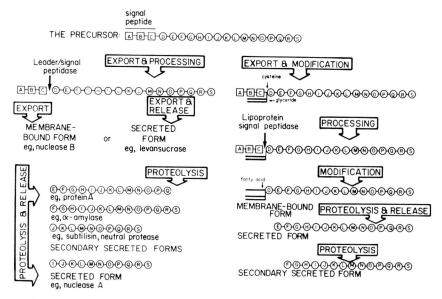

Fig. 3. Pathways for signal peptide-dependent secretion of proteins by gram-positive bacteria.

secretion pathway (Fig. 3). These proteins have a fatty acylated amino-terminal cysteine residue (Regue and Wu, Chapter 14, this volume) which is probably buried in the cytoplasmic membrane, thereby anchoring the polypeptide to the outer surface of the cell. Extracellular penicillinase exists in two major forms, both of which lack the amino-terminal cysteine residue (Izui *et al.*, 1980). This observation led to the proposal that penicillinase is released by the proteolytic removal of the amino terminus of the membrane-bound form of the enzyme, but the possibility that proteolysis occurs only after the polypeptide (lipoprotein) has been released has not been thoroughly investigated.

2. Periplasmic Proteins, and Secreted Proteins of Gram-Positive Bacteria Produced by Gram-Negative Bacteria

The signal peptide pathway for protein export and secretion is illustrated with examples from gram-positive bacteria because they represent the simplest mechanism for secreting proteins. The pathway(s) by which proteins are exported to the gram-negative periplasm is almost identical to the major secretion pathway shown in Fig. 3, except that periplasmic proteins are not known to be subject to secondary proteolysis. It is therefore not surprising that the cloning and expression of genes for secreted proteins of gram-positive bacteria in gram-negative bacteria generally result in the export of the protein into the periplasmic space, rather than in its secretion into the medium (Table I). Lipopenicillinase is exceptional because it is anchored in the outer membrane with the polypeptide chain

TABLE I

Localization of Extracellular Proteins of Gram-Positive Bacteria when Their Structural Genes Are Expressed in *Escherichia coli* or *Pseudomonas aeruginosa*

Protein	Source of gene	New host	Location in new host/other comments	Reference
Penicillinase	*Bacillus licheniformis*	*E. coli*	Outer membrane	Sarvas and Palva (1983)
Penicillinase	*B. licheniformis*	*P. aeruginosa*	Membrane	Nielsen et al. (1983)
Penicillinase	Alkalophilic *Bacillus* 170	*E. coli*	Aspecific leakage into medium	Kobayashi et al. (1986)
α-Amylase	*B. licheniformis*	*E. coli*	Periplasm	Joyet et al. (1984)
α-Amylase	*B. licheniformis*	*P. aeruginosa*	Periplasm	Filloux et al. (1985)
Cellulase	Alkalophilic *Bacillus* N-4	*E. coli*	Periplasm and cell associated	Sashihara et al. (1984)

(continued)

TABLE I (*continued*)

Protein	Source of gene	New host	Location in new host/other comments	Reference
α-Amylase	*Bacillus coagulans*	*E. coli*	Periplasm; aspecific leakage into medium	Cornelis *et al.* (1982), Willemot and Cornelis (1983)
β-Lactamase II	*Bacillus cereus*	*E. coli*	Periplasm and membrane	Hussain *et al.* (1985)
Levansucrase	*Bacillus subtilis*	*E. coli*	Periplasm; cells are fragile	Gay *et al.* (1983)
1,3-1,4-β-D-Glucanase	*B. subtilis*	*E. coli*	Periplasm and cytoplasm; increased in medium after end of exponential growth	Hindcliffe (1984)
β-1,4-Glucanase	*B. subtilis*	*E. coli*	Loosely cell associated and intracellular	Robson and Chambliss (1986)
α-Amylase	*Bacillus stearothermophilus*	*E. coli*	Mainly periplasm	Tsukagoshi *et al.* (1984)
β-Gluconase	*Bacillus amyloliquefaciens*	*E. coli*	Mainly periplasm, some in cytoplasm or membrane associated, some in medium	Borris *et al.* (1985)
Diphtheria toxin	*Corynebacterium diphtheriae*	*E. coli*	Periplasm; proteolytically degraded	Tweten and Collier (1983), Leong *et al.* (1983)
Streptokinase	*Streptococcus equisimilis*	*E. coli*	Periplasm and cytoplasm; release (possibly aspecific) after exponential growth	Malke and Ferretti (1984)
Nuclease B	*Staphylococcus aureus*	*E. coli*	Not secreted, incorrectly processed	Shortle (1983)
Nuclease A	*S. aureus*	*E. coli*	Periplasm; precursor has *E. coli* OmpA protein signal peptide	Takahara *et al.* (1985)
Enterotoxin B	*S. aureus*	*E. coli*	Cytoplasm; signal peptide apparently removed	Ranelli *et al.* (1985)
Staphylokinase	*S. aureus*	*E. coli*	Periplasm	Sako (1985)
Protein A	*S. aureus*	*E. coli*	Periplasm	Lidahl *et al.* (1983)
Lipase	*S. aureus*	*E. coli*	Periplasm	Lee and Iandolo (1985)
Cellulases	*Clostridium thermocellum*	*E. coli*	Periplasm and cytoplasm	Cornet *et al.* (1983)

facing the periplasm (Nielsen *et al.*, 1983; Sarvas and Palva, 1983). Evidence for the "secretion" of some of the other proteins into the growth medium (Table I) was not supported by results showing the selective release of the protein in question [i.e., absence of other, normally cell-associated proteins (Table I)], and must therefore be interpreted with caution. Other proteins may have remained cytoplasmic (Table I) because the precursor could not enter into the protein export route, because the signal peptide was not cleaved correctly, or because the polypeptide formed aggregates in the cytoplasm, as has been reported for other, overproduced proteins in *E. coli* (Cabilly *et al.*, 1984; Sekine *et al.*, 1985; Schoner *et al.*, 1985; Schoemaker *et al.*, 1985).

3. Outer Membrane Proteins

Most integral outer membrane proteins are also synthesized with signal peptides and initially follow the same basic export route(s) as periplasmic proteins (Liss and Oliver, 1986). Proteins are thought to be targeted to the outer membrane by signals in the mature part of the polypeptide (Benson *et al.*, 1985; Pugsley and Schwartz, 1985). These signals may be the unusual transmembrane segments which contribute to the overall protein conformation (Freudl *et al.*, 1985; Bosch *et al.*, 1986) instead of or in addition to any specific targeting sequence such as those proposed by Benson and Silhavy (1983). Signal sequences from outer membrane precursor polypeptides can be replaced by signal sequences from periplasmic or cytoplasmic membrane proteins without affecting the final location of the protein (Tommassen *et al.*, 1983; Jackson *et al.*, 1985), indicating that the signal peptide does not play a direct role in protein targeting.

4. Secreted Proteins

Several well-characterized extracellular proteins of gram-negative bacteria are synthesized with cleavage signal sequences (Table II). These signal peptides (see Table II for references) resemble signal peptides of other secreted or exported proteins in their overall length, number of positively charged amino acids at the extreme amino terminus, length and overall hydrophobicity of the central core region, presence or absence of turn-inducing residues in front of the cleavage site, and amino acids present at the cleavage site (Watson, 1984; von Heijne, 1985). The only feature that might distinguish signal peptides of secreted proteins of gram-negative bacteria from those of their exported counterparts is their somewhat greater length, which results from a higher number of positively charged residues at the extreme amino terminus.

Signal peptides probably initiate translocation of secreted proteins across the cytoplasmic membrane. This could be verified experimentally

TABLE II

Expression of Genes for Extracellular Proteins of Various Gram-Negative Bacteria in *Escherichia coli* K12.

Protein	Source of Gene	Location in E. coli K12/other comments	Reference
Colicins A, D, E1–E8, K, N, DF13	Various E. coli and other enterobacteria (Col plasmid-encoded	Accumulate in cytoplasm, then released into medium; secretion gene (lysis gene) linked to colicin structural gene; no signal peptide	Pugsley (1984a)
α-Hemolysin	Various E. coli strains (Hly plasmid or chromosome)	Released into medium; no signal peptide	Goebel et al. (1984), Mackman et al. (1986), Felmlee et al. (1985a)
Heat-labile enterotoxin I (HLETI)	Various E. coli strains	Similar to cholera toxin; two subunits (A and B); signal peptides in both subunits; mainly in periplasm, inefficient release into medium	Hirst et al. (1984), Dallas and Falkow (1980), Spicer and Noble (1982)
Heat-stable enterotoxin I (HSETI)	Various E. coli strains	~5,000 molecular weight; signal peptide; 70% released, rest in periplasm, largely inactive	Guzmán-Verduzco et al. (1983). So and McCarthy (1980)
HSETII	Various E. coli strains	~5,000 molecular weight; signal peptide; secreted	Lee et al. (1983)
Pullulanase	Klebsiella pneumoniae	Exposed on surface and then released into medium; signal sequence; lipoprotein; secretion genes linked to pullulanase structural gene	Chapon and Raibaud (1985), Pugsley et al. (1986), d'Enfert et al. (1987b)
Chitinase	Serratia marcescens	Secreted; other proteins released; signal peptide	Jones et al. (1986)
Serine protease	S. marcescens	Secreted; signal peptide; possible secondary cleavage site	Yanigida et al. (1986)
Pectate lyase	Erwinia chrysanthemi	Mainly in periplasm	Keen et al. (1984), Collmer et al. (1985)

(continued)

TABLE II (*continued*)

Protein	Source of Gene	Location in E. coli K12/other comments	Reference
Proteases	*E. chrysanthemi*	Released into medium	C. Wandersman (unpublished results), Barras *et al.* (1986)
Pectate lyase	*Erwinia carotovara*	Periplasm, not released or released nonspecifically	Lei *et al.* (1985), Zink and Chatterjee (1985)
Aerolysin	*Aeromonas hydrophila*	Periplasm, not released; signal peptide processed	Howard and Buckley (1986)
Phospholipase C (hemolysin)	*Pseudomonas aeruginosa*	Cytoplasmic, not released into medium; Outer membrane, not released into medium; Signal peptide	Coleman *et al.* (1983); Lory and Tai (1983); Pritchard and Vasil (1986)
Exotoxin A	*P. aeruginosa*	Remains intracellular; signal peptide apparently not cleaved; not released into medium	Gray *et al.* (1984)
Cholera toxin	*Vibrio cholerae*	Mainly periplasmic, not released; two subunits (A and B), both have signal peptides; similar to HLETI	Pearson and Mekalanos (1982) Mekalonos *et al.* (1983)
Hemolysin V	*V. cholerae*	Periplasm, not released into medium	Mercurio and Manning (1985)
DNase	*V. cholerae*	Periplasm, not released into medium; secretion gene(s) tentatively identified	Newland *et al.* (1985), Focareta and Manning (1985)
Thermostable hemolysin	*Vibrio parahaemolyticus*	Periplasm, not released into medium; signal peptide	Nishibuchi and Kaper (1985)
IgA protease	*Haemophilis influenzae*	Periplasm, not released	Bricker *et al.* (1983)
IgA protease	*Neisseria gonorrhoeae*	Released into medium	Koomey *et al.* (1982), Halter *et al.* (1984)
Pertussis toxin	*Bordetella pertussis*	5 subunits, all with signal peptides; periplasm	Nicosia *et al.* (1986)

by replacing the signal sequence with another from an exported protein (see Section III,A,3), which, in most cases, should not affect the release of the protein onto the medium (see also Section V for discussion of a possible exception).

Many of the proteins listed in Table II are normally produced and secreted by bacteria other than *E. coli*. In many cases, expression of the structural gene alone in *E. coli* led to the production of the corresponding enzyme or toxin but not its release into the medium (Table II). *Escherichia coli* thus lacks one or more functions required for protein secretion; in some instances, these functions could be supplied by additional genes cloned from the same source as the structural gene for the secreted protein (Section IV,A). Among the proteins known to be synthesized with signal sequences (Table II), only three, a chitinase and a serine protease normally produced by *Serratia marcescens*, and a *Neisseria gonorrhoea* protease were secreted by *E. coli* in the absence of genes coding for secretion factors (Jones *et al.*, 1986; Yanigida *et al.*, 1986, Pohlner *et al.*, 1987). Yields of secreted chitinase were extremely high, but the medium contained many additional proteins indicative of membrane perturbation similar to that observed when some secreted proteins of gram-positive bacteria were produced by *E. coli* (Section II,A,2 and Table I).

B. Proteins without Cleavable Signal Sequences

Most of the proteins in the gram-negative cytoplasmic membrane are made without cleavable signal sequences. Unlike their outer membrane and periplasmic counterparts, however, inner membrane proteins have one or more hydrophobic, membrane-spanning segments. By analogy with similar proteins produced by eukaryotic cells (Meek *et al.*, 1982; Friedlander and Blobel, 1985), some of these hydrophobic segments may act as membrane insertion sequences, the functional equivalents of cleavable signal peptides, and may, in some cases, interact with components of the signal sequence export route, as defined by the *sec* mutations (Wolfe *et al.*, 1985). Other transmembrane domains of cytoplasmic membrane proteins may serve as stop transfer signals which anchor part of the polypeptide in the membrane. The export process could then be reinitiated by a membrane insertion sequence located further down the polypeptide chain, leading to the complex structure with multiple transmembrane domains which seems to be typical of many cytoplasmic membrane proteins.

Gram-negative bacteria synthesize at least two groups of secreted proteins without cleavable signal sequences: the "secreted" colicins

(Pugsley, 1984a) and the α-hemolysins of *E. coli* (Felmlee *et al.*, 1985a,b; Hess *et al.*, 1986). α-Hemolysin and some colicins contain relatively long stretches of hydrophobic amino acids which could function as membrane insertion sequences, thereby triggering the initial stages of export and leading to eventual release into the medium. Indeed, it has been proposed that one such "signal sequence-like" stretch of amino acids near the carboxy terminus of the colicin E1 polypeptide plays a crucial role in initiating the translocation of this colicin across the cytoplasmic membrane (Yamada *et al.*, 1982a,b; Yamada and Nakazawa, 1986), as well as in the formation by the colicin of pores in the cytoplasmic membranes of target cells (Davidson *et al.*, 1984). Part of the evidence for this was that the imposition of a block on signal sequence-dependent protein export by jamming protein export sites with a hybrid protein containing a normally periplasmic protein coupled to the nonexportable β-galactosidase, or by treatment with globomycin, an inhibitor of some signal peptidases, prevented the colicin E1 export into the periplasm (Yamada *et al.*, 1982b).These conditions, however, actually prevent the processing and export of the product of the "lysis" gene, which is known to be essential for colicin secretion (Section IV,B,1). In particular, the observed effects of globomycin (Yamada *et al.*, 1982b), whose action is restricted to inhibiting the peptidase which cleaves precursors of lipoproteins (Regue and Wu, Chapter 14, this volume), were probably due to the lack of processing of the "prelysis" protein (Section IV,B,1). The second line of evidence for the importance of this region of colicin E1 for its secretion was that sequence changes in this region seemed to affect the ability of the colicin to cross the cytoplasmic membrane (Yamada and Nakazawa, 1984, 1986). These results could also be interpreted as indicating that the integrity of this region is necessary to maintain the colicin polypeptide in a conformation which allows it to cross the membrane, rather than having a direct role in the process (see Section IV,C).

There is no direct evidence that the three major hydrophobic, potential transmembrane segments present in α-hemolysin (Felmlee *et al.*, 1985b; Hess *et al.*, 1986) are involved in hemolysin secretion; indeed these segments of the polypeptide could be involved in the formation of pores in target reticulocyte membranes (Bhakdi *et al.*, 1986). α-Hemolysin secretion is not affected by a *secA* mutation known to prevent protein export by the signal peptide-dependent pathway (I. Holland, personal communication), confirming the separate identity of the α-hemolysin secretion pathway. The potential role of targeting signals in protein secretion will be discussed further in Section IV,C.

One of the characteristic features of signal peptides is the cluster of positively charged amino acids at the extreme amino terminus (von Heijne, 1985). These may be important in facilitating the initial contact between the protein to be exported or secreted and negatively charged head groups on phospholipids in the inner leaflet of the cytoplasmic membrane (Pugsley and Schwartz, 1985). Whether or not the positively charged residues near the amino terminus of the α-hemolysin polypeptide [which actually resembles prepieces or transit peptides present in nuclear-encoded precursors of polypeptides imported into mitochondria or chloroplasts (Hay *et al.*, 1984; Cashmore *et al.*, 1985; Karlin-Neumann and Tobin, 1986)], or, indeed, clusters of positively charged residues elsewhere in hemolysin or colicin polypeptides, have a similar function remains to be determined (see Section IV,C).

IV. LATER STAGES IN THE SECRETION PATHWAY: SECRETION OF PROTEINS AND TARGETING SIGNALS

A. Identification of Secretion Proteins and Secretion Mutants

Three lines of genetic evidence indicate the requirement for envelope components, in addition to those which may be required for protein export, in order to translocate proteins across the outer membrane. First, although *E. coli* can usually express genes coding for secreted proteins from other gram-negative bacteria, the proteins are not always secreted into the medium (Table II). As already mentioned, *E. coli* may secrete only those proteins whose specific protein genes were cloned together with the structural gene for the secreted protein. Second, several classes of non-secreting mutants (Sec⁻ or Out⁻) have been identified in different bacteria, and, third, genes whose products are specifically required to promote protein secretion have been identified.

The known Sec⁻ mutants are affected in the secretion of several different proteins (Table III) which, in many of the mutants, apparently accumulate in the periplasm. Although cell fractionation experiments may not give reliable results with mutants such as these, the data suggest that the mutations specifically affect protein translocation across the outer membrane and imply that the products of secretion genes should be located in the outer membrane. Some of the Sec⁻ mutants do indeed exhibit an altered outer membrane protein composition (Table III). The affected proteins could be the products of the secretion genes themselves or proteins whose localization in the envelope depends on the products of the secretion genes. One of the more intriguing observations made with the Sec⁻ mutants of *Aeromonas hydrophila* is that they can be divided into

TABLE III

Pleiotropic, Nonsecreting Mutants of Gram-Negative Bacteria

Species	Mutant designation	Secreted proteins		Intracellular Accumulation	Membrane proteins affected	Reference
		Affected	Not affected			
Pseudomonas aeruginosa PAO	tox-1a	Exotoxin A, protease, others (un-identified)	Others (uniden-tified)	No (toxin)	Not tested	Gray and Vasil (1981)
P. aeruginosa PAO	Xcp-Ib	Alkaline phos-phatase, phospholipase C, staphylo-lytic enzyme, elastase	Alkaline protease	Variable	Not tested	Wredtlindt and Pavlovskis (1984)
P. aeruginosa PAK	PAKS18	Staphylolytic enzyme, lipase, lecithinase, protease 1	Not specified	Periplasm (protease)	Several	Wredtlindt et al. (1977), Fecycz and Campbell (1985)
Aeromonas hydrophila	—	Hemolysin, protease, acyltransferase	Few (not identified)	Periplasm	Several	Howard and Buckley (1983; 1985)
Erwinia chry-santhemi	Out−	Pectate lyase, polygalacturon-idase, cellu-lases	Proteases	Periplasm	Yes	Andro et al. (1984), Thurn and Chatterjee (1985)

  a The secretion defect in this mutant could be explained by increased proteolytic degradation of certain secreted proteins.

  b Different mutants manifest various defects in protein secretion and in accumulation of normally secreted proteins.

several classes on the basis of their different effects on envelope proteins (Howard and Buckley, 1983; T. Buckley, personal communication), suggesting the existence of multiple secretion steps catalyzed by a secretory complex composed of proteins encoded by different genes. This hypothesis could be verified experimentally by identifying cloned DNA fragments which complement secretion mutations and then testing them for cross-complementation in other Sec⁻ mutants.

   Some of the proteins whose secretion is affected in the Sec⁻ mutants (Table III) are made with signal peptides (Table II; see Section III,A,3). Interestingly, only a subclass of secreted proteins is affected in the *Erwinia chrysanthemi* Sec⁻ mutants (Andro *et al.*, 1984; Thurn and Chatterjee, 1985; Table III), a phenomenon which might be explained by structural differences (e.g., signal peptide) between proteins in these two groups of proteins.

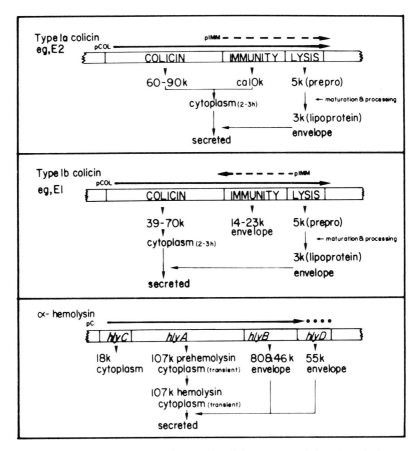

Fig. 4. Structures of two types of secreted colicin operons and the α-hemolysin operon of *E. coli.* Genes are represented by boxes (not to scale) above which the arrows indicate the direction of transcription and location of promoters (pCOL, etc). Note that the colicin immunity gene promoters (denoted by dashed lines) are much less efficient than pCOL. Differences in pC promoter sequences (Felmlee *et al.,* 1985a,b; Hess *et al.,* 1986) may explain why the expression of only some α-hemolysin operons is increased when cells are starved of iron (Lebek and Grüenig, 1985). The products of the different genes [molecular weights $\times 10^{-3}$ (k)] and their locations are indicated below the boxes representing their structural genes.

The *E. coli* genes for α-hemolysin and the secreted colicins are in the same operon as the genes required for protein secretion (Fig. 4). Colicins are produced only following DNA damage (Pugsley, 1984a), and iron availability regulates α-hemolysin production in some *E. coli* strains (Lebek and Grüenig, 1985; see legend to Fig. 4). Thus, these structural and secretion genes may be coregulated in order to avoid producing secretion proteins when there is no protein to secrete. The *Klebsiella pneumoniae* chromosomal genes which, when cloned into *E. coli,* facilitate the specific secretion of pullulanase (α-dextrin endo-1,6-α-glucosidase) are also linked to the pullulanase structural gene, and may be coregulated with it (Chapon and Raibaud, 1985; d'Enfert *et al.,* 1987a,b).

It is tempting to speculate that secretion genes may be linked to the cloned genes for other proteins which are secreted by *E. coli* (e.g., *E. chrysanthemi* proteases) and that their expression is also coordinately regulated. There is, however, no firm evidence for a requirement for secretion proteins in these cases, and we must also, therefore, consider the possibility that secretion of these proteins is an autocatalytic process, as suggested for some low molecular weight secreted polypeptides (Section II). The only potential secretion gene, other than those for pullulanase, which has been cloned into *E. coli* and identified so far is that for *Vibrio cholerae* DNase (Table II; Focareta and Manning, 1985), but possible linkage between the DNase structural and secretion genes and their coregulation were not investigated.

On the other hand, many of the cases of nonsecretion by *E. coli* involve proteins whose structural genes were cloned from bacteria which are known to secrete several different proteins and where genetic evidence suggests that these proteins might be secreted by a common pathway (Table III), even though synthesis of the secreted proteins may be subject to different regulatory factors. In these cases, secretion genes and structural genes for the secreted proteins are unlikely to be linked or coregulated. This also has implications for the cloning of these secretion genes since the possibility that the products of several different genes might be involved in protein secretion in these bacteria makes it unlikely that they could be identified individually by their ability to complement the "secretion defect" in *E. coli.* It would be interesting to see whether the *V. cholerae* DNase secretion gene cloned by Focareta and Manning (1985) can promote cholera toxin or *V. cholerae* hemolysin secretion by *E. coli,* since all three proteins appear to accumulate in the periplasm in this bacterium, as well as in *V. cholerae,* prior to their release into the medium (Table II).

B. Mode of Action of Secretion Proteins

1. Colicin Lysis Proteins

The only case in which any clear idea of how secretion proteins actually function is that of the colicin "lysis" protein, a small, membrane-located lipoprotein which is synthesized with a cleavable signal sequence (Pugsley, 1984a; Cole *et al.*, 1985). It has long been recognized that production of colicins encoded by natural, multiple copy number plasmids is normally a lethal event (Ozeki *et al.*, 1959) and that their release coincides with a decline in culture turbidity. The latter phenomenon was assumed to indicate that colicins were released by cell lysis (Mock and Schwartz, 1978; Jakes and Model, 1979). In 1981, we reported that only periplasmic proteins, and not cytoplasmic proteins, were released along with the colicin, and that extensive cell wall disruption did not occur (Pugsley and Rosenbusch, 1981). As an alternative explanation for colicin release, we proposed that colicins accumulated in very large amounts (see Fig. 2) in the periplasmic space. The observations that colicins could be released together with typical periplasmic proteins by cold osmotic shock or detergent-induced permeabilization of the outer membrane and that they could be labeled by eosin isothiocyanate, which can only penetrate the outer membrane (van Tiel-Menkveld *et al.*, 1979; Pugsley and Rosenbusch, 1981), seemed to support this idea. We further assumed that the massive accumulation of the colicin in the periplasm eventually affected outer membrane integrity, causing periplasmic proteins (mainly colicin) to leak into the medium and altering the light-refracting properties of the cells in such a way that the culture turbidity declined. The hypothesis was supported by the observation that mutations affecting colicin production abolished lysis and leakage of periplasmic proteins (Pugsley and Rosenbusch, 1981).

Studies with cloacin DF13 in the laboratories of Nijkamp and de Graaf (Hakkaart *et al.*, 1981; Oudega *et al.*, 1982) and subsequently with other colicins in our laboratory and elsewhere (reviewed in Pugsley, 1984a) soon indicated the existence of a separate gene on the colicin plasmids (the lysis gene) whose expression was blocked by polar mutations in the colicin structural gene (Fig. 4). Mutations in the lysis gene prevented all colicin secretion-related phenomena, whereas the expression of the lysis gene in the absence of colicin gene expression caused "lysis" and leakage of periplasmic proteins (Oudega *et al.*, 1982; Pugsley and Schwartz, 1983a,b). This in itself did not require a drastic revision of our model for colicin secretion since the lysis protein could (and probably does) cause the outer membrane to become permeable, but elegant immunoelectron microscopic studies of colicin-producing cells by Cavard *et al.* (1984) indicated that colicins actually accumulate in the cytoplasm and not in the periplasm.

Our approach to the study of colicin secretion turned to looking for changes in envelope composition which might explain the increased macropermeability. We thus observed an increase in the amount of lysolipids and free fatty acids in these cells relative to normal cells or to cells which were producing colicin but not lysis protein. The production of these novel products was shown to be due to the action of a normally cryptic envelope phospholipase which was activated when lysis protein was produced (Pugsley and Schwartz, 1984). Phospholipase-deficient mutants secreted colicin very inefficiently and did not "lyse." The products of phospholipase action are known to perturb membranes and, in the case of the lysolipids, to increase the nonbilayer structure of the membrane (Cullis and de Kruijff, 1979). It seems that one of the functions of the lysis protein is to bring about these changes. These observations explain some of the data on which earlier models for colicin release were based, since phospholipase action and, to a lesser extent, lysis protein make cells hyeprsensitive to plasmolysis and osmotic shock (osmotic shock also activates phospholipase) and cause the cytoplasmic membrane to become permeable to normally excluded solutes such as sucrose (used for plasmolysis) and eosin isothiocyanate. Furthermore, detergent actually has the same effect as lysolipids generated by phospholipase action (Pugsley and Schwartz, 1984), which explains why colicin and periplasmic proteins were released by detergent treatment (Pugsley and Rosenbusch, 1981).

The possibility that lysis was required for colicin release was finally disproved by the observation that growth in the presence of high concentrations of Mg^2 or Ca^2 prevented "lysis" without affecting colicin and periplasmic protein release or phospholipase activation (Pugsley and Schwartz, 1984; Luirink et al., 1986). Indeed, loss of viability normally associated with colicin release is almost completely abolished if Mg^2 and K^+ are included in the medium used to grow the cells and to measure viability (A. Pugsley, unpublished results). The main uncertainty now remaining is how the colicins get across the inner membrane so that they can leak across the permeabilized outer membrane. Some results suggest that this stage is independent of phospholipase action (Pugsley and Schwartz, 1984; Luirink et al., 1986) and that lysis protein may have a direct role in translocating colicin across the cytoplasmic membrane.

2. Other Secretion Proteins

The export of a protein to the periplasm, or even into the outer membrane, creates a concentration gradient across the outer membrane. T. Hirst (personal communication) has calculated that the concentration of cholera toxin accumulating in the V. cholerae periplasm is 1800 times greater than that present in the medium. In this and other cases where the protein accumulates in significant amounts in an extracellular compartment prior to its release, the function of the "secretion protein" may be to facilitate the specific diffusion of the secreted protein across the outer membrane. In other cases where no prior accumulation occurs, several secretion proteins or a single, multifunctional secretion protein may ensure the unhindered passage of the protein from the cytoplasm into the medium. In both examples, the secretion protein is assumed to interact directly with the protein being secreted. Secretion proteins could form specific pores, prevent the protein from folding in a conformation which cannot insert into the outer membrane, induce conformational changes in the secreted protein to cause it to partition into or out of the outer membrane, or function as protein shuttle. There is, however, no formal evidence for or against any of these models, all of which predict events for which there is only very limited evidence in other systems of protein export and secretion.

Protein–protein interactions of this type often involve specific recognition sequences in both proteins, although the absence of any strict sequence homology in different signal peptides suggests that this is not the case for the interaction between signal sequences and bacterial proteins involved in the initial stages of protein export (Pugsley and Schwartz, 1985). Nevertheless, it may be instructive to think of protein secretion events in these terms in the hope of defining the modes of action of secretion proteins. Another question which remains to be answered is whether or not secretion proteins function catalytically, as may be the case when the amount of secreted protein synthesized per generation exceeds the amount of secretion protein(s) present in the envelope.

The *hlyB* and *hlyD* gene products (Fig. 4), which are reportedly required for α-hemolysin secretion by *E. coli,* have been localized to the outer and inner membranes (Hartlein *et al.,* 1983; Wagner *et al.,* 1983; Goebel *et al.,* 1984; Mackman *et al.,* 1985). Their absence is reported to lead to the accumulation of α-hemolysin in the periplasm (HlyB) or in the outer membrane (HlyD). It seems, therefore, that the HlyB and HlyD proteins are, respectively, involved in targeting α-hemolsyin to the *E. coli* outer membrane and in catalyzing its release therefrom, as has been proposed by Goebel *et al.* (1984). Translocation of hemolysin across the inner membrane may be a spontaneous event or may be catalyzed by the *hlyB* gene product. These conclusions, however, must be tempered by the fact that the *hlyB* locus seems to encode two different polypeptides (Felmlee *et al.,* 1985a,b; Mackman *et al.,* 1985) and by recent results from R. Welch's laboratory suggesting that the *hlyD* gene product is actually a trans-acting regulator of α-hemolysin production and may not, therefore, play a direct role in hemolysin secretion (R. Welch, personal communication).

The products of the cloned pullulanase secretion gene(s) also seem to be involved in translocating protein across the outer membrane (d'Enfert *et al.,* 1987a,b). Pullulanase secretion is unusual in many respects, including the accumulation of the enzyme on the cell surface prior to its release into the medium (Michaelis *et al.,* 1985; d'Enfert *et al.,* 1987a,b). When *E. coli* is transformed with plasmids carrying the cloned pullulanase structural gene, pullulanase accumulates in the envelope but is neither exposed on the cell surface nor released into the medium. When the pullulanase secretion gene(s) is present, pullulanase is exposed on the cell surface and then released into the medium in the same way as in *K. pneumoniae* (d'Enfert *et al.,* 1987a,b). Thus, at least one of the functions of the pullulanase secretion gene(s) seems to be localization of pullulanase to the outer face of the outer membrane, where it may be anchored by the fatty acyl chains attached to the amino-terminal cysteine residue (see Section V). Since pullulanase is apparently released without the removal of these fatty acyl chains (Pugsley *et al.,* 1986), the final stage of pullulanase secretion could be a spontaneous event which depends solely on the presence of pullulanase on the cell surface, and is therefore only indirectly dependent on the action of the pullulanse secretion protein(s).

The secretion proteins might also be involved in directing the secreted protein to specific areas of the cell surface whence they could be released as specific protein-enriched vesicles (Section V). The specificity of the secretion process is presumably assured by interactions between the secretion protein and the protein being secreted (Section VI,B), and it is therefore not suprising that pullulanase and hemolysin secretion genes are not interchangeable (A. Pugsley, unpublished data), even though their functions are quite similar. It is also worth noting that pullulanse and α-hemolysin release is quite different from colicin release; lysis gene expression does not promote hemolysin or pullulanase release in the absence of their respective specific secretion functions, and changes in envelope lipid composition, phospholipase activation, and outer membrane permeabilization do not occur when cells secrete α-hemolysin or pullulanase.

Proteolytic cleavage of proteins located initially to the outer membrane could result in their specific release into the medium. Yanigida *et al.* (1986) have proposed that proteolysis of the *S. marcescens* serine protease could be involved in its secretion by *E. coli* cells carrying the cloned protease structural gene. The secreted protease (408 residues) was found to lack the 27 residue signal peptide and 646 residues from the carboxy terminus of the predicted prepro form of the enzyme. This carboxy-terminal domain includes a single potential hydrophobic membrane anchor sequence (residues 792 to 807). Thus, if the amino terminus of the membrane-associated form was fully exposed on the cell surface, a single proteolyic cleavage event between residues 408 and 409 could release it from its membrane anchor. Secretion genes were apparently not present in the DNA fragment carrying the cloned serine protease gene. Therefore, production of the mature protease may be autocatalytic or may involve a constitutive *E. coli* cell surface protease (Morona and Reeves, 1984). Immunoglobulin A (IgA) protease of *N. gonorrhoeae* may also be secreted by proteolytic cleavage of an outer membrane-bound proform of the enzyme (Halter *et al.*, 1984; Pohlner *et al.*, 1987).

C. Targeting Signals

Interactions between exported or secreted proteins and proteins specifically involved in the export process are often invoked to explain several aspects of protein insertion and translocation across membranes. Direct evidence for such protein–protein interactions has so far been obtained only in the case of the mammalian prelactalbumin–signal recognition particle (Kurzchalia *et al.*, 1986), although there is strong genetic evidence for the interaction of bacterial signal sequences with the products of the *E. coli sec* and *prl* gene products (Benson *et al.*, 1985). These interactions occur during the early stages of protein translocation across membranes. Thus, the possibility that secreted proteins might interact directly with proteins involved in their translocation across the outer membrane opens up a fascinating new area for research into the movement of proteins across membranes.

1. Sequence Analysis of Secreted Proteins

A genetic approach is likely to prove invaluable in identifying potential secretion or targeting signals in secreted proteins of gram-negative bacteria, as it has in the study of targeting signals in other proteins. Figure 5 summarizes some of the approaches used in this area of research. In those cases where several different proteins might be secreted by a common pathway, as, for example, the extracellular proteins affected by the pleiotropic secretion mutations (Table III), a comparison of the sequences of the secreted proteins may reveal homologies indicative of sites of interaction with secretion proteins. This is already possible in the case of colicins, which may interact directly with lysis proteins (Section IV,B). Lysis proteins encoded by different Col plasmids differ by only a very limited number of conservative amino acid substitutions (Cole *et al.*, 1985; Pugsley, 1988). Furthermore, lysis proteins promote the release of heterologous colicins (Pugsley and Schwartz, 1983a). Therefore, amino acid homologies between different colicins might indicate sites of interaction with lysis protein.

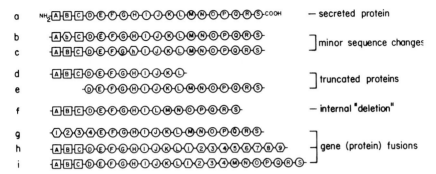

Fig. 5. Strategies for locating targeting signals in a secreted protein. Regions in the sequence of the wild-type precursor polypeptide (a) are denoted A–S, with regions A–C being a potential signal peptide. Minor sequence changes are represented as mutations affecting the sequence of the signal peptide (B → b) or the mature part of the polypeptide (H → h). Truncated polypeptides result from carboxy-terminal deletions or nonsense mutations (d) or amino-terminal deletions (e) in the structural gene for the secreted protein. Line (f) represents the product of an in-frame internal deletion in the structural gene for the secreted protein. Gene fusions are obtained by replacing part of the reading frame by another. In example (g), the 3′ end of the gene, coding for the signal peptide and region 4 of the mature protein, has been replaced by a similar sequence from another protein. In example (h), a similar change affecting the carboxy terminus of the protein has been effected. Example (i) represents an internal gene fusion in which a new reading frame has been inserted into the gene for the secreted protein such that all regions of the original secreted protein are present in the hybrid.

Analyses of published sequences for colicins A, E1, E2, E3, DF13 (Cole *et al.*, 1985; Masaki and Ohta, 1985; Morlon *et al.*, 1983; van den Elzen *et al.*, 1983; Yamada *et al.*, 1982a) reveal the striking feature that about 40% of the amino acids at the amino terminus of these colicins are glycines. It is tempting to speculate that this domain, which is likely to be extremely flexible, could be involved in colicin secretion, although a role in colicin penetration into sensitive cells has traditionally been attributed to this region of the colicin polypeptide (Pugsley, 1984a). Interestingly, colicins B, Ib, and M, which are not efficiently secreted by *E. coli* even when a lysis gene is expressed (ColB, ColI, and ColM plasmids do not have lysis genes; Pugsley, 1983), have considerably less glycine in their amino termini (~20% in colicin B and <10% in colicins Ib and M; Pressler *et al.*, 1986; Varley and Boulnois, 1984; Dreher *et al.*, 1985). It should also be noted, however, that the mechanism by which these colicins penetrate into sensitive cells differs from that of the other, secreted colicins mentioned above (Pugsley, 1984a). In this context, an analysis of the amino-terminal sequence of colicin D, a secreted colicin which is taken up by the same route as colicins Ib, M, and B (with which it shares considerable antigenic homology; Pugsley, 1984a), may prove instructive in assigning a role to the glycine-rich domain.

A region near the carboxy terminus of the α-hemolysin polypeptide contains 13 repeats of the nearly perfectly conserved sequence Leu-X-Gly-Gly-X-Gly-Asn-Asp (Felmlee *et al.*, 1985b; Hess *et al.*, 1986). These repeats may play a role in hemolysin secretion (see Section IV,C,2 and Nicaud *et al.*, 1986a), although they could also be involved in membrane penetration and catalytic action in target cells.

2. Minor Sequence Changes and Truncated Proteins

Secretion signals can be identified genetically by studying the effects of sequence changes affecting the structure of the secreted protein. While this approach is likely to prove useful, it is important to remember that secreted proteins probably undergo several conformational changes during their synthesis and secretion and that even minor sequence changes may affect this process. Thus, changes affecting sequences of the polypeptide outside of those directly involved in secretion need not necessarily be neutral with respect to protein secretion. Minor sequence changes (substitutions, small insertions, or deletions; b and c in Fig. 5) are more likely to prove useful in defining secretion signals because they should induce less drastic conformational changes than major polypeptide rearrangements. However, the only mutations of this type which have been reported so far are those affecting the hydrophobic carboxy-terminal domain of colicin E1 (Yamada and Nakazawa, 1986), which are difficult to interpret because of uncertainty as to the true function of this domain in colicin release (see below and Section III,B), since many of the mutant colicin E1 structural genes encode proteins which differ from the wild-type protein at several sites, and because of uncertainties regarding the method (osmotic shock) used to assess translocation across the cytoplasmic membrane (see Section IV,B,1).

Further searches for mutations affecting protein secretion will probably be based on the tentative localization of secretion signals by rather more drastic alterations to protein structure, such as those represented in d to i in Fig. 5. Analyses of truncated proteins (d) suggest that the carboxy terminus may not always be necessary for protein export by the signal peptide route [reviewed by Pugsley and Schwartz (1985); see Hengge and Boos (1985) and Ohsuye *et al.* (1983) for possible exceptions], although truncated peptides are often unstable (Ito and Beckwith, 1981) or behave aberrantly (Koshland and Botstein, 1982). These problems may be less important in the case of secreted proteins. Nonsense peptides of colicins E2 and N are secreted normally (Pugsley, 1984b; Cole *et al.*, 1985), suggesting that the amino termini of these colicins contain the structural information required for their release from producing cells. However, only some of the similar peptides produced following insertion of a transposon into the structural gene for colicin N were secreted (A. Pugsley, unpublished results). This may indicate a requirement for a specific peptide conformation for colicin N release. Nonsense peptides and peptides produced by transposon-inactivated genes are not directly comparable because the latter contain a limited number of carboxy-terminal residues encoded by sequences in the transposon itself and are therefore hybrid proteins.

Similarly, α-hemolysin remains cytoplasmic when the extreme carboxy terminus is not present (Hartlein *et al.*, 1983; Goebel *et al.*, 1984). Conversely, the carboxy-terminal 23,000 dalton α-hemolysin truncated peptide is secreted (Nicaud *et al.*, 1986a), suggesting that it may contain all of the signals necessary for α-hemolysin secretion.

Some colicins (e.g., E2, E3, D, and DF13) are complexed with immunity protein bound to the carboxy terminus of the colicin polypeptide when they are released (Pugsley, 1984a). This interaction may be necessary for the secretion of full-length colicin since the abundant positively charged residues near the carboxy terminus of the "uncomplexed" colicin polypeptide might impede its translocation through membranes. Similarly, cholera toxin/heat-labile exterotoxin I (HLETI) A subunits are not secreted in the absence of the B subunit (Yamamoto and Yokota, 1982; Hirst *et al.*, 1984). This may again indicate that sequences in the uncomplexed A subunit cannot cross the outer membrane, or, more likely, that the B subunit carries the signals necessary for toxin secretion.

3. Frame-Shift Mutations

A simple approach to obtaining gene fusions, illustrated in line h of Fig. 5, is to change the reading frame near the 3′ end of the structural gene for the secreted protein. This has been done with colicins A (Cavard *et al.*, 1986) and E1 (Yamada and Nakazawa, 1984) and with α-hemolysin (R. Welch, personal communication). None of these hybrid proteins were secreted. Once again, this does not necessarily indicate that the residues missing from the hybrids play a direct role in protein secretion.

4. β-Galactosidase Gene Fusions

Much of the earlier work on protein export in gram-negative bacteria made use of fusions involving the normally cytoplasmic enzyme β-galactosidase, the product of the *E. coli lacZ* gene. Almost all of the evidence accumulated in these studies indicates that β-galactosidase cannot cross the *E. coli* cytoplasmic membrane (Pugsley and Schwartz, 1985), although there are contested reports of β-galactosidase export to the outer membrane when fused to a substantial amino-terminal portion of a normally outer membrane protein (Hall *et al.*, 1982; Tommassen *et al.*, 1985). The use of β-galactosidase fusions in protein secretion studies per se has been restricted to colicins E2 and N. These hybrids were not secreted, despite the fact that colicins follow a protein export/secretion route completely different from that of the previously tested exported proteins (Pugsley and Cole, 1986; A. Pugsley and S. Cole, unpublished results).

We have also constructed a series of β-galactosidase gene fusions with the pullulanase structural gene. These hybrids, like pullulanase itself, have a fatty acylated amino terminus, although we do not know whether all three fatty acyl chains are attached or whether the signal sequence is cleaved from these hybrids (see Regue and Wu, Chapter 14, this volume). Production of these hybrids causes the cells to lyse (d'Enfert and Pugsley, 1987), as does production of hybrids containing β-galactosidase fused to other exported proteins (Ito et al., 1981). Failure to export the latter hybrids is thought to cause lysis by preventing the export of proteins essential for cell wall synthesis, such as penicillin-binding protein 3 (a lipoprotein synthesized as a precursor with a signal sequence similar to that found in other lipoproteins). It seems likely that the pullulanase–β-galactosidase hybrids have a similar effect. In contrast, production of colicin–β-galactosidase hybrids (see above) and a hybrid consisting of almost the entire length of α-hemolysin fused at the carboxy terminus to β-galactosidase (Juarez et al., 1984) was not toxic, presumably because these β-galactosidase hybrids remained in the cytoplasm (Pugsley and Cole, 1986; A. Pugsley and S. Cole, unpublished data; Juarez et al., 1984). It should be noted, however, that the experiments with the hemolysin–β-galactosidase hybrid were performed in the absence of the hemolysin secretion protein(s) (Juarez et al., 1984).

Ding et al. (1985) have constructed a series of gene fusions encoding β-galactosidase fused to the carboxy terminus of varying lengths of the P. aeruginosa secreted polypeptide phospholipase C. These hybrid genes were expressed in E. coli wherein Pseudomonas phospholipase C itself is normally localized to the outer membrane (Lory and Tai, 1983). Although the hybrids were apparently made in large amounts by E. coli carrying the cloned gene fusion, Ding et al. did not remark on their possible toxic effects. This may indicate that these hybrids remain in the cytoplasm because export is not initiated or that phospholipase C export to the outer membrane occurs via a signal peptide-independent mechanism.

5. Other Gene Fusions and Major Gene Rearrangements

Sequences derived from polypeptides which are themselves normally exported out of the cytoplasm are less likely than β-galactosidase to impede the translocation of hybrid proteins through the cell envelope. However, these foreign sequences may contain their own "targeting signals" which could affect the operation of secretion signals in a hybrid polypeptide. Furthermore, the *E. coli* outer membrane is structurally different from the cytoplasmic membrane, and it cannot, therefore, be automatically assumed that sequences introduced into the hybrid would not impede its movement across the outer membrane.

Hybrid polypeptides containing parts of a colicin polypeptide have been the subject of most of the studies on this aspect of protein secretion, largely because the yields of secreted protein can be quite high and because there may be fewer constraints on the sequences which can be fused into the colicin polypeptide without affecting secretion. The results obtained with these hybrids are, for the present, rather limited in scope and difficult to interpret. For example, there were no compelling reasons to expect that hybrid polypeptides containing all but the extreme carboxy terminus of colicin N fused to the entire length of the alkaline phosphatase polypeptide (minus the signal peptide and extreme amino terminus) would not be secreted, since the amino terminus of the colicin can be secreted alone (see above) and alkaline phosphatase, normally a periplasmic protein, can cross lipid bilayers. These hybrids were not secreted, however, apparently because they formed large aggregates in the cytoplasm where they accumulated (Pugsley and Cole, 1986).

Similarly, Cavard *et al.* (1986) have reported that an approximately 18K hybrid polypeptide consisting of the first 70 amino acids of colicin A (the glycine-rich domain; Section IV,C,1) fused to the carboxy-terminal 103 residues of the normally periplasmic β-lactamase was also not secreted. Failure to secrete this hybrid was attributed to the absence of the colicin A carboxy terminus, since the about 32K product of a colicin A structural gene containing a large, internal, in-frame deletion was secreted normally (Cavard *et al.,* 1986; see also Sections III,B and IV,C,2–IV,C,4). A requirement for the carboxy terminus in colicin A secretion would not be in line with our evidence that this region of colicins N and E2 is not required per se for secretion of these colicins (Section IV,C,2). Thus, it could also be that the ability to secrete colicin polypeptides is extremely sensitive to even minor alterations in primary sequence which change the overall conformation of the polypeptide. Some of our recent results from studies with hybrids containing parts of the colicin E2 polypeptide are compatible with this idea, since they suggest that only hybrids containing relatively short stretches of "foreign" amino acids can be secreted (S. Cole, unpublished data).

The only other attempts to secrete hybrid polypeptides concern pullulanase–alkaline phosphatase chimeras studied in our laboratory. Hybrids containing different lengths of the prepullulanase polypeptide fused to alkaline phosphatase without its signal peptide were apparently exported out of the cytoplasm by *E. coli,* since they seemed to be acylated in the same way as pullulanase (d'Enfert *et al.,* 1987). They were not secreted, however, either by *K. pneumoniae* or by *E. coli* carrying the *Klebsiella* secretion genes. This may be because sequences in the pullulanase part of the polypeptide which would normally interact with the secretion proteins either were buried within the tertiary structure of the hybrid polypeptide or are located near the carboxy terminus of the pullulanase polypeptide and therefore absent from the hybrids.

Further studies on the secretion of hybrid polypeptides will almost certainly be conducted, particularly in view of the interest in the potential biotechnological applications of protein secretion (Pugsley and Schwartz, 1985; Nicaud *et al.,* 1986b). However, these applications may be severely restricted by the aforementioned limitations, including the need to fold the polypeptide into a particular tertiary structure, to retain secretion signals, which may be dispersed throughout the length of the secreted protein, and to interact with envelope proteins specifically involved in the secretion process. Gram-negative bacteria may therefore be less attractive for the industrial production of hybrid secreted proteins than gram-positive bacteria, yeast, or mammalian cells currently used for this purpose.

V. RELEASE OF PROTEINS IN VESICLES

The results of many of the early studies on protein secretion by gram-negative bacteria indicated that the proteins were often "contaminated" by typical membrane components such as phospholipids or lipopolysaccharides. This led to the idea that secreted proteins were released in vesicles which were presumed to be "blebbed off" from the cell surface. This aspect of protein secretion has been largely ignored in more recent studies, mainly because protein release was seen to be a specific event occurring without the release of other proteins from the outer membrane (section II). The release of vesicles enriched in particular proteins could, however, represent a highly effective method for secreting proteins into the medium. Secretion or targeting proteins may be required to ensure the correct localization of the proteins in these specialized areas of the cell surface prior to their release (see, e.g., Thompson *et al.,* 1985).

Experimental data suggesting that secreted proteins are released in vesicles include copurification or cofractionation of the secreted protein with membrane components such as phospholipids or lipopolysaccharides in the medium and their presence in the medium as high molecular weight "aggregates." It is difficult to exclude the possibility that secreted proteins are in fact present in "pure" aggregates

(Gonzalez-Carrero *et al.*, 1985) and that their cofractionation with envelope components is coincidental, or that they associate with cell-free membrane components after they have been secreted. On the other hand, the fact that the predominant extracellular form of a secreted protein is not aggregated or present in vesicles need not indicate that it was released from the cell in monomeric form, since specific, limited proteolysis or other covalent modifications could change protein–protein or protein–lipid interactions after the protein has been released.

Recent studies suggest that three of the best-characterized secreted proteins of gram-negative bacteria, α-hemolysin, pullulanase, and cholera toxin/HLETI, may indeed be secreted in membrane vesicles (Bohach and Snyder, 1985; Pugsley *et al.*, 1986; Gankema *et al.*, 1980). It is interesting that these three proteins are secreted by apparently different pathways: α-hemolysin does not accumulate within producing cells (Mackman *et al.*, 1986), whereas pullulanase accumulates on the cell surface (Michaelis *et al.*, 1985) and cholera toxin/HLETI accumulates in the periplasm (Levner *et al.*, 1980; Hirst *et al.*, 1984). It is not known whether the toxin B subunit, which can accumulate in the periplasm and which can be secreted without the A subunit (Section IV,C,1), is also released in vesicles.

The evidence that α-hemolysin or pullulanase secretion do not involve a general blebbing of vesicles from the cell surface is the absence from the medium of significant amounts of major outer membrane proteins, which are made in larger amounts than the secreted proteins (Fig. 2). As already mentioned, this raises the possibility that secretion involves the release of vesicles from specific areas of the cell surface where the secreted protein is either exported across the envelope (α-hemolysin) or accumulates for long periods (pullulanase) prior to its secretion. Lipids or lipopolysaccharides released with the protein may need to form nonbilayer structures or micelles in order to promote vesicularization. Although this may be one of the functions of the protein secretion factors discussed in Section IV,A, such regions of nonbilayer structure may already exist where the outer and inner membranes are naturally in contact [the so-called Bayer patches (Bayer, 1979)] or could be produced through the action of the normally dormant phospholipase A (see Section IV,B).

Long *et al.* (1983) suggested that exotoxin A could be secreted through Bayer patches in the *P. aeruginosa* cell envelope, although the possibility that this protein was released in vesicles was apparently not considered. Bayer reported that regions of contact between the inner and outer membranes are enriched in phospholipase A and C activity (Bayer *et al.*, 1982; Bayer and Bayer, 1985), although the active fraction probably represents only a small proportion of the total amount of largely inactive phospholipase A in the envelope. The absence of phospholipase A activity was apparently without effect on hemolysin secretion by *E. coli* (A. Pugsley, unpublished results), and the absence of detectable changes in phospholipid composition during pullulanase secretion (d'Enfert *et al.,* 1987a) suggests that increased phospholipase activity is not required for pullulanase secretion from *K. pneumoniae.*

Of all the examples of proteins which might be secreted in vesicles, that of pullulanase is perhaps the easiest to visualize mechanistically. We have proposed that fatty acyls attached to the aminoterminal cysteine residue of the mature pullulanase polypeptide might anchor the protein in the outer leaflet of the outer membrane, as well as in vesicles once it is released (Pugsley *et al.,* 1986; d'Enfert *et al.,* 1987a). Indeed, pullulanase might form micelles spontaneously, without the need to interact with lipids or lipopolysaccharides, which would be in line with the observation (d'Enfert *et al.* 1987a) that pullulanese secretion by *K. pneumoniae* does not involve increased lipid or lipopolysaccharide release into the medium. If this is correct, then the replacement of the pullulanase signal sequence and aminoterminal cysteine by another cleavable signal sequence could have either of two distinct effects: (1) the protein may not be secreted because it was not correctly localized on the cell surface or in regions of accentuated nonbilayer structure, or (2) it may be secreted directly into the medium as monomers, without the prior accumulation of a membrane-bound form. In addition to testing these possibilities, we are currently attempting single-step purification of secreted pullulanase by affinity chromatography to see whether other envelope components are indeed associated with it.

VI. ACTIVATORS AND INHIBITORS OF SECRETED PROTEINS

A special feature of certain secreted proteins is that they could be toxic to the producing cells. For example, secreted proteases and other hydrolytic enzymes might degrade macromolecules within the producing cell or from the outside, and some toxins, and in particular colicins, could kill producing cells. Bacteria might circumvent some of these problems by cotranslational secretion of the toxic proteins, by synthesizing an inactive precursor, or by producing inhibitors to prevent the action of the secreted protein, at least while it remains within the producing cell.

A. Activation of Secreted Proteins

Several secreted proteins from gram-negative bacteria, including aerolysin of *A. hydrophila,* hemolysin of *Aeromonas salmonicida,* cholera toxin of *V. cholerae*, and protease of *P. aeruginosa*, are secreted in a partially or totally inactive form. In the first three cases, activation occurs through cleavage by secreted proteases, resulting in the elimination of the carboxy terminus (aerolysin, and possibly hemolysin) or in nicking to produce two polypeptides which remain associated (cholera toxin A subunit) (Howard and Buckley, 1985; Titbank *et al.*, 1985; Booth *et al.*, 1984). *Aeromonas salmonicida* mutants devoid of secreted protease (caseinase) produced an inactive hemolysin precursor (Titbank *et al.*, 1985). It is not known whether aerolysin and hemolysin activation is the sole function of the secreted *Aeromonas* protease(s), but the enzyme which nicks cholera toxin (cholera lectin), is also a polymer-forming hemagglutinin (Finklestein and Hanne, 1982; Booth *et al.*, 1984) and therefore has at least two functions.

Activation of *P. aeruginosa* protease seems to involve a subtle conformational change which occurs spontaneously during the purification of the enzyme or through the action of endogenous proteases, although removal of a protease inhibitor (see Section VI,B) could also be involved in activation (Fecycz and Campbell, 1985). Pleiotropic nonsecreting mutants of *A. hydrophila* and *P. aeruginosa* (Section IV,A) accumulate inactive forms of aerolysin and protease, respectively (Howard and Buckley, 1985; Fecycz and Campbell, 1985). In other cases, conformational differences between the precursor and mature forms of secreted proteins might be such that the precursor is inactive, thereby protecting the producing cell against endogenous attack, at least until the polypeptide is translocated across the cytoplasmic membrane.

The activity of *E. coli* α-hemolysin also depends on the covalent modification of the polypeptide. In this case, the product of the *hlyC* gene, located upstream from the hemolysin structural gene (*hlyA*; Fig. 4), is required for hemolysin activation (Goebel *et al.*, 1984; Nicaud *et al.*, 1985b). Contrary to earlier reports, proteolysis is not needed for hemolysin activation (Goebel *et al.*, 1984); rather, the HlyC protein, which is cytoplasmic, seems to catalyze a minor covalent modification (Nicaud *et al.*, 1985b), probably the addition of a triglyceride which is essential for hemolysin activity (W. Goebel, personal communication). Unmodified hemolysin is secreted normally (Nicaud *et al.*, 1985b). A secreted protease of *S. marcescens* also seems to be covalently modified, but the nature and role of this modification remain to be established (Schmitz and Braun, 1985). [This protease is different from that studied by Yanigida *et al.* (1986) (see below).] Other secreted proteins do not appear to have been examined for covalent modifications, although periplasmic heat-stable enterotoxin I (HSETI) is reportedly inactive unless it is heated (Guzmán-Verduzco *et al.*, 1983).

B. Inhibitors of Secreted Protein Activity

The best-characterized examples of inhibitors of secreted proteins are the colicin immunity proteins. The products of immunity genes which are cotranscribed with the colicin structural genes (Fig. 4) are secreted in stoichiometric amounts with the colicin polypeptide, and remain associated with it at least until the colicin binds to cell surface receptors on target cells (Krone *et al.*, 1986). These immunity proteins must, therefore, be regarded as secreted proteins (Section IV,C,1). The products of immunity genes which are transcribed in the opposite direction to the colicin structural gene (Fig. 4) remain within the producing cell, probably in the cytoplasmic membrane (Weaver *et al.*, 1981; Bishop *et al.*, 1985), where they neutralize the pore-forming activity of these colicins as they attack immune cells from the outside. These bacteria produce considerably less immunity protein than colicin. However, pore-forming colicins do not normally act from within the producing cells, even in the absence of immunity protein, presumably because they cannot penetrate into the cytoplasmic membrane from the inside of the cell or do so only in an inactive conformation (Pugsley, 1984a). Both types of immunity proteins can be produced in the absence of colicin production (the nature of colicin gene regulation is such that only a few cells in the population are producing colicin at any one time), thereby protecting nonproducing cells against the action of colicin secreted by their neighbors (Pugsley, 1984a).

Another example of an inhibitor of secreted proteins is the protease inhibitor of *E. chrysanthemi*. The physiological role of this inhibitor, which is a protein encoded by a gene linked to the protease structural genes, remains to be elucidated. *Escherichia coli* K12 strains bearing plasmids carrying protease structural genes and a mutated inhibitor gene are perfectly viable, indicating that the inhibitor is not required to protect cells against the action of their own proteases (C. Wandersman and P. Delepelaire, personal communication).

VII. SPECIAL CASES

A. Membrane Appendages

Although not normally released into the medium, the protein subunits of pili, fimbriae, and flagellae should perhaps be considered as special examples of secreted proteins because they are localized entirely outside the outer membrane, with which they remain associated through interactions with specific envelope proteins. Translocation of subunit polypeptides, from which these appendages are assembled, across the outer membrane is probably catalyzed by the products of one or more genes which, in many cases, are closely linked to the structural gene for the subunit polypeptide (Mooi *et al.*, 1983; Dougan *et al.*, 1983; Kehoe *et al.*, 1983; Klemm *et al.*, 1985). The failure of *E. coli* to assemble fimbriae from subunit polypeptides encoded by the cloned *Neisseria, Moraxella,* or *Pseudomonas* fimbrilin genes (Meyer *et al.*, 1982; Marrs *et al.*, 1985; Finlay *et al.*, 1986; Strom and Lory, 1986) is almost certainly due to the absence of these assembly and secretion proteins, the genes for which might not, in these cases, be linked to the fimbrilin gene.

Fimbrilins are made as precursors with cleavable signal sequences which seem to be of two different types. Fimbrilins of bacteria in the family Enterobacteriaceae (which includes *E. coli*) are made with classical signal sequences (Klemm, 1984; Orndorff and Falkow, 1985; Rhen *et al.*, 1985; van Die *et al.*, 1985) which, presumably, suffice to engage the polypeptide in the normal signal peptide-dependent export pathway, as has been demonstrated for the type I fimbrilin of *E. coli* (Dodd *et al.*, 1984). The amino termini of fimbrilin precursors produced by other gram-negative bacteria also look like classical signal sequences, but they are cleaved at a site which is much closer to the amino terminus than is normally the case (Marrs *et al.*, 1985; Sastry *et al.*, 1986; Pasloske *et al.*, 1985; Meyer *et al.*, 1982). The predominantly hydrophobic part of the "signal sequence" thus remains intact at the amino terminus of the mature fimbrilin, where it may play a role in later stages of fimbrilin secretion, assembly, or function (adhesion to specific eukaryotic cell surface receptors). The continued presence of the charged aminoterminal amino acids of the signal sequence might have a deleterious effect on these activities, and could actually prevent the release of the polypeptide from the cytoplasmic membrane, as has been suggested for the complete, "normal" signal peptide in precursors of other, exported proteins (Randall and Hardy, 1984; Benson *et al.*, 1985; Pugsley and Schwartz, 1985; Dalbey and Wickner, 1985). Cleavage of these prefimbrilin signal sequences presumably involves a different peptidase from that which acts on most exported proteins [other secreted proteins of *Pseudomonas*, e.g., exotoxin A (Gray *et al.*, 1984) or phospholipase C (Pritchard and Vasil, 1986), have typical signal sequences]. This novel peptidase could be localized in specific areas of the envelope where the subunits were assembled into fimbriae, and its absence from *E. coli* could contribute to the failure of these bacteria to assemble fimbriae from fimbrilin subunits which accumulate in the cytoplasmic membrane (see above).

The signal peptide of pilin encoded by the *E. coli* F sex factor, and by related plasmids, differs from other signal peptides in having a longer positively charged amino-terminal domain (up to and including residue 33

difficulties that are likely to be encountered in attempts to secrete hybrids containing part of the normally secreted protein coupled to another protein (e.g., hormones or proteins useful in the preparation of vaccines); such limitations are likely to restrict further applications of bacterial protein secretion to gram-positive bacteria.

Finally, protein secretion by gram-negative bacteria seems to offer a unique possibility to study protein–protein interactions involved in translocating proteins across bilayer membranes. The structural genes coding for the secretion or translocation of proteins are amenable to refined genetic analysis, and it should be possible to devise an *in vitro* system with purified outer membrane vesicles with which to study events occuring during the final stages of protein secretion.

ACKNOWLEDGMENTS

I am grateful to my colleagues, and in particular to Stewart Cole, Philipe Delepelaire, Christophe d'Enfert, Maxime Schwartz, and Cecile Wandersman, for their material and conceptual contributions to this chapter, and to various authors for permission to quote their unpublished data. Work in the author's laboratory is financed by the Centre National de la Recherche Scientifique, the Ministère de la Recherche et de la Technologie, the Institut Pasteur, and the Foundation pour la Recherche Médicale.

REFERENCES

Andro, T., Chambost, J.-P., Kotoujansky, A., Cattaneo, J., Bertheau, Y., Barras, F., van Gijsegem, F., and Coleno, A. (1984). Mutants of *Erwinia chrysanthemi* defective in secretion of pectinase and cellulase. *J. Bacteriol.* **160,** 1199–1203.

Bakker, E. P., and Randall, L. L. (1984). The requirement for energy during export of β-lactamase in *Escherichia coli* is fulfilled by the total proton motive force. *EMBO J,* **3,** 895–900.

Baquero, F., and Moreno, F. (1984). The microcins. *FEMS Microbiol. Lett.* **23,** 117–124.

Barras, F., Thurn, K. K., and Chatterjee, A. K. (1986). Export of *Erwinia chrysanthemi* (EC16) protease by *Escherichia coli. FEMS Microbiol. Lett.* **34,** 343–348.

Bayer, M. E. (1979). The fusion sites between the outer membrane and the inner membrane: Their role in membrane biogenesis and viral infection. *In* "Bacterial Outer Membranes: Biogenesis and Functions" (M. Inouye, ed.), pp. 67–201. Wiley, New York.

Bayer, M. H., and Bayer, M. E. (1985). Phophoglycerides and phospholipase C in membrane fractions of *Escherichia coli* B. *J. Bacteriol.* **162,** 50–54.

Bayer, M. H., Costello, G. P., and Bayer, M. E. (1982). Isolation and partial characterization of membrane vesicles carrying markers of the membrane adhesion sites. *J. Bacteriol.* **149,** 758–767.

Belland, R. J., and Thrust, T. J. (1985). Synthesis, export and assembly of *Aeromonas salmonicida* A-layer analyzed by transposon mutagenesis. *J. Bacteriol.* **153,** 877–881.

Benson, S. A., and Silhavy, T. J. (1983). Information within mature LamB protein necessary for localization to the outer membrane. *Cell* **32,** 1325–1335.

Benson, S. A., Hall, M. N., and Silhavy, T. J. (1985). Genetic analysis of protein export in *Escherichia coli* K12. *Annu. Rev. Biochem.* **54**, 101–134.

Bhakdi, S., Mackman, N., Nicaud, J.-M., and Holland, I. B. (1986). *Escherichia coli* hemolysin may damage membranes by generating transmembrane pores. *Infect. Immun.* **52**, 63–69.

Bishop, L. J., Bjes, E. S., Davidson, V. L., and Cramer, W. A. (1985). Localization of the immunity protein-reactive domain in unmodified and chemically modified COOH-terminal peptides of colicin E1. *J. Bacteriol.* **164**, 237–244.

Bohach, G. A., and Snyder, I. S. (1985). Chemical and immunological analysis of the complex structure of *Escherichia coli* α-hemolysin. *J. Bacteriol.* **164**, 1071–1080.

Booth, B. A., Boesman-Finklestein, M., and Finklestein, R. A. (1984). *Vibrio cholerae* haemagglutinin/protease nicks cholera enterotoxin. *Infect. Immun.* **45**, 558–560.

Borris, R., Baumlein, H., and Hofemeister, J. (1985). Expression in *Escherichia coli* of a cloned β-gluconase gene from *Bacillus amyloliquefaciens*. *Appl. Microbiol. Biotechnol.* **22**, 63–71.

Bosch, D., Leunissen, J., Verbakel, J., de Jong, M., van Erp, H., and Tommassen, J. (1986). Periplasmic accumulation of outer-membrane PhoE protein of *Escherichia coli* K12. *J. Mol. Biol.* **189**, 449–455.

Bricker, J., Mulks, M. H., Plaut, A. G., Moxan, E. R., and Wright, A. (1983). IgA1 protease of *Haemophilus influenzae:* cloning and characterization in *Escherichia coli*. *Proc. Natl. Acad. Sci. U.S.A.* **80**, 2681–2685.

Cabilly, S., Riggs, A. D., Pande, H., Shively, J. E., Holmes, W. E., Rey, M., Perry, L. J., Wetzel, W., and Heyneker, H. L. (1984). Generation of antibody activity from immunoglobulin polypeptide chains produced in *Escherichia coli*. *Proc. Natl. Acad. Sci. U.S.A.* **81**, 3273–3277.

Cashmore, A., Szabo, L., Timko, A., Van den Broeck, G., Schrier, P., Bohnert, H., Herrera-Estrella, L., Van Montagu, M., and Schell, J. (1985). Import of polypeptides into chloroplasts. *Biotechnol.* **3**, 803–808.

Cavard, D., Bernadac, A., Pagès, J.-M., and Lazdunski, C. (1984). Colicins are not transiently accumulated in the periplasmic space before release from colicinogenic cells. *Biol. Cell* **51**, 79–86.

Cavard, D., Crozel, V., Gorvel, J.-P., Pattus, F., Baty, D., and Lazdunski, C. (1986). A molecular, genetic and immunological approach to the functioning of colicin A, a pore-forming protein. *J. Mol. Biol.* **187**, 449–459.

Chapon, C., and Raibaud, O. (1985). Structure of two divergent promoters located in fron of the gene encoding pullulanase in *Klebsiella pneumoniae* and positively regulated by the *malT* product. *J. Bacteriol.* **164**, 639–645.

Cole, S. T., Saint-Joanis, B., and Pugsley, A. P. (1985). Molecular characterization of the colicin E2 operon and identification of its products. *Mol. Gen. Genet.* **198**, 465–472.

Coleman, K., Dougan, G., and Arbuthnott, J. P. (1983). Cloning and expression in *Escherichia coli* of the chromosomal hemolysin (phospholipase C) determinant of *Pseudomonas aeruginosa*. *J. Bacteriol.* **153**, 909–915.

Collmer, A., Schoedel, C., Roeder, D. L., Ried, J. L., and Rissler, J. F. (1985). Molecular cloning in *Escherichia coli* of *Erwinia chrysanthemi* genes encoding multiple forms of pectate lyase. *J. Bacteriol.* **161**, 913–920.

Cornelis, P., Digneffe, C., and Willemot, K. (1982). Cloning and expression of a *Bacillus coagulans* amylase gene in *Escherichia coli*. *Mol. Gen. Genet.* **186**, 507–511.

Cornet, P., Millet, J., Béguin, P., and Aubert, J.-P. (1983). Characterization of two *cel* (cellulose degradation) genes of *Clostridium thermocellum* coding for endoglucanases. *Biotechnol.* **1**, 589–594.

Cullis, P. R., and de Kruijff, B. (1979). Lipid polymorphism and the functional roles of lipids in biological membranes. *Biochim. Biophys. Acta* **559**, 399–420.

Dalbey, R. E., and Wickner, W. T. (1985). Leader peptidase catalyzes the release of exported proteins from the outer surface of the *Escherichia coli* plasma membrane. *J. Biol. Chem.* **260**, 15925–15931.

Dallas, W. S., and Falkow, S. (1980). Amino acid sequence homology between cholera toxin and *Escherichia coli* heat-labile toxin. *Nature (London)* **288**, 499–501.

Davidson, V. L., Brunden, K. R., Cramer, W. A., and Cohen, F. S. (1984). Studies on the mechanism of action of channel-forming colicins using artificial membranes. *J. Membr. Biol.* **79**, 105–118.

d'Enfert, C. and Pugsley, A. P. (1987). A gene fusion approach to the study of pullulanase export and secretion. *Escherichia coli. Mol. Microbiol.* **1**, 159–168.

d'Enfert, C., Chapon, C., and Puglsey, A. P. (1987a). Export and secretion of the lipoprotein pullulanase by *klebsiella pneumoniae. Mol. Microbiol.* **1**, 107–116.

d'Enfert, C. Ryter, A., and Pugsley, A. P. (1987b). Cloning and expression in *Escherichia coli* of the *klebniella pneumoniue* genes for production surface localization and secretion of the lipoprotein pullulanase. *EMBO. J* **6**, 3531–3538.

Ding, J., Lory, S., and Tai, P. C. (1985). Orientation and expression of the cloned hemolysin gene of *Pseudomonas aeruginosa. Gene* **33**, 313–321.

Dodd, D. C., Bassford, P. J., Jr, and Eisenstein, B. I. (1984). Dependence of secretion and assembly of type 1 fimbrial subunits of *Escherichia coli* on normal protein export. *J. Bacteriol.* **159**, 1077–1079.

Dodt, J., Schmitz, T., Schäfer, T., and Bergmann, C. (1986). Expression, secretion and processing of hirudin in *E. coli* using the alkaline phosphatase signal sequence. *FEBS Lett.* **202**, 373–377.

Dougan, G., Dowd, G., and Kehoe, M. (1983). Organization of K88ac-encoded polypeptides in the *Escherichia coli* cell envelope; use of minicells and outer membrane protein mutants for studying assembly of pili. *J. Bacteriol.* **153**, 364–370.

Dreher, R., Braun, V., and Wittmann-Liebold, B. (1985). Functional domains of colicin M. *Arch. Microbiol.* **140**, 343–346.

Fecycz, I. T., and Campbell, J. N. (1985). Mechanisms of activation and secretion of a cell-associated precursor of an exocellular protease of *Pseudomonas aeruginosa* 34362A. *Eur. J. Biochem.* **146**, 35-42.

Felmlee, T., Pellett, S., Lee, E.-Y., and Welch, R. A. (1985a). *Escherichia coli* hemolysin is released without cleavage of a signal peptide. *J. Bacteriol.* **163**, 88–93.

Felmlee, T., Pellett, S., and Welch, R. (1985b). Nucleotide sequence of an *Escherichia coli* chromosomal hemolysin. *J. Bacteriol.* **163**, 94–105.

Filloux, A., Joyet, P., Murgier, M., and Lazdunski, A. (1985). Cloning and expression of a *Bacillus licheniformis* α-amylase gene in *Psuedomonas aeruginosa. FEMS Microbiol. Lett.* **30**, 203–207.

Finklestein, R. A., and Hanne, L. F. (1982). Purification and characterization of the soluble hemagglutinin (cholera lectin) produced by *Vibrio cholerae. Infect. Immun.* **36**, 1199–1208.

Finlay, B. B., Pasloske, B. L., and Paranchych, W. (1986). Expression of *Pseudomonas aeruginosa* PAK pilin gene in *Escherichia coli. J. Bacteriol.* **165**, 625–630.

Focareta, T., and Manning, P. M. (1985). Molecular cloning of a possible excretion protein of *Vibrio cholerae. FEMS Microbiol. Lett.* **29**, 161–166.

Freudl, R., Schwartz, H., Klose, M., Movva, N. R., and Henning, U. (1985). The nature of information, required for export and sorting, present within the outer membrane protein OmpA of *Escherichia coli* K12. *EMBO J.* **4**, 3595–3598.

Friedlander, M., and Blobel, G. (1985). Bovine opsin has more than one signal sequence. *Nature (London)* **318**, 338–343.

Frost, L. S., Finlay, B. B., Opgenroth, A., Paranchych, W., and Lee, J. S. (1985). Characterization and sequence analysis of pilin from F-like plasmids. *J. Bacteriol.* **164**, 1238–1247.

Gankema, H., Wensink, J., Guinée, P. A. M., Jansen, W. H., and Witholt, B. (1980). Some characteristics of outer membrane material released by growing enterotoxigenic *Escherichia coli. Infect. Immun.* **29**, 704–713.

Gay, P., Le Coq, D., Steinmetz, M., Ferrari, E., and Hoch, J. A. (1983). Cloning structural gene *sacB,* which codes for exoenzyme levansucrase of *Bacillus subtilis:* Expression of the gene in *Escherichia coli. J. Bacteriol.* **153**, 1424–1431.

Geller, B. L., Movva, N. R., and Wickner, W. T. (1986). Both ATP and the electrochemical potential are required for optimal assembly of pro-OmpA into *Escherichia coli* inner membrane vesicles. *Proc. Natl. Acad. Sci. U.S.A.* **83**, 4219–4222.

Goebel, W., Hacker, J., Knapp, S., Then, J., Wagner, W., Hughes, C., and Juarez, A. (1984). Structure, function and regulation of the plasmid-encoded hemolysin determinant of *E. coli. In* "Plasmids in Bacteria" (D. R. Helinski, S. N. Cohen, D. B. Clewell, D. A. Jackson, and A. Hollaender, eds.), pp. 791–805. Plenum, New York.

Gonazles-Carrero, M. I., Zubidu, J. C., de la Cruz, F., and Ortiz, J. M. (1985). Purification of a haemolysin from an overproducing *E. coli* strain. *Mol. Gen. Genet.* **199**, 106–110.

Gould, A. R., May, B. K., and Elliot, W. H. (1974). Release of extracellular enzymes by *Bacillus amyloliquefaciens. J. Bacteriol.* **122**, 34–40.

Gray, G. L., and Vasil, M. L. (1981). Isolation and genetic characterization of toxin-deficient mutants of *Pseudomonas aeruginosa* PAO. *J. Bacteriol.* **147**, 275–281.

Gray, G. L., Smith, D. H., Baldridge, J. S., Harkins, R. N., Vasil, M. L., Chen, E. Y., and Heyneker, H. L. (1984). Cloning, nucleotide sequence and expression in *Escherichia coli* of the exotoxin A structural gene of *Pseudomonas aeruginosa. Proc. Natl. Acad. Sci. U.S.A.* **81**, 2645–2649.

Guzmán-Verduzco, L. M., Fonseca, R., and Kaperszotch-Portnoy, Y. M. (1983). Thermoactivation of a periplasmic heat-stable enterotoxin of *Escherichia coli. J. Bacteriol.* **154**, 146–151.

Hakkaart, M. J. J., Veltkamp, E., and Nijkamp, H. J. J. (1981). Protein H encoded by plasmid CloDF13 involved in lysis of the bacterial host. I. Localization of the gene and identification and subcellular localization of the gene H product. *Mol. Gen. Genet.* **183**, 318–325.

Hall, M. N., Schwartz, M., and Silhavy, T. J. (1982). Sequence information within the *lamB* gene is required for proper routing of the bacteriophage λ receptor to the outer membrane of *Escherichia coli* K12. *J. Mol. Biol.* **156**, 93–112.

Halter, R., Pohlner, J., and Meyer, T. F. (1984). IgA protease of *Neisseria gonorrhoeae:* Isolation and characterization of the gene and its extracellular product. *EMBO J,* **3**, 1595–1601.

Hartlein, M., Schleissl, S., Wagner, W., Rdest, U., Kreft, J., and Goebel, W. (1983). Transport of hemolysin by *Escherichia coli. J. Cell. Biochem.* **22**, 87–97.

Hay, R., Böhni, P., and Gasser, S. (1984). How mitochondria import proteins. *Biochim. Biophys. Acta* **779**, 65–87.

Hengge, R., and Boos, W. (1985). Defective secretion of maltose- and ribose-binding proteins caused by a truncated periplasmic protein in *Escherichia coli. J. Bacteriol.* **162**, 872–978.

Hess, J., Wels, W., Vogel, M., and Goebel, W. (1986). Nucleotide sequence of a plasmid-encoded haemolysin determinant and its comparison with a corresponding chromosomal haemolysin determinant. *FEMS Microbiol. Lett.* **34**, 1–11.

Hindcliffe, E. (1984). Cloning and expression of a *Bacillus subtilis* endo-1-3,1-4-β-D-glucanase gene in *Escherichia coli* K12. *J. Gen. Microbiol.* **130**, 1285–1291.

Hirst, T., Sanchez, J., Kaper, J. B., Hardy, S. J. S., and Holmgren, J. (1984). Mechanism of toxin secretion by *Vibrio cholerae* investigated in strains harboring plasmids that encode heat-labile enterotoxins of *Escherichia coli*. *Proc. Natl. Acad. Sci. U.S.A.* **81**, 7752–7756.

Homma, M., Fujita, H., Yamaguchi, S., and Iino, T. (1984). Excretion of unassembled flagellin by *Salmonella typhimurium* mutants deficient in hook-associated proteins. *J. Bacteriol.* **159**, 1056–1059.

Howard, S. P., and Buckley, J. T. (1983). Intracellular accumulation of extracellular proteins by pleiotropic export mutants of *Aeromonas hydrophila*. *J. Bacteriol.* **154**, 413–418.

Howard, S. P., and Buckley, J. T. (1985). Protein export by a gram-negative bacterium: Production of aerolysin by *Aeromonas hydrophila*. *J. Bacteriol.* **161**, 1118–1124.

Howard, S. P., and Buckley, J. T. (1986). Molecular cloning and expression in *Escherichia coli* of the structural gene for the hemolytic toxin aerolysin from *Aeromonas hydrophila*. *Mol. Gen. Genet.* **204**, 289–295.

Hussain, M., Carlino, A., Madonna, M. J., and Lampen, J. O. (1985). Cloning and sequencing of the metallothioprotein β-lactamase II gene of *Bacillus cereus* 569/H in *Escherichia coli*. *J. Bacteriol.* **164**, 223–229.

Ikeda, T., Kamiya, R., and Yamaguchi, S. (1983). Excretion of flagellin by a short-flagella mutant of *Salmonella typhimurium*. *J. Bacteriol.* **153**, 506–510.

Ito, K., and Beckwith, J. (1981). Role of the mature protein sequence of maltose-binding protein in its secretion across the *E. coli* cytoplasmic membrane. *Cell* **25**, 143–150.

Ito, K., Bassford, P. J., Jr, and Beckwith, J. (1981). Protein localization in *E. coli:* Is there a common step in the secretion of periplasmic and outer membrane proteins? *Cell* **24**, 707–717.

Izui, K., Nielsen, J. B. K., Caulfield, M. P., and Lampen, J. O. (1980). Large exopenicillinase, initial extracellular form detected in cultures of *Bacillus licheniformis*. *Biochemistry* **19**, 1882–1886.

Jackson, M. E., Pratt, J. M., Stoker, N. G., and Holland, I. B. (1985). An inner membrane N-terminal signal sequence is able to promote efficient localization of an outer membrane protein in *Escherichia coli*. *EMBO J*, **4**, 2377–2383.

Jakes, K., and Model, P. (1979). Mechanism of export of colicin E1 and colicin E3. *J. Bacteriol.* **130**, 770–778.

Jones, J. D. G., Grady, K. L., Suslow, T. V., and Bedbrook, J. R. (1986). Isolation and characterization of genes encoding two chitinase enzymes from *Serratia marcescens*. *EMBO J*, **5**, 467–473.

Joyet, P., Guérineau, M., and Heslot, H. (1984). Cloning of a thermostable α-amylase gene from *Bacillus licheniformis* and its expression in *Escherichia coli* and *Bacillus subtilis*. *FEMS Microbiol. Lett* **21**, 253–258.

Juarez, A., Härtlein, M., and Goebel, W. (1984). Study of regulation and transport of hemolysin by using fusion of the β-galactosidase gene (*lacZ*) to hemolysin genes. *J. Bacteriol.* **160**, 161–168.

Karlin-Neumann, G. A., and Tobin, E. M. (1986). Transit peptides of nuclear-encoded chloroplast proteins share a common amino acid framework. *EMBO J*, **5**, 9–13.

Keen, N. T., Dahlbreck, D., Staskawicz, B., and Belser, W. (1984). Molecular cloning of pectate lyase genes from *Erwinia chrysanthemi* and their expression in *Escherichia coli*. *J. Bacteriol.* **159**, 825–831.

Kehoe, M., Winther, M., and Dougan, G. (1983). Expression of a cloned K88ac adhesion antigen determinant: Identification of a new adhesion cistron and role of vector-encoded promoter. *J. Bacteriol.* **155**, 1071–1077.

Klemm, P. (1984). The *fimA* gene encoding the type I-fimbrial subunit of *Escherichia coli:* Nucleotide sequence and primary structure of the protein. *Eur. J. Biochem.* **143**, 395–399.

Klemm, P., Jørgensen, B. J., van Die, I., de Ree, H., and Bergmans, H. (1985). The *fim* genes responsible for synthesis of type I fimbriae in *Escherichia coli:* cloning and genetic analysis. *Mol. Gen. Genet.* **199**, 410–414.

Kobayashi, T., Kato, C., Kado, T., and Horikoshi, H. (1986). Excretion of penicillinase of an alkalophilic *Bacillus* sp, through the *Escherichia coli* outer membrane is caused by insertional activation of the *kil* gene in plasmid pMB9. *J. Bacteriol.* **166**, 728–732.

Koomey, J. M., Gill, R. E., and Falkow, S. (1982). Genetic and biochemical analysis of gonococcal IgA2 protease: Cloning in *Escherichia coli* and construction of mutants of gonococci that fail to produce the activity. *Proc. Natl. Acad. Sci. U.S.A.* **79**, 7881–7885.

Koshland, D., and Botstein, D. (1982). Evidence for posttranslational translocation of β-lactamase across the bacterial inner membrane. *Cell* **30**, 893–902.

Krone, W. J. A., de Vries, P., Koningstein, G., de Jonge, A. J. R., de Graaf, F. K., and Oudega, B. (1986). Uptake of cloacin DF13 by susceptible cells: Removal of immunity protein and fragmentation of cloacin molecules. *J. Bacteriol.* **166**, 260–268.

Kurzchalia, T. V., Wiedmann, M., Girshovich, A. S., Bochkareva, E. S., Bielka, H., and Rapoport, T. A. (1986). The signal sequence of nascent prolactin interacts with the 54K polypeptide of the signal recognition particle. *Nature (London)* **320**, 634–636.

Lebek, G., and Grüenig, H. M. (1985). Relation between the hemolytic property and iron metabolism in *Escherichia coli*. *Infect. Immun.* **50**, 682–686.

Lee, C. Y., and Iandolo, J. J. (1985). Mechanism of bacteriophage conversion of lipase activity in *Staphylococcus aureus*. *J. Bacteriol.* **164**, 288–293.

Lee, C. H., Moseley, S. L., Moon, H. W., Whipp, S. C., Gyles, C. L., and So, M. (1983). Characterization of the gene encoding heat stable entertoxin II and preliminary epidemiological studies of enterotoxigenic *Escherichia coli* heat stable enterotoxin II producers. *Infect. Immun.* **42**, 264–268.

Lei, S.-P., Lin, H.-C., Hefferman, L., and Wilcox, G. (1985). Cloning of the pectate lyase genes from *Erwinia carotovora* and their expression in *Escherichia coli*. *Gene* **35**, 63–70.

Leong, D., Coleman, K. D., and Murphy, J. R. (1983). Cloned diphtheria toxin fragment A expressed from the *tox* promoter and exported to the periplasm by the SecA apparatus of *Escherichia coli* K12. *J. Biol. Chem.* **258**, 15016–15020.

Levner, M., Urbano, C., and Rubin, B. A. (1980). Polymyxin B release of unnicked cholera toxin subunit A. *J. Bacteriol.* **144**, 1203–1204.

Lidahl, S., Guss, B., Uhlen, M., Philipson, L., and Lindberg, M. (1983). Gene for staphylococcal protein A. *Proc. Natl. Acad. Sci. U.S.A.* **80**, 697–701.

Liss, L. R., and Oliver, D. B. (1986). Effects of *secA* mutations on the synthesis and secretion in *Escherichia coli*. Evidence for a major export system for cell envelope proteins. *J. Biol. Chem.* **261**, 2299–2303.

Lopez, J., and Webster, R. E. (1985a). *fipB* and *fipC:* Two bacterial loci required for morphogensis of filamentous bacteriophage f1. *J. Bacteriol.* **163**, 960–905.

Lopez, J., and Webster, R. E. (1985b). Assembly site of bacteriophage f1 corresponds to adhesion zones between the inner and outer membranes of the host cell. *J. Bacteriol.* **163**, 1270–1274.

Lory, S., and Tai, P. C. (1983). Characterization of phospholipase C gene of *Pseudomonas aeruginosa* cloned into *Escherichia coli*. *Gene* **22**, 95–101.

Lory, S., Tai, P. C., and Davis, B. D. (1983). Mechanism of protein excretion by gram-negative bacteria: *Pseudomonas aeruginosa* exotoxin A. *J. Bacteriol.* **156**, 695–702.

Lubitz, W., and Pugsley, A. P. (1985). Changes in host cell phospholipid composition by φX174 gene E product. *FEMS Microbiol. Lett.* **30**, 171–175.

Luirink, J., van der Sande, C., Tommassen, J., Veltkamp, E., de Graaf, F., and Oudega, B. (1986). Effects of divalent cations and phospholipase A on excretion of cloacin DF13 and lysis of host cells. *J. Gen. Microbiol.* **132**, 825–834.

Mackman, N., Nicaud, J.-M., Gray, L., and Holland, I. B. (1985). Identification of polypeptides required for the export of haemolysin 2001 from *E. coli*. *Mol. Gen. Genet.* **201**, 529–536.

Mackman, N., Nicaud, J.-M., Gray, L., and Holland, I. B. (1986). Secretion of haemolysin by *E. coli*. *In* "Current Topics in Microbiology and Immunology. Export and Secretion in Bacteria" (H. Wu and P. C. Tai, eds.), pp. 159–182. Springer Verlag, Berlin and New York.

Malke, H., and Ferretti, J. J. (1984). Streptokinase: Cloning, expression and excretion by *Escherichia coli*. *Proc. Natl. Acad. Sci. U.S.A.* **81**, 3557–3561.

Marrs, C. F., Schoolnik, G., Koomey, J. M., Hardy, J., Rothbard, J., and Falkow, S. (1985). Cloning and sequencing of a *Moraxella bovis* pilin gene. *J. Bacteriol.* **163**, 132–139.

Masaki, H., and Ohta, T. (1985). Colicin E3 and its immunity genes. *J. Mol. Biol.* **182**, 217–227.

Meek, K. L., Walsh, K. A., and Palmitter, R. D. (1982). The signal sequence of ovalbumin is located near the NH_2 terminus. *J. Biol. Chem.* **257**, 12245–12251.

Mekalanos, J. J., Swartz, D. J., Pearson, G. D. N., Harford, N., Groyne, F., and Wilde, M. (1983). Cholera toxin genes: Nucleotide sequence, deletion analysis and vaccine development. *Nature (London)* **306**, 551–557.

Mercurio, A., and Manning, P. M. (1985). Cellular localization and export of soluble haemolysin of *Vibrio cholerae* E1 Tor. *Mol. Gen. Genet.* **200**, 472–475.

Meyer, T., Mlawer, N., and So, M. (1982). Pilus expression in *Neisseria gonorrhoeae* involves chromosome rearrangement. *Cell* **30**, 45–52.

Mézes, P. S. F., and Lampen, J. O. (1985). Secretion of proteins by bacilli. *In* "The Biology of the Bacilli" (D. A. Dubnau, ed.), Vol. 2, pp 151–183. Academic Press, New York.

Michaelis, S. M., Chapon, C., d'Enfert, C., Pugsley, A. P., and Schwartz, M. (1985). Characterization and expression of the structural gene for pullulanase, a maltose-inducible secreted protein of *Klebsiella pneumoniae*. *J. Bacteriol.* **164**, 633–638.

Mock, M., and Schwartz, M. (1978). Mechanisms of colicin E3 production in strains harboring wild-type and mutant plasmids. *J. Bacteriol.* **136**, 700–707.

Mooi, F. R., Wijfjes, A., and de Graaf, F. K. (1983). Identification and characterization of precursors in the biosynthesis of the K88ab fimbria of *Escherichia coli*. *J. Bacteriol.* **154**, 41–49.

Moore, D., Sowa, B. A., and Ippen-Ihler, K. (1981). Location of an F-pilin pool in the inner membrane. *J. Bacteriol.* **146**, 251–259.

Morlon, J., Lloubés, R., Varenne, S., Chartier, M., and Lazdunski, C. (1983). Complete nucleotide sequence of the structural gene for colicin A, a gene translated at nonuniform rate. *J. Mol. Biol.* **170**, 271–285.

Morona, R., and Reeves, P. (1984). Detection of several diisopropylfluorophosphate-binding proteins in the outer membrane of *Escherichia coli* k-12. *FEMS Microbiol. Lett.* **23**, 179–182.

Nagahari, K., Kanaya, S., Munahata, K., Aoyagi, Y., and Mizushima, S. (1985). Secretion into the culture medium of a foreign gene product from *Escherichia coli:* use of the *ompF* gene for secretion of human β-endorphin. *EMBO J*, **3**, 3589–3592.

Newland, J. W., Green, B. A., Foulds, J., and Holmes, R. K. (1985). Cloning of extracellular DNase and construction of a DNase negative strain of *Vibrio cholerae*. *Infect. Immun.* **47**, 691–696.

Nicaud, J.-M., Mackman, N., Gray, L., and Holland, I. B. (1985a). Regulation of haemolysin synthesis by *E. coli* determined by HLY genes of human origin. *Mol. Gen. Genet.* **199**, 111–116.

Nicaud, J.-M., Mackman, N., Gray, L., and Holland, I. B. (1985b). Characterisation of HlyC and mechanism of activation and secretion of haemolysin from *E. coli* 2001. *FEBS Lett.* **187**, 339–344.

Nicaud, J.-M., Mackman, N., Gray, L., and Holland, I. B. (1986a). The C-terminal, 23 kDa peptide of *E. coli* haemolysin 2001 contains all the information necessary for its secretion by the α-haemolysin (Hly) export machinery. *FEBS Lett.* **204**, 331–335.

Nicaud, J. M., Mackman, N., and Holland I. B. (1986b). Current status of secretion of foreign proteins by microorganisms. A review. *J. Biotechnol.* **3**, 255–270.

Nicosia, A., Perugini, M., Franzini, C., Casagli, M. G., Borri, M. G., Antoni, G., Almoni, G., Neri, P., Rutti, G., and Rappaoli, R. (1986). Cloning and sequencing of the pertussis toxin A genes: Operon structure and gene duplication. *Proc. Natl. Acad. Sci. U.S.A.* **83**, 4631–4635.

Nielsen, J. B. K., Mézes, P. S. F., and Lampen, J. O. (1983). Secretion by *Pseudomonas aeruginosa:* Fate of a cloned gram-positive lipoprotein deletion mutant. *J. Bacteriol.* **156**, 559–566.

Nishibuchi, M., and Kaper, J. B. (1985). Nucleotide sequence of the thermostable direct hemolysin gene of *Vibrio parahaemolyticus*. *J. Bacteriol.* **162**, 558–564.

Ohsuye, K., Nomura, M., Tanaka, S., Kubota, I., Nakazato, H., Shinegawa, H., Nakata, A., and Noguchi, T. (1983). Expression of chemically synthesized α-neo-endorphin gene fused to *E. coli* alkaline phosphatase. *Nucleic Acids Res.* **11**, 1283–1294.

Orndorff, P. E., and Falkow, S. (1985). Nucleotide sequence of *pilA,* the gene encoding the structural component of type 1 pili in *Escherichia coli*. *J. Bacteriol.* **162**, 454–457.

Oudega, B., Stegehuis, F., van Tiel-Menkveld, G. J., and de Graaf, F. K. (1982). Protein H encoded by plasmid CloDF13 is involved in excretion of cloacin DF13. *J. Bacteriol.* **150**, 1115–1121.

Ozeki, H., Stocker, B. A. D., and de Margerie, H. (1959). Production of colicine by single bacteria. *Nature (London)* **184**, 337–339.

Pasloske, B. L., Finlay, B. B., and Paranchych, W. (1985). Cloning and sequencing of the *Pseudomonas aeruginosa* PAK pilin gene. *FEBS Lett.* **183**, 408–411.

Pearson, G. D. N., and Mekalanos, J. J. (1982). Molecular cloning of *Vibrio cholerae* enterotoxin genes in *Escherichia coli* K12. *Proc. Natl. Acad. Sci. U.S.A.* **79**, 2976–2980.

Pohlner, J., Halter, R., Beyreuther, K., and Meyer, T. F. (1987). Gene structure and extracellular secretion of *Neisseria* gonorrhoeae IgA protease. *Nature* **325**, 458–462.

Power, S. D., Adams, R. M., and Wells, J. A. (1986). Secretion and autoproteolytic maturation of subtilisin. *Proc. Natl. Acad. Sci. U.S.A.* **83**, 3096–3100.

Pressler, U., Braun, V., Wittman-Liebold, B., and Benz, R. (1986). Structural and functional properties of colicin B. *J. Biol. Chem.* **261**, 2654–2659.

Pritchard, A. E., and Vasil, M. L. (1986). Nucleotide sequence and expression of a phosphate-regulated gene encoding the secreted hemolysin of *Pseudomonas aeruginosa*. *J. Bacteriol.* **167**, 291–298.

Pugsley, A. P. (1983). Obligatory coupling of colicin release and lysis in mitomycin-treated Col⁺ *Escherichia coli*. *J. Gen. Microbiol.* **129**, 1921–1928.

Pugsley, A. P. (1984a). The ins and outs of colicins. *Microbiol. Sci.* **1**, 168–175.

Pugsley, A. P. (1984b). Genetic analysis of ColN plasmid determinants for colicin production, release and immunity. *J. Bacteriol.* **158**, 523–529.

Pugsley, A. P. (1988). The immunity and lysis genes of ColN plasmid pCHAP4. *Mol. Gen. Genet.* (in press).

Pugsley, A. P., and Cole, S. T. (1986). β-Galactosidase and alkaline phosphatase do not become extracellular when fused to the amino-terminal part of colicin N. *J. Gen. Microbiol.* **132**, 2297–2307.

Pugsley, A. P., and Rosenbusch, J. P. (1981). Colicin E2 release from *Escherichia coli* K12. *J. Bacteriol.* **147**, 186–192.

Pugsley, A. P., and Schwartz, M. (1983a). A genetic approach to the study of mitomycin-induced lysis of *Escherichia coli* K12 strains which produce colicin E2. *Mol. Gen. Genet.* **190**, 366–372.

Pugsley, A. P., and Schwartz, M. (1983b). Expression of a gene in a 400 base-pair fragment of the colicin plasmid ColE2-P9 is sufficient to cause host cell lysis. *J. Bacteriol.* **156**, 109–114.

Pugsley, A. P., and Schwartz, M. (1984). Colicin E2 release: Lysis, leakage or secretion? Possible role of a phospholipase. *EMBO J*, **3**, 2393–2397.

Pugsley, A. P., and Schwartz, M. (1985). Export and secretion by bacteria. *FEMS Microbiol. Rev.* **32**, 3–38.

Pugsley, A. P., Chapon, C., and Schwartz, M. (1986). Extracellular pullulanase of *Klebsiella pneumoniae* is a lipoprotein. *J. Bacteriol.* **166**, 1083–1088.

Randall, L. L., and Hardy, S. J. S. (1984). Export of protein by bacteria. *Microbiol. Rev.* **47**, 290–298.

Ranelli, D. M., Jones, C. L., Johns, M. B., Massey, G. J., and Kahn, S. A. (1985). Molecular cloning of staphylococcal enterotoxin B gene in *Escherichia coli* and *Staphylococcus aureus*. *Proc. Natl. Acad. Sci. U.S.A.* **82**, 5850–5854.

Rhen, M., van Die, I., Rhem, V., and Bergmans, H. (1985). Comparison of the nucleotide sequences of the genes encoding the KS71A and F7, fimbrial antigens of uropathogenic *Escherichia coli*. *Eur. J. Biochem.* **151**, 573–577.

Robson, L. M., and Chambliss, G. H. (1986). Cloning of the *Bacillus subtilis* DLG β-1,4-glucanase gene and its expression in *Escherichia coli* and *B. subtilis*. *J. Bacteriol.* **165**, 612–619.

Sako, T. (1985). Overproduction of staphylokinase in *Escherichia coli* and its characterization. *Eur. J. Biochem.* **149**, 557–563.

Sarvas, M., and Palva, I. A. (1983). The penicillinase of *Bacillus licheniformis* is an outer membrane protein in *Escherichia coli*. *J. Bacteriol.* **155**, 657–663.

Sashihara, N., Kudo, T., and Horikoshi, K. (1984). Molecular cloning and expression of cellulase genes of alkalophilic *Bacillus* sp. strain N-4 in *Escherichia coli*. *J. Bacteriol.* **158**, 503–506.

Sastry, P. A., Finlay, B. B., Pasloske, B. L., Paranchych, W., Pearlstone, J. R., and Smillie, L. B. (1986). Cooperative studies of the amino acid and nucleotide sequences of pilin derived from *Pseudomonas aeruginosa* PAK and PAO. *J. Bacteriol.* **164**, 571–577.

Schmitz, G., and Braun, V. (1985). Cell-bound and secreted protease of *Serratia marcescens*. *J. Bacteriol.* **161**, 1002–1009.

Schoemaker, J. M., Brasnett, A. H., and Marston, F. A. D. (1985). Examination of calf prochymosin accumulation in *Escherichia coli*: Disulphide linkages as the structural component of prochymosin-containing inclusion bodies. *EMBO J*, **4**, 1041–1047.

Schoner, R. G., Ellis, L. F., and Schoner, B. E. (1985). Isolation and purification of protein granules from *Escherichia coli* overproducing bovine growth hormone. *Biotechnology* **3,** 151–154.

Sekine, S., Mizukumi, T., Nishi, T., Kuwana, Y., Sato, A., Sato, M., Itoh, S., and Kawauchi, H. (1985). Cloning and expression of cDNA for salmon growth hormone in *Escherichia coli*. *Proc. Natl. Acad. Sci. U.S.A.* **82,** 4306–4310.

Shortle, D. (1983). A genetic system for analysis of staphylococcal nuclease. *Gene* **22,** 181–189.

Spicer, E. J., and Noble, J. A. (1982). *Escherichia coli* heat labile enterotoxin. Nucleotide sequence of the A subunit gene. *J. Biol. Chem.* **257,** 5716–5721.

So, M., and McCarthy, B. T. (1980). Nucleotide sequence of the bacterial transposon Tn*1681* encoding a heat-stable (ST) toxin and its identification in enterotoxigenic *Escherichia coli* strains. *Proc. Natl. Acad. Sci. U.S.A.* **77,** 4011–4015.

Sowa, B. A., Moore, D., and Ippen-Ihler, K. (1983). Physiology of F-pilin synthesis and utilization. *J. Bacteriol.* **153,** 962–968.

Strom, M. S., and Lory, S. (1986). Cloning and expression of the pilin gene of *Pseudomonas aeruginosa* PAK in *Escherichia coli*. *J. Bacteriol.* **165,** 367–372.

Takahara, M., Hibler, D. W., Barr, P. J., Gerit, J. A., and Inouye, M. (1985). The *ompA* signal peptide-directed secretion of staphylococcal nuclease A by *Escherichia coli*. *J. Bio. Chem.* **260,** 2670–2674.

Thompson, S. S., Naidu, Y. M., and Pestka, J. J. (1985). Ultrastructural localization of an extracellular protease in *Pseudomonas fragi* by using the peroxidase–antiperoxidase reaction. *Appli. Environ. Microbiol.* **50,** 1038–1042.

Thurn, K. K., and Chatterjee, A. K. (1985). Single-site chromosomal Tn*5* insertions affect the export of pectolytic and cellulolytic enzymes in *Erwinia chrysanthemi* EC16. *Appl. Environ. Microbiol.* **50,** 894–898.

Titbank, R. W., Bell, A., and Munn, C. B. (1985). Role of caseinase from *Aeromonas salmonicida* in activation of hemolysin. *Infect. Immun.* **49,** 756–759.

Tommassen, J., van Tol, H., and Lugtenberg, B. (1983). The ultimate localization of an outer membrane protein of *Escherichia coli* K12 is not determined by the signal sequence. *EMBO J,* **2,** 1275–1279.

Tommassen, J., Leunissen, J., van Damme-Jongsten, M., and Overduin, P. (1985). Failure of *E. coli* K12 to transport PhoE–LacZ hybrid proteins out of the cytoplasm. *EMBO J,* **4,** 1041–1047.

Tsukagoshi, N., Ihara, H., Yamagata, H., and Udaka, S. (1984). Cloning and expression of a thermophilic α-amylase gene from *Bacillus stearothermophilus* in *Escherichia coli*. *Mol. Gen. Genet.* **193,** 58–63.

Tweten, R. K., and Collier, R. J. (1983). Molecular cloning and expression of gene fragments from corynebacteriophage β encoding enzymically active peptides of diphtheria toxin. *J. Bacteriol.* **156,** 68–685.

van den Elzen, P. J. M., Watters, H. H. B., Veltkamp, E., and Nijkamp, H. J. J. (1983). Molecular structure and function of the bacteriocin gene and bacteriocin protein of plasmid CloDF13. *Nucleic Acids Res.* **11,** 2465–2477.

van Die, I., van Geffen, B., Hoekstra, W., and Bergmans, H. (1985). Type Ic fimbriae of a uropathogenic *Escherichia coli* strain; cloning and characterization of the genes involved in the expression of the Ic antigen and nucleotide sequence of the subunit gene. *Gene* **34,** 187–196.

van Tiel-Menkveld, G. J., Rezee, A., and de Graaf, F. K. (1979). Production and excretion of cloacin DF13 by *Escherichia coli* harboring plasmid CloDF13. *J. Bacteriol.* **140,** 415–423.

Varley, J. M., and Boulnois, G. J. (1984). Analysis of a cloned colicin Ib Gene: Complete nucleotide sequence and implications for regulation of expression. *Nucleic Acids Res.* **12,** 6727–6739.

von Heijne, G. (1985). Signal sequences. The limits of variation. *J. Mol. Biol.* **184,** 99–105.

Wagner, W., Vogel, M., and Goebel, W. (1983). Transport of hemolysin across the outer membrane of *Escherichia coli* requires two functions. *J. Bacteriol.* **154,** 200–210.

Watson, M. E. E. (1984). Compilation of published signal sequences. *Nucleic Acids Res.* **12,** 4155–4174.

Weaver, C. A., Redborg, A. H., and Konisky, J. (1981). Plasmid-determined immunity of *Escherichia coli* K12 to colicin Ia is mediated by a plasmid-encoded membrane protein. *J. Bacteriol.* **148,** 817–828.

Wickner, W. T., and Lodish, H. F. (1985). Multiple mechanisms of protein insertion into and across membranes. *Science* **230,** 400–407.

Willemot, K., and Cornelis, P. (1983). Growth defects of *Escherichia coli* cells which contain the gene for amylase from *Bacillus cereus. J. Gen. Microbiol.* **129,** 311–319.

Wolfe, P. B., Rice, M., and Wickner, W. (1985). Effects of two *sec* genes on protein assembly into the plasma membrane of *Escherichia coli. J. Biol. Chem.* **260,** 1836–1841.

Wredtlindt, B., and Pavlovskis, O. R. (1984). Genetic mapping and characterization of *Pseudomonas aeruginosa* mutants defective in the formation of extracellular proteins. *J. Bacteriol.* **158,** 801–808.

Wredtlindt, B., Sjöberg, L., and Wadström, T. (1977). Protease-deficient mutants of *Pseudomonas aeruginosa:* Pleiotropic changes in activity of other extracellular proteases. *J. Gen. Microbiol.* **103,** 329–336.

Yamada, M., and Nakazawa, A. (1984). Factors necessary for the export process of colicin E1 across cytoplasmic membrane of *Escherichia coli. Eur. J. Biochem.* **140,** 249–255.

Yamada, M., and Nakazawa, A. (1986). Effect of base substitutions in the colicin E1 gene on colicin E1 export and bacteriocin activity. *Mol. Gen. Genet.* **202,** 24–29.

Yamada, M., Ebina, Y., Miyata, T., Nakazawa, T., and Nakazawa, A. (1982a). Nucleotide sequence of the structural gene for colicin E1 and predicted structure of the protein. *Proc. Natl. Acad. Sci. U.S.A.* **79,** 2827–2831.

Yamada, M., Miki, T. and Nakazawa, A. (1982b). Translocation of colicin E1 through cytoplasmic membrane of *Escherichia coli. FEBS Lett.* **150,** 465–468.

Yamamoto, T., and Yokota, T. (1982). Release of heat-labile enterotoxin subunits by *Escherichia coli. J. Bacteriol.* **150,** 1482–1484.

Yanigida, N., Ouozumi, T., and Beppu, T. (1986). Specific excretion of *Serratia marcescens* protease through the outer membrane of *Escherichia coli. J. Bacteriol.* **166,** 937–944.

Zieg, J., and Simon, M. (1980). Analysis of the nucleotide sequence of an invertible controlling element. *Proc. Natl. Acad. Sci. U.S.A.* **77,** 4169–4200.

Zinc, R. T., and Chatterjee, A. K. (1985). Cloning and expression in *Escherichia coli* of pectinase gene of *Erwinia carotovora* subsp. *carotovora. Appl. Environ. Microbiol.* **49,** 714–717.

V

Mitochondrial Assembly

16

Genetic Approaches to the Study of Mitochondrial Protein Import

MARJORIE C. BRANDRISS

I. INTRODUCTION

During the last few years, the field of mitochondrial biogenesis and organelle assembly has witnessed extraordinary growth in the understanding of mechanisms of targeting, delivery, and intraorganelle sorting of mitochondrially imported proteins. Many of the biochemical steps have been elucidated in the yeast *Saccharomyces cerevisiae* and the fungus *Neurospora crassa* where it has been possible to isolate large quantities of purified mitochondria to study the fates of individual proteins. The reader's attention is drawn to several excellent reviews that have appeared in recent years which summarize the remarkable progress made in this area (Hay *et al.*, 1984; Reid, 1985; Douglas *et al.*, 1986) as well as to Nicholson and Neupert, Chapter 17, this volume.

655

PROTEIN TRANSFER AND ORGANELLE BIOGENESIS

A. Biochemistry of Import: A Brief Review

Proteins destined for the mitochondria are encoded by genes in the nucleus and are synthesized in the cytoplasm as precursor polypeptides either longer than, or in a different conformation from, their mature forms found inside the mitochondrion (Hallermayer *et al.*, 1977; Schatz, 1979; Maccecchini *et al.*, 1979; Zimmerman *et al.*, 1979a,b). Precursors do not normally accumulate during steady-state growth, although the use of certain drugs or mutations can lead to their accumulation *in vivo* (Nelson and Schatz, 1979; Suissa and Schatz, 1982). Although posttranslational import has been demonstrated both *in vivo* and *in vitro*, cotranslational import has not yet been ruled out as a mechanism of import *in vivo*.

The proteins carry signal or presequences that target them to the mitochondrial surface where they may bind receptors (Hennig and Neupert, 1981; Riezman *et al.*, 1983a,b; Zimmerman *et al.*, 1981; Zwizinski *et al.*, 1983, 1984). The proteins that are destined for the outer membrane carry signals that are not processed and do not require an electrochemical gradient for their proper localization (Freitag *et al.*, 1982; Gasser and Schatz, 1983). Proteins that reside in the intermembrane space, inner membrane, or matrix space usually carry amino-terminal extensions that are removed some time during their import, a process that requires an electrochemical gradient and ATP (Gasser *et al.*, 1982, and Pfanner *et al.*, 1987). The proteolytic removal of signal sequences is carried out by a matrix-localized processing protease (Böhni *et al.*, 1983; McAda and Douglas, 1982; Mori *et al.*, 1980) and, in some cases, by a protease believed to be located within the intermembrane space that cleaves certain intermembrane space proteins (Ohashi *et al.*, 1982). Signal sequences contain information both to target the proteins to the mitochondrial outer membrane as well as to direct them to the intramitochondrial compartment in which they ultimately reside (Douglas *et al.*, 1984; Hase *et al.*, 1984; Hurt *et al.*, 1985a,b). In Neurospora mitochondrial protein import, Schleyer and Neupert (1985) have shown that a membrane potential is required to initiate import of the amino terminus of the protein across the membrane, but not for completion of translocation of the polypeptide.

B. Mitochondrial Signal Sequences Direct Foreign Proteins into Mitochondria

Hoping to achieve the success that came with the use of gene fusions in the isolation of mutations affecting protein secretion in *Escherichia coli* (for a review, see Silhavy *et al.*, 1983), several laboratories working with *S. cerevisiae* began to construct gene fusions encoding hybrid proteins carrying amino-terminal fragments of mitochondrially imported proteins and carboxy-terminal nonyeast or nonmitochondrial proteins. Geller *et al.* (1983) reported the construction of a gene fusion between *ATP2* (encoding the β subunit of the F_1 ATPase) and *lacZ* (β-galactosidase) of *E. coli.* Douglas *et al.* (1984) demonstrated that this hybrid gene product containing the amino-terminal 382 amino acids of the β subunit and the catalytically active carboxy-terminal portion of β-galactosidase was delivered to the mitochondria *in vivo* and was localized to the inner membrane, the normal location of the F_1 ATPase. Cells carrying this gene fusion on a high copy number plasmid were respiratory deficient [glycerol nonutilizing (Gly$^-$) phenotype] but became respiratory competent when the plasmid was lost from the cells. This phenotype has been exploited in the isolation of new mutations (see Section II,B,1).

In other *in vivo* experiments, Hase *et al.* (1984) fused the gene encoding the major 70 kDa protein of the mitochondrial outer membrane to *lacZ* and showed that the hybrid product was correctly localized in the outer membrane. In the Schatz laboratory, the small murine dihydrofolate reductase protein was delivered to the matrix and processed correctly when fused to the cytochrome oxidase subunit IV amino terminus (Hurt *et al.*, 1984). Similarly, the amino-terminal 48 amino acids of the mitochondrial alcohol dehydrogenase directed its cytoplasmic isozyme to the mitochondrial matrix (van Loon and Young, 1986).

Studies in which the amino-terminal portions of mitochondrial matrix proteins were fused to β-galactosidase resulted in the matrix localization of β-galactosidase in several of the hybrids [δ-aminolevulinate synthase (*HEM1*), Keng *et al.*, 1986; pyrroline-5-carboxylate dehydrogenase (*PUT2*), Brandriss and Krzywicki, 1986]. In spite of the large size of the β-galactosidase moiety and that it was known to remain membrane associated in the protein secretion studies in *E. coli* (Bassford *et al.*, 1979), the mitochondrial import apparatus was able to properly deliver these proteins through both mitochondrial membranes and sort them appropriately. In certain hybrid proteins, or when gene expression was increased by the use of more powerful promoters, the expression of these proteins led to cell death on specific growth media (see Section II,B,1).

Additional studies using the *ATP2* gene (Emr *et al.*, 1986) have demonstrated the importance of the size or conformation of the passenger protein used in these import studies. Only 39 amino acids of the F_1 ATPase β-subunit amino terminus were necessary to deliver invertase (the product of the *SUC2* gene) to the mitochondrion, in contrast to the 169 residues that were required to direct the larger β-galactosidase protein. Interestingly, a β-galactosidase carrying an amino-terminal extension consisting of 90 residues of the *lacI* repressor did not interfere with the targeting ability of the initial 9 amino acids of the δ-aminolevulinate synthase (Keng *et al.*, 1986). It has been hypothesized that the Lac repressor portion separates the signal sequence from the body of the β-galactosidase so that the signal can be seen by the import apparatus (Douglas *et al.*, 1986). Alternatively, the repressor could be providing signaling information.

C. Studies on the Nature of the Signal Sequence

Several recent studies have dissected the information contained in the signal itself to reveal that there are domains responsible for directing the protein to the mitochondrial surface and other regions that cause the protein to take up residence in the appropriate compartment. Cytochrome c_1 is an intermembrane space protein that is anchored to the inner membrane (Li *et al.*, 1981). It is made as a precursor with a 61 amino acid signal sequence that is removed in two steps (Sadler *et al.*, 1984). The first 35 residues of the signal are hydrophilic, and the remainder contains uncharged amino acids that could span the inner membrane. van Loon *et al.* (1986) found that the 61 amino acid presequence of cytochrome c_1 was able to direct murine dihydrofolate reductase to the intermembrane space of yeast mitochondria. When the signal contained only the first 32 amino acids of cytochrome c_1, the dihydrofolate reductase was delivered to the matrix and not the intermembrane space. They suggested that the second half of the presequence served as a stop-transfer signal that prevented the protein from crossing into the matrix space.

The major 70 kDa polypeptide of the mitochondrial outer membrane of yeast is an integral membrane protein with a large domain exposed to the cytoplasm (Riezmann *et al.*, 1983b). The protein is synthesized in the cytoplasm in its mature length, and, although it is not removed, the amino terminus of the protein (as deduced from its DNA sequence, Hase *et al.*, 1983) resembles a mitochondrial signal and membrane anchor region. Hase *et al.* (1984) showed that the amino-terminal 41 residues of this polypeptide contained the information necessary for delivering and anchoring the protein in its proper position. Certain internal deletions in the amino-terminal region of the protein resulted in a small fraction of mislocalization of the protein into the matrix. A recent report (Hase *et al.*, 1986) showed that, while the amino-terminal 61 amino acids of the 70 kDa protein could properly localize β-galactosidase to the outer membrane, a shorter hybrid containing only the first 21 residues directed the β-galactosidase to the matrix.

To delimit the size of the signal sequence, Hurt *et al.* (1985a,b) fused progressively shorter amino-terminal fragments of the cytochrome oxidase subunit IV presequence to dihydrofolate reductase. They found that a signal of fewer than 9 amino acids failed to deliver the protein to the mitochondria. Extensions of 12 or 16 residues allowed the dihydrofolate reductase to be imported into the matrix, but the presequence was not cleaved. Surprisingly, a presequence of 22 residues that allowed proper import was processed (between amino acids 17 and 18), even though the normal cleavage site (between residues 25 and 26) was missing in the construction.

II. ISOLATION OF MUTATIONS AFFECTING MITOCHONDRIAL IMPORT

The goal of ongoing research is aimed at determining the elements involved in the process of mitochondrial protein import. These elements are hypothesized to include proteins that recognize the precursors during or after their translation in the cytoplasm, receptors that recognize the signals that destine the proteins for the mitochondrion, pores through which the proteins pass, the enzymes that process the precursors to their mature forms, among other possibilities. With many of the biochemical and physical aspects of import now understood, it is important to use a genetic approach to complement the biochemical studies and to facilitate the isolation of the genes and gene products that carry out the specific steps that have been demonstrated to occur.

Studies involving a genetic approach to mitochondrial protein import have lagged behind those that have defined the physical elements of the system. This may be attributable to the inability to manipulate with ease the proteins that are imported into the mitochondrion or that are a part of the import apparatus. The lag may also be due to the inability to predict the phenotypes of such import mutations, the difficulty in their isolation, or an inability to appropriately characterize them. With so many of the genes encoding mitochondrially imported proteins now cloned, it has become possible to use them directly, to make the desired mutations *in vitro*, or indirectly, to make gene fusions and hybrid proteins to interfere with the process of import. The focus of the following sections is to review the mutations in mitochondrial protein import that have been reported to date, with particular emphasis on the successful genetic schemes used to isolate and characterize them, and to catalog additional genetic selections and screens that are currently being attempted in order to isolate new mutations which will reveal further aspects of this pathway.

A. Classical Genetic Approaches

1. Pleiotropic Mutations: mas1 and mas2

To date, the only existing mutations that affect the import of many mitochondrial proteins were isolated using a classical genetic approach by Yaffe and Schatz (1984). They made two assumptions: (1) that a defect in mitochondrial protein import would lead to cell death, and (2) that such mutants would accumulate some precursors to mitochondrial proteins in their cytoplasms.

Yaffe and Schatz screened a mutagenized population of yeast cells for temperature-sensitive (*ts*) growth on rich medium (YPD). Both vegetative petite and wild-type strains were used as the starting strains in separate experiments. Cells were spread on YPD plates to grow into colonies at 23°C. They were then replica-plated to two sets of plates and incubated at 23° and 37°C. To allow for delayed expression of the phenotype, the 37°C plates were replica-plated a second time and incubated at 23° and 37°C. Those colonies that failed to grow on the second 37°C plate, but grew at the lower temperature, were examined further.

The temperature-sensitive mutants were screened for accumulation at the nonpermissive temperature of the precursor to the β subunit of the F_1 ATPase, using the Western blotting procedure and antibodies against the β subunit. From 930 *ts* strains isolated from the vegetative petite strain and 270 *ts* strains isolated from the wild-type strain, 29 mutants were identified that accumulated precursor. The mutations fell into two complementation groups, *mas1* and *mas2* (for mitochondrial assembly). Additional tests on these mutant strains indicated that they were also affected in the import of citrate synthase (matrix protein), cytochrome b_2 (intermembrane space protein), and a 90 kDa protein of the matrix. Subsequent screenings of 3000 additional *ts* strains by members of the Schatz laboratory yielded new alleles of *mas1* and *mas2*, but no new complementation groups with a stable precursor accumulation phenotype were discovered (C. R. Bibus, personal communication).

The *mas1* mutant was shown to be fully proficient in the import of precursors but deficient only in their processing. Yaffe *et al.* (1985) demonstrated that the activity of the matrix processing protease that is responsible for maturation of inner membrane, intermembrane space, and matrix proteins was greatly reduced. The gene has been cloned by functional complementation of the *ts* defect in the Schatz laboratory and has been sequenced. Whether *MAS1* is the structural gene for the processing protease or a regulator of its activity will be demonstrated in the near future by the use of antibodies directed toward the product of a *MAS1* gene fusion.

By screening a collection of temperature-sensitive mutants (Hartwell, 1967), M. Yaffe (personal communication) has identified several new alleles of the *mas2* mutation, some with import defects as strong as in the original *mas1* mutant. His laboratory has cloned the *MAS2* gene by functional complementation of the *ts* defect and has sequenced it. The deduced amino acid sequence of this 44.7 kDa protein does not resemble any other known protein, has no recognizable mitochondrial signal sequence and does not appear to have any membrane-spanning regions. It appears to be an average cytoplasmic protein, which is consistent with the finding that mitochondria isolated from *mas2* strains do not show a defect for import of protein *in vitro*. Antibody prepared against *MAS2-lacZ* fusion protein precipitated a 45 kDa protein. If the *mas2* protein does have a cytoplasmic location, it may have a role in translation of mitochondrially imported proteins or as a loosely associated subunit of a receptor in the outer mitochondrial membrane.

2. Specific Mutations: cyc2, cyc3, and pet ts2858

To date, only three cases of nuclear mutations affecting specific mitochondrically imported proteins have been described. These mutations were not isolated as causing defects in protein import, but as affecting the production of a particular protein under study in one case, or as respiratory deficient in the other case. The mutations affect the heme-containing proteins cytochrome c and cytochrome b_2 that are both located in the intermembrane space and have different import pathways.

Cytochrome c is found on the cytoplasmic face of the inner membrane, lacks a cleavable signal sequence, and, although synthesized in the cytoplasm as its mature size, it has a conformation that is different from the mature mitochondrial protein (Korb and Neupert, 1978). A specific receptor component for the binding of apocytochrome c has been identified (Hennig *et al.*, 1983), and the number of binding sites is believed to be limited. Complete transfer of apocytochrome c across the outer mitochondrial membrane depends on covalent attachment of heme to cysteine residues, reactions carried out by proteins in the intermembrane space (Hennig and Neupert, 1981).

Matner and Sherman (1982) have reported the characterization of mutations affecting the import and maturation of cytochrome c. These mutations in the *CYC2* and *CYC3* genes were isolated using selection schemes designed to find strains lacking functional cytochrome c or heme attachment (Sherman *et al.*, 1968, 1974; Rothstein and Sherman, 1980). *cyc2* and *cyc3* mutants were found to accumulate specifically apo-iso-2-cytochrome c (but not apo-iso-1-cytochrome c, perhaps due to differences in protein stability) and lacked the holoenzyme forms. Both genes have now been cloned. The *CYC3* clone in yeast cells leads to the overproduction of cytochrome *c* heme lyase activity that attaches the heme group to cytochrome *c*, and the enzyme activity is associated with the mitochondrion (Dumont *et al.*, 1987). Matner and Sherman (1982) speculated that *CYC2* is a structural protein or a nonenzymatic component of the mitochondria that may participate in import and is a good candidate for the cytochrome *c* receptor. Characterization of the cloned *CYC2* gene is in progress.

Cytochrome b_2 is a soluble intermembrane space protein that is synthesized as a cytoplasmic precursor with an 80 amino acid presequence (Guiard, 1985). The protein undergoes two cleavage reactions, one carried out by the matrix-localized processing protease (Böhni *et al.*, 1983) and the second by an as yet uncharacterized protease presumed to be bound to the inner mitochondrial membrane (Daum *et al.*, 1982). A mutation isolated by E. Schweizer appears to affect this second processing step. The *S. cerevisiae pet ts2858* mutant characterized by Michaelis *et al.* (1982) grew on lactate as the sole carbon source at 22° but not at 37°C. Additional analysis of this mutant showed that it accumulated a longer form of the mitochondrially encoded and synthesized cytochrome oxidase subunit II that contains an amino-terminal presequence (Pratje *et al.*, 1983).

Pratje and Guiard (1986) have recently reported that the *pet ts2858* mutation was found to affect the processing of the nucleus-encoded protein cytochrome b_2. The protein accumulated as its intermediate form, the result of the cell's apparent inability to carry out the second processing step in its maturation. Therefore, a single nuclear mutation appears to affect the processing of two mitochondrial proteins, one encoded in the nuclear genome and the other encoded in the mitochondrial genome. When the processing sites of these two proteins were compared, they appeared to be rather similar: asparagine-15–aspartate-16 in cytochrome oxidase subunit II (Pratje *et al.*, 1983) and asparagine-80–glutamate-81 in cytochrome b_2 (Guiard and Buhler, 1984). The *pet ts2858* mutation had no effect on the two-step processing of another nucleus-encoded intermembrane space protein, cytochrome *c* peroxidase, which lacks the sequence asparagine–glutamate or asparagine–aspartate in its presequence (Kaput *et al.*, 1982). Thus, the gene may encode a sequence-specific protease or a factor activating a protease.

B. Molecular Genetic Approaches

A large number of nuclear genes that encode mitochondrially imported proteins have now been cloned and analyzed. The molecular genetic tools that are currently available allow a determination of the deduced amino acid sequence of the gene product, its *in vitro* production, determination of its processing pathway, the sequence of the mature amino terminus, and its intraorganellar location, all with comparative ease. The cloned genes are being used both to isolate trans-acting mutations that affect mitochondrial import as well as to create cis-acting signal sequence mutations in order to begin to understand the structures and enzymes that are responsible for proper protein localization. The following sections describe the current effort in the isolation of new mutations affecting mitochondrial protein import.

1. Selections for Mutations That Prevent Import of Hybrid Proteins

The ability of the geneticist to isolate useful mutations depends on the strength of the selection and a good prediction of the phenotype of the mutations of interest. The basis of the strategy for the isolation of protein secretion-defective mutants of *E. coli* (Bassford and Beckwith, 1979; reviewed in Silhavy *et al.*, 1983) was that the attempt to export hybrid proteins would jam the secretory apparatus, resulting in cell death under certain conditions. Mutations that suppressed the lethal phenotype were selected by demanding growth under those conditions. Such suppressor mutations were found to lie in the region of the gene fusion encoding the signal sequence or in genes that controlled aspects of the secretion pathway.

Once it was demonstrated that nonmitochondrial proteins could be successfully delivered to the mitochondrion in the form of hybrid proteins carrying mitochondrial signal sequences (see Section I,B), the strategy used in the bacterial systems was applied to eukaryotic cells. A gene fusion is constructed whose hybrid protein, when imported into the mitochondrion, causes cell death under a specific growth condition. Cells that can survive that condition (by exclusion of the toxic protein from the mitochondrion) are candidates for carrying mutations affecting mitochondrial import. This scheme has been applied in the cases described below, and the mutations that have been isolated are currently under intensive investigation. Whether they provide useful insights into the processes of mitochondrial protein import remains to be seen.

a. Reversion of Gly⁻ATP2–lacZ Strains to Gly⁺. *Saccharomyces cerevisiae* strains carrying the *ATP2–lacZ* gene fusion described by Douglas *et al.* (1984) were shown to be respiratory deficient (i.e., unable to grow on a medium containing glycerol as the sole source of carbon). McCammon *et al.* (1986) employed a selection scheme that made two assumptions: (1) mutations affecting the import apparatus would be lethal, and (2) a hybrid protein would be more severely affected in its import compared to the wild-type protein. They isolated yeast strains that were able to grow on glycerol as the sole carbon source and screened among those for ones that were temperature sensitive for growth on all media. They hoped that the *ts* defect would decrease the import of the hybrid protein while not affecting the wild-type protein at the low temperature. At the high temperature, the import defect would be lethal.

They isolated mutants that expressed high levels of the F_1 ATPase β-subunit–β-galactosidase hybrid protein and yet were able to grow on glycerol. Growth on glycerol was poor, and strains that were temperature sensitive for growth on all media were studied further. Four complementation groups have been characterized, and all appear to represent single nuclear genes. One group, defined as the *MAD1* (for mitochondrial assembly or delivery) gene, was shown to accumulate several of the precursors of the F_1 ATPase complex at the nonpermissive temperature. The precursors were associated with the mitochondrion and were membrane localized. Strains carrying the *mad2*, *mad3*, and *mad4* mutations did not accumulate precursors at the nonpermissive temperature, although they expressed the hybrid protein as well as the wild-type strain. These mutations were shown to be different from the *mas1* and *mas2* mutations (Yaffe and Schatz, 1984; Yaffe *et al.*, 1985) by genetic complementation analysis. The *MAD1* gene has been cloned by functional complementation of the *ts* defect and is currently being sequenced (M. G. Douglas, personal communication).

In a similar scheme, Garrett and Emr (1986) independently isolated revertants of the Gly⁻ phenotype of a strain carrying the same *ATP2–lacZ* fusion. Approximately half of the chromosomal Gly⁺ mutants expressing high levels of β-galactosidase activity were temperature sensitive on all media. Recessive mutants were assigned to three complementation groups, called *mtf* (for mitochondrial function), with additional mutants as yet unassigned. J. Garrett (personal communication) characterized the *mtf1* mutant further and found that it showed decreased β subunit–β-galactosidase levels in the mitochondria (from 85 to 60%). Although the *mtf1* mutant grew well on glycerol-containing medium, it has not been possible to demonstrate the accumulation of the precursor to the β subunit of the F_1 ATPase.

J. Garrett has cloned the *MTF1* gene by functional complementation of the temperature-sensitive defect. Since null mutants can grow slowly at 30°C, the *MTF1* gene product is not an essential protein. However, the null mutants also fail to grow at 37°C. The cloned gene hybridized to an abundant mRNA species of 900 bases in length. Based on preliminary studies with a *MTF1–lacZ* gene fusion, it appears as if the *MTF1* gene product is a cytoplasmic protein. The sequencing of the gene is in progress. As an alternative approach to the isolation of mutations affecting import, J. Garrett (personal communication) has reverted the temperature-sensitive defect of the *mtf1* mutant and is looking at the temperature-resistant revertants that have simultaneously become respiratory deficient.

b. Suppression of the Toxic Effects of Colicin E3. C. R. Bibus and G. Schatz (personal communication) have devised several strategies to create hybrid proteins that either have toxic effects when imported into the mitochondria or are needed in the cytoplasm and must be excluded from the mitochondrion in order for the cells to grow under selective conditions. The RNase activity of colicin E3 has been shown to work on chloroplast rRNA as well as on bacterial 16 S rRNA. Bibus and Schatz reasoned that if colicin E3 were introduced into mitochondria, there was a good chance it would destroy mitochondrial rRNA, leaving the cells respiratory deficient (Gly⁻). They constructed a gene fusion between cytochrome oxidase subunit IV and the bacterial colicin E3. The hybrid protein contained the targeting function of the 53 amino-terminal residues of the cytochrome oxidase subunit and the RNase activity of the carboxy-terminal 182 amino acids of the colicin. Although the hybrid protein was successfully imported into yeast mitochondria, the cells remained Gly⁺. Bibus and Schatz concluded that either colicin E3 did not work on mitochondrial rRNA or else this particular hybrid protein was not active for the RNase activity.

c. Selection for Mislocalization of a Mitochondrial Histidine Biosynthetic Enzyme. C. R. Bibus and G. Schatz (personal communication) have also tried an approach developed by Deshaies and Schekman (1987) for the isolation of mutants blocked in translocation of proteins into the endoplasmic reticulum. This scheme involves targeting histidinol dehydrogenase, an enzyme of the histidine biosynthetic pathway, to the mitochondrion, resulting in histidine auxotrophy. Any mutations that excluded the biosynthetic enzyme from the mitochondrion would restore the ability of the cell to convert histidinol to histidine (Hol⁺ phenotype).

Bibus and Schatz made a gene fusion that attached the signal sequence of cytochrome oxidase subunit IV to histidinol dehydrogenase, the product of the *HIS4C* gene, and introduced it on a plasmid into a strain that was deleted for the wild-type *HIS4* gene and carried the *HOL1-1* mutation necessary for growth on histidinol. This gene fusion caused a 20- to 50-fold decrease in growth rate, and the hybrid protein was localized to the mitochondrial matrix and appeared to undergo a maturation step by the processing protease. After mutagenesis with EMS, they selected for growth on histidinol medium at room temperature. Out of 4000 Hol⁺ colonies, approximately 300 were also temperature sensitive. Of these, only two accumulated the precursor to the β subunit of the F_1 ATPase, and that accumulation phenotype was not reproducible.

Bibus and Schatz have concluded that such a scheme which produced so many "false positives" did not yield useful mutants at any greater frequency than the original screen for temperature-sensitive mutants (see Section II,A,1) of Yaffe and Schatz (1984). Interestingly, there were no signal sequence mutations found among the non-temperature-sensitive Hol[+] strains.

d. Import of a Signalless Cytochrome Oxidase Subunit. C. R. Bibus and G. Schatz (personal communication) have also designed a scheme in which a mutated gene encoding a signalless, cytoplasmic cytochrome oxidase subunit IV is placed in a strain otherwise deleted for that subunit. The starting strain is therefore respiratory deficient (Gly[-]), and selection is made for Gly[+] revertants. They expect to find both plasmid-linked mutations that restore a functional signal as well as mutations that alter the translocation components of the import system. Work on these revertants is in progress.

e. Mislocalization of a Nuclear Protein. J. R. Greene and L. Guarente (personal communication) have devised a strategy for isolating mitochondrial import mutants that relies on selection for the proper localization of a protein that normally is found in the nucleus. The *HAP2* gene product is a protein required in the nucleus for the transcription of genes that control growth on nonfermentable carbon sources. J. Pinkham and L. Guarente (personal communication) have constructed a *HAP2–lacZ* gene fusion that encodes a hybrid protein that can still complement a *hap2* mutation and is localized to the nucleus. By fusing to this construction the amino terminus of the mitochondrially imported *HEM8* product (ferrochelatase, the last step in the heme biosynthetic pathway), they have been able to redirect the *HEM8:HAP2:lacZ* protein from the nucleus into the mitochondrion. Such a strain is Lactate[-], and revertants to Lactate[+] may restore the nuclear localization of the *HAP2–lacZ* gene product. Both plasmid-linked and chromosomal Lactate[+] revertants have been isolated and are being analyzed.

f. Suppression of Toxicity of a *HEM1–lacZ* Hybrid Protein. Keng *et al.* (1986) reported the construction of hybrid proteins between δ-aminolevulinate synthase and β-galactosidase that were successfully delivered to the mitochondrial matrix. Under the wild-type *HEM1* promoter, low levels of β-galactosidase were made, and no deleterious effects of the presence of these hybrids within the mitochondrion were detected. When put under the control of the powerful *GAL1–GAL10* promoter and under inducing conditions, however, the gene fusions caused poor growth. By demanding growth on galactose media, Keng *et al.* (1986) could exploit this effect to isolate cis- and trans-acting mutations that prevent the localization of the hybrid protein. With this gene fusion and the *PUT2–lacZ* hybrid described below, the level of expression of the gene appears directly correlated with the poor growth phenotype. The cause of this phenomenon is not understood at present but clearly is critical to our understanding of suppressor mutations that relieve the effect.

g. Suppression of Toxicity of a *PUT2–lacZ* Hybrid Protein. Brandriss and Krzywicki (1986) constructed two hybrid proteins carrying amino-terminal extensions of Δ^1-pyrroline-5-carboxylate dehydrogenase (an enzyme in the mitochondrial proline utilization pathway) attached to β-galactosidase. Although both were delivered to the mitochondrion, one was properly localized to the matrix, while the other (longer) hybrid fractionated partially with the matrix and partially with the inner membrane. Although the presence of the matrix-localized protein somewhat reduced the growth rate when the strain was grown under maximally inducing conditions (proline as the sole source of nitrogen), the longer fusion entirely prevented growth on proline. Mutations that overcome this proline-induced death may affect mitochondrial import.

2. Site-Specific Mutagenesis of the Signal Sequence

Without any preexisting mutations in the presequences of mitochondrially imported proteins to use as guides, several laboratories are currently engineering sequence changes in the cloned genes for various proteins in order to test their effects on import *in vitro* and *in vivo*. Since very little is known about how these signals function to achieve the proper localization, it is, at the present time, a guessing game as to which residues may prove to be important for signal function. Since one characteristic common to most signals is the overall positive charge, however, one obvious type of alteration is the removal of the positively charged residues or the substitution of negatively charged residues.

a. Human Ornithine Transcarbamylase. Horwich *et al.* (1986) are attempting to define the critical regions in the 32 amino acid signal sequence of human ornithine transcarbamylase (OTC). Short amino-terminal and carboxy-terminal deletions in the signal sequence resulted in partial reduction in targeting and processing, while removal of the internal portion (amino acids 8–22) of the signal led to a complete loss of function. Surprisingly, when the amino acids (26–31) immediately adjacent to the cleavage site at position 32 were deleted, the protein was normally delivered and processed. When arginine-6 or arginine-15 was converted to a glycine residue, processing was reduced about 50%. Conversion of arginine-26 to glycine had no effect on function, whereas a change of arginine to glycine at position 23 completely abolished import, and the pre-OTC was not a substrate for the processing protease that normally cleaves it to the mature form. They hypothesized that in this signal sequence, the arginine residue at position 23 has a critical structural role, unlike the other three arginines.

b. F$_1$ ATPase β Subunit. Both the Douglas and Emr laboratories are dissecting the elements of the 19 amino acid signal sequence of the β subunit of the F$_1$ ATPase. Vassarotti *et al.* (1987) have made several internal deletions in the signal of a β subunit–β-galactosidase hybrid protein. Removal of amino acids 17–26 or 10–36 reduce but do not eliminate mitochondrial localization of the protein, and the resulting strains are still Gly$^-$ due to sufficient import of the hybrid protein. Removal of residues 4–36, however, abolished import and restored respiratory competence. By placing these deletions back into an otherwise wild-type *ATP2* gene, they were able to compare their effects on the ATPase itself in a strain deleted for the chromosome *ATP2* copy. The deletions affected the wild-type protein in a manner similar to the hybrid protein.

Two surprises resulted from these studies: (1) deletions that removed the normal processing site for the matrix protease did not prevent the assembly of the unprocessed precursor into a functional enzyme complex, and (2) deletions distal to the processing site by about 20 amino acids were not processed by the protease, in spite of the presence of the correct processing site. Strains carrying deletions of the *ATP2* signal that prevented import will be reverted to Gly$^+$ in order to select for suppressor mutations that improve the quality of signal or adapt the import machinery to recognize the mutant signal sequence.

Deletion analyses by Bedwell and Emr (1987) suggest that signal information may be redundant. When one or both of the arginine residues in the *ATP2* signal were changed to glutamate residues, growth of the yeast strains on a medium containing nonfermentable carbon sources (glycerol–lactate) was unaffected, although the double mutant did show a kinetic defect in import. Strains carrying deletions that removed amino acids 5–12 or 16–19 or both could still grow on glycerol–lactate medium. The double deletion removed all four positively charged residues in the signal sequence. A third deletion (28–34) in the mature portion of the β subunit that removed two positively charged residues at positions 30 and 31 had no affect on growth either singly or when combined with either of the above deletions. However, the triple deletion resulted in lack of growth on glycerol–lactate medium and failure to import the β subunit. Bedwell and Emr have concluded that there are three signaling regions in the F$_1$ ATPase β subunit which are at least partially redundant and individually capable of promoting import. In support of this conclusion, they constructed a strain that carried a fourth deletion which removed codons 2–

19. The strain synthesized a protein containing the initial methionine adjacent to the mature amino terminus that was imported normally and grew on a glycerol–lactate medium.

Bedwell and Emr's strategy for future experiments is to create an *ATP2* gene carrying a single signal region and use that as a substrate for single amino acid changes that affect import of the β subunit. These import-defective strains would then be reverted to isolate both intra- and extragenic suppressor mutations.

c. Alcohol Dehydrogenase III. Pilgrim and Young (1987) have dissected the presequence of the mitochondrial alcohol dehydrogenase (ADH III) by bisulfite-induced mutagenesis and examination of the effects of the amino acid changes on import. The wild-type ADH III protein is synthesized with a 28 amino acid presequence and has a half-time of import of less than 3 min. One class of mutants having single or double amino acid changes was indistinguishable from the wild type. This included an arginine to cysteine change at position 11. A second class of mutants having in common changes at residues 18, 26, or 27 showed slower processing kinetics than the wild-type but with normal cleavage. One mutant enzyme with changes at amino acids, 9, 26, and 27 was imported but showed no detectable processing.

Mutant enzymes with signals of only the first 14 or 16 amino acids still import normally but do not undergo signal processing. When the first 18 residues were deleted and replaced with methionine and aspartic acid residues, only 15–20% of the ADH III was targeted to the mitochondria; it was not imported to the matrix but was membrane associated. These results demonstrate redundancy in this signal as well as suggest that structure rather than sequence determines mitochondrial recognition.

III. CONCLUSIONS

A. Why So Few Mutants?

The yield of mutations affecting steps in the general pathway of mitochondrial protein import has been surprisingly low. In the extensive temperature-sensitive mutant hunt carried out in the Schatz laboratory, only *mas1* and *mas2* mutants fulfilled the two criteria set—temperature sensitivity and stable precursor accumulation. This result leads one to examine the criteria more closely. While it is possible that some genes cannot be mutated to a temperature-sensitive phenotype, this method is generally accepted as the best way to isolate mutations in essential functions. Given the finding that a reduction in processing protease activity leads to cell death, the criterion of essentiality of function would appear to be appropriate. But what about the process is essential? Respiration is clearly not essential for the viability of a yeast cell. Perhaps it is the function of the tricarboxylic acid cycle, or it may be that the mitochondrial import of a *single* unknown protein is essential for cell growth.

Not all gene products involved in import, however, may turn out to be essential for yeast cell viability. Furthermore, if these essential steps are carried out by proteins whose genes are duplicated, then such selections will not be productive. There is no evidence to support this notion at present, other than the difficulty in finding mutations affecting import.

The second criterion, precursor accumulation, deserves careful study. The *mas1* and *mas2* mutants accumulated stable precursors to mitochondrial proteins when the mutations were maintained in their original background. However, in outcrosses to other yeast strains, the precursor accumulation phenotype was frequently lost (J. Garrett, personal communication). Even within the same genetic background, this phenotype in some mutants was variable and not reproducible (C. R. Bibus, personal communication). The ability to maintain the precursors intact in the cytoplasm may be a complex function of genetic background, growth phase, or level of nonspecific cellular proteases. In addition, blocks along the import pathway may occur after the signal sequence has been processed, leaving a mitochondrially associated mature-length polypeptide, perhaps mislocalized inside the mitochondrion. No precursor would be expected to accumulate in the cytoplasm under these circumstances.

The major difficulty with systems using the toxic effects of hybrid proteins is determining the causes of the deleterious effects on cell growth. The strains carrying the *ATP2–lacZ, ATP2–SUC2, HEM1–lacZ, PUT2–lacZ* gene fusions, among others, also had a wild-type copy of the gene in the genome. No attempt was made to determine if an interaction between the two proteins were contributing to the toxicity. If protein–protein interactions are responsible, the mutations that alter such interactions will suppress the toxicity. Likewise, in the cases where the hybrid proteins cause respiratory deficiency, the underlying cause of that effect is not understood. Since respiration involves a large number of protein complexes, restoration of respiratory competence may come through many types of mutational changes, not simply lack of hybrid protein import. In cases where cell death is related to the level of gene expression, subtle alterations in levels of transcription, translation, or protein stability may contribute to cell survival in the selection schemes outlined above.

It would seem that the more specific the isolation scheme, the more successful it will be. However, the strategy described in Section II,A,1 to engineer a mitochondrial histidine biosynthetic enzyme and then select for mutations to histidine prototrophy as a way to restore a cytoplasmic enzyme was unsuccessful in fulfilling the criteria expected of an import mutation.

The large number of mutations that are currently being examined as putative import mutants should be subjected to tests other than precursor accumulation. Although such determinations are not trivial, it may be more useful to examine whether the proteins under study change their localization patterns in measurable ways, for example, total hybrid protein in the cytosol versus total hybrid protein in the mitochondria.

B. Mutations of the Signal Sequence

Based on the work of four laboratories cited in Section II,B,2, several common features of signals have emerged. Given the alterations that the experimenters have decided to make in the DNA sequence of a cloned gene, mitochondrial proteins can tolerate substantial changes in their signals that may slow down, but do not completely interfere with, the delivery and processing pathway. Conformation seems to be important, rather than sequence; this is a finding that supports the earlier observation that mitochondrial presequences do not share sequence homology. In addition, several signals appear to be redundant since deletions of the sequence that remove extensive portions of the signal still import properly. Sequences in the mature portion of a protein may be required for import/ processing and can supply signaling information in the absence of the normal signal. Surprisingly, the signals do not have to be removed in order for a protein to find its appropriate mitochondrial compartment, and their presence does not seem to interfere with enzymatic activity of a multiprotein complex such as the F_1 ATPase.

Studies on signal sequence deletion mutants suffer from the same difficulty plaguing all deletion analyses: the inability to distinguish the effects of changes in amino acid sequence from changes in spacing or protein conformation. Furthermore, the analyses are limited to the types of mutations that are engineered. More useful information will come from letting the cells tell the experimenter which mutational changes result in excluding the import of a toxic protein or allowing a mutant signal to be recognized.

What constitutes the signaling information in a protein destined for the mitochondrion? Work from the Schatz laboratory has shown clearly that the presequence, that portion of the protein which is removed from a polypeptide during the import process, is sufficient to target a foreign, cytosolic protein to the mitochondrion. They and others have shown that the presequence is divisible into smaller units that also carry signaling information. Since Bedwell and Emr (1987) have now found that a mature (i.e., signalless) β subunit of the F_1 ATPase is properly delivered, one is forced to conclude that the presequence may be only a part of the addressing information. Such a result points out that the passenger protein used in such studies may well be playing an important, heretofore unrecognized, role.

C. Prospects for the Future

It is essential that genetic studies continue to be used to complement the elegant biochemical studies that have contributed to our understanding of mitochondrial protein import. Although the number of useful mutations uncovered so far has been smaller than expected, it is anticipated that creative strategies in combination with the increased number of laboratories concentrating on this approach will ultimately provide the tools for isolating elements of the mitochondrial import apparatus.

ACKNOWLEDGMENTS

I am grateful to all of my colleagues who provided preprints, unpublished results, and stimulating discussions.

REFERENCES

Bassford, P., and Beckwith, J. (1979). *Escherichia coli* mutants accumulating the precursor of a secreted protein in the cytoplasm. *Nature (London)* **277**, 538–541.

Bassford, P. J., Jr., Silhavy, T. J., and Beckwith, J. R. (1979). Use of gene fusions to study secretion of maltose-binding protein in *Escherichia coli* periplasm. *J. Bacteriol.* **139**, 19–31.

Bedwell, D. M., Klionsky, D. J., and Emr, S. D. (1987). The yeast F_1-ATPase β subunit precursor contains functionally redundant mitochondrial protein import information. *Mol. Cell. Biol.* **7**, 4038–4047.

Böhni, P., Daum, G., and Schatz, G. (1983). Import of proteins into mitochondria. Partial purification of a matrix-located protease involved in cleavage of mitochondrial precursor polypeptides. *J. Biol. Chem.* **258**, 4937–4943.

Brandriss, M. C., and Krzywicki, K. (1986). Amino-terminal fragments of Δ^1-pyrroline-5-carboxylate dehydrogenase direct β-galactosidase to the mitochondrial matrix in *Saccharomyces cerevisiae*. *Mol. Cell Biol.* **6**, 3502–3512.

Daum, G., Gasser, S., and Schatz, G. (1982). Import of proteins into mitochondria. Energy-dependent two step processing of the intermembrane-space enzyme, cytochrome b_2, by isolated yeast mitochondria. *J. Biol. Chem.* **257**, 13075–13080.

Deshaies, R. J., and Schekman, R. (1987). A yeast mutant defective at an early stage in import of secretory protein precursors into the endoplasmic reticulum. *J. Cell Biol.* **105**, 633–645.

Douglas, M., Geller, B., and Emr, S. (1984). Intracellular targeting and import of an F_1 ATPase β-subunit β-galactosidase hybrid protein into yeast mitochondria. *Proc. Natl. Acad. Sci. U.S.A.* **81**, 3983–3987.

Douglas, M. G., McCammon, M. T., and Vasarotti, A. (1986). Targeting proteins into mitochondria. *Microbiol. Rev.* **50**, 166–178.

Dumont, M. E., Ernst, J. F., Hampsey, D. M., and Sherman, F. (1987). Identification and sequence of the gene encoding cytochrome *c* heme lyase in the yeast *Saccharomyces cerevisiae. EMBO J.* **6**, 235–241.

Emr, S., Vassarotti, A., Garret, J., Geller, B., Takeda, M., and Douglas, M. (1986). The amino terminus of the yeast F_1 ATPase β-subunit precursor functions as a mitochondrial import signal. *J. Cell Biol.* **102**, 523–533.

Freitag, H., Janes, M., and Neupert, W. (1982). Biosynthesis of mitochondrial porin and insertion into the outer mitochondrial membrane of *Neurospora crassa. Eur. J. Biochem.* **126**, 197–202.

Garrett, J. M., and Emr, S. D. (1986). Genetic anlaysis of mitochondrial protein delivery in yeast. *Yeast* **2**, S124.

Gasser, S., and Schatz, G. (1983). Import of proteins into mitochondria. *In vitro* studies on the biogenesis of the outer membrane. *J. Biol. Chem.* **258**, 3427–3430.

Gasser, S. M., Daum, G., and Schatz, G. (1982). Import of proteins into mitochondria. Energy-dependent uptake of precursor by isolated mitochondria. *J. Biol. Chem.* **257**, 13034–13041.

Geller, B., Britten, M., Biggs, C., Douglas, M., and Emr, S. (1983). Import of *ATP2–lacZ* gene fusion proteins into mitochondria. *In* "Mitochondria 1983: Nucleo–Mitochondrial Interactions" (R. Schweyen, K. Wolf, and F. Kaudewitz, eds.), pp. 607–619. Walter de Gruyter, New York.

Guiard, B. (1985). Structure, expression, and regulation of a nuclear gene encoding a mitochondrial protein: the yeast L(+)-lactate cytochrome *c* oxido-reductase (cytochrome b_2). *EMBO J.* **4**, 3265–3272.

Guiard, B., and Buhler, J.-M. (1984). Yeast cytochrome b_2 gene: Isolation with antibody probes. *Biochimie* **66**, 151–158.

Hallermayer, G., Zimmermann, R., and Neupert, W. (1977). Kinetic studies on the transport of cytoplasmically synthesized proteins into the mitochondria in intact cells of *Neurospora crassa. Eur. J. Biochem.* **81**, 523–532.

Hartwell, L. H. (1967). Macromolecule synthesis in ts mutants of yeast. *J. Bacteriol.* **93**, 1662–1670.

Hase, T., Reizman, H., Suda, K., and Schatz, G. (1983). Import of proteins into mitochondria: Nucleotide sequence of the gene for a 70-kd protein of the yeast mitochondrial outer membrane. *EMBO J.* **2**, 2169–2172.

Hase, T., Muller, U., Riezman, H., and Schatz, G. (1984). A 70 kd protein of the yeast mitochondrial outer membrane is targeted and anchored via its extreme amino terminus. *EMBO J.* **3**, 3157–3164.

Hase, T., Nakai, M., and Matsubara, H. (1986). The N-terminal 21 amino acids of a 70 kDa protein of the yeast mitochondrial outer membrane direct *E. coli* β-galactosidase into the mitochondrial matrix space in yeast cells. *FEBS Lett.* **197**, 199–203.

Hay, R., Böhni, P., and Gasser, S. (1984). How mitochondria import proteins. *Biochim. Biophys. Acta* **779**, 65–87.

Hennig, B., and Neupert, W. (1981). Assembly of cytochrome *c*. Apocytochrome *c* is bound to specific sites on mitochondria before its conversion to holocytochrome *c. Eur. J. Biochem.* **121**, 203–212.

Hennig, B., Koehler, H., and Neupert, W. (1983). Receptor sites involved in posttranslational transport of apocytochrome c into mitochondria: specificity affinity and number of sites. *Proc. Natl. Acad. Sci. U.S.A.* **80**, 4963–4967.

Horwich, A., Kalousek, F., Fenton, W., Pollock, R., and Rosenberg, L. (1986). Targeting of pre-ornithine transcarbamylase to mitochondria: Definition of critical regions and residues in the leader peptide. *Cell* **44**, 451–459.

Hurt, E., Pesold-Hurt, B., and Schatz, G. (1984). The amino-terminal region of an imported mitochondrial precursor polypeptide can direct cytoplasmic dihydrofolate reductase into the mitochondrial matrix. *EMBO J.* **3**, 3149–3156.

Hurt, E., Muller, U., and Schatz, G. (1985a). The first twelve amino acids of a yeast mitochondrial outer membrane protein can direct a nuclear encoded cytochrome oxidase subunit to the mitochondrial inner membrane. *EMBO J.* **4**, 3509–3518.

Hurt, E., Pesold-Hurt, B., Suda, K., Oppliger, W., and Schatz, G. (1985b). The first twelve amino acids (less than half of the presequence of an imported mitochondrial protein) can direct mouse cytosolic dihydrofolate reductase into the yeast mitochondrial matrix. *EMBO J* **4**, 2061–2065.

Kaput, J., Goltz, S., and Blobel, G. (1982). Nucleotide sequence of the yeast nuclear gene for cytochrome c peroxidase precursor. Functional implications of the pre-sequence for protein transport into mitochondria. *J. Biol. Chem.* **257**, 15054–15058.

Keng, T., Alani, E., and Guarente, L. (1986). The amino-terminal nine residues of the δ-aminolevulinate synthetase directs β-galactosidase into the mitochondrial matrix. *Mol. Cell. Biol.* **6**, 355–364.

Korb, H., and Neupert, W. (1978). Biogenesis of cytochrome c in *Neurospora crassa:* Synthesis of apocytochrome c, transfer to mitochondria and conversion to holocytochrome c. *Eur. J. Biochem.* **91**, 609–620.

Li, Y., Leonard, K., and Weiss, H. (1981). Membrane-bound and water-soluble cytochrome c_1 from Neurospora mitochondria. *Eur. J. Biochem.* **116**, 199–205.

McAda, P., and Douglas, M. (1982). A neutral metallo endo-protease involved in the processing of an F_1 ATPase subunit precursor in mitochondria. *J. Biol. Chem.* **257**, 3177–3182.

McCammon, M. T., Vallier, L. G., and Douglas, M. G. (1986). A nuclear mutation in *mad1* (mitochondrial assembly or delivery) is associated with accumulation of mitochondrial F_1 ATPase precursors. In "Yeast Genetics and Molecular Biology Meeting Abstracts," p. 2. Cold Spring Harbor Laboratory Press, Cold Spring Harbor, New York.

Maccecchini, M.-L., Rudin, Y., Blobel, G., and Schatz, G. (1979). Import of proteins into mitochondria: Precursor forms of the extramitochondrially made F_1 ATPase subunits in yeast. *Proc. Natl. Acad. Sci. U.S.A.* **76**, 343–347.

Matner, R., and Sherman, F. (1982). Differential accumulation of two apo-isocytochromes c in processing mutants of yeast. *J. Biol. Chem.* **257**, 9811–9821.

Michaelis, G., Mannhaupt, G., Pratje, E., Fischer, E., Naggert, J., and Schweizer, E. (1982). Mitochondrial translation products in nuclear respiration-deficient *pet* mutants of *Saccharomyces cerevisiae.* In "Mitochondrial Genes" (P. P. Slonimski, P. Borst, and G. Attardi, eds.), pp. 311–321. Cold Spring Harbor Laboratory Press, Cold Spring Harbor, New York.

Mori, M., Miura, S., Fatibana, M., and Cohen, P. P. (1980). Characterization of a protease apparently involved in processing of pre-ornithine transcarbamylase of rat liver. *Proc. Natl. Acad. Sci. U.S.A.* **77**, 7044–7048.

Nelson, N., and Schatz, G. (1979). Energy-dependent processing of cytoplasmically made precursors to mitochondrial proteins. *Proc. Natl. Acad. Sci. U.S.A.* **76**, 4365–4369.

Ohashi, A., Gibson, J., Gregor, I., and Schatz, G. (1982). Import of proteins into mitochondria. The precursor of cytochrome c_1 is processed in two steps, one of them heme-dependent. *J. Biol. Chem.* **257**, 13042–13047.

Pfanner, N., Tropschug, M., and Neupert, W. (1987). Mitochondrial protein import: Nucleoside triphosphates are involved in conferring import-competence to precursors. *Cell* **49**, 815–823.

Pilgrim, D., and Young, E. T. (1987). Primary structure requirements for correct sorting of the yeast mitochondrial protein ADHIII to the yeast mitochondrial matrix space. *Mol. Cell. Biol.* **7**, 294–304.

Pratje, E., and Guiard, B. (1986). One nuclear gene controls the removal of transient pre-sequences from two yeast proteins: One encoded by the nuclear, the other by the mitochondrial genome. *EMBO J.* **5**, 1313–1317.

Pratje, E., Mannhaupt, G., Michaelis, G., and Beyreuther, K. (1983). A nuclear mutation prevents processing of a mitochondrially encoded membrane protein in *Saccharomyces cerevisiae*. *EMBO J.* **2**, 1049–1054.

Reid, G. (1985). Transport of proteins into mitochondria. *In* "Current Topics in Membranes and Transport (P. Kanuf and J. Cook, eds.), Vol. 24, pp. 295–336. Academic Press, New York.

Riezman, H., Hase, T., van Loon, A. P. G. M., Grivell, L., Suda, K., and Schatz, G. (1983a). Import of proteins into mitochondria: A 70 kildalton outer membrane protein with a large carboxy-terminal deletion is still transported to the outer membrane. *EMBO J.* **2**, 2161–2168.

Riezman, H., Hay, R., Gasser, S., Daum, G., Schneider, G., Witte, C., and Schatz, G. (1983b). The outer membrane of yeast mitochondria: Isolation of outside-out sealed vesicles. *EMBO J.* **2**, 1105–1111.

Rothstein, R. J., and Sherman, F. (1980). Genes affecting the expression of cytochrome c in yeast. Genetic mapping and genetic interactions. *Genetics* **94**, 871–889.

Sadler, I., Suda, K., Schatz, G., Kaudewitz, F., and Haid, A. (1984). Sequencing of the nuclear gene for the yeast cytochrome c_1 precursor reveals an unusually complex amino-terminal presequence. *EMBO. J.* **3**, 2137–2143.

Schatz, G. (1979). How mitochondria import proteins from the cytoplasm. *FEBS Lett.* **103**, 203–211.

Schleyer, M., and Neupert, W. (1985). Transport of proteins into mitochondria. Translocation intermediates spanning contact sites between outer and inner membranes. *Cell* **43**, 339–350.

Sherman, F., Stewart, J. W., Parker, J. H., Inhaber, E., Shipman, N. A., Putterman, G. J., Gardisky, R. L., and Margoliash, E. (1968). The mutational alteration of the primary structure of yeast iso-1-cytochrome c. *J. Biol. Chem.* **243**, 5446–5456.

Sherman, F., Stewart, J. W., Jackson, M., Gilmore, R. A., and Parker, J. H. (1974). Mutants of yeast defective in iso-1-cytochrome c. *Genetics* **77**, 255–284.

Silhavy, T., Benson, S., and Emr, S. (1983). Mechanisms of protein localization. *Microbiol. Rev.* **47**, 313–344.

Suissa, M., and Schatz, G. (1982). Import of proteins into mitochondria. Translatable mRNAs for imported mitochondrial proteins are present in free as well as mitochondria-bound cytoplasmic polysomes. *J. Biol. Chem.* **257**, 13048–13055.

van Loon, A. P. G. M., and Young, E. T. (1986). Intracellular sorting of alcohol dehydrogenase isoenzymes in yeast: A cytosolic location reflects absence of an amino-terminal targeting sequence for the mitochondrion. *EMBO J.* **5**, 161–165.

van Loon, A. P. G. M., Brändli, A. W., and Schatz, G. (1986). The presequences of two mitochondrial proteins contain information for intracellular and intramitochondrial sorting. *Cell* **44**, 801–812.

Vassarotti, A., Chen, W.-J., Smagula, C., and Douglas, M. G. (1987). Sequences distal to the mitochondrial targeting sequences are necessary for the maturation of the F_1 ATPase β-subunit precursor in mitochondria. *J. Biol. Chem.* **262**, 411–418.

Yaffe, M., and Schatz, G. (1984). Two nuclear mutations that block mitochondrial protein import in yeast. *Proc. Natl. Acad. Sci. U.S.A.* **81**, 4819–4829.

Yaffe, M., Ohta, S., and Schatz, G. (1985). A yeast mutant temperature-sensitive for mitochondrial assembly is deficient in a mitochondrial protease activity that cleaves imported precursor polypeptide. *EMBO J.* **4**, 2069–2074.

Zimmerman, R., Paluch, U., and Neupert, W. (1979a). Cell-free synthesis of cytochrome c. *FEBS Lett.* **108**, 141–146.

Zimmermann, R., Paluch, U., Sprinzl, M., and Neupert, W. (1979b). Cell-free synthesis of the mitochondrial ADP/ATP carrier protein of *Neurospora crassa*. *Eur. J. Biochem.* **99**, 247–252.

Zimmermann, R., Hennig, G., and Neupert, W. (1981). Different transport pathways of individual precursor proteins in mitochondria. *Eur. J. Biochem.* **116**, 455–460.

Zwizinski, C., Schleyer, M., and Neupert, W. (1983). Transfer of proteins into mitochondria. Precursor to the ADP/ATP carrier binds to receptor sites on isolated mitochondria. *J. Biol. Chem.* **258**, 4071–4084.

Zwizinski, C., Schleyer, M., and Neupert, W. (1984). Proteinaceous receptors for the import of mitochondrial precursors. *J. Biol. Chem.* **259**, 7850–7856.

17

Synthesis and Assembly of Mitochondrial Proteins

DONALD W. NICHOLSON AND WALTER NEUPERT

I. INTRODUCTION

Eukaryotic cells can perform a variety of metabolic tasks with high efficiency owing to the compartmentalization of functions within cellular organelles. The organelles are delineated by membrane structures and contain specific subsets of proteins related to their role within the cell. The uniqueness of each type of organelle is maintained by processes in

677

PROTEIN TRANSFER AND ORGANELLE BIOGENESIS

the cell which sort and deliver the individual components in a highly specific manner. In this chapter, we discuss the biogenesis of mitochondria and specifically the processes which are involved in the import of proteins into these organelles.

Mitochondria are not synthesized *de novo*. Instead, control of the number of mitochondria within a cell occurs by division and fusion events. In the steady state there is a constant synthesis and turnover of mitochondrial proteins. Mitochondria grow and are maintained by incorporating newly synthesized material into preexisting organelles. The mitochondrion itself has a genetic apparatus which is discrete from that of the rest of the cell, but because of its relatively small size it can only account for a small percentage of the hundreds of mitochondrial proteins. The remainder are imported from the cytoplasm.

Because mitochondria contain two membranes, the organelle can be spatially divided into four distinct compartments: the outer membrane, the inner membrane, the matrix, and the intermembrane space. These too are specialized in function so that imported proteins must be correctly and specifically sorted within the mitochondrion as well. The pathways by which this occurs can be resolved into a number of discrete steps. In general, most imported mitochondrial proteins are synthesized on free ribosomes in the cytosol as precursors containing amino-terminal extensions. They are released into a cytostolic pool and are rapidly taken up by mitochondria. Receptors on the outer surface of the outer mitochondrial membrane seem to be involved in the initial recognition process. Proteins which must be transported to or through the inner membrane do so via translocation contact sites where the inner and outer membranes come close enough together to be spanned simultaneously. This step is energy dependent and specifically requires an electrochemical potential across the inner membrane. Precursor proteins are proteolytically processed to their mature size by a metal ion-dependent matrix peptidase, in some cases in combination with a second proteolytic event. The imported proteins are sometimes modified further by covalent or noncovalent attachment of cofactors. Many proteins are also assembled into larger complexes composed of several different subunits. In this chapter, we shall examine, in chronological order, each individual step which occurs during the import of mitochondrial proteins.

II. PRECURSOR PROTEINS

A. Properties of Precursor Proteins

Proteins which are imported into mitochondria are synthesized as precursors that differ from their mature counterparts in a number of ways: (1)

most, but not all, are synthesized as higher molecular weight proteins with amino-terminal peptide extensions which are proteolytically removed during or following import; (ii) a number of precursor proteins are covalently modified when imported while some acquire cofactors which are associated with the protein in a tight but noncovalent manner; (iii) precursor proteins differ in conformation from their mature form; (iv) precursor proteins often aggregate in the cytosol, whereas mature proteins in mitochondria are more likely to be found as monomers, dimers, or as subunits in heterologous complexes; (v) precursor proteins accumulated in the cytosol are far less stable than their correctly imported mature equivalent. Changes which accompany the maturation of imported proteins occur for one of two reasons. Some (i.e., removal of amino-terminal extensions) are connected to the import process, while others (i.e., acquisition of cofactors) are required for functional properties once inside the mitochondrion.

Of the more than 60 different proteins examined to date which must be imported into either the matrix or the inner membrane, almost all contain amino-terminal presequences (for an extensive list see Hay *et al.*, 1984, or Harmey and Neupert, 1985). Possible exceptions of the matrix proteins include 2-isopropylmalate synthase (Gasser *et al.*, 1982a; Hampsey *et al.*, 1983; Beltzer *et al.*, 1986), extramitochondrially expressed yeast bI4 maturase (Banroques *et al.*, 1987), and 3-oxoacyl-CoA thiolase from rat (Arakawa *et al.*, 1987), none of which appear to contain an amino-terminal extension. Some proteins of the inner membrane are known to be imported in their mature size, namely, the ADP/ATP carrier (Zimmermann *et al.*, 1979b; Hatalová and Kolarov, 1983), the uncoupling protein of brown adipose tissue (Freeman *et al.*, 1983; Ricquier *et al.*, 1983; Bouillaud *et al.*, 1986; Ridley *et al.*, 1986), two subunits (14 and 11 kDa) of the bc_1 complex (Teintze *et al.*, 1982; van Loon *et al.*, 1983c; Maarse and Grivell, 1987), the ubiquinone-binding protein of the bovine bc_1 complex (Nishikimi *et al.*, 1986), and sweet potato cytochrome oxidase subunit Vc (Nakagawa *et al.*, 1987). Proteins which are imported into the intermembrane space seem to be divided in this regard. Some, like precursors to cytochrome *c* peroxidase (Maccecchini *et al.*, 1979b; Reid *et al.*, 1982), cytochrome b_2 (Reid *et al.*, 1982; Daum *et al.*, 1982b; Gasser *et al.*, 1982b), and sulfite oxidase (Mihara *et al.*, 1982b) carry amino-terminal extensions, while others, like cytochrome *c* (Korb and Neupert, 1978; Zimmermann *et al.*, 1979a; Matsuura *et al.*, 1981) and adenylate kinase (Watanabe and Kubo, 1982), do not.

Although only a few of the proteins which are imported to the outer membrane have been examined, none appears to be synthesized as a higher molecular weight precursor. These include the channel-forming porin protein (Freitag *et al.*, 1982; Mihara *et al.*, 1982a; Gasser & Schatz,

1983) and three outer membrane proteins of unknown function (70, 45, and 14 kDa; Gasser and Schatz, 1983). One possible exception is a 35 kDa protein from rat liver mitochondria (Shore et al., 1981) which appears to migrate on sodium dodecyl sulfate–polyacrylamide gels as a precursor of slightly greater molecular size. Whether this represents a precursor protein containing a peptide extension or is a gel artifact is not entirely clear and will require sequencing data to verify.

The trend that emerges is that the deeper into the mitochondrion a protein must be imported, the more likely it is to be synthesized as a precursor of higher molecular weight. As will be discussed in the following section, the amino-terminal prepiece carries targeting information which is necessary to direct imported proteins to their correct intramitochondrial location. In addition, the prepieces, which are generally hydrophilic, confer different solubility properties on the precursor proteins that may allow for their transfer through the cytosol and are important for subsequent import.

While proteolytic processing is the most prevalent covalent modification of imported mitochondrial proteins, there are a number of other changes which occur in many proteins during or following import. For example, protoheme is covalently attached to cytochromes c and c_1 via cysteine residues in the corresponding apoproteins. Iron–sulfur centers (nonheme iron) are constructed in subunits of complexes I, II, and III of the respiratory chain, again via cysteine residues. In addition, many mitochondrial proteins are modified by tight but noncovalent bonding of coenzymes and cofactors (i.e., NAD, FAD, heme a and b and metal ions such as Cu). These proteins are predominantly constituents of the respiratory chain, and the cofactors which are associated with them are primarily involved in their unique roles as electron carriers. In some cases, however, such modifications also appear to be important for events in the import pathway. Covalent attachment of heme to apocytochrome c (holocytochrome c lacking heme), for instance, initiates conformational changes along the polypeptide chain which pull the protein through the outer membrane to its functional location in the intermembrane space (Hennig and Neupert, 1981; Nicholson et al., 1987). Cytochrome c_1 is proteolytically processed in two distinct steps, the second of which has been suggested to be preceded by covalent attachment of the heme to the intermediate size apoprotein (Gasser et al., 1982b; Ohashi et al., 1982). Likewise, the second processing step of the Fe/S protein of the bc_1 complex may be dependent on formation of the Fe/S cluster (F.-U. Hartl and W. Neupert, unpublished).

Precursor proteins are also distinguishable from their mature counterparts in conformational arrangement. Apocytochrome c (prepared by chemical removal of the heme group from holocytochrome *c*, then renatured to an import-competent form) shows a nearly featureless circular dichroism (CD) pattern. Following interaction with negatively charged lipids in model membranes, which is believed to represent the first event in its import into mitochondria, up to 35% α-helical structure is expressed (Rietveld *et al.*, 1985). The disordered structure of the precursor cytochrome *c* is folded into a highly ordered stable conformation when heme is attached to form holocytochrome *c* (Fisher *et al.*, 1973). The difference between precursor and mature forms of cytochrome *c* is thus dependent on events occurring during its import, namely, interactions with phospholipids and covalent attachment of heme. These conformational changes can be detected *in vitro* with differential antibodies raised against apo- and holocytochrome *c* which do not cross-react (Korb and Neupert, 1978) or by differential proteolytic sensitivity in solution (Basile *et al.*, 1980).

The precursor to the ADP/ATP carrier does not bind carboxyatractyloside whereas the mature form does. This reflects conformational differences in the proteins which can be distinguished by their binding properties to columns of hydroxylapatite in the presence of carboxyatractyloside. *In vitro* imported and correctly assembled ADP/ATP carrier, when solubilized from mitochondria with detergent, behaves like authentic mature ADP/ATP carrier and passes through these columns while the precursor protein does not (Zimmermann and Neupert, 1980; Schleyer and Neupert, 1984). Since ADP/ATP carrier is not proteolytically processed during import, the binding properties to hydroxylapatite serve as useful criteria for establishing whether correct import and assembly has occurred. Such criteria are important for determining whether a protein imported *in vitro* acquires the properties of the mature protein *in vivo*. In some cases, for example, for proteins which are not proteolytically processed, import can only be studied by following these changes.

Precursor conformation is important for import. For example, although CD spectra for apocytochrome *c* show no detectable secondary structure (Rietveld *et al.*, 1985), binding of the protein to mitochondria is sensitive to denaturation by a single freeze–thaw cycle (H. Köhler and W. Neupert, unpublished). Similarly, the apocytochrome *c* produced by a mutant of *Neurospora crassa*, in which the carboxy-terminus is 19 amino acids longer than wild type apocytochrome *c* (the final 27 amino acids being of an unrelated sequence), cannot bind or be imported into mitochondria, most likely because of conformational perturbation (Stuart *et al.*, 1987). On the other hand, the import of a fusion protein containing mouse dihydrofolate reductase linked to the presequence of cytochrome oxidase IV can be blocked by methotrexate, which stabilizes the folding of the dihydrofolate reductase moiety. In this case, the protein must be at least partially unfolded to be imported into mitochondria (Eilers and Schatz, 1986). Recent evidence indicates that most precursor proteins are maintained in, or "defolded" into, an import-competent conformation by the hydrolysis of nucleoside triphosphates (Pfanner and Neupert, 1986; Pfanner et al., 1987; see Section V).

Most newly synthesized precursor proteins tend to aggregate. Though this has not been well characterized in vivo, it is frequently observed in vitro in either homologous or heterologous (i.e., reticulocyte lysate) translation systems. The ADP/ATP carrier, for example, is present in soluble complexes with apparent molecular weights of 120K and 500K (Zimmermann and Neupert, 1980). Aggregation has also been demonstrated for ATPase IX (Schmidt et al., 1983b) and for cytochrome oxidase V (Neupert and Schatz, 1981). In all of these cases it is not clear whether aggregation occurs as homo- or heterooligomers. The precursor to rat ornithine carbamoyltransferase is transported to mitochondria as a 5 S complex (approximately 90 kDa) containing an unidentified import factor (Argan and Shore, 1985). Similarly, the import of the F_1 ATPase β subunit into yeast mitochondria is dependent on a cytosolic factor believed to be a 40 kDa protein which binds to the precursor proteins and enables their correct association with mitochondria (Ohta and Schatz, 1984). These last two examples suggest that aggregation may occur in a heterooligomeric fashion for some imported proteins. On the other hand, the precursor to rat mitochondrial fumarase (fumarate hydratase) was reported to form homooligomeric aggregates containing six to eight molecules (Ono et al., 1985).

Beside the nonspecific interaction of proteins in solution, aggregation of precursor proteins may occur for specific reasons as well. Since many imported mitochondrial proteins have a membrane localization and are therefore at least partially hydrophobic, there probably exists some means for disguising these parts of the molecule and allowing their solubility in the cytosol. The amino-terminal prepiece may confer some solubility in aqueous environments, while aggregation of precursor proteins, presumably via their hydrophobic domains, may also contribute in this respect. The best example of this is ATPase subunit IX [proteolipid or dicyclohexylcarbodiimide (DCCD)-binding protein], one of the most hydrophobic proteins known. It contains a long polar prepiece (66 amino acids in Neurospora crassa) that accounts for the solubility of the highly hydrophobic mature sequence (81 amino acids) in aqueous environments (Viebrock et al., 1982). The protein also forms aggregates, perhaps to further aid its solubility in the cytosol. Cytochrome c, on the other hand, is a soluble protein of the intermembrane space which is not imported with a prepiece and behaves as a monomer or dimer in solution (H. Köhler and W. Neupert, unpublished). Another possible role of aggregation is that it is required for import-competent transport and binding of precursors to mitochondria (Section VI).

Once precursor proteins are synthesized, they are rapidly cleared from the cytosol and imported into mitochondria. Precursors can be accumulated *in vivo* by growing cells in the presence of uncouplers of oxidative phosphorylation. Under these conditions they cannot be imported and are pooled in the cytosol. In contrast to the relative stability of mature mitochondrial proteins, the accumulated precursors (in general) are rapidly degraded. For example, yeast cells grown to early exponential phase and then treated with the uncoupler carbonyl cyanide N-chlorophenylhydrazone (CCCP) accumulated large amounts of some mitochondrial precursor proteins (Reid and Schatz, 1982a). In pulse-labeling experiments, the proteins were degraded at different rates. The precursor of cytochrome c_1 was unstable and was degraded with a half-life of about 10 min. On the other hand, the precursor to F_1 ATPase β subunit was more stable and was degraded with a half-life of 50 min. In similar experiments, the precursor of aspartate aminotransferase, accumulated in chick embryo fibroblast cultures treated with CCCP, was degraded with a half-life of about 5 min (Jaussi *et al.*, 1982), and the precursor of carbamoyl-phosphate synthase in rat liver explants, in which proteolytic processing was blocked, was degraded with a half-life of 2–3 min (Raymond and Shore, 1981). The apparent relative stability comparing precursor with mature proteins is probably also related to the normal subcellular location of the proteins (i.e., cytosol versus mitochondria) and the degradative processes that exist there.

Maturation of imported mitochondrial proteins occurs within the mitochondrion itself and not in the cytosol immediately following synthesis. This occurs in this sequence for two reasons: (1) many maturation events are involved in the import pathway itself and/or can only take place after certain import steps have occurred, and (2) development of functional characteristics within the mitochondrion is consistent with the theme of compartmentation of processes in eukaryotic cells.

B. Cotranslational versus Posttranslational Transport

Two mechanisms exist in eukaryotic cells by which proteins may be synthesized and transported across biological membranes. In cotranslational transport, synthesis begins on soluble cytoplasmic polysomes. As the nascent polypeptide chain appears from the ribosome it is directed, with the accompanying polysome, to the target membrane and is simultaneously inserted into or through the membrane as chain elongation continues. Cotranslational import is the primary means by which proteins are transported across the membranes of the endoplasmic reticulum. It appears that in this case the cotranslational nature, probably with a few exceptions, is obligatory (i.e., a completed polypeptide chain can never enter the transport pathway). It is likely that the polypeptide chain otherwise folds in such a way that it is transport incompetent. The function of the signal recognition particle and docking protein is in arrest and dearrest of the elongation process to guarantee that the nascent polypeptide is kept in a translocation competent state for a sufficient period of time. In posttranslational transport, protein synthesis also begins on soluble cytoplasmic polysomes; however, complete polypeptide synthesis and release into the cytosol occurs before transport into or across membranes begins. Posttranslational import is the mechanism predominant in mitochondria, chloroplasts, microbodies, and probably nuclei. In mitochondria, import may occur with some cotranslational characteristics, though it is clearly not an obligatory process.

In cotranslational compared to posttranslational import, the way in which organelle targeting information is processed is different. For example, in the mechanism for microsomal targeting, the signal recognition particle and docking protein serve as common components for most proteins destined for the endoplasmic reticulum (ER). On the other hand, targeting information for proteins which are posttranslationally transported must be self-contained.

Import of mitochondrial proteins can occur posttranslationally, though early evidence pointed toward a cotranslational mechanism. Cytoplasmic 80 S ribosomes were coisolated with purified yeast mitochondria (Kellems and Butow, 1972). These ribosomes were tightly bound to mitochondria, and only one-third could be released by incubation at high ionic strength. The remaining two-thirds could only be released when their nascent polypeptide chains were dissociated from the ribosome with puromycin (Kellems et al., 1974), suggesting that the ribosomes were anchored via the newly synthesized nascent polypeptide chain which was presumably undergoing cotranslational (vectorial) transport. Furthermore, the mitochondria-associated ribosomes were enriched in mRNAs coding for mitochondrial proteins. When yeast mitochondria were isolated with their accompanying cytoplasmic ribosomes and placed in a readout system to complete polypeptide chain synthesis, over 80% of the products remained associated with the mitochondria and were imported to a location not accessible to externally added proteases (Ades and Butow, 1980a,b). The distribution of total message for mitochondrial proteins, however, was not exclusively associated with these mitochondria-bound ribosomes. Although the mRNA from mitochondria-bound ribosomes was enriched for mitochondrial proteins, indicating that the association was specific, not all mitochondrial proteins were preferentially synthesized on them (Suissa and Schatz, 1982). For example, while 60% of the translatable mRNA for F_1 ATPase β subunit or cytochrome c peroxidase was associated with mitochondria-bound polysomes, more than 95% of the message for cytochrome oxidase V or VI and porin was associated with free polysomes. In no case was the translatable message exclusively associated with mitochondria-bound polysomes. This suggested that cotranslational import was not necessarily the transport mechanism used by all imported mitochondrial proteins.

Evidence for posttranslational import of mitochondrial proteins came from studies both *in vivo* and *in vitro:*

1. Extramitochondrial pools of a number of mitochondrial proteins have been detected in pulse-labeling experiments *in vivo.* The appearance of label in mitochondrial proteins showed a lag compared to total cellular protein in *Neurospora crassa* (Hallermayer *et al.,* 1977). The labeled mitochondrial proteins first appeared in a cytosolic pool and then in mitochondria. Similarly, in yeast, mitochondrial precursor proteins first appeared in a cytosolic pool and were then subsequently imported into mitochondria and converted to their mature forms (Reid and Schatz, 1982b; Schatz, 1979).

2. Posttranslational import in the absence of protein synthesis has been demonstrated both *in vivo* and *in vitro.* When protein synthesis was blocked with cycloheximide immediately following pulse-labeling *in vivo,* the precursor pool of mitochondrial proteins in the cytosol decreased with concomitant posttranslational uptake by mitochondria and conversion to the mature forms during the subsequent chase (Hallermayer *et al.,* 1977; Schatz, 1979; Reid and Schatz, 1982b; Teintze *et al.,* 1982). Similarly, mitochondrial precursor proteins which were accumulated *in vivo* by growing cells in the presence of CCCP could be subsequently chased into mitochondria by removing the inhibitory effects of CCCP with 2-mercaptoethanol (Reid and Schatz, 1982a,b) or cysteamine (Jaussi *et al.,* 1982). This too was unaffected by the presence of cycloheximide and therefore occurred posttranslationally.

3. Perhaps the most convincing evidence that transport is possible in the absence of protein synthesis comes from *in vitro* studies where translation systems were programmed with poly(A)-containing RNA to synthe-

size precursor proteins. If further translation following synthesis was blocked by the addition of cycloheximide or if ribosomes were removed by centrifugation and the resulting supernatants incubated with isolated mitochondria, the precursor proteins were rapidly imported into mitochondria and processed to their mature form (Harmey *et al.*, 1977; Korb and Neupert, 1978; Maccecchini *et al.*, 1979a; Zimmermann and Neupert, 1980). The same effect is observed regardless of whether a heterologous (i.e., reticulocyte lysate) or homologous (from cell cytosol) system is used. Posttranslational transport has been demonstrated *in vitro* for all imported mitochondrial proteins studied so far and in many cases *in vivo* as well.

In mitochondria, protein transport normally occurs posttranslationally although some cotranslational import (which is clearly not an obligatory process) may occur as a consequence of the amino-terminal targeting prepiece being exposed before complete polypeptide synthesis is finished. The prepiece might then initiate the import process before chain elongation is complete. Whether this actually occurs *in vivo* is not clear since the early experiments with mitochondria-bound ribosomes did not distinguish whether the nascent polypeptide chains were concomitantly transported with translation or transported only after complete synthesis. In any case, the enrichment of mRNAs for mitochondrial proteins in mitochondria-bound ribosomes has provided a useful means for isolating and screening for genes for imported proteins. Clones for 16 different genes of imported yeast mitochondrial proteins were identified by selective hybridization to these enriched mRNAs (Suissa *et al.*, 1984).

III. TARGETING AND SORTING SEQUENCES

The intracellular sorting of proteins synthesized on cytoplasmic polysomes is a fairly specific process. The signals which direct these events are contained within the newly synthesized protein itself. In addition, components of the target organelle, such as receptors, recognize these signals and facilitate import and sorting.

A. Fusion Proteins

Delineation of the exact regions in precursor proteins which are responsible for mitochondrial targeting and intramitochondrial sorting has been demonstrated by gene fusion experiments in which parts of a mitochondrial precursor protein can be attached to a nonmitochondrial "passenger" protein. Generally, three observations have been made: (i) the cleavable amino-terminal prepiece of imported mitochondrial proteins fused to nonmitochondrial proteins [such as mouse dihydrofolate reductase (DHFR) or *Escherichia coli* β-galactosidase] can correctly mediate mitochondrial targeting and frequently intramitochondrial sorting of the passenger protein; (ii) mitochondrial proteins in which the prepiece has been removed cannot be imported into mitochondria; and (iii) distinct regions of the polypeptide chain of proteins which are not proteolytically processed contain the information necessary for targeting and sorting. It has been suggested that some mitochondrial precursor proteins also have carboxy-terminal extensions that are removed during import and maturation, though the significance of these findings is as yet unclear (Okamura *et al.*, 1985; Power *et al.*, 1986; Patterson and Poyton, 1986).

Cytochrome oxidase IV (COX IV) from yeast, an inner membrane protein, is synthesized as a precursor with a 25 amino acid amino-terminal prepiece (Maarse *et al.*, 1984). When progressively truncated parts of the COX IV presequence were fused to DHFR, the first 12 amino acids, but no less, directed DHFR to the mitochondrial matrix both *in vivo* and *in vitro* (Hurt *et al.*, 1984b, 1985a). When the entire 25 amino acid presequence or the first 22 amino acids were fused to DHFR, proteolytic processing by the matrix peptidase also took place, albeit at an alternative site in the latter case. Since the fusions were directed to the matrix instead of the inner membrane, even when the first 53 amino acids of the COX IV precursor were fused to DHFR (Hurt *et al.*, 1984a), the intramitochondrial sorting information must be contained within the mature part of COX IV. Mature COX IV, prepared from either pre-COX IV *in vitro* or with fusions *in vivo*, was neither imported nor bound to mitochondria (Hurt *et al.*, 1984a). In fact, removal of only the first 7 amino acids of the prepiece prevented COX IV import (Hurt *et al.*, 1985b).

Similarly, the targeting information for yeast pre-ATPase $F_1\beta$, which contains about a 20 amino acid prepiece and is imported to the matrix, exists within the first 27 amino acids of the precursor protein (Douglas *et al.*, 1984; Emr *et al.*, 1986). In fusion studies, as few as 39 amino acids from the amino terminus of pre-$F_1\beta$ (shorter fusions were not examined) could direct invertase to mitochondria *in vivo*, but 169 amino acids were required to direct β-galactosidase (presumably due to a folding artifact). Internal deletion studies of a fusion between pre-$F_1\beta$ amino acids 1–380 and β-galactosidase narrowed down the portion with the targeting information to the first 27 amino acids of the $F_1\beta$. In all cases, however, the importable fusions were misdirected to the inner membrane. Deletion proteins made from authentic $F_1\beta$ in which amino acids were removed between residues 10 and 36 of the pre-$F_1\beta$ were fully import competent, while deletions within the first 10 amino acids abolished import (Vassarotti et al., 1987a).

The amino-terminal prepiece of both rat and human ornithine carbamoyltransferase (OTC), which is 32 amino acids long and directs the protein to the matrix, can also target nonmitochondrial passenger proteins to the matrix in vitro (Horwich et al., 1985b; Nguyen et al., 1986). In contrast to the case of COX IV, the targeting information does not appear to reside in the very amino terminus of the prepiece since deletion studies in which either the amino or carboxy terminus of the prepiece were removed produced proteins which were still import competent. Instead, the targeting information appears to reside between amino acids 8 and 22 of the 32 amino acid prepiece, while the flanking regions contribute to import efficiency (Horwich et al., 1986). The shortest targeting sequence identified to date is from the prepiece of 5-aminolevulinate synthase, in which the amino-terminal 9 amino acids fused to β-galactosidase correctly directs the fusion protein to the mitochondrial matrix in vivo, though with low efficiency (Keng et al., 1986).

Proteins which do not contain removable prepieces also contain specific targeting information. In yeast ADP/ATP carrier this is contained within the first 115 amino-terminal amino acids but has not been further resolved (Adrian et al., 1986). Since the ADP/ATP carrier is a tripartite protein having three segments (each of approximately 100 amino acids) which have a high degree of homology (Saraste and Walker, 1982), then similar targeting ability might also exist in the other two segments of the protein. [Interestingly, the bovine phosphate carrier protein, which has a high degree of structural homology with the ADP/ATP carrier and would presumably follow a similar import pathway, is synthesized as a precursor with a 49 amino acid amino-terminal extension (Runswick et al., 1987)]. In the 70 kDa outer membrane protein from yeast, the targeting and sorting functions are contained in the first 41 amino-terminal amino acids. Deletion and fusion studies have identified two critical regions: amino acids 1 through 21 are required for mitochondrial targeting while the overlapping amino acid sequence 10 through 37 is necessary for sorting via membrane anchoring (Riezman et al., 1983c; Hase et al., 1984, 1986). As for proteins containing removable prepieces, the information exists in the amino-terminal end of the protein. In fact, the amino-terminal 12 amino acids of the 70 kDa protein fused to mature COX IV could target the fusion protein to mitochondria and restore cytochrome oxidase function in COX IV-deficient mutants in vivo (Hurt et al., 1985b).

That the amino-terminal prepiece carries specific mitochondrial targeting information is confirmed by comparing differences in alcohol dehydrogenase (ADH) isozymes which have different subcellular locations. ADH I and II are located in the cytoplasm while ADH III is a mitochondrial protein. They are 80–90% identical in sequence except for a 27 amino acid prepiece in ADH III. If the presequence of ADH III was removed it could not be imported into mitochondria. If the ADH III prepiece was fused to cytosolic ADH II, then the fusion protein was imported into the mitochondrial matrix. Therefore, ADH II is a cytosolic protein since it lacks an amino-terminal targeting sequence while ADH III is a mitochondrial protein since it contains the targeting sequence (van Loon and Young, 1986).

The mitochondrial targeting function of amino-terminal extensions has also been suggested by studies of other proteins in which isozymes having different subcellular locations are encoded by the same gene. For example, the gene encoding yeast histidyl-tRNA synthase has two in-frame translation starts, both of which are expressed (Natsoulis *et al.*, 1986). Mutations destroying the first start codon resulted in a respiratory deficient (Pet⁻) phenotype without affecting the cytoplasmic isozyme suggesting that the longer mRNA encodes the mitochondrial isozyme while the shorter message encodes the cytoplasmic form. Similarly, the yeast 2-isopropylmalate synthase gene contains multiple in-frame transcription start sites (Beltzer *et al.*, 1986) which produce two related proteins of different sizes in cell-free translation mixtures (Hampsey *et al.*, 1983). Only the larger of these two proteins was imported into mitochondria, suggesting that selection of the appropriate transcription and translation start sites controls the subcellular location of the protein and that mitochondrial targeting information is contained in the amino terminus of the larger precursor protein. This is also the case for a 20 kDa protein from *Neurospora crassa* (M. Tropschug, H. Köhler, R. A. Stuart, and W. Neupert, in preparation). A single gene encodes the mature-size cytosolic form and a larger precursor (24 kDa), which is imported into mitochondria. During import, the amino-terminal extension is removed in two proteolytic steps.

Mitochondrial prepieces appear to be able to direct almost any passenger protein into mitochondria. For example, in addition to fusions with DHFR and β-galactosidase, it has been demonstrated that a mitochondrial presequence (in this case from COX IV) can direct a chloroplast-encoded protein (ribulose-1,5-bisphosphate carboxylase, large subunit) into mitochondria (Hurt *et al.*, 1986b) and that ATPase subunit VIII, normally a mitochondrial gene product, can also be imported back into mitochondria when fused to a mitochondrial-targeting sequence (Gearing and Nagley, 1986). In the latter case, the 66 amino acid prepiece of ATPase subunit IX could mediate import of ATPase VIII, while the shorter 40 amino acid COX VI leader could not. It is clear then that targeting information itself resides in the amino-terminal extension of most precursor proteins. The efficiency of import with a given prepiece, however, is significantly affected by the passenger protein (van Steeg *et al.*, 1986).

B. Properties of Prepiece Sequences

The prepiece sequences for a number of imported mitochondrial proteins have now been determined (Table I). They vary in length between 20 and 80 amino acids. Although the prepieces do not share extensive sequence homology (except perhaps between equivalent proteins in different organisms), they do have several similar characteristics including the following: (i) a high content of positively charged basic amino acids (particularly arginine) which are distributed somewhat randomly throughout the prepiece; (ii) an absence or near absence of negatively charged acidic amino acid residues; (iii) a high content of hydroxylated amino acids (particularly serine); and (iv) a propensity to form amphiphilic α-helical structures. Only a few exceptions exist, the most striking of which is the amino-terminal sequence of the 17 kDa subunit IV of yeast bc_1 complex which contains an extremely high content of acidic amino acids (van Loon *et al.*, 1984); however, import of this protein into mitochondria has not been examined. Another deviation from these general properties occurs in human OTC, which, unlike its counterpart in rat, contains no hydroxylated amino acids. The placement of positively charged amino acids is well conserved between human and rat OTC but the hydroxylated amino acids in rat OTC have been replaced in human OTC by asparagine (instead of threonine) and phenylalanine (instead of serine).

How these amino-terminal prepieces direct proteins to mitochondria and how they facilitate transport across membranes is not entirely clear. One possibility is that they are recognized by specific receptors on the mitochondrial surface. Another is that, because of their amphiphilic helical nature, they can penetrate lipid bilayers. Finally, because they are positively charged, they may be electrophoretically drawn toward the mitochondrial matrix by the potential across the inner membrane (inside negative). It is possible that all of these events are involved. For example, a membrane potential is necessary for the import of all proteins which must be directed to the inner membrane or matrix. Similarly, if positively charged amino acids in the prepiece are replaced by uncharged amino acids, then import is abolished (Horwich *et al.*, 1985a). Specific amino acids, and not just net positive charge, also appear to be important. In human OTC, arginine 23 of the 32 amino acid prepiece is critical for both import and proteolytic processing; however, when it is replaced by an amino acid supporting α-helical structure, the import function is conserved (Horwich *et al.*, 1986).

The ability to form amphiphilic helical structures may be a critical feature of presequences. In a theoretical analysis of many mitochondrial presequences, von Heijne identified regions which can be folded into helices with high hydrophobic moment (von Heijne, 1986a). The segments of highest hydrophobic moment correlated very well with the critical regions identified by fusion studies for yeast COX IV and 70 kDa protein, and human OTC (see Table I). In another study, removal of the basic amphipathic α-helix of the ATPase $F_1\beta$ prepiece prevented import into mitochondria (Vassarotti *et al.*, 1987b). When mutations were selected that restored the correct *in vivo* localization of ATPase $F_1\beta$, it was found that these modifications specifically replaced acidic amino acids at the amino-terminus with basic or neutral amino acids that support amphipathic helix formation.

Some of the physical properties of prepieces have been examined with synthetic prepiece peptides. Synthetic peptides of the COX IV prepiece are soluble in aqueous solutions but were able to penetrate phospholipid monolayers or disrupt liposomes having a diffusion potential negative inside but not of the opposite polarity (Roise *et al.*, 1986). A synthetic peptide equivalent to the first 27 amino acids of pre-OTC assumed an amphiphilic helical conformation that was induced by the presence of anionic phospholipids and could perturb the bilayer of synthetic liposomes (Epand *et al.*, 1986). Synthetic prepiece peptides dissipated the membrane potential in isolated mitochondria (Ito *et al.*, 1985; Gillespie *et al.*, 1985; Roise *et al.*, 1986); however, in the presence of reticulocyte this did not appear to occur (Gillespie *et al.*, 1985). In this case, a synthetic peptide of amino acids 1–27 of the rat OTC prepiece completely blocked the import of pre-OTC, while amino acids 16–27 did not. The peptide (1–27) also blocked the import of pre-malate dehydrogenase (a matrix protein) and pre-thermogenin (uncoupling protein of the inner membrane), suggesting common components are shared in the import pathway of these proteins.

Artificial presequences, encoded by synthetic oligonucleotides, that were fused to the mature part of COX IV were able to mediate import into mitochondria both *in vitro* and *in vivo* if the balance of basic, hydrophobic and hydroxylated amino acids was similar to that of authentic presequences (Allison and Schatz, 1986). In all cases, the import-competent extensions were surface active, indicating that targeting may not depend

TABLE I

Comparison of Amino-Terminal Prepieces of Imported Mitochondrial Proteins

Sequence*	Notes†

OUTER MEMBRANE

```
    ++.   .+ +.        .        .              + ++  .  +--+++
1   MKSFITRNKTAILATVAATGTAIGAYYYYNQLQQQQQRGKKNTINKDEKK...        (a)
```

INTERMEMBRANE SPACE   (b)

```
    +. .   .  +.  .  ++.            .   ++....   .       .   ▼..
2   MTTAVRLLPSLGRTAHKRSLYLFSAAAAAAAAAATFAYSQSHKRSSSSPGGGSNHGWNNWGKAAALASTTPLVHVASV...
    + +  +    +  .+  -  .+_.+.+.,  +  .. +.+.  -.+++.   .    +    ...      - ▼ +  - + +
3   MLKYKPLLKISKNCEAAILRASKTRLNTIRAYGSTVPKSKSFEQDSRKRTQSWTALRVGAILAATSSVAYLNWHNGQIDNEPKLDMNKQK
```

INNER MEMBRANE – C SIDE

```
    +    .   .+      +         + +▼....    ▼ .....   -.
4   MAPVSIVSRAAMRAAAAPARAVRALTTSTALQGSSSSTFESP...               (b)
    +. . .  ++    +.  .+._____+.+. . ..        .       -. .  - ▼ .   -
5   MFSNLSKRWAQRTLSKSFYSTATGAASKSGKLTQKLVTAGVAAAGITASTLLYADSLTAEAMTAAEHGLHA...   (b)
```

– INTRINSIC OR UNKNOWN

```
    +  +  . .+      .+ ...+▼
6   MLRNTFTRAGGLSRITSVRFAQTHALSNAA...
    + +.. . .    +       . .   +▼..
7   MLRTPTVSALVRNVAVRAAKPTMAVRAASTMPISNPT...
    +   .+  .     ++    .     +    ▼.--
8   MLATRVFSLIGRRAISTSVCVRAHGSVVKSED...                         (c)
    +  .+  +  .     .   +  + +    -       .  ..+ +  +
9   MLARGLPLRSALVKACPPILSTVGEGWGHHRVGTGEGAGISTKTPRPYS...
    + .+  .+. ..  .  + + +  ....+ ▼-- ... ...  +
10  MLSRAIFRNPVINRTLLRARPGAYHATRLTKNTFIQSRKYSDAHDEETFE...
    + .+  +. .  .   . + +  +    +  +..    ..+   ..+  ++ ▼ .  -
11  MASTRVLASRLASQMAASAKVARPAVRVAQVSKRTIQTGSPLQTLKRTQMTSIVNATTRQAFQKRAYSSEIAQAMV...  (b)
    +. .  .  +  .+ . . +   .  +-  . .-  .  .+- . . - ▼ -.  .+
12  MQTTGALLISPALIRSCTRGLIRPVSASFLSRPEIQSVQPSYSSGPLQVARREFQTSVVSRDIDTAAKFIG...  (b)
    + . .  .  ++..  .+.  ..+  ++ -. .-- .. .  +  . ...  .+. ..  .▼  . . +
13  MYTCAKFVSTPSLIRRTSTVLSRSLSAVVVRRPETLTDESHSSLAVVPRPLTTSLTPSRSFQTSAISRDIDTAAKFIG...  (b)
```

– M SIDE

```
    +  . + .  + . . +.    ...+  ▼    +   +.
14  MLSLRQSIRFFKPATRTLCSSRYLLQQKPVVKTAQ...
```

MATRIX

```
    +.    ....+_. .+ .,+      +
15  MSAILSTTSKSFLSRGSTRQCQNMQKALFA...
    + .+    .      ++ +.. ..▼.   . ..++.-
16  MLSKLASLQTVAALRRGLRTSVASATSVATKKTEQ...
    +    .+  .  ++  .  +  ▼  .  . -
17  MALLQSRLLLSAPRRAAATARASSWWSHVEMG...
    +   .+.  +    .    .  . +▼.
18  MALLHSGRVLSGVASAFHPGLAAAASARASWWAHVEMGP...
    + +.  .  .++ . . ...   ++▼.+ +  - + -
19  MFAKTAAANLTKKGGLSLLSTTARRTKVTLPDLKWD...
    + . + .  . + + .  +. ++▼ .-  -+ +
20  MNSLRIARAALRVRPTAVRAPLQRRGYAEAVADKIK...                     (d)
    + +.  + .  .    .   +  . + ++ +
21  MQRSIFARFGNSSAAVSTLNRLSTTAAPHAKNGYATA...
    + +.... .++  .     + .   .] ▼ .+
22  MLRTSSLFTRRVQPSLFSRNILRLQSTAAIPKTQKGV...                    (e)
    +   +    +      +  +      + + +      +
23  MLFNLRILLNNAAFRNGHNFMVRNFRCGQPLQNKVQLKGRDL...
    + . +  .    ++ . + + + +     +
24  MLSNLRILLNKAALRKAHTSMVRNFRYGKPVQSQVQLKGRDL...
    + . + . .+.  . .+. . + +   -  -+.+
25  MSALIPRLLTRTAFKASGKLLRLSSVISRTFSQTTTSYAAAFDRSKP...          (f)
    +.+ .  .+  .+.  .  .+ . . + . + .
26  MTRILTACKVVKTLKSGFGLANVTSKRQWDFSRPGIRLLSVKAQTAHI...         (g)
    +   +    . ..+   +_     ...  ++
27  MVLPRLYTATSRAAFKAAQSAPLLSTSWKRCMASA...                      (h)
    + .++ .   + . .+.        . .  .+ + .+  +  ▼.
28  MASRRLLASLLRQSAQRGGGLISRSLGNSIPKSASRASSRASPKGFLLNRAVQYATSAAAPASQP...  (i)
    +.. . +   +  + +     ++ . .   .  + +++. +- .    ▼.  +
29  MTSLWGKGTGCKLFKFRVAAAPASGALRRLTPSASLPPAQLLLRAVRRRSHPVRDYAAQTSPSPKAGAA...  (h,j)
    + . +  .   .  ++ + ++ + .   +.   .+. . .+ --.+ .
30  MAARLLRVASAALGDTAGRYRLLVRPRAGAGGLRGSRGPGLGGGAVATRTLSVSGRAQSSSEDKITVH...
```

TABLE I (*continued*)

Sequence*	Notes†

OTHER SEQUENCES OF INTEREST (k)

```
      +       +...  .    +  +.. +       +    +                    .          +
31  MAAVIAKSSVSAAVARPARSSVRPMAALKPAVKAAPVAAPAQANQMMVWTPVNNK...
    +-   - - - - +  .                 --- -  -  --+---  ---
32  MDMLELVGEYWEQLKITVVPVVAAAEDDDNEQHEEKAAEGEEKEEENGDE...
    +    .    +.  - +.     .+                  ++    +  --
33  MPQSFTSIARIGDYILKSPVLSKLCVPVANQFINLAGYKKLGLKFDDLIA...
    ++   +    .+       +    ..++  .+   .  .   .   .  +  -
34  MKIQLVRWHCSRNALWNRAFYSTRKATKNASSATPATMTSMVSQRQDLFM...
    +. +    +    .++   ..   + - +    .-      .   .  .
35  MSNKQAVLKLISKRWISTVQRADFKLNSEALHSNATVFSMIQPTGCFHLG...
    + .+.  +   ..  +...   +  ...   ..    .     + .+  ++ +
36  MLSRSLNKVVTSIKSSSIIRMSSATAAATSAPTANAANALKASKAPKKGK...
    +  +-.    -    .+ .+     +    +     +- ..+ +    + .+
37  MVKESIIALAEHAASRASRVIPPVKLAYKNMLKDPSSKYKPFNAPKLSNR...
    +.        .  .      ++.     +       .  ..++     +    -
38  MTVLYAPSGATQLYFHLLRKSPHNRLVVSHQTRRHLMGFVRNALGLDPPP...   (l)
    +     .    .  ++  ...+ ++     .  + + ++.    .  -
39  MVNWQTLFMVSLRRQGSSSRYRYKFNMENITHQVFPRCKQAFKKTNLSYE...
    +...    +-.  ++    -+ +-+    .  +-   + ++       -
40  MSSSQVVRDSAKKLVNLLEKYPKDRIHHLVSFRDVQIARFRRVAGLPNVD...
```

* The amino-terminal prepiece sequences of imported mitochondrial precursor proteins are listed (by the single letter amino acid code) in groups according to the intramitochondrial location of the mature protein. Above the primary sequence, basic (lysine and arginine), acidic (aspartate and glutamate), and hydroxylated amino acids (serine and threonine) are identified (+, −, and ·, respectively). Proteolytic cleavage sites which yield intermediate or mature proteins are indicated by an arrow or by a bracket above the sequence where only the general region of proteolytic processing is known (i.e., based on apparent molecular weight differences between precursor and intermediate or mature proteins). The segment of the prepiece identified by von Heijne (1986a) as having the highest hydrophobic moment (18 residue window, Eisenberg et al., 1984) when plotted in a helical wheel projection (Schiffer and Edmundson, 1967) is indicated by a box. Proteins: (1) 70 kDa protein (yeast), Hase et al. (1983); (2) cytochrome c peroxidase (yeast), Kaput et al. (1982), Reid et al. (1982); (3) cytochrome b_2 (yeast), Guiard (1985), Gasser et al. (1982b); (4) Rieske Fe/S of bc_1 complex (N. crassa), Harnisch et al. (1985), Hartl et al. (1986); (5) cytochrome c_1 (yeast), Sadler et al. (1984), Gasser et al. (1982b); (6) cytochrome oxidase V (yeast), Koerner et al. (1985); (7) cytochrome oxidase V (N. crassa), Sachs et al. (1986); (8) cytochrome oxidase IV (bovine), Lomax et al. (1984); (9) cytochrome P-450 (SCC) (bovine), Morohashi et al. (1984); (10) cytochrome oxidase VI (yeast), Wright et al. (1984); (11) ATPase IX (N. crassa), Viebrock et al. (1982), Schmidt et al. (1984); (12) ATPase IX-P1 (bovine), Gay and Walker (1985); (13) ATPase IX-P2 (bovine), Gay and Walker (1985); (14) cytochrome oxidase IV (yeast), Maarse et al. (1984); (15) citrate synthase (yeast), Suissa et al. (1984); (16) ornithine aminotransferase (rat), Mueckler and Pitot (1985), Simmaco et al. (1986); (17) aspartate aminotransferase (chicken), Jaussi et al. (1985); (18) aspartate aminotransferase (porcine), Joh et al. (1985); (19) Mn-superoxide dismutase (yeast), Marres et al. (1985); (20) ATPase F_1 δ subunit (N. crassa), Kruse and Sebald (1984); (21) 5-aminolevulinate synthase (yeast), Keng et al. (1986), Urban-Grimal et al. (1986); (22) alcohol dehydrogenase iso-III (yeast), Young and Pilgrim (1985); (23) ornithine carbamoyltransferase (human), Horwich et al. (1984); (24) ornithine carbamoyltransferase (rat), McIntyre et al. (1984), Takiguchi et al. (1984), Kraus et al. (1985); (25) mEF-Tu (yeast), Nagata et al. (1983); (26) carbamoyl-phosphate synthase I (rat), Nyunoya et al. (1985); (27) ATPase $F_1\beta$ subunit (yeast), Takeda et al. (1985); (28) ATPase F_1 β (Nicotiana plumbaginifolia), Boutry and Chua (1985), Kobayashi et al. (1986); (29)

on specific amino acid sequences but rather on the overall composition of the prepiece and its amphipathic nature. Sequences with these characteristics may reside at the amino-terminus of mitochondrial precursor proteins, where they are exposed and can be active, but they also appear to exist within nonmitochondrial proteins, although they are apparently masked such that they do not normally direct these proteins to mitochondria (Hurt and Schatz, 1987).

C. The Stop Transport Model

Based on the sequence characteristics of mitochondrial precursor proteins, Hurt and van Loon (1986) proposed a model for intracellular target-

---

ATPase $F_1\beta$ (human), Ohta and Kagawa (1986); (30) adrenodoxin (bovine), Okamura et al. (1985); (31) ribulose-1,5-bisphosphate carboxylase small subunit (Chlamydomonas reinhardtii), Schmidt et al. (1979), Hurt et al. (1986a); (32) subunit VI of bc_1 complex (yeast), van Loon et al. (1984); (33) 14 kDa subunit of bc_1 complex (yeast), De Haan et al. (1984); (34) threonyl-tRNA synthase (yeast), Pape et al. (1985); (35) tryptophanyl-tRNA synthase (yeast), Myers and Tzagoloff (1985); (36) histidyl-tRNA synthase (yeast), Natsoulis et al. (1986); (37) 2-isopropylmalate synthase (yeast), Beltzer et al. (1986); (38) MSS51 (COX I pre-mRNA maturase, yeast), Faye and Simon (1983); (39) CBP2 (cytochrome b pre-mRNA maturase, yeast), McGraw and Tzagoloff (1983); (40) CBP6 (yeast), Dieckmann and Tzagoloff (1985).

† Notes: (a) Not an actual prepiece (i.e., not proteolytically removed) but amino terminus is able to specifically target β-galactosidae to mitochondria (Hase et al., 1984, 1986). (b) Precursors are processed in two proteolytic steps. This has not yet been demonstrated for bovine ATPase IX (P1 or P2) but it may occur in two steps like N. crassa ATPase IX. (c) Bovine COX IV is equivalent to COX V in yeast or N. crassa (Gregor and Tsugita, 1982). (d) Possible cleavage site based on homology with N. crassa ATPase IX processing sites. Grouped with matrix proteins since it is probably imported as a matrix protein and then assembled into ATPase. (e) Possible cleavage site based on alignment with cytosolic ADH II. (f) Possible cleavage site based on alignment with bacterial mature sequence. (g) An alternate region with similar hydrophobic moment exists between residues 22 and 39. (h) ATPase $F_1\beta$ (a subunit of ATPase at the inner face of the inner membrane) is grouped with matrix proteins since it is probably imported as a matrix protein and then assembled into ATPase. (i) Possible cleavage site based on homology with amino terminus of mature $F_1\beta$ from Ipomoea batatas (Kobayashi et al., 1986). (j) Possible cleavage site based on homology with bovine mature sequence. (k) The presequence of chloroplast RuBPCase (31) is presented since the first 35 amino-terminal amino acids can direct DHFR or mature COX IV to mitochondria (Hurt et al., 1986). The first 50 amino acids of the 17K subunit VI of the bc_1 complex (32; determined from the nuclear gene sequence)—which contains an unusually high content of acidic amino acids—are presented, though import has not been examined. Sequences 33–40 are amino-terminal sequences (determined from the nuclear gene sequence) for putative mitochondrial proteins or proteins for which import into mitochondria has not been characterized. The first 50 amino acids from the presumed start site are presented. (l) Apparently imported into mitochondrial matrix without proteolytic processing (Gasser et al., 1982a; Hampsey et al., 1983).

ing and intramitochondrial sorting of imported mitochondrial proteins. In this model, mitochondrial presequences (or amino-terminal sequences in noncleavable proteins) can be divided into distinct domains: (1) matrix targeting domains; (2) stop transport domains; and (3) proteolytic cleavage sites. The presence or absence and arrangement of these domains determines the targeting and sorting of the accompanying mature piece of the protein. The matrix targeting domain (the amino-terminal positively charged region of the prepiece) directs the attached protein to mitochondria and mediates its transfer across both mitochondrial membranes to the matrix. If this domain is followed by a stop transport domain (a long uninterrupted sequence of uncharged amino acids capable of forming a transmembrane anchor), then the transfer of the protein is halted at either the outer or inner membrane, depending on subtle differences in the size and strength of the stop transport region. Further sorting or simply removal of the prepiece can then occur by proteolytic processing.

For example, in this model ADH III is a matrix protein because it contains the amino-terminal matrix targeting domain but no subsequent stop transport sequence. Inner membrane proteins also contain the matrix targeting domain but are halted at the inner membrane because of stop transfer domains within the mature part of the sequence. The matrix targeting sequence is then removed by proteolytic processing to prevent further import. For intermembrane space proteins, such as cytochrome b_2, the stop transport domain is in the prepiece so that when transport is halted at the inner membrane and proteolytic cleavage occurs at the outer face of the inner membrane the mature part of the protein is released into the intermembrane space. Cytochrome c_1 is proposed to follow a similar mechanism but remains associated with the inner membrane because of a carboxy-terminal anchor. Indeed, when the first 64 amino acids of cytochrome c_1 (containing the entire 61 amino acid presequence) was fused to DHFR, the fusion protein was directed to the inner membrane and DHFR was released into the intermembrane space following processing (van Loon et al., 1986). Identical results were observed when the cytochrome c_1 prepiece was fused to the mature part of COX IV (van Loon et al., 1987). Outer membrane proteins are simply halted at the outer membrane by a stop transport domain with presumably different characteristics from the inner membrane stop transport region.

The stop transport model accounts for the intramitochondrial sorting of many imported proteins. For the most part, it assumes a relatively simple linear arrangement of targeting and sorting domains, which may not be the case for all proteins. It does not, however, account for the import mechanism of intermembrane space proteins which do not contain presequences, such as cytochrome c or adenylate kinase. Other movements may also be possible which the data supporting the stop-transport model cannot yet exclude. For example, the Fe/S protein of the bc_1 complex, which like cytochrome c_1 is also synthesized with a long prepiece (that is processed in two steps), is fully transported into the matrix, where it is processed by the matrix peptidase, and is then redirected back to the intermembrane space (Hartl *et al.*, 1986; see Section XIB). In *Neurospora crassa,* cytochrome c_1 follows a similar transport pathway through the matrix (F.-U. Hartl, J. Ostermann, and W. Neupert, in preparation). In yeast, however, the imported cytochrome c_1 was never found on the matrix side of the inner membrane (van Loon and Schatz, 1987).

D. Heterologous Import

The precursor proteins and import machinery of mitochondria from different organisms share many common features. This can be demonstrated in heterologous import experiments in which the mitochondrial precursor protein from one organism can be imported into the mitochondria of another. For example, the Fe/S protein from *Neurospora crassa* bc_1 complex could be imported into yeast mitochondria (Teintze *et al.*, 1982), and, similarly, either ADP/ATP carrier or ATPase subunit IX could be imported into rat liver mitochondria (Schleyer *et al.*, 1982). This is not at all surprising since conservation of complicated import machinery is likely to be high during evolution. What is interesting, however, is that the precursor to *N. crassa* ATPase subunit IX could be imported into yeast mitochondria and be correctly processed (Schmidt *et al.*, 1983a; Fig. 1). In *Neurospora,* ATPase IX is a nuclear gene product which is synthesized on cytoplasmic polysomes (Jackl and Sebald, 1975; van den Boogaart *et al.*, 1982b) while in yeast it is coded for by mitochondrial DNA and synthesized inside the mitochondrion (Tzagoloff and Meagher, 1972; Macino and Tzagoloff, 1979; Hensgens *et al.*, 1979). The mature proteins are similar, and their sequences are 53% homologous in overlapping regions; however, the *Neurospora* pre-ATPase IX, which must be imported from the cytosol, contains a long 66 amino acid amino-terminal prepiece which presumably directs it to the mitochondrial inner membrane. A similar prepiece is absent in the yeast mitochondrial gene product.

Heterologous import has also been demonstrated between different subcellular organelles. When the prepiece (transit peptide) of chloroplast ribulose-1,5-bisphosphate carboxylase (small subunit) was fused to either DHFR or the mature part of cytochrome oxidase IV, the resulting fusion proteins were imported into mitochondria *in vivo*, albeit with lower efficiency than with authentic mitochondrial prepieces. In the latter case, the fusion protein could restore cytochrome oxidase activity *in vivo* when

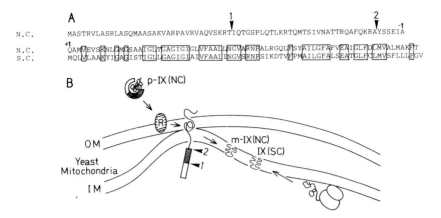

Fig. 1. *Neurospora crassa* pre-ATPase subunit IX is imported into yeast mitochondria. (A) Comparison of amino acid sequences of ATPase IX from *N. crassa* (N.C.) and *Saccharomyces cerevisiae* (S.C.) beginning with the amino terminus. The two sites of proteolytic processing of the *N. crassa* ATPase IX prepiece are indicated by arrows. Identical amino acid residues are indicated by boxes. (See text for references.) (B) *Neurospora crassa* ATPase IX is a nuclear gene product synthesized in the cytosol with a long prepiece. It can be imported into and processed by yeast mitochondria (*in vitro*) where ATPase IX is a mitochondrial gene product. Left: Sequence of events in the heterologous import of *N. crassa* ATPase IX. Right: Synthesis of yeast ATPase IX on membrane-bound mitochondrial ribosomes and insertion into the inner membrane. OM, Outer membrane; IM, inner membrane; R, receptor; p, precursor; m, mature.

used to transform a COX IV-deficient mutant of yeast (Hurt *et al.*, 1986a,b). When compared to mitochondrial prepiece sequences, the chloroplast transit peptide shares many common features (e.g., high content of basic and hydroxylated amino acids with no acidic amino acids; Table I). What subtle differences exist in the targeting prepieces that correctly direct proteins to either chloroplasts or mitochondria in photosynthetic cells are unknown.

IV. RECEPTORS

While much information is obviously contained within specific regions of mitochondrial precursor proteins, there must exist complementary structures within mitochondria themselves to decode and process it. One obvious possibility is that receptors are involved in the initial recognition of proteins imported from the cytosol.

Proteinaceous components on the outer surface of mitochondrial membranes were first demonstrated by shaving isolated mitochondria with low concentrations of protease which did not penetrate or destroy the outer membrane (Gasser *et al.*, 1982a; Zwizinski *et al.*, 1984; Pfaller and Neupert, 1987). Following this treatment, the specific binding of precursor proteins to the outer membrane was blocked and import was abolished.

The binding of precursor proteins to mitochondria is independent of import and precedes all other events in the import pathway. For example, mature outer membrane porin is in a location which is protected from externally added proteases. If the insertion of precursor porin into the lipid bilayer of the outer membrane was inhibited by importing at 0°C, the binding of the protein to mitochondria was unaffected and it remained in a protease-susceptible location at the outer face of the outer membrane. When mitochondria containing the bound porin were reisolated and warmed to 25°C, the protein was subsequently imported from its receptor sites to the protease-protected location (Freitag *et al.*, 1982). Similarly, the import of cytochrome *c* could be blocked with the heme analog deuterohemin which prevents covalent attachment of heme to the precursor apocytochrome *c* and subsequent translocation across the outer membrane. Under these conditions, apocytochrome *c* could still bind to mitochondria independently of import. When the inhibition by deuterohemin was reversed by adding excess amounts of hemin, cytochrome *c* was subsequently imported from its receptor sites into the intermembrane space (Hennig and Neupert, 1981).

Proteins such as the ADP/ATP carrier, which require a membrane potential for import, could be stalled at their receptor sites by disrupting the potential with CCCP. On reestablishing the membrane potential, the bound precursor could be imported (Zwizinski *et al.*, 1983; Pfanner and Neupert, 1985). In these cases, however, the binding to deenergized mitochondria was slow, though clearly independent of a membrane potential. Binding of pre-ATPase $F_1\beta$ to deenergized mitochondria did not occur at all (Zwizinski *et al.*, 1984), though for unknown reasons. The binding of membrane potential-dependent proteins to deenergized mitochondria may be poor because of the low capacity of specific sites which may be occupied by precursors with higher affinity.

Receptors for imported mitochondrial proteins are specific in many regards as was demonstrated by the receptors for precursors to cytochrome b_2 and citrate synthase in yeast (Riezman *et al.*, 1983b). First, binding was specific to the outer mitochondrial membrane. These precursor proteins were able to bind to whole mitochondria or to isolated outer membrane vesicles (Riezman *et al.*, 1983a) but not to inner membrane preparations. Binding was also specific for precursor proteins. When the partially processed intermediate form of the cytochrome b_2 precursor was generated by treatment with an extract of the mitochondrial matrix (containing the processing peptidase), the intermediate size cytochrome b_2 did not bind to mitochondria. Similarly, binding of the mature form did not occur either. The same observation has been made with cytochrome c where mature holocytochrome c did not recognize the binding sites in *Neurospora* mitochondria (Hennig *et al.*, 1983) and even large excesses could not displace prebound apocytochrome c (H. Köhler and W. Neupert, unpublished). Last, binding was specific for mitochondrial proteins. For example, the binding of cytosolic hexokinase and glyceraldehyde-3-phosphate dehydrogenase to whole yeast mitochondria or isolated outer membrane vesicles was negligible (Riezman *et al.*, 1983b).

How many different receptors exist to mediate the recognition and binding of imported mitochondrial proteins? It is clear from the large number of mitochondrial proteins and the limited number associated with the outer membrane that it would be impossible, if not just simply impractical, for a different receptor to exist for each protein. Nevertheless, experiments have shown that many mitochondrial proteins do not share common binding sites so that a single receptor for all imported proteins is not the case either. For example, the binding of porin to yeast mitochondria was not affected when the mitochondria were pretreated with low concentrations of trypsin that were sufficient to abolish the binding of F_1 ATPase β subunit and cytochrome b_2 (Gasser and Schatz, 1983). This suggests that they are bound by different receptors having different trypsin sensitivities. Similarly, shaving *Neurospora* mitochondria with elastase inhibited the binding and import of ADP/ATP carrier and porin but had no effect on ATPase $F_1\beta$ (Zwizinski *et al.*, 1984).

By this criterion it is difficult to assign possible common binding sites for the few mitochondrial proteins that have been studied to date. In addition, proper competition studies are limited by the difficulties of preparing sufficient amounts of mitochondrial precursor proteins. In three cases this has been overcome: (1) apocytochrome c, the precursor to holocytochrome c, can be prepared by chemical cleavage of the heme group and renaturation of the protein by dialysis from urea (Ambler and Wynn, 1973); (2) a water-soluble form of porin can be made by subjecting isolated porin to acid precipitation and then resolubilizing the protein at alkaline pH in the absence of detergent (Pfaller *et al.*, 1985); and (3) the precursor to F_1 ATPase β subunit can be accumulated then purified from yeast *rho*$^-$ mutants grown in the presence of CCCP (Ohta and Schatz, 1984). In all three cases, the resulting precursor proteins were import competent and retained characteristics of their precursors synthesized in cell-free systems. Apocytochrome c does not compete for the binding of any mitochondrial proteins tested so far, including cytochrome c_1, the Fe/S protein of bc_1 complex, ATPase $F_1\beta$, ATPase IX, or ADP/ATP carrier

(Zimmermann *et al.*, 1981; Teintze *et al.*, 1982). It appears to have a unique receptor. Porin, however, is able to compete for the binding and import of the ADP/ATP carrier, suggesting a common receptor (Pfaller and Neupert, 1987). Binding competition has not yet been examined using the ATPase $F_1\beta$ precursor purified from yeast.

The binding of cytochrome *c* to *Neurospora* mitochondria is the best characterized of imported mitochondrial proteins so far. When radiolabeled apocytochrome *c* was bound to mitochondria in the presence of deuterohemin (to prevent subsequent import), it could be completely displaced by adding unlabeled apocytochrome *c* but not holocytochrome *c* (Hennig *et al.*, 1983). Apocytochrome *c* from different species could also displace the precursor protein to varying degrees. Titration of the binding sites on mitochondria by Scatchard analysis indicated that there were 90 pmol of high affinity binding sites for apocytochrome *c* per milligram of mitochondrial protein. These sites had an association constant (K_a) of 2.2 $\times$ 10^7 M^{-1} (Table II).

When mitochondria were solubilized with octyl glucoside and the resulting extract was reconstituted into liposomes, the high affinity binding sites were also reconstituted (Köhler *et al.*, 1987). Using this procedure to identify high affinity binding for apocytochrome *c*, a putative receptor protein was identified and purified to homogeneity. The protein, when reconstituted into liposomes, bound apocytochrome *c* with similar affinity compared to whole mitochondria. In addition, the purified protein could compete with mitochondria for the binding of apocytochrome *c*. The unusual feature of the apocytochrome *c* binding protein, however, is that it is a soluble protein of the intermembrane space. This explains why more than a 10-fold higher concentration of proteases are required to abolish apocytochrome *c* binding to mitochondria compared to other imported proteins. Although the binding of apocytochrome *c* to the protein meets all criteria for specific interaction of a protein with a receptor—namely, (i)

TABLE II

Mitochondrial Receptors for Imported Precursor Proteins

Precursor protein	Number of binding sites (pmol/mg mitochondrial protein)	Association constant (K_a)	Reference
Apocytochrome *c*	90	2.2 $\times$ 10^7 M^{-1}	Hennig *et al.* (1983)
Porin	5–10	2 $\times$ 10^8 M^{-1}	Pfaller *et al.* (1985, 1987)
ADP/ATP carrier	1.7	1.1 $\times$ 10^9 M^{-1}	Schmidt *et al.* (1985)

rapid and reversible binding, (ii) saturable and limited in number, and (iii) specificity—the location in the intermembrane space is unexpected. Cytochrome c appears to have an unusual import pathway which is not representative of the mechanism for other imported proteins. Apocytochrome c is able to spontaneously insert into lipid bilayers in a nonspecific manner with low affinity (Rietveld *et al.*, 1983, 1985, 1986a,b; Rietveld and Kruijff, 1984; Dumont and Richards, 1984). A possible explanation of the binding protein's location is that it recognizes the partially inserted apocytochrome c and then binds it from the inner face of the outer membrane, perhaps to mediate exposure of the cysteine sulfhydryl groups to the heme attaching enzyme. The apocytochrome c binding protein and the heme attaching enzyme are distinct proteins (Nicholson *et al.*, 1987).

The high affinity binding of porin to mitochondria was sensitive to treatment of whole mitochondria with very low concentrations of trypsin, indicating that the receptor is exposed to the outer surface of the outer membrane (Pfaller and Neupert, 1987). As for cytochrome c, the high affinity binding sites for porin could be solubilized with detergent and reconstituted into liposomes. Using this approach, it should be possible to purify the porin receptor.

The identification of putative receptor proteins mediating import into mitochondria has also been successful using other methods. For example, antibodies raised against the total outer mitochondrial membrane of yeast blocked the import of the preCOX IV-DHFR fusion protein (Ohba and Schatz, 1987a). Antibodies raised against a 45 kDa outer membrane protein also blocked import while sera against other outer membrane proteins did not. The Fab fragments of the anti-45 kDa antibodies alone also blocked import. In another study, a 30 kDa protein from the outer membrane of rat mitochondria was specifically cross-linked to the synthetic prepiece (amino acids 1–27) of pre-OTC (Gillespie, 1987). Mild pretreatment of mitochondria with trypsin reduced both cross-linking of the prepiece peptide to the 30 kDa protein and the import of pre-OTC into mitochondria.

V. ENERGY REQUIREMENTS

Precursor proteins bind to mitochondria independently of other events in the import pathway; however, subsequent translocation into the mitochondrion is dependent on energy in most cases. Generally, energy is required for (i) all proteins having an amino-terminal extension, (ii) all proteins which are inserted into or translocated across the inner membrane, and (iii) only for import to the first processing stage for those proteins which are proteolytically processed in two steps. Where import requires energy, it is needed in the form of an energized inner membrane, specifically, the membrane potential component ($\Delta\psi$) of the total proton-motive force (Δp).

The primary energy source required for import was first thought to be ATP (Nelson and Schatz, 1979). It was later shown, however, that an energized inner membrane was the necessary energy form. Even a low potential of 20–40 mV, about one-tenth the normal value of 230 mV (Mitchell and Moyle, 1969), can drive the import of mitochondrial proteins (Pfanner and Neupert, 1985).

The requirement for an electrochemical potential and not ATP as the immediate energy source was clearly demonstrated *in vitro* in a series of experiments by Schleyer *et al.*, (1982) in which high intramitochondrial ATP plus a dissipated membrane potential versus low ATP plus a normal membrane potential were compared in *Neurospora crassa* mitochondria. In the first case (high ATP/no membrane potential), the membrane potential was dissipated by the protonophore CCCP, and intramitochondrial ATP was elevated via the ADP/ATP carrier by adding ATP externally. Oligomycin was added to inhibit the F_0F_1 ATPase and prevent the hydrolsis of ATP which would generate a small potential. Under these conditions, high internal concentrations of ATP in the absence of an energized inner membrane could not drive import. In the opposite case (low ATP/normal membrane potential), intramitochondrial ATP was depleted by preincubating mitochondria in the presence of oligomycin (to inhibit the membrane potential-driven synthesis of ATP) and carboxyatractyloside (to block the ADP/ATP carrier and prevent the translocation of external ATP into mitochondria). Under these circumstances, the electrochemical potential in the absence of ATP was sufficient to power import. In yeast mitochondria, import was blocked when the membrane potential was dissipated with CCCP but not when ATP synthesis was inhibited by oligomycin (Gasser *et al.*, 1982a). Furthermore, ATP-supported import in cyanide-inhibited mitochondria was blocked by either carboxyatractyloside (so ATP could not enter the matrix) or oligomycin (which would prevent ATP hydrolysis), demonstrating that ATP-stimulated import was a consequence of the small electrochemical gradient generated by the hydrolysis of ATP by the F_0F_1 ATPase.

The electrochemical potential (or total protonmotive force) Δp is the sum of the membrane potential ($\Delta\psi$) from the separation of charged species across the inner membrane and the force exerted by the pH gradient (ΔpH) ($\Delta p = \Delta\psi - Z\,\Delta pH$). It is the membrane potential component of the total protonmotive force which is responsible for powering import. This was first suggested by experiments where the ionophore nigericin, which exchanges K^+ for H^+ in a stoichiometric way across the inner membrane, did not affect the import of ADP/ATP carrier or ATPase IX into *Neurospora* mitochondria (Schleyer *et al.*, 1982). Since nigericin leads to the breakdown of the proton gradient without affecting the membrane potential, it is the latter which drives import of precursor proteins. This was substantiated in experiments where the physiological membrane potential was inhibited by antimycin A/oligomycin and then a valinomycin-induced potassium diffusion potential could drive the import of the ADP/ATP carrier into *Neurospora* mitochondria (Pfanner and Neupert, 1985). This import was not abolished by protonophores like CCCP which dissipate the proton gradient and, in the absence of a K^+ diffusion potential, would normally dissipate the accompanying membrane potential. Furthermore, the establishment of a ΔpH did not power import. The importance of the membrane potential component of the total protonmotive force is not surprising since ΔpH makes only a relatively small contribution to Δp under physiological conditions.

The energy necessary for import is not needed for proteolytic processing itself. For example, proteins that are transported to the inner membrane or matrix, but which are not proteolytically processed, require an energized inner membrane. This has been demonstrated for the ADP/ATP carrier (Schleyer *et al.*, 1982; Pfanner and Neupert, 1985) and for the matrix-localized 2-isopropylmalate synthase (Hampsey *et al.*, 1983). In addition, the purified matrix processing peptidase does not require any form of energy to process protein outside of whole mitochondria (G. Hawlitschek and W. Neupert, unpublished). Instead, energy is required for interaction with the mitochondrial inner membrane. Evidence for this is 3-fold. First, all precursor proteins which must be translocated into or across the inner membrane require a membrane potential for import. This is also true for intermembrane space proteins which require proteolytic processing (e.g., cytochrome b_2, cytochrome c_1, and cytochrome c peroxidase). In these cases, the precursor protein must at least partially penetrate the inner membrane to reach the matrix peptidase for the first processing event (Daum *et al.*, 1982b; Gasser *et al.*, 1982b; Reid *et al.*, 1982; Teintze *et al.*, 1982). Second, proteins that are imported into the intermembrane space but do not come in contact with the inner membrane, such as cytochrome c (Zimmermann *et al.*, 1981), and those imported into the outer membrane, such as porin (Freitag *et al.*, 1982; Mihara *et al.*, 1982a) and the major outer membrane polypeptides (Gasser and Schatz, 1983), do not require an energized inner membrane for import. Finally, when import of cytochrome c_1 and F_1 ATPase β subunit was performed at 7°C, an intermediate could be trapped in which the amino-terminal prepiece could be processed by the matrix peptidase, but the major portion of the precursor was still outside the mitochondrion and could be digested by externally added proteases (Schleyer and Neupert, 1985). Import to this stage was dependent on energy; however, subsequent translocation of the proteins completely into mitochondria was independent of a membrane potential when chased at 25°C. Therefore, only import of the amino terminus through the inner membrane required a membrane potential while transport of the rest of the protein did not.

Proteolytic processing, however, is not obligatory for import past the energy-dependent step. Precursors to F_1 ATPase β and IX subunits could be partially imported into mitochondria without processing in the presence of o-phenanthroline, which blocks the matrix processing peptidase. The precursors could then be chased to the mature size by adding Mn^{2+} in a step that did not require a membrane potential (Zwizinski and Neupert, 1983). The energy-dependent step precedes and is independent of proteolytic processing.

Exactly why the membrane potential is required for import is not clear. One possibility is that it produces an electrophoretic driving force (negative inside) on the positively charged prepieces which mediates pentration into or through the inner membrane (Fig. 2). Another is that it induces transient conformational changes in lipid and protein organization which allow the initial entry of the precursor protein into or through the membrane barrier. Both effects probably contribute to energy-dependent import.

Not all energy requirements are strictly for an energized inner mem-

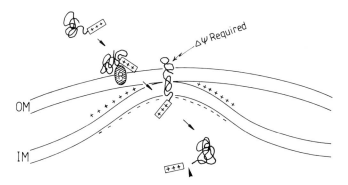

Fig. 2. A membrane potential ($\Delta\psi$) is required for the import of proteins which must go to or through the inner membrane. Energy is not required for binding to receptors but is necessary for penetration of the positively charged amino-terminal prepiece through the inner membrane. Subsequent translocation of the remainder of the protein through the inner membrane, or movement away from contact sites (i.e., inner membrane proteins), does not require the continued presence of $\Delta\psi$. Proteolytic processing is also independent of energy. OM, Outer membrane; IM, inner membrane; R, receptor.

brane. For example, degradation of endogenous ATP, by the enzyme apyrase, reduced the import of ATPase $F_1\beta$ into *Neurospora crassa* mitochondria even in the presence of a membrane potential, whereas subsequent addition of ATP or GTP restored import (Pfanner and Neupert, 1986). Nonhydrolyzable ATP analogues also blocked the import of mitochondrial precursor proteins (Pfanner and Neupert, 1986; Eilers *et al.*, 1987; Chen and Douglas, 1987), indicating that cleavage of the nucleoside triphosphate (NTP) phosphodiester bond is necessary. It has been suggested that NTPs maintain or confer an import-competent conformation in mitochondrial precursor proteins. This is supported by experiments in which the proteolytic sensitivity of precursor proteins is greater in the presence of NTPs (Pfanner *et al.*, 1987; Verner and Schatz, 1987), indicating that a less folded conformation is sustained by NTP hydrolysis and that such a conformation is necessary for import. The requirement for NTPs depends primarily on the mature part of the precursor protein. For example, precursors having identical presequences but different mature polypeptides required different concentrations of NTPs for optimal import (Pfanner *et al.*, 1987). It appears that NTPs are necessary for confering import-competence during all steps that precede and include the interaction of the precursor with the outer membrane (Pfanner *et al.*, 1987; Eilers *et al.*, 1987). In addition, maturation steps for some imported proteins require NADH. For example, the covalent attachment of heme to apocytochrome *c* and concomitant translocation across the outer membrane is dependent on NADPH in yeast (Basile *et al.*, 1980) and NADH in *N. crassa* (Nicholson *et al.*, 1987). Similarly, the second processing step of cytochrome c_1, which is accompanied by covalent heme attachment, is also dependent on NADH (Teintze *et al.*, 1982; Schleyer and Neupert, 1985). The second processing step of the Fe/S protein of the bc_1 complex requires NADH (F.-U. Hartl and W. Neupert, unpublished), presumably for the Fe/S cluster formation. Maturation of cytochrome oxidase II, a mitochondrial translation product, also requires NADH (W. Driever, R. Cook, and W. Neupert, unpublished). These proteins all have the common feature of having iron, either in heme or as nonheme Fe/S clusters.

VI. CYTOSOLIC COFACTORS

By virtue of the elaborate mixtures which are required for the cell-free synthesis of precursor proteins, import reactions *in vitro* contain many components. In the few cases where mitochondrial precursor proteins could be purified or where minimal amounts of cell-free translation mixtures have been used, a number of potential "cytosolic" cofactors which are required for import have been identified. These cofactors are present both in reticulocyte lysates and in homologous postribosomal cytosol preparations. They fall into three broad classes: (i) soluble low molecular weight components; (ii) proteins; and (iii) RNA.

When a reticulocyte lysate translation mixture containing the newly synthesized precursor to rat ornithine carbamoyltransferase (OTC) was passed over a Sephadex G-25 column, the precursor, recovered in the excluded fraction, could no longer by imported into mitochondria. Import of the precursor could be restored by fresh unlabeled reticulocyte lysate mixture but not by its individual components, including Mg^{2+}, K^+, or ATP (Argan et al., 1983). The reticulocyte lysate itself, without additives for translation, completely restored import. In an independent report, the postribosomal supernatant of the reticulocyte lysate, which had been dialyzed, stimulated the import of pre-OTC severalfold (Miura et al., 1983). The stimulating activity of the added dialyzed lysate was inactivated by either pretreatment with trypsin or heat denaturation, suggesting that the cofactor is a protein. The cofactor was further characterized by examining requirements for the import of pre-OTC, which had been purified by immunoaffinity chromatography, into mitochondria (Argan and Shore, 1985). When pre-OTC was mixed with the untreated lysate, the cofactor bound the precursor to form a 5 S complex. The pre-OTC in the isolated 5S complex was imported into mitochondria without a requirement for extra lysate. When mitochondria were pretreated with reticulocyte lysate and then reisolated, import of purified pre-OTC did not occur. However, when pre-OTC plus reticulocyte lysate was mixed with mitochondria at 4°C, 50% of the pre-OTC bound to the mitochondrial surface and could be imported when the mitochondria were isolated and incubated at 30°C. The import factor appears to be necessary for the import-competent delivery and binding of pre-OTC to mitochondria.

The purified precursor to the F_1 ATPase β subunit is poorly imported into yeast mitochondria. In the presence of the cytosolic fraction from yeast or reticulocytes, however, import and processing was stimulated 4- to 8-fold (Ohta and Schatz, 1984). The cofactor was nondialyzable, protease sensitive, and had an apparent molecular mass of 40 kDa. Similar stimulation of import of the in vitro synthesized cytochrome b_2 precursor by unlabeled reticulocyte lysate indicated that cofactors may be necessary for other proteins as well. Again, the cofactor appeared to aid the correct association of the precursor with the mitochondrial surface.

A cytoplasmic RNA component has also been implicated in the import of several proteins into rat liver mitochondria. Posttranslational treatment with RNase of reticulocyte lysates containing newly synthesized precursor proteins inhibited subsequent import into mitochondria (Firgaira et al., 1984). The cofactor and OTC precursor fractionated as a 400 kDa complex with characteristics of a ribonucleoprotein. On the other hand, the cofactor which stimulated the import of purified OTC precursors was not sensitive to RNase (Argan and Shore, 1985). The import of F_1 ATPase β subunit and citrate synthase precursors into yeast mitochondria was also inhibited by pretreatment of the reticulocyte lysate (in which they were synthesized) with RNase; however, removal of the ribosomes by centrifugation abolished the sensitivity of import to RNase, suggesting that the RNA cofactor is not a specific component necessary for import but that degraded ribosomes inhibit import (Burns and Lewin, 1986). In contrast, RNase treatment of reticulocyte lysate postribosomal supernatants containing newly synthesized ATPase $F_1\beta$ inhibited binding and import into *Neurospora crassa* mitochondria (N. Pfanner and W. Neupert, unpublished). The role of putative RNA cofactors requires further investigation.

Protein and RNA cofactors present in cytosolic fractions may serve a number of possible functions during import. For example, they may stabilize precursor proteins against premature proteolytic digestion in the cytosol or mediate conformational arrangements which are necessary for import competence. These cofactors appear to be necessary for the specific binding of precursors to mitochondria but not for subsequent stages of import. Specific cofactors have not yet been isolated so that it cannot be ruled out that they may act in a nonspecific manner by, for instance, preventing the small amounts of precursor proteins from aggregating. Whether such cofactors are important *in vivo* is unknown.

The import of apocytochome *c* into mitochondria is also dependent on a cytosolic or reticulocyte lysate cofactor. In this case, however, the stimulatory component is a low molecular weight, heat-stable factor which is not sensitive to proteases. In yeast, the cofactor can be substituted by an NADPH-regenerating system (Taniuchi *et al.*, 1982), but in *N. crassa* the cofactor appears to serve some other function (Nicholson *et al.*, 1987). It is not involved in the binding of apocytochrome c to mitochondria, but is necessary for enzymatic attachment of heme and subsequent translocation across the outer membrane.

VII. TRANSLOCATION CONTACT SITES

Proteins which are imported into the mitochondrial matrix or inner membrane must cross two membrane barriers to reach their final location. The question arises as to whether transport across the outer and inner membrane occurs in two distinct steps, with a soluble intermediate in the intermembrane space, or whether the inner and outer membranes come close enough together to be spanned and crossed in a single event. For proteins imported into the inner membrane or matrix of *Neurospora crassa* mitochondria, the latter mechanism appears to be the case. These translocation contact sites have been demonstrated for the import of the F_1 ATPase β subunit and cytochrome c_1 (Schleyer and Neupert, 1985), the Fe/S protein of bc_1 complex (Hartl *et al.*, 1986), and the ADP/ATP carrier (Pfanner and Neupert, 1987).

Membrane-spanning intermediates could be detected when translocation was impeded by importing the precursors of ATPase $F_1\beta$ or cytochrome c_1 at low temperatures (4–12°C). Under these conditions, translocation intermediates were accumulated which had penetrated far enough through the inner membrane to be processed by the matrix-localized processing peptidase but which still had the major portion of the polypeptide exposed to the outer surface of the outer membrane since they could be digested with externally added proteases (Schleyer and Neupert, 1985). Import to this stage was dependent on a membrane potential. When translocation intermediates spanning contact sites were accumulated this way, they could be subsequently chased into a protease-insensitive location by raising the temperature to 25°C in a step that was independent of the membrane potential. Similarly, complete transport across the mitochondrial membranes could be blocked by prebinding the precursor proteins to specific antibodies generated against the mature size proteins. When incubated with mitochondria, the amino terminus of the antibody-bound precursors penetrated far enough into the matrix to be proteolytically processed while the major part of the precursor was retained by the antibodies outside the mitochondria in a location susceptible to externally added proteases. These experiments indicate that (i) the ATPase $F_1\beta$ and cytochrome c_1 precursors first entered the matrix by their amino termini in a step that is dependent on a membrane potential; (ii) the membrane potential is required only for import of the amino-terminal prepiece which is then sufficient to trigger the transmembrane movement of the remainder of the polypeptide chain; and (iii) transport across the mitochondrial membranes occurs via translocation contact sites in which the outer and inner membranes come close enough together to be simultaneously spanned by the imported polypeptide.

Sites of contact between the outer and inner mitochondrial membranes have been visualized by transmission electron microscopy (Hackenbrock, 1968). They appear to be stable structures since they are visible in mitochondria in both the condensed and orthodox conformations. It was estimated that there were 115 of these sites in a 1 μm diameter mitochondrion from rat liver. Similar regions of contact have been observed in mitochondria that were prepared for electron microscopy by freeze-etching (van

Venetie and Verkleij, 1982). Interpretation of the fracture plane, which jumped back and forth between the outer and inner limiting membranes, suggested that semifusion of the membranes had occurred in which nonbilayer lipids (specifically hexagonal II phase lipids) were involved.

Recently, the sites where mitochondrial precursor proteins form translocation intermediates spanning both membranes were correlated to the morphological contact regions (Schwaiger et al., 1987). The precursor of the F_1 ATPase β subunit, synthesized in reticulocyte lysate, was prebound to a specific antibody directed against the mature protein and then incubated with mitochondria. The antibody prevented the protein from being completely imported into mitochondria, but the amino-terminal prepiece was removed by the matrix-localized processing peptidase. When the antibody-bound pre-$F_1\beta$ was tagged by protein A–gold and visualized by electron microscopy, the gold particles were exclusively localized in regions of contact between inner and outer membranes. These contact sites had three distinct characteristics. First, they appeared to be stable structures. Mitochondria which had been treated with low concentrations of digitonin to disrupt the outer membrane still contained regions of contact between the inner membrane and outer membrane fragments. This agrees with the persistent occurrence of contact sites in condensed versus orthodox states. Second, contact site formation did not require precursor proteins. Mitochondria which had been pretreated with digitonin were able to mediate protein import. Therefore, contact sites seem to be preformed to facilitate import as opposed to being formed as a consequence of the presence of precursor proteins. And third, the outer membrane was necessary for contact site-mediated import. When mitochondria were pretreated with digitonin (to create mitoplasts), then incubated with antibody-bound $F_1\beta$, none of the protein A–gold particles was associated with the inner membrane, but they were associated with the remaining contact regions. Similarly, intact mitochondria which were shaved with trypsin then treated with digitonin did not import pre-$F_1\beta$ while the nonprotease-treated mitochondria (also treated with digitonin) did. While this indicates that second sites on the inner membrane that recognize precursor proteins do not exist, it has been demonstrated in yeast that import into trypsin-inactivated mitochondria can be restored if the outer membrane is disrupted by osmotic shock (Ohba and Schatz, 1987b).

How contact sites mediate transmembrane transport of proteins is unclear, particularly for proteins directed to the inner membrane. One possibility is that the formation and dissipation of contact sites is in constant flux so that proteins are translocated into the sites and then pulled into the mitochondrion with the inner membrane as they are dissipated. Another possibility is that the imported proteins migrate laterally from the contact regions. Finally, inner membrane proteins may be completely transported through the contact sites to the matrix and then redirected back to the inner membrane. Whether physical contact between the two membranes is necessary for import is unknown. The shortest precursor protein shown to be imported via contact sites so far is the Fe/S protein of the bc_1 complex (231 amino acids) (Hartl *et al.*, 1986). Calculation of the minimal distance which the membranes must come together to be spanned by pre-Fe/S indicates that they do not necessarily have to make contact, but they must come very close together (i.e., 10–20 nm, Fig. 3). The constituents and events occurring at contact sites remain unclear.

VIII. PROTEOLYTIC PROCESSING

During or shortly following the translocation step, the amino-terminal prepiece of many proteins directed to the inner membrane or matrix is removed by a specific protease which is located in the matrix (Böhni *et al.*, 1980; Mori *et al.*, 1980; Schmidt *et al.*, 1984). This occurs very rapidly *in vivo*. For example, in pulse–chase experiments in rat liver explants, the precursor for carbamoyl-phosphate synthase was imported into mitochondria and processed with a half-life of 2 min (Raymond and Shore, 1981). Similarly, in yeast, the F_1 ATPase β subunit precursor was imported and processed with a half-life of 0.5 min (Reid and Schatz, 1982b). Proteolytic processing must normally occur immediately on exposure of the precursor proteins to the matrix protease since precursor proteins cannot normally be detected in mitochondria. Processing, however, is not obligatory for import since the precursors to F_1 ATPase β and IX subunits could be imported into mitochondria when proteolytic processing was blocked with *o*-phenanthroline (Zwizinski and Neupert, 1983). Similarly, the precursor of the bc_1 complex Fe/S protein could be imported and did accumulate in the mitochondrial matrix when processing was blocked (Hartl *et al.*, 1986).

The matrix proteases identified and characterized in yeast, rat, and *Neurospora* mitochondria share similar properties: (1) the protein is a soluble component of the mitochondrial matrix; (2) it has a neutral pH optimum; (3) it is not affected by inhibitors of serine proteases [i.e., 4-nitrophenyl phosphate (PMSF)]; (4) it is sensitive to divalent cation chelators (i.e., *o*-phenanthroline); and (5) activity is stimulated by divalent cations such as Co^{2+}, Mn^{2+}, and Zn^{2+} (which can also reverse chelator inhibition; Böhni *et al.*, 1980, 1983; McAda and Douglas, 1982; Mori *et al.*, 1980; Miura *et al.*, 1982; Conboy *et al.*, 1982; Schmidt *et al.*, 1984). In whole mitochondria, only *o*-phenanthroline inhibits processing activity, whereas chelators such as ethylenediaminetetraacetic acid (EDTA) or bathophenanthroline, which cannot penetrate the inner membrane, do not. Partially purified preparations of the protease are inhibited by all these chelators and also by nucleoside triphosphates (Böhni *et al.*, 1983). In addition, activity was reported to be specifically inhibited in whole cells (isolated hepatocytes) by rhodamine 123 and 6G, and it was described to occur without affecting the membrane potential (Morita *et al.*, 1982; Kolarov and Nelson, 1984; Kuzela *et al.*, 1986). The underlying mechanism is not understood, but inhibition could not be overcome with excess divalent cations.

The matrix-located peptidase has been partially purified from yeast mitochondria. In one report, activity was purified 200-fold over whole mitochondria (McAda and Douglas, 1982). When analyzed by gel filtration, the protease had an apparent molecular weight of 150,000, and activity correlated best with a 59 kDa protein subunit identified by sodium dodecyl sulfate–polyacrylamide gel electrophoresis (SDS–PAGE). In another study where the protease activity was enriched 100-fold over whole mitochondria, the partially purified enzyme migrated with an apparent molecular weight of 115,000 but displayed 10 protein bands on SDS–PAGE, none of which corresponded to the 59 kDa band identified in the former study (Böhni *et al.*, 1983). The matrix protease was found to be a nuclear gene product which itself must be imported into mitochondria.

The matrix-localized peptidase has been purified to homogeneity from *N. crassa* mitochondria. Activity was enriched 5000-fold over whole mitochondria by a series of purification steps and displayed two bands on SDS–PAGE (52 kDa and 57 kDa; Hawlitschek *et al.*, 1988). When the two subunits were separated, neither was able to process precursor proteins. Activity was fully restored, however, when they were recombined. Activity of the matrix-localized processing peptidase depends on the presence of both proteins, though the contribution of each is as yet unclear.

Some imported mitochondrial proteins are proteolytically processed in two distinct steps. This has been demonstrated in yeast cytochrome b_2 (Gasser *et al.*, 1982b; Reid *et al.*, 1982; Daum *et al.*, 1982b), yeast and *Neurospora* cytochrome c_1 (Gasser *et al.*, 1982b; Ohashi *et al.*, 1982; Teintze *et al.*, 1982; Schleyer and Neupert, 1985), yeast cytochrome *c* peroxidase (Reid *et al.*, 1982), *Neurospora* ATPase subunit IX (Schmidt *et al.*, 1984), and the Fe/S protein of bc_1 complex in yeast (Sidhu and Beattie, 1983) and *Neurospora* (Hartl *et al.*, 1986). The precursor for rat liver OTC was originally thought to be processed in two steps as well, with the transient formation of an intermediate size protein (Mori *et al.*, 1980; Kraus *et al.*, 1981; Miura *et al.*, 1982; Conboy *et al.*, 1982); how-

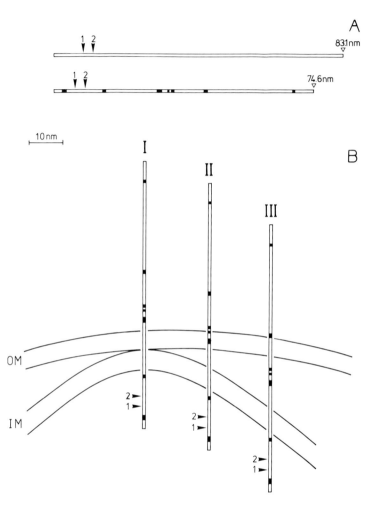

Fig. 3. Proximity of inner and outer membranes is required for simultaneous crossing by imported precursor proteins. Components are drawn approximately to scale. (A) The shortest protein demonstrated to be imported via translocation contact sites to date is the Fe/S protein of the bc_1 complex (231 amino acids). The approximate linear length (at 0.36 nm/residue) is 83.1 nm. When compression owing to a helical structure is included (yielding 0.15 nm/residue), as predicted from the primary sequence (Chou and Fasman, 1974; Argos et al., 1978), the contour length is 74.6 nm. Helical regions are indicated by shaded boxes. Sites of proteolytic processing are indicated by arrowheads. (B) Translocation intermediates are shown where the amino terminus protrudes far enough into the matrix to be processed by the chelator-sensitive matrix peptidase but the protein still has a major part outside the mitochondrion where it is accessible to externally added proteases. Three contact proximities are illustrated: (I) membranes in contact, (II) a 5 nm space between membranes as observed by electron microscopy (Schwaiger et al., 1987), and (III) estimated maximum

ever, evidence now suggests otherwise since: (i) "intermediate" OTC can be found outside of mitochondria while the processing peptidase is located in the mitochondrial matrix (Kolansky *et al.*, 1982); (ii) intermediate size OTC cannot be detected *in vivo* (Mori *et al.*, 1981; Morita *et al.*, 1982); and (iii) apparent "intermediate" OTC is found in the absence of mitochondria and does not associate with mitochondria *in vitro* (Argan *et al.*, 1983).

The first and second processing events appear to be performed by distinct proteases in different submitochondrial locations, with the possible exception of *N. crassa* ATPase IX processing. The ATPase IX precursor is imported to the inner membrane where its amino-terminal extension protrudes into the matrix and is processed in two steps by the chelator-sensitive matrix peptidase (Schmidt *et al.*, 1984). Evidence that the same enzyme is responsible for both cleavage steps includes the following: (1) both processing steps take place in the matrix, (2) both steps have the same sensitivity to chelating agents, (3) the two cleavage sites share amino acid sequence homology (hydrophobic-polar-Lys-Arg-small/bulky hydrophobic), and (5) the purified processing peptidase performs both cleavage steps (G. Hawlitschek and W. Neupert, unpublished).

Other than ATPase IX, precursor proteins which are processed in two steps share common features (Table III): (i) their respective mature forms are exposed to the intermembrane space (cytochrome b_2 and cytochrome c peroxidase are soluble components of the intermembrane space, whereas cytochrome c_1 and Fe/S of bc_1 complex are attached to the inner membrane but with their bulk protruding into the intermembrane space); (ii) they all make membrane-potential dependent contact with the inner membrane; (iii) the first proteolytic processing event occurs by the chelator-sensitive matrix peptidase; and (iv) the second processing event is catalyzed by a different protease.

Evidence indicates that different proteases are involved in two-step processing. First, whereas the first processing step is clearly performed by the matrix protease, the second step of cytochrome b_2 maturation is not sensitive to chelators (Daum *et al.*, 1982b). The sensitivity of the second step of bc_1 complex Fe/S protein processing to *o*-phenanthroline is believed to be due to inhibition of the Fe/S cluster formation (Hartl *et al.*, 1986), which, like attachment of heme to intermediate cytochrome c_1

---

separation which could still account for experimental observations (Schleyer and Neupert, 1985). Distances greater than 10–20 nm would probably produce detectable fragments from the imported part of the protein following external protease digestion. Mitochondrial membranes are assumed to be 5 nm thick. Insertion of the first processing site 10 nm past the inner face of the inner membrane is allowed for accessibility by the matrix peptidase.

TABLE III
Topology and Requirements for Mitochondrial Precursor Proteins Processed in Two Steps[a]

Precursor protein	Final location[b]	Two steps in vivo?	First step Energy[c]	First step o-Phe-sensitive	Intermediate location[b]	Second step Energy[c]	Second step o-Phe-sensitive[d]	Reference
Cytochrome b_2 (yeast)	IMS	+	+	+	O-IM	–	–	Gasser et al. (1982b), Reid et al. (1982), Daum et al. (1982a,b), Reid et al. (1982), Maccecchini et al. (1979b)
Cytochrome c Peroxidase (yeast)	IMS	?	+	+	O-IM			
Cytochrome c_1 (yeast)	O-IM	+	+	+	O-IM			Gasser et al. (1982b), Ohashi et al. (1982)
Cytochrome c_1 (Neurospora)	O-IM.	+	+	+	O-IM	+(NADH)		Teintze et al. (1982), Schleyer and Neupert (1985), B. Schmidt and W. Neupert (unpublished) Hartl et al. (1987)
Fe/S of bc_1 complex (Neurospora)	O-IM	+	+	+	M	+(NADH)	+	Hartl et al. (1986)
ATPase IX (Neurospora)	IM	+	+	+			+	Schmidt et al. (1984)

[a] Features of import for proteins which are processed in two steps are summarized. The energy for the first processing step is required for transport of the precursor to or through the inner membrane and not for proteolytic activity itself. Features of the second processing step are described only where conversion of the intermediate to the mature form has been examined independently of the first processing step. Areas left blank were not determined.

[b] IMS, intermembrane space; O-IM, outer surface of inner membrane; M, matrix.

[c] Energy required is a membrane potential unless otherwise indicated.

[d] o-Phe, o-phenanthroline.

(Ohashi *et al.*, 1982), might precede the second proteolytic processing step. Second, the topological arrangement of the intermembrane cytochromes b_2 and c_1 (and probably cytochrome c peroxidase) presumably exposes the second processing site to the intermembrane space (Daum *et al.*, 1982b; Ohashi *et al.*, 1982; Reid *et al.*, 1982). Third, the partially purified matrix peptidase cleaves cytochome b_2, c_1, and cytochrome c peroxidase precursors only to their intermediate size (Gasser *et al.*, 1982b; Reid *et al.*, 1982). Finally, a mutation in yeast blocks the second processing step of cytochrome b_2 without affecting the first step (Pratje and Guiard, 1986).

Two-step processing of these proteins has been demonstrated *in vivo*, with the exception of cytochrome c peroxidase (Reid *et al.*, 1982; Maccecchini *et al.*, 1979b) in which an intermediate size protein cannot be detected (presumably for kinetic reasons). For some proteins, NADH is required for the second processing step. As discussed in Section V, this may be necessary for heme attachment or Fe/S cluster formation and not for proteolytic processing itself. In these cases, such modifications appear to be obligatory and precede the second processing step.

Mutants have been isolated in yeast which are defective in proteolytic processing activity. In an attempt to identify essential components in the pathway of protein import, two complementation groups of temperature-sensitive mutants, defective in the import of mitochondrial proteins, were identified (Yaffe and Schatz, 1984). Termed *mas1* and *mas2* (mitochondrial assembly), the mutants were normal at the permissive temperature (23°C) but accumulated F_1 ATPase β subunit precursor at 37°C and stopped growing after 2–3 generations. The *mas1* mutants were deficient in the matrix-localized protease activity and could not process a number of mitochondrial precursor proteins (Yaffe *et al.*, 1985). The phenotype suggested that processing of imported proteins is essential for mitochondrial function (though processing is not necessary for import). Another temperature-sensitive mutation in yeast has also been identified which was defective in the second processing step of cytochrome b_2 but not cytochrome c peroxidase (Pratje *et al.*, 1983; Pratje and Guiard, 1986). In addition, the mutant was unable to process the precursor of cytochrome oxidase subunit II, a mitochondrial gene product. Assuming that the mutation is in a structural gene coding for a processing enzyme, this suggests that (i) the second processing protease differs from the first step protease; (ii) the processing of the mitochondrial gene product COX II is catalyzed by the same protease, and probably in the same location, as a second step protease; and (iii) the second processing step for cytochrome b_2 and cytochrome c peroxidase occurs by different proteases. All evidence taken together, the minimum number of distinct processing peptidases in mitochondria now appears to be three.

TABLE IV

Comparison of Sites of Proteolytic Processing of Imported Mitochondrial Precursor Proteins[a]

One-step Processing	−10	+10
1. Cytochrome oxidase IV (N. crassa)	...RSIATTVVRC|NAGTKPVPPH...	
2. Cytochrome oxidase V (N. crassa)	AAKPTMAVRA|ASTMPISNPT	
3. Cytochrome oxidase IV (yeast)	RTLCSSRYLL|QQKPVVKTAQ	
4. Cytochrome oxidase V (yeast)	GLSRITSVRF|AQTHALSNAA	
5. Cytochrome oxidase VI (yeast)	KNTFIQSRKY|SDAHDEETFE	
6. Mn-superoxide dismutase (yeast)	SLLSTTARRT|KVTLPDLKWD	
7. Aspartate aminotransferase (chicken)	PRRAAATARA|SSWWSHVEMG	
8. Ornithine aminotransferase (rat)	RGLRTSVASA|TSVATKKTEQ	
9. Ornithine carbamoyltransferase (rat)	RNFRYGKPVQ|SQVQLKGRDL	
10. Carbamoyl-phosphate synthase I (rat)	WDFSRPGIRL|LSVKAQTAHI	
11. Aspartate aminotransferase (porcine)	GLAAAASARA|SWWAHVEMGP	
12. Adrenodoxin (bovine)	RTLSVSGRAQ|SSSEDKITVH	
13. Cytochrome oxidase IV (bovine)	RAISTSVCVR|AHGSVVKSED	
14. Cytochrome P-450 (SCC) (bovine)	HRVGTGEGAG|ISTKTPRPYS	
15. Ornithine carbamoyltransferase (Human)	RNFRCGQPLQ|NKVQLKGRDL	

Two-step processing +10	First Site −10	Second Site +10	−10
16. ATPase IX (N. crassa)	VRVAQVSKRT|IQTGSPLQTL	TTRQAFQKRA|YSSEIAQAMV	b
17. Rieske Fe/S of bc_1 complex (N. crassa)	AAAPARAVRA|LTTSTALQGS	LTTSTALQGS|SSSTFESP	c
18. Cytochrome c peroxidase (yeast)	?	NWGKAAALAS|TTPLVHVASV	—
19. Cytochrome b_2 (yeast)	?	LNWHNGQIDN|EPKLDMNKQK	—
20. Cytochrome c_1 (yeast)	?	LYADSLTAEA|MTAAEHGLHA	—
21. ATPase IX P1 (bovine)	?	REFQTSVVSR|DIDTAAKFIG	d
22. ATPase IX P2 (bovine)	?	RSFQTSAISR|DIDTAAKFIG	d

[a] The amino acid sequences flanking the sites of proteolytic processing of imported mitochondrial proteins are listed by the single letter amino acid code (left to right, amino to carboxy terminus). The vertical lines indicate the sites of cleavage. Basic, acidic, and hydroxylated amino acids are indicated by the symbols +, −, and ·, respectively, above the primary sequence. The proteins are grouped by the organism in which they occur and then according to rough evolutionary order. Only sequences in which the cleavage sites have been positively identified are presented. References are the same as those indicated in Table I. In addition, the processing site for cytochrome oxidase IV from Neurospora crassa was determined by Sachs et al. (1986).

[b] The second processing step of N. crassa ATPase IX is catalyzed by the same protease as the first step (Schmidt et al., 1984). Comparison of the second processing site with first site sequences might be more appropriate.

[c] Only eight amino acids preceding the second processing site are presented since the first processing step occurs at this position.

[d] Homology of bovine ATPase IX presequences (P1 and P2) with ATPase IX from N. crassa suggests that the bovine prepieces might also be processed in two steps. This has not been demonstrated yet.

What are the recognition sites for proteolytic processing? A comparison of the amino acid sequences flanking the cleavage sites (Table IV) does not indicate a clear consensus sequence. There may be a number of reasons why very few trends are apparent. (1) the site specificity may vary between different organisms; however, a certain degree of conservation must exist since heterologous import and processing can occur. (2) Processing could be catalyzed by more than one chelator-sensitive matrix protease with different specificities. First site processing and processing of proteins in which the prepiece is removed in a single step, however, all appear to be catalyzed by the same chelator-sensitive matrix peptidase. The partially purified matrix peptidase from yeast processed precursors to F_1 ATPase β and α subunits, cytochrome oxidase IV, citrate synthase (Böhni *et al.*, 1983), and cytochrome oxidase V (Cerletti *et al.*, 1983). Similarly, the purified *Neurospora* matrix peptidase processed all precursor proteins tested to date (G. Hawlitschek and W. Neupert, unpublished). (3) The specificity of the matrix peptidase may depend on regions within the prepiece or mature part of the protein distal from the actual cleavage site. For example, cytochrome oxidase IV is normally processed between amino acids 25 and 26 to remove the 25 amino acid prepiece. When only the first 22 amino acids of the COX IV prepiece were fused to DHFR, thus removing the normal cleavage site, the fusion protein was still processed, but between amino acids 17 and 18 instead (Hurt *et al.*, 1985a). The sequence of amino acids flanking the new cleavage site show no homology to the authentic site, suggesting that other elements in the prepiece may contribute to the signal for processing. Processing of a fusion protein consisting of the presequence of pre-COX IV and DHFR by the solubilized matrix peptidase was blocked when even small deletions were made at the very amino terminus of the prepiece (Hurt *et al.*, 1987).

In some cases, the specificity of processing has requirements in the mature sequence. Correct processing of the rat OTC prepiece occurred when the first 60 amino-terminal amino acids of the pre-OTC, containing the 32 amino acid prepiece, were fused to asparagine synthase; however, incorrect processing at a site 14 amino acids closer to the amino terminus occurred when only the first 37 amino acids (still containing the full 32 amino acid prepiece) were fused (Nguyen *et al.*, 1986). The processing of human pre-OTC, also containing a 32 amino acid prepiece, was inhibited when deletions or substitutions were made between amino acids 8 and 22 of the prepiece. In addition, the glutamine at position -1 was also critical for proteolytic cleavage (Horwich *et al.*, 1986). Similarly, small deletions up to 17 amino acids away from the ATPase $F_1\beta$ cleavage site also prevented processing (Vassarotti *et al.*, 1987a).

How processing peptidases recognize the correct sites of cleavage is unclear. Undoubtedly, conformation plays an important role in the recognition of cleavage sites. The alkali-denatured OTC precursor, for example, was not processed by a partially purified form of the matrix peptidase (Miura *et al.*, 1986). Sequences around the cleavage sites show only minor similarities (Table IV, sequences 1–17). (1) Position -1 rarely has a charged amino acid. In lower eukaryotes it is usually a hydrophobic amino acid while in higher eukaryotes it is more often an uncharged polar amino acid. (2) In position -2, 10 of 17 amino acids are positively charged, mostly in lower eukaryotes. When not positively charged, the -2 amino acid is usually hydrophobic with an aliphatic side chain. (3) Charged amino acids are also rare in the $+1$ and $+2$ positions, but hydroxylated amino acids (particularly serine) frequently occur. Despite no obvious consensus sequence, however, proteolytic processing is highly specific. Neither mature mitochondrial proteins nor nonmitochondrial proteins are cleaved by the matrix-located peptidase (Böhni *et al.*, 1983).

IX. MITOCHONDRIAL GENE PRODUCTS

Not all mitochondrial proteins are nuclear gene products which are imported from their site of synthesis in the cytosol. A small number ($<10\%$) are coded for by the mitochondrial genome and are synthesized on 70 S mitochondrial ribosomes which are associated with the inner face of the inner membrane. The mitochondrial genomes in a number of species have been completely or partially sequenced. They vary in size from 17 kilobases in humans (Anderson *et al.*, 1981) to over 200 kilobases in plants (Palmer and Shields, 1984). Despite this variation, however, they code for a similar complement of proteins. These usually include apocytochrome *b,* cytochrome oxidase subunits I, II, and III, ATPase subunits VI and VIII, and several subunits from the NADH dehydrogenase complex (for review, see Breitenberger and RajBhandary, 1985). Yeast, fungi, and plant mitochondrial genomes also code for other proteins. For example, yeast and fungi mitochondria contain genes for the S5 protein of the small ribosomal subunit, ATPase subunit IX (which is dormant in *Neurospora crassa*), and intron-coded proteins involved in RNA maturation. Plant mitochondrial genomes also encode ATPase $F_1\alpha$. Finally, information coding for mitochondrial transfer RNAs and for ribosomal RNAs are contained within the genome.

Why mitochondria need a distinct genome at all and why the specific proteins that they encode are so highly conserved is unknown. One theory is that the hydrophobic nature of the mitochondrial gene products necessitates their synthesis within the organelle. This, however, seems unlikely to be the only reason since ATPase IX, a very hydrophobic protein, is synthesized in the cytosol of *N. crassa* containing a hydrophilic amino-terminal prepiece to mask the hydrophobic mature part (Jackl and Sebald, 1975; van den Boogaart *et al.*, 1982b). In fact, the ATPase IX prepiece has been shown to be able to mediate the import of ATPase VIII, an authentic mitochondrial gene product, back into mitochondria (Gearing and Nagley, 1986), although the shorter COX VI prepiece did not. On the other hand, the S5 ribosomal protein is water soluble but is still a mitochondrial gene product in some species. A second possibility is that the mitochondrial genome is an evolutionary remnant and that most of the genes from the bacterial endosymbiont were transferred to the nucleus of the host cell, but the process was incomplete. A third possibility is that the mitochondrial gene products form nucleating points around which the remainder of respiratory complexes are built. This cannot be absolutely critical though, since ATPase, for example, is assembled into a functional, albeit somewhat less efficient, complex when the subunits encoded by the mitochondrial genome are absent (Schatz, 1968; De Jong *et al.*, 1979; Marzuki and Linnane, 1985).

A final hypothesis is that the cotranslational protein export mechanism existed prior to the evolutionary endosymbiotic event. Accordingly, the genes for proteins which had amino acid sequences within them that resembled the export signal sequence were retained in the mitochondrial genome so that they would not be mistakenly exported. Indeed, it has been determined that most mitochondrial gene products from *Xenopus* and yeast seem to contain signal sequence-like segments near the amino terminus (von Heijne, 1986b). Eukaryotic cells maintain a distinct genome in the mitochondrion at great expense, since a large number of proteins must be imported simply to accommodate a separate protein synthetic system.

Mitochondrial protein synthesis occurs on ribosomes which are associated with the inner mitochondrial membrane, suggesting that translation is coupled to the insertion of the newly synthesized proteins into the membrane. Even the soluble S5 ribosomal protein, however, is synthesized on membrane-bound ribosomes (Marzuki and Hibbs, 1986). Cytochrome oxidase II is synthesized as a larger precursor in *N. crassa* (Machleidt and Werner, 1979; van den Boogaart *et al.*, 1982a) and in yeast (Sevarino and Poyton, 1980). In both cases, the higher molecular weight precursor can be chased to the mature size protein in the absence of protein synthesis (Sevarino and Poyton, 1980; Driever *et al.*, 1987). Furthermore, the newly synthesized COX II precursor from *Neurospora* was not integrated into the inner membrane, demonstrating that assembly can

occur posttranslationally. Processing of the COX II precursor was catalyzed by a protease which is located in the intermembrane space, and
formation of the mature size protein is dependent on NADH but not on a
membrane potential. Cytochrome oxidase II is probably not representative of the other mitochondrial gene products since it is synthesized with
an amino-terminal prepiece while others from the mitochondrial genome
are not (with the possible exception of COX I; Burger *et al.*, 1982).

X. ASSEMBLY AND COORDINATION

Many of the proteins which are imported into mitochondria are subunits of respiratory complexes. Therefore, assembly into functional complexes represents the final step in the import pathway. What sequence of
events occurs during assembly, and how is the supply of subunits coordinated and regulated?

Since two genetic systems are responsible for the synthesis of proteins
for most respiratory complexes, it would seem logical that they would be
synchronized in some way. No doubt there is some long-term regulation
of the mitchondrial genome by the nucleus since most of the proteins
comprising the mitochondrial transcription/translation system are nuclear
gene products. In the short term, however, the two systems do not appear
to be tightly coupled. For example, when synthesis of mitochondrial gene
products was blocked *in vivo* by growth of *Neurospora crassa* on chloramphenicol, thereby inhibiting the synthesis of cytochrome *b* of the bc_1
complex, normal amounts of cytochrome c_1 were synthesized in the cytosol (Weiss and Kolb, 1979). Furthermore, cytochrome c_1 was imported
normally and assembled into a cytochrome *b*-deficient complex. Similarly, when overexpression of the 11 kDa subunit (a nuclear gene product)
of the bc_1 complex was induced in yeast cells by transformation, it did not
affect the rate of synthesis or degradation of the other subunits of the
complex (van Loon *et al.*, 1983a). The same was observed when individual subunits of the bc_1 complex (also nuclear gene products) were over-
expressed, suggesting that stringent coupling does not exist (van Loon *et
al.*, 1983b).

Stringent coordination may not occur at the level of translation for the
nuclear gene products either. Cytoplasmically made subunits for the
ATPase, the bc_1 complex, and cytochrome oxidase are synthesized as
individual subunits, not as polyproteins (Lewin *et al.*, 1980; Mihara and
Blobel, 1980; van Loon *et al.*, 1983c). Notwithstanding this apparent lack
of coordination, excess unincorporated subunits do not accumulate in
mitochondria. One suggestion has been that surplus amounts of unas-

sembled subunits are simply eliminated by proteolytic digestion (Luzikov, 1986) regardless of the waste of cellular resources by such a process.

Despite the apparent lack of coordination of synthesis of subunits in these experiments, nuclear genes are involved in regulating the expression of the mitochondrial genome. This appears to occur at three levels: (i) control of mRNA processing, (ii) control of translation, and (iii) post-translational modification (e.g., proteolytic processing of pre-COX II by a nuclear-coded protease; Pratje *et al.*, 1983; Pratje and Guiard, 1986). These systems have been studied best in *Saccharomyces cerevisiae*. Yeast mitochondrial genes contain introns, separating the exons of a gene, which must be spliced out prior to translation. These intervening sequences are removed from pre-mRNAs by self-splicing mechanisms and by protein-assisted splicing. In the latter case, these proteins (termed mRNA maturases) are encoded by either mitochondrial intron open reading frames (Weiss-Brummer *et al.*, 1982; Carignani *et al.*, 1983; Guiso *et al.*, 1984; Jacq *et al.*, 1984) or are nuclear gene products (Faye and Simon, 1983; Pillar *et al.*, 1983; McGraw and Tzagoloff, 1983; Dieckmann *et al.*, 1984) and have been shown to be necessary for processing of COX I and apocytochrome *b* pre-mRNAs. The interesting feature of these studies is that the maturation proteins appear to be specific for individual mitochondrial gene transcripts so that the expression of mitochondrial gene products can be selectively controlled by the activity of the different nuclear-encoded mRNA maturases.

Nuclear gene products are also required for the specific translation of mitochondrial mRNAs (Fox, 1986). For example, the yeast *pet494* mutant has normal levels of fully processed COX III mRNA. A nuclear gene product was found to be required to promote the translation of the COX III mRNA that appeared to interact with the 5'-untranslated leader (Müller *et al.*, 1984). Similar nuclear gene products were also required for the specific translation of apocytochrome *b* (CPB6, Dieckmann and Tzagoloff, 1985; and MK2, Rödel *et al.*, 1985) and for COX II (PET111; Fox, 1986). Though they serve similar functions, there seems to be little homology between the PET494, CBP6, and PET111 gene products. They all appear, however, to act on the 5'-untranslated mRNA leader sequence in a specific way so that translation of mitochondrial-encoded proteins can be independently controlled.

Compared to the loose coordination between nuclear and mitochondrial genomes, the assembly of individual subunits into functional complexes is a more ordered process. The best studied examples of complex assembly to date are the F_0F_1 ATPase and cytochrome oxidase. When the synthesis of yeast mitochondrial gene products was blocked by growth in the presence of chloramphenicol (De Jong *et al.*, 1979) or as in *rho*⁻ mutants

(Schatz, 1968), a correctly assembled functional ATPase was still produced which was loosely associated with the inner membrane but no longer sensitive to oligomycin. This suggests that the mitochondrial gene products (subunits VI, VIII, and IX) are not essential for either assembly or function but are required for stability of the complex and confer oligomycin sensitivity. Sequential assembly of the ATPase subunits was demonstrated in yeast mutants lacking each of the mitochondrially encoded proteins (Marzuki and Linnane, 1985). The mutant lacking subunit IX was deficient in the assembly of both VI and VIII, while the mutant lacking subunit VIII could assemble IX but not VI. This indicated that the relative order of assembly of the mitochondrial gene products was IX then VIII then VI and that sequential assembly was necessary. Each of these mutants had a functional but unstable ATPase.

A large number of components are required for the functional assembly of cytochrome oxidase. In a *N. crassa* mutant lacking COX I (COX I, II, and III are mitochondrial gene products) only the assembly of subunits V and VI occurred (Nargang *et al.*, 1978). In pulse–chase experiments in rats, COX II and III were immediately assembled while COX I arrived only after a long chase (Wielburski *et al.*, 1982). Taken together, this suggests that COX II and III are assembled first, followed by COX I, which in turn is necessary for the subsequent assembly of COX V and VI. In a yeast mutant lacking COX IV, no cytochrome oxidase activity was observed, although the mutant still contained the other mitochondrial COX subunits. This suggested that COX IV is necessary for the proper assembly of cytochrome oxidase (Dowhan *et al.*, 1985).

Cofactors such as heme, oxygen, and copper are also necessary for cytochrome oxidase assembly. In yeast, a mutant lacking heme contained no cytochrome oxidase activity. Although the mitochondria still contained subunits II, III, and IV, it had only low amounts of COX I and IV and no V or VI. The residual subunits were not assembled (Saltzgaber-Müller and Schatz, 1978). In rats, heme was shown to be necessary for the assembly of subunit I with the preassembled COX II and III (Wielburski and Nelson, 1984). COX I was predominantly associated with COX III. Yeast cells grown anaerobically did not contain assembled cytochrome oxidase. When shifted to an oxygen environment, however, subunits I and II were immediately assembled with VI and VII (Woodrow and Schatz, 1979). Cytochrome oxidase is not assembled in copper-depleted *N. crassa,* although both nuclear and mitochondrial gene products are synthesized normally (Werner *et al.*, 1974).

The coordination and assembly of functional mitochondrial complexes occurs by a series of events which are dependent on components other than the mitochondrial subunits alone. Whether assembly is controlled by factors other than the simple stoichiometric availability of components is unknown.

XI. OVERVIEW

A. Import Pathway Models

Owing to the diversity of mitochondrial proteins and their topological locations, a variety of different pathways exist to faciliate their import. We summarize what is known about the import pathways of representative proteins and what can be hypothesized.

Porin: Outer Membrane (Fig. 4). The precursor to porin differs from the mature form only in conformational arrangement, thereby allowing its solubility in the cytosol. It does not contain an amino-terminal prepiece and does not require an energized inner membrane for import. Preporin binds to its receptor at the outer face of the outer membrane and is inserted into the outer membrane where it is protected from externally added proteases. It then forms dimers and trimers. The porin pathway is probably representative of most outer membrane proteins (Pfaller *et al.*, 1985; Pfaller and Neupert, 1987, and references therein).

Cytochrome c: Intermembrane Space (Fig. 5). Cytochrome *c* is synthesized as apocytochrome *c* which does not contain covalently attached heme and has a loosely ordered conformation. Apocytochrome *c* spontaneously inserts partway through the lipid bilayer of the outer membrane. It is then sequestered at the inner face of the outer membrane by an apocytochrome *c* binding protein which in turn exposes the cysteine

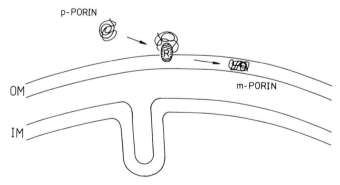

Fig. 4. Import pathway of porin into mitochondria. OM, Outer membrane; IM, inner membrane; R, receptor; p, precursor; m, mature.

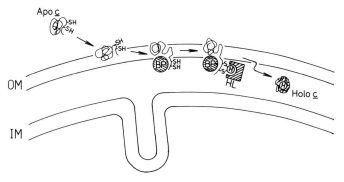

Fig. 5. Import pathway of cytochrome c into mitochondria. OM, Outer membrane; IM, inner membrane; Apo c, apocytochrome c; Holo c, holocytochrome c; BP, apocytochrome c binding protein; H, heme; HL, cytochrome c heme lyase.

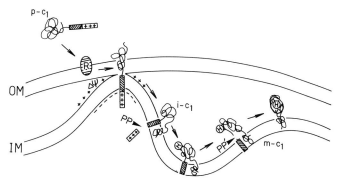

Fig. 6. Import pathway of cytochrome c_1 into mitochondria (according to Hurt and van Loon, 1986). OM, Outer membrane; IM, inner membrane; p, precursor; i, intermediate; m, mature; R, receptor; $\Delta\psi$, membrane potential; PP, processing peptidase; H, heme. Boxes represent prepiece segments.

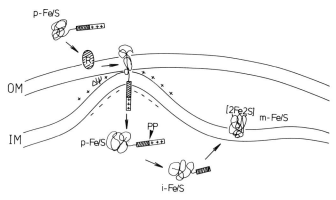

Fig. 7. Import pathway of the Rieske Fe/S protein of bc_1 complex into mitochondria. Abbreviations are as in Fig. 6.

thiols to the intermembrane space. Heme is enzymatically attached by cytochrome c heme lyase, and the resulting conformational change pulls the protein through the outer membrane. Holocytochrome c then migrates to its functional location in association with cytochrome c reductase and cytochrome c oxidase at the outer face of the inner membrane (Hennig and Neupert, 1981; Nicholson *et al.*, 1987, and references therein).

Cytochrome c_1: Intermembrane Space, Inner Membrane (Fig. 6). The cytochrome c_1 precursor is synthesized with a two-domain prepiece. Pre-c_1 binds to its receptor at the outer face of the outer membrane. The amino terminus is then transloc ted through the inner membrane via contact sites in a step which is dependent on a membrane potential. The first part of the prepiece is removed by the chelator-sensitive matrix peptidase to generate intermediate c_1. The carboxy terminus, which eventually anchors the mature protein to the inner membrane, is embedded into the inner membrane at some point following translocation across the outer membrane. Heme is covalently attached to intermediate c_1, and the second part of the prepiece is then removed yielding the mature c_1 which is then assembled into the bc_1 complex (Ohashi *et al.*, 1982; Schleyer and Neupert, 1985; Hurt and van Loon, 1986, and references therein.) The same basic pathway may be followed by cytochrome c peroxidase, which is released as a soluble protein into the intermembrane space following the second proteolytic step because it does not contain a carboxy-terminal anchoring segment. This pathway demonstrates the principles of the stop transfer model suggested by Hurt and van Loon (1986). Another possibility is that pre-c_1 is completely translocated into the matrix, then redirected back to the inner membrane like the Fe/S protein of the bc_1 complex (Fig. 7). This has been demonstrated to be the case in *Neurospora crassa* and also for the import of cytochrome b_2 into yeast mitochondria (Hartl *et al.*, 1987).

Fe/S Protein of bc_1 Complex: Inner Membrane (Fig. 7). The precursor to the Fe/S protein is synthesized with a two-part prepiece. Pre-Fe/S binds to its receptor on the outer membrane and is then completely translocated into the mitochondrial matrix via translocation contact sites in a step which is dependent on the membrane potential. The matrix-located pre-Fe/S is then processed to the intermediate size protein by the chelator-sensitive processing peptidase. The Fe/S protein is then directed back across the inner membrane to the outer face of the inner membrane. The protein is processed to its mature form by removal of the second half of the prepiece and formation of the Fe/S cluster, but when and where this occurs are unknown (Hartl *et al.*, 1986).

ADP/ATP Carrier: Inner Membrane (Fig. 8). The ADP/ATP carrier is synthesized without a prepiece but contains stretches of positively

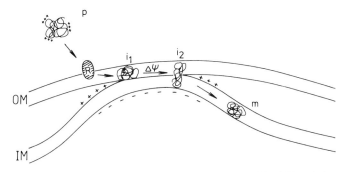

Fig. 8. Import pathway of ADP/ATP carrier into mitochondria. Abbreviations are as in Fig. 6.

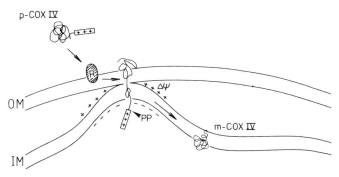

Fig. 9. Import pathway of cytochrome oxidase subunit IV into mitochondria. Abbreviations are as is Fig. 6.

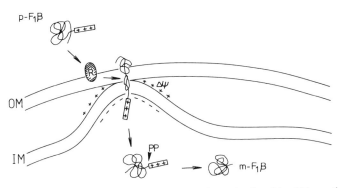

Fig. 10. Import pathway of F_1 ATPase β subunit into mitochondria. Abbrevations are as in Fig. 6.

charged amino acids which resemble mitochondrial targeting sequences. The precursor binds to its receptor and is inserted into the mitochondrial membrane where it is protected from externally added proteases. The first intermediate (i_1) is past the receptor stage but before the membrane potential-dependent stage of import. The second intermediate (i_2) is found in the presence of a membrane potential and is transported to the inner membrane via translocation contact sites. The imported ADP/ATP carrier then undergoes a conformational change, in which it acquires properties of the mature protein, followed by the formation of dimers (Schleyer and Neupert, 1984; Pfanner and Neupert, 1987). The uncoupling protein from brown fat mitochondria has similar sequence and folding characteristics (Aquila *et al.*, 1985) and is probably imported in a similar manner.

Cytochrome Oxidase IV: Inner Membrane (Fig. 9). Pre-COX IV is synthesized with an amino-terminal prepiece in the cytosol. It binds to its receptor on the outer membrane and is subsequently transported to the inner membrane by translocation contact sites in a membrane potential-dependent step. The amino-terminal prepiece, which protrudes into the matrix, is removed by the chelator-sensitive peptidase, and the mature part of the protein remains integrated in the inner membrane where it is assembled into cytochrome oxidase (Hurt and van Loon, 1986, and references therein). This mechanism is probably representative of most imported inner membrane proteins which are processed in a single step.

F_oF_1 ATPase $F_1\beta$: Matrix (Fig. 10). Pre-$F_1\beta$ binds to its receptor and is transported completely into the matrix. This occurs via translocation contact sites and is dependent on a membrane potential. During or shortly following translocation, the mature $F_1\beta$ is generated by removal of the prepiece by the matrix peptidase. Mature $F_1\beta$ is then assembled with the other F_1 ATPase subunits at the inner face of the inner membrane. Most imported matrix proteins probably follow a similar pathway (Schleyer and Neupert, 1985, and references therein).

Cytochrome Oxidase II: A Mitochondrial Gene Product (Fig. 11). Most mitochondrial gene products are probably inserted directly into the inner membrane as they are synthesized on membrane-bound ribosomes (except the S5 ribosomal protein which is released into the matrix). COX II, on the other hand, is synthesized with a prepiece (in lower eukaryotes) and can be posttranslationally inserted into the inner membrane (B). Cotranslational insertion may occur *in vivo* (A) but is not obligatory. Pre-COX II is converted to mature COX II by removal of the prepiece by a protease located in the intermembrane space and noncovalent binding of heme *a* and copper. The sequence in which these changes occur is unknown, but is dependent on NADH. Mature COX II is then assembled into cytochrome oxidase (Driever *et al.*, 1987).

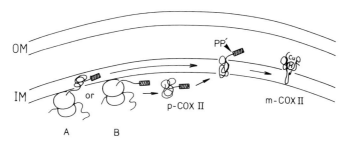

Fig. 11. Insertion pathway of cytochrome oxidase subunit II. Abbreviations are as in Fig. 6.

B. Evolutionary Considerations

A particularly interesting finding which emerged while examining the import of the Fe/S protein of bc_1 complex into *Neurospora crassa* mitochondria is that the precursor protein was completely translocated into the matrix where it was partially processed and then redirected back across the inner membrane to its functional location on the other side facing the intermembrane space. (Hartl *et al.*, 1986; see Fig. 7). This seemingly complex assembly pathway was explained in terms of a rerouting mechanism by which the Fe/S protein is returned to remnants of its "ancestral" assembly pathway. Following evolutionary gene transfer from the bacterial endosymbiont to the host cell nucleus, mechanisms had to evolve to return the gene product to its functional location. Rather than completely rebuild the means by which the protein was folded, assembled, and acquired the correct topology, the import pathway of the Fe/S protein has evolved to make use of preexisting mechanisms. To do this, the Fe/S protein had to be completely transported across both mitochondrial membranes back into the matrix. This was accommodated by adding a matrix targeting prepiece to the Fe/S protein (i.e., the first part of the pre-Fe/S prepiece) which allows it to enter the matrix by components which had evolved to mediate the import of other mitochondrial proteins. At this point, "ancestral" transport pathways take over to correctly insert the Fe/S protein into the inner membrane.

Indeed, the equivalent Fe/S protein from the photosynthetic bacteria *Rhodopseudomonas sphaeroides* is synthesized in the bacterial cytoplasm (comparable to the mitochondrial matrix) and transferred across the photosynthetic membrane to the side opposite the cytoplasm where, as in mitochondria, it is topologically opposed to the F_1 part of ATPase. This transfer is accompanied by a reduction in molecular weight (Gabellini *et al.*, 1985) which may be equivalent to processing of the mitochon-

Table V

Comparison of Amino-Terminal Sequences of Mitochondrial and Bacterial Proteins[a]

A. Rieske Fe/S protein of bc_1 complex

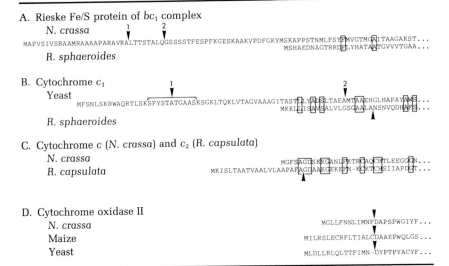

B. Cytochrome c_1

C. Cytochrome c (N. crassa) and c_2 (R. capsulata)

D. Cytochrome oxidase II

[a] (A – C) Amino acid sequences of the amino terminus of mitochondrial precursor proteins are compared to their equivalent in bacteria. Arrows indicate the sites of proteolytic processing. The estimated region containing the first processing site of yeast cytochrome c_1 is indicated by a bracket. The processing site for R. sphaeroides Fe/S protein is not known. Segments having identical amino acid sequences are marked by a box. Alignment inserts are indicated by (−). (D) The prepiece sequences of mitochondrial cytochrome oxidase subunit II. Sequences were taken from Harnisch et al. (1985) (N. crassa Fe/S protein of bc_1 complex), Gabellini and Sebald (1986) (R. sphaeroides Fe/S protein of bc_1 complex and cytochrome c_1), Sadler et al. (1984) (yeast cytochrome c_1), Lederer and Simon (1974) (N. crassa cytochrome c), Daldal et al. (1986) (R. capsulata cytochrome c_2), van den Boogaart et al. (1982a) (N. crassa COX II), Fox and Leaver (1981) (maize COX II), and Coruzzi and Tzagoloff (1979) (yeast COX II).

drial intermediate Fe/S to mature Fe/S. The bacterial Fe/S protein is homologous to the mature *Neurospora* Fe/S protein except in the targeting prepieces (Gabellini and Sebald, 1986). The amino terminus of the bacterial Fe/S protein has characteristics of a leader sequence for export through the cytoplasmic membrane, but the second part of the mitochondrial Fe/S protein prepiece is not comparable as might be predicted from this model (Table VA). Information to redirect the mitochondrial Fe/S protein from the matrix back across the inner membrane might not be contained in sequences analogous to bacterial leader sequences. In this regard, the second part of the cytochrome c_1 prepiece bears a close resemblance to the export leader sequence of the equivalent bacterial protein (Table VB).

The Fe/S protein of the bc_1 complex may have retained this elaborate import pathway because of stringent assembly requirements which necessitate its insertion into the inner membrane from the matrix side. On the other hand, cytochrome c is a soluble protein of the intermembrane space and is less likely to be so demanding. This is reflected in the way in which import pathway for cytochrome c has evolved. In the bacterium *Rhodopseudomonas capsulata,* the equivalent protein (cytochrome c_2) is synthesized as a precursor protein in the bacterial cytosol and is processed as it is secreted into the periplasmic space (comparable to the mitochondrial intermembrane space). Rather than via a redirected import pathway, mitochondrial cytochrome c is transported across just the outer membrane (Fig. 5). On evolutionary gene transfer to the host cell nucleus, the bacterial leader prepiece was removed (Table VC).

Some bacterial-like leader sequences remain in mitochondria. In lower eukaryotes such as *Neurospora,* yeast, and maize, for example, cytochrome oxidase II (a mitochondrial gene product) is synthesized with a prepiece having some characteristics of bacterial leader sequences (e.g., a stretch of hydrophobic amino acids following a charged amino terminus; Table VD). On the other hand, proteins which have no equivalent in bacteria, and therefore presumably did not evolve from the bacterial endosymbiont, may be imported by novel mechanisms which differ from most mitochondrial proteins. For example, the ADP/ATP carrier does not have an amino-terminal prepiece, but specific targeting information is contained within internal sequences that have accompanied the structural part of the protein during evolutionary formation.

Evolutionary remnants of the bacterial endosymbiont may still be present in the import pathways of mitochondrial proteins. The differing pathways could reflect specific assembly requirements of the various proteins and, in some cases, the relative time at which they were introduced as mitochondrial constituents.

C. Summary

Nearly the entire complement of mitochondrial proteins must be imported from the cytosol. To do this correctly, mitochondrial precursor proteins must be specifically targeted to mitochondria and then properly sorted to their functional submitochondrial location. In addition, the proper topological arrangement and assembly with other subunits must occur. Information to enable these processes is embodied within the precursor proteins themselves, and components required to facilitate these events are molecules and structures within the mitochondrion. The import pathways for a variety of mitochondrial proteins have been resolved into a sequence of discrete but contiguous steps. In doing so, the components necessary for each step have frequently been identified at a molecular level. These can be divided into three broad classes, namely, components necessary for (i) specific targeting to mitochondria, (ii) sorting and translocation processes, and (iii) processing events. In the first group, targeting signals, usually contained within the amino-terminal prepiece of the precursor protein, mediate recognition of the remainder of the protein by mitochondria. Receptors on the surface of the mitochondrion bind the precursor protein and initiate its entry into mitochondria. In the second group, translocation contact sites facilitate transmembrane movement of most imported proteins. The constituents of these contact regions are unknown but probably contain specific proteins which are necessary for import. The membrane potential supplies the energy or the circumstances to initiate translocation. Features within the precursor proteins and perhaps processes which exist at the site of translocation guide the imported protein to its correct submitochondrial location. Finally, processing occurs in many forms. Prepieces are proteolytically removed by specific proteases, and some proteins are covalently or noncovalently modified. Refolding into an active conformation often occurs, and then assembly into a functional location marks the end of the import process.

The general sequence of events by which import occurs has been well-characterized for a number of proteins having different submitochondrial locations and different roles in mitochondrial function. The molecular mechanisms are now of greater interest and are the focus of current research.

ACKNOWLEDGMENTS

We would like to thank M. E. Clement for typing and Dr. M. Schwaiger for critically reviewing the manuscript. We would also like to thank our colleagues for their many contributions to this review.

REFERENCES

Ades, I. Z., and Butow, R. A. (1980a). The products of mitochondria-bound cytoplasmic polysomes in yeast. *J. Biol. Chem.* **255,** 9918–9924.

Ades, I. Z., and Butow, R. A. (1980b). The transport of proteins into yeast mitochondria: Kinetics and pools. *J. Biol. Chem.* **255,** 9925–9935.

Adrian, G. S., McCammon, M. T., Montgomery, D. L., and Douglas, M. G. (1986). Sequences required for delivery and localization of the ADP/ATP translocator to the mitochondrial inner membrane. *Mol. Cell. Biol.* **6**, 626–634.

Allison, D. S., and Schatz, G. (1986). Artificial mitochondrial presequences. *Proc. Natl. Acad. Sci. U.S.A.* **83**, 9011–9015.

Ambler, R. P., and Wynn, M. (1973). The amino acid sequences of cytochromes *c*-551 from three species of *Pseudomonas*. *Biochem. J.* **131**, 485–498.

Anderson, S., Bankier, A. T., Barrell, B. G., de Bruijn, M. H. L., Coulson, A. R., Drouin, J., Eperon, I. C., Nierlich, D. P., Roe, B. A., Sanger, F., Schreier, P. H., Smith, A. J. H., Staden, R., and Young, I. G. (1981). Sequence and organization of the human mitochondrial genome. *Nature (London)* **290**, 457–465.

Arakawa, H., Takiguchi, M., Amaya, Y., Nagata, S., Hayashi, H., and Mori, M. (1987). cDNA-derived amino acid sequence of rat mitochondrial 3-oxyacyl-CoA thiolase with no transient presequence: Structural relationship with peroxisomal isozyme. *EMBO J.* **6**, 1361–1366.

Aquila, H., Link, T. A., and Klingenberg, M. (1985). The uncoupling protein from brown fat mitochondria is related to the mitochondrial ADP/ATP carrier. Analysis of sequence homologies and of folding of the protein in the membrane. *EMBO J.* **4**, 2369–2376.

Argan, C., and Shore, G. C. (1985). The precursor to ornithine carbamoyltransferase is transported to mitochondria as a 5S complex containing an import factor. *Biochem. Biophys. Res. Commun.* **131**, 289–298.

Argan, C., Lusty, C. J., and Shore, G. C. (1983). Membrane and cytosolic components affecting transport of the precursor for ornithine carbamoyltransferase into mitochondria. *J. Biol. Chem.* **258**, 6667–6670.

Argos, P., Hanei, M., and Garavito, R. M. (1978). The Chou–Fasman secondary structure prediction method with an extended data base. *FEBS Lett.* **93**, 19–24.

Banroques, J., Perea, J., and Jacq, C. (1987). Efficient splicing of two yeast mitochondrial introns controlled by a nuclear-encoded maturase. *EMBO J.* **6**, 1085–1091.

Basile, G., Di Bello, C., and Taniuchi, H. (1980). Formation of an iso-1-cytochrome *c*-like species containing a covalently bonded heme group from the apoprotein by a yeast cell-free system in the presence of hemin. *J. Biol. Chem.* **255**, 7181–7191.

Beltzer, J. P., Chang, L.-F. L., Hinkkanen, A. E., and Kohlhaw, G. B. (1986). Structure of yeast *LEU1*: The 5′ flanking region contains features that predict two modes of control and two productive translation starts. *J. Biol. Chem.* **261**, 5160–5167.

Böhni, P., Gasser, S., Leaver, C., and Schatz, G. (1980). A matrix-localized mitochondrial protease processing cytoplasmically made precursors to mitochondrial proteins. *In* "The Organization and Expression of the Mitochondrial Genome" (A. M. Kroon, and C. Saccone, eds.), pp. 423–433. Elsevier/North-Holland, Amsterdam.

Böhni, P. C., Daum, G., and Schatz, G. (1983). Import of proteins into mitochondria: Partial purification of a matrix-located protease involved in cleavage of mitochondrial precursor polypeptides. *J. Biol. Chem.* **258**, 4937–4943.

Bouillaud, F., Weissenbach, J., and Ricquier, D. (1986). Complete cDNA-derived amino acid sequence of rat brown fat uncoupling protein. *J. Biol. Chem.* **261**, 1487–1490.

Boutry, M., and Chua, N.-H. (1985). A nuclear gene encoding the β subunit of the mitochondrial ATP synthase in *Nicotiana plumbaginifolia*. *EMBO J.* **4**, 2159–2165.

Breitenberger, C. A., and RajBhandary, U. L. (1985). Some highlights of mitochondrial research based on analyses of *Neurospora crassa* mitochondrial DNA. *Trends Biochem. Sci.* **10**, 478–483.

Burger, G., Scriven, C., Machleidt, W., and Werner, S. (1982). Subunit 1 of cytochrome oxidase from *Neurospora crassa*: Nucleotide sequence of the coding gene and partial amino acid sequence of the protein. *EMBO J.* **1**, 1385–1391.

Burns, D., and Lewin, A. (1986). Inhibition of the import of mitochondrial proteins by RNase. *J. Biol. Chem.* **261**, 6153–6155.

Carignani, G., Groudinsky, O., Frezza, D., Schiavon, E., Bergantino, E., and Slonimski, P. P. (1983). An mRNA maturase is encoded by the first intron of the mitochondrial gene for the subunit I of cytochrome oxidase in *S. cerevisiae*. *Cell* **35**, 733–742.

Cerletti, N., Böhni, P. C., and Suda, K. (1983). Import of proteins into mitochondria: Isolated yeast mitochondria and a solubilized matrix protease correctly process cytochrome *c* oxidase subunit V at the NH$_2$ terminus. *J. Biol. Chem.* **258**, 4944–4949.

Chen, W.-J., and Douglas, M. G. (1987). Phosphodiester bond cleavage outside mitochondria is required for the completion of protein import into the mitochondrial matrix. *Cell* **49**, 651–658.

Chou, P. Y., and Fasman, G. D. (1974). Prediction of protein conformation. *Biochemistry* **13**, 222–245.

Conboy, J. G., Fenton, W. A., and Rosenberg, L. E. (1982). Processing of pre-ornithine transcarbamylase requires a zinc-dependent protease localized to the mitochondrial matrix. *Biochem. Biophys. Res. Commun.* **105**, 1–7.

Coruzzi, G., and Tzagoloff, A. (1979). Assembly of the mitochondrial membrane system: DNA sequence of subunit 2 of yeast cytochrome oxidase. *J. Biol. Chem.* **254**, 9324–9330.

Daldal, F., Cheng, S., Applebaum, J., Davidson, E., and Prince, R. C. (1986). Cytochrome c$_2$ is not essential for photosynthetic growth of *Rhodopseudomonas capsulata*. *Proc. Natl. Acad. Sci. U.S.A.* **83**, 2012–2016.

Daum, G., Böhni, P. C., and Schatz, G. (1982a). Import of proteins into mitochondria: Cytochrome b$_2$ and cytochrome *c* peroxidase are located in the intermembrane space. *J. Biol. Chem.* **257**, 13028–13033.

Daum, G., Gasser, S. M., and Schatz, G. (1982b). Import of proteins into mitochondria: Energy-dependent, two-step processing of the intermembrane space enzyme cytochrome b$_2$ by isolated yeast mitochondria. *J. Biol. Chem.* **257**, 13075–13080.

De Haan, M., van Loon, A. P. G. M., Kreike, J., Vaessen, R. T. M. J., and Grivell, L. A. (1984). The biosynthesis of the ubiquinol–cytochrome *c* reductase complex in yeast: DNA sequence analysis of the nuclear gene coding for the 14-kDa subunit. *Eur. J. Biochem.* **138**, 169–177.

De Jong, L., Holtrop, M., and Kroon, A. M. (1979). The biogenesis of rat liver mitochondrial ATPase: Subunit composition of the normal ATPase complex and of the deficient complex formed when mitochondrial protein synthesis is blocked. *Biochim. Biophys. Acta* **548**, 48–62.

Dieckmann, C. L., and Tzagoloff, A. (1985). Assembly of the mitochondrial membrane system: *CBP6*, a yeast nuclear gene necessary for synthesis of cytochrome *b*. *J. Biol. Chem.* **260**, 1513–1520.

Dieckmann, C. L., Koerner, T. J., and Tzagoloff, A. (1984). Assembly of the mitochondrial membrane system: *CBPI*, a yeast nuclear gene involved in 5′ processing of cytochrome *b* pre-mRNA. *J. Biol. Chem.* **259**, 4722–4731.

Douglas, M. G., Geller, B. L., and Emr, S. D. (1984). Intracellular targeting and import of an F$_1$-ATPase β-subunit–β-galactosidase hybrid protein into yeast mitochondria. *Proc. Natl. Acad. Sci. U.S.A.* **81**, 3983–3987.

Dowhan, W., Bibus, C. R., and Schatz, G. (1985). The cytoplasmically made subunit IV is necessary for assembly of cytochrome *c* oxidase in yeast. *EMBO J.* **4**, 179–184.

Driever, W., Cook, R. A., and Neupert, W. (1987). In preparation.

Dumont, M. E., and Richards, F. M. (1984). Insertion of apocytochrome *c* into lipid vesicles. *J. Biol. Chem.* **259**, 4147–4156.

Eilers, M., and Schatz, G. (1986). Binding of a specific ligand inhibits import of a purified precursor protein into mitochondria. *Nature (London)* **322,** 228–232.

Eilers, M., Oppliger, W., and Schatz, G. (1987). Both ATP and an energized inner membrane are required to import a purified precursor protein into mitochondria. *EMBO J.* **6,** 1073–1077.

Eisenberg, D., Schwarz, E., Komaromy, M., and Wall, R. (1984). Analysis of membrane surface protein sequences with the hydrophobic moment plot. *J. Mol. Biol.* **179,** 125–142.

Emr, S. D., Vassarotti, A., Garrett, J., Geller, B. L., Takeda, M., and Douglas, M. G. (1986). The amino terminus of the yeast F_1-ATPase β-subunit precursor functions as a mitochondrial import signal. *J. Cell Biol.* **102,** 523–533.

Epand, R. M., Hui, S.-W., Argan, C., Gillespie, L. L., and Shore, G. C. (1986). Structural analysis and amphiphilic properties of a chemically synthesized mitochondrial signal peptide. *J. Biol. Chem.* **261,** 10017–10020.

Faye, G., and Simon, M. (1983). Analysis of a yeast nuclear gene involved in the maturation of mitochondrial pre-messenger RNA of the cytochrome oxidase subunit I. *Cell* **32,** 77–87.

Firgaira, F. A., Hendrick, J. P., Kalousek, F., Kraus, J. P., and Rosenberg, L. E. (1984). RNA required for import of precursor proteins into mitochondria. *Science* **226,** 1319–1322.

Fisher, W. R., Taniuchi, H., and Anfinsen, C. B. (1973). On the role of heme in the formation of the structure of cytochrome c. *J. Biol. Chem.* **248,** 3188–3195.

Fox, T. D. (1986). Nuclear gene products required for translation of specific mitochondrially coded mRNAs in yeast. *Trends Genet.* **2,** 97–100.

Fox, T. D., and Leaver, C. J. (1981). The *Zea mays* mitochondrial gene coding cytochrome oxidase subunit II has an intervening sequence and does not contain TGA codons. *Cell* **26,** 315–323.

Freeman, K. B., Chien, S.-M., Lichtfield, D., and Patel, H. V. (1983). Synthesis *in vitro* of rat brown adipose tissue 32000 M_r protein. *FEBS Lett.* **158,** 325–330.

Freitag, H., Janes, M., and Neupert, W. (1982). Biosynthesis of mitochondrial porin and insertion into the outer mitochondrial membrane of *Neurospora crassa. Eur. J. Biochem.* **126,** 197–202.

Gabellini, N., and Sebald, W. (1986). Nucleotide sequence and transcription of the *fbc* operon from *Rhodopseudomonas sphaeroides:* Evaluation of the deduced amino acid sequences of the FeS protein, cytochrome *b* and cytochrome c_1. *Eur. J. Biochem.* **154,** 569–579.

Gabellini, N., Harnish, U., McCartny, J. E. G, Hauska, G., and Sebald, W. (1985). Cloning and expression of the *fbc* operon encoding the FeS protein, cytochrome *b* and cytochrome c_1 from the *Rhodopseudomonas sphaeroides* b/c_1 complex. *EMBO J.* **4,** 549–553.

Gasser, S. M., and Schatz, G. (1983). Import of proteins into mitochondria: *In vitro* studies on the biogenesis of the outer membrane. *J. Biol. Chem.* **258,** 3427–3430.

Gasser, S. M., Daum, G., and Schatz, G. (1982a). Import of proteins into mitochondria: Energy-dependent uptake of precursors by isolated mitochondria. *J. Biol. Chem.* **257,** 13034–13041.

Gasser, S. M., Ohashi, A., Daum, G., Böhni, P. C., Gibson, J., Reid, G. A., Yonetani, T., and Schatz, G. (1982b). Imported mitochondrial proteins cytochrome b_2 and cytochrome c_1 are processed in two steps. *Proc. Natl. Acad. Sci. U.S.A.* **79,** 267–271.

Gay, N. J., and Walker, J. E. (1985). Two genes encoding the bovine mitochondrial ATP synthase proteolipid specify precursors with different import sequences and are expressed in a tissue-specific manner. *EMBO J.* **4,** 3519–3524.

Gearing, D. P., and Nagley, P. (1986). Yeast mitochondrial ATPase subunit 8, normally a mitochondrial gene product, expressed *in vitro* and imported back into the organelle. *EMBO J.* **5**, 3651–3655.

Gillespie, L. L. (1987). Identification of an outer mitochondrial membrane protein that interacts with a synthetic signal peptide. *J. Biol. Chem.* **262**, 7939–7942.

Gillepsie, L. L., Argan, C., Taneja, A. T., Hodges, R. S., Freeman, K. B., and Shore, G. C. (1985). A synthetic signal peptide blocks import of precursor proteins destined for the mitochondrial inner membrane or matrix. *J. Biol. Chem.* **260**, 16045–16048.

Gregor, I., and Tsugita, A. (1982). The amino acid sequence of cytochrome *c* oxidase subunit VI from *Saccharomyces cerevisiae. J. Biol. Chem.* **257**, 13081–13087.

Guiard, B. (1985). Structure, expression and regulation of a nuclear gene encoding a mitochondrial protein: The yeast L(+)-lactate cytochrome *c* oxidoreductase (cytochrome b_2). *EMBO J.* **4**, 3265–3272.

Guiso, N., Dreyfus, M., Siffert, O., Danchin, A., Spyridakis, A., Gargouri, A., Claisse, M., and Slonimski, P. P. (1984). Antibodies against synthetic oligopeptides allow identification of the mRNA-maturase encoded by the second intron of the yeast *cob-box* gene. *EMBO J.* **8**, 1769–1772.

Hackenbrock, C. R. (1968). Chemical and physical fixation of isolated mitochondria in low-energy and high-energy states. *Proc. Natl. Acad. Sci. U.S.A.* **61**, 589–602.

Hallermayer, G., Zimmermann, R., and Neupert, W. (1977). Kinetic studies of the transport of cytoplasmically synthesized proteins into the mitochondria in intact cells of *Neurospora crassa. Eur. J. Biochem.* **81**, 523–532.

Hampsey, D. M., Lewin, A. S., and Kohlhaw, G. B. (1983). Submitochondrial localization, cell-free synthesis, and mitochondrial import of 2-isopropylmalate synthase of yeast. *Proc. Natl. Acad. Sci. U.S.A.* **80**, 1270–1274.

Harmey, M. A., and Neupert, W. (1985). Synthesis and intracellular transport of mitochondrial proteins. *In* "The Enzymes of Biological Membranes" (A. Martonosi, ed.), Vol. 4, pp. 431–464. Plenum, New York.

Harmey, M. A., Hallermayer, G., Korb, H., and Neupert, W. (1977). Transport of cytoplasmically synthesized proteins into the mitochondria in a cell free system from *Neurospora crassa. Eur. J. Biochem.* **81**, 533–544.

Harnisch, U., Weiss, H., and Sebald, W. (1985). The primary structure of the iron–sulfur subunit of ubiquinol–cytochrome *c* reductase from *Neurospora,* determined by cDNA and gene sequencing. *Eur. J. Biochem.* **149**, 95–99.

Hartl, F.-U., Schmidt, B., Weiss, H., Wachter, E., and Neupert, W. (1986). Transport into mitochondria and intramitochondrial sorting of the Fe/S protein of ubiquinol cytochrome *c* reductase. *Cell* **47**, 939–951.

Hartl, F.-U., Ostermann, J., Guiard, B., and Neupert, W. (1987). Successive translocation into and out of the mitochondrial matrix: targeting of proteins to the intermembrane space by a bipartite signal peptide. *Cell* (in press).

Hase, T., Riezman, H., Suda, K., and Schatz, G. (1983). Import of proteins into mitochondria: Nucleotide sequence of the gene for a 70-kd protein of the yeast mitochondrial outer membrane. *EMBO J.* **2**, 2169–2172.

Hase, T., Müller, U., Riezman, H., and Schatz, G. (1984). A 70-kd protein of the yeast mitochondrial outer membrane is targeted and anchored via its extreme amino terminus. *EMBO J.* **3**, 3157–3164.

Hase, T., Nakai, M., and Matsubara, H. (1986). The N-terminal 21 amino acids of a 70 kDa protein of the yeast mitochondrial outer membrane direct *E. coli* β-galactosidase into the mitochondrial matrix space in yeast cells. *FEBS Lett.* **197**, 199–203.

Hatalová, I., and Kolarov, J. (1983). Synthesis and intracellular transport of cytochrome oxidase subunit IV and ADP/ATP translocator protein in intact hepatoma cells. *Biochem. Biophys. Res. Commun.* **110**, *132–139.*

Hay, R., Böhni, P., and Gasser, S. (1984). How mitochondria import proteins. *Biochim. Biophys. Acta* **779**, 65–87.

Hawlitschek, G., Hartl, F.-U., Schmidt, B., and Neupert, W. (1988). The processing peptidase of mitochondria: Two different polypeptides cooperate to cleave imported precursor proteins. *Cell* (submitted).

Hennig, B., and Neupert, W. (1981). Assembly of cytochrome *c*. Apocytochrome *c* is bound to specific sites on mitochondria before its conversion to holocytochrome *c*. *Eur. J. Biochem.* **81**, 533–544.

Hennig, B., Köhler, H., and Neupert, W. (1983). Receptor sites involved in posttranslational transport of apocytochrome *c* into mitochondria: Specificity, affinity and number of sites. *Proc. Natl. Acad. Sci. U.S.A.* **80**, 4963–4967.

Hensgens, L. A. M., Grivell, L. A., Borst, P., and Bos, J. L. (1979). Nucleotide sequence of the mitochondrial structural gene for subunit 9 of yeast ATPase complex. *Proc. Natl. Acad. Sci. U.S.A.* **76**, 1663–1667.

Horwich, A. L., Fenton, W. A., Williams, K. R., Kalousek, F., Kraus, J. P., Doolittle, R. F., Konigsberg, W., and Rosenberg, L. E. (1984). Structure and expression of a complementary DNA for the nuclear coded precursor of human mitochondrial ornithine transcarbamylase. *Science* **224**, 1068–1074.

Horwich, A. L., Kalousek, F., and Rosenberg, L. E. (1985a). Arginine in the leader peptide is required for both import and proteolytic cleavage of a mitochondrial precursor. *Proc. Natl. Acad. Sci. U.S.A.* **82**, 4930–4933.

Horwich, A. L., Kalousek, F., Mellman, I., and Rosenberg, L. E. (1985b). A leader peptide is sufficient to direct mitochondrial import of a chimeric protein. *EMBO J.* **4**, 1129–1135.

Horwich, A. L., Kalousek, F., Fenton, W. A., Pollock, R. A., and Rosenberg, L. E. (1986). Targeting of pre-ornithine transcarbamylase to mitochondria: Definition of critical regions and residues in the leader peptide. *Cell* **44**, 451–459.

Hurt, E. C., and Schatz, G. (1987). A cytosolic protein contains a cryptic mitochondrial targeting signal. *Nature (London)* **325**, 499–503.

Hurt, E. C., and van Loon, A. P. G. M. (1986). How proteins find mitochondria and intramitochondrial compartments. *Trends Biochem. Sci.* **11**, 204–207.

Hurt, E. C., Pesold-Hurt, B., and Schatz, G. (1984a). The amino-terminal region of an imported mitochondrial precursor polypeptide can direct cytoplasmic dihydrofolate reductase into the mitochondrial matrix. *EMBO J.* **3**, 3149–3156.

Hurt, E. C., Pesold-Hurt, B., and Schatz, G. (1984b). The cleavable prepiece of an imported mitochondrial protein is sufficient to direct cytosolic dihydrofolate reductase into the mitochondrial matrix. *FEBS Lett.* **178**, 306–310.

Hurt, E. C., Pesold-Hurt, B., Suda, K., Oppliger, W., and Schatz, G. (1985a). The first twelve amino acids (less than half of the pre-sequence) of an imported mitochondrial protein can direct mouse cytosolic dihydrofolate reductase into the yeast mitochondrial matrix. *EMBO J.* **4**, 2061–2068.

Hurt, E. C., Müller, U., and Schatz, G. (1985b). The first twelve amino acids of a yeast mitochondrial outer membrane protein can direct a nuclear-encoded cytochrome oxidase subunit to the mitochondrial inner membrane. *EMBO J.* **4**, 3509–3518.

Hurt, E. C., Soltanifar, N., Goldschmidt-Clermont, M., Rochaix, J.-D., and Schatz, G. (1986a). The cleavable pre-sequence of an imported chloroplast protein directs attached polypeptides into yeast mitochondria. *EMBO J.* **5**, 1343–1350.

Hurt, E. C., Goldschmidt-Clermont, M., Peshold-Hurt, B., Rochaix, J.-D., and Schatz, G.

(1986b). A mitochondrial presequence can transport a chloroplast-encoded protein into yeast mitochondria. *J. Biol. Chem.* **261,** 11440–11443.

Hurt, E. C., Allison, D. S., Müller, U., and Schatz, G. (1987). Amino-terminal deletions in the presequence of an imported mitochondrial protein block the targeting function and proteolytic cleavage of the presequence at the carboxy terminus. *J. Biol. Chem.* **262,** 1420–1424.

Ito, A., Ogishima, T., Ou, W., Omura, T., Aoyagi, H., Lee, S., Mihara, H., and Izumiya, N. (1985). Effects of synthetic model peptides resembling the extension peptides of mitochondrial enzyme precursors on import of the precursors into mitochondria. *J. Biochem.* **98,** 1571–1582.

Jackl, G., and Sebald, W. (1975). Identification of two products of mitochondrial protein synthesis associated with mitochondrial adenosine triphosphatase from *Neurospora crassa*. *Eur. J. Biochem.* **54,** 97–106.

Jacq, C., Banroques, J., Becam, A. M., Slonimski, P. P., Guiso, N., and Danchin, A. (1984). Antibodies against a fused '*lacZ*–yeast mitochondrial intron' gene product allow identification of the mRNA maturase encoded by the fourth intron of the yeast *cob-box* gene. *EMBO J.* **3,** 1567–1572.

Jaussi, R., Sonderegger, P., Flückiger, J., and Christen, P. (1982). Biosynthesis and topogenesis of aspartate aminotransferase isoenzymes in chicken embryo fibroblasts: The precursor of the mitochondrial isoenzyme is either imported into mitochondria or degraded in the cytosol. *J. Biol. Chem.* **257,** 13334–13340.

Jaussi, R., Cotton, B., Juretic, N., Christen, P., and Schümperli, D. (1985). The primary structure of the precursor of chicken mitochondrial aspartate aminotransferase. *J. Biol. Chem.* **260,** 16060–16063.

Joh, T., Nomiyama, H., Maeda, S., Shimada, K., and Morino, Y. (1985). Cloning and sequence analysis of a cDNA encoding porcine mitochondrial aspartate aminotransferase precursor. *Proc. Natl. Acad. Sci. U.S.A.* **82,** 6065–6069.

Kaput, J., Goltz, S., and Blobel, G. (1982). Nucleotide sequence of the yeast nuclear gene for cytochrome *c* peroxidase precursor: Functional implications of the pre-sequence for protein transport into mitochondria. *J. Biol. Chem.* **257,** 15054–15058.

Kellems, R. E., and Butow, R. A. (1972). Cytoplasmic-type 80 S ribosomes associated with yeast mitochondria: I. Evidence for ribosome binding sites on yeast mitochondria. *J. Biol. Chem.* **247,** 8043–8050.

Kellems, R. E., Allison, V. F., and Butow, R. A. (1974). Cytoplasmic-type 80 S ribosomes associated with yeast mitochondria: II. Evidence for the association of cytoplasmic ribosomes with the outer mitochondrial membrane *in situ*. *J. Biol. Chem.* **249,** 3297–3303.

Keng, T., Alani, E., and Guarente, L. (1986). The nine amino-terminal residues of δ-aminolevulinate synthase direct β-galactosidase into the mitochondrial matrix. *Mol. Cell. Biol.* **6,** 355–364.

Kobayashi, K., Iwasaki, Y., Sasaki, T., Nakamura, K., and Asahi, T. (1986). Putative amino-terminal presequence for β-subunit of plant mitochondrial F_1 ATPase deduced from the amino-terminal sequence of the mature subunit. *FEBS Lett.* **203,** 144–148.

Koerner, T. J., Hill, J., and Tzagoloff, A. (1985). Cloning and characterization of the yeast nuclear gene for subunit 5 of cytochrome oxidase. *J. Biol. Chem.* **260,** 9513–9515.

Köhler, H., Stuart, R. A., and Neupert, W. (1987). In preparation.

Kolansky, D. M., Conboy, J. G., Fenton, W. A., and Rosenberg, L. E. (1982). Energy-dependent translocation of the precursor of ornithine transcarbamylase by isolated rat liver mitochondria. *J. Biol. Chem.* **257,** 8467–8471.

Kolarov, J., and Nelson, B. D. (1984). Import and processing of cytochrome b–c_1 complex subunits in isolated hepatoma ascites cells: Inhibition by rhodamine 6G. *Eur. J. Biochem.* **144**, 387–392.

Korb, H., and Neupert, W. (1978). Biogenesis of cytochrome *c* in *Neurospora crassa:* Synthesis of apocytochrome *c*, transfer to mitochondria and conversion to holocytochrome *c*. *Eur. J. Biochem.* **91**, 609–620.

Kraus, J. P., Conboy, J. G., and Rosenberg, L. E. (1981). Pre-ornithine transcarbamylase: Properties of the cytoplasmic precursor of a mitochondrial matrix enzyme. *J. Biol. Chem.* **256**, 10739–10742.

Kraus, J. P., Hodges, P. E., Williamson, C. L., Horwich, A. L., Kalousek, F., Williams, K. R., and Rosenberg, L. E. (1985). A cDNA clone for the precursor of rat mitochondrial ornithine transcarbamylase: Comparison of rat and human leader sequences and conservation of catalytic sites. *Nucleic Acids Res.* **13**, 943–952.

Kruse, B., and Sebald, W. (1984). Nucleotide sequences of the nuclear genes for the proteolipid and δ subunit of the mitochondrial ATP synthase from *Neurospora crassa. In* "H⁺-ATPase (ATP Synthase): Structure, Function, Biogenesis of the F_0 F_1 Complex of Coupling Membranes" (S. Papa, K. Altendorf, L. Ernster, and L. Packer, eds.), pp. 67–75. ICSU Press/Adriatica. Editrice, Bari, Italy.

Kuzela, S., Joste, V., and Nelson, B. D. (1986). Rhodamine 6G inhibits the matrix-catalyzed processing of precursors of rat-liver mitochondrial proteins. *Eur. J. Biochem.* **154**, 553–557.

Lederer, F., and Simon, A. M. (1974). *Neurospora crassa* and *Humicola lanuginosa* cytochromes *c*: More homology in the heme region. *Biochem. Biophys. Res. Commun.* **56**, 317–323.

Lewin, A. S., Gregor, I., Mason, T. L., Nelson, N., and Schatz, G. (1980). Cytoplasmically made subunits of yeast mitochondrial F_1-ATPase and cytochrome *c* oxidase are synthesized as individual precursors, not as polyproteins. *Proc. Natl. Acad. Sci. U.S.A.* **77**, 3998–4002.

Lomax, M. I., Bachman, N. J., Nasoff, M. S., Caruthers, M. H., and Grossman, L. I. (1984). Isolation and characterization of a cDNA clone for bovine cytochrome *c* oxidase subunit IV. *Proc. Natl. Acad. Sci. U.S.A.* **81**, 6295–6299.

Luzikov, V. N. (1986). Proteolytic control over topogenesis of membrane proteins. *FEBS Lett.* **200**, 259–264.

Maarse, A. C., and Grivell, L. A. (1987). Nucleotide sequence of the gene encoding the 11-kDa subunit of the ubiquinol-cytochrome-*c* oxidoreductase in *Saccharomyces cerevisiae. Eur. J. Biochem.* **165**, 419–425.

Maarse, A. C., van Loon, A. P. G. M., Riezman, H., Gregor, J., Schatz, G., and Grivell, L. A. (1984). Subunit IV of yeast cytochrome *c* oxidase: Cloning and nucleotide sequencing of the gene and partial amino acid sequencing of the mature protein. *EMBO J.* **3**, 2831–2837.

McAda, P. C., and Douglas, M. G. (1982). A neutral metallo endoprotease involved in the processing of an F_1-ATPase subunit precursor in mitochondria. *J. Biol. Chem.* **257**, 3177–3182.

Maccecchini, M.-L., Rudin, Y., Blobel, G., and Schatz, G. (1979a). Import of proteins into mitochondria: Precursor forms of the extramitochondrially made F_1 ATPase subunits in yeast. *Proc. Natl. Acad. Sci. U.S.A.* **76**, 343–347.

Maccecchini, M.-L., Rudin, Y., and Schatz, G. (1979b). Transport of proteins across the mitochondrial outer membrane: A precursor form of the cytoplasmically made intermembrane enzyme cytochrome *c* peroxidase. *J. Biol. Chem.* **254**, 7468–7471.

McGraw, P., and Tzagoloff, A. (1983). Assembly of the mitochondrial membrane system: Characterization of a yeast nuclear gene involved in the processing of the cytochrome *b* pre-mRNA. *J. Biol. Chem.* **258**, 9459–9468.

Machleidt, W., and Werner, S. (1979). Is the mitochondrially made subunit 2 of cytochrome oxidase synthesized as a precursor in *Neurospora crassa? FEBS Lett.* **107,** 327–330.

Macino, G., and Tzagoloff, A. (1979). Assembly of the mitochondrial membrane system: The DNA sequence of a mitochondrial ATPase gene in *Saccharomyces cerevisiae. J. Biol. Chem.* **254,** 4617–4623.

McIntyre, P., Graf, L., Mercer, J., Peterson, G., Hudson, P., and Hoogenraad, N. (1984). A highly basic N-terminal extension of the mitochondrial matrix enzyme ornithine transcarbamylase from rat liver. *FEBS Lett.* **177,** 41–46.

Marres, C. A. M., van Loon, A. P. G. M., Oudshoorn, P., van Steeg, H., Grivell, L. A., and Slater, E. C. (1985). Nucleotide sequence analysis of the nuclear gene coding for manganese superoxide dismutase of yeast mitochondria, a gene previously assumed to code for the Rieske iron–sulphur protein. *Eur. J. Biochem.* **147,** 153–161.

Marzuki, S., and Hibbs, A. R. (1986). Are all mitochondrial translation products synthesized on membrane-bound ribosomes? *Biochim. Biophys. Acta* **866,** 120–124.

Marzuki, S., and Linnane, A. W. (1985). Biosynthesis of the yeast mitochondrial H^+-ATPase complex. *In* "The Enzymes of Biological Membranes" (A. N. Martonosi, ed.), Vol. **4,** pp. 415–430. Plenum, New York.

Matsuura, S., Arpin, M., Hannum, C., Margoliash, E., Sabatini, D. D., and Morimoto, T. (1981). *In vitro* synthesis and posttranslational uptake of cytochrome *c* into isolated mitochondria: Role of a specific addressing signal in the apocytochrome. *Proc. Natl. Acad. Sci. U.S.A.* **78,** 4368–4372.

Mihara, K., and Blobel, G. (1980). The four cytoplasmically made subunits of yeast mitochondrial cytochrome *c* oxidase are synthesized individually and not as a polyprotein. *Proc. Natl. Acad. Sci. U.S.A.* **77,** 4160–4164.

Mihara, K., Blobel, G., and Sato, R. (1982a). *In vitro* synthesis and integration into mitochondria of porin, a major protein of the outer mitochondrial membrane of *Saccharomyces cerevisiae. Proc. Natl. Acad. Sci. U.S.A.* **79,** 7102–7106.

Mihara, K., Omura, T., Harano, T., Brenner, S., Fleischer, S., and Rajagopalan, K. V. (1982b). Rat liver L-glutamate dehydrogenase, malate dehydrogenase, D-β-hydroxybutyrate dehydrogenase, and sulfite oxidase are each synthesized as larger precursors by cytoplasmic free polysomes. *J. Biol. Chem.* **257,** 3355–3358.

Mitchell, P., and Moyle, J. (1969). Estimation of membrane potential and pH difference across the cristae membrane of rat liver mitochondria. *Eur. J. Biochem.* **7,** 471–484.

Miura, S., Mori, M., Amaya, Y., and Tatibana, M. (1982). A mitochondrial protease that cleaves the precursor of ornithine carbamoyltransferase: Purification and properties. *Eur. J. Biochem.* **122,** 641–647.

Miura, S., Mori, M., and Tatibana, M. (1983). Transport of ornithine carbamoyltransferase precursor into mitochondria: Stimulation by potassium ion, magnesium ion, and a reticulocyte cytosolic protein(s). *J. Biol. Chem.* **258,** 6671–6674.

Miura, S., Amaya, Y., and Mori, M. (1986). A metalloprotease involved in the processing of mitochondrial precursor proteins. *Biochem. Biophys. Res. Commun.* **134,** 1151–1159.

Mori, M., Miura, S., Tatibana, M., and Cohen, P. P. (1980). Characterization of a protease apparently involved in processing of pre-ornithine transcarbamylase of rat liver. *Proc. Natl. Acad. Sci. U.S.A.* **77,** 7044–7048.

Mori, M., Morita, T., Ikeda, F., Amaya, Y., Tatibana, M., and Cohen, P. P. (1981). Synthesis, intracellular transport, and processing of the precursors for mitochondrial ornithine transcarbamylase and carbamoylphosphate synthetase I in isolated hepatocytes. *Proc. Natl. Acad. Sci. U.S.A.* **78,** 6056–6060.

Morita, T., Mori, M., Ikeda, F., and Tatibana, M. (1982). Transport of carbamoyl phosphate synthetase I and ornithine transcarbamylase into mitochondria: Inhibition by rhodamine 123 and accumulation of enzyme precursors in isolated hepatocytes. *J. Biol. Chem.* **257**, 10547–10550.

Morohashi, K., Fujii-Kuriyama, Y., Okada, Y., Sogawa, K., Hirose, T., Inayama, S., and Omura, T. (1984). Molecular cloning and nucleotide sequence of cDNA for mRNA of mitochondrial cytochrome *P*-450 (SCC) of bovine adrenal cortex. *Proc. Natl. Acad. Sci. U.S.A.* **81**, 4647–4651.

Mueckler, M. M., and Pitot, H. C. (1985). Sequence of the precursor to rat ornithine aminotransferase deduced from a cDNA clone. *J. Biol. Chem.* **260**, 12993–12997.

Müller, P. P., Reif, M. K., Zonghou, S., Sengstag, C., Mason, T. L., and Fox, T. D. (1984). A nuclear mutation that posttranscriptionally blocks accumulation of a yeast mitochondrial gene product can be suppressed by a mitochondrial gene rearrangement. *J. Mol. Biol.* **175**, 431–452.

Myers, A. M., and Tzagoloff, A. (1985). *MSW*, a yeast gene coding for mitochondrial tryptophanyl-tRNA synthetase. *J. Biol. Chem.* **260**, 15371–15377.

Nagata, S., Tseunetsugu-Yokata, Y., Naito, A., and Kaziro, Y. (1983). Molecular cloning and sequence determination of the nuclear gene coding for mitochondrial elongation factor Tu of *Saccharomyces cerevisiae*. *Proc. Natl. Acad. Sci. U.S.A.* **80**, 6192–6196.

Nakagawa, T., Maeshima, M., Muto, H., Kajiura, H., Hattori, H., and Asahi, T. (1987). Separation, amino-terminal sequence and cell-free synthesis of the smallest subunit of sweet potato cytochrome *c* oxidase. *Eur. J. Biochem.* **165**, 303–307.

Nargang, F. E., Bertrand, H., and Werner, S. (1978). A nuclear mutant of *Neurospora crassa* lacking subunit 1 of cytochrome *c* oxidase. *J. Biol. Chem.* **253**, 6364–6369.

Natsoulis, G., Hilger, F., and Fink, G. R. (1986). The *HST1* gene encodes both the cytoplasmic and mitochondrial histidine tRNA synthetases of *S. cerevisiae*. *Cell* **46**, 235–243.

Nelson, N., and Schatz, G. (1979). Energy-dependent processing of cytoplasmically made precursors to mitochondrial proteins. *Proc. Natl. Acad. Sci. U.S.A.* **76**, 4365–4369.

Neupert, W., and Schatz, G. (1981). How proteins are transported into mitochondria. *Trends Biochem. Sci.* **6**, 1–4.

Nguyen, M., Argan, C., Lusty, C. J., and Shore, G. C. (1986). Import and processing of hybrid proteins by mammalian mitochondria *in vitro*. *J. Biol. Chem.* **261**, 800–805.

Nicholson, D. W., Köhler, H., and Neupert, W. (1987). Import of cytochrome *c* into mitochondria: Cytochrome *c* heme lyase. *Eur. J. Biochem.* **164**, 147–157.

Nishikimi, M., Shimomura, Y., and Ozawa, T. (1986). Cell-free synthesis of ubiquinone-binding protein of mitochondrial cytochrome bc_1 complex. *Biochem. Biophys. Res. Commun.* **138**, 1291–1297.

Nyunoya, H., Broglie, K. E., Widgren, E. E., and Lusty, C. J. (1985). Characterization and derivation of the gene coding for mitochondrial carbamoyl phosphate synthase I of rat. *J. Biol. Chem.* **260**, 9346–9356.

Ohashi, A., Gibson, J., Gregor, I., and Schatz, G. (1982). Import of proteins into mitochondria: The precursor of cytochrome c_1 is processed in two steps, one of them heme-dependent. *J. Biol. Chem.* **257**, 13042–13047.

Ohba, M., and Schatz, G. (1987a). Protein import into yeast mitochondria is inhibited by antibodies raised against 45-kd proteins of the outer membrane. *EMBO J.* **6**, 2109–2115.

Ohba, M., and Schatz, G. (1987b). Disruption of the outer membrane restores protein import to trypsin-treated yeast mitochondria. *EMBO J.* **6**, 2117–2122.

Ohta, S., and Kagawa, Y. (1986). Human F_1-ATPase: Molecular cloning of cDNA for the β subunit. *J. Biochem.* **99**, 135–141.

Ohta, S., and Schatz, G. (1984). A purified precursor polypeptide requires a cytosolic protein fraction for import into mitochondria. *EMBO J.* **3**, 651–657.

Okamura, T., John, M. E., Zuber, M. X., Simpson, E. R., and Waterman, M. R. (1985).

Molecular cloning and amino acid sequence of the precursor form of bovine adrenodoxin: Evidence for a previously unidentified COOH-terminal peptide. *Proc. Natl.
Acad. Sci. U.S.A.* **82,** 5705–5709.

Ono, H., Yoshimura, N., Sato, M., and Tuboi, S. (1985). Translocation of proteins into rat
liver mitochondria: Existence of two different precursor polypeptides of liver fumarase and import of the precursor into mitochondria. *J. Biol. Chem.* **260,** 3402–3407.

Palmer, J. D., and Shields, C. R. (1984). Tripartite structure of the *Brassica campestris*
mitochondrial genome. *Nature (London)* **307,** 437–440.

Pape, L. K., Koerner, T. J., and Tzagoloff, A. (1985). Characterization of a yeast nuclear
gene (*MST1*) coding for the mitochondrial threonyl-tRNA$_1$ synthetase. *J. Biol. Chem.*
260, 15362–15370.

Patterson, T., and Poyton, R. O. (1986). COX8, the structural gene for yeast cytochrome *c*
oxidase subunit VIII. *J. Biol. Chem.* **261,** 17192–17197.

Pfaller, R., and Neupert, W. (1987). High-affinity binding sites involved in the import of
porin into mitochondria. *EMBO J.* **6,** 2635–2642.

Pfaller, R., Freitag, H., Harmey, M. A., Benz, R., and Neupert, W. (1985). A water-soluble
form of porin from the mitochondrial outer membrane of *Neurospora crassa*. *J. Biol.
Chem.* **260,** 8188–8193.

Pfanner, N., and Neupert, W. (1985). Transport of proteins into mitochondria: A potassium
diffusion potential is able to drive the import of ADP/ATP carrier. *EMBO J.* **4,** 2819–
2825.

Pfanner, N., and Neupert, W. (1986). Transport of F$_1$-ATPase subunit β into mitochondria
depends on both a membrane potential and nucleoside triphosphates. *FEBS Lett.* **209,**
152–156.

Pfanner, N., and Neupert, W. (1987). Distinct steps in the import of ADP/ATP carrier into
mitochondria. *J. Biol. Chem.* **262,** 7528–7536.

Pfanner, N., Tropschug, M., and Neupert, W. (1987). Mitochondrial protein import: Nucleoside triphosphates are involved in conferring import-competence to precursors. *Cell*
49, 815–823.

Pillar, T., Lang, B. F., Steinberger, I., Vogt, B., and Kaudewitz, F. (1983). Expression of
the "split gene" *cob* in yeast mtDNA: Nuclear mutations specifically block the excision of different introns from its primary transcript. *J. Biol. Chem.* **258,** 7954–7959.

Power, S. D., Lochrie, M. A., and Poyton, R. O. (1986). The nuclear-coded subunits of
yeast cytochrome *c* oxidase: The amino acid sequences of subunits VII and VIIa,
structural similarities between the three smallest polypeptides of the holoenzyme, and
implications for biogenesis. *J. Biol. Chem.* **261,** 9206–9209.

Pratje, E., and Guiard, B. (1986). One nuclear gene controls the removal of transient presequences from two yeast proteins: One encoded by the nuclear the other by the
mitochondrial genome. *EMBO J.* **5,** 1313–1317.

Pratje, E., Mannhaupt, G., Michaelis, G., and Beyreuther, K. (1983). A nuclear mutation
prevents processing of a mitochondrially encoded membrane protein in *Saccharomyces cerevisiae*. *EMBO J.* **2,** 1049–1054.

Raymond, Y., and Shore, G. C. (1981). Processing of the precursor for the mitochondrial
enzyme, carbamoyl phosphate synthetase. *J. Biol. Chem.* **256,** 2087–2090.

Reid, G. A., and Schatz, G. (1982a). Import of proteins into mitochondria: Yeast cells grown
in the presence of carbonyl cyanide *m*-chlorophenylhydrazone accumulate massive
amounts of some mitochondrial precursor polypeptides. *J. Biol Chem.* **257,** 13056–
13061.

Reid, G. A., and Schatz, G. (1982b). Import of proteins into mitochondria: Extramitochondrial pools and posttranslational import of mitochondrial protein precursors *in vivo*. *J.
Biol. Chem.* **257,** 13062–13067.

Reid, G. A., Yonetani, T., and Schatz, G. (1982). Import of proteins into mitochondria: Import and maturation of the mitochondrial intermembrane space enzymes cytochrome b_2 and cytochrome c peroxidase in intact yeast cells. *J. Biol. Chem.* **257**, 13068–13074.

Ricquier, D., Thibault, J., Bouillaud, F., and Kuster, Y. (1983). Molecular approach to thermogenesis in brown adipose tissue. *J. Biol. Chem.* **258**, 6675–6677.

Ridley, R. G., Patel, H. V., Gerber, G. E., Morton, R. C., and Freeman, K. B. (1986). Complete nucleotide and derived amino acid sequence of cDNA encoding the mitochondrial uncoupling protein of rat brown adipose tissue: Lack of a mitochondrial targeting presequence. *Nucleic Acids Res.* **14**, 4025–4035.

Rietveld, A., and de Kruijff, B. (1984). Is the mitochondrial precursor protein apocytochrome c able to pass a lipid barrier? *J. Biol. Chem.* **259**, 6704–6707.

Rietveld, A., Sijens, P., Verkleij, A. J., and de Kruijff, B. (1983). Interaction of cytochrome c and its precursor apocytochrome c with various phospholipids. *EMBO J.* **2**, 907–913.

Rietveld, A., Ponjee, G. A. E., Schiffers, P., Jordi, W., Van De Coolwijk, P. J. F. M., Demel, R. A., Marsh, D., and de Kruijff, B. (1985). Investigations on the insertion of the mitochondrial precursor protein apocytochrome c into model membranes. *Biochim. Biophys. Acta* **818**, 398–409.

Rietveld, A., Jordi, W., and de Kruijff, B. (1986a). Studies on the lipid dependency and mechanism of the translocation of the mitochondrial precursor protein apocytochrome c across model membranes. *J. Biol. Chem.* **261**, 3846–3856.

Rietveld, A., Berkhout, T. A., Roenhorst, A., Marsh, D., and de Kruijff, B. (1986b). Preferential association of apocytochrome c with negatively charged phospholipids in mixed model membranes. *Biochem. Biophys. Acta* **858**, 38–46.

Riezman, H., Hay, R., Gasser, S., Daum, G., Schneider, G., Witte, C., and Schatz, G. (1983a). The outer membrane of yeast mitochondria: Isolation of outside-out sealed vesicles. *EMBO J.* **2**, 1105–1111.

Riezman, H., Hay, R., Witte, C., Nelson, N., and Schatz, G. (1983b). Yeast mitochondrial outer membrane specifically binds cytoplasmically synthesized precursors of mitochondrial proteins. *EMBO J.* **2**, 1113–1118.

Riezman, H., Hase, T., van Loon, A. P. G. M., Grivell, L. A., Suda, K., and Schatz, G. (1983c). Import of proteins into mitochondria: A 70 kilodalton outer membrane protein with a large carboxy-terminal deletion is still transported to the outer membrane. *EMBO J.* **2**, 2161–2168.

Rödel, G., Körte, A., and Kaudewitz, F. (1985). Mitochondrial suppression of a yeast nuclear mutation which affects the translation of the mitochondrial apocytochrome b transcript. *Curr. Genet.* **9**, 641–648.

Roise, D., Horvath, S. J., Tomich, J. M., Richards, J. H., and Schatz, G. (1986). A chemically synthesized pre-sequence of an imported mitochondrial protein can form an amphiphilic helix and perturb natural and artifical phospholipid bilayers. *EMBO J.* **5**, 1327–1334.

Runswick, M. J., Powell, S. J., Nyren, P., and Walker, J. E. (1987). Sequence of the bovine mitochondrial phosphate carrier protein: Structural relationship to ADP/ATP translocase and the brown fat mitochondria uncoupling protein. *EMBO J.* **6**, 1367–1373.

Sachs, M. S., David, M., Werner, S., and RajBhandary, U. L. (1986). Nuclear genes for cytochrome c oxidase subunits of *Neurospora crassa*. *J. Biol. Chem.* **261**, 869–873.

Sadler, I., Suda, K., Schatz, G., Kaudewitz, F., and Haid, A. (1984). Sequencing of the nuclear gene for the yeast cytochrome c_1 precursor reveals an unusually complex amino-terminal presequence. *EMBO J.* **3**, 2137–2143.

Saltzgaber-Müller, J., and Schatz, G. (1978). Heme is necessary for the accumulation and assembly of cytochrome c oxidase subunits in *Saccharomyces cerevisiae*. *J. Biol. Chem.* **253**, 305–310.

Saraste, M., and Walker, J. E. (1982). Internal sequence repeats and the path of polypeptide in mitochondrial ADP/ATP translocase. *FEBS Lett.* **144**, 250–254.

Schatz, G. (1968). Impaired binding of mitochondrial adenosine triphosphatase in the cytoplasmic "petite" mutant of *Saccharomyces cerevisiae*. *J. Biol. Chem.* **243**, 2192–2199.

Schatz, G. (1979). How mitochondria import proteins from the cytoplasm. *FEBS Lett.* **103**, 203–211.

Schiffer, M., and Edmundson, A. B. (1967). Use of helical wheels to represent the structures of proteins and to identify segments with helical properties. *Biophys. J.* **7**, 121–135.

Schleyer, M., and Neupert, W. (1984). Transport of ADP/ATP carrier into mitochondria: Precursor imported *in vitro* acquires functional properties of the mature protein. *J. Biol. Chem.* **259**, 3487–3491.

Schleyer, M., and Neupert, W. (1985). Transport of proteins into mitochondria: Translocational intermediates spanning contact sites between outer and inner membranes. *Cell* **43**, 339–350.

Schleyer, M., Schmidt, B., and Neupert, W. (1982). Requirement of a membrane potential for the posttranslational transfer of proteins into mitochondria. *Eur. J. Biochem.* **125**, 109–116.

Schmidt, G. W., DeVillers-Thiery, A., Desruisseaux, H., Blobel, G., and Chua, N.-H. (1979). NH$_2$-Terminal amino acid sequences of precursor and mature forms of the ribulose-1,5-bisphosphate carboxylase small subunit from *Chlamydomonas reinhardtii*. *J. Cell Biol.* **83**, 615–622.

Schmidt, B., Hennig, B., Köhler, H., and Neupert, W. (1983a). Transport of the precursor to *Neurospora* ATPase subunit 9 into yeast mitochondria: Implications on the diversity of the transport mechanism. *J. Biol. Chem.* **258**, 4687–4689.

Schmidt, B., Hennig, B., Zimmermann, R., and Neupert, W. (1983b). Biosynthetic pathway of mitochondrial ATPase subunit 9 in *Neurospora crassa*. *J. Cell Biol.* **96**, 248–255.

Schmidt, B., Wachter, E., Sebald, W., and Neupert, W. (1984). Processing peptidase of *Neurospora* mitochondria: Two-step cleavage of imported ATPase subunit 9. *Eur. J. Biochem.* **144**, 581–588.

Schmidt, B., Pfaller, R., Pfanner, N., Schleyer, M., and Neupert, W. (1985). Transport of proteins into mitochondria: Receptors, recognition and transmembrane movement of precursors. *In* "Achievements and Perspectives in Mitochondrial Research" (E. Quagliarello, E. C. Slater, F. Palmieri, C. Saccone, and A. M. Kroon eds.), pp. 389–396. Elsevier, Amsterdam.

Schwaiger, M., Herzog, V., and Neupert, W. (1987). Characterization of translocation sites involved in the import of mitochondrial proteins. *J. Cell Biol.* **105**, 235–246.

Sevarino, K. A., and Poyton, R. O. (1980). Mitochondrial membrane biogenesis: Identification of a precursor to yeast cytochrome c oxidase subunit II, an integral polypeptide. *Proc. Natl. Acad. Sci. U.S.A.* **77**, 142-146.

Shore, G. C., Power, F., Bendayan, M., and Carignan, P. (1981). Biogenesis of a 35-kilodalton protein associated with outer mitochondrial membrane in rat liver. *J. Biol. Chem.* **256**, 8761–8766.

Sidhu, A., and Beattie, D. S. (1983). Kinetics of assembly of complex III into the yeast mitochondrial membrane. *J. Biol. Chem.* **258**, 10649–10656.

Simmaco, M., John, R. A., Barra, D., and Bossa, F. (1986). The primary structure of ornithine aminotransferase: Identification of active-site sequence and site of posttranslational proteolysis. *FEBS Lett.* **199**, 39–42.

Stuart, R. A., Neupert, W., and Tropschug, M.(1987). Deficiency in mRNA splicing in a cytochrome c mutant of *Neurospora crassa:* Importance of carboxy terminus for import of apocytochrome c into mitochondria. *EMBO J.* **6,** 2131–2137.

Suissa, M., and Schatz, G. (1982). Import of proteins into mitochondria: Translatable mRNAs for imported mitochondrial proteins are present in free as well as mitochondria-bound cytoplasmic polysomes. *J. Biol. Chem.* **257,** 13048–13055.

Suissa, M., Suda, K., and Schatz, G. (1984). Isolation of the nuclear yeast genes for citrate synthase and fifteen other mitochondrial proteins by a new screening method. *EMBO J.* **3,** 1773–1781.

Takeda, M., Vassarotti, A., and Douglas, M. G. (1985). Nuclear genes coding the yeast mitochondrial adenosine triphosphatase complex: Primary sequence analysis of *ATP2* encoding the F_1-ATPase β-subunit precursor. *J. Biol. Chem.* **260,** 15458–15465. (Also see correction: *J. Biol. Chem.* **261,** 10466.)

Takiguchi, M., Miura, S., Mori, M., Tatibana, M., Nagata, S., and Kaziro, Y. (1984). Molecular cloning and nucleotide sequence of cDNA for rat ornithine carbamoyltransferase precursor. *Proc. Natl. Acad. Sci. U.S.A.* **81,** 7412–7416.

Taniuchi, H., Basile, G., Taniuchi, M., and Veloso, D. (1982). Solubilization of yeast mitochondrial cytochrome c synthetase. *Fed. Proc., Fed. Am. Soc. Exp. Biol.* **41,** 1209.

Teintze, M., Slaughter, M., Weiss, H., and Neupert, W. (1982). Biogenesis of mitochondrial ubiquinol:cytochrome c reductase (cytochrome bc_1 complex). *J. Biol. Chem.* **257,** 10364–10371.

Tzagoloff, A., and Meagher, P. (1972). Assembly of the mitochondrial membrane system: VI. Mitochondrial synthesis of subunit proteins of the rutamycin-sensitive adenosine triphosphatase. *J. Biol. Chem.* **247,** 594–603.

Urban-Grimal, D., Volland, C., Garnier, T., DeHoux, P., and Labbe-Bois, R., (1986). The nucleotide sequence of the *HEM1* gene and evidence for a precursor form of the mitochondrial 5-aminolevulinate synthase in *Saccharomyces cerevisiae. J. Biochem.* **156,** 511–519.

van den Boogaart, P., van Dijk, S., and Agsteribbe, E. (1982a). The mitochondrially made subunit 2 of *Neurospora crassa* cytochrome aa_3 is synthesized as a precursor protein. *FEBS Lett.* **147,** 97–100.

van den Boogaart, P., Samallo, J., and Agsteribbe, E. (1982b). Similar genes for a mitochondrial ATPase subunit in the nuclear and mitochondrial genomes of *Neurospora crassa. Nature (London)* **298,** 187–189.

van Loon, A. P. G. M., and Schatz, G. (1987). Transport of proteins to the mitochondrial intermembrane space: the "sorting" domain of the cytochrome c_1 presequence is a stop-transfer sequence specific for the mitochondrial inner membrane. *EMBO J.* **6,** 2441–2448.

van Loon, A. P. G. M., and Young, E. T. (1986). Intracellular sorting of alcohol dehydrogenase isozymes in yeast: A cytosolic location reflects absence of an amino-terminal targeting sequence for the mitochondrion. *EMBO J.* **5,** 161–165.

van Loon, A. P. G. M., van Eijk, E., and Grivell, L. A. (1983a). Biosynthesis of the ubiquinol–cytochrome c reductase complex in yeast. Discoordinate synthesis of the 11-kd subunit in response to increased gene copy number. *EMBO J.* **2,** 1765–1770.

van Loon, A. P. G. M., Maarse, A. C., Riezman, H., and Grivell, L. A. (1983b). Isolation, characterization and regulation of expression of the nuclear genes of the core II and Rieske iron–sulfur proteins of the yeast ubiquinol–cytochrome c reductase. *Gene* **26,** 261–272.

van Loon, A. P. G. M., Kreike, J., de Ronde, A., van der Horst, G. T. J., Gasser, S. M., and Grivell, L. A. (1983c). Biosynthesis of the ubiquinol–cytochrome c reductase complex in yeast: Characterization of precursor forms of the 44-kDa, 40-kDa and 17-kDa subunits and identification of individual messenger RNAs for these and other imported subunits of the complex. *Eur. J. Biochem.* **135**, 457–463.

van Loon, A. P. G. M., De Groot, R. J., De Haan, M., Dekker, A., and Grivell, L. A. (1984). The DNA sequence of the nuclear gene coding for the 17-kd subunit VI of the yeast ubiquinol–cytochrome *c* reductase: A protein with an extremely high content of acidic amino acids. *EMBO J.* **3**, 1039–1043.

van Loon, A. P. G. M., Brändli, A. W., and Schatz, G. (1986). The presequences of two imported mitochondrial proteins contain information for intracellular and intramitochondrial sorting. *Cell* **44**, 801–812.

van Loon, A. P. G. M., Brändli, A. W., Peshold-Hurt, B., Blank, D., and Schatz, G. (1987). Transport of proteins to the mitochondrial intermembrane space: The "matrix-targeting" and the "sorting" domains in the cytochrome c_1 presequence. *EMBO J.* **6**, 2433–2439.

van Steeg, H., Oudshoorn, P., van Hell, B., Polman, J. E. M., and Grivell, L. A. (1986). Targeting efficiency of a mitochondrial pre-sequence is dependent on the passenger protein. *EMBO J.* **5**, 3643–3650.

van Venetie, R., and Verkleij, A. K. (1982). Possible role of nonbilayer lipids in the structure of mitochondria: A freeze–fracture electron microscopy analysis. *Biochem. Biophys. Acta* **692**, 397–405.

Vassarotti, A., Chen, W.-J., Smagula, C., and Douglas, M. G. (1987a). Sequences distal to the mitochondrial targeting sequences are necessary for the maturation of the F_1-ATPase β-subunit precursor in mitochondria. *J. Biol. Chem.* **262**, 411–418.

Vassarotti, A., Stroud, R., and Douglas, M. G. (1987b). Independent mutations at the amino terminus of a protein act as surrogate signals for mitochondrial import. *EMBO J.* **6**, 705–711.

Verner, K., and Schatz, G. (1987). Import of an incompletely folded precursor protein into isolated mitochondria requires an energized inner membrane, but no added ATP. *EMBO J.* **6**, 2449–2456.

Viebrock, A., Perz, A., and Sebald, W. (1982). The imported preprotein of the proteolipid subunit of the mitochondrial ATP synthase from *Neurospora crassa*. Molecular cloning and sequencing of the mRNA. *EMBO J.* **1**, 565–571.

von Heijne, G. (1986a). Mitochondrial targeting sequences may form amphiphilic helices. *EMBO J.* **5**, 1335–1342.

von Heijne, G. (1986b). Why mitochondria need a genome. *FEBS Lett.* **198**, 1–4.

Watanabe, K., and Kubo, S. (1982). Mitochondrial adenylate kinase from chicken liver: Purification, characterization and its cell-free synthesis. *Eur. J. Biochem.* **123**, 587–592.

Weiss, H., and Kolb, H. J. (1979). Isolation of mitochondrial succinate:ubiquinone reductase, cytochrome *c* reductase and cytochrome *c* oxidase from *Neurospora crassa* using nonionic detergents. *Eur. J. Biochem.* **99**, 139–149.

Weiss-Brummer, B., Rödel, G., Schweyen, R. J., and Kaudewitz, F. (1982). Expression of the split gene *cob* in yeast: Evidence for a precursor of a "maturase" protein translated from inton 4 and preceeding exons. *Cell* **29**, 527–536.

Werner, S., Schwab, A. J., and Neupert, W. (1974). Precursors of cytochome oxidase in cytochrome-oxidase-deficient cells of *Neurospora crassa:* Comparison of the nuclear mutant *cni-1*, the cytoplasmic mutant *mi-1*, and copper-depleted wild type. *Eur. J. Biochem.* **49**, 607–617.

Wielburski, A., and Nelson, B. D. (1984). Heme *a* induces assembly of rat liver cytochrome *c* oxidase subunits I–III in isolated mitochondria. *FEBS Lett.* **177**, 291–294.

Wielburski, A., Kuzela, S., and Nelson, B. D. (1982). Studies on the assembly of cytochrome oxidase in isolated rat hepatocytes. *Biochem. J.* **204**, 239–245.

Woodrow, G., and Schatz, G. (1979). The role of oxygen in the biosynthesis of cytochrome *c* oxidase of yeast mitochondria. *J. Biol. Chem.* **254,** 6088–6093.

Wright, R. M., Ko, C., Cumsky, M. G., and Poyton, R. O. (1984). Isolation and sequence of the structural gene for cytochrome *c* oxidase subunit VI from *Saccharomyces cerevisiae. J. Biol. Chem.* **259,** 15401–15407.

Yaffe, M. P., and Schatz, G. (1984). Two nuclear mutations that block mitochondrial protein import in yeast. *Proc. Natl. Acad. Sci. U.S.A.* **81,** 4819–4823.

Yaffe, M. P., Ohta, S., and Schatz, G. (1985). A yeast mutant temperature-sensitive for mitochondrial assembly is deficient in a mitochondrial protease activity that cleaves imported precursor polypeptides. *EMBO J.* **4,** 2069–2074.

Young, E. T., and Pilgrim, D. B. (1985). Isolation and DNA sequence of *ADH3*, a nuclear gene encoding the mitochondrial isozyme of alcohol dehydrogenase in *Saccharomyces cerevisiae. Mol. Cell. Biol.* **5,** 3024–3034.

Zimmermann, R., and Neupert, W. (1980). Transport of proteins into mitochondria: Post-translational transfer of ADP/ATP carrier into mitochondria *in vitro. Eur. J. Biochem.* **109,** 217–229.

Zimmermann, R., Paluch, U., and Neupert, W. (1979a). Cell-free synthesis of cytochrome *c. FEBS Lett.* **108,** 141–146.

Zimmermann, R., Paluch, U., Sprinzl, M., and Neupert, W. (1979b). Cell-free synthesis of the mitochondrial ADP/ATP carrier protein of *Neurospora crassa. Eur. J. Biochem.* **99,** 247–252.

Zimmermann, R., Hennig, B., and Neupert, W. (1981). Different transport pathways of individual precursor proteins in mitochondria. *Eur. J. Biochem.* **116,** 455–460.

Zwizinski, C., and Neupert, W. (1983). Precursor proteins are transported into mitochondria in the absence of proteolytic cleavage of the additional sequences. *J. Biol. Chem.* **258,** 13340–13346.

Zwizinski, C., Schleyer, M., and Neupert, W. (1983). Transfer of proteins into mitochondria: Precursor to the ADP/ATP carrier binds to receptor sites on isolated mitochondria. *J. Biol. Chem.* **258,** 4071–4074.

Zwizinski, C., Schleyer, M., and Neupert, W. (1984). Proteinaceous receptors for the import of mitochondrial precursor proteins. *J. Biol. Chem.* **259,** 7850–7856.

VI

Nuclear Transport

18

Transport of Proteins into the Nucleus

PAMELA A. SILVER AND MICHAEL N. HALL

I. INTRODUCTION

In this chapter, we summarize what is known about how certain proteins accumulate specifically in the cell nucleus and the way in which these proteins are transported across the double membrane that divides the nuclear interior from the cytoplasm. In the second part of the chapter, we describe in more detail results from studies of nuclear protein localization in the yeast, *Saccharomyces cerevisiae*.

749

PROTEIN TRANSFER AND ORGANELLE BIOGENESIS

II. NUCLEAR PORE

The nucleus is surrounded by an envelope composed of two membrane bilayers. The presence of pores which traverse the envelope distinguishes the nuclear envelope from its counterpart in other organelles. These nuclear structures have been identified in all eukaryotes examined. A possible arrangement of the proteins composing the pores has been proposed (Franke *et al.*, 1981), although only a very few pore-associated proteins have been identified (Gerace *et al.*, 1982; Davis and Blobel, 1986). The pore complex is a grommetlike structure with a diameter of approximately 1200 Å (Unwin and Milligan, 1982). The actual pore, or aqueous channel, is thought to have a diameter of approximately 90 Å (Paine *et al.*, 1975). The mere presence of the nuclear pore has led to numerous proposals that it is the site of entry into and exit from the nucleus of proteins and nucleic acids.

Why does the nucleus require specialized pores as sites of nucleocytoplasmic exchange? As in all organelles, nuclear proteins must cross the membranous barrier presented by the nuclear envelope. Other organelles have apparently acquired the property that proteins may pass through their surrounding membranes in the absence of structures as large and distinct as the nuclear pore. Although pores have been postulated to play a role in transit of proteins into the lumen of the endoplasmic reticulum (reviewed in Walter *et al.*, 1984) very little experimental evidence exists to support the existence of such structures. The nuclear pore, on the other hand, is clearly visible by electron microscopy and is present in large numbers on most nuclei. Perhaps the nuclear pore evolved to accommodate the heavy two-way traffic of very large, water-soluble molecules across the nuclear envelope (Bonner, 1978). A rapidly growing cell must transport approximately 100 histone molecules into the nucleus per minute per pore (Alberts *et al.*, 1983). The same cell must transport about three newly assembled ribosomes out of the nucleus per minute per pore. This nucleocytoplasmic exchange, not even considering transport of RNA, is a small fraction of the total traffic across the nuclear envelope.

Experimental evidence has recently demonstrated that proteins do actually utilize the nuclear pore to gain access to the nuclear interior. Feldherr *et al.* (1984) prepared nucleoplasmin, one of the major proteins of the *Xenopus* oocyte, conjugated with colloidal gold. After injection into the *Xenopus* oocyte cytoplasm, the fate of the gold-conjugated nucleoplasmin was determined by electron microscopy. The gold particles could be seen in the nucleus and at the nuclear envelope in the pores. However, this result together with other data on the physical properties of the nuclear pore provides us with a paradox. The functional pore diameter has been estimated to be about 90 Å, and the molecular weight cutoff for particle diffusion into the nucleus has been estimated to be about 70 K. Some of the gold-conjugated nucleoplasmin particles were shown to be as large as 170 Å in diameter. In addition, we know that the nucleus contains many very large proteins (greater that 100 kDa). How do these proteins enter the nucleus through a channel that, at face value, appears to be too small? In the following section, we discuss possible mechanisms for specific nuclear protein localization and some of the experiments which support these models.

III. POSSIBLE MECHANISMS FOR SPECIFIC NUCLEAR PROTEIN LOCALIZATION

A. Passive Retention versus Active Transport

Several mechanisms have been proposed for the specific localization of proteins to the cell nucleus (see also Bonner, 1978; DeRobertis, 1983). According to one model, nuclear association is the result of selective retention of a subset of proteins within the nucleus. Entry into the nucleus is by simple diffusion through the nuclear pore. Subsequent nuclear stabilization could be the consequence of binding to a nondiffusible nuclear substrate such as DNA. Increased stability of certain proteins in the nuclear interior could also add to the apparent accumulation of these proteins in the nucleus. Thus, if a normally nuclear protein is not sequestered within the nucleus, it may be selectively degraded in the cytoplasm. However, this possible mechanism is not sufficient to explain how very large proteins such as RNA polymerase are transported across the nuclear envelope to be retained in the nucleus. As discussed above, proteins of very high molecular weight (greater than 70 K) would not be expected to diffuse readily through the nuclear pore.

By an alternative mechanism, receptors on the nuclear envelope may recognize specific amino acid sequences or tertiary structures of certain proteins and thereby allow passage of only these proteins across the nuclear envelope. The receptor could even be a cytoplasmic component that interacts transiently with a newly synthesized nuclear protein in a manner similar to that by which signal recognition particle (SRP) (Walter et al., 1984) targets nascent secreted proteins to the endoplasmic reticulum. Specific information for nuclear localization would then reside in the transported protein as well as in the transport apparatus. A specific transporter might require energy for transport across the nuclear envelope. Consistent with this suggestion, depletion of ATP blocks nuclear protein accumulation in *Xenopus* oocytes (Newmeyer *et al.*, 1986). Similarly, ATP enhances the uptake of dextrans (64K apparent molecular weight) by isolated nuclei (Schindler and Jiang, 1986). A mechanism proposing a specific transport system is distinguished from the first model discussed because it suggests selective translocation into the nucleus rather than passive retention of a subset of proteins in the nuclear interior.

B. Nuclear Protein Localization Is Not Necessarily Cotranslational

Much of what we know about translocation of proteins into the nucleus is the result of experiments with *Xenopus* oocytes. When injected into the *Xenopus* oocyte cytoplasm, nuclear proteins such as histones move to the nucleus (Bonner, 1975). This indicates that nuclear association of these particular proteins is not necessarily coupled to protein synthesis. In addition, no obvious posttranslational processing is required for nuclear protein uptake. Four nuclear proteins from *Xenopus* oocytes, when synthesized *in vitro*, are identical to the same proteins present in nuclei from living oocytes (Dabauvalle and Franke, 1982). These proteins synthesized *in vitro* readily accumulate in nuclei following injection into the cytoplasm of the *Xenopus* oocyte. From a practical standpoint, the apparent lack of protein modification as a measureable event complicates the study of nuclear protein localization. With no processing event to follow, nuclear localization studies must rely more on criteria such as relative rate and extent of nuclear uptake of a particular protein.

Like histones, intact nucleoplasmin will localize to the nucleus after injection into the *Xenopus* oocyte cytoplasm. Dingwall *et al.* (1982) have further demonstrated that nucleoplasmin lacking its carboxy-terminal "tail" does not enter the nucleus from the cytoplasm, yet remains nuclear when injected directly into the nucleus. Moreoever, the carboxy-terminal peptide alone is efficiently tranported into the nucleus. Thus, the carboxy terminus of nucleoplasmin is necessary and probably sufficient to direct passage of the complete protein into the nucleus, indicating that specific amino acid sequences may act as nuclear determinants. Thus, selective nuclear retention as the sole mechanism of nuclear accumulation of nucleoplasmin can be ruled out because the localization-defective amino-terminal peptide of nucleoplasmin remains in the nucleus if it is injected into the nuclear interior.

Similarly, proteolytic fragments of histone H1 were examined for intracellular localization following microinjection into the *Xenopus* oocyte cytoplasm (Dingwall and Allan, 1984). There are three structural domains of H1; an amino-terminal extension of 36 amino acids, a globular core extending from amino acid 36 to 121, and a carboxy-terminal region extending from amino acid 122 to 210. As in the case of nucleoplasmin, only a fragment containing the carboxy-terminal domain of H1 was capable of moving from the cytoplasm to the nucleus following injection into the cytoplasm.

C. Short Amino Acid Sequences Act as Nuclear Targeting Signals

The experiments described above suggest that nuclear proteins contain specific targeting signals. Two types of experiments have further defined regions of proteins involved in their nuclear localization. In one case, mutations that cause single amino acid changes have delineated stretches of amino acids necessary for nuclear targeting. In the other case, short stretches of amino acids from a nuclear protein have been shown to be sufficient to target a normally nonnuclear protein to the nucleus.

Although the finding that a short amino acid sequence is capable of mediating nuclear localization was first described in the studies on the yeast nuclear protein $\alpha2$ (see Section IV,B), the two types of experiments discussed above are best exemplified by work done on SV40 T antigen. The large T antigen of SV40 is normally found predominantly in the nucleus of infected mammalian cells (Pope and Rowe, 1964). The nuclear location of T antigen is consistent with its role in viral gene expression and DNA replication (Tooze, 1980). Its size (M_r 94,000) should render it too large to enter the nucleus simply by diffusion through the nuclear pore, suggesting that a specific transport system may be at work. Lanford and Butel (1984) and Kalderon et al. (1984a) found that a single amino acid change at lysine-128 of T antigen renders the protein nonnuclear in vivo. This mutant protein has specific DNA binding ability in vitro, indicating that it retains at least some regions of normal conformation (Paucha et al., 1985). Mutations resulting in conversion of lysine-128 to either a threonine (Kalderon et al., 1984a) or an asparagine residue (Lanford and Butel, 1984) result in nonnuclear localization of SV40 T antigen, as determined by indirect immunofluorescence. Mutations at a neighboring lysine and arginine also cause cytoplasmic T antigen to accumulate in the cytoplasm but do not prevent entry into the nucleus. Interestingly, the mutations at lysine-128 confer a dominant negative phenotype by preventing nuclear uptake of wild-type T antigen (Landford and Butel, 1984). Possibly the mutant T antigen oligomerizes with the wild-type protein in the cytoplasm and inhibits further transport into the nucleus.

The findings from analysis of mutant T antigens led to the suggestion that amino acids 126 through 132 with the sequence Pro-Lys-Lys-Lys-Arg-Lys-Val direct the protein to the nucleus. This prediction was confirmed by the demonstration that this amino acid sequence is sufficient to cause nuclear localization of normally nonnuclear proteins in mammalian cells (Kalderon et al., 1984b). These experiments were performed by constructing gene fusions between the portion of the T antigen gene encoding the nuclear targeting sequence and the gene encoding pyruvate kinase or *Escherichia coli* β-galactosidase.

Thus, by two criteria, mutational analysis and gene fusions, amino acids 126 through 132 of T antigen (see Fig. 1) are both necessary and sufficient as a localization signal. However, Welsh et al. (1986) have reported that changes at amino acids 185, 186, and 199 of T antigen result in mutant T antigen that accumulated in both the nucleus and the cytoplasm. This second region of T antigen which also appears to be implicated in nuclear localization does not resemble in its primary sequence the signal at amino acids 126 through 132. Why T antigen that is altered at positions 185, 186, or 199 is improperly localized despite having an intact localization signal at amino acids 126 through 132 which has been shown to be sufficient for localization is not clear. Perhaps T antigen has two signals but the second, defined by the changes at positions 185, 186, and 199, is not essential under all conditions.

The findings of Lanford and Butel (1984) and Kalderon et al. (1984a,b) have led to searches in known nuclear proteins for sequences homologous to the T antigen sequence that may act to target these proteins to the nucleus. Polyoma large T antigen, 785 amino acids, contains a similar sequence around amino acid 282 (Richardson et al., 1986). However, fragments of polyoma large T containing this sequence are unable to confer a nuclear location on pyruvate kinase. A second stretch of amino acids in polyoma T antigen around amino acid 192 is capable of conferring a nuclear location to pyruvate kinase. Deleting both of these sequences results in nonnuclear polyoma large T; deletion of either one has only a partial effect on the nuclear localization of large T. Apparently, unlike SV40 T antigen, polyoma T antigen may contain two nuclear localization signals.

Goldfarb et al. (1986) have examined the role of a nuclear targeting peptide in the rate of nuclear transport. They synthesized a peptide that contains the sequence shown to be sufficient to target SV40 T antigen to

$$\text{Pro}^{126} - \text{Lys} - \text{Lys} - \text{Lys} - \text{Arg} - \text{Lys} - \text{Val}^{132}$$

$$\downarrow$$

Thr or Asn

Fig. 1. Proposed nuclear localization sequence from SV40 T antigen with amino acid changes at Lys[128] which alter T antigen nuclear localization.

the nucleus (Fig. 1). The peptide was cross-linked to bovine serum albumin (BSA) and immunoglobulin G (IgG). When microinjected into *Xenopus* oocytes, the conjugated proteins accumulated in the nucleus. The nuclear uptake of the BSA conjugated with this peptide was saturable, and the rate of uptake was reduced by coinjection of the free peptide. Furthermore, similar conjugates with a peptide containing a threonine at the position equivalent to lysine-128 in SV40 T antigen also led to nuclear uptake of BSA and IgG, but at a rate six times lower than that of the wild-type peptide sequence. In a similar series of experiments, Lanford *et al.* (1986) made conjugates between a similar peptide and proteins as large as ferritin (M_r 465,000) and injected them into mammalian cells. Even the ferritin conjugate was efficiently transported into the nucleus. The rate of nuclear uptake, but not the final nuclear protein concentration, was influenced by the number of nuclear target peptides conjugated to the protein. Certainly, these experiments lend support to the idea of a specific nuclear uptake apparatus rather than passive retention of proteins in the nuclear interior.

IV. NUCLEAR PROTEIN LOCALIZATION
IN THE YEAST *Saccharomyces cerevisiae*

The yeast *S. cerevisiae* has been successfully used to study protein localization. We now understand in great detail how proteins assemble into the yeast mitochondria from the studies of Schatz and co-workers (reviewed in Schatz and Butow, 1983). They were able to reconstruct *in vitro* the process of protein import into isolated mitochondria and to generate yeast mutants defective in mitochondrial protein import (Yaffe and Schatz, 1984). Schekman and co-workers (Novick *et al.*, 1980, 1981) have successfully dissected some of the steps involved in protein secretion from yeast. Both of these studies demonstrate that yeast is the organism of choice for identifying genes essential for protein transport systems. More recently, studies have begun on the mechanism of nuclear protein localization in yeast. Because of the obvious potential in using the genetics of yeast to study this problem, we shall use the remainder of this chapter to review some of the lessons learned, thus far, from these studies.

The nuclear envelope of *S. cerevisiae* is indistinguishable from its mammilian counterpart during interphase. It is composed of a double membrane that is visible by electron microscopy (Fig. 2). Nuclear pores are present and are best observed by freeze–fracture electron microscopy (Willison and Johnston, 1978). For the purposes of the studies described

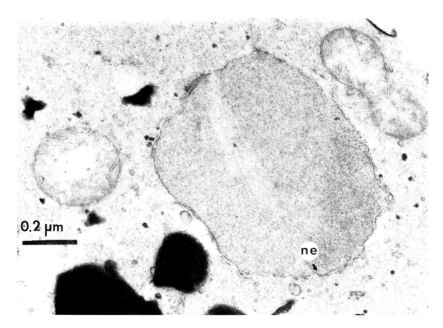

Fig. 2. Thin section from *Saccharomyces cerevisiae* examined by electron microscopy. ne, nuclear envelope.

here, it is probably safe to assume that there is little difference between transport of proteins into the yeast nucleus and into the nucleus of higher eukaryotes. Interestingly, the yeast nuclear envelope (as well as the nuclear envelope of most other fungi) remains visibly intact throughout mitosis. Thus, yeast nuclear proteins must always traverse the barrier of the nuclear envelope in order to gain access to the nuclear interior. We know very little about the composition of the yeast nuclear envelope because few, if any, of its proteins have been identified.

The experiments described below employ fusions between a gene encoding a yeast protein and the *E. coli lacZ* gene. It is worth discussing why the product of the *E. coli lacZ* gene, β-galactosidase, has been chosen as a nonnuclear component of the chimeric proteins used in these and many other studies of this type. It is a soluble protein of the *E. coli* cytoplasm, and is not specifically localized to any subcellular organelle in yeast. Its hydrophilicity should prevent direct passage through membranes, and its size (116 kDa) might limit its ability to diffuse through the nuclear pore. It has been demonstrated that fusions between yeast genes and *lacZ*, when introduced into yeast, produce active β-galactosidase (Rose *et al.*, 1981). The presence of β-galactosidase then provides both a convenient biochemical and immunological tag.

The approach of using protein fusions to study protein movement was initiated by Beckwith and colleagues to study protein secretion in *E. coli* (reviewed in Silhavy and Beckwith, 1985). They demonstrated that protein fusions containing β-galactosidase and the signal peptide of an exported protein are directed out of the *E. coli* cytoplasm. This finding has enabled definition of the signals in certain exported proteins which are sufficent for conferring extracytoplasmic localization on the normally cytoplasmic β-galactosidase. In addition, phenotypes unique to cells containing aberrant hybrid proteins have been exploited to isolate export-defective mutants (Emr and Silhavy, 1980; Oliver and Beckwith, 1981). The same kind of approach has been used in studies of nuclear protein localization in yeast.

The analyses discussed in the following sections examine a specific function, nuclear localization, of mutant hybrid proteins containing different portions of a nuclear protein fused to β-galactosidase. Loss of function as a result of a particular deletion or as a result of the position of a fusion joint, could therefore be due to indirect effects transmitted through protein conformation, and might not be due to removal of a sequence mediating the function being assayed. Although this problem is partly circumvented by a systematic analysis, any conclusions must be tempered by the caveat that loss of function could be due to improper protein folding. This problem is perhaps more severe in studying nuclear localization than in studying protein export by *E. coli* since nuclear targeting signals appear to be an integral part of mature proteins.

A. Yeast GAL4 Protein Contains a Nuclear Determinant

As a positive activator protein of the yeast genes *GAL1*, *GAL7*, and *GAL10*, necessary for galactose metabolism in *S. cerevisiae* (Douglas and Hawthorne, 1966), the *GAL4* gene product interacts directly with DNA (Giniger *et al.*, 1985) and hence is a nuclear protein. The *GAL4* gene has been cloned and shown to encode a 99K protein (Laughon and Gesteland, 1984). The protein mediates positive control by binding to specific sites on the DNA upstream of the starting site of transcription, and thereby stimulating RNA polymerase II-dependent transcription (Keegan *et al.*, 1986).

The 74 amino-terminal amino acids of the *GAL4* gene product are sufficient to localize a GAL4–β-galactosidase chimeric protein to the yeast nucleus (Silver *et al.*, 1984). Gene fusions with various portions of the *GAL4* gene fused to the 5' end of the *E. coli lacZ* gene were constructed and introduced into yeast cells. The resulting gene fusions, when placed in yeast, yield chimeric proteins of the predicted size that have β-galactosidase activity. The intracellular location of the chimeric proteins was determined by indirect immunofluorescence using anti-β-galactosidase antibody and cell fractionation. Intact β-galactosidase is not localized to any intracellular compartment when produced in yeast. On the other hand, when portions of the GAL4 amino terminus are fused to β-galactosidase, the β-galactosidase is predominantly associated with the cell nucleus (Fig. 3). By this analysis, the 74 amino-terminal amino acids of the total 881 GAL4 amino acids are sufficient to redirect the intracellular location of the β-galactosidase to the nucleus. Chimeric proteins lacking this nuclear determinant and beginning at amino acid 79 of GAL4 are apparently excluded from the nucleus. Although it is very lysine rich, the amino acid sequence of the GAL4 amino terminus does not agree precisely with the previously proposed consensus sequence derived from SV40 (Fig. 1) and polyoma T antigen.

An unexpected finding from the analysis of the GAL4–β-galactosidase fusion proteins in yeast was that the chimeric proteins lacking the 78 amino-terminal amino acids of GAL4, which do not accumulate in the nucleus, were produced at higher levels than the appropriately transported GAL4–β-galactosidase fusion protein. It is possible that nuclear transport somehow limits the amount of protein that is made. For example, there may be a limited number of nuclear envelope-associated receptors that recognize a nuclear determinant in the 74 GAL4 amino-terminal amino acids. Once these sites are saturated, excess synthesized protein that can no longer bind to or translocate into the nucleus may inhibit further synthesis or simply be degraded.

Mutations in the GAL4 portion of the *GAL4–lacZ* gene fusion have been isolated that result in increased production of GAL4–β-galactosidase (Silver *et al.*, 1987). The mutations are single base substitutions that result in single amino acid changes in the GAL4 portion of the protein. Two of these mutations at Ser[6] and Lys[20], cause significant nonnuclear accumulation of the normally nuclear protein. Because these mutations cause changes in amino acids 6 and 20, it was possible that fewer than 74 amino-terminal amino acids were sufficient for nuclear location of GAL4–β-galactosidase hybrids. A fusion between only the first 62 amino-terminal amino acids of GAL4 and β-galactosidase is not nuclear localized, however, but rather appears distributed throughout the cell.

Do these three mutations define one or more nuclear targeting sequences in GAL4? The effect of these mutations is to cause nonnuclear protein to accumulate. This result could be explained by a mutant protein that fails to efficiently interact with a transport apparatus. Alternatively, the mutant proteins may be taken up into the nucleus normally but fail to be retained within the nuclear interior. Finally, the mutant proteins could be transported to the nucleus more slowly than wild-type protein. In any

A

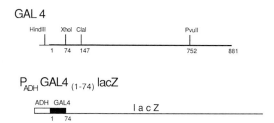

GAL 4

HindIII   XhoI ClaI   PvuII

1   74   147   752   881

P_{ADH} GAL4 $_{(1-74)}$ lacZ

ADH GAL4

l a c Z

1   74

B

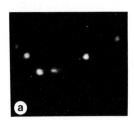
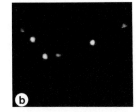
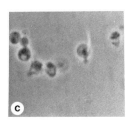

a   b   c

Fig. 3. Immunofluorescence of yeast cells producing nuclear-associated GAL4–β-galactosidase proteins, from a gene fusion P_{ADH}–*GAL4*(1–74)–*lacZ*. (A) Restriction map of the *GAL4* gene and of the *GAL4–lacZ* gene fusion. ADH, Promoter from the yeast alcohol dehydrogenase gene. The number below each restriction site represents the amino acid codon at which it occurs. (B) Cells were prepared for indirect immunofluorescence and treated with either mouse anit-β-galactosidase antibody, followed by fluorescein isothiocyante (FITC)-conjugated anti-mouse IgG (a) to identify the chimeric proteins, or 4',6-diamidino-2-phenylindole (DAPI) (b) to identify the cell nuclei. (c) The same cells photographed with phase contrast.

case, the mutations are spread throughout the 74 amino acid region rather than clustered around a certain stretch of amino acids. Either there is an as yet unidentified shorter nuclear target sequence in GAL4 or the entire 74 amino-terminal amino acids are necessary and sufficient for nuclear localization.

B. α2 Protein Contains Two Nuclear Localization Sequences

The α2 protein, the product of the *MATα2* gene, is a DNA-binding protein which regulates cell type in the yeast *S. cerevisiae*. It represses transcription of a group of cell type-specific genes by binding specifically to an operator located upstream of each target gene (Johnson and Herskowitz, 1985). The α2 monomer, 210 amino acids, has a molecular weight of approximately 25K. A cell that is defective for α2 is viable, but it is unable to mate as can a wild-type cell (Strathern *et al.*, 1981). The gene fusion approach described above has been employed to define nuclear localization signals in α2 (Hall *et al.*, 1984). An extensive series of gene fusions encoding hybrid proteins containing different portions of the α2 protein fused to β-galactosidase was constructed (see Fig. 4).

The cellular location of the different hybrid proteins was determined by indirect immunofluorescence on whole fixed cells. The results of this study are summarized in Fig. 4. According to these results, the α2 protein has two localization determinants, either one of which is sufficient for directing β-galactosidase to the nucleus. One of the signals is in the amino-terminal 13 amino acids. This follows from the observation that a hybrid protein containing only these 13 amino acids of α2 fused to β-galactosidase is targeted to the nucleus. A second signal resides, at least in part, in amino acids 141 through 152. This is based on the following two observations (M. N. Hall, C. Craik, G. Mullenbach, and Y. Hiraoka,

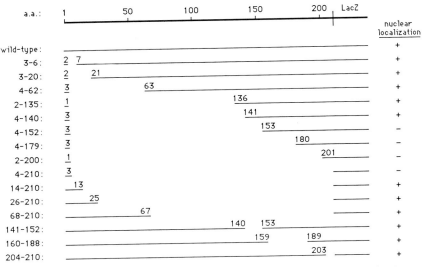

Fig. 4. Schematic of the α2 portion of hybrid proteins encoded by *MATα2–lacZ* gene fusions. The hybrid proteins contain different portions of α2 protein at the amino terminus and a constant, enzymatically active portion of β-galactosidase (not drawn to scale) at the carboxy terminus. The wild-type hybrid protein contains all 210 amino acids of α2. The numbers in the left-hand column refer to the amino acids of α2 that are missing. The number at each deletion end point refers to the α2 amino acid at that position.

unpublished). (1) A hybrid protein lacking amino acids 4 through 140 is localized to the nucleus whereas a protein missing 12 more amino acids, amino acids 4 through 152, is dispersed throughout the cell. (2) A protein missing only amino acids 141 through 152, a deletion of the interval defined by the two deletions mentioned above, exhibits altered localization properties. Hybrid proteins missing these 12 amino acids appear to accumulate at discrete sites, presumably pores, in the nuclear envelope (although this is not wild-type localization it is still considered localization for the purpose of Fig. 4). Why this phenotype is uniquely conferred by a deletion of amino acids 141 through 152 is not understood. However, the observation does suggest that the deleted region is implicated in the localization process. The accumulation in the envelope in this case was determined by a recently developed computer-assisted optical sectioning technique on whole cells in which the hybrid protein is visualized by indirect immunofluoresence (Y. Hiraoka and M. N. Hall, unpublished).

Neither one of the two regions of α2 shown to contain a localization determinant exhibits any homology with the SV40 T antigen signal (see Section III,C), nor do the two signals of α2 resemble each other. However, comparison of amino acid sequences of other yeast proteins with the amino-terminal 13 residues of α2 reveals a sequence that might be important for nuclear targeting, Lys³-Ile-Pro-Lys⁷. A similar sequence of two positively charged amino acids flanking three hydrophobic residues, one of which is proline, is present in several other yeast nuclear proteins (see Fig. 5) (Hall *et al.*, 1984). This sequence is not present in any yeast cytoplasmic proteins currently known. A region of influenza virus nucleoprotein which has been shown to mediate nuclear accumulation contains a similar sequence (Davey *et al.*, 1985).

How do the two signals in α2 mediate the observed nuclear localization? Formal possibilities for the function of either one or both of the signals are (1) a DNA binding domain, (2) a multimerization domain, or (3) a "true" localization signal. These three possibilities are considered individually below.

According to the passive diffusion and selective retention model of nuclear localization (see Section III,A), the function of the signals would be to bind DNA, the intranuclear substrate of α2. This prediction was tested by assaying the different hybrid proteins for nonspecific DNA binding (Hall and Johnson, 1986; M. N. Hall and A. Johnson, unpublished). Determination of nonspecific DNA binding is appropriate for this study, even though α2 is a specific DNA binding protein, since specific binding alone would not be sufficient to retain α2 in the nucleus because of the relative scarcity of specific binding sites. The results of the test reveal that there is no correlation between nuclear localization of the hybrid proteins

```
                          +    HYDROPHOBIC    +

       MET ASN   | LYS   ILE PRO ILE   LYS |  ASP LEU LEU ASN PRO GLN    α2 (210 A.A.)
                 |  3                    7 |
   MET PHE THR SER | LYS   PRO ALA PHE   LYS |  ILE LYS ASN LYS ALA SER    α1 (175 A.A.)
                 |  5                    9 |
   PRO ALA GLU LYS | LYS   PRO ALA ALA   LYS |  LYS THR SER THR SER THR    H2B-1 (131 A.A.)
                 | 18                   22 |
   GLN GLY ILE THR | LYS   PRO ALA ILE   ARG |  ARG LEU ALA ARG ARG GLY    H4 (103 A.A.)
                 | 32                   36 |
   ASN LYS LYS THR | ARG   ILE ILE PRO   ARG |  HIS LEU GLN LEU ALA ILE    H2A (132 A.A.)
                 | 79                   83 |
   SER VAL TYR GLU | LYS   PHE ALA PRO   LYS |  GLY LYS GLN LEU SER MET    RAD52 (504 A.A.)
                 |411                  415 |
   HIS ALA PRO LEU | LYS   PRO VAL VAL   ARG |  LYS LYS PRO GLU PRO ILE    SPT2 (333 A.A.)
                 |112                  116 |
```

Fig. 5. Sequence present within the amino-terminal 13 residues of the $\alpha2$ protein and in the other known or presumed nuclear proteins $\alpha1$ (Astell *et al.,* 1981), H2B-1 (Wallis *et al.,* 1980), H4 (Smith and Andresson, 1983), H2A (Choe *et al.,* 1982), RAD52 (Adzuma *et al.,* 1984), and SPT2 (Roeder *et al.,* 1985). Numbers below an amino acid indicate the position of the amino acid within the respective protein. Numbers in parentheses indicate the total number of amino acids in the designated proteins.

and an ability to bind DNA *in vitro,* indicating that DNA binding is neither necessary nor sufficient for nuclear localization (see Section IV,D). Assuming that $\alpha2$ does not have a second nondiffusible intranuclear substrate which could retain $\alpha2$ in the absence of DNA binding, the above results argue against passive diffusion and retention as the mechanism of nuclear localization.

Multimeric nuclear proteins are thought to assemble in the cytoplasm before entering the nucleus (Dingwall *et al.,* 1982; Kauffman and Ginsberg, 1976). Furthermore, studies on *Xenopus* nucleoplasmin suggest that a localization-defective subunit can piggyback into the nucleus by multimerizing with a wild-type subunit (Dingwall *et al.,* 1982). Accordingly, one of the localization signals in $\alpha2$ could be a multimerization domain which enables a mutant subunit to piggyback into the nucleus. This possibility can be ruled out in the case of $\alpha2$ since none of the hybrid proteins containing only one of the two signals requires the presence of wild-type $\alpha2$ to be delivered to the nucleus (M. N. Hall, unpublished).

Both signals in $\alpha2$ appear to be, but only for lack of evidence to the contrary, "true" localization signals. Presumably, they interact with a receptor in the localization pathway (see Section III,A). Why are there two signals? The two signals could (1) act independently and redundantly,

(2) act at different steps in a localization pathway, or (3) act cooperatively at the same step(s). Determining which of these possibilities is correct will require identification of the cellular components with which the signals interact. The observation that a hybrid protein which is missing amino acids 141 through 152 of α2 appears to be trapped in the nuclear envelope at what might normally be an intermediate step in nuclear entry, as reviewed above, is consistent with the last two possibilities.

The α2 protein is not unique in having two localization signals. As discussed above, polyoma T antigen (Richardson *et al.*, 1986), SV40 T antigen (Welsh *et al.*, 1986), and influenza virus nucleoprotein (Davey *et al.*, 1985) are also known or suspected to have two localization determinants. Two localization determinants could be a common feature of the process by which proteins are delivered to and accumulate in the nucleus.

C. Ribosomal Proteins Move in and out of the Nucleus

The first 64 amino acids of the 135 amino acid yeast ribosomal RP51A protein are sufficient to direct β-galactosidase to the yeast nucleus (Gritz *et al.*, 1985). When the *RP51A–lacZ* gene fusion encoding this protein is expressed in cells containing a wild-type copy of RP51A, low levels of only nuclear associated RP51A–β-galactosidase are produced. When the same gene fusion is expressed in cells missing the intact RP51A gene, most of the fusion protein is cytoplasmic. About 30% of the β-galactosidase activity cofractionates with polysomes. Because the wild-type gene product is absent, the fusion gene product assembles into ribosomes instead. It appears that the first 64 amino acids of RP51A are sufficient not only to direct the chimeric protein to the nucleus but, once there, to allow it to interact with ribosomal RNA and move back out of the nucleus. Moreover, some of the fusion protein maintains the capacity to assemble into ribosomes. A fusion between only the first 7 RP51A amino acids and β-galactosidase appears to be both nuclear and cytoplasmic, but not associated with ribosomes.

The first 21 amino acids of the 387 amino acid yeast ribosomal protein L3 are sufficient for nuclear localization when fused to β-galactosidase (Moreland *et al.*, 1985). A protein having only the first 5 L3 amino acids fused to β-galactosidase is primarily cytoplasmic. A protein with almost the entire L3 protein (373 amino acids) fused to β-galactosidase is both nuclear and cytoplasmic. In the latter case, the cytoplasmic protein is associated with polysomes like the longer RP51A–β-galactosidase protein. The amino terminus of L3 contains a sequence Pro[18]-Arg-Lys-Arg[21] that has been proposed to be the important element for nuclear localization. Moreland *et al.* (1985) propose that the information is not so much contained in the primary sequence as in the general motif of a proline residue followed by several basic amino acids.

Once in the nucleus, what causes these ribosomal proteins to move back out into the cytoplasm? One possibility is that, once protein is bound to the rRNA, a new signal is revealed that causes the protein to be transported out of the nucleus. This kind of mechanism has been proposed for the localization of snRNPs to the nucleus (Mattaj and DeRobertis, 1985). Proteins normally associated with snRNPs are cytoplasmic and excluded from the nucleus when not associated with RNA in *Xenopus* oocytes. Once bound to RNA, the protein–RNA complex moves into the nucleus. It is proposed that by formation of the protein–RNA complex a targeting signal is exposed. A similar type of mechanism, but in reverse, could explain the ability of ribosomal proteins to move in and out of the nucleus.

D. Prokaryotic DNA Binding Proteins Enter the Yeast Nucleus

Several genes encoding bacterial DNA binding proteins have been expressed in yeast. These proteins have been shown to enter the nucleus; however, in all cases examined, these proteins do not accumulate in the nucleus. This is additional evidence that DNA binding is not sufficient to mediate nuclear accumulation.

Barnes and Rine (1985) have shown that the restriction endonuclease *Eco*RI, when produced in high amounts in yeast, degrades the chromosomal DNA. *Eco*RI shares no significant sequence homology with any of the other identified nuclear target sequences. It is not known what fraction of the *Eco*RI synthesized in yeast actually enters the nucleus. Because the *Eco*RI was produced at fairly high levels in yeast cells, it may be that it enters the nucleus by a nonphysiological mechanism, if not by diffusion through nuclear pores.

LexA is a small (24 kDa) protein that represses the *E. coli* genes necessary for the response to DNA damage (SOS response) by binding to a 22 base pair operator (Brent and Ptashne, 1981). The LexA protein, when produced in yeast, is capable of repressing transcription by binding to a synthetic operator placed appropriately in a yeast promoter (Brent and Ptashne, 1984). Thus, LexA apparently can gain access to the nuclear interior.

A larger hybrid protein (99 kDa) that contains the LexA DNA binding domain at its amino terminus followed by the carboxy-terminal 91% of GAL4 is capable of activating transcription of yeast genes that have the

LexA operator near the start of transcription (Brent and Ptashne, 1985). Thus, the LexA–GAL4 hybrid protein must enter the nucleus.

When yeast cells producing LexA are examined by indirect immunofluorescence, the LexA appears distributed throughout the cell and not localized to any specific compartment (Silver *et al.*, 1986). Because of its small size, LexA may freely diffuse through the nuclear pore. However, interaction with nuclear DNA is not sufficient to cause LexA to be concentrated in the nucleus. As in the case of LexA alone, the LexA–GAL4 hybrid protein is distributed throughout the cell and not concentrated only at the nucleus. This LexA–GAL4 protein lacks the GAL4 DNA binding region, the same portion that has been shown to be sufficient to localize β-galactosidase to the nucleus.

Is it reasonable to propose that DNA binding alone is sufficient to retain proteins in the nucleus? Several indirect observations argue against this idea. SV40 T antigen molecules defective in their nuclear localization *in vivo* still bind to the SV40 origin of replication *in vitro* (Paucha *et al.*, 1985). Similarly, among several mutations in amino acids 1–74 of GAL4, some fail to turn on *GAL1* expression but are still nuclear-localized, while others are not concentrated in the nucleus but can still turn on *GAL1* (P. Silver, unpublished results).

These experiments rule out the hypothesis that specific DNA binding is responsible for the nuclear localization of certain proteins. However, it remains possible that nonspecific DNA binding, which was not directly assayed, accounts for the localization of mutant T antigen and GAL4 proteins described above. The experiments with $\alpha2$ (see Section IV,B), LexA, and the LexA–GAL4 hybrid protein, when considered more closely, bear on this latter point. In the case of $\alpha2$, there is no correlation between ability of mutant proteins to bind DNA nonspecifically and an ability to accumulate in the nucleus. Certain mutant $\alpha2$ proteins are localized to the nucleus despite not being able to bind DNA; others fail to accumulate in the nucleus although they are capable of DNA binding. In the case of LexA and LexA–GAL4, versions of these proteins which contain a DNA binding domain are not localized to the yeast nucleus. Possibly, the nucleus provides enough of a permeability barrier so that the localization-defective but DNA binding-proficient $\alpha2$ and prokaryotic DNA binding proteins cannot freely associate with DNA in the nucleus. Some LexA does apparently enter the nucleus (see above). In both the LexA and the LexA–GAL4 cases, proteins are expressed at very high levels, far beyond that of the intact *GAL4* gene. It may be that if LexA were expressed at lower levels, not enough of it would get into the nucleus to bind specifically to its site.

V. CONCLUSION

It is clear that we still do not completely understand how a certain group of proteins are specifically localized to the nucleus. However, we now know that nucleus-destined proteins contain specific amino acid sequences which are necessary and sufficient to localize the proteins to the nucleus. The evidence for this comes from analysis of genes for nuclear-localized proteins in cells from many organisms, especially the yeast *S. cerevisiae*. Mutations in short, discrete sequences can block or reduce localization of a protein to the nucleus without affecting other activities of the protein. Attachment of such a nuclear target sequence to a cytoplasmic protein results in that protein being localized to the nucleus. However, we still do not know precisely how these sequences are used by the protein for transport to the nucleus. Future research will focus on what cellular component(s) the nuclear targeting sequence may interact with. A genetic approach with *S. cerevisiae* should be particularly useful in this regard.

The nuclear pore provides a channel through which nucleus-destined proteins may pass. The evidence for this comes from direct observations of proteins passing through the pores. The nuclear pore is a large, complex structure which is poorly characterized. Only a few of its component proteins have been identified. No mutation affecting its function has been identified in any organism. Clearly, much future research will focus on the assembly and function of the nuclear pore complex. One particular outstanding problem is how proteins of different sizes pass through the pore. Also, it is not at all clear that all macromolecules must pass through this structure to enter the nucleus. The mode of action of the nuclear targeting sequences and what they interact with will further clarify the role of the nuclear pore.

ACKNOWLEDGMENTS

We would like to thank John Bonner, Ralph Greenspan, Mark Samuels, and Jeff Way for critical reading of the manuscript and Barbara Baggaley for help with its preparation. Parts of this work were supported by National Institutes of Health Grant GM3673-01 awarded to P.A.S. and NIH Grant GM35284 to M.N.H. P.S. also wishes to acknowledge support of M. Ptashne, in whose lab portions of this work were done.

REFERENCES

Adzuma, K., Ogawa, T., and Ogawa, H. (1984). Primary structure of the *RAD52* gene in *Saccharomyces cerevisiae*. *Mol. Cell. Biol.* **4,** 2735–2744.

Alberts, A., Bray, D., Lewis, J., Raff, M., Roberts, K., and Watson, J. (1983). "Molecular Biology of the Cell." Garland, New York.

Astell, A. R., Ahlstrom-Jonasson, L., Smith, M., Tatchell, K., Nasmyth, K. A., and Hall, B. D. (1981). The sequence of the DNAs coding for the mating-type loci of *Saccharomyces cerevisiae*. *Cell* **27**, 15–23.

Barnes, G., and Rine, J. (1985). Regulated expression of endonuclease *Eco*RI in *Saccharomyces cerevisiae*: Nuclear entry and biological consequences. *Proc. Natl. Acad. Sci. U.S.A.* **82**, 1354–1358.

Bonner, W. M. (1975). Protein migration into nuclei. I. Frog oocyte nuclei *in vivo* accumulate microinjected histones, allow entry to small proteins, and exclude large proteins. *J. Cell Biol.* **64**, 421–430.

Bonner, W. M. (1978). Protein migration and accumulation in nuclei. *In* "The Cell Nucleus" (H. Busch, ed.), pp. 97–148. Academic Press, New York.

Brent, R., and Ptashne, M. (1981). Mechanism of action of the *lexA* gene product. *Proc. Natl. Acad. Sci. U.S.A.* **78**, 4204–4208.

Brent, R., and Ptashne, M. (1984). A bacterial repressor protein on a yeast transcriptional terminator can block upstream activation of a yeast gene. *Nature (London)* **312**, 612–615.

Brent, R., and Ptashne, M. (1985). A eukaryotic transcriptional activator bearing the DNA specificity of a prokaryotic repressor. *Cell* **43**, 729–736.

Choe, J., Kolodrubetz, D., and Grunstein, M. (1982). The two yeast histone H2A genes encode similar protein subtypes. *Proc. Natl. Acad. Sci. U.S.A.* **79**, 1484–1487.

Dabauvalle, M.-C., and Franke, W. W. (1982). Karyophilic proteins: Polypeptides synthesized *in vitro* accumulate in the nucleus on microinjection into the cytoplasm of amphibian oocytes. *Proc. Natl. Acad. Sci. U.S.A.* **79**, 5302–5306.

Davey, J., Dimmock, N. J., and Colman, A. (1985). Identification of the sequence responsible for the nuclear accumulation of the influenza virus nucleoprotein in *Xenopus* oocytes. *Cell* **40**, 667–675.

Davis, L. I., and Blobel, G. (1986). Identification and characterization of a nuclear pore complex protein. *Cell* **45**, 699–709.

DeRobertis, E. M. (1983). Nucleocytoplasmic segregation of proteins and RNA. *Cell* **32**, 1021–1025.

Dingwall, C., and Allan, J. (1984). Accumulation of the isolated carboxyl-terminal domain of histone H1 in the *Xenopus* oocyte nucleus. *EMBO J.* **3**, 1933–1937.

Dingwall, C., Sharnick, S. V., and Laskey, R. A. (1982). A polypeptide domain that specifies migration of nucleoplasmin into the nucleus. *Cell* **30**, 449–458.

Douglas, H. C., and Hawthorne, D. C. (1966). Regulation of genes controlling synthesis of the galactose pathway enzymes in yeast. *Genetics* **54**, 911–916.

Emr, S. D., and Silhavy, T. J. (1980). Mutations affecting localization of an *Escherichia coli* outer membrane protein, the bacteriophage λ receptor. *J. Mol. Biol.* **141**, 63–90.

Feldherr, C. M., Kallenbach, E., and Schultz, N. (1984). Movement of a karyophilic protein through the nuclear pores of oocytes. *J. Cell Biol.* **999**, 2216–2222.

Franke, W. W., Scheer, V., Krohne, G., and Jarasch, E. D. (1981). The nuclear envelope and the architecture of the nuclear periphery. *J. Cell Biol.* **91**, 39–50.

Gerace, L., Ottaviano, Y., and Kondor-Koch, C. (1982). Identification of a major polypeptide of the nuclear pore complex. *J. Cell Biol.* **95**, 826–837.

Giniger, E., Varnum, S. M., and Ptashne, M. (1985). Specific DNA binding of GAL4, a positive regulatory protein of yeast. *Cell* **40**, 767–774.

Goldfarb, D. S., Gariepy, J., Schoolnik, G., and Kornberg, R. D. (1986). Synthetic peptides as nuclear location signals. *Nature (London)* **322**, 641–644.

Gritz, L., Abovich, N., Teem, J. L., and Rosbash, M. (1985). Posttranscriptional regulation and assembly into ribosomes of a *S. cerevisiae* ribosomal protein–β-galactosidase fusion. *Mol. Cell. Biol.* **5,** 3436–3442.

Hall, M. N., and Johnson, A. D. (1987). The homeo-domain of the yeast repressor α2 is a sequence specific DNA domain but is not sufficient for repression. *Science* **238,** 1007–1012.

Hall, M. N., Hereford, L., and Herskowitz, I. (1984). Targeting of *E. coli* β-galactosidase to the nucleus in yeast. *Cell* **36,** 1057–1065.

Johnson, A. D., and Herskowitz, I. (1985). A repressor (*MATα2* product) and its operator control expression of a set of cell type specific genes in yeast. *Cell* **42,** 237–247.

Kalderon, D., Richardson, W. D., Markham, A. F., and Smith, A. E. (1984a). Sequence requirements for nuclear location of simian virus 40 large-T antigen. *Nature (London)* **311,** 33–38.

Kalderon, D., Roberts, B. L., Richardson, W. D., and Smith, A. E. (1984b). A short amino acid sequence able to specify nuclear location. *Cell* **39,** 499–509.

Kauffman, R. S., and Ginsberg, H. S. (1976). Characterization of a temperature-sensitive, hexon transport mutant of type 5 adenovirus. *J. Virol.* **19,** 643–658.

Keegan, L., Gill, G., and Ptashne, M. (1986). Separation of DNA binding from the transcription-activating function of a eukaryotic regulatory protein. *Science* **231,** 699–704.

Lanford, R. E., and Butel, J. S. (1984). Construction and characterization of an SV40 mutant defective in nuclear transport of T antigen. *Cell* **37,** 801–813.

Lanford, R. E., Kanda, P., and Kennedy, R. C. (1986). Induction of nuclear transport with a synthetic peptide homologous to the SV40 T antigen transport signal. *Cell* **46,** 575–582.

Laughon, A., and Gesteland, R. F. (1984). Primary structure of the *Saccharomyces cerevisiae GAL4* gene. *Mol. Cell. Biol.* **4,** 260–267.

Mattaj, I. W., and DeRobertis, E. M. (1985). Nuclear segregation of U2 snRNA requires binding of specific proteins. *Cell* **40,** 111–118.

Moreland, R. B., Nam, H. G., Hereford, L. M., and Fried, H. M. (1985). Identification of a nuclear signal of a yeast ribosomal protein. *Proc. Natl. Acad. Sci. U.S.A.* **82,** 6561–6565.

Newmeyer, D. D., Lucocq, J. M., Burglin, T. R., and DeRobertis, E. M. (1986). Assembly *in vitro* of nuclei active in nuclear protein transport: ATP is required for nucleoplasmin accumulation. *EMBO J.* **5,** 501–510.

Novick, P., Field, C., and Schekman, R. (1980). Identification of 23 complementation groups required for posttranslational events in the yeast secretory pathway. *Cell* **21,** 205–215.

Novick, P., Ferro, S., and Schekman, R. (1981). Order of events in the yeast secretory pathway. *Cell* **25,** 461–469.

Oliver, D. B., and Beckwith, J. (1981). *E. coli* mutant pleiotropically defective in the export of secreted proteins. *Cell* **25,** 765–772.

Paine, P. L., Moore, L. C., and Horowitz, S. B. (1975). Nuclear envelope permeability. *Nature (London)* **254,** 109–114.

Paucha, E., Kalderon, D., Richardson, W. D., Harvey, R. W., and Smith, A. E. (1985). The abnormal location of cytoplasmic SV40 large T is not caused by failure to bind to DNA or to p53. *EMBO J.* **4,** 3235–3240.

Pope, J. H., and Rowe, W. P. (1964). Detection of specific antigen in SV40-transformed cells by immunofluorescence. *J. Exp. Med.* **120,** 121–128.

Richardson, W. D., Roberts, B. L., and Smith, A. E. (1986). Nuclear location signals in polyoma virus large-T. *Cell* **44,** 77–85.

Roeder, G. S., Beard, C., Smith, M., and Keranen, S. (1985). Isolation and characterization of the SPT2 gene, a negative regulator of Ty-controlled yeast gene expression. *Mol. Cell. Biol.* **5,** 1543–1553.

Rose, M., Casadaban, M. J., and Botstein, D. (1981). Yeast genes fused to β-galactosidase in *Escherichia coli* can be expressed normally in yeast. *Proc. Natl. Acad. Sci. U.S.A.* **78,** 2460–2464.

Schatz, G., and Butow, R. A. (1983). How are proteins imported into mitochondria? *Cell* **32,** 316–318.

Schindler, M., and Jiang, L.-W. (1986). Nuclear actin and myosin as control elements in nucleocytoplasmic transport. *J. Cell Biol.* **102,** 859–862.

Silhavy, T. J., and Beckwith, J. R. (1985). Uses of *lac* fusions for the study of biological problems. *Microbiological Rev.* **49,** 398–418.

Silver, P. A., Keegan, L. P., and Ptashne, M. (1984). Amino terminus of the yeast *GAL4* gene product is sufficient for nuclear localization. *Proc. Natl. Acad. Sci. U.S.A.* **81,** 5951–5955.

Silver, P. A., Brent, R., and Ptashne, M. (1986). DNA binding is not sufficient for nuclear localization of regulatory proteins in *Saccharomyces cerevisiae*. *Mol. Cell. Biol.* **6,** 4763–4766.

Silver, P. A., Chiang, A., and Sadler, I. (1987). Mutations which simultaneously alter localization and expression of a yeast nuclear protein: Support for a receptor model of nuclear localization. (Submitted).

Smith, M. M., and Andresson, O. S. (1983). The DNA sequences of yeast H3 and H4 histone genes from two non-allelic gene sets encode identical H3 and H4 proteins. *J. Mol. Biol.* **169,** 663–690.

Strathern, J., Hicks, J., and Herskowitz, I. (1981). Control of cell type in yeast by the mating type locus, the α1–α2 hypothesis. *J. Mol. Biol.* **147,** 357–372.

Tooze, J. (ed.) (1980). "Molecular Biology of Tumor Viruses," 2nd Ed., Part 2 (DNA Tumor Viruses), pp. 61–338. Cold Spring Harbor Press, Cold Spring Harbor, New York.

Unwin, N. T., and Milligan, R. A. (1982). A large particle associated with the perimeter of the nuclear pore complex. *J. Cell Biol.* **93,** 63–75.

Wallis, J. W., Hereford, L., and Grunstein, M. (1980). Histone H2B genes of yeast encode two different proteins. *Cell* **22,** 799–805.

Walter, P., Gilmore, R., and Blobel, G. (1984). Protein translocation across the endoplasmic reticulum. *Cell* **38,** 5–8.

Welsh, J. D., Swimmer, C., Cocke, T., and Shenk, T. (1986). A second domain of simian virus 40 T antigen in which mutations can alter the cellular localization of the antigen. *Mol. Cell. Biol.* **6,** 2207–2212.

Willison, J. H. M., and Johnston, G. C. (1978). Altered nuclear pore diameters in G_1-arrested cells of the yeast *Saccharomyces cerevisiae*. *J. Bacteriol.* **136,** 318–323.

Yaffe, M. P., and Schatz, G. (1984). Two nuclear mutations that block mitochondrial protein import in yeast. *Proc. Natl. Acad. Sci. U.S.A.* **81,** 4819–4823.

VII

Applied Research

19

Secretion Research in Industrial Mycology

RAMUNAS BIGELIS AND RATHINDRA C. DAS

I. INTRODUCTION

Applied mycology is an ancient science, as is also the study of fungal secretion. Undoubtedly, the earliest applications of fungi were unintentional. Later investigations were deliberate and yielded a variety of foods and beverages produced by fermentation or processed by secreted microbial enzymes. Quite likely the enzymes played a role in improving the palatability, preservation, and nutrition of the foods.

These ancient origins are the very basic foundation for today's applied mycology. But contemporary applied mycology also serves a great many other human needs. It plays a role not only in the production of foods and beverages but also in the manufacture of fuel, commodity chemicals, industrial enzymes, and diverse pharmaceuticals (Demain, 1981; Bigelis,

773

1985, 1988; Bennett, 1985; Bigelis and Lasure, 1987). And, in the future, it promises to furnish a variety of new specialty products that will play an important role in industry, agriculture, and medicine.

This chapter focuses on the exploitation of secretion processes for the production of substances beneficial to human needs. Primarily, it discusses the production of commercial enzymes of fungal origin and emphasizes the impact of modern molecular biology on the secretion of these proteins (Table I). It also assembles recent information on the current status of research on the secretion of heterologous proteins with potential as therapeutic agents. Other reviews have discussed the fundamentals of secretion processes in *Saccharomyces cerevisiae* (Schekman, 1982; Schekman and Novick, 1982), fungal extracellular enzymes (Aunstrup, 1979; Fogarty and Kelly, 1981; Priest, 1984), and the use of genetically engineered fungi to produce pharmaceuticals (Bitter, 1984; Hollenberg *et al.*, 1985; Primrose, 1986; Nicaud *et al.*, 1986).

II. COMMERCIALLY IMPORTANT ENZYMES OF FUNGAL ORIGIN

A. Amylases

α-Amylases degrade polysaccharides such as starch by randomly hydrolyzing internal α-1,4 linkages (Fogarty and Kelly, 1980; Fogarty, 1983). Glucoamylases (glucan 1, 4-α-glucosidase), on the other hand, can also digest starch but cleave α-1,4 linkages consecutively from the nonreducing end. The enzyme can cleave α-1,3 and α1,6 bonds, but at slower rates (Kulp, 1975; Fogarty and Kelly, 1979; MacAllister, 1979). α-Amylases from species of *Aspergillus* are used in the baking and brewing industries to elevate the levels of fermentable sugars in dough and malt (Taylor and Richardson, 1979; Schwimmer, 1981). Glucoamylases from improved strains of *Rhizopus* or, more commonly, *Aspergillus* are used for commercial starch processing (Meyrath and Bayer, 1980).

1. Expression of α-Amylases in Yeast

The development of *S. cerevisiae* strains able to secrete α-amylase would have significant industrial impact and would be potentially useful to the brewing and fuel alcohol industries (Stewart and Russell, 1986). Rothstein *et al.* (1984) combined the wheat α-amylase gene (Baulcombe and Buffard, 1983) with a yeast phosphoglycerate kinase promoter still bearing the sequence for the first 12 amino acids of the protein. The α-amylase protein produced by *S. cerevisiae* strains with this hybrid α-amylase gene

TABLE I

Enzymes of Importance in Industrial Mycology and Secretion Research Which Have Been Examined by Recombinant DNA Methods

Industrial enzyme	Source	Recombinant fungal host	Primary reference for secretion studies
α-Amylase	Mouse salivary	Saccharomyces cerevisiae	Thomsen (1983)
	Mouse pancreas	Saccharomyces cerevisiae	Filho et al. (1986)
	Human salivary	Saccharomyces cerevisiae	Nakamura et al. (1986)
	Saccharomycopsis fibuligera	Saccharomyces cerevisiae	Yamashita et al. (1985a)
	Wheat	Saccharomyces cerevisiae	Rothstein et al. (1984)
Glucoamylase	Aspergillus awamori	Saccharomyces cerevisiae	Innis et al. (1985)
	Aspergillus niger	Aspergillus niger, Aspergillus awamori	Finklestein et al.. 1986
	Rhizopus sp.	Saccharomyces cerevisiae	Ashikari et al. (1985)
	Saccharomyces diastaticus	Saccharomyces cerevisiae	Yamashita and Fukui (1983)
	Saccharomyces diastaticus	Schizosaccharomyces pombe	Yamashita and Fukui (1984)
	Saccharomycopsis fibuligera	Saccharomyces cerevisiae	Yamashita et al. (1985b)
Protease (rennet)	Calf	Saccharomyces cerevisiae	Smith et al. (1985)
	Calf	Aspergillus nidulans	Hayenga et al. (1986)
	Mucor miehei	Aspergillus nidulans	Gray et al. (1987)
Cellulases			
Cellobiohydrolase	Trichoderma reesei	Saccharomyces cerevisiae	Shoemaker (1984)
Endoglucanase	Trichoderma reesei	Saccharomyces cerevisiae	van Arsdell et al. (1987)
	Cellulomonas fimi	Saccharomyces cerevisiae	Skipper et al. (1985)
β-Glucosidase	Aspergillus niger	Saccharomyces cerevisiae	Penttilä et al. (1984)
	Kluyveromyces fragilis	Saccharomyces cerevisiae	Raynal and Guerineau (1984)
	Candida pelliculosa	Saccharomyces cerevisiae	Kohchi and Toh-e (1986)
Lactase	Kluyveromyces marxianus var. lactis	Saccharomyces cerevisiae	Dickson (1980)
Invertase	Saccharomyces cerevisiae	Saccharomyces cerevisiae	Perlman et al. (1982), Kaiser and Botstein (1986)
β-Glucanase	Bacillus subtilis	Saccharomyces cerevisiae	Hinchliffe and Box (1984)
Ligninase	Phanerochaete chysosporium	Saccharomyces cerevisiae	Shoemaker et al. (1984)

was secreted in a fully active form into the yeast medium. The signal sequence of this fused gene, which was located internally, was correctly processed even though it appears to be somewhat buried (Rothstein *et al.*, 1984). Interestingly, the high level of α-amylase secretion was obtained only when the cells were grown in rich medium. Furthermore, this is probably one of the rare examples where a heterologous protein of considerable size (molecular weight 44K) has been secreted into the grown medium by *S. cerevisiae*.

Mouse salivary (Thomsen, 1983) and pancreatic (Filho *et al.*, 1986) α-amylases have also been expressed and secreted by *S. cerevisiae*. Pre-amylase cDNA behind the yeast *ADC1* promotor was used in the former case, while preproleader of α-factor was used to direct secretion of the latter enzyme. Pancreatic α-amylase is an extremely efficient enzyme capable of digesting raw starch, producing maltose and maltotriose substrates easily assimilated by *MAL* yeast strains. Thus, the derivation of yeast strains that secrete functional pancreatic α-amylase should permit ethanolic fermentation directly from starchy materials. Such fermentations have already been achieved with an *S. cerevisiae* strain transformed with a plasmid containing the cDNA for human salivary α-amylase, a yeast *PH05* promoter, and a mammalian signal sequence (Nakamura *et al.*, 1986; Sato *et al.*, 1986). The recombinant yeast secreted glycosylated, processed enzyme that permitted growth on starch and ethanol formation.

Saccharomyces cerevisiae has also been a host for cloned fungal and bacterial α-amylases. A fungal gene from *Saccharomycopsis fibuligera* has been introduced into *S. cerevisiae* and expression and secretion of α-amylase has been achieved (Yamashita *et al.*, 1985a). Expression and secretion of prokaryotic α-amylases could also be extremely useful, as in the case of the *Bacillus amyloliquefaciens* α-amylase gene which has been expressed in *S. cerevisiae* (Gorozhankina *et al.*, 1985).

2. *Aspergillus Glucoamylase*

Emerging information on the nature of *Aspergillus* glucoamylase has revealed its molecular structure (Matsuura *et al.*, 1980) and nucleotide sequence (Boel *et al.*, 1984a; Nunberg *et al.*, 1984), and now should clarify pathways for its formation and export. *Aspergillus niger* secretes two glycosylated forms of glucoamylase. Only form G1 can bind to and digest raw starch, while both G1 and the smaller form G2 act on soluble polysaccharide and oligosaccharide substrates. Boel *et al.* (1984a,b) propose that the multiple forms of glucoamylase are the result of differential mRNA splicing of glucoamylase G1 mRNA. Molecular cloning of cDNA and the use of synthetic glucoamylase gene fragments have permitted the determination of the primary structure of G1 mRNA. Apparently, the precursor of G1 glucoamylase is a protein of 640 amino acids and contains a putative signal peptide of 18 residues and a propeptide of 6 residues. Cell-free translation of glucoamylase message followed by immunoprecipitation with antibody raised to the purified enzyme reveals two glucoamylase polypeptides. The larger form, G1, has a molecular weight of 71,000, while the smaller species, G2, has a molecular weight of 61,000. Nucleotide sequence data suggest that a 169 base pair intron may be removed from G1 mRNA to make G2 mRNA (Boel *et al.*, 1984b).

Recent data of Svensson and co-workers (1986), however, indicate that glucoamylase G2 consists of two molecular species which share residues 1–512 and 1–514 with glucoamylase G1. In addition, homologous peptides of 70 amino acids from the G1 and G2 forms contain identical O-linked carbohydrates. Thus, these alternative results indicate that the G2 forms are quite likely generated posttranslationally from G1 by limited proteolysis. Experiments with G1 modified with proteinase VIII from *Bacillus subtilis* tend to support these results (Bartoszewicz, 1986). The influence of growth conditions on proteolysis or differential splicing of the glucoamylase message may reconcile the divergent observations on glucoamylases G1 and G2.

Analysis of the O-glycosidically linked carbohydrate chains of *A. niger* glucoamylase G1 indicates that it contains 19% (w/w) neutral carbohydrates. Most of the sugars are linked to threonine or serine via a mannose moiety, a glycosylation pattern which is rare in glycoproteins. Most of the O-linked carbohydrates are attached to a 70 amino acid segment of the protein approximately 100 residues from the carboxy-terminal end of the molecule. Of the few glycosylated amino acids outside of this region, two are N-glycosylated asparagine residues (Gunnarsson, *et al.*, 1984).

3. Expression of an Aspergillus Glucoamylase Gene in Saccharomyces cerevisiae

Cloned glucoamylase genes in yeast have been studied extensively since the availability of strains able to completely hydrolyze starch would eliminate pretreatment steps and improve industrial processes and fermentations (Tubb, 1986). Recently, glucoamylase from *Aspergillus awamori* has been expressed, glycosylated, and secreted in *S. cerevisiae* by Innis *et al.* (1985). In their initial studies the authors could detect neither initiation or termination of transcription, nor correct splicing of the four intervening sequences of the *A. awamori* G1 gene in yeast. Subsequently, the introns were deliberately excised, and the promotor and terminator regions of the yeast enolase gene were added to flank the glucoamylase gene. Transformation with the resultant plasmid indicated the presence of glucoamylase transcripts in yeast. Furthermore, some transformants were able to grow on starch as the sole carbon source in liquid medium. Secreted glucoamylase, which constituted 25–30% of extracellular protein, was detectable in the culture medium immunologically. Only 10% of the activity remained within cells.

Secreted glucoamylase had an electrophoretic mobility similar to that of the native glycosylated *A. awamori* enzyme but appeared to have a lower mobility than the intracellular recombinant protein. Purified recombinant glucoamylase preparations shared a 30 amino acid amino-terminal amino acid sequence with native *A. awamori* glucoamylase and revealed no heterogeneity at the amino terminus. Recombinant enzyme contained only slightly more glycosylated serine and threonine residues than did the native enzyme, and N-glycosylation patterns were similar. Treatment with endo-β-N-acetylglucoaminidase H (endo H) followed by analysis on denaturing gels revealed a small decrease in mobility of both enzymes. Thus, the overall degree of glycosylation of native and recombinant enzymes appeared quite similar. However, the amount of glucoamylase secreted into the yeast growth medium was considerably lower than that secreted by the original *A. awamori* strain.

Innis and co-workers (1985) have also characterized the leader sequence of *A. awamori* glucoamylase expressed in yeast. Their data indicate that signal peptidase cleaves an 18-amino acid peptide and that an endopeptidase liberates 6 additional amino acids from the amino-terminal end of the protein. This secondary endopeptidase processing reaction is thought to be carried out by the *KEX2* gene product, which cleaves the protein after Lys[23]-Arg[24]. Similar endopeptidase cleavage of Lys-Arg residues in the α-factor mating hormone maturation process is discussed in more detail in Section III and in Fig. 1. The function of the endopeptidase cleavage of glucoamylase is not understood. However, mutants bearing the *KEX2* mutation still secrete fully active glucoamylase that is larger by six amino-terminal amino acids.

In essence, these results indicate that in spite of a close phylogenetic relationship between *Saccharomyces* and *Aspergillus,* the promoter and terminator recognition machinery of *Saccharomyces* is not compatible with that of *Aspergillus*. Nevertheless, the protein glycosylation pathway and the secretory mechanisms appear to be quite similar in these two organisms.

4. Expression of a Rhizopus Glucoamylase Gene in Saccharomyces cerevisiae

Glucoamylase from certain species of *Rhizopus* has attracted attention because it has greater activity than other such enzymes that have been examined. The glucoamylase gene from *Rhizopus* sp. has been cloned and expressed in yeast using a cDNA library of *Rhizopus* total poly(A) RNA.

Sequence analysis of the genomic *Rhizopus* glucoamylase gene has been performed and reveals four intervening sequences. The glucoamylase itself consists of 604 amino acids and possesses a putative signal peptide. A lysine is present at the 9 position, and the cleavage point for the signal peptide lies between two alanines in the 25 and 26 position (Ashikari *et al.*, 1986).

Expression of the chromosomal glucoamylase gene in yeast has been unsuccessful. The studies by Ashikari *et al.* (1985, 1986) have demonstrated that yeast bearing the cloned *Rhizopus* gene secretes active glucoamylase, and they later showed that recombinant glucoamylase resembled Gluc 1, one of three forms of *Rhizopus* glucoamylase (Tanaka *et al.*, 1986). Thus, yeast transformants grow normally on starch as the sole carbon source and are able to produce ethanol efficiently under these conditions. Surprisingly, recombinant enzyme degraded raw starch more efficiently than native enzyme though both enzymes share the same amino acid sequence.

5. Yeast Glucoamylases

While *S. cerevisiae* lacks the ability to utilize starch, a number of other yeasts can produce glucoamylases which permit growth on this carbon source. *Saccharomyces diastaticus*, a yeast so closely related to *S. cerevisiae* that it is now considered the same species, has been studied extensively because it has the capacity to produce and secrete this enzyme. *S. diastaticus* carries three unlinked genes, *STA1*, *STA2*, and *STA3* (allelic to *DEX2* and *MAL5*, *DEX1*, and *DEX3*, respectively, according to Pretorius *et al.*, 1986b), which encode three extracellular isozymes of glucoamylase. The polymorphic glucoamylases have similar amino acid sequences, but the glycosylation patterns appear to be strain specific (Yamashita *et al.*, 1984).

The *STA1* gene has been cloned in *S. cerevisiae* (Yamashita and Fukui, 1983) and *Schizosaccharomyces pombe* (Yamashita and Fukui, 1984). Both yeast hosts transformed with the plasmid bearing the glucoamylase gene gained the ability to express the enzyme, secrete it into the medium, and ferment starch. The data suggest that these two related yeasts share similar elements of the secretory pathway.

The *STA2* gene has been expressed in *S. cerevisiae* (Meaden *et al.*, 1985; Pretorius *et al.* 1986a; Pardo *et al.*, 1986), and a putative sporulation-specific glucoamylase gene, also from *S. diastaticus*, has been cloned in *S. cerevisiae* and *S. pombe* (Erratt and Nasim, 1986). In both cases, the *S. cerevisiae* transformants were able to produce extracellular glucoamylase activity and grow on starch as the sole carbon source.

Recent studies with the *STA1* and *STA3* genes cloned in *S. cerevisiae* also reveal that transformants secrete glucoamylases (Yamashita *et al.*,

1985c,d) immunologically indistinguishable from those of *S. diastaticus*. Furthermore, the *STA1* and *STA3* genes show homology with one another and also with a *S. cerevisiae* sporulation-specific glucoamylase gene sequence. It is proposed that the *S. diastaticus* extracellular glucoamylase genes arose from an intracellular *S. cerevisiae* sporulation-specific glucoamylase (Colonna and Magee, 1978) that acquired the promoter and signal sequence of the extracellular enzyme.

Nucleotide sequence data for the *STA1* gene indicate that the 778 amino acid protein has a putative hydrophobic signal of about 30 amino acids and 13 potential N-glycosylation sites. A threonine-rich tract is followed by a pair of lysine residues which may be signals for proteolytic processing. The carboxy-terminal portion of *S. diastaticus* glucoamylase shows distinct homology with α-amylases from *Aspergillus oryzae, B. subtilis*, and mouse. In fact, the three *S. diastaticus* genes for extracellular glucoamylase are highly homologous and probably are derived from the same ancestral gene (Yamashita *et al.*, 1985c,d).

The α-amylase gene (Yamashita *et al.*, 1985a), as mentioned before in Section II,A,1, and the glucoamylase gene (Yamashita *et al.*, 1985b) from another yeast, *S. fibuligera*, have also been cloned and expressed in *S. cerevisiae*. As in the cases reported above, both enzymes are secreted by the transformants.

B. Proteases

Extracellular proteolytic enzymes from fungi, primarily species of *Aspergillus, Mucor, Rhizopus,* or *Endothia* are used in the baking, dairy, and soybean processing industries or in food fermentation (Aunstrup, 1974, 1977, 1980; Yamamoto, 1975; Ward, 1983). *Aspergillus* spp. of the *A. flavus–A. oryzae* and *A. niger* groups typically produce acid, neutral, and alkaline proteases in characteristic proportions, while *Mucor miehei* and *Mucor pusillus* produce very stable milk coagulating enzymes with a pH optimum around pH 3.5. Acid protease preparations from *Endothia parasitica (Priphonectria parasitica)* are also used in cheesemaking (Sardinas, 1976).

Very little modern secretion research has been performed with fungal species used for production of commercial proteases. However, advances in heterologous gene expression in *S. cerevisiae* (Kingsman *et al.*, 1985), the recent application of recombinant DNA technology to the filamentous fungi (Hynes, 1986; Upshall, 1986), and an increasing interest in recombinant calf rennet (Beppu, 1983; Nishimori, 1986) are generating new information on mechanisms of extracellular protease production by fungi.

1. Expression of the Prochymosin Gene in Saccharomyces cerevisiae

Much attention is being directed to prochymosin owing to its significant role in the dairy industry. Prochymosin, the inactive precursor of chymosin (rennin), is converted to an acid protease used in cheesemaking that acts on κ-casein and causes milk coagulation (Olson, 1979; Higgins *et al.*, 1985). Native calf prochymosin is activated autocatalytically to chymosin at acid pH by removal of 42 amino acids from the amino terminus (Pedersen and Foltmann, 1975) after it is secreted into the stomach using a 16 residue amino-terminal secretion signal (Moir *et al.*, 1982). Since the availability of chymosin depends on the supply of fourth stomachs from unweaned calves, alternative sources have long been sought. The possibility of producing chymosin by fermentation of *S. cerevisiae* or *A. niger* able to secrete this protein offers several advantages, primarily, lowered cost of production, reliable and continuous supply, ease of purification and processing, and a favorable food regulatory status.

Since the entire preprochymosin gene was cloned in *Escherichia coli* and sequenced (Harris *et al.*, 1982; Moir *et al.*, 1982; Emtage *et al.*, 1983), the yeast *S. cerevisiae* has been studied as a convenient and food-approved host for expression and secretion of recombinant calf rennet. In 1983 a strain of *S. cerevisiae* was reported that has the ability to synthesize an activatable prochymosin as 5% of its total cell protein (Mellor *et al.*, 1983). An expression vector with the preprochymosin gene inserted between the 5′ and 3′ control regions of the yeast phosphoglycerate kinase gene was able to direct the expression of chymosin and its two precursors, preprochymosin and prochymosin. The chymosin polypeptides were not secreted, however, but appeared to be associated with insoluble cell wall material. Nevertheless, whole cell extracts could be activated by acidification, yielding enzyme that clotted milk.

Duncan *et al.* (1984) have linked the prochymosin gene to the promoter, signal sequence, and part of the structural gene of yeast invertase encoded by the *SUC2* gene in attempts to achieve correct folding and deter aggregation. Approximately 10% of the invertase–prochymosin fusion protein was secreted. The invertase portion was glycosylated, and the entire fusion product was regulated by the same growth conditions as external invertase. Acidic growth conditions converted all of the extracellular protein to its active form, but most remained internal and unactivatable (Smith *et al.*, 1985).

Goff *et al.* (1984) combined the prochymosin gene with the highly regulated *GAL1* yeast promoter and then made constructions with and without the *SUC2* transcription terminator. The two constructions yielded equal amounts of prochymosin-specific mRNA and acid activatable prochymosin. Still, approximately 80% of the prochymosin in these yeast strains remained insoluble, even after cell breakage. The exact localization of the insoluble fraction remains unknown. It appears that aberrant disulfide bonding may be the cause for the inability to solubilize or activate prochymosin made in the yeast cytoplasm (Smith *et al.*, 1985).

Other investigations (Moir *et al.*, 1985; Smith *et al.*, 1985) compared the efficiency of natural calf prochymosin secretion signal (Moir *et al.*, 1982), yeast invertase leader (Carlson *et al.*, 1983), and yeast prepro-α-factor sequence (Kurjan and Herskowitz, 1982) on prochymosin secretion from recombinant *S. cerevisiae* strains. Whereas in one construction the yeast invertase (*SUC2*) promoter was used as the site of transcription initiation, *GAL1, URA3, MFα1,* and triose phosphate isomerase promoters were utilized in others. As noted before, along with the *SUC2* signal sequence, codons for the first 11 amino acid residues of mature invertase were fused to the prochymosin gene sequence. This segment possesses one N-glycosylation site. Likewise, the yeast α-factor pro segment contains as many as three oligosaccharide attachment sequences. About 10% of the prochymosin bearing an invertase signal sequence was secreted, while slightly less of the prochymosin with α-factor signal appeared in culture medium. A *PH05* signal sequence (Arima *et al.*, 1983) likewise resulted in less efficient secretion. The natural calf secretion signal proved to be totally ineffective in directing export of the protein from yeast. Secreted recombinant prochymosin could be fully activated to chymosin with the same procedure used for conversion of natural calf prochymosin to its active form. Identical milk clotting activity was observed in both cases. In contrast, prochymosin which remained in the yeast cytoplasm could not be activated. Therefore, it appears that passage through the secretion pathway may be an obligatory process for correct folding of prochymosin which allows biological activity (Moir *et al.*, 1985; Smith *et al.*, 1985).

The processing and glycosylation patterns of the invertase–prochymosin fusion protein were investigated by Smith *et al.* (1985). The precise secretion signal processing was not determined, although gel electrophoresis data indicated that the 19 amino acid residue invertase signal was processed. Furthermore, the glycosylation pattern of the fusion protein was investigated using endo H and tunicamycin, an antibiotic which inhibits the addition of asparagine-linked oligosaccharide. The data indicated that of the three potential asparagine-linked glycosylation sites in the fusion protein, one on the invertase gene segment and two on the prochymosin sequence, only the invertase segment of the protein was glycosylated. Likewise, amino acid residues 310 and 349 of prochymosin are not glycosylated in the calf cells. Since protein folding plays a role in susceptibility to glycosylation, it seems that the folding patterns of the fusion protein in yeast are similar to those of the native protein in calf cells.

The studies of Smith et al. (1985) show that despite channeling of recombinant prochymosin to secretion pathways in yeast, only about one-tenth of the total is actually secreted. This pattern is observed with two different controlling elements, the *SUC2* promoter, and even the triose phosphate isomerase promoter which elevates prochymosin levels to 0.25% of the total cell protein. The inability to export all of the prochymosin may be caused by a rate-limiting step in the secretion pathway involving the endoplasmic reticulum or the Golgi apparatus. Intracellular, insoluble prochymosin bears only core oligosaccharide added in the endoplasmic reticulum, while extracellular, prochymosin contains this moiety and the outer chain carbohydrate added in the Golgi apparatus. Thus, possible rate-limiting steps may involve transfer of the partially glycosylated protein from the endoplasmic reticulum to the Golgi or completion of the modification process within the Golgi itself. Alternatively, since indirect immunofluorescence indicates that much of the core glycosylated invertase–prochymosin fusion protein is vacuolar, partially modified prochymosin may be diverted to this cellular compartment and then be unavailable for export (Smith and Gill, 1985). Further study of the secretion pathway in yeast should permit identification of possible rate-limiting steps that influence the solubility properties of recombinant proteins and regulate their exit from the cell.

Mutagenesis and screening has been applied to the derivation of super-secreting mutants of *S. cerevisiae,* strains able to secrete a larger proportion of prochymosin (Smith and Gill, 1985; Smith et al., 1985). After examination of 120,000 colonies, 39 strains bearing partially dominant or recessive mutations were identified as supersecretors, some being able to produce eight to ten times more extracellular prochymosin than the parent strain. The distribution of prochymosin between the vacuole and medium was changed, while absolute levels were unaffected. Four recessive genes were studied most extensively, and two, *SSC1* and *SSC2*, were found to be strongest and most easily manipulated. Haploid mutant strains with the *ssc1* and *ssc2* mutations were more efficient secretors than single mutants. It is evident that such additive effects leading to secretion of up to 30% of the total prochymosin made will be desirable features in new hosts constructed for the secretion of other proteins. Thus, mutations *ssc1* and *ssc2,* and quite likely other genetic loci, together or singly, will be applicable to other systems. For example, it has already been shown that the *ssc1* mutation elevates the secretion of pre-bovine growth hormone in yeast at least 10-fold and that *ssc2* increases the yield also, though to a lesser extent (Smith *et al.*, 1985).

Several of the complementation groups related to supersecretion have been associated with pleiotropic effects. Indications are that they have a deleterious influence on sporulation, thus linking the sporulation process with secretion pathways in *S. cerevisiae* (Smith and Gill, 1985).

2. Expression of the Prochymosin Gene in Aspergillus nidulans

Recent work by Hayenga *et al.* (1986) has demonstrated the expression of the prochymosin gene in *Aspergillus nidulans,* a member of a class of filamentous fungi long used for the manufacture of food-grade products. A construct of the *A. nidulans* vector (Balance and Turner, 1987) employed for the transformation contained the presumptive upstream activation sequence, promoter, and terminator of the *A. niger* glucoamylase gene. *Aspergillus nidulans* transformants, which were selected by *pyr4* complementation, secreted active prochymosin when grown in the presence of starch or xylose, indicating that the *A. niger* promoter is not regulated in *A. nidulans*. This achievement represents the first time a mammalian gene has been expressed in filamentous fungi. It also raises the possibility that other recombinant proteins can be made in these organisms which naturally secrete proteins so well (Esser and Mohr, 1986; Upshall, 1986; Van Brunt, 1986a,b).

3. Expression of Mucor miehei Protease in Aspergillus nidulans

Mucor miehei secretes an acid protease which is used in cheese manufacturing. Recently Gray *et al.* (1987) have cloned and expressed the protease gene in *A. nidulans*. The genomic clone of the protein with its flanking control regions was first incorporated in an *A. nidulans pyr4*-based vector. *Aspergillus nidulans* transformants, which were detected by *pyr4* complementation as described above, secreted the acid protease when grown in submerged culture. The yield of the protein could roughly be correlated with the number of integrated copies of the gene.

4. Other Fungal Proteases

It is likely that new secreted proteinases with industrial applications will be discovered in the future. Ahearn *et al.* (1968) tested 800 yeasts and other fungi for extracellular caseinolysis. Significant proteolytic activity was found with isolates of various genera, primarily *Kluyveromyces, Endomycopsis, Aureobasidium,* and *Cephalosporium*. Fungal species with the highest levels were *Aureobasidium pullulans,* species of *Cephalosporium, Candida punicea,* and *Candida lipolytica.* Potent extracellular proteolytic activity has also been found with members of the genera *Debaryomyces, Rhodotorula,* and *Metschnikowia* (Foda and Badr El-Din, 1979).

A screening for extracellular proteolytic activity by Nelson and Young (1986) examined 119 yeast strains for their ability to produce proteases. The most significant producers belonged to the genera *Candida, Kluyveromyces,* and *Hansenula.* Among 50 strains of *Saccharomyces* examined, none were identified as protease secreters. The extracellular yeast proteases were tested for chill-proofing activity, that is, their ability to reduce a haze of polypeptide material in beer that complexes with polyphenols and precipitates at low pH and temperature. Proteases able to reduce this chill haze were generally those secreted by species of the genus *Candida,* though one strain of *Debaryomyces* and one of *Pichia* also had chill-proofing activity. The secreted protease from *Candida olea* strain 148 was an especially promising enzyme with potential as a chill-proofing protease. Other recent studies report the applicability of proteases from *Saccharomycopsis fibuligera* and *Torulopsis magnoliae* in degrading haze-forming proteins in brewer's wort (Bilinski *et al.,* 1987).

In recent years the genetics and biochemistry of extracellular alkaline protease production by *Yarrowia (Saccharomycopsis) lipolytica* has been examined in detail, and at least 16 genes that influence protease production have been found (Ogrydziak and Mortimer, 1977; Mehta and von Borstel 1979; Ogrydziak *et al.,* 1982). The cloning of the acid protease from *Saccharomycopsis fibuligera,* a related yeast, in *S. cerevisiae* has already been achieved (Yamashita *et al.,* 1986). Transformants of *S. cerevisiae* which secrete the protease have projected use in the food processing and brewing industries.

Owing to the additional ability of *Y. lipolytica* to secrete acid proteases, lipases, and RNases, transformation systems have been developed for this yeast that will lead to a better understanding of its secretion processes (Davidow *et al.,* 1986; Gaillardin *et al.,* 1986). *Yarrowia lipolytica* may serve not only as a source of digestive enzymes but may also become a convenient organism for the high level expression of heterologous genes and the secretion of valuable specialty proteins.

C. Cellulases

Fungal cellulases, which are usually extracellular, are multienzyme complexes that act synergistically to hydrolyze cellulose, cellodextrin, cellobiose, or chemically modified forms of these sugar polymers. Cellulases have not seen broad use in industry but have great potential for biomass conversion to glucose syrups. Nevertheless, enzyme preparations are used in food processing, the extraction of natural products, and digestive aids (Ghose and Pathak, 1973; Goksøyr and Eriksen, 1980; Enari, 1983; Mandels, 1985; Coughlan, 1985). Most cellulase manufacturing takes place in Japan and the United States from filamentous fungi; enzyme from *A. niger* is produced mainly for food use, and *Trichoderma viride* enzyme is prepared primarily for nonfood use (Scott, 1978). A potential new use may involve washing products; cellulase from the mold *Humicola insolens* may find application in detergent formulations (Stinson, 1987).

Other fungi, some with commercial potential, also produce significant amounts of cellulase (Mandels and Andreotti, 1978; Mandels, 1982; Vanbelle *et al.*, 1982; Reese and Mandels, 1984). For example, *Talaromyces emersonii* produces a potent extracellular cellulolytic system which includes a thermostable β-glucosidase of broad specificity and insensitive to substrate inhibition (McHale and Coughlan, 1981). Because cellulase activity is inducible by lactose in this thermophilic fungus, cheese whey may someday serve as a medium for cellulase and single cell protein production (McHale and Morrison, 1986).

1. Trichoderma Cellulase

Of the cellulolytic fungi, *Trichoderma reesei* and *T. viride* have been studied most extensively (Gong and Tsao, 1979; Gritzali and Brown, 1979; Ryu and Mandels, 1980; Montenecourt, 1983; Montenecourt and Eveleigh, 1985; Wood, 1985; Merivuori *et al.*, 1985a,b). *Trichoderma* cellulolytic activity, which consists of endo-1,4-β-glucanase (cellulase), cellobiohydrolase, and β-glucosidase, is induced late in fermentation by cellulose, mixed β-1,4-glucans, and certain oligosaccharides. Synergistic interaction of cellobiohydrolase and endoglucanase on cellulose yields cellodextrins and cellobiose. Endoglucanase randomly cleaves β-1,4-glucosidic linkages, nicking chains which lack strong hydrogen bonds. Cellobiohydrolase, an exo-β-glucanase which removes cellobiose units from the nonreducing end of shortened chains, appears to be essential for the degradation of crystalline cellulose, while β-glucosidase is needed for the breakdown of cellobiose and, thus, for total hydrolysis of cellulose to glucose. The β-glucosidase activity in *T. viride* is represented by at least two extracellular enzymes, one of which is glycosylated (Wilhelm and Sahm, 1986), and is of special importance since it removes cellobiose inhibition of cellulase. Therefore, in industrial applications additional β-glucosidase from one of several *Aspergillus* species is often supplemented to increase the rate and extent of saccharification. It is conceivable that future processes may employ mutants with enhanced β-glucosidase, such as the *T. reesei* strains selected by Kawamori *et al.* (1986) that have elevated levels of this extracellular enzyme.

Multiple forms of these three cellulolytic enzymes have been isolated and investigated. Most of these isoenzymes are glycoproteins with a high-mannose structure, but they exhibit variability in carbohydrate content, usually from 1 to 10%. One cellulase component even consists of as much as 50% carbohydrate (McHale and Coughlan, 1981). The origin of the isozymes remains unclear and could be based on differential glycosylation, partial proteolysis, or multiple genes (Enari, 1985; Wood, 1985).

The primary structure of *T. reesei* cellobiohydrolase I, which cleaves cellobiosyl units from the nonreducing end of cellulose and is the major secreted protein of this organism, has been determined (Fägerstam *et al.*, 1984), and the gene encoding this enzyme has been cloned in *E. coli* (Shoemaker *et al.*, 1983; Teeri *et al.*, 1983). The protein sequence data indicate that the enzyme consists of approximately 490 amino acid residues and reveal the presence of a 17 amino acid hydrophobic signal sequence (Shoemaker *et al.*, 1983). The enzyme bears carbohydrate bound to a short region near the carboxy terminus and possesses three attachment sites for *N*-acetylglucosamine residues. Attached carbohydrates consist of O-glycosidic glycans which are mostly mannose and N-glycosidic glycans which are $Man_5GlcNAc_2$ and $Man_9GlcNAc_2$ (Salovuori *et al.*, 1987). Recently, the gene for cellobiohydrolase II from *T. reesei* has also been cloned and the primary structure of the enzyme deduced from the nucleotide sequence (Chen *et al.*, 1987; Teeri *et al.*, 1987).

The full-length cDNA for the major endoglucanase of *T. reesei* has also been cloned in *E. coli* and its nucleotide sequence determined (Penttilä *et al.*, 1986). The deduced protein sequence data revealed 437 amino acids and a 22 amino acid signal peptide which shows 45% homology to the signal sequence of cellobiohydrolase I isozyme. Endoglucanase I isozyme contains six putative N-glycosylation sites and a putative O-glycosylation region. Since the carboxy-terminal region shows about 70% homology to a sequence in cellobiohydrolase I, Penttilä *et al.* (1986) have suggested that the two enzymes may have evolved from a common ancestor by gene duplication.

2. Expression of Trichoderma Cellobiohydrolase and Endoglucanase Genes in Saccharomyces cerevisiae

Cellulase-secreting *S. cerevisiae* would be applicable to ethanolic fermentations and food processing technologies employing yeast. Consequently, a number of recombinant DNA programs have centered on this organism as a host for fungal and bacterial cellulase genes. Active, glycosylated cellobiohydrolase and endo-1,4-β-glucanase of *T. reesei* have been expressed and secreted in *S. cerevisiae* using plasmids bearing the intronless genes flanked by yeast enolase promoter and downstream regulatory sequences. Transformed yeast cells secreted into the medium 70–80% of the enzyme, which constituted about 0.5–1% of the total protein in the culture. Both native and recombinant yeast-secreted cellobiohydrolase share the same amino-terminal sequence indicating correct processing of the secretory leader peptide of *T. reesei* in yeast. Enzymes of both origins possess O-glycosylated serine and threonine residues and an amino terminus blocked by a pyroglutamyl residue, but, unlike native enzyme, recombinant cellobiohydrolase is extremely N-glycosylated. Although the specific activity of this hyperglycosylated enzyme is essentially equivalent to that of the native protein (Shoemaker *et al.*, 1984), it appears that some of the oligosaccharide processing events in yeast differ from those in *T. reesei* (Shoemaker *et al.*, 1984).

Information is emerging on the endoglucanase from *T. reesei* cloned in *S. cerevisiae*. Available data indicate that the glucanase secreted by yeast has a higher molecular weight than enzyme secreted by *T. reesei* (Shoemaker, 1984). Other studies reveal that yeast plasmids bearing the coding sequence for mature protein with its signal peptide and yeast transcription control sequences yield transformants which secrete enzymatically active but hyperglycosylated endoglucanase isozyme I into culture medium (Van Arsdell *et al.*, 1987). This glycosylation pattern does not alter the specific activity of the enzyme, but it seems to protect the enzyme from inactivation by heat. Other processing steps appear to be similar to those in *T. reesei*. Cleavage of the signal peptide at a glutamic acid residue forms a pyrrolidine carboxylic acid, leading to a blocked amino terminus. In addition, O-linked sugars are added at threonine and serine sites of both the recombinant endoglucanase and the native enzyme.

3. Expression of β-Glucosidase Genes in Saccharomyces cerevisiae

In attempts to construct yeast strains that can ferment cellulose to ethanol, cloned β-glucosidase genes from several sources have been introduced into *S. cerevisiae*, which has low levels of this enzyme and cannot grow on cellobiose. Plasmids bearing the gene from *A. niger* (Penttilä *et al.*, 1984), *Kluyveromyces fragilis* (Raynal and Guerineau, 1984), and *Candida pelliculosa* (Kohchi and Toh-e, 1986) have been used to transform *S. cerevisiae*. None of the genes conferred on yeast the ability to grow on cellobiose. The *K. fragilis* enzyme, which is intracellular, is expressed efficiently in *S. cerevisiae* but not secreted. Only permeabilized mutants of the transformant can grow on cellobiose. The β-glucosidase from *A. niger* is expressed poorly in *S. cerevisiae*, also preventing growth on cellobiose. Glucosidase activity associated with a gene from *C. pellicu-*

losa is produced in *S. cerevisiae* and secreted into the periplasmic space, though most transformants do not grow on cellobiose. It is likely that future efforts to develop yeast strains with the capacity to metabolize cellobiose will focus on cellobiase expression levels and secretion or the ability of the transformant strains to take up the disaccharide (Montenecourt and Eveleigh, 1985).

4. Expression of a Bacterial Cellulase Gene in Saccharomyces cerevisiae

A bacterial cellulase gene has also been cloned in *S. cerevisiae*. A *S. cerevisiae* strain bearing a plasmid encoding an extracellular β-1,4-glucanase gene from *Cellulomonas fimi* was used to secrete active enzyme into culture medium. Secretion was significantly increased when the leader of the secreted yeast protein, K1 toxin, was inserted upstream from the bacterial gene. It is likely that such novel recombinant yeast strains may permit industrial fermentations that use cellulose as a carbon source (Skipper *et al.*, 1985).

5. Secretory Mutants of Trichoderma reesei

It is known that during growth on cellulose all *Trichoderma* extracellular proteins are glycosylated and are localized in the endoplasmic reticulum (ER) and in secretory vesicles (Chapman *et al.*, 1983). To elucidate the cellulase secretion process, the physiology of cellulase production is being investigated in wild-type and mutant strains of *T. reesei*. A major goal has been to develop strains with improved cellulase yield and decreased sensitivity to catabolite repression. Rut-C30, a hypercellulolytic and hypersecretory mutant isolated on the basis of its resistance to catabolite repression, has been shown to possess 6- to 7-fold greater ER content. The mutant also has ER with wider cisternae and ER-associated saccules instead of typical Golgi bodies. In addition, this mutant synthesizes more α-amylase and acid phosphatase, and it and another mutant, RL-P37, show increased levels of cellulase, endoglucanase, and two acid proteases (Ghosh *et al.*, 1982, 1984; Sheir-Neiss and Montenecourt, 1984). Study of RL-P37 and its temperature-sensitive derivative which shows significantly reduced secretion of active cellulases at a nonpermissive temperature (Suh *et al.*, 1986) should help to elucidate the secretory pathway of this cellulolytic fungus.

Mutants Rut-C30 and RL-P37 display altered profiles of secreted proteins. Since both mutants were isolated by selection with 2-deoxyglucose, an agent that intereferes with protein glycosylation, a defect in carbohydrate transfer mechanisms has been proposed. Such a defect would explain the altered profiles of extracellular proteins of both mutants (Sheir-Neiss and Montenecourt, 1984).

Since most fungal cellulases are glycoproteins, the function of the carbohydrate moieties is of interest. Though the role of O-linked oligosaccharides remains to be clarified, studies by Merivuori *et al.* (1985b) with
cellulase from strain RL-P37 and wild-type *T. reesei* show that N-linked
oligosaccharides do not appear to be essential for secretion or activity but
do appear to play a role in enzyme stabilization. Tunicamycin did not
lower the levels of secreted cellulase or total extracellular protein. However, endoglucanase from mycelia grown in the presence of tunicamycin
was more sensitive to heat treatment and proteolysis. An identical observation has been made with exoglucanase from *S. cerevisiae* (Sanchez *et
al.*, 1982). Thus, the role of glycosylation remains uncertain, and its involvement in cellulase secretion, tertiary structure determination, substrate interaction, or stabilization is yet to be determined (Montenecourt
and Eveleigh, 1985).

Mutants of *T. reesei* that overproduce cellulases will be of great industrial importance. A process for cellulase production has already been
tested in a 3000-liter French pilot plant that uses a hypersecreting mutant
of *T. reesei*. Of the 35 grams of secreted protein produced per liter,
cellulase is the major fraction (Anonymous, 1987).

D. Lactases

Lactases, or β-D-galactosidases, are produced mostly from the yeast
Kluyveromyces marxianus var. *lactis* (*Torula cremoris*) and species of
Aspergillus for use in the dairy industry (Miller and Brand, 1980; Richmond *et al.*, 1981; Gekas and López-Leiva, 1985). The yeast enzyme has
a pH optimum of 6.8–7.0 and poor stability at low pH, while the *A. niger*
lactase has a pH optimum of 4.0–4.5 making it suitable for use in acid
wheys to break down lactose (Woychik and Holsinger, 1977; Nijpels,
1983). The *Aspergillus* enzyme is known to be extracellular (Borglum and
Sternberg, 1972), though little information has been published on specific
aspects of lactase secretion. The yeast enzyme is intracellular.

Interest in recent years has focused on the β-galactosidase gene *LAC4*
from the yeast *K. marxianus* var. *lactis*. The gene has been cloned in *E.
coli* (Dickson and Markin, 1978) and *S. cerevisiae* (Dickson, 1980; Sreekrishna and Dickson, 1985) and its expression studied in the latter organism (Velati-Bellini *et al.*, 1986). Secretion has not been achieved. Deduced amino acid sequence analysis of the *LAC4* gene product indicates
significant homology to *E. coli* β-galactosidase (Breunig *et al.*, 1984).
Secretion of *E. coli* β-galactosidase in *S. cerevisiae* was attempted using
either the invertase signal or prepro-α-factor leader sequence (Emr *et al.*,
1984; Das and Shultz, 1987). The invertase signal sequence transported

the protein to the secretory pathway where it was localized in the endoplasmic reticulum. On the other hand, α-factor leader sequence could direct β-galactosidase through the yeast endoplasmic reticulum to the periplasm.

Another filamentous fungus, *T. reesei* mutant Rut-C30, produces not only elevated levels of secreted cellulase but also lactase (Castillo *et al.*, 1984). Information on secretion processes in this mutant may provide insight into lactase secretion in fungi.

E. Invertases

For many years invertase (β-fructofuranosidase) has served as a model system for the study of secretion in *S. cerevisiae*. Its examination continues to provide substantial information on the genetics and molecular biology of the *SUC* genes in yeast. The application of the invertase secretion signal to other recombinant proteins has promoted their secretion in *S. cerevisiae*, as in the case of calf chymosin (Smith *et al.*, 1985) and human interferon (Chang *et al.*, 1986).

Baker's yeast invertase is used commercially to break down sucrose, degrade di- and trisaccharides and also fructans, and make artificial honey. A major application is in the manufacture of a wide variety of desserts and confections. These and other industrial applications have been reviewed by Wiseman (1981). Further basic research involving invertase should generate superior yeast strains for commercial purposes.

Six different, unlinked loci for the structural genes of invertase (*SUC1–SUC5, SUC7*) have been described. The presence of only one active *SUC* gene permits *S. cerevisiae* to ferment raffinose or sucrose, though some strains possess more than one *SUC* gene. Five of the *SUC* genes have been cloned and studied in *S. cerevisiae*. All five genes (*SUC1–SUC5*) have even been expressed on a single multicopy vector in yeast (Hohmann and Zimmerman, 1986). The different *SUC* genes have similar nucleotide sequences, and the enzymes display similar biochemical properties. Of the six yeast invertases, the *SUC2* gene product has been studied in the greatest detail.

The *SUC2* gene is known to code for two forms of invertase, one form of which is secreted, glycosylated, and regulated by glucose repression. The other form is known to be intracellular, nonglycosylated, and expressed constitutively. Investigations with the cloned *SUC2* gene reveal that the clones encode two mRNAs (1.8 and 1.9 kilobases) that differ at their 5' ends (Perlman and Halvorson, 1981; Carlson and Botstein, 1982; Taussig and Carlson, 1983). Only the larger, glucose-regulated mRNA encodes a signal peptide which enables the enzyme to be exported via the endoplasmic reticulum and Golgi membranes to the periplasmic space (Perlman *et al.*, 1982). About 3% of total invertase is normally found in the culture medium. Invertase generally forms an octameric structure in the endoplasmic reticulum which is retained during its transport to the periplasm. A dimeric molecule, however, is found in the growth medium. It is possible that the octamer is required for protein processing or for packaging prior to transport of the protein between intracellular compartments (Esmon *et al.*, 1986).

The study of mutations that alter synthesis and secretion of the *SUC2* invertase has provided valuable details on the mechanism of protein secretion by this organism. Certain signal sequence mutations that substitute or extensively delete bases result in the formation of active internal invertase that is unable to enter the secretory pathway. Some short deletions and several extensive substitutions do not prevent normal secretion, however (Kaiser and Bostein, 1986). Similarly, Perlman *et al.* (1986) have reported that while some insertion mutations in the proximal portion of the signal sequence do not interfere with the processing or export of invertase, large deletions in the hydrophobic region of the signal do prevent secretion and glycosylation. The recent results of Kaiser *et al.* (1987) shed more light on these observations and show that the specificity of the secretion signal for yeast *SUC2* invertase is very low. Their data demonstrate that about one-fifth of a number of random sequences are able to functionally replace the export signal for yeast invertase. Whether such broad specificity is characteristic of signal sequences for other secreted proteins remains to be determined.

F. Lipases

Commercial lipases are used mainly in food processing and as digestive aids (Seitz, 1974; Shahani, 1975; Arnold *et al.*, 1975; Macrae, 1983a,b). Future uses may include the production of desirable food lipids, synthesis and interesterification of fats, and addition to washing products (Posorske, 1984). The last application is attracting considerable attention from manufacturers of commodity enzymes since lipases are expected to be the next frontier of detergent enzymes. Their adaptation to washing products will be a challenge to industrial enzymologists and genetic and protein engineers (Cannon *et al.*, 1987).

A large number of yeasts and molds produce extracellular lipases. Though the structure, properties, and modes of action of some of these lipases may be well understood (Brockman, 1984), little is known about the processes involved in the export of these enzymes from cells. Classical approaches with species of *Aspergillus, Mucor, Rhizopus, Penicillium, Candida,* and *Geotrichum* have demonstrated that fungi of these groups produce lipases, often multiform enzymes, of industrial significance. Lipase synthesis, in many cases, appears to be induced by lipid material in the medium, and production generally requires somewhat higher levels of nitrogen sources (Böing, 1982). Extracellular lipases often have a high carbohydrate content, which is lowered during prolonged storage, and typically they are rapidly degraded by coexistent secreted proteases (Iwai and Tsujisaka, 1984).

G. Pectic Enzymes

Fungal pectic enzyme mixtures containing primarily polygalacturonase, or pectinase, are used to clarify fruit juices and wines and to macerate fruit pulps (Neubeck, 1975; Rombouts and Pilnik, 1980; Whitaker, 1984). Commercial preparations in the United States usually consist of extracellular material from *A. niger* cultures with pectinase and pectin methylesterase, or pectase, activity (Fogarty and Kelly, 1983; Pilnik and Rombouts, 1983). Research has centered on strain improvement, medium optimization, and applications. Secretion mechanisms for fungal pectic enzymes remain unknown, and to date no fungal genes related to pectin degradation have been cloned.

H. Glucanases

High concentrations of barley β-glucans in brewers wort impede beer filtration and can cause the formation of undesirable cloudiness or gelatinous precipitates in beer. β-Glucanases are added during the fermentation process or during beer storage. The supplemented β-1,3-β-1,4-glucanase is a concomitant product of *Bacillus amyloliquefaciens* fermentation for α-amylase or *A. niger* fermentation for pectinase. Commercial β-glucanases from strains of *Penicillium emersonii* are also produced. Fermentation conditions are adjusted to yield high levels of extracellular enzyme (Ward, 1985).

The endo-1,3-1,4-β-glucanase gene from *Bacillus subtilis* has been cloned in a yeast replicative plasmid containing a *LEU2* gene and used to transform *S. cerevisiae*. The recombinant gene was expressed in yeast, but low levels of β-glucanase activity could be detected only in crude cell extract (Hinchliffe and Box, 1984). Subsequent studies removed the 5'-nontranslated region of the *B. subtilis* β-glucanase gene, placed the gene under control of the yeast *ADH1* promoter, and achieved significantly higher levels of β-glucanase in yeast (Cantwell *et al.*, 1986). Though not yet available, strains of brewers yeast able to secrete β-glucanase will be extremely useful to degrade β-glucans in wort and prevent the formation of gels during the beer fermentation process.

III. SECRETION OF THERAPEUTIC MAMMALIAN
POLYPEPTIDES IN *Saccharomyces cerevisiae*

Over the past several years the yeast *S. cerevisiae* has been developed as an organism for production of a variety of mammalian polypeptides of pharmaceutical importance. Most such proteins or peptides with therapeutic activity are secretory proteins. However, some of these proteins can be produced intracellularly in yeast as biologically active molecules (e.g., hepatitis B surface antigen) while others are synthesized as inactive products. The inactivity of some intracellular recombinant proteins can be explained by the absence of disulfide bonds or oligosaccharides which often permit a protein to attain a correct and active conformation. Since the secretory organelles are the sites of the addition of oligosaccharides and the formation of disulfide bonds in a protein, much recent effort has been dedicated to the secretion of these recombinant proteins rather than their intracellular accumulation in yeast.

Secretion of a foreign protein in yeast has been achieved either by using the protein's own signal sequence in association with a yeast promoter or by using a yeast signal or leader sequence under the control of a yeast promoter. In fact, efficient processing of the signal peptide prior to protein secretion has been consistently obtained only when the yeast leader sequence was exploited. Signal sequences from three yeast secretory proteins have been tested so far, for example, acid phosphatase (*PHO5*), invertase (*SUC2*), and α-factor (*MFα1*).

Both acid phosphatase and invertase possess a "consensus" signal sequence (Perlman and Halvorson, 1983) having three domains, namely, an amino terminus with charged amino acids, a central core region consisting of hydrophobic amino acids, and a classic signal peptidase cleavage site of Ala-X where cleavage occurs between Ala and the adjacent amino acid. Acid phosphatase and invertase are normally localized in the periplasmic space, and less than 5% of the total protein is secreted into the yeast growth medium. In fact, attempts at secretion of significant amounts of foreign proteins into the yeast growth medium using the signal sequence of acid phosphatase or invertase have generally yielded poor results. For example, Hinnen *et al.* (1983) could observe secretion of human leukocyte interferon only at a 10% level when the acid phosphatase promoter and the first 14 amino acids of the acid phosphatase signal sequence were used. Likewise, by using the signal sequence and part of the structural gene of invertase, Smith *et al.* (1985) achieved secretion of a heterologous protein (calf prochymosin) only to a level of about 4% unless a hypersecreting mutant strain was utilized.

Potentially more promising results were obtained recently by Chang *et al.* (1986) when they attempted to achieve secretion of human interferon by utilizing the invertase signal sequence. In contrast to the system of Smith *et al.* (1985), the invertase signal sequence used by these workers contained no mature invertase sequence. As much as 30% of the expressed interferon was secreted into the growth medium of yeast cells transformed with such an invertase–human interferon hybrid gene when placed under the control of the 3-phosphoglycerate kinase promoter.

The leader sequence of the yeast mating pheromone, α-factor, has been used most successfully for the secretion of heterologous proteins into yeast culture medium. α-Factor, a peptide of 13 amino acids, is synthesized as a precursor protein consisting of 165 amino acids of which an 81 amino acid amino-terminal segment constitutes the leader sequence. The leader sequence is composed of a signal sequence of approximately 20 hydrophobic amino acids followed by a pro segment of 61 amino acids (Kurjan and Herskowitz, 1982; Julius *et al.*, 1983). The pro segment contains three potential N-glycosylation sites beside a spacer peptide with the carboxy-terminal sequence of Lys-Arg-Glu-Ala-Glu-Ala. The yeast *KEX2* gene product cleaves the pro sequence at the carboxyl side of Lys-Arg, while dipeptidyl aminopeptidase A, a product of the *STE13* gene, removes the Glu-Ala sequence. A processing scheme for α-factor production is shown in Fig. 1. Further details of the cloning, biosynthesis, processing and secretion of α-factor can be found elsewhere (Kurjan and Herskowitz, 1982; Julius *et al.*, 1983, 1984).

Table II shows a list of foreign proteins or peptides of therapeutic importance that have been secreted into yeast growth medium by utilizing the prepro-α-factor sequence. In most cases, the protein of interest was found to be cleaved out of the α-factor leader peptide during its transit through the secretory pathway. However, incomplete processing of the pro-α-factor spacer peptide was observed when the proteins were produced at a high level using a multicopy vector (Bitter *et al.*, 1984). This was caused by the limited availability of dipeptidyl aminopeptidase A in yeast cells, resulting in the production of foreign proteins with amino-terminal Glu-Ala dipeptide extensions. Such problems of incomplete processing of the Glu-Ala dipeptide sequence were overcome by utilizing a pro-α-factor sequence having only Lys-Arg as the spacer sequence (Brake *et al.*, 1984; Bitter *et al.*, 1984; Miyajima *et al.*, 1985; Zsebo *et al.*, 1986; Vlasuk *et al.*, 1986; Mullenbach *et al.*, 1986).

Indeed, the α-factor leader sequence has been judiciously utilized to secrete a significant number of heterologous proteins or peptides having biological activity in yeast growth medium. Nevertheless, in most cases the total yield of the protein was far too low to be commercially viable. It is expected that future research will focus on obtaining a better understanding of the α-factor secretion system, exploitation of the yeast secretory organelles, and further manipulation of the genetic background of

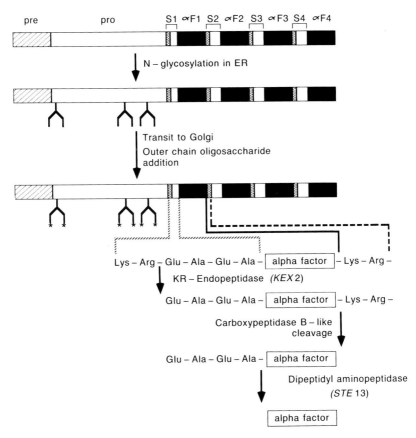

Fig. 1. Processing pathway of prepro-α-factor. αF1–αF4 are the four copies of α-factor peptides present within the precursor protein. Similarly, S1–S4 are the four copies of spacer peptides. The amino acid sequence of the spacer peptides are as follows: S1, Lys-Arg-(Glu-Ala)₂; S2, Lys-Arg-(Glu-Ala)₃; S3 and S4, Lys-Arg-Glu-Ala-Asp-Ala-Glu-Ala. ⋏ Symbols denote N-linked core glycosylation, while asterisks stand for outer chain oligosaccharide addition. Further details of the biosynthesis of α-factor have been provided in the text and in Julius *et al.* (1983, 1984).

yeast to improve product yield and secretion. This research could extend to immobilized yeast systems, such as those developed for α-factor-secreting *S. cerevisiae* (Okada *et al.*, 1987), and lead to methods for continuous heterologous protein production. Other approaches may involve the expression of silent genes in yeast strains after alcohol fermentation and recovery of the ethanol. The heterologous protein would not be expressed until an inducer has been added. Such a yeast fermentation process using the *GAL10–CYC1* hybrid promoter for the production of β-lactamase, β-

TABLE II

Prepro-α-Factor-Directed Secretion of Therapeutic Proteins into the Growth Medium of *Saccharomyces cerevisiae*

Protein/peptide	Source for the DNA	Amount secreted	Reference
Atrial natriuretic peptide	Synthetic[a]	570 μg/liter	Vlasuk et al. (1986)
Calcitonin	Synthetic[a]	12.0 mg/liter	Zsebo et al. (1986)
Consensus α interferon	Synthetic [a]	100 mg/liter	Zsebo et al. (1986)
Connective tissue activating peptide III	Synthetic[a]	2.5×10^4 units/liter	Mullenbach et al. (1986)
β-Endorphin	Synthetic[a]	7.0 mg/liter	Zsebo et al. (1986)
Epidermal growth factor	Synthetic[a]	4.6 mg/liter	Brake et al. (1984)
Granulocyte–macrophage colony-stimulating factor	Human	NK[b]	Miyajima et al. (1986)
Granulocyte–macrophage colony-stimulating factor	Murine	1×10^8 units/liter	Miyajima et al. (1986)
Growth hormone releasing factor	Human	NK	Brake et al. (1984)
Insulin	Human	0.18–2 mM	Thim et al. (1986)
Insulin-like growth factors	Human	NK	Brake et al. (1984)
Interleukin 2	Mouse	10 μg/liter	Miyajima et al. (1985)
Interleukin 2	Human	NK	Brake et al. (1984)
Leukocyte interferon D (IFN-α1)	Human	1×10^8 units/liter	Singh et al. (1984)
Somatomedin C	Synthetic[a]	12 mg/liter	Ernst (1986)
Somatostatin	Angler fish	200 ng/10⁷ cells	Green et al. (1986)

[a] The DNA sequence was based on the amino acid sequence of the native peptide.
[b] NK, Not known.

glucanase, β-galactosidase, human serum albumin, and human immunoglobulin has already been proposed by Hinchliffe and Kenny (1986).

Other yeast systems may also be important in industrial secretion research. For example, the development of the methylotrophic yeast, *Pichia pastoris,* as a host for the production of *S. cerevisiae* invertase has been reported and may be extended to the large-scale production of heterologous proteins in the future (Cregg and Madden, 1988).

Research will also examine the filamentous fungi as secretion sytems.

Rapid advances are being made with these organisms as evidenced by results discussed in earlier sections of this chapter and by the expression of tissue plasminogen activator in *Aspergillus* using the promoter and terminator sequences of alcohol dehydrogenase (Upshall *et al.*, 1986; Anonymous, 1986). Low levels of peptide were secreted into culture medium. Other filamentous fungi beside species of *Aspergillus,* such as *Caldariomyces fumago* (Nuell *et al.,* 1986) and *T. reesei* (Salovuori *et al.,* 1987), may also be excellent secretion systems.

IV. APPLIED SECRETION RESEARCH: PROSPECTS FOR THE FUTURE

Fungal secretion research is a rapidly advancing field as evidenced by the investigations cited in this chapter. Most published information has appeared only in the last five years or so, and already much of it is being applied to industrial microbiology. Successes with yeast and filamentous fungal systems make it likely that this pace will continue and even accelerate (Esser and Mohr, 1986; Esser and Meinhardt, 1986; Cullen and Leong, 1986; Hynes, 1986; Saunders *et al.,* 1986; Upshall, 1986; Van Brunt, 1986a,b; Stewart and Russell, 1986).

Perhaps future directions in fungal secretion research will follow already familar paths. Attempts to improve both the yield of secreted products and their rates of export will continue, as will the development of superior organisms, expression vectors, and secretion-related nucleotide sequences. However, it is conceivable that, in the future, applied secretion researchers not only will be concerned with molecular genetic aspects, but will also be responsive to the complexities of fermentation engineering, process development, pharmacology, immunology, and even economics. A narrow focus will probably widen, broadening the scope of modern secretion research. Secretion scientists in industrial mycology will play the expanded roles indicated below in the development of new products and processes.

Applied secretion researchers will play a role in the development of fermentation systems optimized for production of valued commodity or specialty proteins. They will work closely with fermentation engineers to develop fermentations that ideally will use the least expensive raw materials and generate extracellular product with few undesirable by-products and a minimum of excess biomass. The fermentation systems could involve submerged culture with recycled cells, continuous culture, immobilized cells, or novel technologies adapted to the particular organism. Volumes could range from 50 liters for specialty proteins to 500,000 liters or more for commodity enzymes. The production of marketable intracellular or extracellular coproducts could influence the economic feasibility of some fermentation systems, as could the availability of fungi containing silent genes that can be turned on for production of a second product.

Applied secretion researchers will play a role in the development of efficient purification strategies designed with the final use of the product in mind. Having an intimate understanding of the sequences of peptides and thereby some of their physical and chemical properties, they will be advisers to biochemical engineers. A major goal of both will be to maximize purity using scaled-up purification processes. The extent of purification will sometimes depend on the yield and type of product. For example, detergent enzymes will be essentially preparations derived from cell-free culture fluid, while pharmacological peptides will be extensively purified preparations devoid of contaminants, especially immunogenic substances.

Applied secretion researchers will modify the final protein product, as needed, to meet the stringent requirements of its applications. They will be key members of protein engineering teams. The teams will engineer peptide sequences to enhance therapeutic properties or improve stability. Also, glycosylation patterns will be modified by inducing the fungal production organisms to attach sugars only at certain sites. Thus, glycoproteins that are not immunogenic to humans will be developed for medical applications.

Secretion researchers will work closely with specialists in industry, medicine, and agriculture, and with business managers, to propose commercial targets. New ideas will result from the close interaction of experts in diverse fields. New products will follow. In future years, the emergence of a valuable peptide into the culture medium could very likely represent the first step of its journey to the marketplace.

The impact of secretion research on industrial mycology will continue to increase, as will the role of secretion researchers in the conception of new products of biotechnology. The reasons are many, but one is primary: like today, most commodity enzymes and recombinant specialty peptides of the future will be *secreted* proteins.

ACKNOWLEDGMENTS

The authors thank Dr. Douglas A. Campbell for his critical comments and Ms. Phyllis L. Vandegrift for typing the manuscript.

REFERENCES

Ahearn, D. G., Meyers, S. P., and Nichols, R. A. (1968). Extracellular proteinases of yeasts and yeast-like fungi. *Appl. Microbiol.* **16,** 1370–1374.

Anonymus. (1986). Fungi become more useful for rDNA products. *Genet. Technol. News* **6**(9), 9.

Anonymous. (1987). Enzymatic hydrolysis at preindustrial scale. *Process. Technol.* **9**(2), 4.

Arima, K., Oshima, T., Kubota, I., Nakamura, N., Mizunaga, T., and Toh-e, A. (1983). The nucleotide sequence of the yeast *PHO5* gene: A putative precursor of repressible acid phosphatase contains a signal peptide. *Nucleic Acids Res.* **11,** 1657–1672.

Arnold, R. G., Shahani, K. M., and Dwivedi, B. K. (1975). Application of lipolytic enzymes to flavor development in dairy products. *J. Dairy Sci.* **58,** 1127–1143.

Ashikari, T., Nakamura, N., Tanaka, Y., Kiuchi, N., Shibano, Y., Takaharu, T., Amachi, T., and Yoshizumi, H. (1985). Cloning and expression of the *Rhizopus* glucoamylase gene in yeast. *Agric. Biol. Chem.* **49,** 2521–2523.

Ashikari, T., Nakamura, N., Tanaka, Y., Kiuchi, N., Shibano, Y., Tanaka, T., Amachi, T., and Yoshizumi, H. (1986). *Rhizopus* raw-starch-degrading glucoamylase: Its cloning and expression in yeast. *Agric. Biol. Chem.* **50,** 957–964.

Aunstrup, K. (1974). Industrial production of proteolytic enzymes. *In* "Industrial Aspects of Biochemistry" (B. Spencer, ed.), pp. 23–46. Fed. Eur. Biochem. Soc., Amsterdam.

Aunstrup, K. (1977). Enzymes of industrial interest. Traditional products. *Annu. Rep. Ferment. Process.* **1,** 181–204.

Aunstrup, K. (1979). Production, isolation, and economics of extracellular enzymes. *In* "Applied Biochemistry and Bioengineering" (L. B. Wingaard, E. Katchalski-Katzir, and L. Goldstein, eds.), pp. 27–69. Vol. 2, Academic Press, New York.

Aunstrup, K. (1980). Proteinases. *In* "Economic Microbiology. Vol. 5. Microbial Enzymes and Bioconversions" (A. H. Rose, ed.), pp. 49–114. Academic Press, New York.

Balance, D. J., and Turner, G. (1987). *Gene* (in press).

Bartoszewicz, K. (1986) Glucoamylase of *Aspergillus niger. Acta Biochim. Polon.* **33,** 2–29.

Baulcombe, D. C., and Buffard, D. (1983). Gibberellic acid-regulated expression of α-amylase and 6 other genes in wheat aleurone layers. *Planta* **157,** 493–501.

Bennett, J. W. (1985). Molds, manufacturing, and molecular genetics. *In* "Molecular Genetics of Filamentous Fungi" (W. E. Timberlake, ed.), pp. 345–366. Alan R. Liss, New York.

Beppu, T. (1983). The cloning and expression of chymosin (rennin) genes in microorganisms. *Trends Biotechnol.* **1,** 85–89.

Bigelis, R. (1985). Primary metabolism and industrial fermentations. *In* "Gene Manipulations in Fungi" (J. W. Bennett and L. L. Lasure, eds.), pp. 357–401. Academic Press, New York.

Bigelis, R. (1988). Industrial products of biotechnology: Production, uses, and application of gene technology. *In* "Biotechnology. Gene Technology" (S. Jolly and G. Jacobson, eds.), Vol. 7b, VCH Verlagsgesellschaft, Weinheim (in press).

Bigelis, R., and Lasure, L. L. (1987). Fungal enzymes and primary metabolites used in food processing. *In* "Food and Beverage Mycology" (L. R. Beuchat, ed.), pp. 473–516. Avi/van Nostrand, New York.

Bilinski, C. A., Russell, I., and Stewart, G. G. (1987). Applicability of yeast extracellular proteinases in brewing: Physiological and biochemical aspects. *Appl. Environ Microbiol.* **53,** 495–499.

Bitter, G. A. (1984). Development of yeast as a host for the production of recombinant DNA derived pharmaceuticals. *In* "The Work Biotech Report 1984," Vol. 2, pp. 381–391. Online Publications, Middlesex.

Bitter, G. A., Chen, K. K., Banks, A. R., and Lai, P-H. (1984). Secretion of foreign proteins from *Saccharomyces cerevisiae* directed by α-factor gene fusions. *Proc. Natl. Acad. Sci. U.S.A.* **81,** 5330–5339.

Boel, E., Hjort, I., Svensson, B., Norris, F., Norris, K. E., and Fiil, N. P. (1984a). Glucoamylases G1 and G2 from *Aspergillus niger* are synthesized from two different but closely related mRNAs. *EMBO J.* **3,** 1097–1102.

Boel, E., Hansen, M. T., Hjort, I., Hoegh, I., and Fiil, N. P. (1984b). Two different types of intervening sequences in the glucoamylase gene from *Aspergillus niger. EMBO J.* **3,** 1581–1585.

Böing, J. T. P. (1982). Enzyme production. *In* "Prescott & Dunn's Industrial Microbiology" (G. Reed, ed.), pp. 634–708. Avi, Westport, Connecticut.

Borglum, G. B., and Sternberg, M. (1972). Properties of a fungal lactase. *J. Food Sci.* **37,** 619–623.

Brake, A. J., Merryweather, J. P., Coit, D. G., Heberlein, U. A., Masiarz, F. R., Mullenbach, G. T., Urdea, M. S., Valenzuela, P., and Barr, P. J. (1984). α-Factor directed synthesis and secretion of mature foreign proteins in *Saccharomyces cerevisiae. Proc. Natl. Acad. Sci. U.S.A.* **81,** 4642–4646.

Breunig, K. D., Dahlems, U., Das, S., and Hollenberg, C. P. (1984). Analysis of a eukaryotic β-galactosidase gene: the N-terminal end of the *Kluyveromyces lactis* protein shows homology to the *Escherichia coli* lacZ gene product. *Nucleic Acids Res.* **12,** 2327–2341.

Brockman, H. L. (1984). General features of lipolysis: Reaction scheme, interfacial structure and experimental approaches. *In* "Lipases" (B. Borgstrom and H. L. Brockman, eds.), pp. 3–46. Elsevier, Amsterdam.

Cannon, D. R., Savage, P., Watzman, A., Lazorko, L., Bishop, J., Shafers, T., Kemezis, P., and Slakter, A. (1987). Soaps and detergents: A basketful of high-tech products. *Chem. Week* **140,** 22, 24, 28, 33, 39, 42, 44, 48, 53.

Cantwell, B. A., Brazil, G., Murphy, N., and McConnell, D. J. (1986). Comparison of expression of the endo-β-1,3-1,4-glucanase gene from *Bacillus subtilis* in *Saccharomyces cerevisiae* from the *CYC1* and *ADH1* promoters. *Curr. Genet.* **11,** 65–70.

Carlson, M., and Botstein, D. (1982). Two differentially regulated mRNAs with different 5′ ends encode secreted and intracellular forms of yeast invertase. *Cell* **28,** 145–154.

Carlson, M., Taussig, R., Kustu, S., and Botstein, D. (1983). The secreted form of invertase in *Saccharomyces cerevisiae* is synthesized from mRNA encoding a signal sequence. *Mol. Cell. Biol.* **3,** 439–447.

Castillo, F. J., Blanch, H. W., and Wilke, C. R. (1984). Lactase production in continuous culture by *Trichoderma reesei* RUT-C30. *Biotechnol. Lett.* **6,** 593–596.

Chang, C. N., Matteucci, M., Perry, L. J., Wulf, J. J., Chen, C. Y. and Hitzeman, R. A. (1986). *Saccharomyces cerevisiae* secretes and correctly processes human interferon hybrid proteins containing yeast invertase signal peptides. *Mol. Cell. Biol.* **6,** 1812–1819.

Chapman, C. M., Lowenberg, J. R., Schaller, M. J., and Peichura, J. E. (1983). Ultrastructural localization of cellulase in *Trichoderma reesei* using immunocytochemistry and enzyme cytochemistry. *J. Histochem. Cytochem.* **31,** 1363–1368.

Chen, C. M., Gritzali, M., and Stafford, D. W. (1987). Nucleotide sequence and deduced primary structure of cellobiohydrolase II from *Trichoderma reesei. Bio/Technology* **5,** 274–278.

Colonna, W. J., and Magee, P. T. (1978) Glycogenolytic enzymes in sporulating yeast. *J. Bacteriol.* **134,** 844–853

Coughlan, M. P. (1985). The properties of fungal and bacterial cellulases with comment on their production and application. *Biotechnol. Genet. Eng. Rev.* **3,** 39–111.

Cregg, J. M., and Madden, K. R. (1988). Development of the methylotrophic yeast, *Pichia pastoris*, as a host system for the production of foreign proteins. *Gene* (in press).

Cullen, D., and Leong, S. (1986). Recent advances in the molecular genetics of industrial filamentous fungi. *Trends Biotechnol.* **4**, 285–288.

Das, R. C., and Shultz, J. L. (1987). Secretion of heterologous proteins from *Saccharomyces cerevisiae*. *Biotechnol. Prog.* **3**, 43–48.

Davidow, L. S., Apostolakos, D., O'Donnell, M. M., Proctor, A. R., Ogrydziak, D. M., Wing, R. A., Stasko, I., and DeZeeuw, J. R. (1986). Integrative transformation of the yeast *Yarrowia lipolytica*. *Curr. Genet.* **10**, 39–48.

Demain, A. L. (1981). Industrial microbiology. *Science* **214**, 987–995.

Dickson, R. C. (1980). Expression of a foreign eukaryotic gene in *Saccharomyces cerevisiae: β*-Galactosidase from *Kluyveromyces lactis*. *Gene* **10**, 347–356.

Dickson, R. C., and Markin, J. S. (1978). Molecular cloning and expression in *E. coli* of a yeast gene coding for *β*-galactosidase. *Cell* **15**, 123–130.

Duncan, M., Kohno, T., Stashenko, K., and Yamasaki, E. (1984). Secretion of prochymosin by *Saccharomyces cerevisiae*. *Int. Conf. Yeast Genet. Mol. Biol., 12th, Edinburgh, Scotland*, p. 296.

Emr, S. D., Schauer, I., Hansen, W., Esmon, P., and Schekman, R. (1984). Invertase *β*-galactosidase hybrid proteins fail to be transported from the endoplasmic reticulum in *Saccharomyces cerevisiae*. *Mol. Cell. Biol.* **4**, 2347–2355.

Emtage, J. S., Angal, S., Doel, M. T., Harris, T. J. R., Jenkins, B., Lilley, G., and Lowe, P. A. (1983). Synthesis of calf prochymosin (prorennin) in *Escherichia coli*. *Proc. Natl. Acad. Sci. U.S.A.* **80**, 3671–3675.

Enari, T.-M. (1983). Microbial cellulases. *In* "Microbial Enzymes and Biotechnology" (W. M. Fogarty, ed.), pp. 183–223. Applied Science, London.

Ernst, J. F. (1986). Improved secretion of heterologous proteins by *Saccharomyces cerevisiae:* Effects of promoter substitution in *α*-factor fusions. *DNA* **5**, 483–491.

Erratt, J. A., and Nasim, A. (1986). Cloning and expression of a *Saccharomyces diastaticus* glucoamylase gene in *Saccharomyces cerevisiae* and *Schizosaccharomyces pombe*. *J. Bacteriol.* **166**, 484–490.

Esmon, P. C., Esmon, B. E., and Schekman, R. W. (1986). Invertase forms octamers during secretion. *In* "Microbiology–1986" (L. Leive, ed.), pp. 317–319. American Society for Microbiology, Washington, D.C.

Esser, K., and Meinhardt, F. (1986). Genetics of strain improvement in filamentous fungi with respect to biotechnology. *In* "Overproduction of Microbial Metabolites" (Z. Vaněk and Z. Hošťalek, eds.), Vol. 7, pp. 143–164. Butterworths, Stoneham, Massachusetts.

Esser, K., and Mohr, G. (1986). Integrative transformation of filamentous fungi with respect to biotechnological application. *Proc. Biochem.* **21**(5), 153–159.

Fägerstam, L. G., Pettersson, L. G., and Engström, J. A. (1984). The primary structure of a *β*-1,4-glucan cellobiohydrolase from the fungus *Trichoderma reesei* QM 9414. *FEBS Lett.* **167**, 309–315.

Filho, S. A., Galembeck, E. V., Faria, J. B., and Frascino, A. C. S. (1986). Stable yeast transformants that secrete functional *α*-amylase encoded by cloned mouse pancreatic DNA. *Bio/Technology* **4**, 311–315.

Finkelstein, D. B., Rambosek, J. A., Leach, J., Wilson, R. E., Larson, A. E., Soliday, C. L., and McAda, P. C. (1986). Transformation of *Aspergillus niger* to give enhanced protein secretion. *Abstr. Annu. Meet. Soc. Ind. Microbiol.,* S55.

Foda, M., and Badr El-Din, S. M. (1979). Distribution of extracellular proteolytic activities among various yeasts. *Zbl. Bakt. II. Abt.* **134**, 89–93.

Fogarty, W. M. (1983). Microbial amylases. *In* "Microbial Enzymes and Biotechnology" (W. M. Fogarty, ed.), pp. 1–92. Applied Science, London.

Fogarty, W. M., and Kelly, C. T. (1979). Starch-degrading enzymes of microbial origin part 1. *Prog. Ind. Microbiol.* **15**, 89–150.

Fogarty, W. M., and Kelly, C. T. (1980). Amylases, amyloglucosidases, and related glucanases. *In* "Economic Microbiology. Vol. 5. Microbial Enzymes and Bioconversions" (A. H. Rose, ed.), pp. 115–170. Academic Press, London.

Fogarty, W. M., and Kelly, C. T. (1981). Developments in microbial extracellular enzymes, *Topics Enzyme Ferment. Biotechnol.* **3**, 45–102.

Fogarty, W. M., and Kelly, C. T. (1983). Pectic enzymes. *In* "Microbial Enzymes and Biotechnology" (W. M. Fogarty, ed.), pp. 131–182. Applied Science, London.

Gaillardin, C., Ribet, A. M., and Heslot, H. (1986). Integrative transformation of the yeast *Yarrowia lipolytica. Curr. Genet.* **10**, 49–58.

Gekas, V., and López-Leiva, M. (1985). Hydrolysis of lactose: A literature review. *Proc. Biochem.* **20**(2), 2–12.

Ghose, T. K., and Pathak, A. N. (1973). Cellulases–2: Applications. *Proc. Biochem.* **8**(5), 20–21, 24.

Ghosh, A., Al-Rabiai, S., and Ghosh, B. K., Trimino-Vazquez, H., Eveleigh, D. E., and Montenecourt, B. S. (1982). Increased endoplasmic reticulum content of a mutant of *Trichoderma reesei* (RUT-C30) in relation to cellulase synthesis. *Enzyme Microb. Technol.* **4**, 110–113.

Ghosh, A., Ghosh, B. K., Trimino-Vazquez, H., Eveleigh, D. E., and Montenecourt, B. S. (1984). Cellulase secretion from a hypercellulolytic mutant of *Trichoderma reesei* RUT-C30. *Arch. Microbiol.* **140**, 126–133.

Goff, C. G., Moir, D. T., Kohno, T., Gravius, T. C., Smith, R. A., Yamasaki, E., and Taunton-Rigby, A. (1984). Expression of calf prochymosin in *Saccharomyces cerevisiae. Gene* **27**, 35–46.

Goksøyr, J., and Eriksen, J. (1980). Cellulases. *In* "Economic Microbiology. Vol. 5. Microbial Enzymes and Bioconversions" (A. H. Rose, ed.), pp. 283–330. Academic Press, London.

Gong, C.-S., and Tsao, G. T. (1979). Cellulase and biosynthesis regulation. *Annu. Rep. Ferment. Process.* **3**, 111–140.

Gorozhankina, T. F., Beburov, M. Yu., Sorokin, A. V., and Stepanov, A. I. (1985). Expression of the α-amylase gene of *Bacillus amyloliquefaciens* in the yeast *Saccharomyces cerevisiae. Dokl. Akad. Nauk SSSR* **285**, 717–720.

Gray, G. L., Hayenga, K., Cullen, D., Wilson, L., and Norton, S. (1987). *Gene* (in press).

Green, R., Schaber, M. D., Shields, D., and Kramer, R. (1986). Secretion of somatostatin by *Saccharomyces cerevisiae. J. Biol. Chem.* **261**, 7558–7565.

Gritzali, M., and Brown, R. D., Jr. (1979). The cellulase system of *Trichoderma. Adv. Chem. Ser.* **181**, 237–260.

Gunnarsson, A., Svensson, B., and Svensson, S. (1984). Structural studies on the 0-glycosidically linked carbohydrate chains of glucoamylase G1 from *Aspergillus niger. Eur. J. Biochem.* **145**, 463–467.

Gwynne, D. I., Buxton, F. P., Williams, S. A., Garven, S., and Davies, R. W. (1987). Genetically engineered secretion of active human interferon and a bacterial endoglucanase from *Aspergillus nidulans. Bio/Technology* **5**, 713–719.

Harris, T. J. R., Lowe, P. A., Eaton, M. A. W., Millican, T. A., Patel, T. P., Bose, C. C., Carey, N. H., and Doel, M. T. (1982). Molecular cloning and nucleotide sequence of the cDNA coding for calf preprochymosin. *Nucleic Acids Res.* **10**, 2177–2187.

Hayenga, K., Berka, R., Cullen, D., Gray, G., Norton, S., Rey, M., and Wilson, L. (1986). Expression and secretion of bovine calf chymosin by *Aspergillus nidulans*. *J. Cell, Biochem. Suppl.* **10A**, 274.

Higgins, I. J., Best, D. J., and Jones, J. (1985). "Biotechnology Principles and Applications" pp. 78–79. Blackwell, Oxford.

Hinchliffe, E., and Box, W. G. (1984). Expression of the cloned endo-1,3-1,4-β-glucanase gene of *Bacillus subtilis* in *Saccharomyces cerevisiae*. *Curr. Genet.* **8**, 471–475.

Hinchliffe, E., and Kenny, E. (1986). European Patent Application 201239.

Hinnen, A., Meyhack, B., and Tsapis, R. (1983). High expression and secretion of foreign proteins in yeast. *In* "Gene Expression in Yeast, Foundation for Biotechnical and Industrial Fermentation Research" (M. Korhola and E. Vaisanen, eds.), Vol. 1, pp. 157–166. Kauppakirjapaino Oy, Helsinki.

Hohmann, S., and Zimmerman, F. K. (1986). Cloning and expression on a multicopy vector of five invertase genes of *Saccharomyces cerevisiae*. *Curr. Genet.* **11**, 217–225.

Hollenberg, C. P., Roggenkamp, R., Reipen, G., and Bielefeld, M. (1985). Synthesis and secretion of foreign gene products in the yeast *Saccharomyces cerevisiae*. Therapeutic agents produced by gene engineering. *"Quo Vadis" Symp., Sanofi Group, May 29–30, Toulouse-Labege, France,* pp. 65–78.

Hynes, M. J. (1986). Transformation in filamentous fungi. *Exp. Mycol.* **10**, 1–8.

Innis, M. A., Holland, M. J., McCabe, P. C., Cole, G. E., Wittman, V. P., Tal, R., Watt, K. W. K., Gelfand, D. H., Holland, J. P., and Meade, J. H. (1985). Expression, glycosylation, and secretion of an *Aspergillus* glucoamylase by *Saccharomyces cerevisiae*. *Science* **228**, 21–26.

Iwai, M., and Tsujisaka, Y. (1984). Fungal lipase. *In* "Lipases" (B. Borgstrom and H. L. Brockman, eds.), pp. 443–469. Elsevier, Amsterdam.

Julius, D., Blair, L., Brake, A., Sprague, G., and Thorner, J. (1983). Yeast α-factor is processed from a larger precursor polypeptide: The essential role of a membrane bound dipeptidyl aminopeptidase. *Cell* **32**, 839–852.

Julius, D., Schekman, R., and Thorner, J. (1984). Glycosylation and processing of prepro-α-factor through the yeast secretory pathway. *Cell* **36**, 309–318.

Kaiser, C. A., and Botstein, D. (1986). Secretion defective mutations in signal sequence for *Saccharomyces cerevisiae* invertase. *Mol. Cell. Biol.* **6**, 2382–2391.

Kaiser, C. A., Preuss, D., and Botstein, D. (1987). Many random sequences functionally replace the secretion signal sequence of yeast invertase. *Science* **235**, 312–317.

Kawamori, M., Ado, Y., and Takasawa, S. (1986). Preparation and application of *Trichoderma reesei* mutants with enhanced β-glucosidase. *Agric. Biol. Chem.* **50**, 2477–2482.

Kingsman, S. M., Kingsman, A. J., Dobson, M. J., Mellor, J., and Roberts, N. A. (1985). Heterologous gene expression in *Saccharomyces cerevisiae*. *Biotechnol. Genet. Eng. Rev.* **3**, 377–416.

Kohchi, C. and Toh-e, A. (1986). Cloning of *Candida pelliculosa* β-glucosidase gene and its expression in *Saccharomyces cerevisiae*. *Mol. Gen. Genet.* **203**, 89–94.

Kulp. K. (1975). Carbohydrases. *In* "Enzymes in Food Processing" (G. Reed, ed.), 2nd Ed., pp. 53–122. Academic Press, New York.

Kurjan, J., and Herskowitz, I. (1982). Structure of a yeast pheromone gene (MFα): A putative α-factor precursor contains four tandem copies of mature α-factor. *Cell* **30**, 933–943.

MacAllister, R. V. (1979). Nutritive sweeteners made from starch. *Adv. Carbohydr. Chem.* **36**, 15–56.

McHale, A., and Coughlan, M. P. (1981). The cellulolytic system of *Talaromyces emersonii*. Identification of the various components produced during growth on cellulosic media. *Biochim. Biophys. Acta* **662**, 145–151.

McHale, A. P., and Morrison, J. (1986). Cellulase production during growth of *Talaromyces emersonii* CBS 814.70 on lactose containing media. *Enzyme Microb. Technol.* **8**, 749–754.

Macrae, A. R. (1983a). Extracellular microbial lipases. *In* "Microbial Enzymes and Biotechnology" (W. M. Fogarty, ed.), pp. 225–250. Applied Science, London.

Macrae, A. R. (1983b). Lipase-catalyzed interesterification of oils and fats. *J. Am. Oil Chem. Soc.* **60**, 291–294.

Mandels, M. (1982). Cellulases. *Annu. Rep. Ferment. Process.* **5**, 35–78.

Mandels, N. (1985). Applications of cellulases. *Biochem. Soc. Trans.* **13**, 414–416.

Mandels, M., and Andreotti, R. E. (1978). Problems and challenges in the cellulose to cellulase fermentation. *Proc. Biochem.* **13(5)**, 6–13.

Matsuura, Y., Kusunoki, N., Harada, W., Tanaka, N., Iga, Y., Yasuoka, N., Toda, H., Narita, K., and Kakudo, M. (1980). Molecular structure of Taka-amylase A I. Backbone chain folding at 3 Å resolution. *J. Biochem.* **87**, 1555–1558.

Meaden, P., Ogden, K., Bussey, H., and Tubb, R. S. (1985). A *DEX* gene conferring production of extracellular amyloglucosidase on yeast. *Gene* **34**, 325–334.

Mehta, R. D., and von Borstel, R. C. (1979). Genetic regulation of extracellular protease and ribonuclease production in *Saccharomycopsis lipolytica*. *Genetics* **91**, 580–581.

Mellor, J., Dobson, M. J., Roberts, N. A., Tuite, M. F., Emtage, J. S., White, S., Lowe, P. A., Patel, T., Kingsman, A. J., and Kingsman, S. M. (1983). Efficient synthesis of enzymatically active calf chymosin in *Saccharomyces cerevisiae*. *Gene* **24**, 1–4.

Merivuori, H., Siegler, K. M., Sands, J. A., and Montenecourt, B. S. (1985a). Regulation of cellulase biosynthesis and secretion in fungi. *Biochem. Soc. Trans.* **13**, 411–414.

Merivuori, H., Sands, A., and Montenecourt, B. S. (1985b). Effects of tunicamycin on secretion and enzymatic activities of cellulase from *Trichoderma reesei*. *Appl. Microbiol. Biotechnol.* **23**, 60–66.

Meyrath, J., and Bayer, G. (1980). Environmental factors and cultivation techniques in fungal α-amylase production. *Proc. FEBS Mtg.* **61**, 331–338.

Miller, J. J., and Brand, J. C. (1980). Enzymic lactose hydrolysis. *Food Technol. Aust.* **32**, 144–146.

Miyajima, A., Bond, M. W., Otsu, K., Arai, K.-I., and Arai, N. (1985). Secretion of mature mouse interleukin-2 by *Saccharomyces cerevisiae:* Use of a general secretion vector containing promoter and leader sequences of the mating pheromone α-factor. *Gene* **37**, 155–161.

Miyajima, A., Otsu, K., Schreurs, J., Bond, M. W., Abrams, J. S., and Arai, K. (1986). Expresssion of murine and human granulocyte–macrophage colony stimulating factors in *S. cerevisiae:* Mutagenesis of the potential glycosylation sites. *EMBO J.* **5**, 1193–1197.

Moir, D., Mao, J., Schumm, J. W., Vovis, G. F., Alford, B. L., and Taunton-Rigby, A. (1982). Molecular cloning and characterization of double-stranded cDNA coding for bovine chymosin. *Gene* **19**, 127–138.

Moir, D. T., Mao, J., Duncan, M. J., Smith, R. A., and Kohno, T. (1985). Production of calf chymosin by the yeast *S. cerevisiae*. *Dev. Ind. Microbiol.* **26**, 75–85.

Montenecourt, B. S. (1983). *Trichoderma reesei* cellulases. *Trends Biotechnol.* **5**, 156–161.

Montenecourt, B. S., and Eveleigh, D. E. (1985). Fungal carbohydrases: Amylases and cellulases. *In* "Gene Manipulations in Fungi" (J. W. Bennett and L. L. Lasure, eds.), pp. 491–512. Academic Press, New York.

Moody, A. J., Norris, F., Norris, K., Hanse, M. T., and Thim, L. (1987). The secretion of glucagon by transformed yeast strains. *FEBS Lett.* **212**, 302–306.

Mullenbach, G. T., Tabrizi, A., Blacher, R. W., and Steimer, K. S. (1986). Chemical synthesis and expression in yeast of a gene encoding connective tissue activating peptide III *J. Biol. Chem.* **261**, 719–722.

Nakamura, Y., Sato, T., Emi, M., Miyanohara, A., Nishide, T., and Matsubara, K. (1986). Expression of human salivary α-amylase gene in *Saccharomyces cerevisiae* and its secretion using the mammalian signal sequence. *Gene* **50**, 239–245.

Nelson, G., and Young, T. W. (1986). Yeast extracellular proteolytic enzymes for chill-proofing beer. *J. Inst. Brew.* **92**, 599–603.

Neubeck, C. E. (1975). Fruits, fruit products. In "Enzymes in Food Processing" (C. Reed, ed.), pp. 397–442. Academic Press, New York.

Nicaud, J. M., Mackman, N., and Holland, I. B. (1986). Current status of secretion of foreign proteins by microorganisms. *J. Biotechnol.* **3**, 255–270.

Nijpels, H. H. (1983). Lactases and their applications. In "Enzymes and Food processing" (G. G. Birch, N. Blakebrough, and K. J. Parker, eds.), pp. 89–104 Applied Science, London.

Nishimori, K. (1986). Molecular cloning of calf prochymosin cDNA and its expression in microorganisms. *J. Agric. Chem. Soc. Jpn.* **60**, 119–127.

Nuell, M., Axley, M., Kenigsberg, P., Fang, G. H., and Hager, L. P. (1986). Isolation and sequence of the chloroperoxidase gene of *Caldariomyces fumago* and studies of its regulation. *Fed. Proc., Fed. Am. Soc. Exp. Biol.* **45**, 1700.

Nunberg, J. H., Meade, J. H., Cole, G., Lawyer, F. C., McCabe, P., Schweickart, V., Tal. R., Wittman, V. P., Flatgaard, J. E., and Innis, M. A. (1984). Molecular cloning and characterization of the glucoamylase gene of *Aspergillus awamori. Mol. Cell. Biol.* **4**, 2306–2315.

Ogrydziak, D. M., and Mortimer, R. K. (1977). Genetics of extracellular protease production in *Saccharomycopsis lipolytica. Genetics* **87**, 621–632.

Ogrydziak, D. M., Cheng, S.-C., and Scharf, S. J. (1982). Characterization of *Saccharomycopsis lipolytica* mutants producing lowered levels of alkaline protease. *J. Gen. Microbiol.* **128**, 2271–2280.

Okada, T., Sonomoto, K., and Tanaka, A. (1987). Application of entrapped growing yeast cells to peptide secretion system. *Eur. J. Appl. Microbiol. Biotechnol.* **26**, 112–116.

Olson, N. F. (1979). Cheese. In "Microbial Technology. Vol. 2. Fermentation Technology" (H. J. Peppler and D. Perlman, eds.), 2nd Ed., pp. 39–77. Academic Press, New York.

Pardo, J. M., Polaina, J., and Jimenez, A. (1986). Cloning of the *STA2* and *SGA* genes encoding glucoamylases in yeast and regulation of their expression by the *STA10* gene of *Saccharomyces cerevisiae. Nucleic Acids Res.* **14**, 4701–4718.

Pedersen, V. B., and Foltmann, B. (1975). Amino-acid sequence of the peptide segment liberated during activation of prochymosin (prorennin). *Eur. J. Biochem.* **55**, 95–103.

Penttilä, M. E., Nevalainen, K. M. H., Raynal, A., and Knowles, J. K. C. (1984). Cloning of *Aspergillus niger* genes in yeast. Expression of the gene coding *Aspergillus* β-glucosidase. *Mol. Gen. Genet.* **194**, 494–499.

Penttilä, M., Legtovaara, P., Nevalainen, H., Bhikhabhai, R., and Knowles, J. (1986). Homology between cellulase genes of *Trichoderma reesei:* Complete nucleotide sequence of the endoglucanase I gene. *Gene* **45**, 253–263.

Perlman, D., and Halvorson, H. O. (1981). Distinct repressible mRNAs for cytoplasmic and secreted yeast invertase are encoded by a single gene. *Cell* **25**, 525–536.

Perlman, D., and Halvorson, H. O. (1983). A putative signal peptidase recognition site in eukaryotic and prokaryotic signal peptides. *J. Mol. Biol.* **167**, 391–409.

Perlman, D., Halvorson, H. O., and Cannon, L. E. (1982). Presecretory and cytoplasmic invertase polypeptides encoded by distinct mRNAs derived from the same structural gene differ by a signal sequence. *Proc. Natl. Acad. Sci. U.S.A.* **79**, 781–785.

Perlman, D., Raney, P., and Halvorson, H. O. (1986). Mutations affecting the signal sequence alter synthesis and secretion of yeast invertase. *Proc. Natl. Acad. Sci. U.S.A.* **83**, 5033–5037.

Pilnik, W., and Rombouts, F. M. (1983). Pectic enzymes. In "Enzymes and Food Processing" (G. G. Birch, N. Blakebrough, and K. J. Parker, eds.), pp. 105–128. Applied Science, London.

Posorske, L. H. (1984). Industrial-scale application of enzymes to the fats and oil industry. *J. Am. Oil Chem. Soc.* **61**, 1758–1760.

Pretorius, I. S., Chow, T., Modena, D., and Marmur, J. (1986a). Molecular cloning and characterization of the *STA2* gene of *Saccharomyces diastaticus. Mol. Gen. Genet.* **203**, 29–35.

Pretorius, I. S., Chow, T., and Marmur, J. (1986b). Identification and physical characterization of yeast glucoamylase structural genes. *Mol. Gen. Genet.* **203**, 36–41.

Priest, F. G. (1984). "Extracellular Enzymes." American Society for Microbiology, Washington, D.C.

Primrose, S. B. (1986). The application of genetically engineered micro-organisms in the production of drugs. *J. Appl. Bacteriol.* **61**, 99–116.

Raynal, A., and Guerineau, M. (1984). Cloning and expression of the structural gene for β-glucosidase of *Kluyveromyces fragilis* in *Escherichia coli* and *Saccharomyces cerevisiae. Mol. Gen. Genet.* **195**, 108–115.

Reese, E. T., and Mandels, M. (1984). Rolling with the times: Production and applications of *Trichoderma reesei* cellulase. *Annu. Rep. Ferment. Process.* **7**, 1–20.

Richmond, M. L., Gray, J. I., and Stine, C. M. (1981). β-Galactosidase: Review of recent research related to technological application, nutritional concerns, and immobilization. *J. Dairy Sci.* **64**, 1759–1771.

Rombouts, F. M., and Pilnik, E. (1980). Pectic enzymes. In "Economic Microbiology. Vol. 5. Microbial Enzymes and Bioconversions" (A. H. Rose, ed.), pp. 227–282. Academic Press, London.

Rothstein, S. J., Lazarus, C. M., Smith, W. E., Baulcombe, D. C., and Gatenby, A. A. (1984). Secretion of a wheat α-amylase expressed in yeast. *Nature (London)* **308**, 662–665.

Ryu, D. D. Y., and Mandels, M. (1980). Cellulases: Biosynthesis and applications. *Enzyme Microb. Technol.* **2**, 91–102.

Salovuori, I., Makarow, M., Rauvala, H., Knowles, J., and Kaariainen, L. (1987). Low molecular weight high-mannose type glycans in a secreted protein of the filamentous fungus *Trichoderma reesei. Bio/Technology* **5**, 152–156.

Sanchez, A., Villanueva, J. R., and Villa, T. G. (1982). Effect of tunicamycin an exo-1,3-β-D-glucanase synthesis and secretion by cells and protoplasts of *Saccharomyces cerevisiae. J. Gen. Microbiol.* **128**, 3051–3060.

Saunders, G., Tuite, M. F., and Holt, G. (1986). Fungal cloning vectors. *Trends Biotechnol.* **4**, 93–98.

Sardinas, J. L. (1976). Calf rennet substitutes. *Proc. Biochem.* **11**, (4), 10–17.

Sato, T., Tsunasawa, S., Nakamura, Y., Emi, M., Sakiyama, F., and Matsubara, K. (1986). Expression of the human salivary α-amylase gene in yeast and characterization of the secreted protein. *Gene* **50**, 247–257.

Schekman, R. (1982). The secretory pathway in yeast. *Trends Biochem. Sci.* **7**, 243–246.

Schekman, R., and Novick, P. (1982). The secretory process and yeast cell-surface assembly. In "The Molecular Biology of the Yeast *Saccharomyces:* Metabolism and Gene Expression" (J. N. Strathern, E. W. Jones, and J. R. Broach, eds.), pp. 361–393. Cold Spring Harbor Laboratory, Cold Spring Harbor, New York.

Schwimmer, S. (1981). "Source Book of Food Enzymology." Avi, Westport, Connecticut.

Scott, D. (1978). Enzymes, industrial. *In* "Kirk–Othmer Encyclopedia of Chemical Technology" (M. Grayson and D. Ekroth, eds.), pp. 173–224. Wiley, New York.

Seitz, E. W. (1974). Industrial applications of microbial lipases: A review. *J. Am. Oil Chem. Soc.* **51**, 12–15.

Shahani, K. M. (1975). Lipases and esterases. *In* "Enzymes in Food Processing" (G. Reed, ed.), 2nd Ed., pp. 181–217. Academic Press, New York.

Sheir-Neiss, G., and Montenecourt, B. S. (1984). Characterization of the secreted cellulases of *Trichoderma reesei* wild type and mutants during controlled fermentations. *Appl. Microbiol. Biotechnol.* **20**, 46–53.

Shoemaker, S. P. (1984). The cellulase system of *Trichoderma reesei: Trichoderma* strain improvement and expression of *Trichoderma* cellulases in yeast. *In* "The World Biotech Report 1984," Vol. 2, pp. 593–600. Online Publications, Middlesex.

Shoemaker, S., Schweickart, V., Ladner, M., Gelfand, D., Kwok, S., Myambo, K., and Innis, M. (1983)., Molecular cloning of exo-cellobiohydrolase I derived from *Trichoderma reesei* strain L27. *Bio/Technology* **1**, 691–696.

Shoemaker, S. P., Gelfand, D. H., Innis, M. A., Kwok, S. Y., Ladner, M. B., and Schweickart, V. (1984). European Patent Application 84110305.4.

Singh, A., Chen, E.-Y., Lugovy, J. M., Chang, C. N., Hitzeman, R. A., and Seeburg, P. (1984). *Saccharomyces cerevisiae* contains two discrete genes coding for the α-factor pheromone. *Nucleic Acids Res.* **11**, 4049–4063.

Skipper, N., Sutherland, M., Davies, R. W., Kilburn, D., Miller, Jr., R. C., Warren, A., and Wong, R. (1985). Secretion of a bacterial cellulase in yeast. *Science* **230**, 958–960.

Smith, R. A., and Gill, T. (1985). Yeast mutants that have improved secretion efficiency for calf prochymosin. *J. Cell. Biochem. Suppl.* **9C**, 157.

Smith, R. A., Duncan, M. J., and Moir, D. T. (1985). Heterologous protein secretion from yeast. *Science* **229**, 1219–1223.

Sreekrishna, K., and Dickson, R. C. (1985). Construction of strains of *Saccharomyces cerevisiae* that grow on lactose. *Proc. Natl. Acad. Sci. U.S.A.* **82**, 7909–7913.

Stewart, G. G., and Russell, I. (1986). One hundred years of yeast research and development in the brewing industry. *J. Inst. Brew.* **92**, 537–558.

Stinson, S. C. (1987). Consumer preferences spur innovation in detergents. *Chem. Eng. News* **65**(4), 21–22, 24, 30–32, 37, 39–40, 42, 46.

Suh, D. H., Sands, J. A., and Montenecourt, B. S. (1986). Temperature sensitive mutants of *Trichoderma reesei* defective in secretion of cellulase. *Appl. Microbiol. Biotechnol.* **25**, 277–284.

Svensson, B., Larsen, K., and Gunnarsson, A. (1986). Characterization of a glucoamylase G2 from *Aspergillus niger*. *Eur. J. Biochem.* **154**, 497–502.

Tanaka, Y., Ashikari, T., Nakamura, N., Kiuchi, N., Shibano, Y., Amachi, T., and Yoshizumi, H. (1986). Glucoamylase produced by *Rhizopus* and by a recombinant yeast containing the *Rhizopus* glucoamylase gene. *Agric. Biol. Chem.* **50**, 1737–1742.

Taussig, R., and Carlson, M. (1983). Nucleotide sequence of the yeast *SUC2* gene for invertase. *Nucleic Acids Res.* **11**, 1943–1954.

Taylor, M. J., and Richardson, T. (1979). Applications of microbial enzymes in food systems and in biotechnology. *Adv. Appl. Microbiol.* **25**, 7–35.

Teeri, T., Salovuori, I., and Knowles, J. (1983). The molecular cloning of the major cellulase gene from *Trichoderma reesei*. *Bio/Technology* **1**, 696–699.

Teeri, T. T., Lehtovaara, P., Kauppinen, S., Saluvuori, I., and Knowles, J. (1987). Homologous domains in *Trichoderma reesei* cellulolytic enzymes: Gene sequence and expression of cellobio-hydrolase II. *Gene* **51**, 43–52.

Thim, L., Hansen, M. T., Norris, K., Hoegh, I., Boel, E., Forstrom, J., Ammerer, G., and Fiil, N. P. (1986). Secretion and processing of insulin precursors in yeast. *Proc. Natl. Acad. Sci. U.S.A.* **83**, 6766–6770.

Thomsen, K. K. (1983). Mouse α-amylase synthesized by *Saccharomyces cerevisiae* is released into culture medium. *Carlsberg Res. Commun.* **48**, 545–555.

Tubb, R. S. (1986). Amylolytic yeasts for commercial applications. *Trends Biotechnol.* **4**, 98–104.

Upshall, A. (1986). Filamentous fungi in biotechnology. *BioTechniques* **4**, 158–166.

Van Arsdell, J. N., Kwok, S., Schweickart, V., Ladner, M. B., Gelfand, D. H., and Innis, M. A. (1987). Cloning, characterization, and expression in *Saccharomyces cerevisiae* of endoglucanase I from *Trichoderma reesei*. *Bio/Technology* **5**, 60–64.

Vanbelle, M., Meurens, M., and Crichton, R. R. (1982). Enzymes in foods and feeds. *Rev. Ferment Industries Alimentaires* **37**, 124–135.

Van Brunt, J. (1986a). Filamentous fungi join the production ranks. *Bio/Technology* **4**, 385–386.

Van Brunt, J. (1986b). Fungi: The perfect hosts. *Bio/Technology* **4**, 1057–1062.

Velati-Bellini, A., Pedroni, P., Martegani, E., and Alberghina, L. (1986). High levels of inducible expression of cloned β-galactosidase of *Kluyveromyces lactis* in *Saccharomyces cerevisiae*. *Appl. Microbiol. Biotechnol.* **25**, 124–131.

Vlasuk, G. P., Bencen, G. H., Scarborough, R. M., Tsai, P.-K., Whang, J. L., Maack, T., Camargo, M. J. F., Kirsher, S. W., and Abraham, J. A. (1986). Expression and secretion of biologically active human atrial natriuretic peptide in *Saccharomyces cerevisiae*. *J. Biol. Chem.* **261**, 4789–4796.

Ward, O. P. (1983). Proteinases. *In* "Microbial Enzymes and Biotechnology" (W. M. Fogarty, ed.), pp. 251–317. Applied Science, London.

Ward, O. P. (1985). Hydrolytic enzymes. *In* "Comprehensive Biotechnology. The Practice of Biotechnology: Current Commodity Products" (H. W. Blanch, S. Drew, and D. I. C. Wang, eds.), Vol. 3, pp. 819–835. Pergamon, Oxford.

Whitaker, J. R. (1984). Pectic substances, pectic enzymes and haze formation in fruit juices. *Enzyme Microb. Technol.* **6**, 341–349.

Wilhelm, M., and Sahm, H. (1986). Purification and characterization of two extracellular β-glucosidases from *Trichoderma viride* ITCC 1433. *Acta Biotechnol.* **6**, 115–121.

Wiseman, A. (1981). New and modified invertases—and their applications. *Topics Enzyme Ferment. Biotechnol.* **3**, 267—288.

Wood, T. M. (1985). Properties of cellulolytic enzyme systems. *Biochem. Soc. Trans.* **13**, 407–410.

Woychik, J. H., and Holsinger, V. H. (1977). Use of lactase in the manufacture of dairy products. *In* "Enzymes in Food and Beverage Processing" (R. L. Ory and A. J. St. Angelo, eds.), pp. 67–79. Am. Chem. Soc., Washington, D.C.

Yamamoto, A. (1975). Proteolytic enzymes. *In* "Enzymes in Food Processing" (G. Reed, ed.), pp. 123–179. Academic Press, New York.

Yamashita, I., and Fukui, S. (1983). Molecular cloning of a glucoamylase-producing gene in the yeast *Saccharomyces*. *Agric. Biol. Chem.* **47**, 2689–2692.

Yamashita, I., and Fukui, S. (1984). Secretion of *Saccharomyces diastaticus* glucoamylase from *Schizosaccharomyces pombe*. *Agric. Biol. Chem.* **48**, 1931–1932.

Yamashita, I., Hatano, T., and Fukui, S. (1984). Subunit structure of glucoamylase of *Saccharomyces diastaticus*. *Agric. Biol. Chem.* **48**, 1611–1616.

Yamashita, I., Itoh, T., and Fukui, S. (1985a). Cloning and expression of the *Saccharomycopsis fibuligera* α-amylase gene in *Saccharomyces cerevisiae*. *Agric. Biol. Chem.* **49**, 3089–3091.

Yamashita, I., Itoh, T., and Fukui, S. (1985b). Cloning and expression of the *Saccharomycopsis fibuligera* glucoamylase gene in *Saccharomyces cerevisiae*. *Appl. Microbiol. Biotechnol.* **23,** 130–133.

Yamashita, I., Suzuki, K., and Fukui, S. (1985c). Nucleotide sequence of the extracellular glucoamylase gene *STA1* in the yeast *Saccharomyces diastaticus*. *J. Bacteriol.* **161,** 567–573.

Yamashita, I., Maemura, T., Hatano, T., and Fukui, S. (1985d). Polymorphic extracellular glucoamylase genes and their evolutionary origin in the yeast *Saccharomyces diastaticus*. *J. Bacteriol.* **161,** 574–582.

Yamashita, I., Hirata, D., Machida, M., and Fukui, S. (1986). Cloning and expression in *Saccharomyces cerevisiae* of the secretable acid protease gene from *Saccharomycopsis fibuligera*. *Agric. Biol. Chem.* **50,** 109–113.

Zsebo, K. M., Lu, H.-S., Fieschko, J. C., Goldstein, L., Davis, J., Duker, K., Suggs, S. V., Lai, P.-H., and Bitter, G. A. (1986). Protein secretion from *Saccharomyces cerevisiae* directed by the prepro-α-factor leader region. *J. Biol. Chem.* **261,** 5858–5865.

Index